Trigonometric Concepts and Formulas

Circular Functions

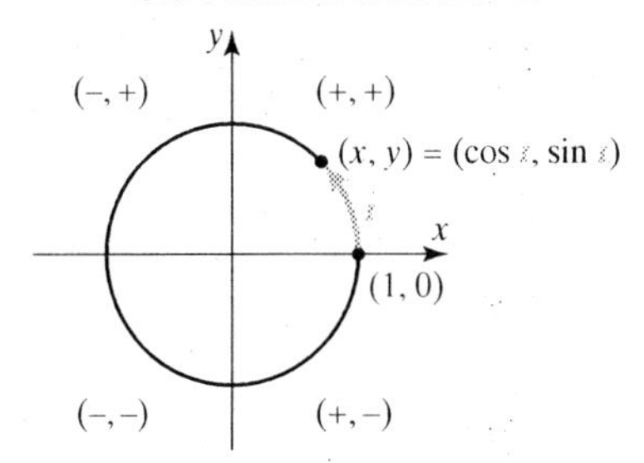

Pythagorean identity:

$$\cos^2 t + \sin^2 t = 1$$

x	$\cos x$	$\sin x$	$\tan x$
0	1	0	0
$\frac{\pi}{6}$	$\frac{\sqrt{3}}{2}$	$\frac{1}{2}$	$\frac{1}{\sqrt{3}}$
$\frac{\pi}{4}$	$\frac{1}{\sqrt{2}}$	$\frac{1}{\sqrt{2}}$	1
$\frac{\pi}{3}$	$\frac{1}{2}$	$\frac{\sqrt{3}}{2}$	$\sqrt{3}$
$\frac{\pi}{2}$	0	1	undefined
π	-1	0	0
$\frac{3\pi}{2}$	0	-1	undefined

Definitions/Identities

$$\tan x = \frac{\sin x}{\cos x}$$

$$\sec x = \frac{1}{\cos x}, \qquad \cos x = \frac{1}{\sec x}$$

$$\csc x = \frac{1}{\sin x}, \qquad \sin x = \frac{1}{\csc x}$$

$$\cot x = \frac{1}{\tan x} = \frac{\cos x}{\sin x}, \quad \tan x = \frac{1}{\cot x}$$

$$\cos(-x) = \cos x$$

$$\sin(-x) = -\sin x$$

$$\tan(-x) = -\tan x$$

Graphs of Circular Functions

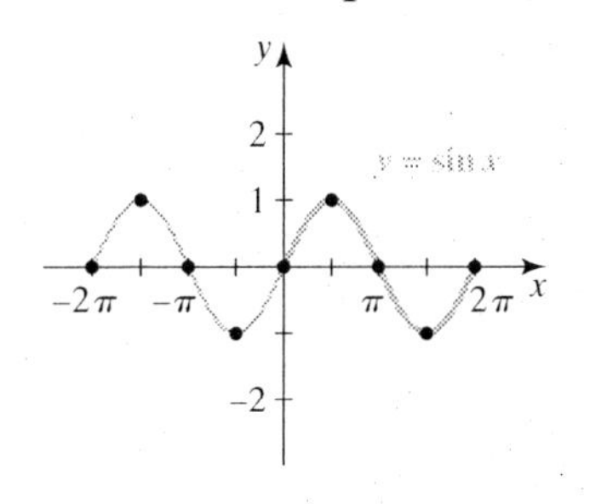

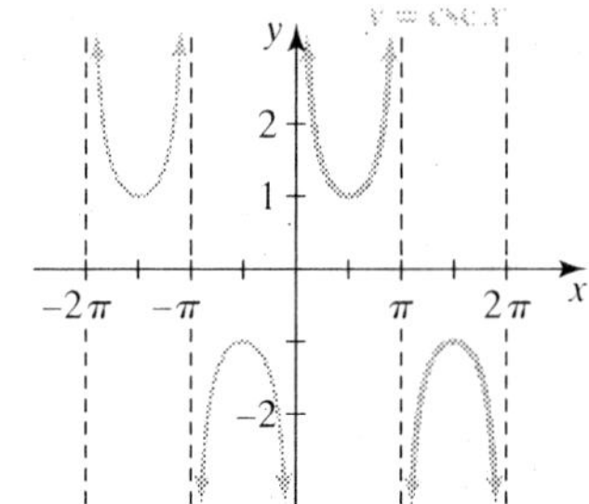

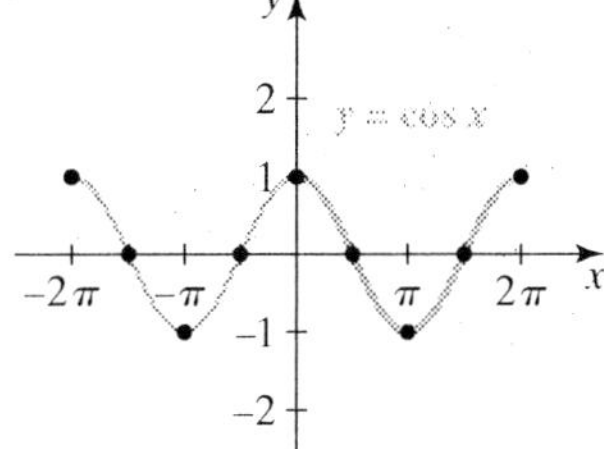

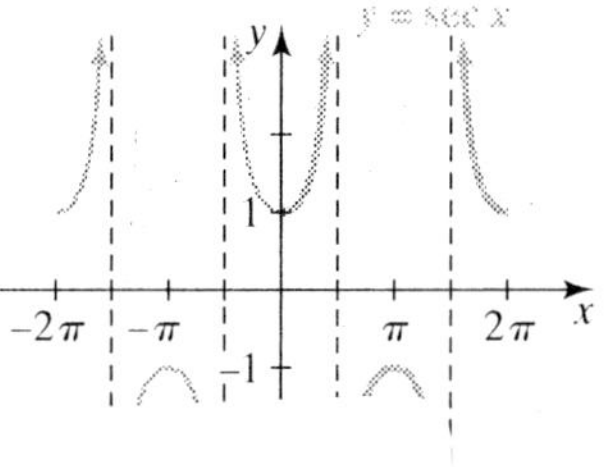

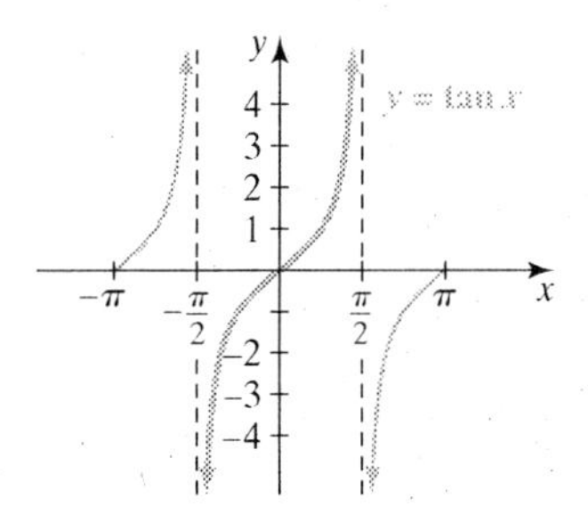

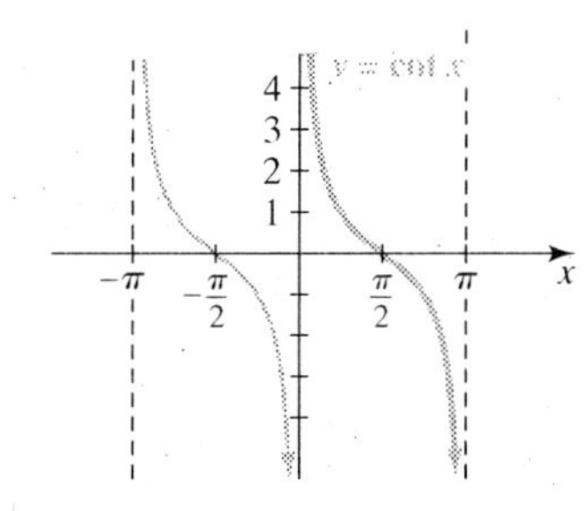

Inverse Functions

$$y = \sin^{-1} x \leftrightarrow \sin y = x, \quad -\frac{\pi}{2} \leq y \leq \frac{\pi}{2}$$

$$y = \tan^{-1} x \leftrightarrow \tan y = x, \quad -\frac{\pi}{2} < y < \frac{\pi}{2}$$

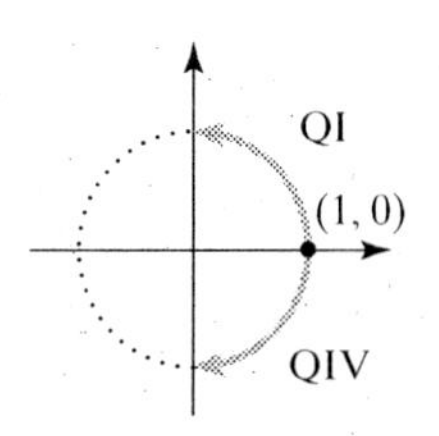

$$y = \cos^{-1} x \leftrightarrow \cos y = x, \quad 0 \leq y \leq \pi$$

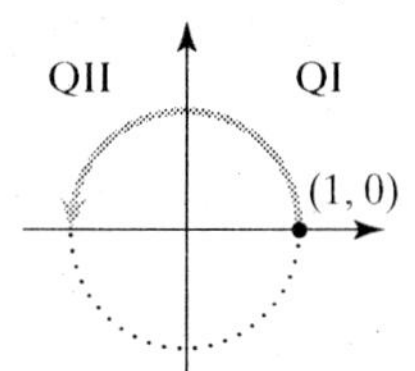

Exact Values of Common Angles with Common Reference Angles

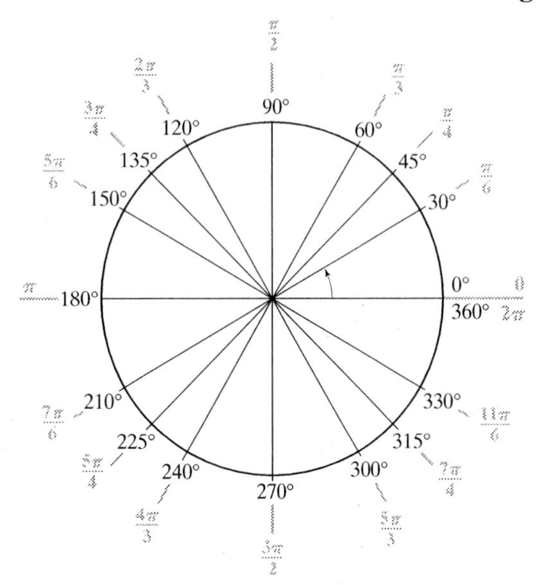

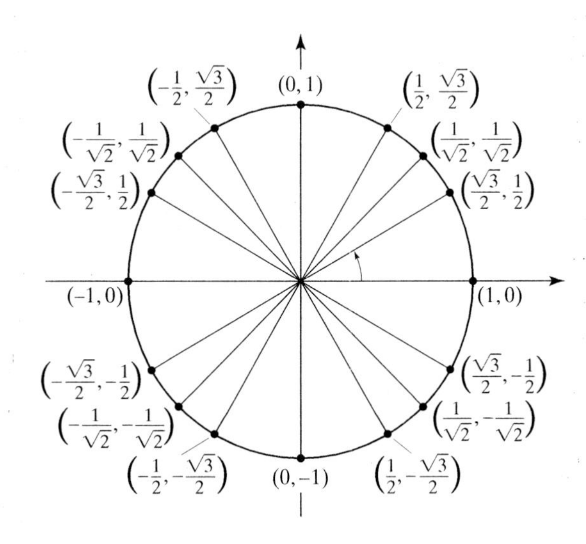

Conversion Formulas

Radians to Degrees

$$\alpha \text{ rad} = \alpha \cdot \frac{180^\circ}{\pi}$$

Degrees to Radians

$$\alpha^\circ = \alpha \cdot \frac{\pi}{180}$$

Arc Length and Velocity

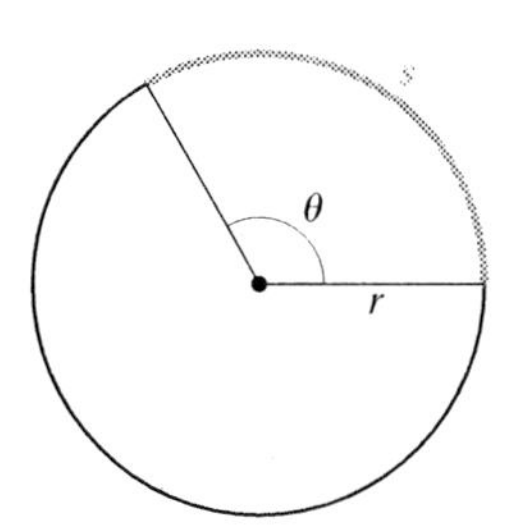

θ is measured in radians.

arc length s:

$$s = r\theta$$

linear velocity v:

$$v = \frac{s}{t} = \frac{r\theta}{t}$$

angular velocity ω:

$$\omega = \frac{\theta}{t}$$

Trigonometric Functions

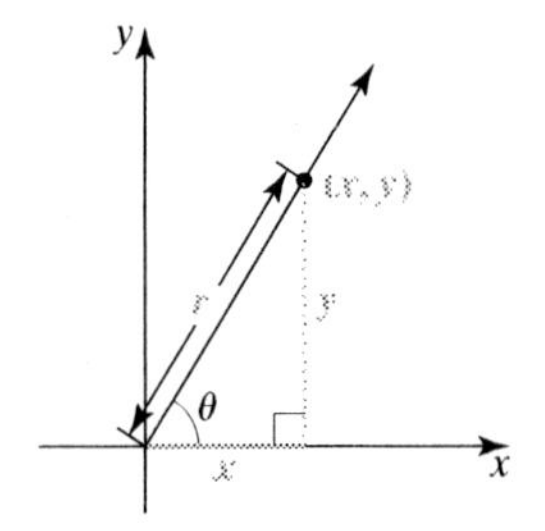

$$\cos\theta = \frac{x}{r} \qquad \sin\theta = \frac{y}{r} \qquad \tan\theta = \frac{y}{x}, \quad x \neq 0$$

$$\sec\theta = \frac{r}{x}, \quad x \neq 0 \qquad \csc\theta = \frac{r}{y}, \quad y \neq 0 \qquad \cot\theta = \frac{x}{y}, \quad y \neq 0$$

$$r = \sqrt{x^2 + y^2} \text{ and } r > 0$$

PEARSON ALWAYS LEARNING

Marie Aratari

Trigonometry
A Circular Function Approach

Taken from:
Trigonometry: A Circular Function Approach
by Marie Aratari

Cover Art: Courtesy of Glow Images, Photodisc, Stockbyte/Getty Images.

Taken from:

Trigonometry: A Circular Function Approach
by Marie Aratari

Published by Addison Wesley
Boston, Massachusetts 02116

This special edition published in cooperation with Pearson Learning Solutions.

For permission to use copyrighted material, grateful acknowledgment is made to the following copyright holders: Page 1 © PhotoDisc Blue. Page 73 © Corbis RF. Page 159 © National Geographic / Barry C. Bishop 1963. Page 235 © Brand X Pioctures. Page 291 © resexcellence.com. Page 327 © NASA Dryden Flight Research Center.

Pearson Learning Solutions, 501 Boylston Street, Suite 900, Boston, MA 02116
A Pearson Education Company
www.pearsoned.com

Printed in the United States of America

1 2 3 4 5 6 7 8 9 10 V036 16 15 14 13 12 11

000200010271271371

SP

ISBN 10: 1-256-46990-4
ISBN 13: 978-1-256-46990-2

Dedication

To John, Nicole, Johnny, Thomas,
and the memory of Donald and Gladys

Trigonometry

Contents

Preface

I hope that posterity will judge me kindly, not only as to the things which I have explained, but also to those which I have intentionally omitted so as to leave to others the pleasure of discovery.

René Descartes
French mathematician, scientist, and philosopher, 1596–1650

This text presents a circular function approach to trigonometry by demonstrating connections between the familiar language of algebra and the new language of trigonometry. This method of providing students with a comfortable base of learning something new from something old or familiar is used throughout the text. (See Section 1.2, Examples 5, 6, 7, 8, and 9.) With just a few connections to algebra, students have the tools to understand the circular functions, their domains and ranges, and the relationship between the circular functions and the functional values. The approach immediately launches the student into the concept of periodic functions and their applications, graphs, and use in modeling many periodic phenomena. Beginning the study of trigonometry with the circular function approach provides the student with a common thread that can be used to discover, connect, and understand the remaining concepts of trigonometry.

Teaching the student to understand trigonometry not as segmented topics, but as topics that are all related and that flow from prior learning, is the goal of this text. For this reason, the text is designed to be read and taught in sequence. In addition, the content is delivered in a student-friendly format, and therefore, asking students to read the sections as part of their assignment is highly recommended. Section 2.3, which is designed to have the student understand how and which sinusoidal function models a particular periodic phenomena, can be considered optional. Numerous graphics are provided to help students visualize concepts. Students are told (and reminded) when memorization is necessary, cautioned to avoid typical mistakes, and presented with ways to organize their written work (Section 1.3, Example 2; Section 3.3, Example 4.) While students are asked to memorize certain items, such as basic identities, they are also encouraged to develop others on the basis of their understanding of the unit circle. This should improve their ability to use trigonometry in future courses. Many examples are designed to lead the student through the comprehension of the topic, as well as to teach the student successful problem solving. Since many trigonometry problems can be solved in several different ways, some examples in this text provide two different methods for the solution.

The topics of algebra and geometry that are necessary for the understanding of a trigonometry concept are reviewed as they are needed (Section 5.2, Examples 1–4). A more detailed presentation of these algebra and geometry topics is available in the appendices for students who need extra review.

Chapter Openers and Objectives

Each chapter begins with a real-world application designed to motivate study of the topics of trigonometry discussed in the chapter. A list of the important concepts of the chapter is also given.

Exercises

Exercise Sets

Each exercise set is designed to provide the student with conceptual understanding, variety, practice with memorization where necessary, and graded problems, from the basic to the more challenging, that require synthesis and conjecture. Many exercise sets contain specific examples in order to get students started on the concept *and* their homework. Several of the exercises, especially those designated as requiring a graphing utility, are designed to have the student discover a concept prior to its presentation in the text (Section 2.1, Exercises 41–47). In addition, many of these graphing calculator exercises are intended to enhance the student's understanding of the topic.

Exercises identified by the accompanying icon involve graphics that can be duplicated by going to www.aw-bc.com/aratari to download a copy of the table, geometric figure, or graph.

Explore Patterns and Relationships

One way to engage students in the memory process is to have them discover patterns or relationships that will help them understand or memorize specific items, like the functional values of the common arcs and angles of the cosine and sine. Discussion exercises are provided to emphasize key concepts and can also be used as group or individual assignments.

GC/CAS—Graphing Calculators or Computer Algebra Systems

Several sections provide examples of graphing utility technology to visualize or calculate the mathematics of trigonometry. Students are cautioned to avoid common input errors during calculation and certain pitfalls of the technology. While graphing calculators are not required for the text, their use will provide the student with the necessary visual support to explore topics, discover patterns, and reinforce answers or concepts. The "Graphing Calculator/CAS Explorations" exercises require the use of a graphing calculator or other graphing utility. The text uses the TI-83 Plus for screen demonstrations. However, any graphing calculator or CAS system that has the indicated functions can be used in the exercise sets.

Web Activities

Several exercise sets contain web activities to involve the trigonometry student with current technology *and finding mathematics on the web*!

Chapter Summary, Chapter Review, and Chapter Test

Each chapter has a detailed summary, a set of review questions, and a chapter test. In order to provide the student with a study tool, the review exercises are identified by section.

Accuracy of Solutions and Appropriate Units

In the real world, the level of accuracy may be tied to the specific application. When possible and unless indicated otherwise, exact values are used in the body of the problem as well as in the answers. In general, we round to four decimal places when approximating answers. Specific instructions about rounding and which decimal place to use to approximate a solution are given in detail in the text. To reduce confusion while working many of the solutions requiring trigonometry formulas, units are often omitted in the steps leading up to the answer, unless they can be incorporated easily into the body of the problem. The final solutions to applications have appropriate units.

Supplements

Student Study Guide and Solutions Manual

ISBN 0-321-17244-2

- **Worksheet**
  Each section is provided with a worksheet that can be used as an outline or a guide for the classroom presentation of the major concepts in the section. Alternatively, the worksheet can act as an end-of-class assessment, a preview to the exercise sets, a quiz, a take-home assignment dealing with the important concepts of the section, or a review of the material covered in the section.

Because there is a lot of material to memorize in trigonometry, the worksheets can be an invaluable tool for learning the formulas and definitions, as well as for preparing for the homework and tests.

- **Appendix Review Exercises**
  Exercise sets are included to review algebraic and geometric concepts that coordinate with the review material presented in the appendices of the text.
- **Solutions**
  Detailed explanations of all the odd problems in the text are included.

Addison-Wesley Math Tutor Center

Free tutoring is available to students who purchase a new copy of the text. The Addison–Wesley Math Tutor Center is staffed by qualified mathematics and statistics instructors who provide students with tutoring on examples and odd-numbered exercises from the textbook. Tutoring is available via toll-free telephone, fax, email, or the Internet, and White Board technology allows tutors and students to actually see the problems worked while they "talk" in real time over the Internet during tutoring sessions. An access code is required. For more information, go to www.aw.com/tutorcenter.

Instructor's Solutions Manual

ISBN 0-321-17245-0

Detailed explanations of all the problems are available in the instructor's solutions manual. Most odd-numbered problems duplicate the concept of the following even-numbered problem.

Instructor's Testing Manual

ISBN 0-321-17425-9

The testing manual contains quizzes and chapter tests with answer keys.

- **Quizzes**
  Each chapter has a set of quizzes for each section that are cumulative to the chapter. There are two forms of each quiz, and both forms can be used as a 5–10-minute quiz, a collaborative group project, or a study guide. It is highly

recommended that they be used as quizzes, as they provide the student with ongoing assessment. The quizzes are not intended to be high in point value; rather, they are intended to provide an immediate indication of which concepts the student understands and which he or she still needs to master before moving ahead. Since many students have difficulty using any calculator, for example, to find reciprocal or inverse trigonometric functional values, quizzes specifically devoted to the calculator are provided to assess these skills.

- **Tests**
  There are four alternative forms of tests per chapter.

TestGen-EQ with QuizMaster-EQ

ISBN 0-321-17426-7

TestGen enables instructors to build, edit, print, and administer tests using a computerized bank of questions developed to cover all the objectives of the text. TestGen is algorithmically based, so that multiple, yet equal, versions of the same question or test can be generated at the click of a button. Instructors can also modify test bank questions or add new questions by using the built-in question editor, which allows users to create graphs, import graphics, insert mathematical notation, and insert variable numbers or text. Tests can be printed or administered on-line via the Web or another network. Many questions in TestGen can be expressed in a short-answer or multiple-choice form, giving instructors greater flexibility in their test preparation. TestGen comes packaged with QuizMaster, which allows students to take tests on a local area network. The software is available on a dual-platform Windows/Macintosh CD-ROM.

Web Site (url: http://www.aw-bc.com/aratari)

The free-access website has additional resources available to the student and the instructor. Student resources include downloadable graphics and homework assignments.

Acknowledgments

I am extremely grateful to the following people who reviewed the manuscript at various stages of its development:

Kent Aeschliman	Oakland Community College
Douglas B. Aichele	Oklahoma State University
Tim Britt	Jackson State Community College
Kathy Chiasson	Oakland Community College
Janis Cimperman	St. Cloud State University
Terry Cremeans	Oakland Community College
Art Fruhling	Yuba College
Jim Fryxell	College of Lake County
Tuesday J. Johnson	Broome Community College
Raja Khoury	Collin County Community College
Richard Leedy	Polk Community College
Stephen J. Nicoloff	Paradise Valley Community College

Shirley Pereira	Grossmont College
Susan Schibel	New Mexico State University
Ron Smith	Edison Community College
Fran Smith	Oakland Community College
Rob Wylie	Carl Albert State College

All of these individuals, as well as the students who have worked with this text in progress, had valuable suggestions and comments that have been incorporated into the text.

I am also extremely privileged to have worked with Elka Block, developmental editor of the book. Her talent, knowledge, support, and encouragement, along with the understanding and love of my family and friends, made this book possible. In addition, I would like to thank Julie LaChance, Joe Vetere, Barbara Atkinson, Cecilia Stashwick, Jaime Bailey, and Becky Anderson of Addison Wesley as well as the accuracy checkers Steve Ouellette and Cathy Ferrer.

Chapter 1

Circular Functions

What do a snowboarding half-pipe rider, your heart, a pendulum, a vibrating guitar string, an earthquake tremor, and a microwave have in common? Each one is an example of rhythmic motion that involves the circular functions, which we study in this chapter.

Important Concepts

- arcs on the unit circle
- the quadrant of an arc
- cosine and sine functions
  - —domain and range of cosine
  - —domain and range of sine
- Pythagorean identity
- common arcs
- functional values of common arcs
- functional values of noncommon arcs
- reference arcs
- ratio and reciprocal functions
  - —tangent, secant, cosecant, and cotangent
- periodic property
- negative identities
- harmonic motion

1.1 Fundamentals for Trigonometry

As you begin the study of trigonometry, there are specific areas of algebra and geometry that are critical for your readiness, comprehension, and success in learning trigonometry. We will review these concepts as they are needed. We begin with a few questions involving three important algebraic ideas. For more review on these topics of algebra, see Appendix A.

Are the following algebraic ideas and questions familiar?

The Coordinate Plane

QUESTION 1 Are the coordinates of the points A, H, S, T, and C in the accompanying figure positive, negative, or zero? What quadrants are they in?

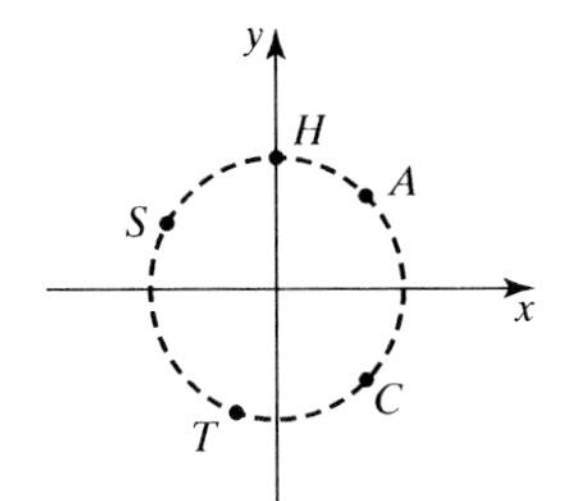

SOLUTION
Both the x- and y-coordinates of point A are positive, which can be denoted by

$A = (+, +)$. Point A is in Quadrant I (QI).
$H = (0, +)$ Since point H lies on the y-axis, it is not in any quadrant.
$S = (-, +)$ Point S is in QII.
$T = (-, -)$ Point T is in QIII.
$C = (+, -)$ Point C is in QIV. ■

QUESTION 2 Name the quadrant in which the y-coordinate is negative $(y < 0)$ and the x-coordinate is positive $(x > 0)$.

SOLUTION In QIV, each point has the form $(+, -)$.

x-coordinate y-coordinate ■

Graphing

QUESTION 3 What does the graph of $x^2 + y^2 = 1$ represent?

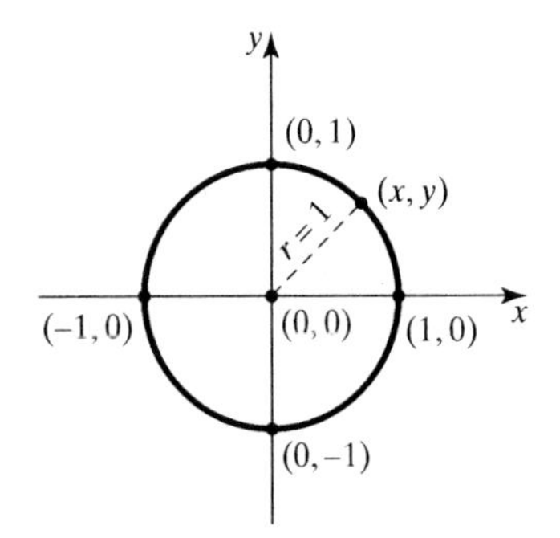

SOLUTION The graph represents a circle whose center is at the origin $(0, 0)$ and whose radius is $r = 1$. This circle is called the **unit circle**. Recall that a circle is the set of all points in a plane at the same distance r from a center. If $r = 1$, and center is $(0, 0)$, then using the distance formula we get:

$$\sqrt{(x_2 - x_1)^2 + (y_2 - y_1)^2} = d$$
$$\sqrt{(x - 0)^2 + (y - 0)^2} = 1 \quad d = r = 1$$
$$x^2 + y^2 = 1 \quad \text{Square both sides.}$$
■

QUESTION 4 What is the range of numerical values for the x-coordinates on the unit circle? What are the largest and smallest of these values? Answer the same questions for the y-coordinates.

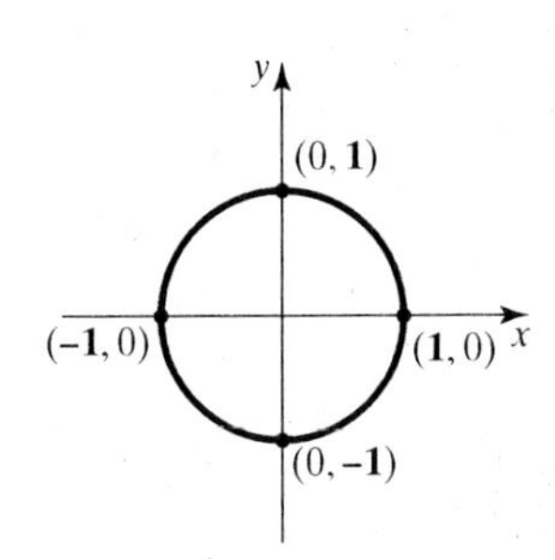

SOLUTION

$-1 \leq x \leq 1$, or $|x| \leq 1$ The largest value for the x-coordinates is 1 and the smallest is -1.

$-1 \leq y \leq 1$, or $|y| \leq 1$ The largest value for the y-coordinates is 1 and the smallest is -1. ■

Solving Equations

QUESTION 5 If P is a point on the unit circle $(x^2 + y^2 = 1)$ and you are given one of its coordinates, can you find the other coordinate? For example, if the x-coordinate of P is $\frac{1}{2}$, what is the y-coordinate of P if P is in QI?

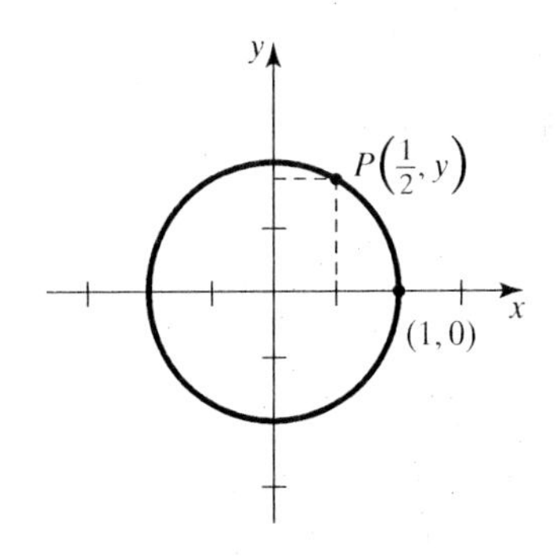

SOLUTION Since P is a point on the circle, the coordinates of P satisfy the circle's equation.

$$x^2 + y^2 = 1$$

$$\left(\frac{1}{2}\right)^2 + y^2 = 1 \quad \text{Substitute } x = \tfrac{1}{2}.$$

$$\frac{1}{4} + y^2 = 1 \quad \text{Simplify.}$$

$$y^2 = \frac{3}{4} \quad \text{Subtract } \tfrac{1}{4} \text{ from both sides.}$$

$$y = \pm\sqrt{\frac{3}{4}} = \pm\frac{\sqrt{3}}{2} \quad \text{Take the square root of both sides.}$$

Since P is a point in QI, the y-coordinate is positive $(y > 0)$. Therefore, $y = \sqrt{3}/2$. ■

Each one of these preceding familiar questions, which we refer to as "algebra questions," plays an important role in your understanding as you learn trigonometry. You will be asked questions just like these again and again, but in a new context that uses the language of trigonometry. Everything we need to know about trigonometry—our new language—will be derived from diagrams of the unit circle similar to the one shown below. If you are able to understand and answer these familiar questions in algebra, *let's get started understanding them in trigonometry!*

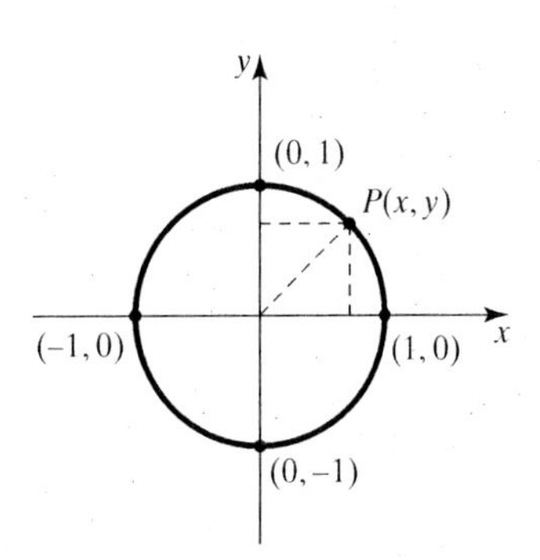

1.2 Circular Functions: The Cosine and Sine Functions

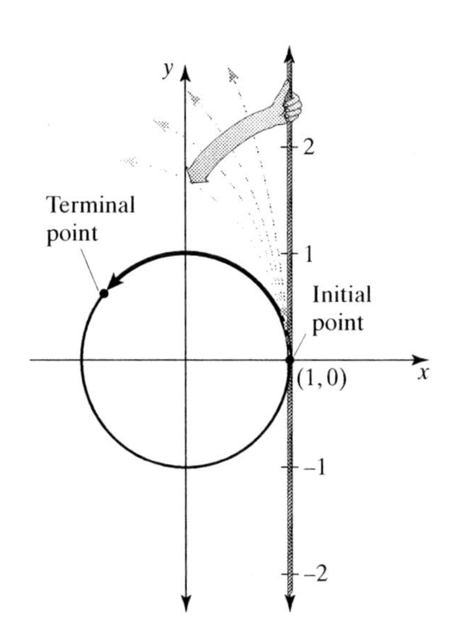

Let's again look at the unit circle, $x^2 + y^2 = 1$. Notice a real number line has been drawn tangent to the circle at $(1, 0)$. We imagine that this line is a string that can be wrapped around the circle. If the string is cut at a given length and wrapped around the circle, it will take on the shape of an arc whose length is equal to the length of the string. Each string that we wrap around the circle defines a specific arc whose initial point is $(1, 0)$ and whose terminal point will be somewhere on the circle, depending on its length.

The positive portion of the number line wraps the circle *counterclockwise* and defines arcs that have *positive direction.*

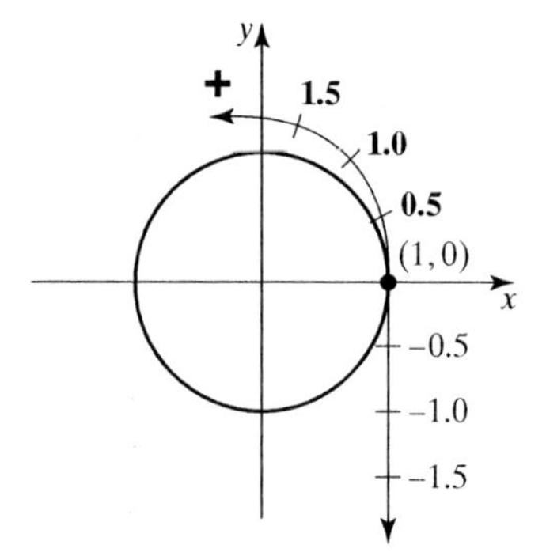

The negative portion of the number line wraps the circle *clockwise* and defines arcs that have *negative direction.*

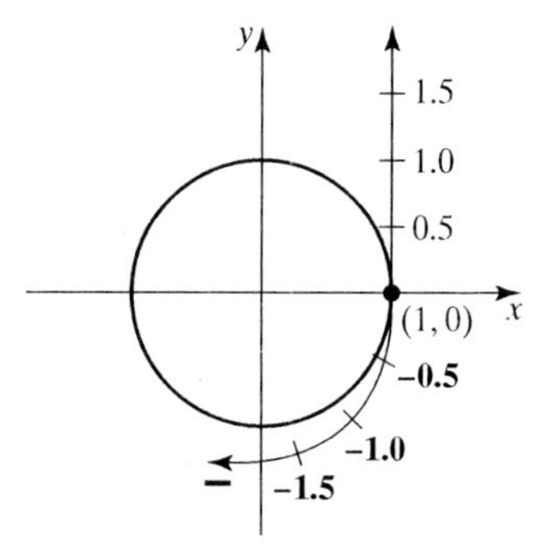

EXAMPLE 1 What would be the length of the string if it wrapped the unit circle exactly once?

SOLUTION

Length for one wrap of the circle = circumference $= 2\pi r = 2\pi(1) = 2\pi \approx 6.28$

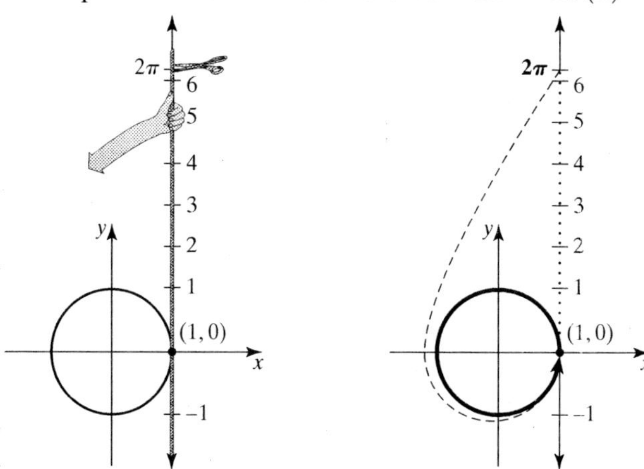

We could have chosen to wrap the string around the circle clockwise. The length of the string would still be 2π because length is always nonnegative, but the arc would correspond to -2π on the number line, indicating a wrap in a clockwise (negative) direction.

NOTE: If the radius of the unit circle was in inches, then the circumference (one wrap) would be in inches. In the examples we will omit the unit until we have an application with a specific unit measurement.

EXAMPLE 2 What would be the length of the string if it wrapped the unit circle twice?

SOLUTION Length of one wrap $= 2\pi$

Length of two wraps $= 2(2\pi) = 4\pi \approx 12.57$

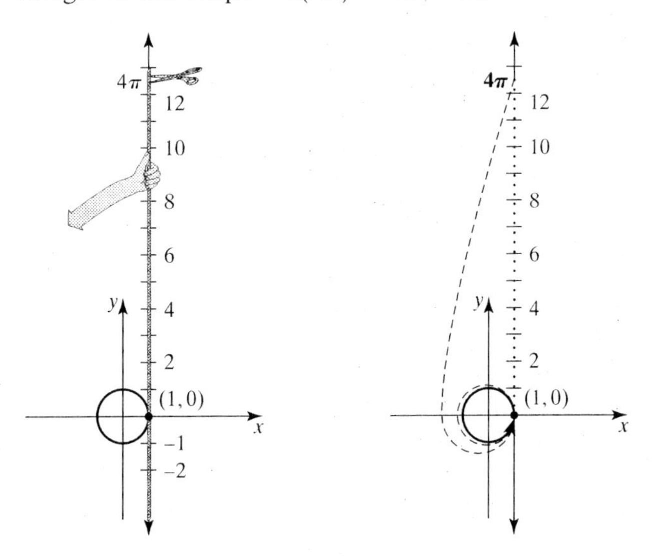

EXAMPLE 3 What would be the length of the string if it wrapped $\frac{1}{4}$ the unit circle? $\frac{1}{2}$ the circle?

SOLUTION Length of one wrap $= 2\pi$

$$\text{Length of } \frac{1}{4} \text{ wrap} = \frac{1}{4}(2\pi) = \frac{\pi}{2} \approx 1.57$$

$$\text{Length of } \frac{1}{2} \text{ wrap} = \frac{1}{2}(2\pi) = \pi \approx 3.14$$

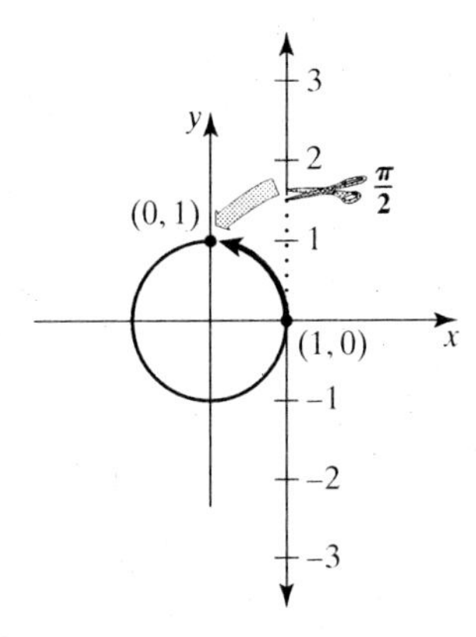

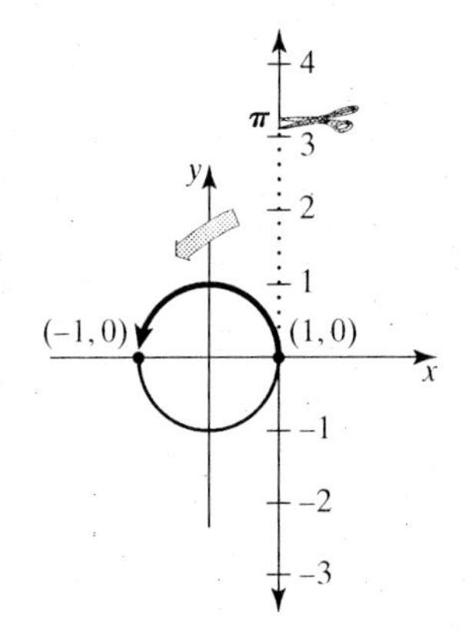

We could cut the string at a place corresponding to any real number, and wrap the unit circle to create an arc. Therefore, we have a correspondence between the real numbers on the number line and arcs that wrap the circle. We are interested in the length and direction of an arc represented by t and the quadrant that contains its terminal point. It is often convenient to use multiples of π to represent the arc lengths.

Special Arcs on the Unit Circle

The following diagrams demonstrate arcs with lengths that represent multiples of $\pi/6$ and $\pi/4$. We may also use an approximation of π (≈ 3.14) when the arcs are not in terms of π. *Notice we use an arrow at the terminal point of the arc to indicate a positive or negative direction.*

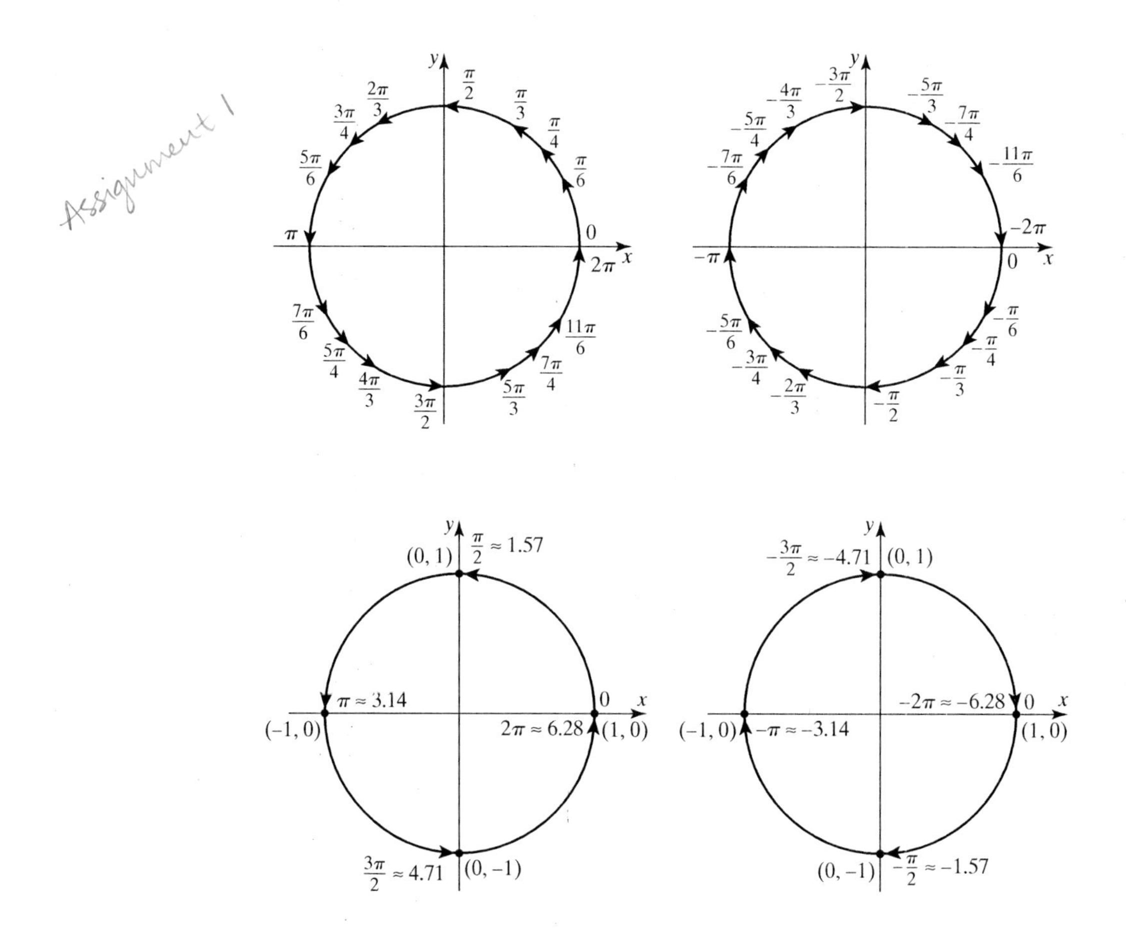

EXAMPLE 4 Name the quadrant that contains the terminal point of each arc on the unit circle.

a. $t = \dfrac{7\pi}{8}$ **b.** $t = \dfrac{11\pi}{6}$ **c.** $t = 4$ **d.** $t = -6$ **e.** $t = \dfrac{13\pi}{4}$

Solution

a. As a reference, we use the arcs 0, $\pi/2$, π, $3\pi/2$, or 2π. Look at the denominator of t and rewrite all the arcs in terms of this common denominator. We locate the terminal point in QII.

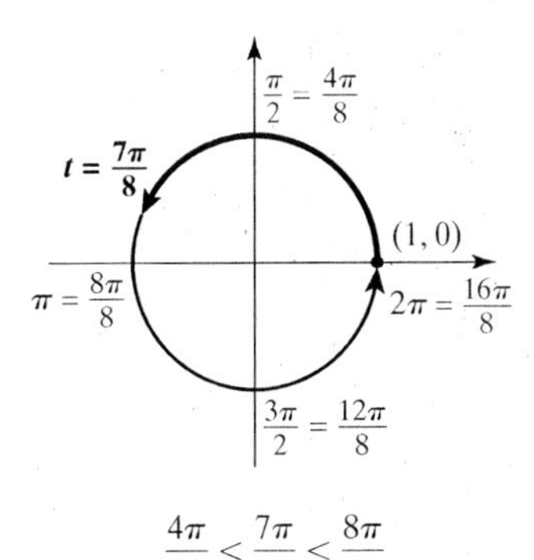

$$\frac{4\pi}{8} < \frac{7\pi}{8} < \frac{8\pi}{8}$$

b. Using the common denominator 6, we locate the terminal point in QIV.

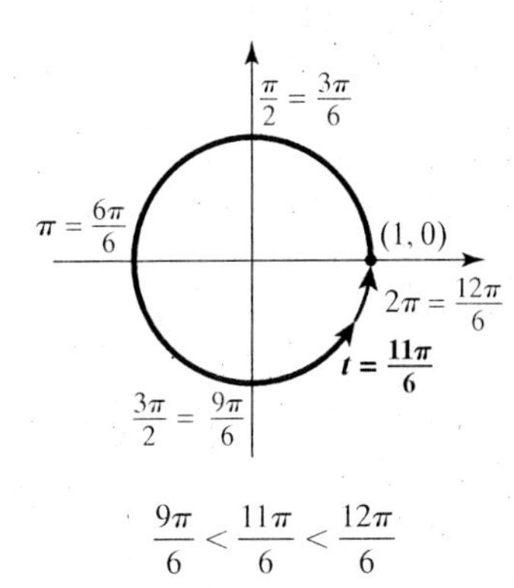

$$\frac{9\pi}{6} < \frac{11\pi}{6} < \frac{12\pi}{6}$$

c. Since t is not in terms of π, we use $\pi \approx 3.14$ to find that the terminal point of the arc is in QIII.

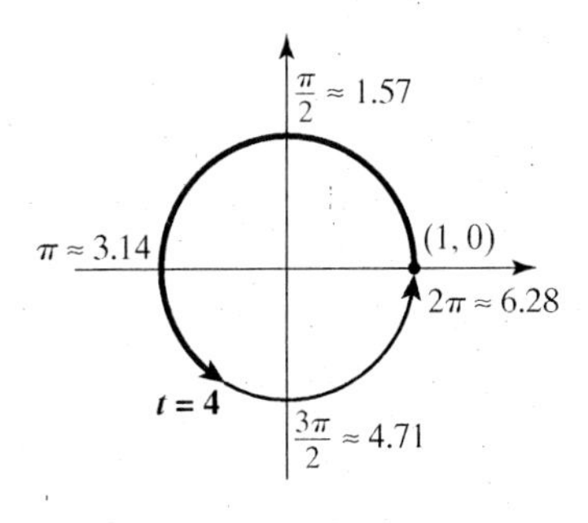

$$3.14 < 4 < 4.71$$

d. Again t is not in terms of π. Because t is negative, the wrap is clockwise. We find that the terminal point of the arc is in QI.

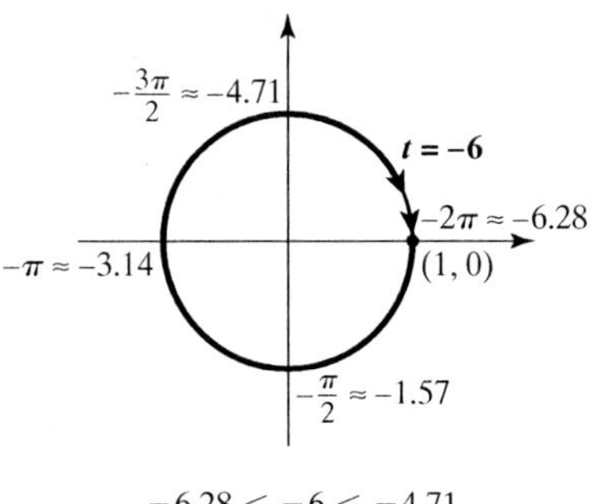

$$-6.28 < -6 < -4.71$$

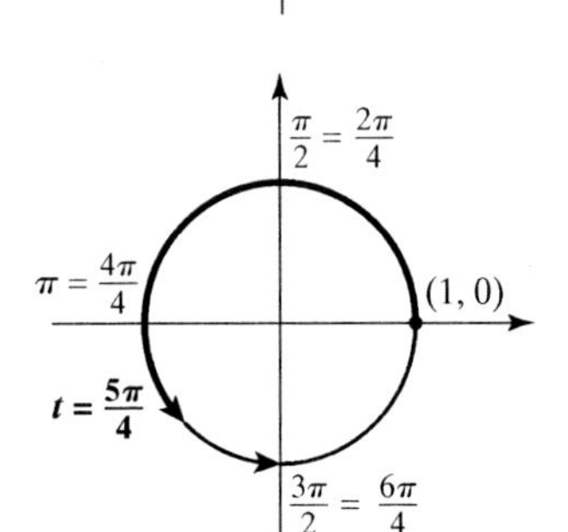

e. Here the arc is in terms of π but larger than one wrap of the circle since $2\pi = 8\pi/4$. (Refer back to Example 1.) To find the quadrant that contains the terminal point of this arc, we subtract multiples of 2π until we have a number between 0 and 2π:

$$\frac{13\pi}{4} - \frac{8\pi}{4} = \frac{5\pi}{4},$$

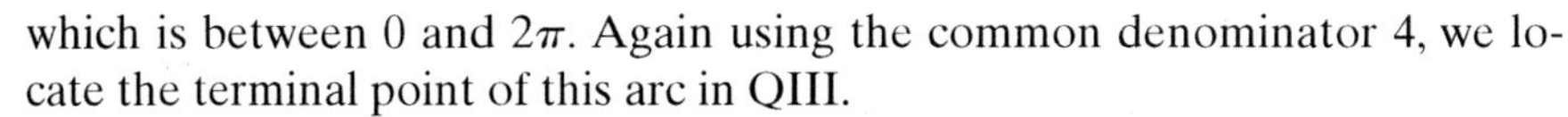

which is between 0 and 2π. Again using the common denominator 4, we locate the terminal point of this arc in QIII.

Notice that the arcs $13\pi/4$ and $5\pi/4$ have the same terminal point. ■

Arcs, such as $t = 13\pi/4$ and $5\pi/4$, that have the same initial point and terminal point are called **coterminal arcs**. Coterminal arcs are useful and will be discussed in Section 1.4.

Circular Functions

Now that we can find the location of the terminal point of an arc on the unit circle with initial point $(1, 0)$, we want to know about the coordinates of this terminal point. But instead of saying "on the unit circle the x-coordinate of the terminal point of an arc t with initial point $(1, 0)$," we simply say " cosine t." And instead of saying "on the unit circle the y-coordinate of the terminal point of an arc t with initial point $(1, 0)$," we simply say "sine t." We begin using this new language for these algebraic ideas in the following definition.

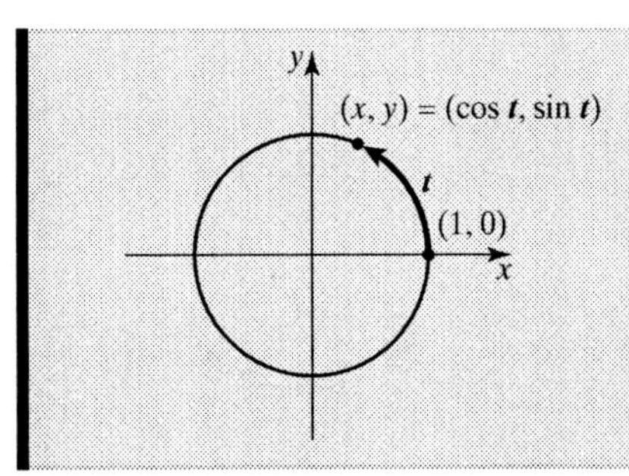

Cosine t and Sine t

On the unit circle $x^2 + y^2 = 1$ if t represents the length and direction of an arc with initial point $(1, 0)$ and terminal point (x, y), where t, x, and y are real numbers $(t, x, y \in \mathbb{R})$, then:

$$\text{cosine } t = x \quad \text{and} \quad \text{sine } t = y$$

or $\cos t = x$ and $\sin t = y$ common abbreviations

Since each arc t has only one terminal point, it has *only one* x-coordinate at that terminal point. Recall from algebra that a set of ordered pairs is a function if each first component is associated with exactly one second component. Pairing each arc t with the x-coordinate, which we now call $\cos t$, produces the set of ordered pairs called the **cosine function:** $\{(t, \cos t)\}$.

Since each arc t has only one terminal point, it has *only one* y-coordinate at that terminal point. Likewise, pairing each arc t with the y-coordinate, or $\sin t$, produces the set of ordered pairs called the **sine function:** $\{(t, \sin t)\}$.

These functions are called **circular functions** because of their correspondence to coordinates of points on the unit circle. Considering the usual functional notation, we can also write $\cos t$ as $\cos(t)$ and $\sin t$ as $\sin(t)$. Even though we do not always write the arc in parentheses they are always implied. As you will see, there are times when we must insert them around the arc.

In algebra we learn that the set of first components in a set of ordered pairs is called the *domain* and the set of second components is called the *range*. Recall that an arc t corresponds to any real number, or $t \in \mathbb{R}$. And since t is the first component in the ordered pair, the domain of both the cosine function and sine function is the real numbers, or $\mathbb{R}$.

With the definitions of the cosine and sine functions, we are now ready to begin the connections to our familiar questions of algebra.

EXAMPLE 5 What are the ranges for the cosine function and sine function? What is the largest and smallest value in the range of these functions? Keep in mind that $\cos t$ and $\sin t$ are the x- and y-coordinates of points on the unit circle, respectively.

ALGEBRA QUESTION: What is the range of numerical values for the x- and y-coordinates on the unit circle? What are the largest and smallest of these numerical values?

SOLUTION

$-1 \le x \le 1$, or $|x| \le 1$

and $-1 \le y \le 1$, or $|y| \le 1$.

The largest x-coordinate is 1, smallest -1.

The largest y-coordinate is 1, smallest -1.

ALGEBRA QUESTION IN TRIGONOMETRY: What are the range values for the cosine function and sine function? What is the largest and smallest value in the range of these functions?

SOLUTION

$-1 \le \cos t \le 1$, $|\cos t| \le 1$

and $-1 \le \sin t \le 1$, $|\sin t| \le 1$.

The largest $\cos t$ value is 1, smallest is -1.

The largest $\sin t$ value is 1, smallest is -1.

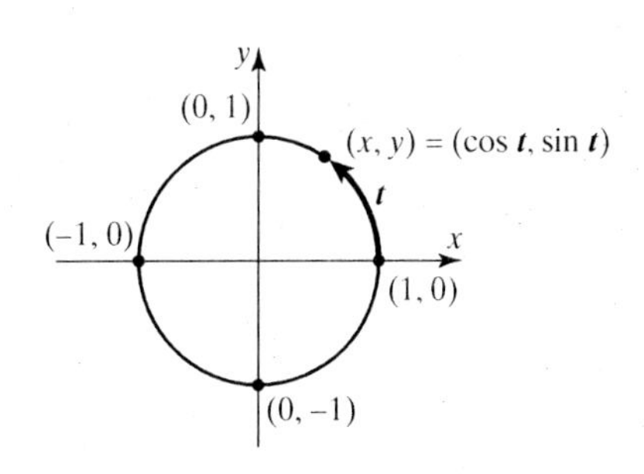

Let's summarize this important information.

Function	Domain	Range
$\{(t, \cos t)\}$	$t \in \mathbb{R}$	$-1 \le \cos t \le 1$
$\{(t, \sin t)\}$	$t \in \mathbb{R}$	$-1 \le \sin t \le 1$

Positive and Negative Values of cos *t* and sin *t*

Notice that the values of $\cos t$ and $\sin t$ (the x- and y-coordinates of the terminal point of an arc t on the unit circle) are either positive or negative depending on the quadrant that contains the terminal point of arc t.

EXAMPLE 6 Name the quadrant in which $\sin t < 0$ and $\cos t > 0$.

ALGEBRA QUESTION: Name the quadrant in which the y-coordinate is negative and the x-coordinate positive.

SOLUTION In QIV, we have

$$(x, y) = (+, -).$$

ALGEBRA QUESTION IN TRIGONOMETRY: Name the quadrant in which $\sin t < 0$ and $\cos t > 0$.

SOLUTION In QIV, we have

$$(\cos t, \sin t) = (+, -).$$

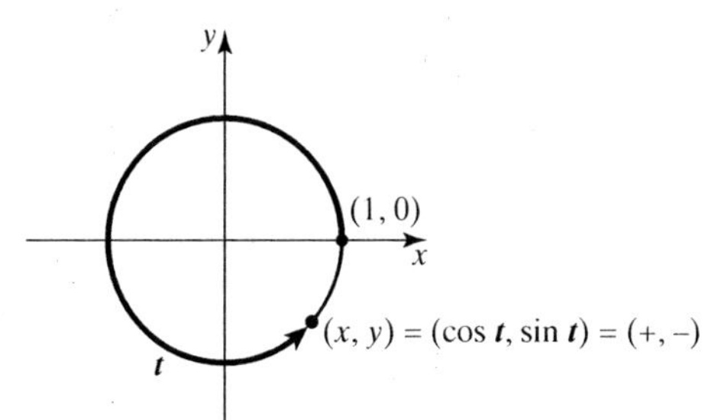

■

EXAMPLE 7 Complete the given table with + or − and fill in each blank with < or > to answer the algebra question using the new language.

ALGEBRA QUESTION: Are the coordinates of the points A, S, T, and C positive or negative? What quadrants are they in?

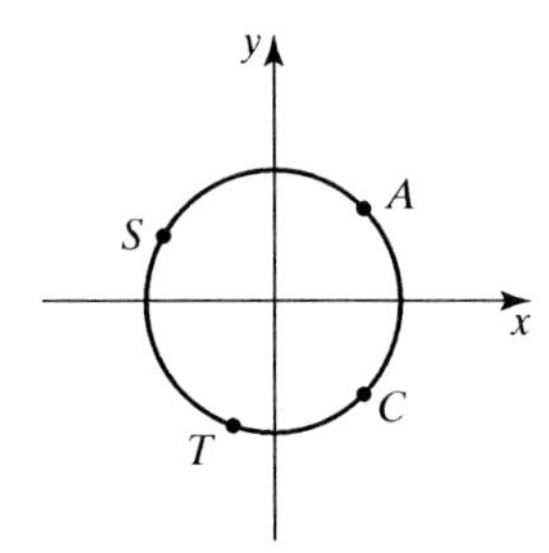

ALGEBRA QUESTION IN TRIGONOMETRY: Are the $\cos t$ and $\sin t$ values positive or negative when the terminal point of t is in the indicated quadrant?

Quadrant	$\cos t$	$\sin t$
I		
II		
III		
IV		

$\cos t$ ___ 0 $\cos t$ ___ 0

$\sin t$ ___ 0 $\sin t$ ___ 0

y

(1, 0)

x

$\cos t$ ___ 0 $\cos t$ ___ 0

$\sin t$ ___ 0 $\sin t$ ___ 0

SOLUTION

	Q	x	y
A	I	+	+
S	II	−	+
T	III	−	−
C	IV	+	−

SOLUTION

Quadrant	$\cos t$	$\sin t$
I	+	+
II	−	+
III	−	−
IV	+	−

$\cos t \le 0$, $\sin t \ge 0$ $\quad$ $\cos t \ge 0$, $\sin t \ge 0$

(1, 0)

$\cos t \le 0$, $\sin t \le 0$ $\quad$ $\cos t \ge 0$, $\sin t \le 0$

EXAMPLE 8 Determine whether $\cos 7\pi/8$ and $\sin 7\pi/8$ are positive or negative.

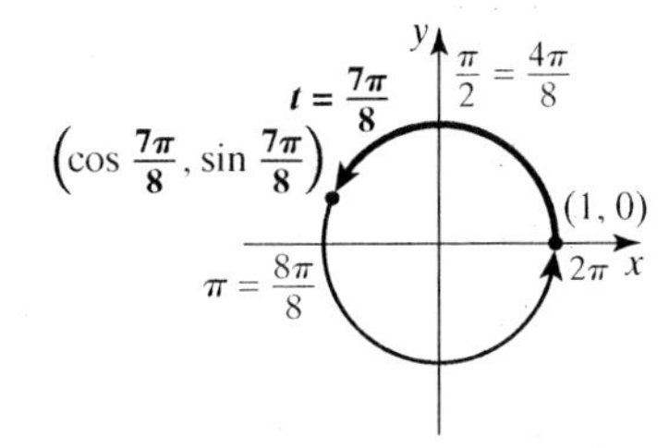

SOLUTION From Example 4(a), arc $t = 7\pi/8$ terminates in QII. Since $\cos 7\pi/8$ and $\sin 7\pi/8$ are the respective x- and y-coordinates at the terminal point of arc $t = 7\pi/8$, this is the familiar algebra question "Are the x- and y-coordinates in QII positive or negative?" Therefore, with the new language we say

$$\cos 7\pi/8 < 0 \text{ (negative) and } \sin 7\pi/8 > 0 \text{ (positive).}$$

Basic Identity

An identity is a statement that is true for all values for which it is defined. For example, $2x + 3x = 5x$ is an identity since it is true for all values of x. ($2x + 3x = 5$ is not an identity since it is true only for a specific value of x, $x = 1$. It is not true for all values of x.)

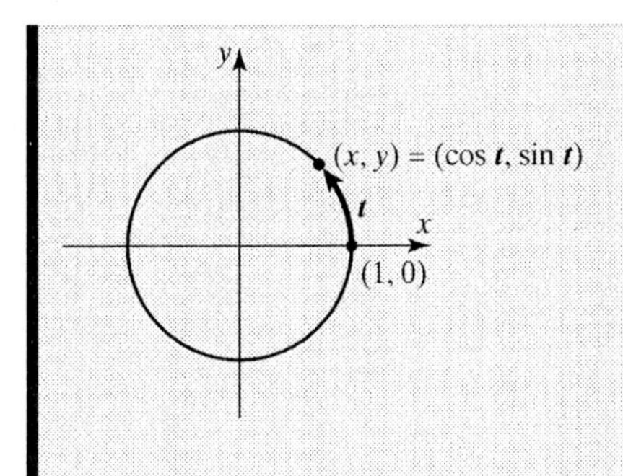

Pythagorean Identity

Using the definitions $x = \cos t$ and $y = \sin t$, along with the equation of the unit circle $x^2 + y^2 = 1$, we obtain one of the most important trigonometric identities. It is called the **Pythagorean identity**:

$$(\cos t)^2 + (\sin t)^2 = 1$$

or

$$\cos^2 t + \sin^2 t = 1,$$

for all real numbers t.

We now use the circular function definitions and the Pythagorean identity to answer another old question with the new language.

EXAMPLE 9 If $\cos t = \frac{1}{2}$ and $\sin t > 0$, find $\sin t$.

*A1

ALGEBRA QUESTION: If P is a point on the unit circle and you are given that its x-coordinate is $\frac{1}{2}$, find the y-coordinate of P if P is in QI.

SOLUTION Since P is a point on the circle, the coordinates of P satisfy the circle's equation $x^2 + y^2 = 1$.

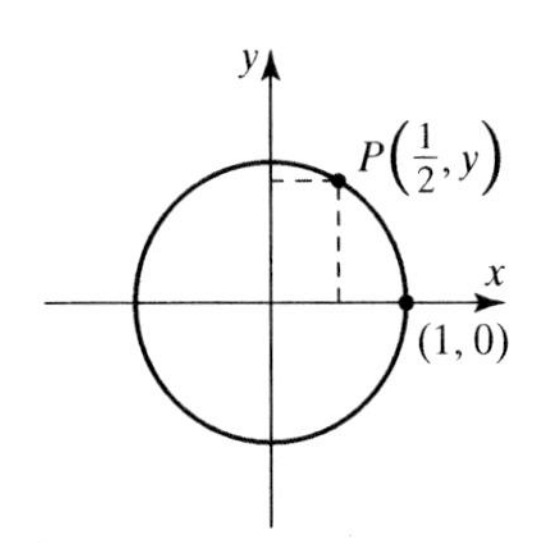

$$x^2 + y^2 = 1$$

$$\left(\frac{1}{2}\right)^2 + y^2 = 1$$

$$\frac{1}{4} + y^2 = 1$$

$$y^2 = \frac{3}{4}$$

$$y = \pm\sqrt{\frac{3}{4}} = \pm\frac{\sqrt{3}}{2}$$

Since P is a point in QI, the y-coordinate is *positive*, that is, $y > 0$. Therefore $y = \sqrt{3}/2$.

ALGEBRA QUESTION IN TRIGONOMETRY: If $\cos t = \frac{1}{2}$, and $\sin t > 0$, find $\sin t$.

SOLUTION If both $\cos t$ and $\sin t$ are positive, then t must be an arc that terminates in QI. Use the Pythagorean identity $\cos^2 t + \sin^2 t = 1$.

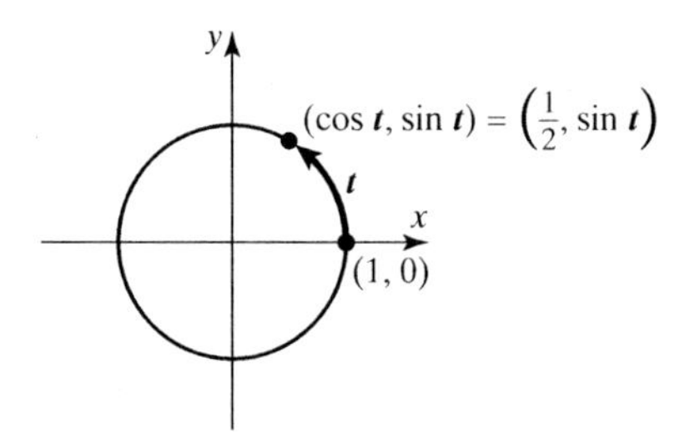

$$\cos^2 t + \sin^2 t = 1$$

$$\left(\frac{1}{2}\right)^2 + \sin^2 t = 1$$

$$\frac{1}{4} + \sin^2 t = 1$$

$$\sin^2 t = \frac{3}{4}$$

$$\sin t = \pm\sqrt{\frac{3}{4}} = \pm\frac{\sqrt{3}}{2}$$

Since t is an arc in QI, we know $\sin t > 0$. So, $\sin t = \sqrt{3}/2$. ■

Example 9 should demonstrate to you that if you can locate the terminal point of the arc t and you know the value of either $\cos t$ or $\sin t$, then you can find the other one.

Let's practice finding the quadrant that contains the terminal point of an arc t, determining if the functional values are positive or negative and finding a missing functional value at the terminal point of the arc. Continue to translate the familiar ideas to this new language. At first you will say "$\cos t$ is the x-coordinate at the terminal point of an arc t" and "$\sin t$ is the y-coordinate at the terminal point of an arc t." This is all a part of learning any new language. Are you ready?

NOTE: Since the new sine function sounds the same as the word "sign," we will refrain from using the word "sign" when asking about the positive or negative nature of coordinates. This will help avoid confusion when you are speaking about these two ideas in trigonometry.

Exercise Set 1.2

1–6. *Select the appropriate item from Column A to make each statement a true statement. Some items may be used more than once or not at all.*

1. A portion of a real **(a)**______ was wrapped around the **(b)**______ circle to create an arc whose length and **(c)**______ can be represented by any real number t. Arcs, with initial point **(d)**______, that are wrapped **(e)**______ are considered to be positive, and those that are wrapped clockwise are considered to be **(f)**______.
2. The functional value cosine t is defined to be the **(a)**______ at the **(b)**______ point of an arc t whose initial point is **(c)**______ on the unit circle. The largest value for $\cos t$ is **(d)**______ and the smallest value for $\cos t$ is **(e)**______.
3. The functional value sine t is defined to be the **(a)**______ at the terminal point of an arc **(b)**______ whose initial point is $(1, 0)$ on the unit circle. The largest value for $\sin t$ is **(c)**______ and the smallest value for $\sin t$ is **(d)**______.
4. To use the new language for the familiar question, "On the unit circle, name the quadrant in which the x-coordinate is negative and the y-coordinate is positive," we would ask, "Where is **(a)**______ < 0 and **(b)**______ > 0?"
5. The Pythagorean identity is (**(a)**______$)^2 + (\sin t)^2 = 1$, or $\cos^2 t +$ **(b)**______ $= 1$.
6. The domain of both the cosine function and sine function is ______.

Column A

negative
number line
x-coordinate
y-coordinate
$(1, 0)$
unit
1
-1
direction
$\sin^2 t$
counterclockwise
clockwise
positive
$\cos t$
$\sin t$
terminal
t
$\mathbb{R}$

7–8. *Select the appropriate arc from Column B that describes t— the direction and length of the arc on the unit circle.*

7.

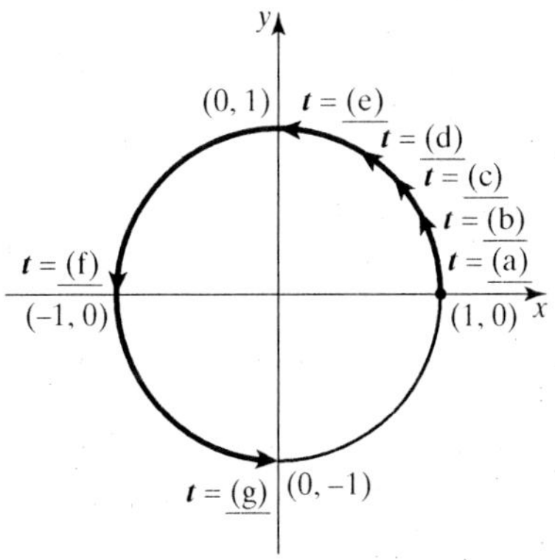

y
$(0, 1)$ $t = \underline{(e)}$
$t = \underline{(d)}$
$t = \underline{(c)}$
$t = \underline{(b)}$
$t = \underline{(a)}$
$t = \underline{(f)}$
$(-1, 0)$
$(1, 0)$ x
$t = \underline{(g)}$ $(0, -1)$

8.

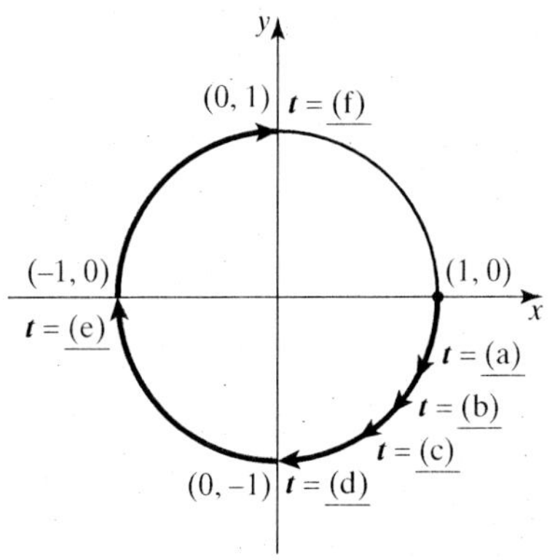

y
$(0, 1)$ $t = \underline{(f)}$
$(-1, 0)$
$t = \underline{(e)}$
$(1, 0)$
x
$t = \underline{(a)}$
$t = \underline{(b)}$
$t = \underline{(c)}$
$(0, -1)$ $t = \underline{(d)}$

Column B

$\frac{\pi}{3}$	$-\frac{\pi}{6}$
π	$-\frac{\pi}{4}$
$\frac{\pi}{2}$	$-\frac{3\pi}{2}$
$-\frac{\pi}{3}$	$\frac{\pi}{4}$
$-\pi$	$\frac{3\pi}{2}$
0	$-\frac{\pi}{2}$
$\frac{\pi}{6}$	

9–30. *Name the quadrant that contains the terminal point of each given arc with initial point $(1, 0)$ on the unit circle.*

EXAMPLE **a.** $t = \dfrac{2\pi}{3}$ **b.** $t = 1$

SOLUTION

a. Using the common denominator 6, $t = 2\pi/3 = 4\pi/6$ is in QII because $3\pi/6 < 4\pi/6 < 6\pi/6$.

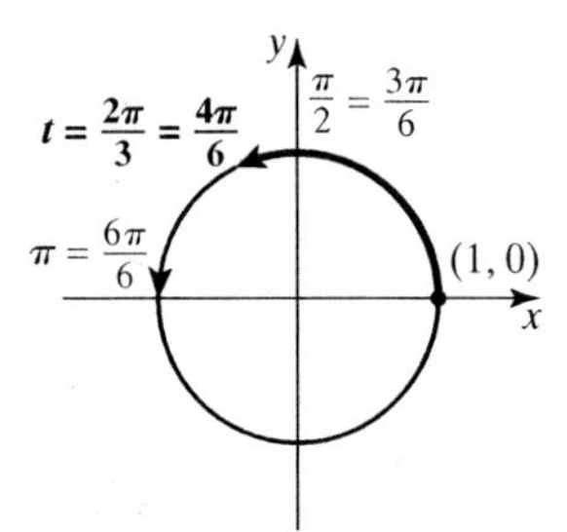

b. Using $\pi/2 \approx 1.57$, the terminal point lies in QI because $0 < 1 < 1.57$.

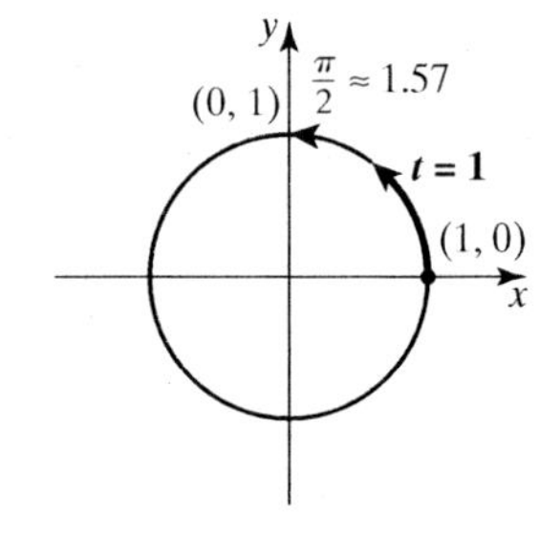

■

9. $\dfrac{7\pi}{6}$ **10.** $\dfrac{5\pi}{3}$ **11.** $\dfrac{4\pi}{3}$ **12.** $\dfrac{3\pi}{4}$ **13.** $\dfrac{7\pi}{4}$

14. $\dfrac{5\pi}{6}$ **15.** $\dfrac{17\pi}{6}$ **16.** $\dfrac{8\pi}{3}$ **17.** $-\dfrac{5\pi}{3}$ **18.** $-\dfrac{2\pi}{3}$

19. $-\dfrac{5\pi}{4}$ **20.** $-\dfrac{9\pi}{4}$ **21.** $-\dfrac{15\pi}{4}$ **22.** $-\dfrac{23\pi}{6}$ **23.** 2

24. 5 **25.** 7.1 **26.** 8.3 **27.** -1.82 **28.** -2.9

29. -6.5 **30.** -12.28

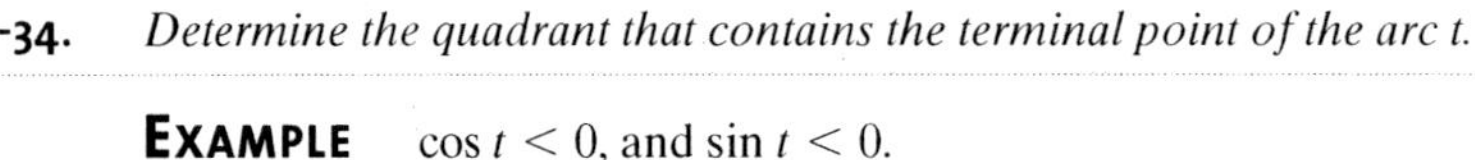

31–34. *Determine the quadrant that contains the terminal point of the arc t.*

y
(1, 0)
x
t
cos $t < 0$
sin $t < 0$
(−, −)

EXAMPLE $\cos t < 0$, and $\sin t < 0$.

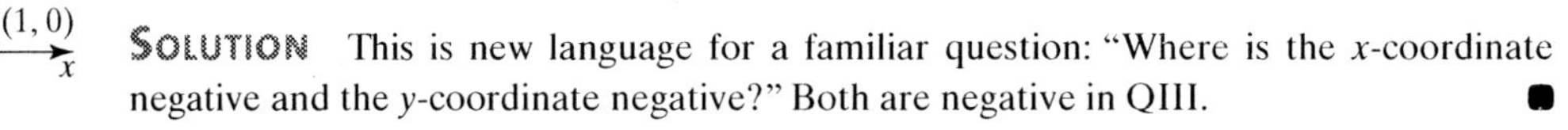

SOLUTION This is new language for a familiar question: "Where is the x-coordinate negative and the y-coordinate negative?" Both are negative in QIII. ■

31. $\cos t > 0$ and $\sin t > 0$ **32.** $\sin t > 0$ and $\cos t < 0$

33. $\sin t < 0$ and $\cos t > 0$ **34.** $\sin t < 0$ and $\cos t < 0$

35–44. *Find the indicated functional value(s).*

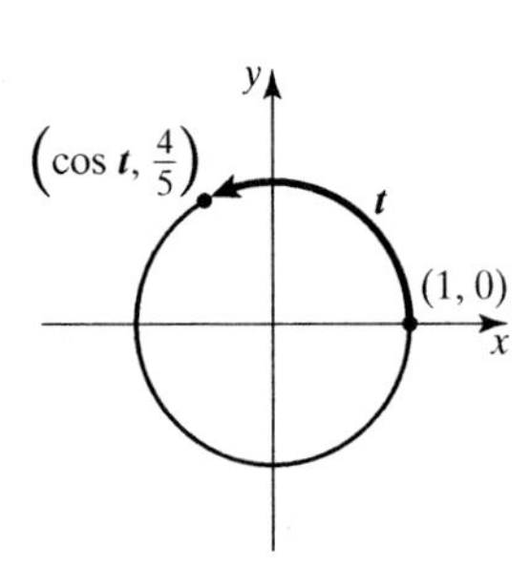

EXAMPLE If $\sin t = \dfrac{4}{5}$ and $\cos t < 0$, find $\cos t$.

SOLUTION Since $\cos t < 0$ and $\sin t > 0$, t is in QII. Use the Pythagorean identity.

$$\cos^2 t + \sin^2 t = 1$$

$$\cos^2 t + \left(\frac{4}{5}\right)^2 = 1$$

$$\cos^2 t = 1 - \frac{16}{25} = \frac{9}{25}$$

$$\cos t = \pm\sqrt{\frac{9}{25}} = \pm\frac{3}{5}$$

Since $\cos t < 0$, $\cos t = -3/5$. ■

35. If $\sin t = \frac{5}{13}$ and $\cos t > 0$, find $\cos t$.

36. If $\cos t = -\frac{4}{5}$ and $\sin t > 0$, find $\sin t$.

37. If $\cos t = -\frac{\sqrt{2}}{2}$ and $\sin t < 0$, find $\sin t$.

38. If $\sin t = -\frac{1}{2}$ and t is in QIII, find $\cos t$.

39. If $\cos t = -\frac{8}{17}$ and $\frac{\pi}{2} < t < \pi$, find each of the following.

a. The quadrant that contains the terminal point of t

b. $\sin t$ **c.** $\frac{1}{\cos t}$ **d.** $\frac{1}{\sin t}$ **e.** $\frac{\sin t}{\cos t}$ **f.** $\frac{\cos t}{\sin t}$

40. If $\cos t = \frac{12}{13}$ and $0 < t < \frac{\pi}{2}$, find each of the following.

a. The quadrant that contains the terminal point of t

b. $\sin t$ **c.** $\frac{1}{\cos t}$ **d.** $\frac{1}{\sin t}$ **e.** $\frac{\sin t}{\cos t}$ **f.** $\frac{\cos t}{\sin t}$

41. If $\sin t = -\frac{\sqrt{3}}{2}$ and $\cos t > 0$, find each of the following.

a. The quadrant that contains the terminal point of t

b. $\cos t$ **c.** $\frac{1}{\cos t}$ **d.** $\frac{1}{\sin t}$ **e.** $\frac{\sin t}{\cos t}$ **f.** $\frac{\cos t}{\sin t}$

42. If $\sin t = -\frac{8}{9}$ and $\cos t > 0$, find each of the following.

a. The quadrant that contains the terminal point of t

b. $\cos t$ **c.** $\frac{1}{\cos t}$ **d.** $\frac{1}{\sin t}$ **e.** $\frac{\sin t}{\cos t}$ **f.** $\frac{\cos t}{\sin t}$

43. If $\cos t = x$ and $\sin t < 0$, find $\sin t$.

44. If $\sin t = y$ and $\cos t < 0$, find $\cos t$.

Graphing Calculator/CAS Exploration

45. The definitions for the set of points (x, y) on the unit circle are defined as functions of t: $x = \cos(t) = f(t)$, and $y = \sin(t) = g(t)$. In Chapter 6 we will discuss that these points are said to be defined parametrically and the equations $x = \cos(t)$ and $y = \sin(t)$ are called *parametric equations.* Use a graphing utility with the following settings:

MODE	radian
	parametric (par, para, param)
WINDOW	Tmin $= 0$ Tmax $= 2\pi$
	T step $= \frac{\pi}{24}$
	Xmin $= -3$ Xmax $= 3$
	Ymin $= -2$ Ymax $= 2$

Input $x_1 = \cos(t)$ and $y_1 = \sin(t)$, and graph these parametric equations.

a. Describe the graph. Is the graph what you would expect?

b. Trace the curve and describe what's happening.

c. Why are you only able to trace the curve once?

d. What do you think will happen if you change Tmax to 4π (≈ 12.56)? Test your thinking by again tracing the curve with Tmax as 4π.

(continued)

e. Trace the curve again and stop at different points on the unit circle. Notice that each time you stop, the screen displays values. A graphing calculator will display three values below the graph, as indicated in the accompanying screen. In the context of the circular functions, what does each value represent?

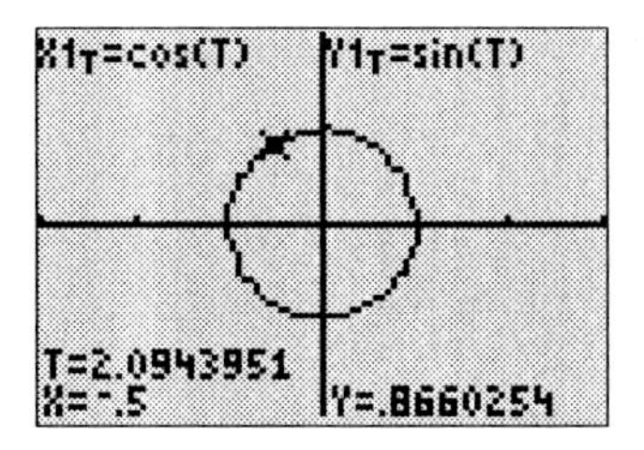

Discussion

46. Bill finished his homework and didn't check all his answers. For one problem, where he was asked to find the functional value sin t, his answer is

$$\sin t = 8.$$

Without even knowing which question he is answering, how can you tell his answer is wrong?

47. If you have the following multiple choice question on a test, discuss how you could immediately determine three incorrect answers.

If $\sin t = -\frac{8}{17}$ and $\cos t < 0$, find $\cos t$.

a. $-\frac{17}{8}$ **b.** $-\frac{15}{17}$ **c.** -1.5 **d.** -2

Web Activity

48. A Web search of Animation of the Circular Functions produces several sites that demonstrate the definitions of the cosine and sine circular functions and that show terminal points of the arcs as they move along the unit circle.

a. Visit at least two of these sites and describe the similarities or differences in their demonstrations. You will notice that most of the sites provide you with additional trigonometric material with which you may be unfamiliar.

b. Make a list of the Web sites. You may want to visit them again as you progress through trigonometry to see if they assist you in reviewing important information.

1.3 Functional Values for Common Arcs

In Section 1.2 we were given the location of the terminal point of an unknown arc t on the unit circle and either $\cos t$ (the x-coordinate at the terminal point of the arc t) or $\sin t$ (the y-coordinate). Then we were asked to find the other coordinate. Now we do something different—we start with an arc on the unit circle in which t is known, and then find *both* $\cos t$ and $\sin t$.

Circular Functional Values for $t = 0, \frac{\pi}{2}, \pi, \frac{3\pi}{2}$

The cosine and sine functional values can be found when the given arc is one of a group of arcs considered to be *special arcs*. For example, any multiple of $\pi/2$ is a special arc. We look at these arcs first.

EXAMPLE 1 Find $\cos t$ and $\sin t$ for each given arc.

a. $t = 0$ **b.** $t = \dfrac{\pi}{2}$ **c.** $t = \pi$ **d.** $t = \dfrac{3\pi}{2}$

SOLUTION

a. Since $t = 0$, the terminal point of the arc is the same as the initial point $(1, 0)$. Thus, $\cos 0 = 1$ and $\sin 0 = 0$.

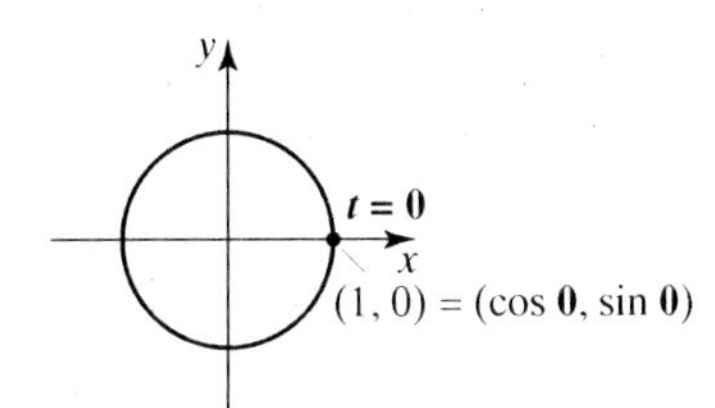

b. $\cos \dfrac{\pi}{2} = 0$ and $\sin \dfrac{\pi}{2} = 1$.

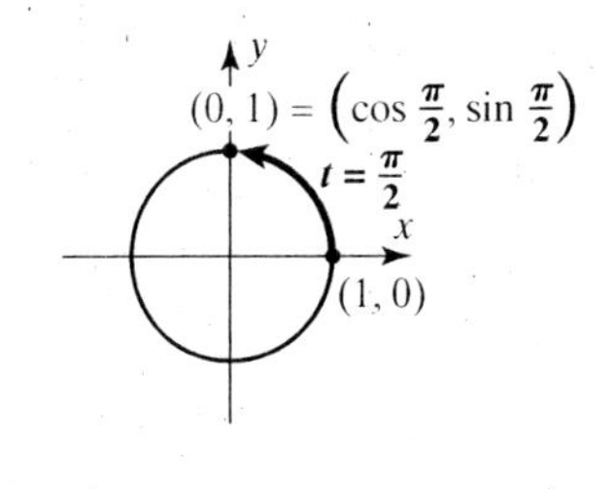

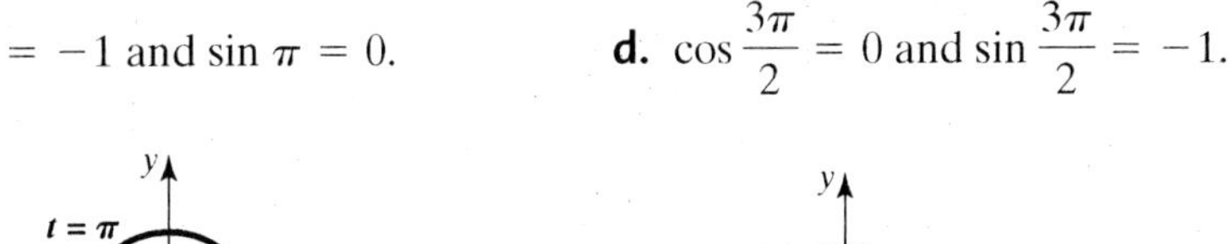

c. $\cos \pi = -1$ and $\sin \pi = 0$.

d. $\cos \dfrac{3\pi}{2} = 0$ and $\sin \dfrac{3\pi}{2} = -1$.

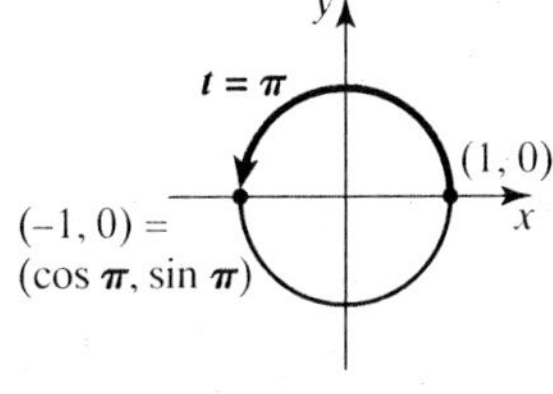

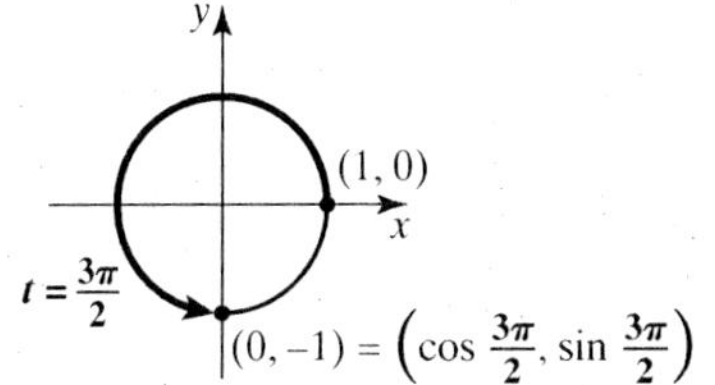

Since these arcs terminate at either an x-intercept or y-intercept of the unit circle, the cosine and sine values are readily known. The next arcs we discuss all terminate in QI and are special multiples of $\pi/6$ and $\pi/4$. Finding the functional values of these arcs requires a little more work. We need to use some geometry and algebra.

Circular Functional Values for $t = \dfrac{\pi}{4}$

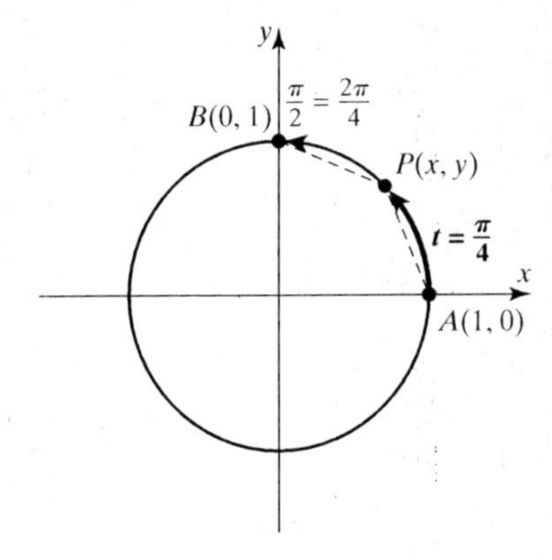

If $t = \pi/4$, we know that the terminal point bisects the arc whose length is $\pi/2$. In a circle if the arcs are equal, the distances between the endpoints of each arc are equal. As a result, we know the distance from P to A is the same as the distance from P to B, or $PA = PB$. Using the distance formula, we get:

$$PA = PB$$

$$\sqrt{(x-1)^2 + (y-0)^2} = \sqrt{(x-0)^2 + (y-1)^2}$$

$$(x-1)^2 + y^2 = x^2 + (y-1)^2 \qquad \text{Square both sides.}$$

$$x^2 - 2x + 1 + y^2 = x^2 + y^2 - 2y + 1 \qquad \text{Recall } (a+b)^2 = a^2 + 2ab + b^2.$$

$$-2x = -2y \qquad \text{Simplify by subtracting } x^2, y^2, \text{ and 1 from both sides.}$$

$$x = y \qquad \text{Divide both sides by } -2.$$

This tells us the two coordinates are equal, but *we still do not know what they are.* Since $x = y$, let's substitute x for y in the equation of the unit circle:

$$x^2 + y^2 = 1$$
$$x^2 + (x)^2 = 1$$
$$2x^2 = 1$$
$$x^2 = \frac{1}{2}$$
$$x = \pm\sqrt{\frac{1}{2}} = \pm\frac{1}{\sqrt{2}}, \quad \text{or} \quad \pm\frac{\sqrt{2}}{2}$$

Point P is in QI $(x, y > 0)$, therefore $x = 1/\sqrt{2}$. Since $x = y$, we have $y = 1/\sqrt{2}$.

Now we know both coordinates at the terminal point of arc $t = \pi/4$. And using the definitions of the circular functions, we get for $t = \pi/4$:

$$\boxed{\cos\frac{\pi}{4} = \frac{1}{\sqrt{2}} \quad \text{and} \quad \sin\frac{\pi}{4} = \frac{1}{\sqrt{2}}}$$

NOTE: For this text, we can use either $1/\sqrt{2}$ or $\sqrt{2}/2$. However, we may find it easier to use fractions in which the denominator contains the radical. Answers for exercises should be in proper radical form (see Appendix A).

Circular Functional Values for $t = \dfrac{\pi}{6}$

For $t = \pi/6$ we can find another pair of arcs that are equal and again use the distance formula.

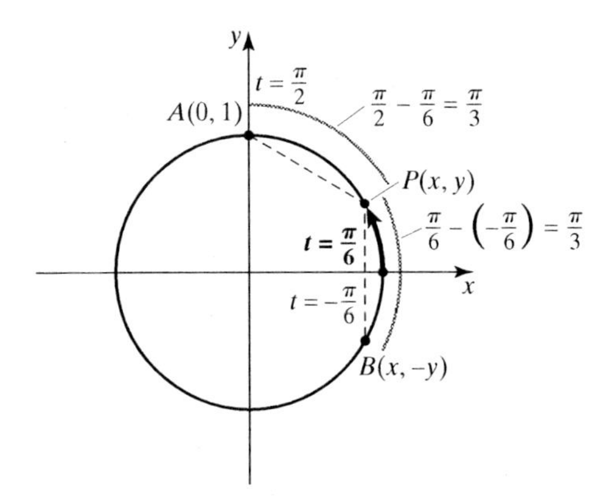

Let $P = (x, y)$ be the terminal point of an arc on the unit circle. Using the symmetry of the circle, we notice that if $P = (x, y)$ is at the terminal point of $t = \pi/6$, then $B = (x, -y)$ is the terminal point of $t = -\pi/6$. The arc length from P to B is

$$\frac{\pi}{6} + \frac{\pi}{6} = \frac{\pi}{3}.$$

The arc length from P to A is

$$\frac{\pi}{2} - \frac{\pi}{6} = \frac{\pi}{3}.$$

Therefore, $PA = PB$.

$$
\begin{aligned}
\sqrt{(x-0)^2 + (y-1)^2} &= \sqrt{(x-x)^2 + (y-(-y))^2} \\
x^2 + (y-1)^2 &= (2y)^2 && \text{Square both sides.} \\
x^2 + y^2 - 2y + 1 &= 4y^2 && \text{Multiply.} \\
1 - 2y + 1 &= 4y^2 && \text{Substitute } x^2 + y^2 = 1. \\
0 &= 4y^2 + 2y - 2 && \text{Write the quadratic equation in standard form.} \\
0 &= 2(2y-1)(y+1) && \text{Factor.} \\
y = \frac{1}{2} \quad &\text{or} \quad y = -1 && \text{Set each factor equal to 0 and solve for } y.
\end{aligned}
$$

Because point P is in QI, $y > 0$, and so $y = 1/2$.

This means we only know one of the coordinates. Can we find the other? (Recognize this question?) Using the unit circle equation, we get:

$$
\begin{aligned}
x^2 + y^2 &= 1 \\
x^2 + \left(\frac{1}{2}\right)^2 &= 1 \\
x^2 &= \frac{3}{4} \\
x &= \pm\frac{\sqrt{3}}{2}
\end{aligned}
$$

Since P is a point in QI, $x = \sqrt{3}/2$.

Now we know both coordinates at the terminal point of this arc. And using the definitions of the circular functions, we get for $t = \pi/6$:

$$\boxed{\cos\frac{\pi}{6} = \frac{\sqrt{3}}{2} \quad \text{and} \quad \sin\frac{\pi}{6} = \frac{1}{2}}$$

Circular Functional Values for $t = \frac{\pi}{3}$

Since we know the functional values for $t = \pi/6$, we can use the same procedure to arrive at the functional values for $t = \pi/3$:

$$\boxed{\cos\frac{\pi}{3} = \frac{1}{2} \quad \text{and} \quad \sin\frac{\pi}{3} = \frac{\sqrt{3}}{2}}$$

We leave the algebraic calculations for you to do in Exercise 43.

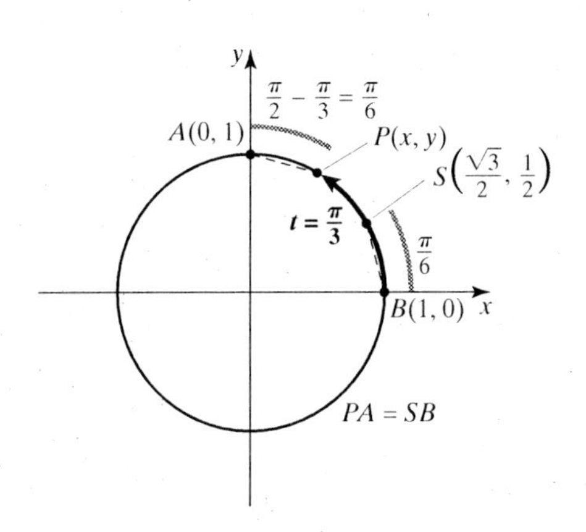

Let's summarize the special arc values $t \in \{0, \pi/6, \pi/4, \pi/3, \pi/2, \pi, 3\pi/2, 2\pi\}$ graphically and in a table. These special arcs are usually called **common arcs**.

Common Arcs and Their Functional Values

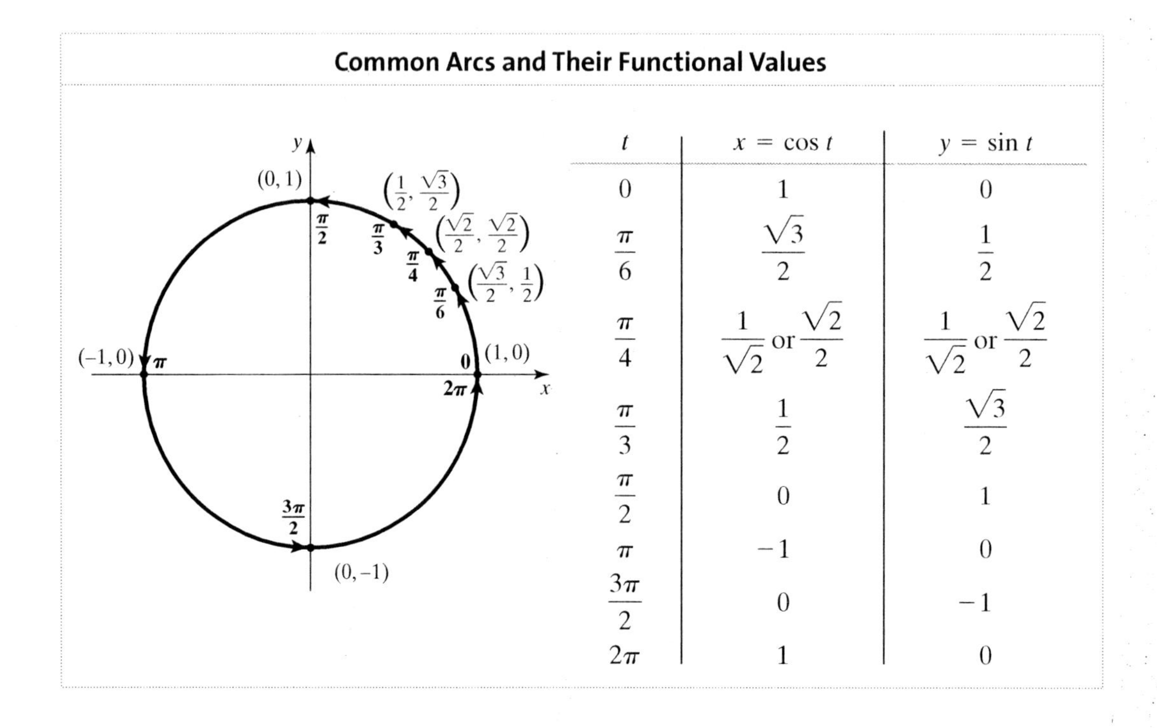

t	$x = \cos t$	$y = \sin t$
0	1	0
$\frac{\pi}{6}$	$\frac{\sqrt{3}}{2}$	$\frac{1}{2}$
$\frac{\pi}{4}$	$\frac{1}{\sqrt{2}}$ or $\frac{\sqrt{2}}{2}$	$\frac{1}{\sqrt{2}}$ or $\frac{\sqrt{2}}{2}$
$\frac{\pi}{3}$	$\frac{1}{2}$	$\frac{\sqrt{3}}{2}$
$\frac{\pi}{2}$	0	1
π	-1	0
$\frac{3\pi}{2}$	0	-1
2π	1	0

This summary should be memorized. When you learn any language, there is a great deal of memorization required. Trigonometry is no exception and it would be in your very best interest to memorize these values now!

Notice that $\sqrt{3}/2$ and $1/\sqrt{2}$ involve radicals and are irrational numbers. These numbers are *exact functional values,* and we refer to them as **exact values**. We will use exact values rather than decimal approximations until we have a need to approximate them, in which case, we will use a calculator.

```
cos(π/6)
         .8660254038   ← ≈ √3/2
sin(π/4)
         .7071067812   ← ≈ 1/√2
sin(π/3)
         .8660254038   ← ≈ √3/2
```

Noncommon Arcs

```
cos(π/11)
         .9594929736
sin(π/9)
         .3420201433
```

Finding exact functional values for many other first quadrant arcs, such as $\pi/11$ or $\pi/9$, would range from difficult to impossible with the same techniques we used for the common arcs. The functional values for these and other arcs, which are mostly irrational values, are found using advanced methods of calculus. The calculator will approximate these functional values for any domain value, $t \in \mathbb{R}$. Calculators may vary in the number of digits displayed. However, unless otherwise indicated, we will round answers for the exercises to *four decimal places.*

Using a Calculator to Find Functional Values

We use a calculator when we cannot find the exact functional values of arcs or when we need to approximate known exact irrational values. The circular function values we have defined can be found by setting any calculator to **radian mode** and using the [COS] or [SIN] key. We will define the term *radian* in Chapter 3.

CAUTION: It is common when writing circular functions to leave parentheses off the arc when it involves multiplication or division by a positive real number. For example, $\cos(t) = \cos t$, $\sin(2x) = \sin 2x$ or $\cos(x/2) = \cos x/2$. Parentheses should not be omitted when the arc involves other operations. For example, parentheses are not omitted when writing $\cos(2 + x)$, $\sin(-3x^2)$, but notice the *need for parentheses around the arc and any operations involving the arc* when using most technology to find functional values. On the calculator we also use parentheses around the functional value if it is to be raised to any power.

EXAMPLE 2 Use a calculator to find the approximate functional value rounded to four decimal places.

a. $\cos \dfrac{\pi}{8}$ **b.** $\sin 1.5 + \cos 3.6$ **c.** $\sin(1 + \pi)$

d. $\cos^2 2.34$ **e.** $\dfrac{1}{\cos \frac{1}{2}}$

SOLUTION Use a graphing calculator in radian mode.

MODE

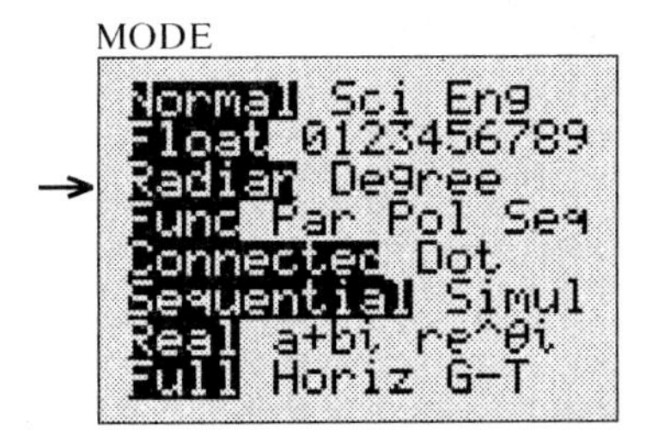

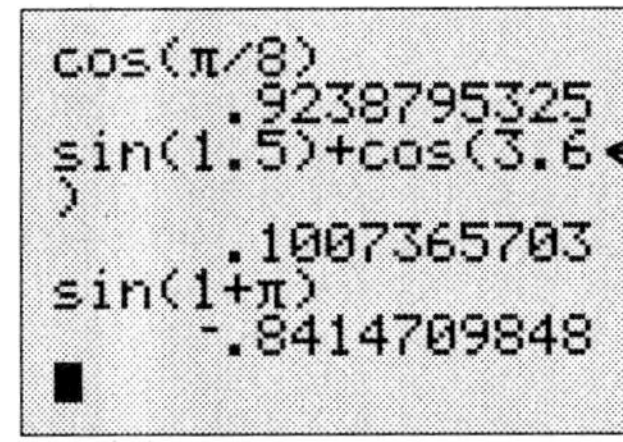

Be careful to use appropriate parentheses (both right and left) when finding functional values on the calculator.

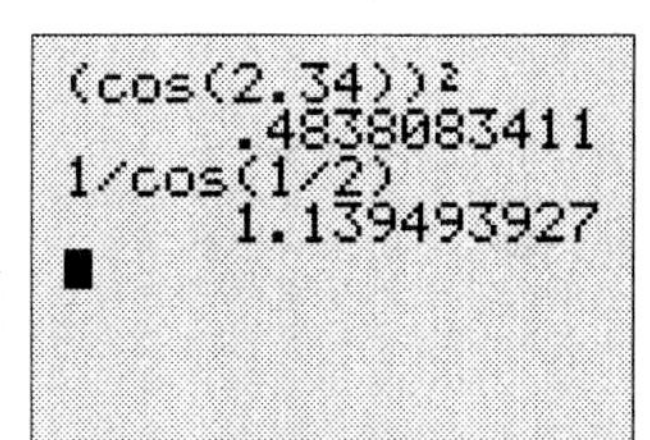

We obtain the following functional values:

a. $\cos \dfrac{\pi}{8} \approx 0.9239$ **b.** $\sin 1.5 + \cos 3.6 \approx 0.1007$

c. $\sin(1 + \pi) \approx -0.8415$ **d.** $\cos^2 2.34 \approx 0.4838$ **e.** $\dfrac{1}{\cos \frac{1}{2}} \approx 1.1395$ ■

EXAMPLE 3 Find the *exact value* for the expression

$$\cos^2 \frac{\pi}{3} + \sin\left(\frac{\pi}{2} - \frac{\pi}{4}\right).$$

Then, use a calculator to check your answer.

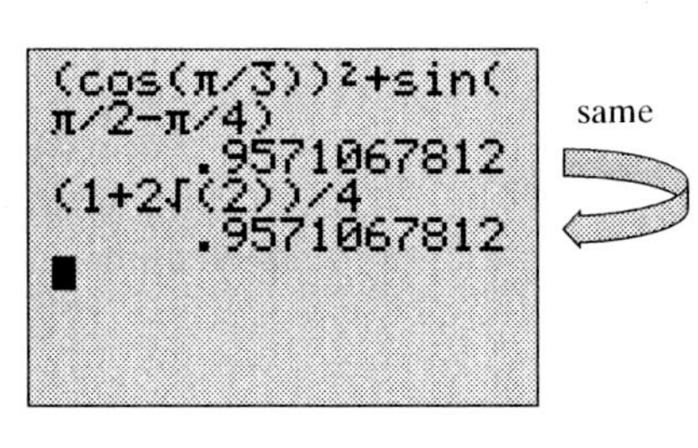

SOLUTION $$\cos^2\frac{\pi}{3}+\sin\left(\frac{\pi}{2}-\frac{\pi}{4}\right)=\left(\cos\frac{\pi}{3}\right)^2+\sin\frac{\pi}{4}$$
$$=\left(\frac{1}{2}\right)^2+\left(\frac{\sqrt{2}}{2}\right)$$
$$=\frac{1}{4}+\frac{\sqrt{2}}{2}\cdot\frac{2}{2}$$
$$=\frac{1+2\sqrt{2}}{4}$$

NOTE: The left bracket key [and the right bracket key] are not used on the calculator. Instead, a second set of parentheses are used because the calculator brackets can indicate a different mathematical concept.

Choice of Variable

We have been using the variable t to represent the length and direction of an arc, which is the domain of the circular functions. We could essentially use any symbol or variable without disturbing the definitions. For example, instead of saying $\cos t$, we could say $\cos s$ or $\cos x$ where s and x, like t, represent the length and direction of the arc on the unit circle. And $\cos s$ or $\cos x$ still represent the x-coordinate at the terminal point of the particular arc. Likewise, $\sin t$, $\sin s$, and $\sin x$ each represent the y-coordinate at the terminal point of the arc. Using x in this way may be a little confusing at first since we are using the variable in two different ways. Don't be confused—just go back to the ideas in the original definition of the circular functions.

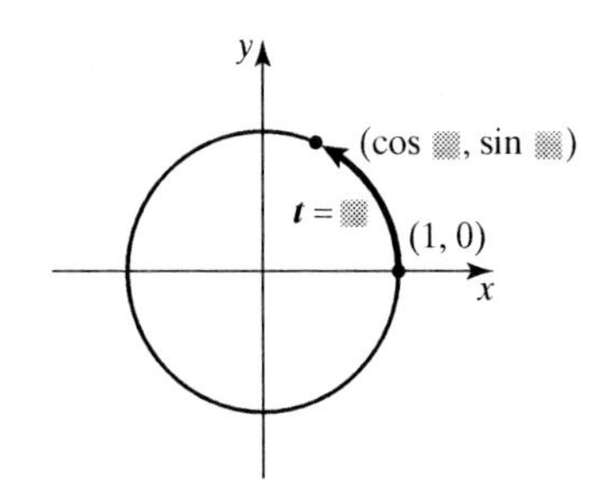

For each arc $\square \in \mathbb{R}$ on the unit circle, the

cosine function: $\{(\square, \cos\square)\}$, where $\cos\square$ = x-coordinate at the terminal point of arc $\square$,

and

sine function: $\{(\square, \sin\square)\}$, where $\sin\square$ = y-coordinate at the terminal point of arc $\square$,

are such that $\square$ can be replaced with any variable or expression, as long as you use the same variable or expression in each ordered pair. For example, $(m, \cos m)$ is valid, but $(x, \cos t)$ is not.

We could also use any same variable or expression for the arc $\square$ in the Pythagorean identity.

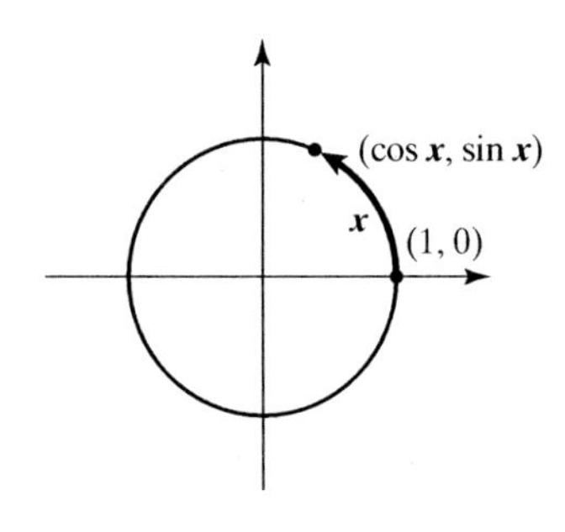

$$(\cos\square)^2+(\sin\square)^2=1, \quad \text{or} \quad \cos^2\square+\sin^2\square=1$$

For example, if we let x represent the arc in the Pythagorean identity, we have

$$\cos^2 x+\sin^2 x=1.$$

Exercise Set 1.3

1–4. *Using the information shown on each unit circle, fill in the blanks with exact values.*

1. **a.** $t =$ ____
b. $\sin t =$ ____

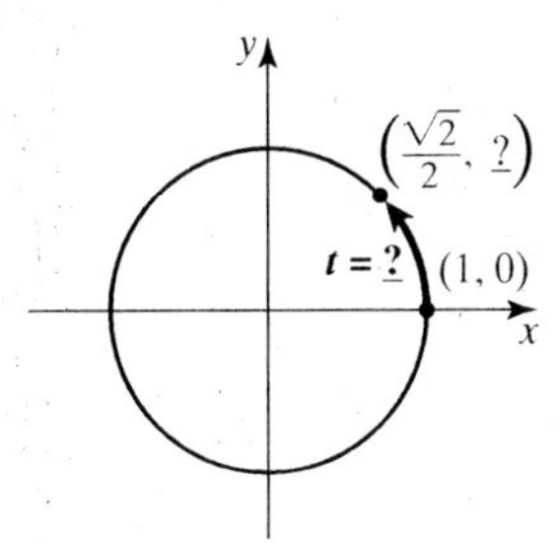

2. **a.** $\cos t =$ ____
b. $\sin t =$ ____

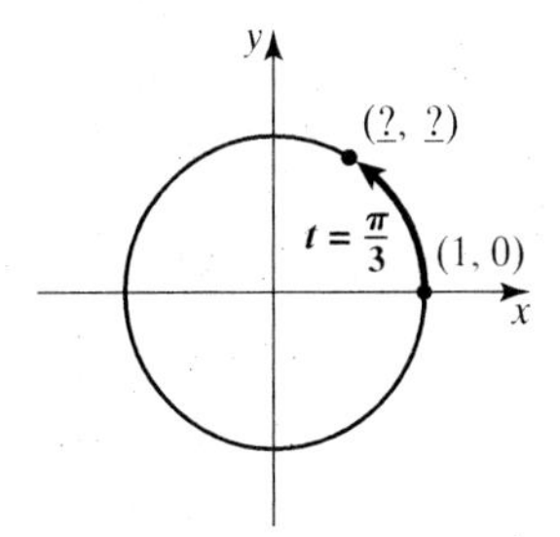

3. **a.** $\sin t =$ ____
b. $\cos t =$ ____

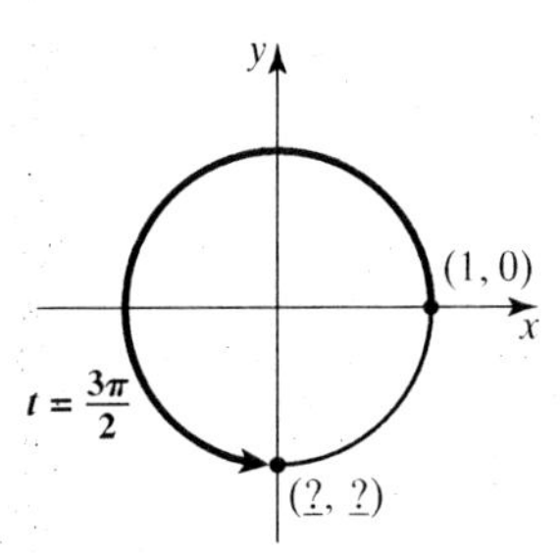

4. **a.** $t =$ ____
b. $\cos t =$ ____

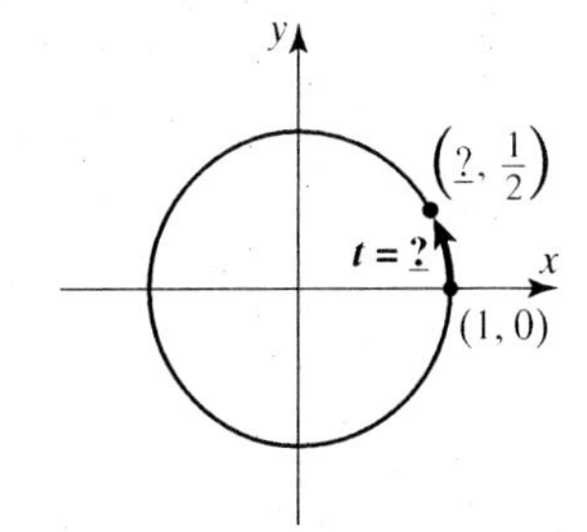

5–6. *Complete the table with exact functional values for arc x.*

5.

x	$\cos x$
0	
$\frac{\pi}{4}$	
π	
$\frac{\pi}{3}$	
$\frac{3\pi}{2}$	
$\frac{\pi}{2}$	
$\frac{\pi}{6}$	
2π	

6.

x	$\sin x$
π	
0	
$\frac{\pi}{6}$	
$\frac{\pi}{3}$	
$\frac{\pi}{4}$	
$\frac{\pi}{2}$	
2π	
$\frac{3\pi}{2}$	

7–18. *Evaluate each expression, if it is defined. Give the exact value for the answer. (Do not give a calculator approximation.)*

EXAMPLE $\cos \pi + \sin^2 \frac{\pi}{6}$

SOLUTION $\cos \pi + \left(\sin \frac{\pi}{6}\right)^2 = (-1) + \left(\frac{1}{2}\right)^2 = -\frac{4}{4} + \frac{1}{4} = -\frac{3}{4}$ ■

7. $\cos^2 \frac{\pi}{6}$

8. $\sin^2 \frac{\pi}{2} + \cos \frac{3\pi}{2}$

9. $\dfrac{\sin\left(\frac{\pi}{2} - \frac{\pi}{4}\right)}{\cos 0 - \cos \pi}$

10. $2 \sin^2 \frac{\pi}{4} + \sin 2\pi$

11. $3 \sin \frac{\pi}{2} - \sin \frac{\pi}{3}$

12. $\cos \frac{\pi}{4} + \sin\left(\frac{\pi}{2} - \frac{\pi}{6}\right)$

13. $\left(\cos \frac{\pi}{3}\right)^2 + \sin^2 \frac{\pi}{3}$

14. $\dfrac{2 \cos \frac{\pi}{6} - 4 \cos \frac{\pi}{3}}{\sin \frac{3\pi}{2}}$

15. $\dfrac{\sin \frac{\pi}{6}}{\cos \frac{\pi}{6}}$

16. $\dfrac{\cos \frac{\pi}{6}}{\sin \frac{\pi}{6}} + (\sin 0)\left(\cos \frac{\pi}{2}\right)$

17. $\dfrac{\cos^2 2\pi + \sin \frac{\pi}{3}}{\cos \frac{3\pi}{2} + \sin 0}$

18. $\dfrac{9 \cos \frac{\pi}{3} + \cos \pi}{(\sin \pi)(\cos^2 0)}$

19–22. *Find the exact value for each expression, and then use your calculator to check your answer. (Make sure your calculator is in radian mode.)*

EXAMPLE $\cos 0 + \cos \frac{\pi}{6}$

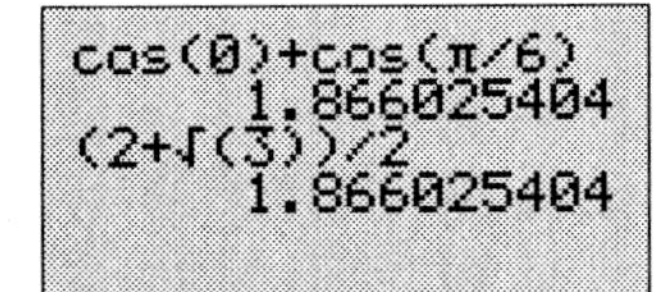

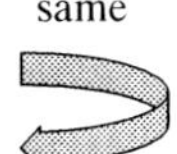

SOLUTION
$$\cos 0 + \cos \frac{\pi}{6} = 1 + \frac{\sqrt{3}}{2}$$
$$= \frac{2}{2} + \frac{\sqrt{3}}{2}$$
$$= \frac{2 + \sqrt{3}}{2}$$ ■

19. $\cos^2 \frac{\pi}{4} - \sin \frac{3\pi}{2}$

20. $\dfrac{5 \cos \pi - \sin 0}{\sin \frac{\pi}{2}}$

21. $\dfrac{\sin \frac{\pi}{6} + \cos \frac{\pi}{3}}{\sin^2 \frac{\pi}{4} + \sin \pi}$

22. $\left(\cos^3 \frac{\pi}{6}\right)\left(\sin \frac{\pi}{4}\right)$

23–30. *Compare the exact values to determine if each statement is true or false.*

EXAMPLE $\cos 0 = \cos^2 \pi$

SOLUTION
$$\cos 0 = \cos^2 \pi$$
$$1 = (-1)^2$$
$$1 = 1 \qquad \text{True}$$ ■

23. $\sin\left(\frac{\pi}{2}+\frac{\pi}{2}\right)=\sin\frac{\pi}{2}\cos\frac{\pi}{2}+\cos\frac{\pi}{2}\sin\frac{\pi}{2}$

24. $\cos\left(\frac{\pi}{4}+\frac{\pi}{4}\right)=\cos\frac{\pi}{4}\cos\frac{\pi}{4}-\sin\frac{\pi}{4}\sin\frac{\pi}{4}$

25. $\dfrac{\cos\frac{\pi}{3}}{\sin\frac{\pi}{3}}=\sqrt{3}$

26. $\dfrac{\cos\frac{\pi}{2}}{\sin\frac{\pi}{2}}=0$

27. $\cos\frac{\pi}{6}=\sin\frac{\pi}{3}$

28. $\dfrac{1}{\sin\pi}=-1$

29. $\cos^2\frac{\pi}{4}+\sin^2 0=1$

30. $\cos^2\frac{\pi}{6}+\sin^2\frac{\pi}{6}=1$

31–42. *Use a calculator to find an approximation rounded to four decimal places for each expression.*

```
sin(1.885-.11)
      .9792227801
cos(π/7)
      .9009688679
```

EXAMPLES **a.** $\sin(1.885-0.11)$ Make sure your calculator is in radian mode.

b. $\cos\frac{\pi}{7}$

SOLUTION

a. $\sin(1.885-0.11)\approx 0.9792$ **b.** $\cos\frac{\pi}{7}\approx 0.9010$ ■

31. $\sin 1.2$

32. $\cos 0.002$

33. $\cos(\pi+0.14)$

34. $\sin(\pi-0.2257)$

35. $\sin\frac{11\pi}{13}$

36. $\cos\frac{8\pi}{15}$

37. $\sin 2(4.3)$

38. $\cos 4(1.9)$

39. $\dfrac{1}{\sin 0.287}$

40. $\dfrac{1}{\cos 0.765}$

41. $\dfrac{\cos 1.56-\sin 1.1}{\sin 0.1}$

42. $\dfrac{\cos 2+\cos 5}{\sin 5}$

43. Derive the cosine and sine functional values for $t=\pi/3$.

44. Verify that every common arc x satisfies the Pythagorean identity $\cos^2 x+\sin^2 x=1$.

Graphing Calculator/CAS Exploration

Do you know what happens inside your calculator when you push the COS or SIN key? Many students think their calculator has the cosine and sine functional values for any arc stored in memory. If you think about this, it is impossible. (Why?) Since we have infinitely many arcs that could be wrapped around the circle, what do you think the calculator is doing to produce the values if it doesn't have them stored?

The formulas the calculator could be using are derived in calculus, where it can be shown that

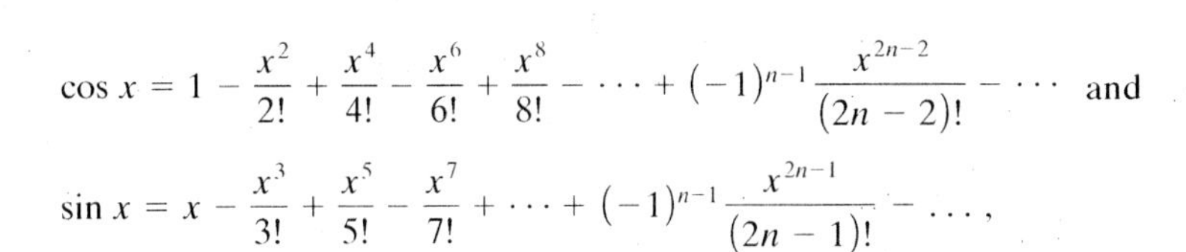

$$\cos x = 1-\frac{x^2}{2!}+\frac{x^4}{4!}-\frac{x^6}{6!}+\frac{x^8}{8!}-\cdots+(-1)^{n-1}\frac{x^{2n-2}}{(2n-2)!}-\cdots \quad\text{and}$$

$$\sin x = x-\frac{x^3}{3!}+\frac{x^5}{5!}-\frac{x^7}{7!}+\cdots+(-1)^{n-1}\frac{x^{2n-1}}{(2n-1)!}-\cdots,$$

for $x\in\mathbb{R}$, $n\in\{1,2,3,\dots\}$, where n represents the term in the sum ($n=1$ for first term, $n=2$ for second term, and so on) and $n!=n\cdot(n-1)\cdot\cdots\cdot 3\cdot 2\cdot 1$, read "$n$ factorial." The right side of each equation is called an *infinite power series.* If we add a *finite* number of terms in the power series, we find an approximation for the functional value.

45–46. *Substitute the given* arc *for* x *into the first four terms of the right side of the appropriate equation on the previous page given for* $\cos x$ *and* $\sin x$ *and, with the help of a calculator, find the sum of the four terms. Next, use the calculator cosine or sine key to find the required functional value. Is your answer for the sum of the four terms fairly close to the calculator functional value? If they are significantly different, how could we make the values closer?*

EXAMPLE $\cos 0.25$

Substituting 0.25 for x in the first four terms of the given $\cos x$ infinite series, we get:

$$\cos 0.25 = 1 - \frac{0.25^2}{2 \cdot 1} + \frac{0.25^4}{4 \cdot 3 \cdot 2 \cdot 1} - \frac{0.25^6}{6 \cdot 5 \cdot 4 \cdot 3 \cdot 2 \cdot 1} + \cdots$$

If we add the first four terms, we get:

$$\cos 0.25 \approx 1 - \frac{0.0625}{2} + \frac{0.00390625}{24} - \frac{0.000244140625}{720}$$

$$\approx 0.968912421332$$

Using a calculator, $\cos(0.25) \approx 0.968912421711$. We see the values are the same to nine decimal places.

45. a. $\cos 0.8$ **b.** $\sin 1$

46. a. $\cos 0.1$ **b.** $\sin \frac{1}{2}$

47. Refer to Exercise 45 in Section 1.2, and set up a graphing utility in the same manner.

MODE radian and parametric (param)

WINDOW $\text{Tmin} = 0$ $\text{Tmax} = 2\pi$

$\text{Tstep} = \frac{\pi}{24}$

$\text{Xmin} = -3$ $\text{Xmax} = 3$

$\text{Ymin} = -2$ $\text{Ymax} = 2$

Input $x_1 = \cos(t)$, $y_1 = \sin(t)$, and graph these parametric equations, which will again graph the unit circle. Trace around the circle and watch *just the cosine values.*

As the length of the arc increases from 0 to 1.57, what is happening to the cosine values? Continue to trace all around the circle. Complete the table that follows by filling in the appropriate description of the functional values—increasing (getting larger) or decreasing (getting smaller)—in the indicated intervals.

Repeat the preceding instructions, only this time pay attention to what is happening to *the sine values.* Complete the following table for the sine values on the indicated intervals.

x	0 to $\pi/2$	$\pi/2$ to π	π to $3\pi/2$	$3\pi/2$ to 2π
$\cos x$				

x	0 to $\pi/2$	$\pi/2$ to π	π to $3\pi/2$	$3\pi/2$ to 2π
$\sin x$				

Explore the Pattern

48. **a.** Complete the table that follows by matching each cosine value with the correct one from the column on the right. Leave the values in the form that they appear in the right column.

x	$\cos x$
0	
$\frac{\pi}{6}$	
$\frac{\pi}{4}$	
$\frac{\pi}{3}$	
$\frac{\pi}{2}$	

$\frac{\sqrt{4}}{2}$

$\frac{\sqrt{2}}{2}$

$\frac{\sqrt{0}}{2}$

$\frac{\sqrt{3}}{2}$

$\frac{\sqrt{1}}{2}$

b. Do you notice a pattern with the cosine functional values? Describe the pattern.

c. Do you think the pattern could help you remember the functional values of the common arcs?

d. Complete the table by matching each sine value with the correct one from the column on the right. Leave the values in the form that they appear in the right column.

x	$\sin x$
0	
$\frac{\pi}{6}$	
$\frac{\pi}{4}$	
$\frac{\pi}{3}$	
$\frac{\pi}{2}$	

$\frac{\sqrt{4}}{2}$

$\frac{\sqrt{2}}{2}$

$\frac{\sqrt{0}}{2}$

$\frac{\sqrt{3}}{2}$

$\frac{\sqrt{1}}{2}$

e. Do you notice a pattern with the sine functional values? Describe the pattern.

f. How is the pattern for the sine function different from the pattern for the cosine function?

g. Does this contradict the results you found in Exercise 47 when you described the functional values of $\cos x$ and $\sin x$ in the interval from 0 to $\pi/2$?

1.4 Reference Arcs and Functional Values for cos x and sin x, $x \in \mathbb{R}$

In Section 1.3 we obtained functional values for the common arcs. Now we use these arcs to find functional values of additional arcs. Doing so extends our knowledge of arcs that have exact functional values. We begin with a new definition that depends on our ability to locate terminal points of arcs.

Reference Arc

We define the **reference arc $\hat{x}$** as an arc with *positive direction* formed by the terminal point of the arc x and the nearest x-intercept on the unit circle. The length of the reference arc $\hat{x}$ is therefore the length of the smallest arc along the circle from the terminal point of arc x to the x-axis.

For example, the arc $x = 3\pi/4$ terminates in QII. So, its reference arc (the arc from the terminal point of $3\pi/4$ to the x-axis) must be $\pi - (3\pi/4) = \pi/4$. The arc $x = 11\pi/6$ terminates in QIV. So, its reference arc (the arc from the terminal point of $11\pi/6$ to the x-axis) must be $2\pi - (11\pi/6) = (12\pi/6) - (11\pi/6) = \pi/6$.

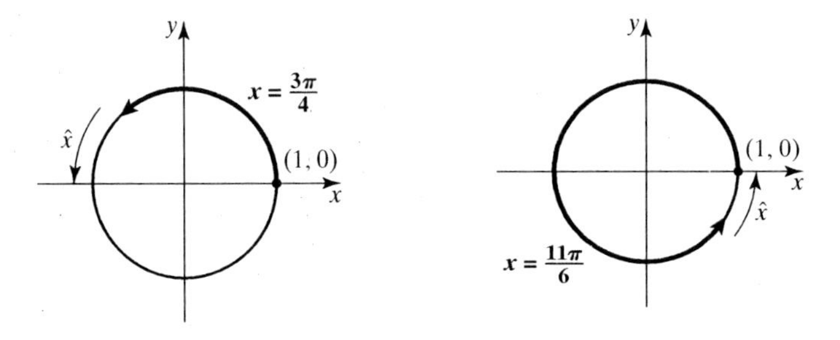

The following figures show reference arcs for arcs x within one wrap of the circle $(0 < x < 2\pi)$.

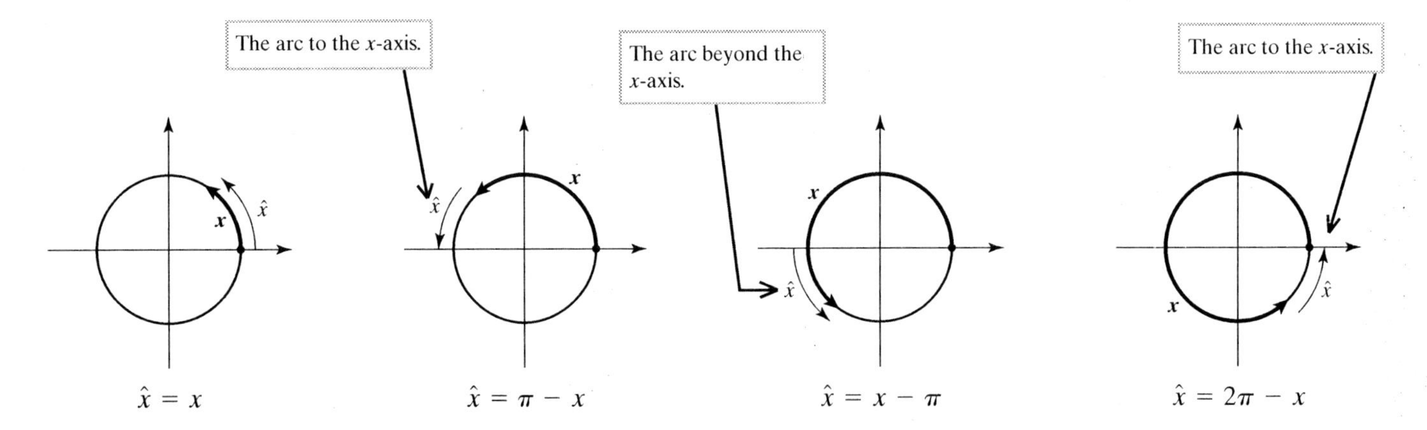

As a result of the definition, the length of any reference arc is always less than $\pi/2$; that is, $0 < \hat{x} < \pi/2$.

CAUTION: Be careful to remember that we find reference arcs with respect to the x-axis, *not the y-axis.*

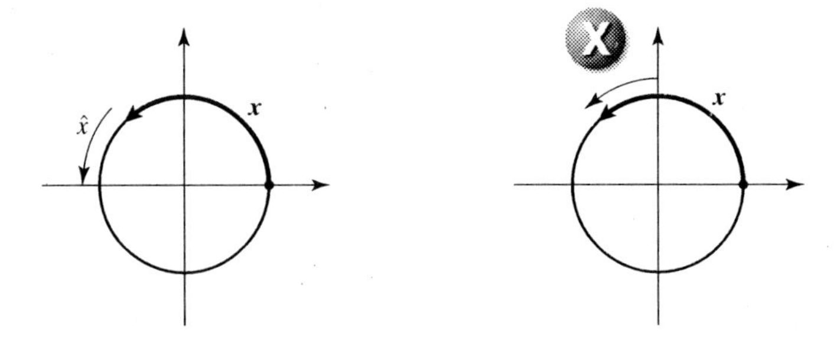

EXAMPLE 1 Find the reference arc $\hat{x}$ for each given arc. (Determining the quadrant that contains the terminal point was discussed in Sections 1.2 and 1.3.)

a. $4\pi/3$ **b.** $5\pi/6$ **c.** $7\pi/4$ **d.** $7\pi/3$

SOLUTION

a. $4\pi/3$ terminates in QIII.

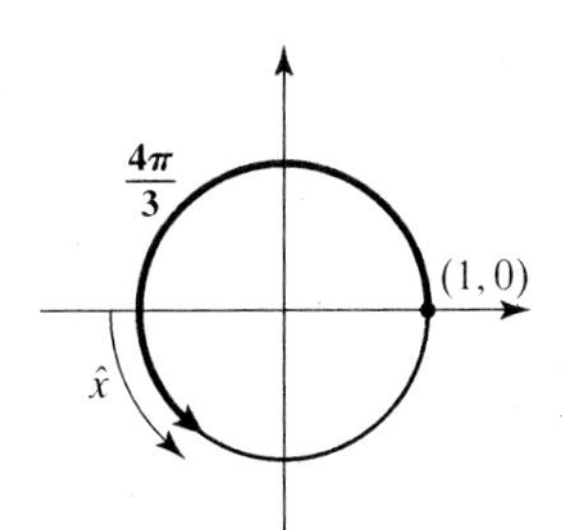

$$\hat{x} = \frac{4\pi}{3} - \pi = \frac{\pi}{3}$$

b. $5\pi/6$ terminates in QII.

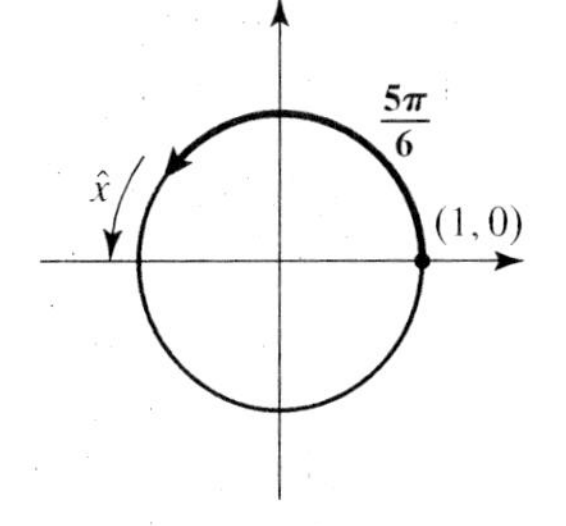

$$\hat{x} = \pi - \frac{5\pi}{6} = \frac{\pi}{6}$$

c. $7\pi/4$ terminates in QIV.

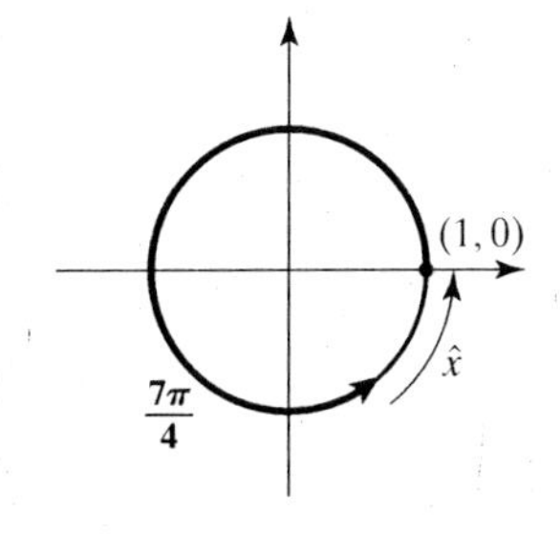

$$\hat{x} = 2\pi - \frac{7\pi}{4} = \frac{\pi}{4}$$

d. $7\pi/3$ wraps the circle more than once and terminates in QI.

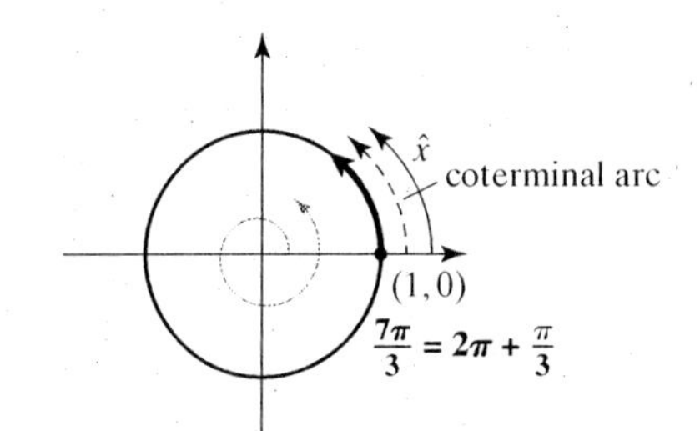

$$\hat{x} = \frac{7\pi}{3} - 2\pi = \frac{\pi}{3}$$

Notice $7\pi/3$ and $\pi/3$ are coterminal. ■

Least Positive Coterminal Arc

For arcs that wrap the circle more than once (Example 1d) or arcs that wrap clockwise, it is advisable to find their least positive coterminal arcs first and then find the reference arcs for those arcs. Coterminal arcs were defined in Section 1.2 as arcs that have the same initial and terminal points. To find the least positive coterminal arc, we either subtract multiples of 2π from positive arcs or add multiples of 2π to negative arcs to get a positive arc within one wrap of the circle. Let's practice finding least positive coterminal arcs before we find more reference arcs.

EXAMPLE 2 Sketch each arc and find its least positive coterminal arc.

a. $\dfrac{15\pi}{4}$ **b.** $-\dfrac{3\pi}{4}$ **c.** $-\dfrac{7\pi}{6}$ **d.** $-\dfrac{5\pi}{3}$ **e.** $\dfrac{14\pi}{3}$

SOLUTION

a. $\dfrac{15\pi}{4} - 2\pi = \dfrac{7\pi}{4}$

Least positive coterminal arc is $7\pi/4$.

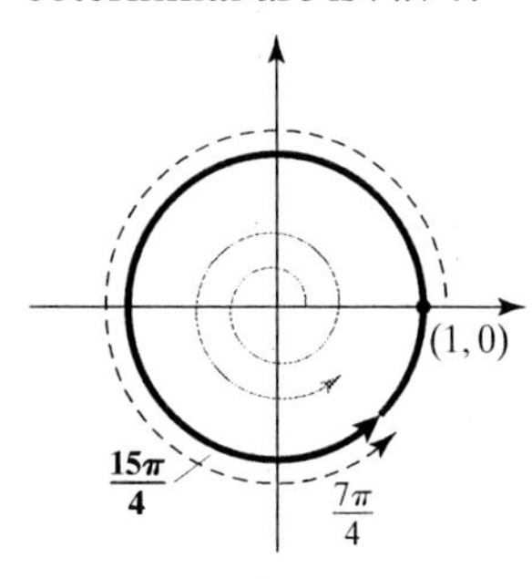

b. $-\dfrac{3\pi}{4} + 2\pi = \dfrac{5\pi}{4}$

Least positive coterminal arc is $5\pi/4$.

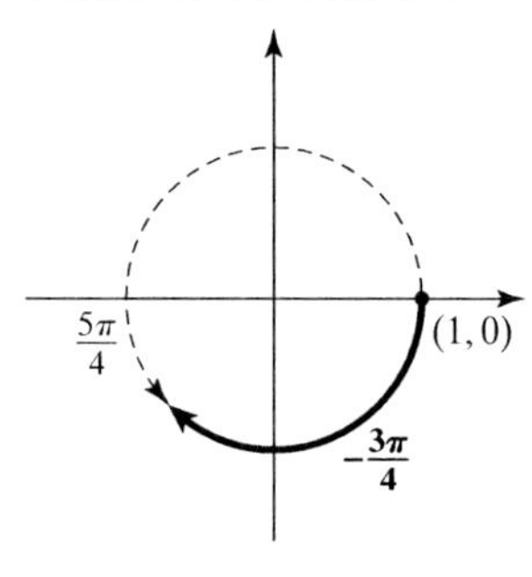

c. $-\dfrac{7\pi}{6} + 2\pi = \dfrac{5\pi}{6}$

Least positive coterminal arc is $5\pi/6$.

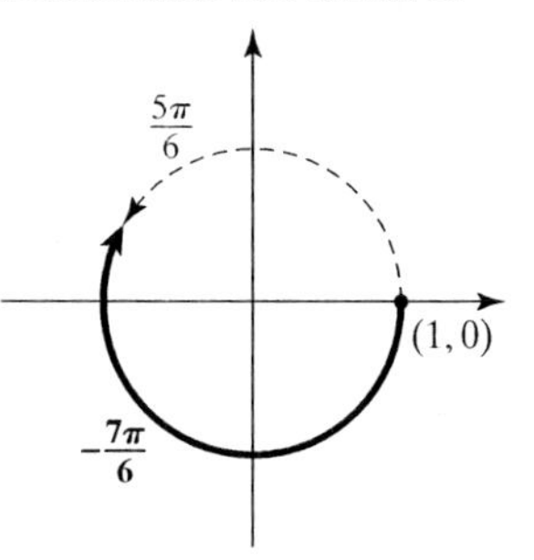

d. $-\dfrac{5\pi}{3} + 2\pi = \dfrac{\pi}{3}$

Least positive coterminal arc is $\pi/3$.

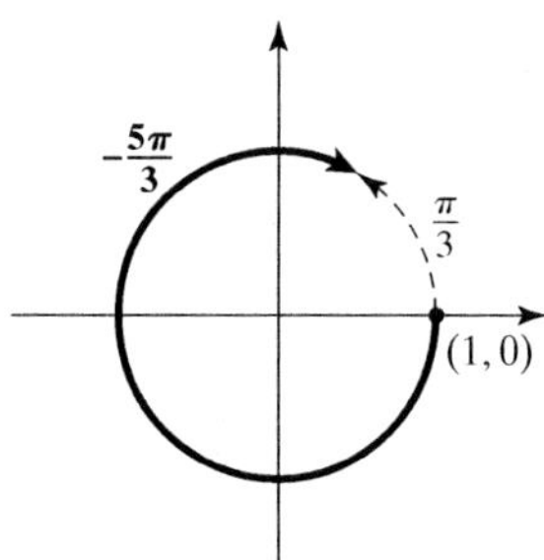

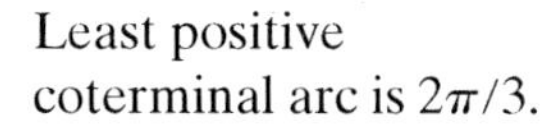

e. $\dfrac{14\pi}{3} - 4\pi = \dfrac{2\pi}{3}$

Least positive coterminal arc is $2\pi/3$.

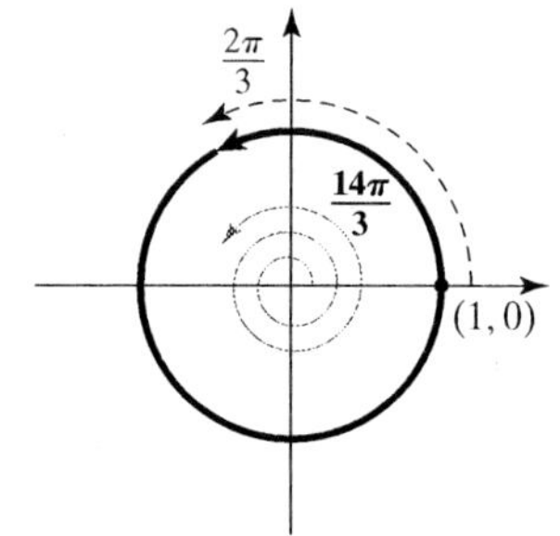

EXAMPLE 3 Find the reference arc $\hat{x}$ for each given arc.

a. $\dfrac{15\pi}{4}$ **b.** $-\dfrac{3\pi}{4}$ **c.** $-\dfrac{7\pi}{6}$ **d.** $-\dfrac{5\pi}{3}$ **e.** $\dfrac{14\pi}{3}$

SOLUTION Refer to Example 2 for the least positive coterminal arc.

a.

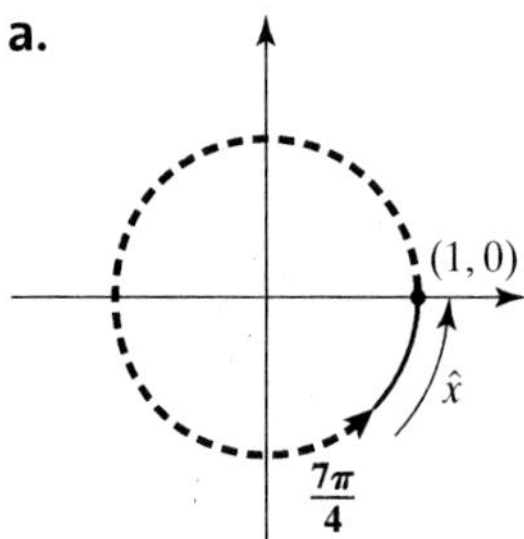

$$\hat{x} = 2\pi - \frac{7\pi}{4} = \frac{\pi}{4}$$

b.

$$\hat{x} = \frac{5\pi}{4} - \pi = \frac{\pi}{4}$$

c.

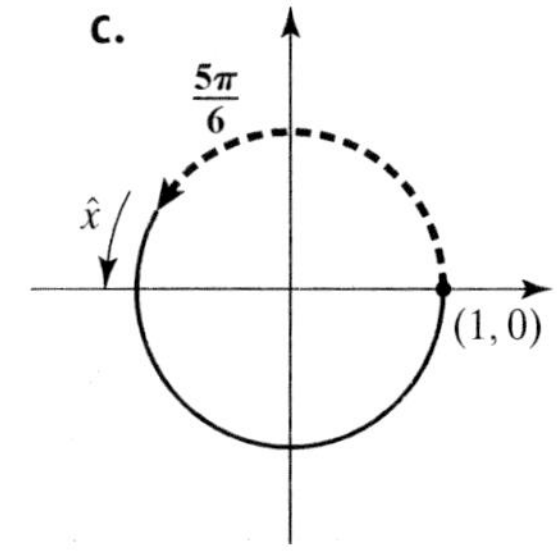

$$\hat{x} = \pi - \frac{5\pi}{6} = \frac{\pi}{6}$$

d.

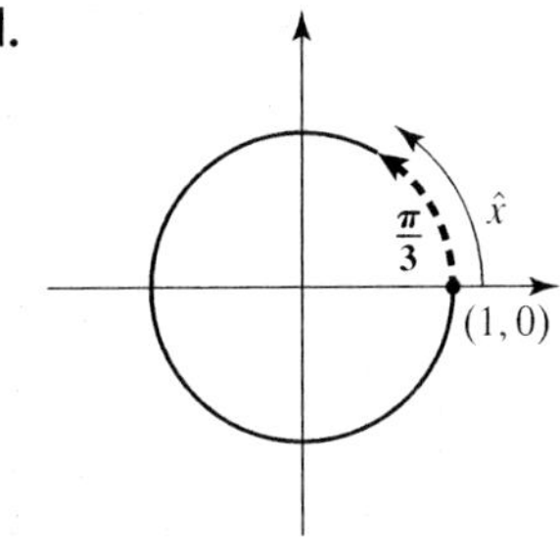

$$\hat{x} = \frac{\pi}{3}$$

e.

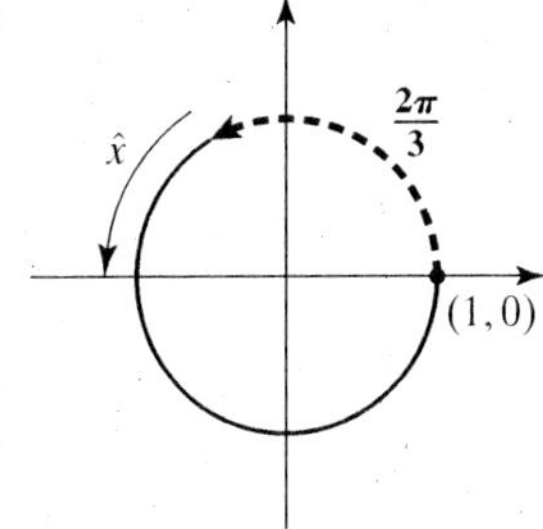

$$\hat{x} = \pi - \frac{2\pi}{3} = \frac{\pi}{3}$$

Integer Multiples of Common Arcs

Why do we want to know the reference arc? Let's consider two arcs, a common arc and a multiple of the common arc: $\pi/4$ and $3\pi/4$.

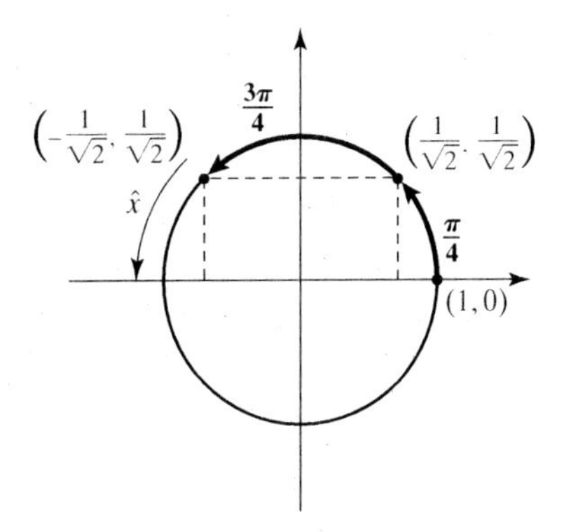

The reference arc for $3\pi/4$ is $\hat{x} = \pi/4$, which happens to be a common arc. We know the functional values for ALL the common arcs from the previous section. (You've memorized them, right?) The coordinates of the common arc are $(1/\sqrt{2}, 1/\sqrt{2})$. Using the symmetry of the circle, we notice that the coordinates at the terminal point of $3\pi/4$ and $\pi/4$ are symmetric with respect to the y-axis. Since $3\pi/4$ is in QII, the coordinates are $(-1/\sqrt{2}, 1/\sqrt{2})$. Therefore, the functional values are:

$$\cos\frac{3\pi}{4} = -\frac{1}{\sqrt{2}} = -\frac{\sqrt{2}}{2} \quad \text{and} \quad \sin\frac{3\pi}{4} = \frac{1}{\sqrt{2}} = \frac{\sqrt{2}}{2}$$

So, the benefit of knowing the reference arc—if it is a common arc—is that it allows us to know exact functional values of integer multiples of these arcs. In other words, if you know the functional values of the common arcs $\pi/6$, $\pi/4$, $\pi/3$, $\pi/2$, π and $3\pi/2$, and you know whether the coordinates are positive or negative in the different quadrants, then you automatically know the functional values of any integer multiples of these common arcs. Because of this fact, you can see how important it is for you to know the common-arc functional values.

NOTE: The numbers $k \in \{\ldots, -2, -1, 0, 1, 2, \ldots\}$ are called integers. Values such as

$$\ldots, -\frac{2\pi}{6}, -\frac{\pi}{6}, 0, \frac{\pi}{6}, \frac{2\pi}{6}, \ldots$$

are referred to as integer multiples of $\pi/6$.

Functional Values for Arcs with Common Reference Arcs

We give the following procedure for finding the functional values for any arc that has a common reference arc.

Step 1. Sketch the arc (x) to determine the quadrant of the terminal point.

Step 2. Find the reference arc ($\hat{x}$).

Step 3. Since the reference arc is a common arc, recall its functional values.

Step 4. Determine the positive or negative nature for the desired functional value.

EXAMPLE 4 Find $\sin \dfrac{4\pi}{3}$.

SOLUTION

We want to find the $\sin 4\pi/3$ (the y-coordinate on the unit circle at the terminal point of the arc $4\pi/3$).

1. Sketch the arc.
2. Find the reference arc.

$$\hat{x} = \frac{4\pi}{3} - \pi = \frac{\pi}{3}$$

3. Recall the functional value: $\sin \dfrac{\pi}{3} = \dfrac{\sqrt{3}}{2}$.
4. Since $4\pi/3$ is in QIII, $\sin x$ is negative. Therefore,

$$\sin \frac{4\pi}{3} = -\sin \frac{\pi}{3} = -\frac{\sqrt{3}}{2}.$$

EXAMPLE 5 Find the exact functional value.

a. $\cos \frac{11\pi}{6}$ **b.** $\cos \frac{2\pi}{3}$ **c.** $\sin\left(-\frac{7\pi}{4}\right)$ **d.** $\sin \frac{7\pi}{3}$

SOLUTION We use the same procedure shown in Example 4 to find the exact values.

		Sketch arc	Reference arc	Functional value
a.	$\cos \frac{11\pi}{6}$	(1, 0); $\hat{x}$; $\frac{11\pi}{6}$	$\hat{x} = 2\pi - \frac{11\pi}{6}$ $= \frac{\pi}{6}$	$11\pi/6$ is in QIV, so cos x is positive. $\cos \frac{11\pi}{6} = \cos \frac{\pi}{6}$ $= \frac{\sqrt{3}}{2}$
b.	$\cos \frac{2\pi}{3}$	$\frac{2\pi}{3}$; $\hat{x}$; (1, 0)	$\hat{x} = \pi - \frac{2\pi}{3}$ $= \frac{\pi}{3}$	$2\pi/3$ is in QII, so cos x is negative. $\cos \frac{2\pi}{3} = -\cos \frac{\pi}{3}$ $= -\frac{1}{2}$
c.	$\sin\left(-\frac{7\pi}{4}\right)$	$-\frac{7\pi}{4}$; $\hat{x}$; (1, 0)	$\hat{x} = -\frac{7\pi}{4} + 2\pi$ $= \frac{\pi}{4}$	$-7\pi/4$ is in QI, so sin x is positive. $\sin\left(-\frac{7\pi}{4}\right) = \sin \frac{\pi}{4}$ $= \frac{1}{\sqrt{2}} = \frac{\sqrt{2}}{2}$
d.	$\sin \frac{7\pi}{3}$	$\frac{7\pi}{3}$; $\hat{x}$; (1, 0)	$\hat{x} = \frac{7\pi}{3} - 2\pi$ $= \frac{\pi}{3}$	$7\pi/3$ is in QI, so the sin x is positive. $\sin \frac{7\pi}{3} = \sin \frac{\pi}{3}$ $= \frac{\sqrt{3}}{2}$

■

In the last example we found that $\sin 7\pi/3 = \sin \pi/3 = \sqrt{3}/2$. You may have already noticed that when arcs are coterminal, they have the same functional values. For example,

$$\sin\frac{\pi}{3} = \sin\left(\frac{\pi}{3} + 2\pi\right) = \sin\left(\frac{\pi}{3} + 2 \cdot 2\pi\right) = \sin\left(\frac{\pi}{3} + 3 \cdot 2\pi\right), \text{ and so on.}$$

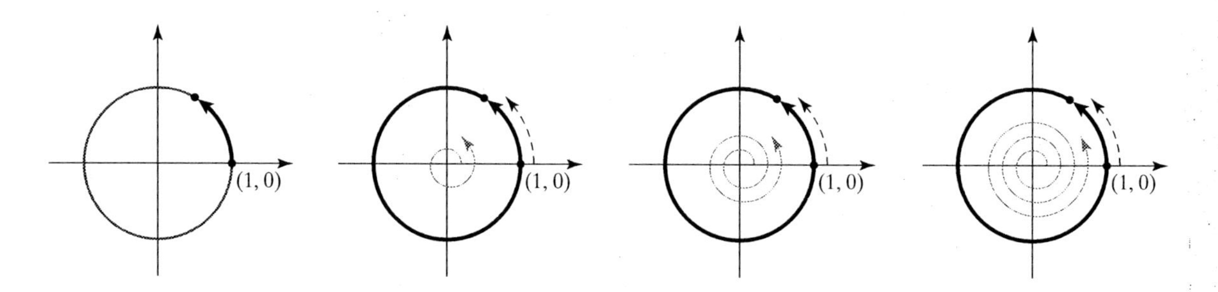

Thus, for arcs that involve multiple wraps,

$$\sin(x + k \cdot 2\pi) = \sin x,$$

where k is any integer $(k \in \{\ldots, -2, -1, 0, 1, 2, \ldots\})$.

The formula suggests that the sine function repeats its values, meaning it is **periodic**. The cosine function is also periodic.

> $\sin(x + k \cdot 2\pi) = \sin x$ and $\cos(x + k \cdot 2\pi) = \cos x$,
>
> where k is any integer, or $k \in \{\ldots, -2, -1, 0, 1, 2, \ldots\}$.

We will discuss these results again in Section 1.6.

The periodic property is a very important property of circular functions. We will see how important it is in our applications, since these functions model behavior that is repetitious. You may not realize how much you are surrounded by repetitious behavior: each day repeats every 24 hours; leap year repeats every four years; summer, fall, winter, and spring repeat each year; the number of daylight hours in a year repeats; your heartbeat, your breathing, the motion of the pendulum of a clock, and so on, all repeat. As a result, we will use the periodic property of the cosine and sine functions in many applications.

Functional Values for Arcs without Common Reference Arcs

What about arcs that do not have a common reference arc?

EXAMPLE 6 Find the functional value.

a. $\sin\frac{7\pi}{8}$ **b.** $\cos 6$

SOLUTION We begin by finding the reference arc $\hat{x}$ for each given arc.

a.

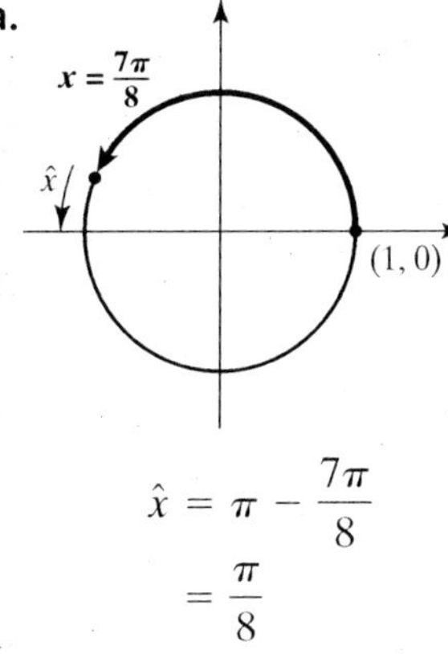

$$\hat{x} = \pi - \frac{7\pi}{8} = \frac{\pi}{8}$$

b.

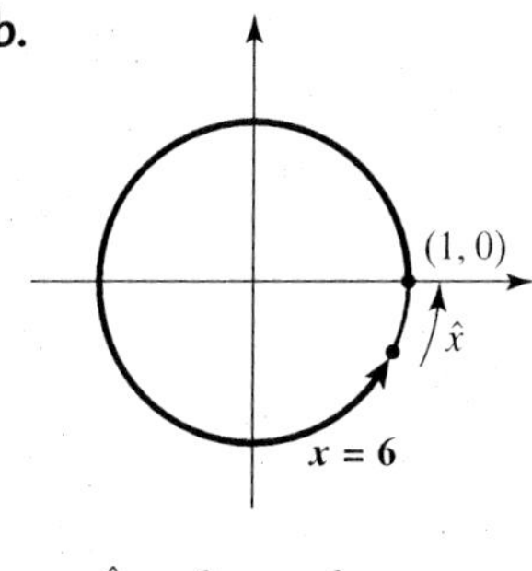

$$\hat{x} = 2\pi - 6 \approx 6.28 - 6 \approx 0.28$$

```
sin(7π/8)
     .3826834324
cos(6)
     .9601702867
```

Since neither $\hat{x} = \pi/8$ nor $\hat{x} = 0.28$ are common arcs, we do not know the exact functional values of these reference arcs. So finding the reference arcs is not helpful. Therefore, we use a calculator to find approximations of these functional values. Remember to set your calculator to radian mode and watch the use of parentheses!

a. $\sin \frac{7\pi}{8} \approx 0.3827$

b. $\cos 6 \approx 0.9602$ ■

We have extended our knowledge of exact functional values for integer multiples of common arcs. As Example 6 demonstrates, if an arc is not an integer multiple of a common arc, we find an approximation for the functional value using a calculator.

Now let's turn the question around. Suppose you know the functional values, can you find the arc?

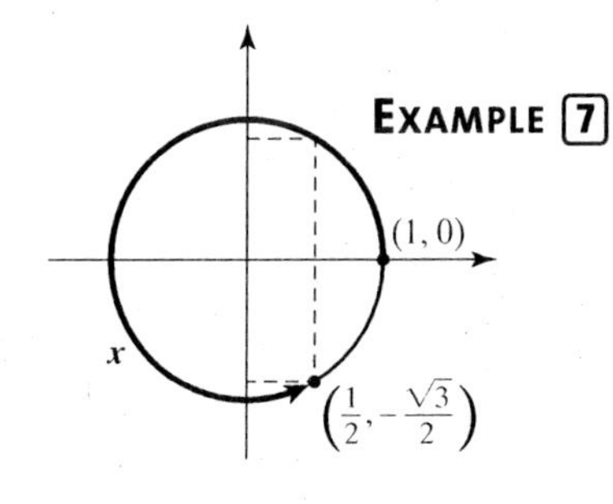

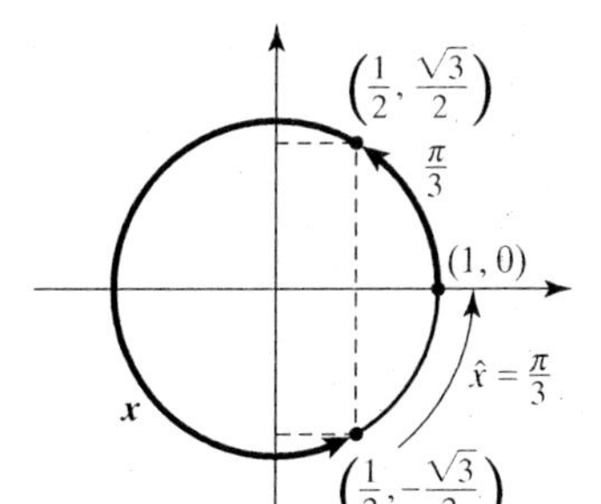

EXAMPLE 7 If $\cos x = \frac{1}{2}$ and $\sin x = -\frac{\sqrt{3}}{2}$, where $0 \le x < 2\pi$, find x.

SOLUTION This problem contains a previous question, "Where is the cosine positive and the sine negative?" The answer to this question and the fact that $0 \le x < 2\pi$ tells us that x is a positive arc within one wrap of the circle terminating in QIV. Since we have memorized the common arc values, we notice that if we ignore the negative on the value for $\sin x$, the values would be $\cos x = 1/2$ and $\sin x = \sqrt{3}/2$, the same as the functional values for $x = \pi/3$. So now the question becomes, "Which positive arc in QIV has $\pi/3$ as a reference arc?"

$$\hat{x} = 2\pi - x$$

$$\frac{\pi}{3} = 2\pi - x$$

$$\frac{\pi}{3} = \frac{6\pi}{3} - x$$

$$x = \frac{6\pi}{3} - \frac{\pi}{3} = \frac{5\pi}{3}$$ ■

Example 7 is a little challenging. Problems like this will become easier for you with more practice finding reference arcs, but they require you to be *very familiar with the common arc functional values.* We recommend that if you are having difficulty at any point in the text, go back and read sections over again. When you are learning a new language, it is always good to review. In the next section, we will ask you to go back and look at work you have done so you can make new connections with old problems. This process should become standard practice.

As a result of our work in this section, we know how to find exact functional values when the arc is an integer multiple of common arcs. (We can always use a calculator to find approximations for the functional value of any arc.) And we know how to find an integer multiple of a common arc when we are given its exact functional values.

Exercise Set 1.4

1–4. *Copy the diagram and find the exact cosine and sine values for the indicated arcs on the unit circle.*

1.

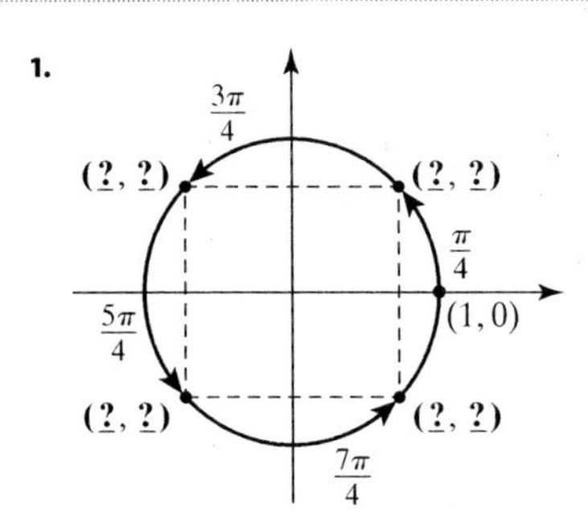

2.

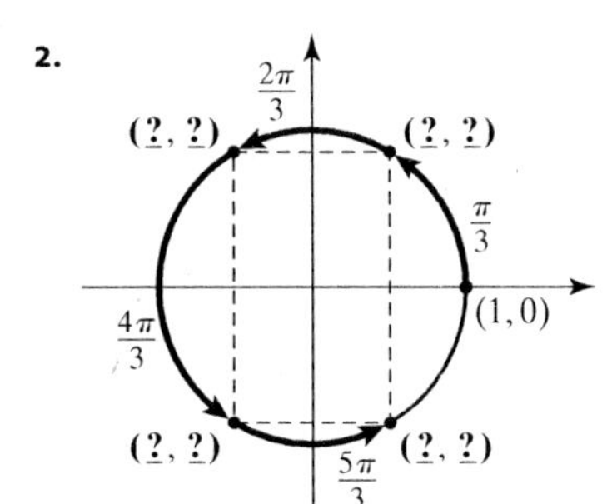

3.

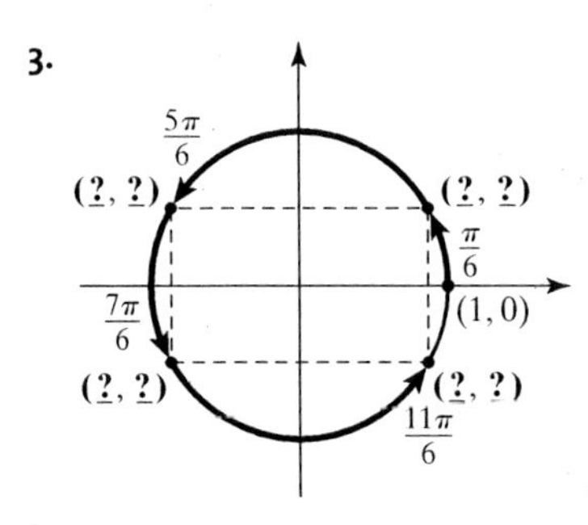

4.

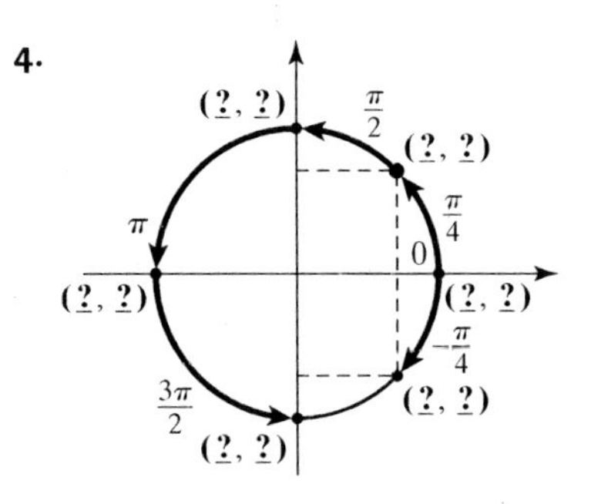

5–8. *Find each arc x on the unit circle. (Be careful of the direction of the arc.)*

5. $x =$ ______

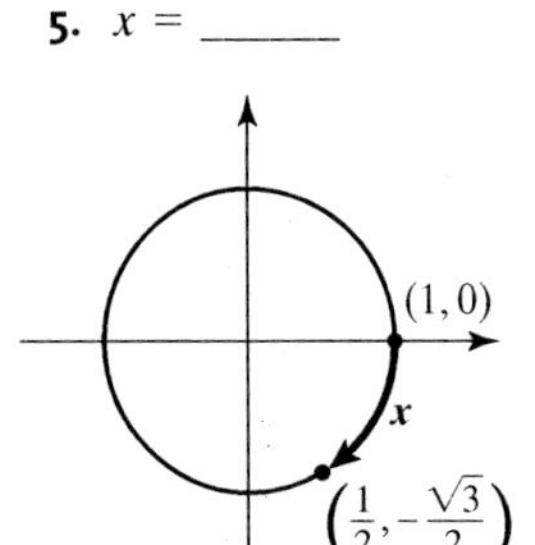

6. $x =$ ______

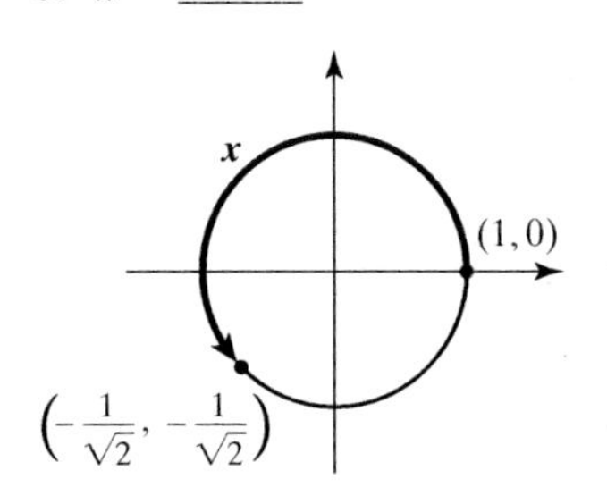

7. $x =$ ______

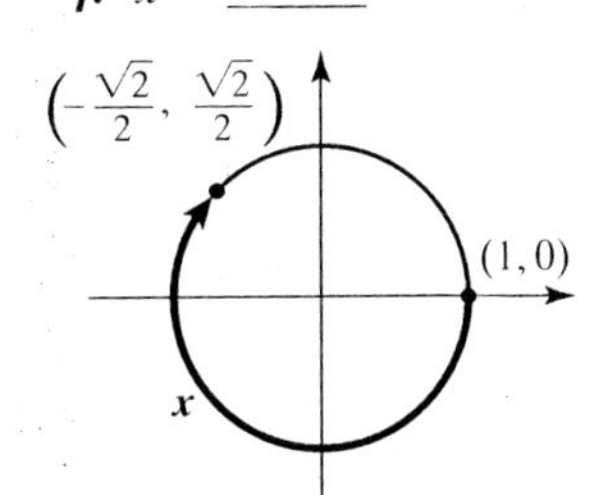

8. $x =$ ______

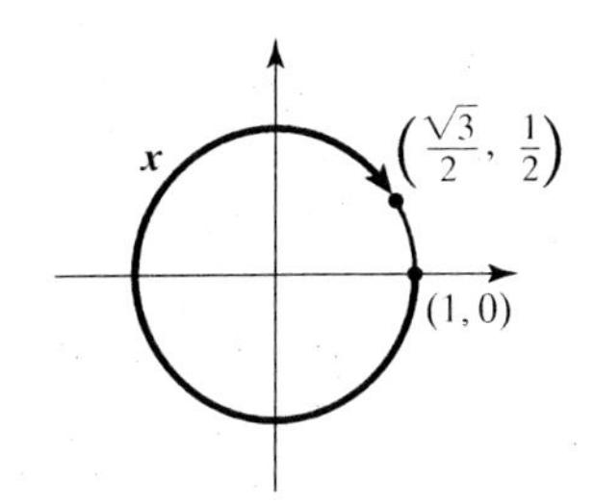

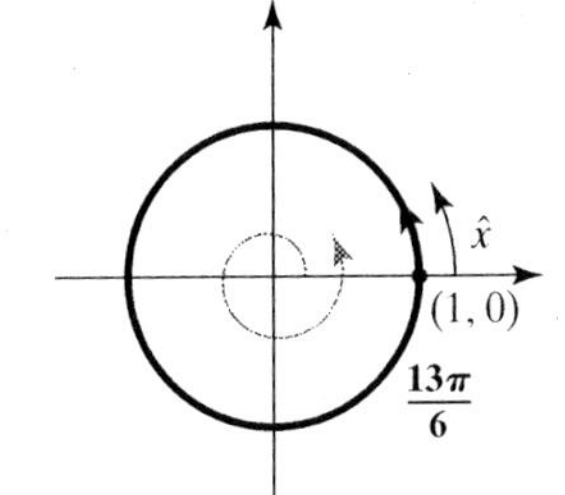

9–20. *Find the reference arc for each given arc. Sketch the arc and the reference arc.*

Example

$\frac{13\pi}{6}$

Solution

$\hat{x} = \frac{13\pi}{6} - 2\pi = \frac{\pi}{6}$

9. $\frac{5\pi}{3}$ **10.** $\frac{5\pi}{4}$ **11.** $\frac{7\pi}{6}$ **12.** $\frac{2\pi}{3}$

13. $\frac{9\pi}{4}$ **14.** $\frac{17\pi}{6}$ **15.** $-\frac{4\pi}{3}$ **16.** $-\frac{11\pi}{6}$

17. $-\frac{13\pi}{6}$ **18.** $-\frac{9\pi}{4}$ **19.** $-\frac{25\pi}{6}$ **20.** $-\frac{13\pi}{3}$

21–34. *For each expression sketch the given arc and state the reference arc. Then find the exact functional value.*

Example

Solution

	Sketch arc	Reference arc	Functional value
$\sin\frac{5\pi}{4}$		$\hat{x} = \frac{\pi}{4}$	$\sin\frac{5\pi}{4} = -\sin\frac{\pi}{4} = -\frac{\sqrt{2}}{2}$

21. $\sin\frac{2\pi}{3}$ **22.** $\sin\frac{3\pi}{4}$ **23.** $\cos\frac{7\pi}{6}$ **24.** $\cos\frac{4\pi}{3}$

25. $\cos\frac{7\pi}{4}$ **26.** $\sin\frac{11\pi}{6}$ **27.** $\sin\frac{13\pi}{6}$ **28.** $\sin\frac{9\pi}{4}$

29. $\sin\left(-\frac{3\pi}{4}\right)$ **30.** $\cos\left(-\frac{2\pi}{3}\right)$ **31.** $\cos\left(-\frac{5\pi}{3}\right)$ **32.** $\sin\left(-\frac{7\pi}{6}\right)$

33. $\cos\left(-\frac{13\pi}{6}\right)$ **34.** $\cos\left(-\frac{9\pi}{4}\right)$

35–46. *If possible, find the exact functional values. If not, use a calculator to find an approximate value rounded to four decimal places.*

35. $\dfrac{\cos^2 \pi}{\sin \frac{5\pi}{4}}$

36. $\sin^2 \dfrac{5\pi}{4}$

37. $\cos 21.23$

38. $\sin 7.81$

39. $15 \cos(-0.98)$

40. $12 \cos 2 + \sin 5.68$

41. $\sin \dfrac{\pi}{11}$

42. $\cos\left(-\dfrac{10\pi}{13}\right)$

43. $\sin\left(-\dfrac{5\pi}{6}\right) + \cos \dfrac{7\pi}{4}$

44. $\cos \dfrac{7\pi}{6} + \sin \dfrac{7\pi}{3}$

45. $\dfrac{-5}{\cos 2.763215 + 13 \sin 0}$

46. $\dfrac{\sin 5.8}{\cos 5.8}$

47–62. *Are the following statements true or false? If a statement is false, explain why or give an example that shows why it is false.*

47. We are able to find the exact functional values of every arc.

48. We are able to find the exact functional values of integer multiples of common arcs.

49. Reference arcs have positive direction.

50. $\cos 2.76 = \cos(-2.76)$

51. $\sin 59.1 = \sin(-59.1)$

52. $\sin\left(\dfrac{\pi}{4} + 2\pi\right) = \sin \dfrac{\pi}{4}$

53. $\cos\left(\dfrac{\pi}{6} + 2\pi\right) = \cos \dfrac{\pi}{6}$

54. Arcs $\dfrac{3\pi}{4}$ and $-\dfrac{5\pi}{4}$ are coterminal.

55. $\cos^2 2x + \sin^2 2t = 1$

56. The x-coordinate at the terminal point of an arc t whose initial point is $(1, 0)$ is called $\sin t$.

57. $\cos \dfrac{\pi}{3} = \dfrac{\sqrt{3}}{2}$

58. $\sin^2 x = (\sin x)^2 = (\sin x)(\sin x)$

59. $\cos x = 5$ is not possible for any arc x $(x \in \mathbb{R})$.

60. The largest value that $\cos x$ or $\sin x$ can be is 1.

61. $\cos^2 x = 1 - \sin^2 x$

62. $\cos^2 = 1 - \sin^2$

63–70. *Find the arc x with initial point $(1, 0)$, either within one wrap of the unit circle $(0 \leq x < 2\pi)$ or in the indicated interval, that makes each statement true.*

EXAMPLE $\cos x = -\dfrac{\sqrt{3}}{2}$ and $\sin x = \dfrac{1}{2}$

SOLUTION Since $\cos x < 0$ and $\sin x > 0$, arc x is in QII.

If we ignore the negative on the value for $\cos x$, the functional values are recognized as the functional values of the common arc $\pi/6$. Therefore, we need an arc in QII within one wrap that has $\pi/6$ as a reference arc.

$$\hat{x} = \pi - x$$

$$\frac{\pi}{6} = \frac{6\pi}{6} - x$$

$$x = \frac{6\pi}{6} - \frac{\pi}{6} = \frac{5\pi}{6}$$

63. $\cos x = \frac{1}{2}$ and $\sin x = -\frac{\sqrt{3}}{2}$

64. $\cos x = -\frac{\sqrt{2}}{2}$ and $\sin x = -\frac{\sqrt{2}}{2}$

65. $\sin x = -1$ and $\cos x = 0$

66. $\cos x = 0$ and $\sin x = 1$

67. $\sin x = -\frac{\sqrt{2}}{2},\ -\frac{\pi}{2} \le x \le \frac{\pi}{2}$

68. $\sin x = -\frac{1}{2},\ -\frac{\pi}{2} \le x \le \frac{\pi}{2}$

69. $\cos x = -\frac{\sqrt{3}}{2},\ 0 \le x \le \pi$

70. $\cos x = -\frac{1}{2},\ 0 \le x \le \pi$

Graphing Calculator/CAS Exploration

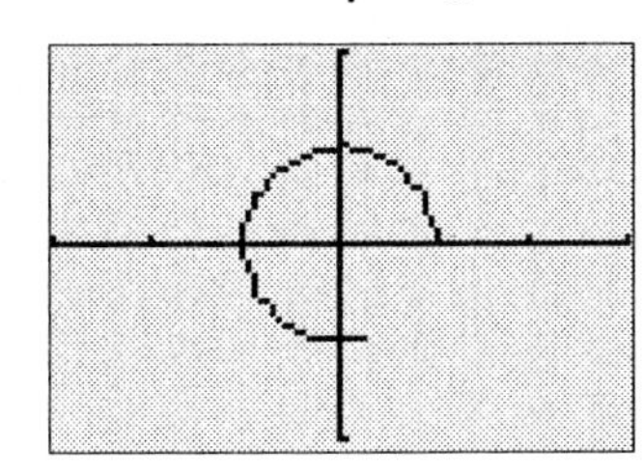

71. Dennis is having trouble locating the terminal point of the arc $x = 5$. He decides to use a graphing calculator. He wants his arc to stop at 5 so that he can see the terminal point. He uses the same setup as that indicated in Exercise 45 of Section 1.2 (graphing calculator/CAS exploration problem) with the exception of one change in the window setting. His graph looks like the one shown, which answers the question as to where the terminal point is located.

    a. What change did Dennis make? b. When won't this method help Dennis?

Explore the Pattern/Relationships

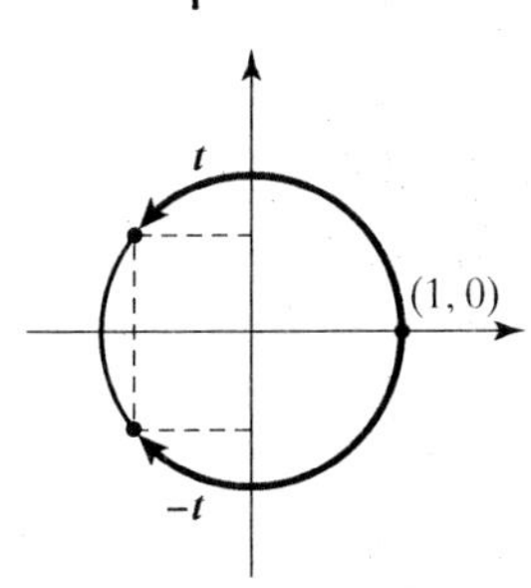

72. Consider arc t and arc$(-t)$ on the diagram shown. Discuss what you know about:
    a. The reference arc for each.
    b. The relationship between the cosine values of each arc.
    c. The relationship between the sine values of each arc.
    d. Is $\cos t = \cos(-t)$? e. Is $\sin t = \sin(-t)$?

73. a. Write out each statement and fill in the missing common arc $(0, \pi/6, \pi/4, \pi/3, \pi/2)$ to make a statement true.

    $\cos 0 = \sin \pi/2$ $\cos$_____ $= \sin \pi/6$

    $\cos \pi/6 = \sin$_____ $\cos$_____ $= \sin 0$

    $\cos \pi/4 = \sin$_____

    b. List in pairs the two arcs used in each true statement. (*Example*: For $\cos 0 = \sin \pi/2$, the two arcs are 0 and $\pi/2$.)

    c. Can you find the relationship that exists for each pair of arcs listed in part (b)? (*Hint*: Consider the sum of each pair of arcs.) Describe this relationship and how it affects sin and cos values.

    d. Does the relationship indicate that $\cos \pi/8 = \sin 3\pi/8$ is a true statement? Verify your answer with a calculator.

    e. Using the relationship you found, what arc x would make the following statement true? Verify your answer with a calculator.

    $$\sin 5\pi/12 = \cos x$$

    f. Fill in the missing function to make the statement true.

    _____$9\pi/20 = \sin \pi/20$

1.5 Four Additional Circular Functions: Tangent, Secant, Cosecant, and Cotangent Functions

In the last three sections we found functional values for cosine and sine. In the corresponding exercise sets you added, subtracted, multiplied, and divided these values and even found their reciprocals. For example, Exercises 39–42 in Section 1.2 asked for reciprocals and ratios of cosine and sine values. Actually these reciprocals and ratios, which are defined in terms $\cos x$ and $\sin x$, are used so often they have special names.

Tangent Function

If we consider the ratio $\frac{\sin x}{\cos x}$, it is defined only when the denominator is not equal to zero. That is, the ratio

$$\frac{\sin x}{\cos x}$$

is defined if $\cos x \neq 0$. Recall $\cos \pi/2 = 0$, $\cos 3\pi/2 = 0$, $\cos 5\pi/2 = 0$, and so on. Therefore the domain values of $\frac{\sin x}{\cos x}$ are all real numbers x such that $x \neq \pi/2, 3\pi/2, 5\pi/2, 7\pi/2, \ldots$.

We now give the ratio $\frac{\sin x}{\cos x}$ a name and describe the domain of this new function using a formula.

$$\text{tangent } x = \frac{\sin x}{\cos x},\ x \neq \frac{\pi}{2} + k\pi, \text{ where } k \in \{\ldots, -2, -1, 0, 1, 2, \ldots\}$$

$$\tan x = \frac{\sin x}{\cos x} \qquad \text{common abbreviation}$$

Pairing each arc x with the ratio

$$\tan x = \frac{\sin x}{\cos x}$$

produces the

tangent function: $\{(x, \tan x)\}$,

whose domain is $x \in \mathbb{R}$ such that $x \neq (\pi/2) + k\pi$, where $k \in \{\ldots, -2, -1, 0, 1, 2, \ldots\}$.

Let's extend the table in Example 7 from Section 1.2 that shows in which quadrants $\cos x$ and $\sin x$ are positive or negative to include $\tan x = \frac{\sin x}{\cos x}$. For example, in QIII $\tan x$ is positive because $\sin x$ and $\cos x$ are both negative, and the quotient of two negative numbers is always positive.

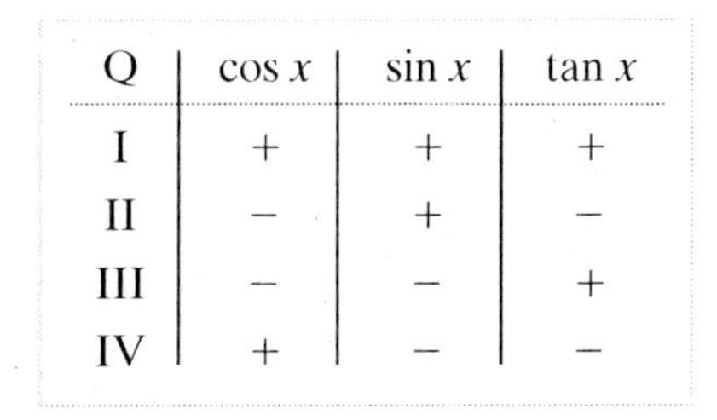

Q	$\cos x$	$\sin x$	$\tan x$
I	+	+	+
II	−	+	−
III	−	−	+
IV	+	−	−

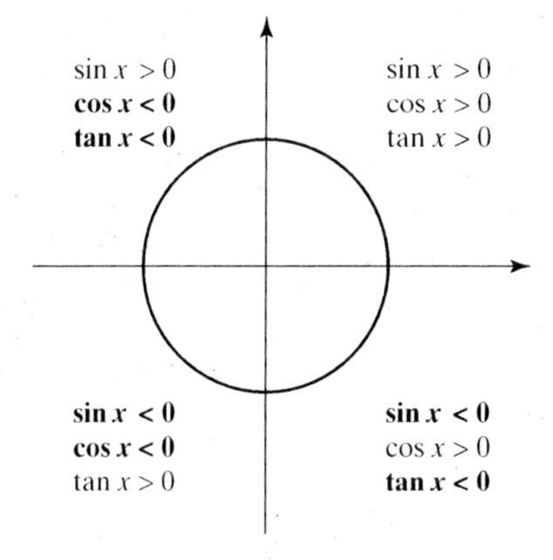

Since we know the cosine and sine exact functional values for common arcs, we can find the corresponding exact functional values for the tangent by taking the ratios. Let's extend the table from Section 1.2 to summarize these values.

EXAMPLE 1 Complete the table for functional values of the common arcs to include $\tan x$.

SOLUTION

x	$\cos x$	$\sin x$	$\tan x$
0	1	0	0
$\frac{\pi}{6}$	$\frac{\sqrt{3}}{2}$	$\frac{1}{2}$	$\frac{1}{\sqrt{3}}$ or $\frac{\sqrt{3}}{3}$
$\frac{\pi}{4}$	$\frac{1}{\sqrt{2}}$ or $\frac{\sqrt{2}}{2}$	$\frac{1}{\sqrt{2}}$ or $\frac{\sqrt{2}}{2}$	1
$\frac{\pi}{3}$	$\frac{1}{2}$	$\frac{\sqrt{3}}{2}$	$\sqrt{3}$
$\frac{\pi}{2}$	0	1	undefined
π	-1	0	0
$\frac{3\pi}{2}$	0	-1	undefined

$$\tan 0 = \frac{\sin 0}{\cos 0} = \frac{0}{1} = 0$$

$$\tan \frac{\pi}{6} = \frac{\sin \pi/6}{\cos \pi/6} = \frac{1/2}{\sqrt{3}/2} = \frac{1}{2} \cdot \frac{2}{\sqrt{3}} = \frac{1}{\sqrt{3}}$$

$$\tan \frac{\pi}{4} = \frac{\sin \pi/4}{\cos \pi/4} = \frac{1/\sqrt{2}}{1/\sqrt{2}} = 1$$

$$\tan \frac{\pi}{3} = \frac{\sin \pi/3}{\cos \pi/3} = \frac{\sqrt{3}/2}{1/2} = \frac{\sqrt{3}}{2} \cdot \frac{2}{1} = \sqrt{3}$$

$$\tan \frac{\pi}{2} = \frac{\sin \pi/2}{\cos \pi/2} = \frac{1}{0} \rightarrow \text{undefined}$$

$$\tan \pi = \frac{\sin \pi}{\cos \pi} = \frac{0}{-1} = 0$$

$$\tan \frac{3\pi}{2} = \frac{\sin 3\pi/2}{\cos 3\pi/2} = \frac{-1}{0} \rightarrow \text{undefined}$$ ■

Notice that our table contains values $\pi/2$ and $3\pi/2$ for which the tangent function is not defined.

You already know that this is an important table, and you now have *one more* column to memorize. Of course, if you know the cosine and sine values, you could always stop and take their ratio. But it may save you time to memorize this complete table.

Using a calculator we can look at other values for $\tan x = \dfrac{\sin x}{\cos x}$. We notice that when x is getting closer to $\pi/2 \approx 1.57$, $\tan x$ values become very large in absolute value. But when x gets close to zero, so does $\tan x$.

The range of the tangent function is $\mathbb{R}$, the real numbers.

x	$\tan x$
1.5	≈ 14.1
1.55	≈ 48.1
1.56	≈ 92.6
1.565	≈ 172.5
1.569	≈ 556.7
1.58	≈ -108.6

x	$\tan x$
0.2	≈ 0.203
0.1	≈ 0.100
0.05	≈ 0.050
0.01	≈ 0.010
0.001	≈ 0.001
0.0001	≈ 0.0001

Definitions of the Reciprocal Functions

We also give special names to the reciprocals of the cosine, sine, and tangent functions. Notice that the domains of these functions consist of the values for which the ratios are defined. Again, each domain is expressed by a formula that eliminates the values that make the denominator zero.

Secant—the Reciprocal of the Cosine

$$\text{secant } x = \frac{1}{\cos x}, \quad x \neq \frac{\pi}{2} + k\pi, \text{ where } k \in \{\ldots, -2, -1, 0, 1, 2, \ldots\}$$

$$\sec x = \frac{1}{\cos x} \quad \text{common abbreviation}$$

Like the tangent function, $\sec x$ has $\cos x$ as the denominator. Its domain also consists of real numbers x such that $\cos x \neq 0$.

Pairing each arc x with $\sec x = \dfrac{1}{\cos x}$ produces the

secant function: $\{(x, \sec x)\}$,

whose domain is $x \in \mathbb{R}$ such that $x \neq (\pi/2) + k\pi$.

If $\cos x = -\frac{1}{10}$, $\sec x = -\frac{10}{1}$.

If $\cos x = \frac{1}{5}$, $\sec x = 5$.

If $\cos x = \frac{1}{2}$, $\sec x = 2$.

If $\cos x = 1$, $\sec x = 1$.

Since the secant function is the reciprocal of the cosine and $|\cos x| \le 1$, then $|\sec x| \ge 1$. Therefore, the range of the secant function is $|\sec x| \ge 1$ $(\sec x \le -1$ or $\sec x \ge 1)$.

Cosecant—The Reciprocal of the Sine

$$\text{cosecant } x = \frac{1}{\sin x}, \quad x \ne k\pi, \text{ where } k \in \{\ldots, -2, -1, 0, 1, 2, \ldots\}$$

$$\csc x = \frac{1}{\sin x} \quad \text{common abbreviation}$$

Since $\csc x$ has $\sin x$ as the denominator, its domain consists of values for x such that $\sin x \ne 0$. (Recall that $\sin 0 = \sin \pi = \sin 2\pi = \sin(-\pi) = 0$, and so on.)

The pairing of each arc x with $\csc x = \frac{1}{\sin x}$ produces the

cosecant function: $\{(x, \csc x)\}$,

whose domain is $x \in \mathbb{R}$ such that $x \ne k\pi$. Since cosecant is the reciprocal of the sine and $|\sin x| \le 1$, then $|\csc x| \ge 1$. Therefore, the range of the cosecant function is $|\csc x| \ge 1$ $(\csc x \le -1$ or $\csc x \ge 1)$.

Cotangent—The Reciprocal of the Tangent

$$\text{cotangent } x = \frac{1}{\tan x} = \frac{1}{\frac{\sin x}{\cos x}} = \frac{\cos x}{\sin x}$$

$$\cot x = \frac{\cos x}{\sin x} \quad \text{or} \quad \text{ctn } x = \frac{\cos x}{\sin x}, \quad \text{common abbreviations}$$

$$x \ne k\pi, \text{ where } k \in \{\ldots, -2, -1, 0, 1, 2, \ldots\}$$

Like the cosecant function, $\cot x$ has $\sin x$ as the denominator. The domain also consists of values for x such that $\sin x \ne 0$.

The pairing of each arc x with $\cot x = \frac{\cos x}{\sin x}$ produces the

cotangent function: $\{(x, \cot x)\}$,

whose domain is $x \in \mathbb{R}$ such that $x \ne k\pi$.

Since the range of the tangent function is the real numbers, so is the range of the cotangent function. We summarize the domain and range of these new circular functions. The definitions of these four new functions are also called identities.

Function	Domain	Range
$\{(x, \tan x)\}$	$x \neq \frac{\pi}{2} + k\pi$	$\mathbb{R}$
$\{(x, \sec x)\}$	$x \neq \frac{\pi}{2} + k\pi$	$\lvert\sec x\rvert \geq 1$
$\{(x, \csc x)\}$	$x \neq k\pi$	$\lvert\csc x\rvert \geq 1$
$\{(x, \cot x)\}$	$x \neq k\pi$	$\mathbb{R}$

We now extend our table of exact functional values for common arcs to include these three new functions.

x	$\cos x$	$\sin x$	$\tan x$	$\sec x$	$\csc x$	$\cot x$
0	1	0	0	1	undefined	undefined
$\frac{\pi}{6}$	$\frac{\sqrt{3}}{2}$	$\frac{1}{2}$	$\frac{1}{\sqrt{3}}$	$\frac{2}{\sqrt{3}}$	2	$\sqrt{3}$
$\frac{\pi}{4}$	$\frac{1}{\sqrt{2}}$	$\frac{1}{\sqrt{2}}$	1	$\sqrt{2}$	$\sqrt{2}$	1
$\frac{\pi}{3}$	$\frac{1}{2}$	$\frac{\sqrt{3}}{2}$	$\sqrt{3}$	2	$\frac{2}{\sqrt{3}}$	$\frac{1}{\sqrt{3}}$
$\frac{\pi}{2}$	0	1	undefined	undefined	1	0
π	-1	0	0	-1	undefined	undefined
$\frac{3\pi}{2}$	0	-1	undefined	undefined	-1	0

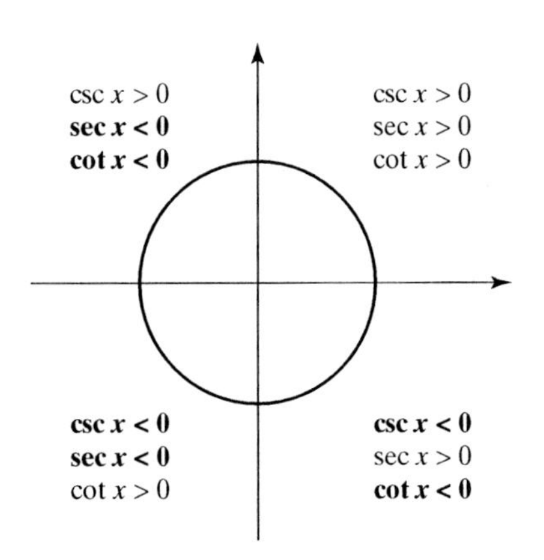

It is not necessary to memorize the entire table. If you know the first three columns, the remaining three columns are just the respective reciprocals. It is also not necessary to extend the summary diagram for positive and negative values with the three additional functions since the reciprocal of a positive number is positive and the reciprocal of a negative number is negative.

You do need to memorize the definitions of the new functions being careful to take the correct ratio or reciprocal.

NOTE: $\csc x = \dfrac{1}{\sin x}$ and $\sec x = \dfrac{1}{\cos x}$.

What are we going to do with these new functions? The same ideas from the previous sections can now be used with these new functions. First let's take a moment and look back at what we have done:

1. We located the quadrant that contains the terminal point of a given arc.
2. Given one coordinate at the terminal point of a given arc t, we found the other coordinate $(x = \cos t \quad \text{or} \quad y = \sin t)$.
3. We found (and memorized) the exact functional values for common arcs.
4. We found reference arcs.
5. We found exact functional values of arcs that are integer multiples of common arcs.
6. We used a calculator to find approximations of functional values for any noncommon arcs.

Now that we have four new functions, we will apply these six ideas to each of them. And, in many cases, we use a lot of these ideas in the same problem—all at once!

For Examples 2–5, do the following:

a. Locate the terminal point of the arc.
b. Find the reference arc.
c. Find the functional value.

EXAMPLE 2 Find $\tan \dfrac{3\pi}{4}$.

The tangent is negative in QII since it is the ratio of a positive and a negative value.

SOLUTION $3\pi/4$ terminates in QII in which the tangent values are negative.

$$\tan \frac{3\pi}{4} = -\tan \frac{\pi}{4} = -1$$

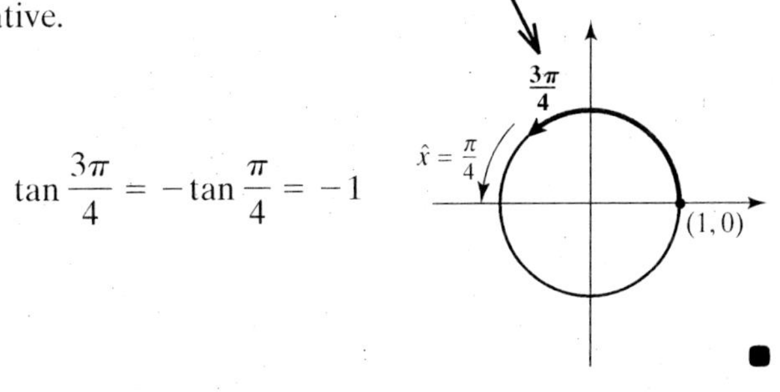

EXAMPLE 3 Find $\sec \dfrac{7\pi}{6}$.

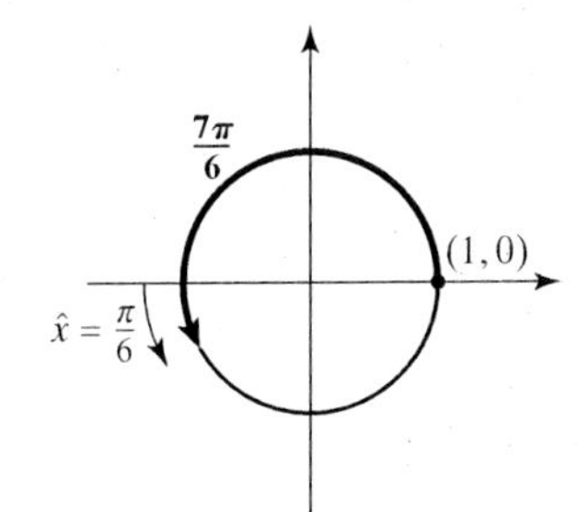

SOLUTION $7\pi/6$ terminates in QIII in which cosine values are negative.

$$\sec \frac{7\pi}{6} = \frac{1}{\cos \frac{7\pi}{6}} = \frac{1}{-\cos \frac{\pi}{6}} = \frac{1}{-\frac{\sqrt{3}}{2}} = -\frac{2}{\sqrt{3}} = -\frac{2\sqrt{3}}{3}$$

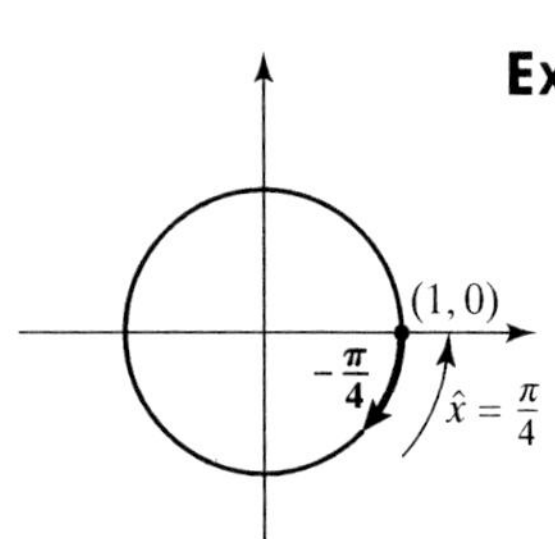

EXAMPLE 4 Find $\csc\left(-\frac{\pi}{4}\right)$.

SOLUTION $-\pi/4$ terminates in QIV in which sine values are negative.

$$\csc\left(-\frac{\pi}{4}\right) = \frac{1}{\sin\left(-\frac{\pi}{4}\right)} = \frac{1}{-\sin\frac{\pi}{4}} = \frac{1}{-\frac{1}{\sqrt{2}}} = -\sqrt{2}$$

■

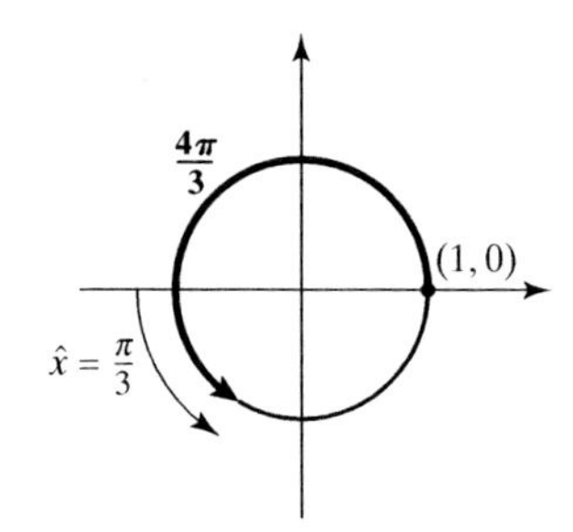

EXAMPLE 5 Find $\cot\frac{4\pi}{3}$.

SOLUTION $4\pi/3$ terminates in QIII in which tangent values are positive.

$$\cot\left(\frac{4\pi}{3}\right) = \frac{1}{\tan\left(\frac{4\pi}{3}\right)} = \frac{1}{\tan\frac{\pi}{3}} = \frac{1}{\sqrt{3}} = \frac{\sqrt{3}}{3}$$

■

As we found approximations to cosine and sine functional values of arcs that were not multiples of common arcs, we can do the same with these four new circular functions. You may have already noticed that your calculator has a key for the tangent function (TAN key), but may not have keys for the reciprocal functions (COT, CSC, or SEC). If your calculator does not have these reciprocal keys, you can use the ratio definitions to find these functional values. Again, make sure your calculator is set in radians and be careful to use appropriate parentheses.

CAUTION: [SIN⁻¹] is not a reciprocal key on the calculator. It represents a special function that we will discuss in Chapter 2. In other words, $\frac{1}{\sin x} \neq \text{SIN}^{-1}x$.

EXAMPLE 6 Use a calculator to find an approximation to four decimal places.

a. $\csc 2$ **b.** $\sec\left(-\frac{\pi}{8}\right)$ **c.** $\tan^2 5.76$ **d.** $\cot 36.5612$ **e.** $\sec 3/\cot 5$

SOLUTION

a. $\csc 2 \approx 1.0998$

b. $\sec\left(-\frac{\pi}{8}\right) \approx 1.0824$

c. $\tan^2 5.76 \approx 0.3327$

d. $\cot 36.5612 \approx -0.4621$

e. $\frac{\sec 3}{\cot 5} \approx 3.4147$

```
1/sin(2)
        1.09975017
1/cos(-π/8)
         1.0823922
(tan(5.76))²
       .3326972127
```

```
1/tan(36.5612)
       -.4621168443
(1/cos(3))/(1/ta
n(5))
        3.414687503
```

■

We have used almost all of the ideas from our previous sections with these four new functions. We have one more idea to look at.

EXAMPLE 7 If $\cos x = \frac{1}{2}$ and $\sin x > 0$, find $\tan x$.

ALGEBRA QUESTION: If P is the point graphed below on the unit circle and given one of its coordinates, what is the other coordinate and the ratio of y/x?

SOLUTION Since P is a point on the circle, the coordinates of P satisfy the circle's equation $x^2 + y^2 = 1$.

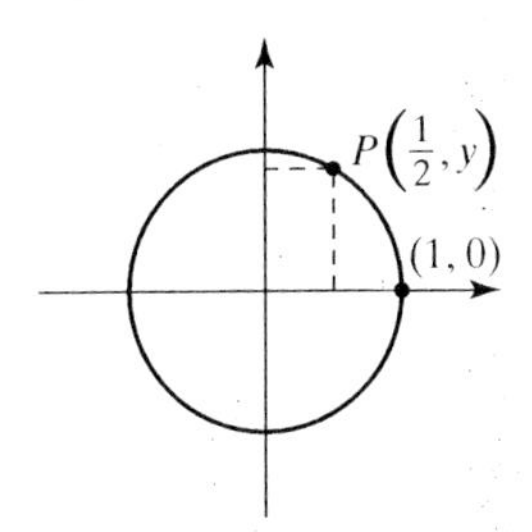

$$x^2 + y^2 = 1$$
$$\left(\frac{1}{2}\right)^2 + y^2 = 1$$
$$\frac{1}{4} + y^2 = 1$$
$$y^2 = \frac{3}{4}$$
$$y = \pm\sqrt{\frac{3}{4}} = \pm\frac{\sqrt{3}}{2}$$

Since P is a point in QI, the y coordinate is positive $(y > 0)$. Therefore $y = \sqrt{3}/2$. The ratio we want is

$$\frac{y}{x} = \frac{\frac{\sqrt{3}}{2}}{\frac{1}{2}} = \frac{\sqrt{3}}{\not{2}} \cdot \frac{\not{2}}{1} = \sqrt{3}.$$

ALGEBRA QUESTION IN TRIGONOMETRY: If $\cos x = \frac{1}{2}$, and $\sin x > 0$, find $\tan x$.

SOLUTION If both $\cos x$ and $\sin x$ are positive, then x must be an arc in QI. Use the Pythagorean identity $\cos^2 x + \sin^2 x = 1$.

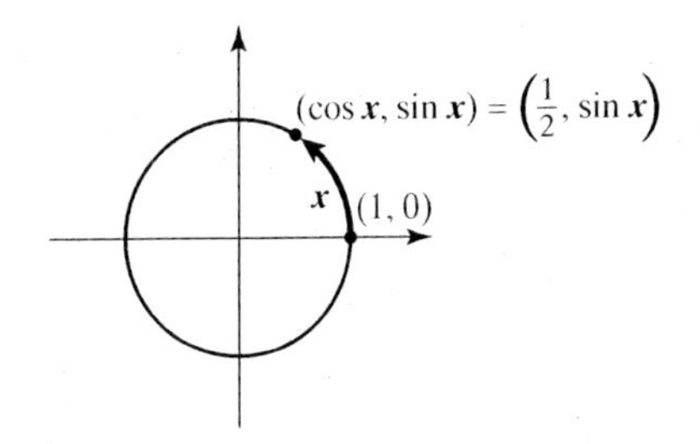

$$\cos^2 x + \sin^2 x = 1$$
$$\left(\frac{1}{2}\right)^2 + \sin^2 x = 1$$
$$\frac{1}{4} + \sin^2 x = 1$$
$$\sin^2 x = \frac{3}{4}$$
$$\sin x = \pm\sqrt{\frac{3}{4}} = \pm\frac{\sqrt{3}}{2}$$

Since x is an arc in QI, and we know $\sin x > 0$, $\sin x = \sqrt{3}/2$. But we were asked to find $\tan x$. Using the definition of $\tan x$, we find the ratio of the $\sin x$ and $\cos x$ values:

$$\tan x = \frac{\sin x}{\cos x} = \frac{\frac{\sqrt{3}}{2}}{\frac{1}{2}} = \frac{\sqrt{3}}{\not{2}} \cdot \frac{\not{2}}{1} = \sqrt{3}.$$

■

NOTE: In Example 7 you may have recognized the cosine functional value for the common arc $\pi/3$. In doing so, you would know that if $\cos x = 1/2$, then $\sin x = \sqrt{3}/2$ or $\sin x = -\sqrt{3}/2$, depending on the quadrant of x. It certainly saves time if you have carefully memorized all the common arc functional values. If you missed recognizing this common arc, you can always do the question using the Pythagorean identity as demonstrated in the solution.

Now that we have six circular function definitions, if we are given one functional value and the location of the terminal point of an arc, *we can find any of the remaining five functional values.*

EXAMPLE 8 If $\cos x = \frac{1}{2}$ and $\sin x > 0$, find $\sec x$, $\csc x$ and $\cot x$.

SOLUTION From Example 7, we found that $\sin x = \sqrt{3}/2$ and $\tan x = \sqrt{3}$. Therefore,

$$\sec x = \frac{1}{\cos x} = 2, \quad \csc x = \frac{1}{\sin x} = \frac{2}{\sqrt{3}} = \frac{2\sqrt{3}}{3}, \quad \text{and} \quad \cot x = \frac{1}{\tan x} = \frac{1}{\sqrt{3}} = \frac{\sqrt{3}}{3}.$$

■

We have a language that has two ways to write the reciprocal of functions (for example, $\frac{1}{\cos x}$ or $\sec x$). We actually have a language that has many ways to express the same idea.

EXAMPLE 9 What is another way to write $\tan x \cdot \csc x$?

POSSIBLE SOLUTION

We use the definitions of the new functions to rewrite this expression in terms of $\sin x$ and $\cos x$.

> Notice the original expression is defined where $\cos x \neq 0$ and $\sin x \neq 0$. (We can cancel $\sin x$ only if this is the case.) Consequently, the equivalent expression is only valid for the values where the original expression was defined.

$$\tan x \cdot \csc x = \frac{\sin x}{\cos x} \cdot \frac{1}{\sin x} = \frac{\cancel{\sin x}}{\cos x} \cdot \frac{1}{\cancel{\sin x}} = \frac{1}{\cos x}$$

Or, we could write $\tan x \cdot \csc x = \sec x$. ■

You may be wondering if there are more circular functions. Actually, we have all of them. But sometimes the new functions are written in a slightly different form. The following identities are all based on the original definitions.

$$\sec x = \frac{1}{\cos x} \qquad \cos x = \frac{1}{\sec x}$$

$$\csc x = \frac{1}{\sin x} \qquad \sin x = \frac{1}{\csc x}$$

$$\tan x = \frac{\sin x}{\cos x} = \frac{1}{\cot x} \qquad \cot x = \frac{\cos x}{\sin x} = \frac{1}{\tan x}$$

Now you need to practice all our ideas with the new circular functions.

Exercise Set 1.5

1. Find the tangent values, if they are defined, for each indicated arc.

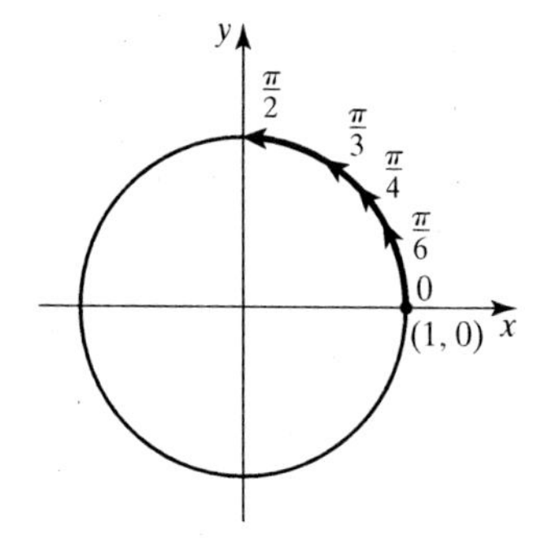

2. Find the cotangent values, if they are defined, for each indicated arc.

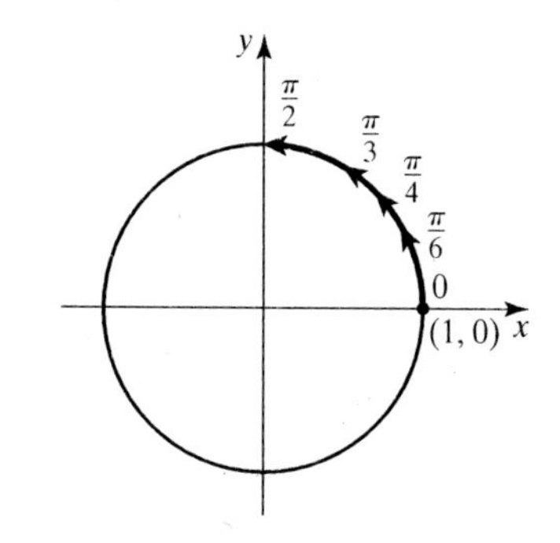

3. Find the secant values, if they are defined, for each indicated arc.

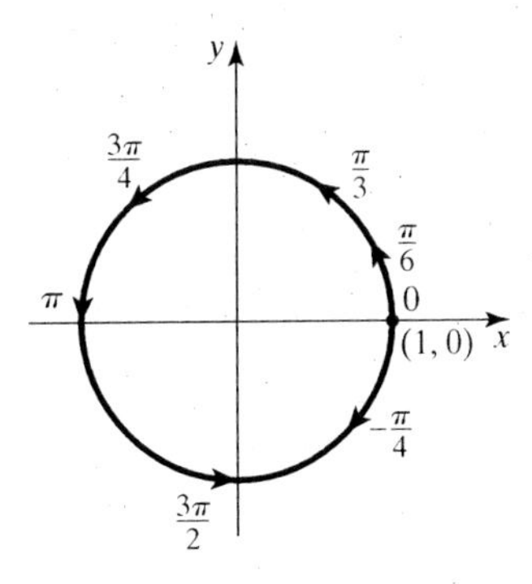

4. Find the cosecant values, if they are defined, for each indicated arc.

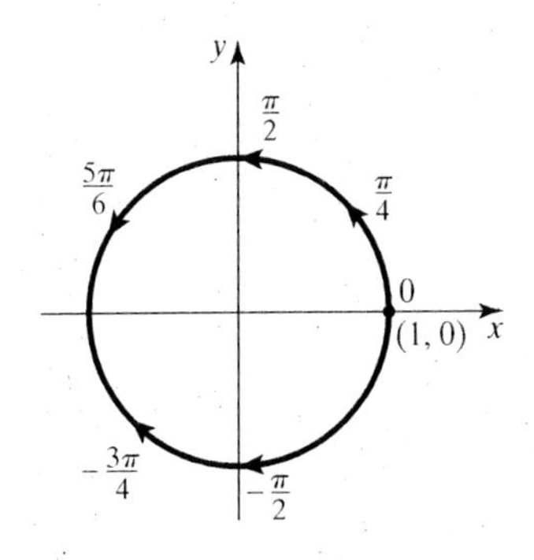

5–8. *If x is a positive arc in the first quadrant $(0 < x < \pi/2)$ with the given functional value, find x.*

EXAMPLE $\csc x = 2$

SOLUTION Using the definition $\csc x = \dfrac{1}{\sin x}$, if $\csc x = 2$, then $\sin x = \dfrac{1}{2}$. Therefore, $x = \dfrac{\pi}{6}$.

5. $\sec x = \sqrt{2}$
6. $\tan x = 1$
7. $\cot x = \sqrt{3}$
8. $\csc x = \dfrac{2}{\sqrt{3}}$

9–16. *Determine the quadrant in which the arc x has the following functional values.*

EXAMPLE $\sin x > 0$ and $\tan x < 0$

SOLUTION The $\sin x > 0$ in QI and QII, the $\tan x < 0$ in QII and QIV. The quadrant that satisfies both conditions is QII.

This is like the algebra questions about the positive or negative nature of the coordinates in the quadrants.

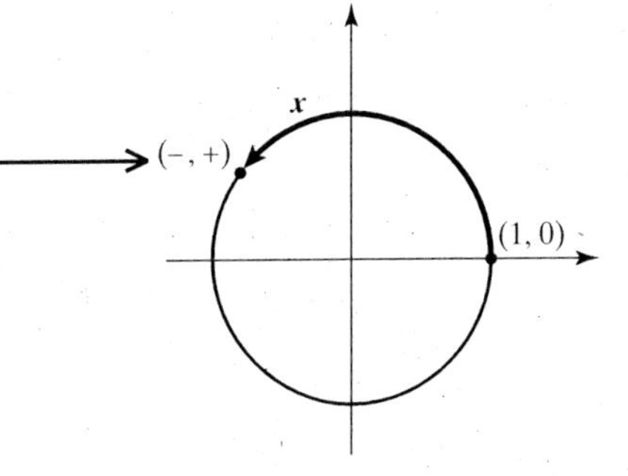

9. $\cos x > 0$ and $\tan x < 0$
10. $\sec x > 0$ and $\csc x < 0$
11. $\tan x > 0$ and $\sec x > 0$
12. $\sin x > 0$ and $\cot x > 0$
13. $\cot x > 0$ and $\csc x < 0$
14. $\cos x < 0$ and $\tan x < 0$
15. $\sin x > 0$ and $\sec x < 0$
16. $\csc x < 0$ and $\sec x < 0$

17–40. *Find the exact functional value for each expression if it is defined.*

EXAMPLE Find $\sec \dfrac{5\pi}{4}$.

SOLUTION Graph arc $5\pi/4$. It terminates in QIII and the reference arc is $\hat{x} = \pi/4$.

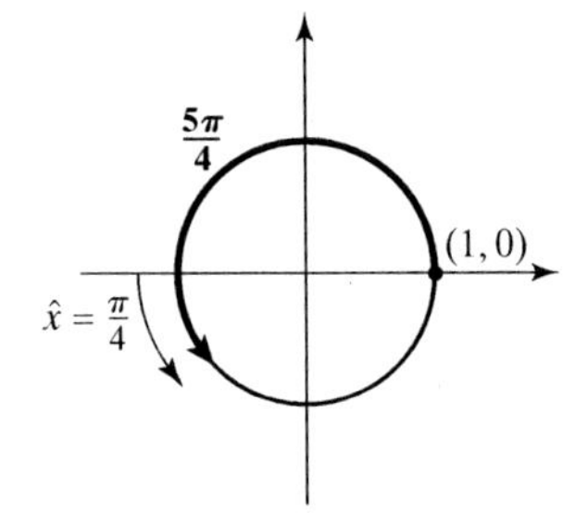

$$\sec \frac{5\pi}{4} = \frac{1}{\cos \frac{5\pi}{4}}$$

$$= \frac{1}{-\cos \frac{\pi}{4}} = \frac{1}{-\frac{1}{\sqrt{2}}} = -\sqrt{2}$$

17. $\tan \dfrac{5\pi}{4}$
18. $\csc \dfrac{2\pi}{3}$
19. $\sec \dfrac{5\pi}{6}$
20. $\sec \dfrac{11\pi}{6}$
21. $\csc \dfrac{\pi}{6}$
22. $\cot \dfrac{3\pi}{4}$
23. $\csc \dfrac{9\pi}{4}$
24. $\tan \dfrac{7\pi}{6}$
25. $\cot\left(-\dfrac{4\pi}{3}\right)$
26. $\sec \dfrac{7\pi}{3}$
27. $\tan\left(-\dfrac{7\pi}{6}\right)$
28. $\csc\left(-\dfrac{11\pi}{6}\right)$
29. $\sec\left(-\dfrac{\pi}{4}\right)$
30. $\cot\left(-\dfrac{7\pi}{3}\right)$
31. $\tan \dfrac{5\pi}{3}$
32. $\cot(-2\pi)$
33. $\sec\left(-\dfrac{3\pi}{2}\right)$
34. $\tan \pi$
35. $\csc\left(-\dfrac{2\pi}{3}\right)$
36. $\sec 3\pi$
37. $\csc^2 \dfrac{7\pi}{6} + (\sec 0)\left(\csc \dfrac{\pi}{6}\right)$
38. $\tan \dfrac{2\pi}{3} + \left(\cot \dfrac{5\pi}{6}\right)\left(\sin \dfrac{\pi}{2}\right)$
39. $\dfrac{\sin \frac{\pi}{2} + \tan \frac{2\pi}{3}}{\sec \frac{5\pi}{3}}$
40. $\dfrac{\cot \frac{3\pi}{4} + 2 \sin \frac{5\pi}{6}}{\csc \frac{\pi}{6}}$

41–52. *Use a calculator to find an approximation rounded to four decimal places for the following functional values. Use the radian setting and be careful of the parentheses.*

EXAMPLE $4 \cot 5.27$

SOLUTION

$4 \cot 5.27 \approx -2.4945$

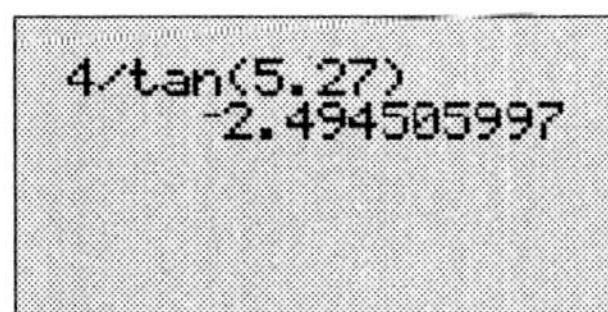

41. $\tan \dfrac{7\pi}{9}$
42. $\sec \dfrac{8\pi}{5}$
43. $\csc\left(\dfrac{5}{9} + \dfrac{2}{3}\right)$
44. $\cot(-0.689)$
45. $\sec 89.09$
46. $\csc(-3.25)$
47. $\tan^2 1.56689$
48. $\cot^2 100$
49. $15 \csc 5 + 2.2 \cot 1$
50. $\dfrac{\sec 2 - 5 \sin 8}{\tan 7.46}$
51. $\cot \dfrac{\pi}{5} - \dfrac{3 \sin(-9)}{\tan 33}$
52. $15.69 \sin(17.69 + 8) - 44.6 \sec 15.69$

53–58. *Find the indicated exact functional values.*

EXAMPLE If $\cos x = \frac{3}{5}$ and $\sin x < 0$, find $\cot x$.

SOLUTION Since $\cos x > 0$ and $\sin x < 0$, the arc must terminate in QIV. Use the Pythagorean identity:

$$\cos^2 x + \sin^2 x = 1$$

$$\left(\frac{3}{5}\right)^2 + \sin^2 x = 1$$

$$\frac{9}{25} + \sin^2 x = 1$$

$$\sin^2 x = 1 - \frac{9}{25} = \frac{25}{25} - \frac{9}{25} = \frac{16}{25}$$

$$\sin x = \pm\sqrt{\frac{16}{25}} = \pm\frac{4}{5}$$

Since $\sin x < 0$, $\sin x = -\frac{4}{5}$. Therefore,

$$\cot x = \frac{\cos x}{\sin x} = \frac{\frac{3}{5}}{-\frac{4}{5}} = \frac{3}{5} \cdot \left(-\frac{5}{4}\right) = -\frac{3}{4}.$$

■

53. If $\sin x = -\dfrac{5}{13}$ and $\cos x < 0$, find $\tan x$.

54. If $\cos x = -\dfrac{1}{2}$ and $\sin x > 0$, find $\csc x$.

55. If $\cos x = -\dfrac{7}{25}$ and $\sin x > 0$, find the following.

a. $\sin x$ **b.** $\sec x$ **c.** $\tan x$ **d.** $\csc x$ **e.** $\cot x$

56. If $\sin x = \dfrac{\sqrt{2}}{2}$ and $\dfrac{\pi}{2} < x < \pi$, find the following.

a. $\cos x$ **b.** $\sec x$ **c.** $\tan x$ **d.** $\csc x$ **e.** $\cot x$

57. If $\cos x = -\dfrac{5}{\sqrt{26}}$ and x terminates in QIII, find $\cot x$.

58. If $\sin x = \dfrac{1}{4}$ and $\cos x > 0$, find $\sec x$.

59–70. *Are the following statements true or false? If a statement is false, explain why or give an example that shows why it is false.*

59. $\tan x = \dfrac{\cos x}{\sin x}$

60. $\cot x = \dfrac{\cos x}{\sin x}$

61. $\tan^2 x = (\tan x)^2 = \left(\dfrac{\sin x}{\cos x}\right)^2 = \dfrac{\sin^2 x}{\cos^2 x}$

62. $\cos^2 x + \sin^2 x = 1$

63. $\sin^2 x = 1 - \cos^2 x$

64. $\sin^2 x = 1 - \cos^2 t$

65. $\tan \pi = -1$

66. If $\cos x = \dfrac{1}{2}$, then $\sin x = \dfrac{\sqrt{3}}{2}$.

67. $\cos x = \dfrac{1}{\csc x}$

68. $\tan 0 = \tan \pi = \tan 2\pi = \tan 3\pi = 0$

69. The tangent function is undefined at $x = \dfrac{\pi}{2}$.

70. The secant function is undefined at $x = \pi$.

71–74. *Use the definitions of the circular functions to rewrite the right side of the following equations in terms of* $\cos x$ *and* $\sin x$. *Then simplify to determine if the equation is an identity.*

EXAMPLE $\tan x = \sin x \cdot \sec x$

SOLUTION

$$\begin{aligned} \tan x &= \sin x \cdot \sec x \\ &= \sin x \cdot \frac{1}{\cos x} \\ &= \frac{\sin x}{\cos x} = \tan x \\ \tan x &= \tan x \end{aligned}$$

This is an identity.

71. $\cot x = \cos x \csc x$

72. $\csc x = \cot x \csc x$

73. $\sec^2 x = \dfrac{\cos^2 x + \sin^2 x}{\cos^2 x}$

74. $\tan^2 x = \dfrac{1 - \cos^2 x}{\cos^2 x}$

75–80. *Find the arc x that satisfies the given conditions.*

75. $\tan x = 1, \cos x = -\dfrac{1}{\sqrt{2}},\ 0 < x < 2\pi$

76. $\sin x = \dfrac{1}{\sqrt{2}}, \sec x = \sqrt{2},\ 0 < x < 2\pi$

77. $\cos x = -\dfrac{1}{2}, \dfrac{\pi}{2} \le x \le \pi$

78. $\tan x = -\sqrt{3},\ -\dfrac{\pi}{2} \le x \le 0$

79. $\csc x = -1,\ 0 < x < 2\pi$

80. $\sec x = -\dfrac{2}{\sqrt{3}},\ \pi \le x \le \dfrac{3\pi}{2}$

81. The domain of both the tangent and secant functions is

$$x \ne \frac{\pi}{2} + k\pi, \text{ where } k \in \{\ldots, -2, -1, 0, 1, 2, \ldots\}.$$

Use this formula to list several values of x that are *not* in their domain.

82. The domain of both the cotangent and cosecant functions is

$$x \ne k\pi, \text{ where } k \in \{\ldots, -2, -1, 0, 1, 2, \ldots\}.$$

Use this formula to list several values of x that are *not* in their domain.

Explore the Pattern

83. In Section 1.3, you were asked to discover a pattern for the cosine and sine functional values of common arcs between and including 0 and $\pi/2$. Similarly, a pattern can be found for the tangent function.

Complete the table on the next page for $\tan x$ using the given patterns of $\sin x$ and $\cos x$ functional values of the common arcs to find a pattern for $\tan x$. (*Hint:* Leave the values expressed as a radical of a fraction.) Carefully explain the pattern.

	0	$\frac{\pi}{6}$	$\frac{\pi}{4}$	$\frac{\pi}{3}$	$\frac{\pi}{2}$
$\sin x$	$\frac{\sqrt{0}}{2}$	$\frac{\sqrt{1}}{2}$	$\frac{\sqrt{2}}{2}$	$\frac{\sqrt{3}}{2}$	$\frac{\sqrt{4}}{2}$
$\cos x$	$\frac{\sqrt{4}}{2}$	$\frac{\sqrt{3}}{2}$	$\frac{\sqrt{2}}{2}$	$\frac{\sqrt{1}}{2}$	$\frac{\sqrt{0}}{2}$
$\tan x$	$\sqrt{\frac{0}{4}}$	$\sqrt{\frac{1}{?}}$	$\sqrt{\frac{?}{?}}$	$\sqrt{\frac{?}{?}}$	$\sqrt{\frac{?}{?}}$

Discussion

84. **a.** State the domain for the tangent and secant functions.

b. Discuss why the following diagram could be used to visually represent the domain of the tangent and secant function.

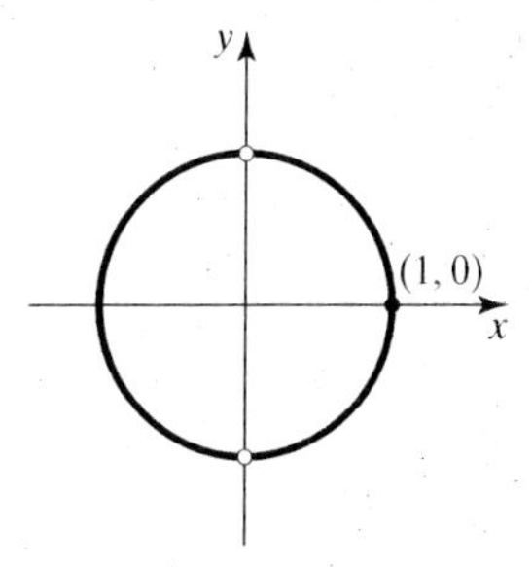

85. **a.** State the domain for the cotangent and cosecant functions.

b. Discuss why the following diagram could be used to visually represent the domain of the cotangent and cosecant functions.

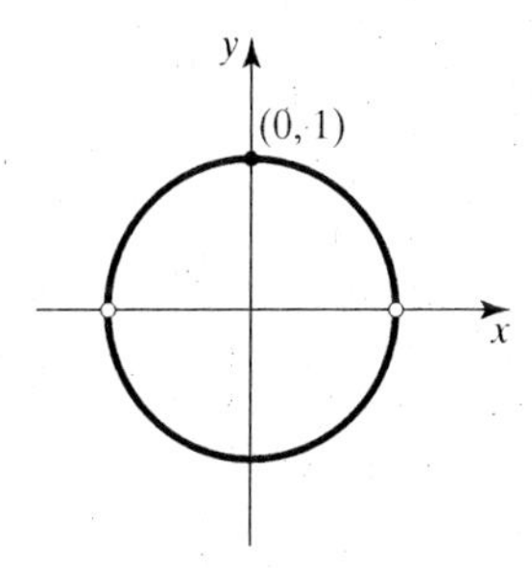

1.6 Negative Identities and Periods for the Circular Functions

As a result of our work in this chapter, you may have discovered some interesting relationships that exist between different arcs on the unit circle. Using the symmetry of the circle again, let's find the relationship between the functional values of positive arcs and the corresponding negative arcs.

EXAMPLE 1 Consider the arcs $\frac{3\pi}{4}$ and $-\frac{3\pi}{4}$.

a. Compare the cosine values of these arcs.

b. Compare the sine values of these arcs.

c. Compare the tangent values of these arcs.

SOLUTION

a. $\cos\frac{3\pi}{4} = -\frac{1}{\sqrt{2}}$ and $\cos\left(-\frac{3\pi}{4}\right) = -\frac{1}{\sqrt{2}}$. The cosine values are the same.

b. $\sin\frac{3\pi}{4} = \frac{1}{\sqrt{2}}$ and $\sin\left(-\frac{3\pi}{4}\right) = -\frac{1}{\sqrt{2}}$. The sine values are opposites.

c. $\tan\frac{3\pi}{4} = -1$ and $\tan\left(-\frac{3\pi}{4}\right) = 1$. The tangent values are opposites. ■

EXAMPLE 2 Consider the arcs $\frac{\pi}{3}$ and $-\frac{\pi}{3}$.

a. Compare the cosine values of these arcs.

b. Compare the sine values of these arcs.

c. Compare the tangent values of these arcs.

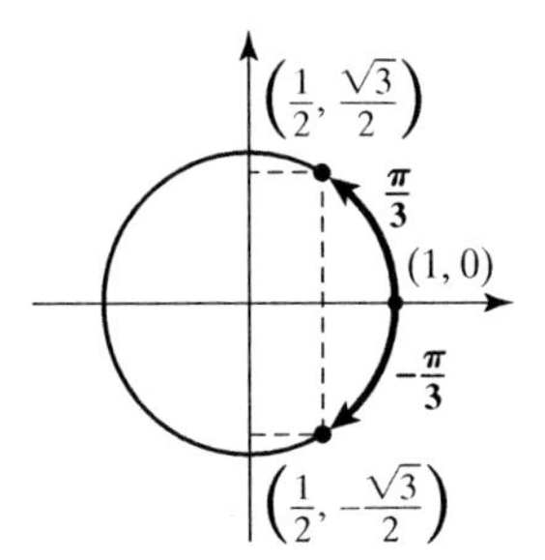

SOLUTION

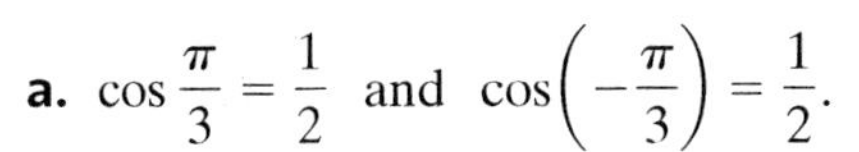

a. $\cos\frac{\pi}{3} = \frac{1}{2}$ and $\cos\left(-\frac{\pi}{3}\right) = \frac{1}{2}$. The cosine values are the same.

b. $\sin\frac{\pi}{3} = \frac{\sqrt{3}}{2}$ and $\sin\left(-\frac{\pi}{3}\right) = -\frac{\sqrt{3}}{2}$. The sine values are opposites.

c. $\tan\frac{\pi}{3} = \sqrt{3}$ and $\tan\left(-\frac{\pi}{3}\right) = -\sqrt{3}$. The tangent values are opposites. ■

The relationship in general between the functional values of any positive arc and its corresponding negative arc, which is found in Examples 1 and 2, is illustrated in the following diagrams.

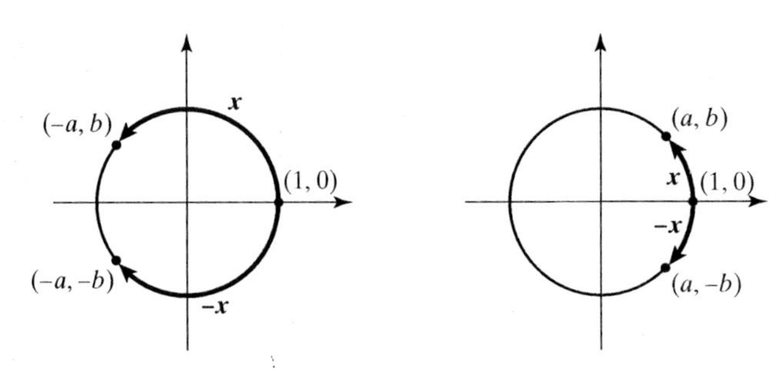

These comparisons hold for any two arcs x and $-x$. If we write the relationships in a more general form, we get the following three new identities.

Negative Identities

For all $x \in \mathbb{R}$ in the domain of the function,

a. $\cos(-x) = \cos x$

b. $\sin(-x) = -\sin x$

c. $\tan(-x) = -\tan x$

Periodic Property of the Circular Functions

Another very important result we discussed in Section 1.4 occurs with coterminal arcs (arcs that have both the same initial point and same terminal point).

$$\cos\frac{\pi}{4} = \cos\left(\frac{\pi}{4} + 2\pi\right) = \cos\left(\frac{\pi}{4} + 2 \cdot 2\pi\right) = \cos\left(\frac{\pi}{4} + 3 \cdot 2\pi\right)$$

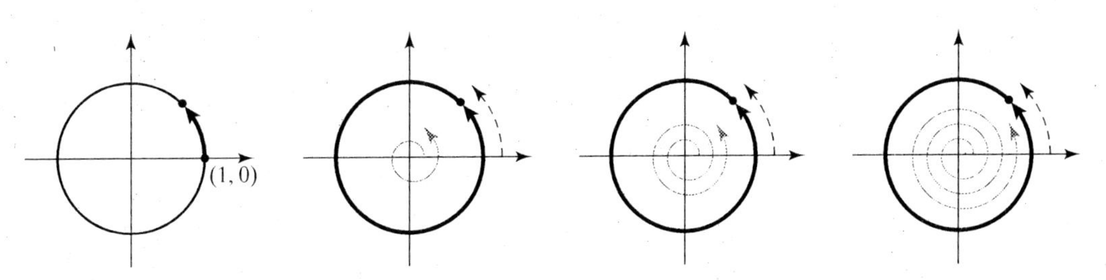

Every time we add another complete wrap (2π) of the circle from the terminal point of any arc, we arrive at the same terminal point and, therefore, the same functional values. As a result, we introduced the formulas that reflect this repetitious or periodic property, where $k \in \{\ldots, -2, -1, 0, 1, 2, \ldots\}$:

$$\cos(x + k \cdot 2\pi) = \cos x$$
$$\sin(x + k \cdot 2\pi) = \sin x$$

We now define a *periodic function* more formally.

Periodic Function

If f is a function and there is a number $p > 0$ that is the smallest value such that $f(x + p) = f(x + k \cdot p) = f(x)$ for all x in the domain of f, where $k \in \{\ldots, -2, -1, 0, 1, 2, \ldots\}$, then f is a **periodic function** and p is the **period** of f.

Period of cos x and sin x

Using this definition, $f(x + p) = f(x + k \cdot p) = f(x)$, and knowing that

$$\cos(x + 2\pi) = \cos(x + k \cdot 2\pi) = \cos(x)$$

and

$$\sin(x + 2\pi) = \sin(x + k \cdot 2\pi) = \sin(x),$$

where $k \in \{\ldots, -2, -1, 0, 1, 2, \ldots\}$, we see that the cosine and sine functions are periodic, and the period of these functions is $p = 2\pi$.

What about the remaining four circular functions? Are they periodic? If so, what is the period? Let's see.

Period of sec x and csc x

It seems reasonable that if $\cos x$ is periodic with period $p = 2\pi$, the reciprocal function, $\sec x$, is also periodic with period $p = 2\pi$. In other words, if the values of $\cos x$ repeat every wrap of the circle (2π), then so do the reciprocals of these values. The same can be said of $\sin x$ and its reciprocal function, $\csc x$. Therefore, $\sec x$ and $\csc x$ are periodic with period $p = 2\pi$ for all values of x in their respective domains.

Period of tan x and cot x

The $\tan x$ and $\cot x$ are periodic as well, since they are also the ratios of periodic functions. As a result, $\tan x$ and $\cot x$ values repeat every 2π. A closer look at these functions indicates their periods are not as large as 2π. Let's look at a brief table of the tangent function to see where it begins to repeat values. Recall from the definition of a periodic function, p should be the *smallest value* such that $\tan(x + p) = \tan(x)$.

x	$\tan x$
0	0
$\frac{\pi}{6}$	$\frac{1}{\sqrt{3}}$
$\frac{\pi}{4}$	1
$\frac{\pi}{3}$	$\sqrt{3}$
$\frac{\pi}{2}$	undefined
$\frac{2\pi}{3}$	$-\sqrt{3}$
$\frac{3\pi}{4}$	-1
$\frac{5\pi}{6}$	$-\frac{1}{\sqrt{3}}$

x	$\tan x$
π	0
$\frac{7\pi}{6}$	$\frac{1}{\sqrt{3}}$
$\frac{5\pi}{4}$	1
$\frac{4\pi}{3}$	$\sqrt{3}$
$\frac{3\pi}{2}$	undefined
$\frac{5\pi}{3}$	$-\sqrt{3}$
$\frac{7\pi}{4}$	-1
$\frac{11\pi}{6}$	$-\frac{1}{\sqrt{3}}$

$$\tan 0 = \tan(0 + \pi)$$

$$\tan\frac{\pi}{6} = \tan\frac{7\pi}{6} = \tan\left(\frac{\pi}{6} + \pi\right)$$

$$\tan\frac{\pi}{4} = \tan\frac{5\pi}{4} = \tan\left(\frac{\pi}{4} + \pi\right)$$

$$\vdots$$

$$\tan(x) = \tan(x + \pi)$$

Therefore $\tan x$ is periodic with period $p = \pi$. Using the same argument, $\cot x$ is also periodic with period $p = \pi$.

Let's summarize the new identities and the periods of our circular functions. The period of a function is also referred to as the length of the *cycle*. The negative identities for the secant, cosecant, and cotangent are included and left as an exercise for you to verify.

Negative Identities

$\cos(-x) = \cos x$ $\quad$ $\sec(-x) = \sec x$

$\sin(-x) = -\sin x$ $\quad$ $\csc(-x) = -\csc x$

$\tan(-x) = -\tan x$ $\quad$ $\cot(-x) = -\cot x$

Function	Period
$\cos x$	2π
$\sin x$	
$\sec x$	
$\csc x$	
$\tan x$	π
$\cot x$	

EXAMPLE 3 Find all values of x within one wrap $(0 \le x < 2\pi)$ of the circle where $\sin x = \sqrt{2}/2$. Then use the periodic property to find an expression for all values of x.

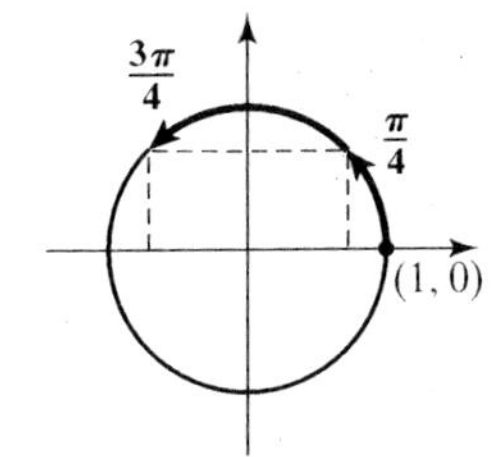

SOLUTION We know that if $\sin x = \sqrt{2}/2$, then

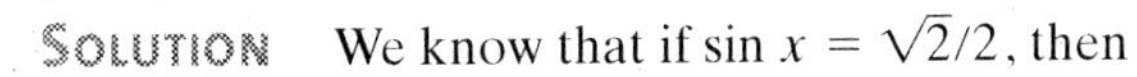

$$x = \frac{\pi}{4} \quad \text{and} \quad x = \frac{3\pi}{4}$$

within one wrap of the circle. Using the periodic property $\sin x = \sin(x + k \cdot 2\pi)$, all values of x are expressed as

$$x = (\pi/4) + k \cdot 2\pi, \text{ and } x = (3\pi/4) + k \cdot 2\pi.$$

Application of Circular Functions

Since the circular functions are periodic, they are often used as mathematical models to describe repetitious behavior. Repetitious situations can be observed with a swinging pendulum, a bobbing spring, a vibrating violin string, and with alternating electrical current, to name just a few examples.

Simple Harmonic Motion

A point on a pendulum that is swinging back and forth, or the bobbing up and down of a weight attached to a spring, repeats the movement in equal intervals of time. This systematic motion about a rest position (equilibrium point) is periodic.

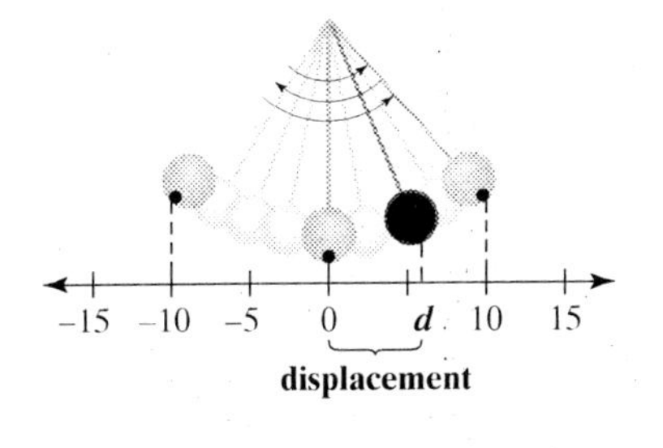

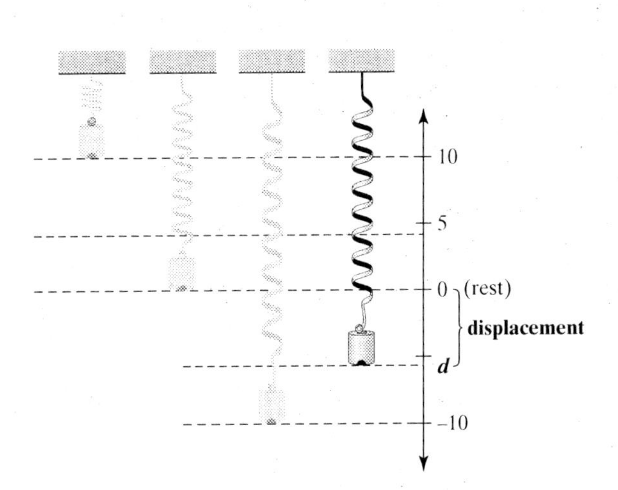

The motion is often defined as **simple harmonic motion**. For simple harmonic motion, the displacement of a point about its rest position can be determined using a mathematical model that uses the cosine or sine function. These two models may take on the following form:

$$d(t) = A\cos(Bt) \qquad \text{or} \qquad d(t) = A\sin(Bt)$$

- $d(t)$ is the displacement determined at a specific measure of time t, where $d(t)$ is usually positive when the position is up or to the right of the rest position, and negative when down or to the left of rest.
- A and B are constants where A represents the initial displacement of the point from the rest position, and B is a constant value that is determined by the speed or frequency of the motion.

For the problems in this section you will be given all necessary constants to determine the displacement. We will discuss these values in more detail in the next chapter. Be careful with the evaluations of these problems. In other words, watch the parentheses. If these problems are presented without parentheses around Bt, don't forget to insert them for evaluation.

EXAMPLE 4 Determine the displacement of a point on a pendulum with initial displacement of 6 centimeters (cm) to the right that has the simple harmonic motion form $d(t) = 6\cos(2\pi t)$, where t is in seconds and $d(t)$ is in centimeters. If necessary, approximate the displacement to the nearest tenth of a centimeter. Draw the location of the pendulum at each given time.

a. $t = 0$ **b.** $t = \frac{1}{4}$ **c.** $t = \frac{1}{2}$ **d.** $t = 1$ **e.** $t = 1.2$

SOLUTION

a. If $t = 0$,

$$d(0) = 6\cos(2\pi \cdot 0) = 6\cos(0) = 6(1) = 6.$$

The displacement is 6 cm to the right, which is what we would expect since we were told this was the initial displacement $(t = 0)$.

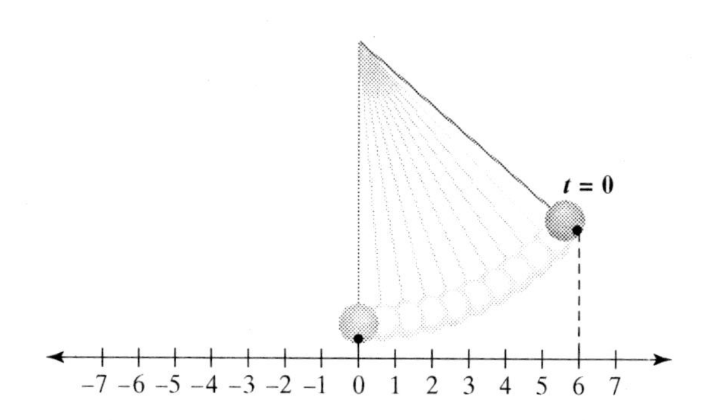

b. If $t = \frac{1}{4}$,

$$d\left(\frac{1}{4}\right) = 6\cos\left(2\pi \cdot \frac{1}{4}\right) = 6\cos\left(\frac{\pi}{2}\right) = 6(0) = 0.$$

The point is in the rest position since $d = 0$. ($d = 0$ does not require units.)

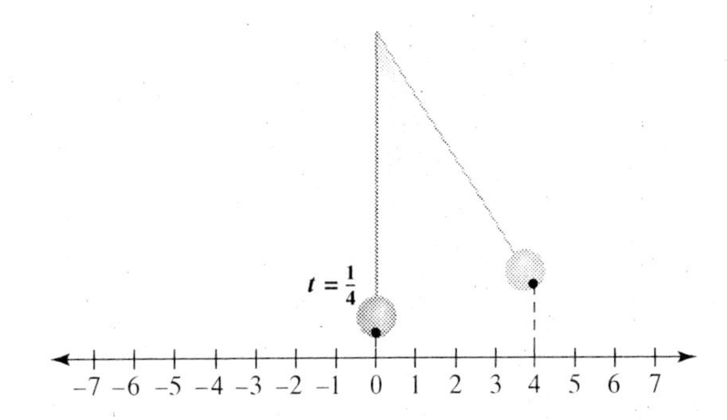

c. If $t = \frac{1}{2}$,

$$d\left(\frac{1}{2}\right) = 6\cos\left(2\pi \cdot \frac{1}{2}\right) = 6\cos(\pi) = 6 \cdot (-1) = -6.$$

The point is displaced 6 cm to the left.

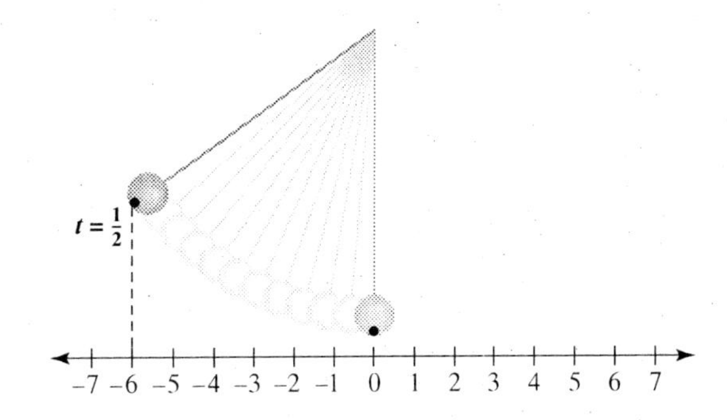

d. If $t = 1$,

$$d(1) = 6\cos(2\pi \cdot 1) = 6(1) = 6.$$

The point is again displaced 6 cm to the right.

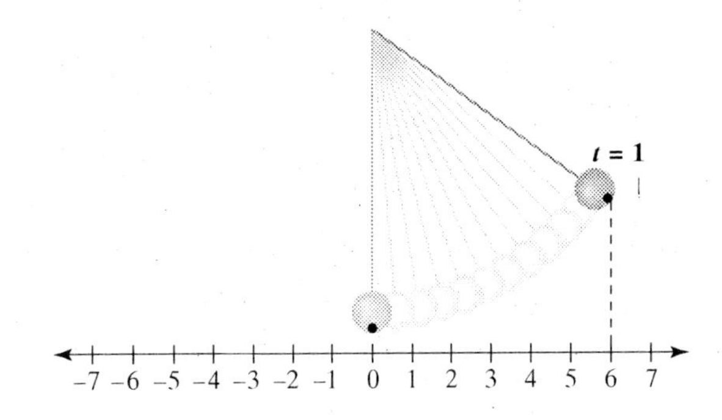

e. If $t = 1.2$,

$$d(1.2) = 6\cos(2\pi \cdot 1.2) = 6\cos(2.4\pi) \approx 1.9.$$

The point is displaced approximately 1.9 cm to the right.

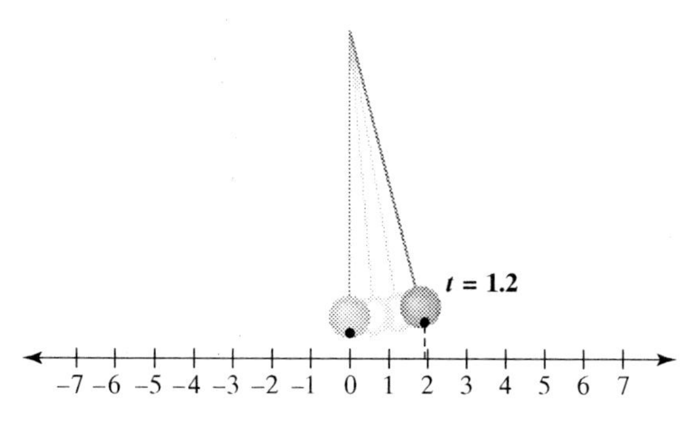

EXAMPLE 5 If the spring shown is weighted on one end and is bouncing up and down every 2 seconds with an initial displacement of 5.2 inches below the rest position, then the displacement is given by the equation $d(t) = -5.2\cos(\pi t)$, where t is in seconds and the displacement is in inches (in.). Find the displacement (to the nearest tenth of an inch where necessary) for the following times.

+5.2 in.
0 (rest)
−5.2 in.

a. $t = 0$ **b.** $t = \frac{1}{2}$ **c.** $t = 1\frac{1}{4}$

SOLUTION

a. $d(0) = -5.2\cos(\pi \cdot 0) = -5.2\cos(0) = -5.2 \cdot 1 = -5.2$

The spring is displaced -5.2 in., or down 5.2 in.

b. $d\left(\frac{1}{2}\right) = -5.2\cos\left(\pi \cdot \frac{1}{2}\right) = -5.2\cos\left(\frac{\pi}{2}\right) = 0$

The spring is displaced 0, or it is at the rest position.

c. $d(1\frac{1}{4}) = -5.2\cos(\pi \cdot 1\frac{1}{4}) = -5.2\cos\left(\frac{5\pi}{4}\right) \approx 3.676955\ldots \approx 3.7$

The spring is displaced up approximately 3.7 in.

EXAMPLE 6 For an electrical circuit, the voltage at a time t is given by the equation $E(t) = 200\sin 2250t$, where t is expressed in seconds and $E(t)$ is expressed in volts.

a. Find the voltage at time $t = 0.005$ to the nearest tenth of a volt.

b. Find the maximum voltage for this circuit.

SOLUTION

a. Letting $t = 0.005$,

$$E(0.005) = 200\sin(2250 \cdot 0.005) = 200\sin(11.25)$$
$$E(0.005) \approx -193.6 \text{ volts}$$

b. Since the maximum value of the sine function is 1, $E(t) = 200(1)$. Therefore, 200 volts will be the maximum voltage for this circuit.

CAUTION: Just as with other functions whose notation requires careful reading and evaluating, the circular functions are no exception. For $\cos Bt$ or $\cos B(t+C)$, you need to see it as $\cos(Bt)$ or $\cos(B(t+C))$. Recall that parentheses are always implied around the arc. Remember to evaluate Bt or $B(t+C)$ first, then find the functional value. (Do not read $\cos Bt$ as $(\cos B)t$, which finds the functional value $\cos B$ first, then multiplies the answer by t. As a reminder, many calculators automatically insert the left parenthesis before the arc.) The same would be true when you use any of the other circular functions.

Exercise Set 1.6

1–2. *Fill in the blank to make a true statement.*

1. The circular functions with period 2π are ________.
2. The circular functions with period π are ________.

3–12. *Are the following statements true or false? If a statement is false, explain why or give an example that shows why it is false.*

3. All the circular functions are periodic functions.
4. $\cos(-x) = \cos x$
5. $\csc(-x) = -\csc x$
6. $\cot(-x) = \cot x$
7. $\tan\left(\pi + \frac{\pi}{8}\right) = \tan\frac{\pi}{8}$
8. $\cos\left(\frac{\pi}{3} + \pi\right) = \cos\frac{\pi}{3}$
9. $\sin 3x = (\sin 3)x$
10. $\cos 5x = \cos(5x)$
11. $\sec(x + 2\pi) = \sec x$
12. $\sin(-x) = -\sin x$

13–16. *Match each value with a value from the right column so that the functional values are equal. A value may be used more than once or not at all. More than one answer is possible.*

13. $\sin\frac{2\pi}{3} =$ ________
14. $\tan\frac{3\pi}{4} =$ ________
15. $\cos\frac{7\pi}{6} =$ ________
16. $\sin\left(2\pi + \frac{2\pi}{3}\right) =$ ________

i. $\tan\left(-\frac{\pi}{4}\right)$

ii. $\tan\frac{7\pi}{4}$

iii. $\sin\frac{8\pi}{3}$

iv. $\cos\left(-\frac{7\pi}{6}\right)$

v. $\cos\frac{19\pi}{6}$

vi. $\sin\left(-\frac{2\pi}{3}\right)$

vii. $-\sin\left(-\frac{2\pi}{3}\right)$

17–18. *Find all values of x within one wrap of the circle $(0 \le x < 2\pi)$ that make the statement true. Then use the periodic property to find all values that make it true.*

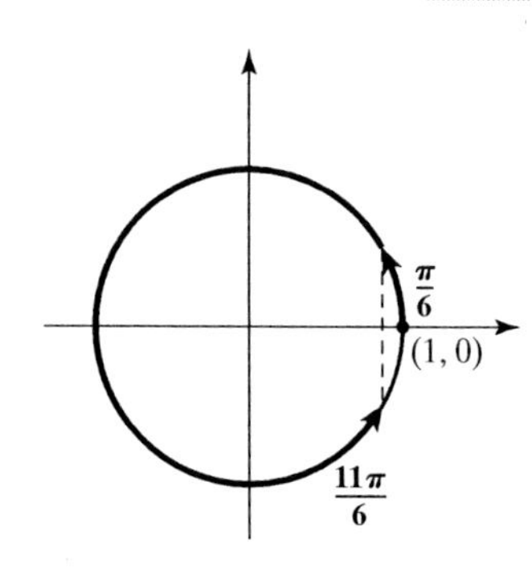

EXAMPLE $\cos x = \frac{\sqrt{3}}{2}$

SOLUTION $\cos x = \frac{\sqrt{3}}{2}$ when $x = \frac{\pi}{6}$ and $\frac{11\pi}{6}$ (within one wrap).

Since the period of the cosine is 2π,

$$x = \frac{\pi}{6} + k \cdot 2\pi \quad \text{and} \quad x = \frac{11\pi}{6} + k \cdot 2\pi.$$

17. **a.** $\cos x = \frac{1}{2}$ **b.** $\sin x = -\frac{\sqrt{2}}{2}$ **c.** $\tan x = 1$

18. **a.** $\sin x = \frac{1}{2}$ **b.** $\cos x = -\frac{\sqrt{2}}{2}$ **c.** $\tan x = -1$

19–22. *A pendulum is moving back and forth (oscillating) and the displacement of its periodic motion is described as $d(t) = 4\cos(\pi t)$, where t is in seconds and the displacement is in inches. Determine the displacement to the nearest tenth (if necessary) of an inch for the following times, and draw the location of the pendulum.*

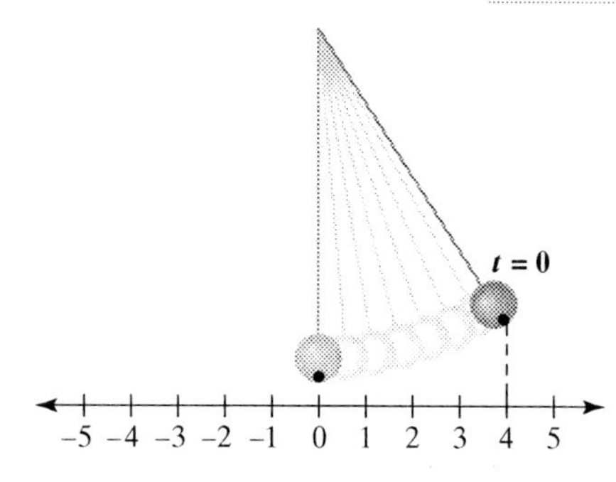

EXAMPLE $t = 0$

SOLUTION $d(0) = 4\cos(\pi \cdot 0) = 4\cos(0) = 4 \cdot 1 = 4$. So, $d(0) = 4$ in.

19. $t = \frac{1}{2}$ **20.** $t = 1$

21. $t = \frac{3}{2}$ **22.** $t = 2.4$

23–26. *A spring weighted at one end is bouncing up and down with initial displacement of 15 cm. The displacement is given by the equation $d(t) = 15\cos(5t)$, where t is in seconds and $d(t)$ is in centimeters. Find the displacement to the nearest tenth of a centimeter for the following times.*

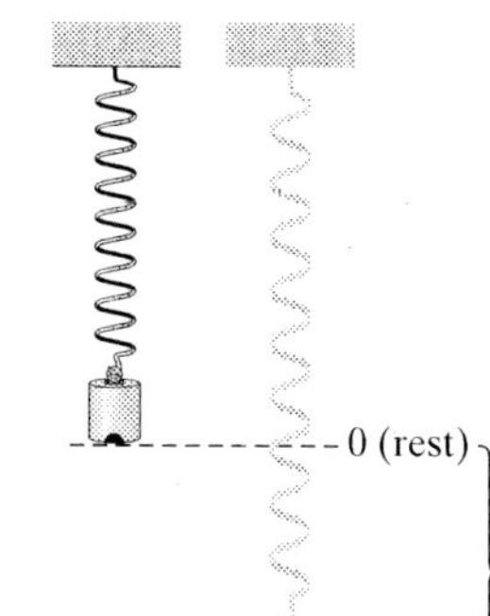

23. $t = 0.002$

24. $t = 0.5$

25. $t = 2$

26. $t = 4.6$

27–30. *The voltage E in a circuit is given by $E(t) = 170\sin 376t$, where t is in seconds and $E(t)$ is in volts. Find the voltage to the nearest tenth of a volt at the following times.*

27. $t = 0.005$ **28.** $t = 0.017$ **29.** $t = \frac{1}{4}$ **30.** $t = 1$

31–34. *The change in pressure P on an eardrum from a pure tone at time t is given by the equation: $P(t) = 0.001\sin(1882\pi t + (\pi/4))$, where t is expressed in seconds, and P is in pounds per square foot. Find P to the nearest thousandth of a pound for the following times t.*

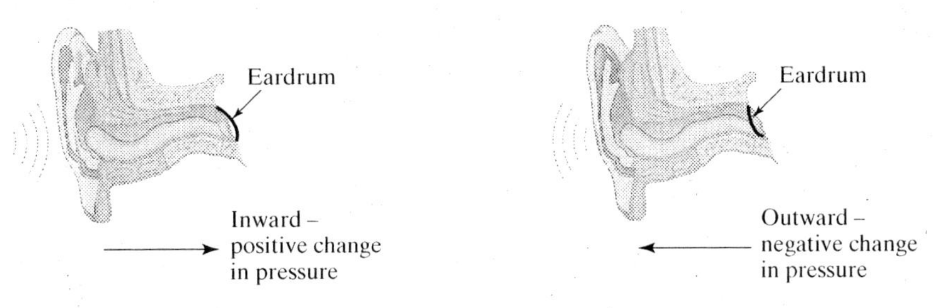

31. $t = 0.01$ **32.** $t = 1.01$ **33.** $t = 4$ **34.** $t = 6.5$

35. Explain why in Example 5 the spring whose displacement is given by

$$d(t) = -5.2\cos(\pi t)$$

will not bounce more than 5.2 inches above or below the rest position.

36. Using the definitions of the reciprocal functions, verify the following negative identities:

a. $\sec(-x) = \sec x$ **b.** $\csc(-x) = -\csc x$ **c.** $\cot(-x) = -\cot x$

37. On the first page in Section 1.1, points labeled A, S, T, and C in the four quadrants are given. Explain how each letter designates the positive nature of the functional values in those quadrants. (Some students remember this relationship by saying "*A*ll *S*tudents *T*ake *C*alculus!")

Web Activity

38. A Web search of Trigonometry and Harmonic Motion for simple harmonic motion applet demonstrations can provide you with some interesting examples. Find a demonstration of simple harmonic motion that includes examples other than pendulums and springs. Carefully describe the simple harmonic motion demonstration and provide the web address.

Chapter 1 Summary

1.2 If we wrap the real number line that is tangent to the unit circle at $(1, 0)$ around the circle, we create arcs that have length and direction and that correspond to the real numbers. If the line is wrapped in a counterclockwise direction, the arc corresponds to a positive real number. If the line is wrapped in a clockwise direction, the arc corresponds to a negative real number. The terminal point $P(x,y)$ of each arc t with initial point $(1, 0)$ lies on the unit circle. The coordinates of P are functions of t (called circular functions):

$$x = \cos t$$
$$y = \sin t,$$

where $t \in \mathbb{R}$.

Using familiar ideas of algebra, we have the following connections and definitions given on the next page.

ALGEBRA IDEAS:

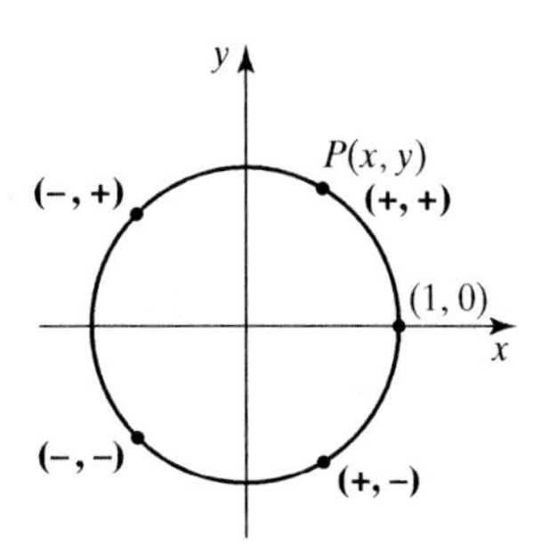

$$x^2 + y^2 = 1$$

$$-1 \le x \le 1$$
$$-1 \le y \le 1$$

ALGEBRA IDEAS IN TRIGONOMETRY:

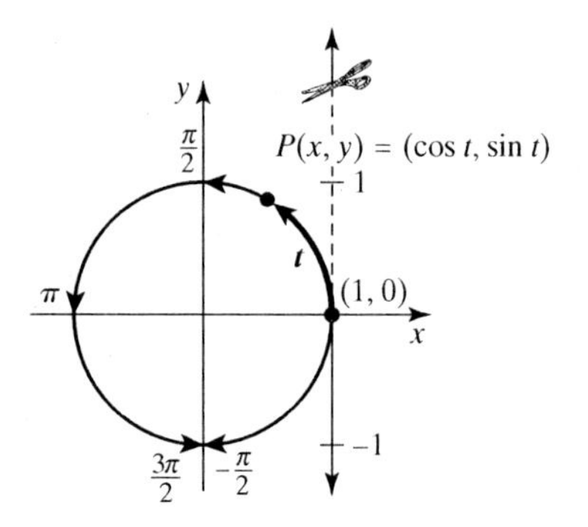

$$(\cos t)^2 + (\sin t)^2 = 1 \qquad \text{Pythagorean Identity}$$
$$\text{or} \quad \cos^2 t + \sin^2 t = 1$$
$$-1 \le \cos t \le 1$$
$$-1 \le \sin t \le 1$$

Therefore, the domain of $\cos t$ or $\sin t$ is all real numbers, or $t \in \mathbb{R}$. The corresponding range is all real numbers between and including -1 and 1; that is, $|\cos t| \le 1$ and $|\sin t| \le 1$.

Given any arc t, we are able to locate the quadrant that contains its terminal point. And when we are given the quadrant of t and a functional value of t, we are able to find the other functional value by using the Pythagorean identity.

1.3 For common arcs we have exact functional values for the cosine and sine functions. *The following table for functional values of the common arcs should be memorized.* For arcs that are not common, the functional values are approximated using a calculator, as shown.

x	$\cos x$	$\sin x$
0	1	0
$\frac{\pi}{6}$	$\frac{\sqrt{3}}{2}$	$\frac{1}{2}$
$\frac{\pi}{4}$	$\frac{1}{\sqrt{2}}$ or $\frac{\sqrt{2}}{2}$	$\frac{1}{\sqrt{2}}$ or $\frac{\sqrt{2}}{2}$
$\frac{\pi}{3}$	$\frac{1}{2}$	$\frac{\sqrt{3}}{2}$
$\frac{\pi}{2}$	0	1
π	-1	0
$\frac{3\pi}{2}$	0	-1
2π	1	0

```
(cos(1.3))^2
         .0715556233
sin(2π/5)
         .9510565163
cos(5(π+2))
         .8390715291
```

1.4 Arcs that are integer multiples of common arcs can use reference arcs to find exact values for the cosine and sine functions.

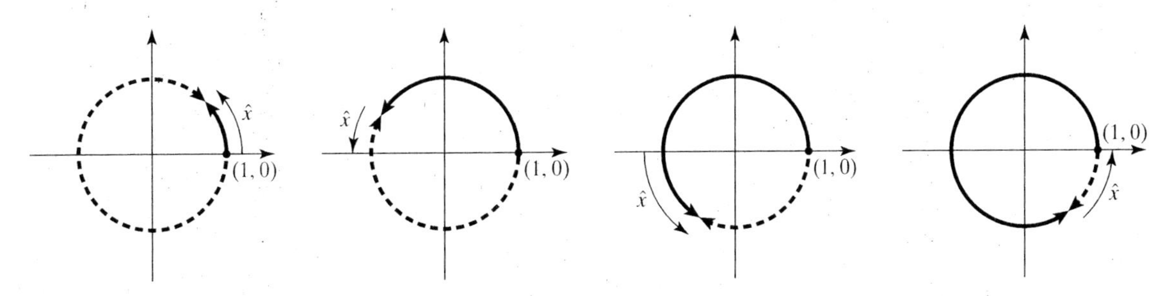

Functional values of arcs that are not integer multiples of common arcs can be approximated using a calculator (in radian mode).

1.5 Four additional circular functions are defined by finding the quotients and reciprocals of $\cos x$ and $\sin x$, as follows:

Function	Domain	Range
tangent $x = \tan x = \dfrac{\sin x}{\cos x}$	$x \neq \dfrac{\pi}{2} + k\pi$	$\mathbb{R}$
secant $x = \sec x = \dfrac{1}{\cos x}$	$x \neq \dfrac{\pi}{2} + k\pi$	$\lvert \sec x \rvert \geq 1$
cosecant $x = \csc x = \dfrac{1}{\sin x}$	$x \neq k\pi$	$\lvert \csc x \rvert \geq 1$
cotangent $x = \cot x = \dfrac{\cos x}{\sin x}$	$x \neq k\pi$	$\mathbb{R}$

x	$\cos x$	$\sin x$	$\tan x$
0	1	0	0
$\dfrac{\pi}{6}$	$\dfrac{\sqrt{3}}{2}$	$\dfrac{1}{2}$	$\dfrac{1}{\sqrt{3}}$ or $\dfrac{\sqrt{3}}{3}$
$\dfrac{\pi}{4}$	$\dfrac{1}{\sqrt{2}}$ or $\dfrac{\sqrt{2}}{2}$	$\dfrac{1}{\sqrt{2}}$ or $\dfrac{\sqrt{2}}{2}$	1
$\dfrac{\pi}{3}$	$\dfrac{1}{2}$	$\dfrac{\sqrt{3}}{2}$	$\sqrt{3}$
$\dfrac{\pi}{2}$	0	1	undefined
π	-1	0	0
$\dfrac{3\pi}{2}$	0	-1	undefined
2π	1	0	0

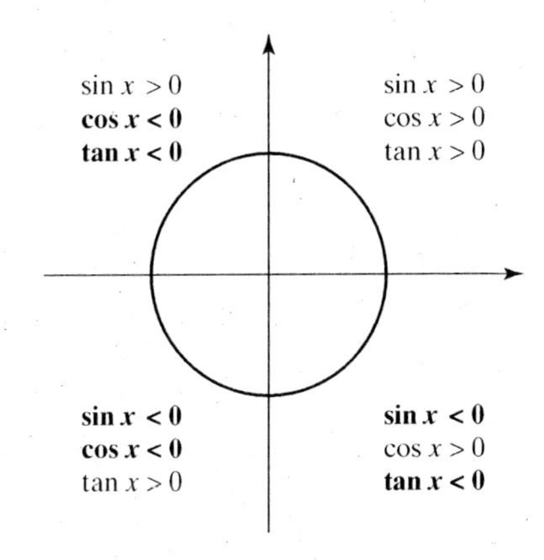

If we are given one circular function value, and if we know the location of the terminal point of the arc, we can find all the remaining five circular functional values.

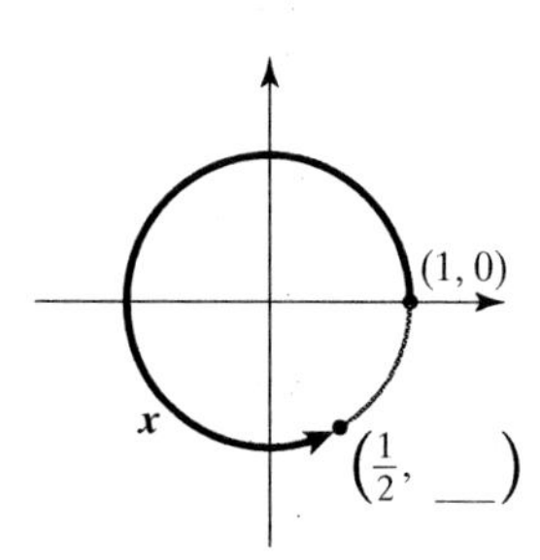

If $\cos x = \frac{1}{2}, \frac{3\pi}{2} < x < 2\pi$ (QIV)

then:

$$\sin x = -\frac{\sqrt{3}}{2} \qquad \sec x = 2$$

$$\csc x = -\frac{2}{\sqrt{3}} = -\frac{2\sqrt{3}}{3}$$

$$\tan x = -\sqrt{3} \qquad \cot x = -\frac{1}{\sqrt{3}} = -\frac{\sqrt{3}}{3}$$

1.6 The negative identities, the relationships between the circular functional values of arcs x and $-x$, are as follows.

$$\cos(-x) = \cos(x) \qquad \sin(-x) = -\sin x$$
$$\sec(-x) = \sec(x) \qquad \csc(-x) = -\csc x$$
$$\tan(-x) = -\tan x$$
$$\cot(-x) = -\cot x$$

The six circular functions are periodic.

Function	Period
$\cos x$, $\sec x$ $\sin x$, $\csc x$	2π
$\tan x$, $\cot x$	π

Among the mathematical models that can be used to describe many types of periodic behavior are the following:

$$d(t) = A\cos(Bt) \qquad d(t) = A\sin(Bt)$$
$$P = A\cos(B(t + C)) \qquad P = A\sin(B(t + C))$$

Chapter 1 Review Exercises

1.2 **1–6.** *Fill in the blanks to make a true statement.*

1. The cos t is defined to be the **(a)**______-coordinate at the **(b)**______ point of arc **(c)**______ with initial point **(d)**______ on the **(e)**______ circle. The **(f)**______ and direction of the arc is represented by t.

2. The largest possible numerical value in the range of $\cos t$ and $\sin t$ is ______.
3. The smallest possible numerical value in the range of $\cos t$ and $\sin t$ is ______.
4. Arcs wrapped around the unit circle counterclockwise are considered **(a)**______, while arcs wrapped **(b)**______ are considered negative.
5. The Pythagorean identity is ______ = 1.
6. The domain of both the cosine and sine functions is ______.
7. Name the quadrant that contains the terminal point of each arc with initial point $(1, 0)$ on the unit circle.

   a. $\dfrac{7\pi}{4}$ **b.** $\dfrac{7\pi}{3}$ **c.** $-\dfrac{5\pi}{6}$ **d.** $\dfrac{11\pi}{4}$ **e.** 4.9 **f.** -3 **g.** $-\dfrac{3\pi}{4}$ **h.** $-\dfrac{4\pi}{3}$

8. Determine the quadrant that contains the terminal point of the arc t.

   a. $\sin t < 0$ and $\cos t > 0$ **b.** $\cos t < 0$ and $\sin t < 0$

9. If $\sin t = \dfrac{\sqrt{3}}{2}$ and $\cos t < 0$, find:

   a. the quadrant that contains the terminal point of t

   b. $\cos t$ **c.** $\dfrac{1}{\sin t}$ **d.** $\dfrac{1}{\cos t}$ **e.** $\dfrac{\sin t}{\cos t}$ **f.** $\dfrac{\cos t}{\sin t}$

10. If $\cos t = \frac{3}{5}$ and $\sin t < 0$, find:

    a. the quadrant that contains the terminal point of t

    b. $\sin t$ **c.** $\dfrac{1}{\sin t}$ **d.** $\dfrac{1}{\cos t}$ **e.** $\dfrac{\sin t}{\cos t}$ **f.** $\dfrac{\cos t}{\sin t}$

1.3

11. Using the information shown on each unit circle, fill in the blanks with exact values.

    a. $\cos t =$ ______ **b.** $t =$ ______ **c.** $t =$ ______

    $\sin t =$ ______ $\sin t =$ ______ $\cos t =$ ______

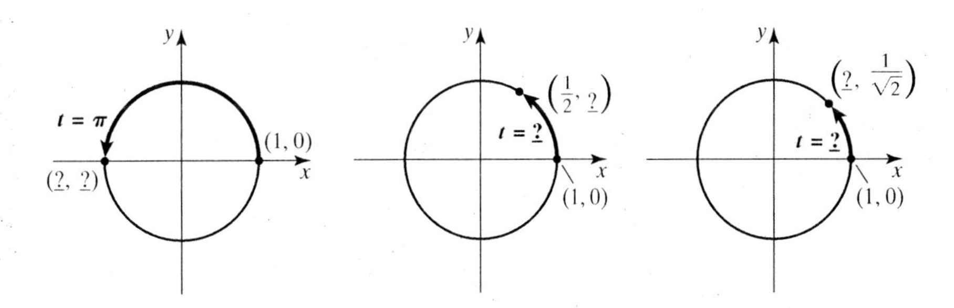

12. Find the exact value for the following expressions.

    a. $\sin\dfrac{\pi}{3} + \cos 0$ **b.** $2\cos^2\dfrac{\pi}{6} + \left(\sin\dfrac{\pi}{4}\right)^2$ **c.** $\cos^2\dfrac{\pi}{6} - \sin^2\dfrac{\pi}{6}$

    d. $4\sin\dfrac{\pi}{3} \cdot \cos\dfrac{\pi}{3}$ **e.** $\dfrac{\cos\dfrac{\pi}{2} + \sin\dfrac{\pi}{6}}{-2\cos^2\pi}$ **f.** $\dfrac{\sin^2\left(\dfrac{\pi}{4} + \dfrac{\pi}{4}\right)}{\sin\dfrac{\pi}{4}}$

13. Use a calculator (radian setting) to find an approximation rounded to four decimal places for each expression.

a. $2\cos\frac{\pi}{4} + \sin\frac{\pi}{8}$ **b.** $4\cos 2(1.423 + 0.12)$ **c.** $\sin\frac{5\pi}{9} + \frac{1}{\cos 1}$

d. $\dfrac{\cos\frac{2}{3}}{\sin\frac{1}{5}}$ **e.** $\dfrac{1}{\sin^2 0.0001}$

14. Compare the exact values to determine whether each statement is true or false.

a. $\cos\frac{\pi}{4} = \sin\frac{\pi}{4}$ **b.** $\dfrac{1}{\sin\frac{3\pi}{2}} = 0$

c. $\dfrac{\sin\frac{\pi}{6}}{\cos\frac{\pi}{3}} = 1$ **d.** $\sin\left(\frac{\pi}{2} - \frac{\pi}{4}\right) = \sin\frac{\pi}{2}\cos\frac{\pi}{4} - \cos\frac{\pi}{2}\sin\frac{\pi}{4}$

e. $\cos^2\frac{\pi}{2} = 1 - \sin^2\frac{\pi}{2}$ **f.** $\sin\left(\frac{\pi}{2} + \frac{\pi}{2}\right) = \sin\frac{\pi}{2} + \sin\frac{\pi}{2}$

g. $\cos^2\left(\frac{\pi}{6}\right) + \sin^2\left(\frac{\pi}{6}\right) = 1$ **h.** $\cos\left(\frac{\pi}{4} + \frac{\pi}{4}\right) = \cos\frac{\pi}{4} + \cos\frac{\pi}{4}$

i. $\sin\left(2 \cdot \frac{\pi}{2}\right) = 2\sin\frac{\pi}{2}$ **j.** $\sec\frac{\pi}{2} = 0$

1.4

15. Find the reference arc for each given arc. Sketch the arc and the reference arc.

a. $\frac{5\pi}{3}$ **b.** $\frac{9\pi}{4}$ **c.** $-\frac{7\pi}{6}$ **d.** $-\frac{3\pi}{4}$

16. Find the exact functional value.

a. $\sin\frac{5\pi}{3}$ **b.** $\cos\frac{9\pi}{4}$ **c.** $\cos\left(-\frac{7\pi}{6}\right)$ **d.** $\sin\left(-\frac{3\pi}{4}\right)$

17. If possible, find the exact functional values. If not, use a calculator to find an approximation rounded to four decimal places for each expression.

a. $\dfrac{\sin\frac{7\pi}{4}}{\cos\left(-\frac{3\pi}{4}\right)}$ **b.** $\cos^2 2.4 + \sin^2 2.4$ **c.** $\cos(\pi + 2) \cdot \sin(2(-6.2))$

d. $\cos\pi + \sin\frac{3\pi}{2} + \sin 4.2 + \sin(-4.2)$

18. Find the arc x with initial point $(1, 0)$ and within one counterclockwise wrap of the unit circle that makes each statement true.

a. $\cos x = -\frac{\sqrt{3}}{2}$ and $\sin x = -\frac{1}{2}$ **b.** $\cos x = \frac{-\sqrt{2}}{2}$ and $\sin x = \frac{\sqrt{2}}{2}$

c. $\sin x = -1$ and $\cos x = 0$ **d.** $\sin x = \frac{\sqrt{3}}{2}$ and $\cos x = \frac{1}{2}$

1.5

19. Using the information shown on each unit circle, fill in the blanks with exact values.

a. $x =$ ______ $\csc x =$ ______

b. $x =$ ______ $\tan x =$ ______

c. $x =$ ______ $\sec x =$ ______

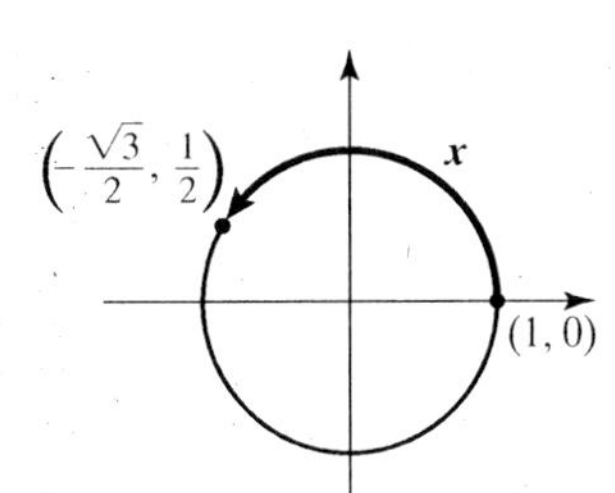

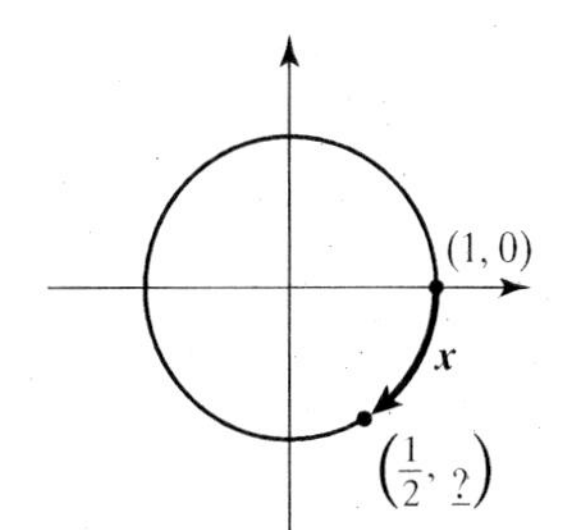

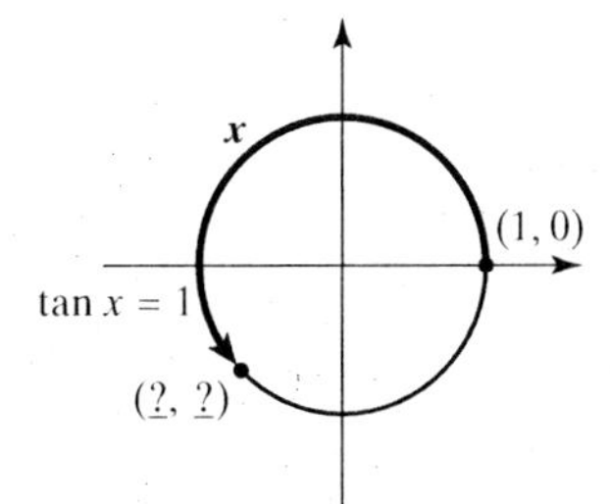

20. Determine the quadrant, if possible, in which the arc x has the following functional values.

a. $\tan x < 0$ and $\sec x > 0$

b. $\cot x > 0$ and $\sin x < 0$

c. $\sec x > 0$ and $\cos x < 0$

d. $\csc x < 0$ and $\sec x > 0$

e. $\cot x > 0$ and $\sec x > 0$

f. $\csc x > 0$ and $\cos x < 0$

g. $\cot x > 0$ and $\tan x > 0$

21. Find the exact functional value of each expression, if defined.

a. $\tan \frac{2\pi}{3}$

b. $\csc \frac{5\pi}{4}$

c. $\sec \frac{19\pi}{6}$

d. $\cot \frac{5\pi}{3}$

e. $\tan^2\left(-\frac{11\pi}{6}\right)$

f. $\sec\left(-\frac{3\pi}{2}\right)$

g. $\csc^2\left(-\frac{7\pi}{6}\right)$

h. $\cot 0$

i. $\dfrac{\sec 5\pi + \csc \frac{2\pi}{3}}{\tan \frac{11\pi}{4}}$

22. Use a calculator, along with the definitions of the circular functions, to find an approximation (rounded to four decimal places) for each expression.

a. $3 \cot 5.1 + \sec\left(-\frac{7\pi}{5}\right)$

b. $\dfrac{\csc 5}{\cot(-9.1)}$

c. $\cot(-7.62) + \tan\left(\frac{5}{3}\right)$

d. $\dfrac{\sec 1.5}{\tan \frac{\pi}{12}}$

23. Find the indicated exact functional values.

a. If $\sin x = \frac{8}{17}$ and $\cos x < 0$, find $\cos x$ and $\tan x$.

b. If $\cos x = \frac{\sqrt{2}}{2}$ and $\frac{3\pi}{2} < x < 2\pi$, find $\sin x$ and $\cot x$.

c. If $\cos x = -\frac{12}{13}$ and the terminal point of x is in QIII, find $\sin x$, $\csc x$, and $\sec x$.

d. If $\sin x = -\frac{2}{3}$ and $-\frac{\pi}{2} \le x \le 0$, find $\cos x$ and $\tan x$.

24. Find the positive arc in the indicated quadrant that would have the given functional value.

a. $x =$ ______ **b.** $x =$ ______ **c.** $x =$ ______

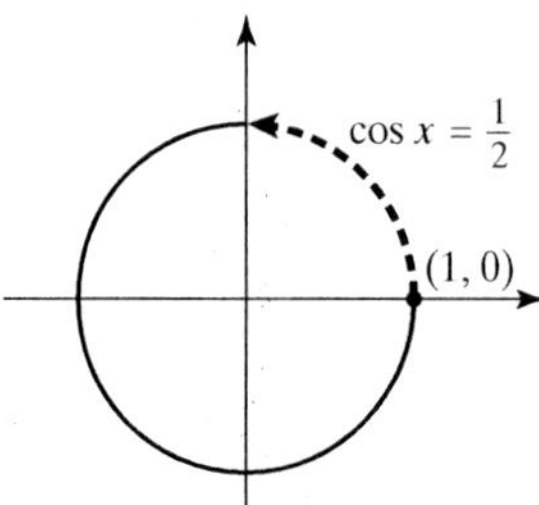

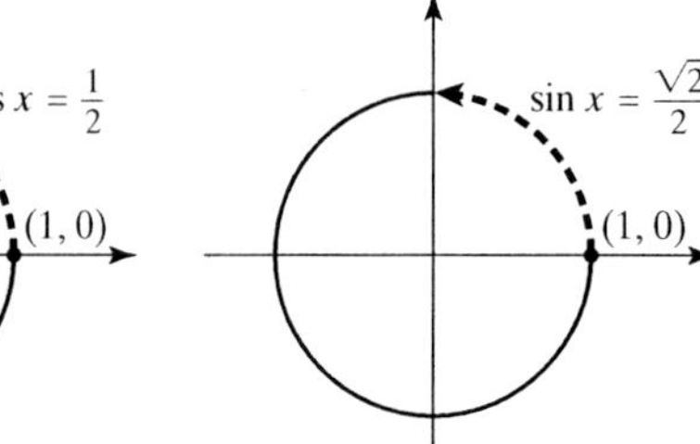

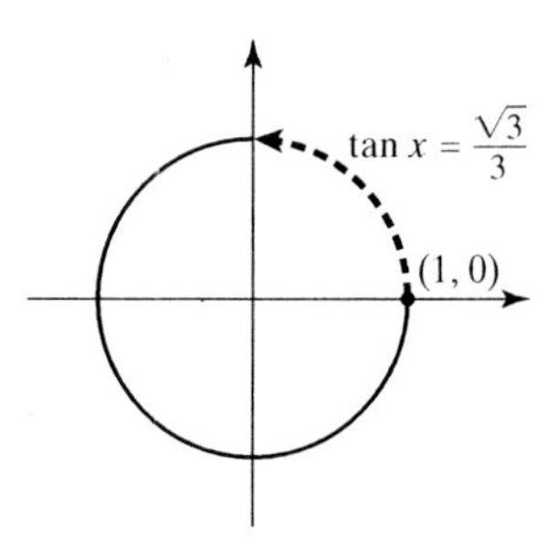

25. Find the positive or negative arc that lies in the indicated quadrant and has the given functional value.

a. $x =$ ______ **b.** $x =$ ______ **c.** $x =$ ______

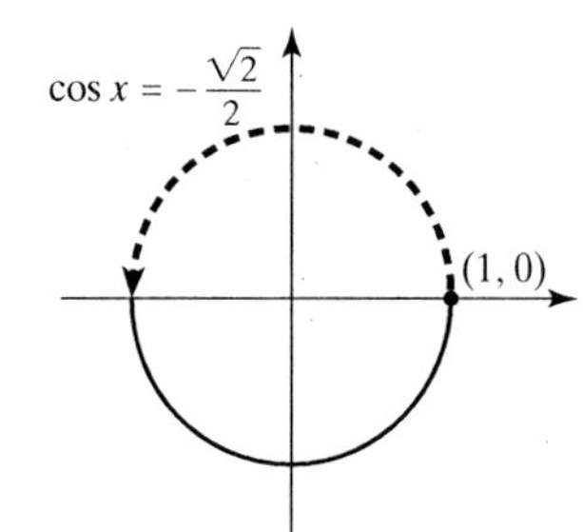

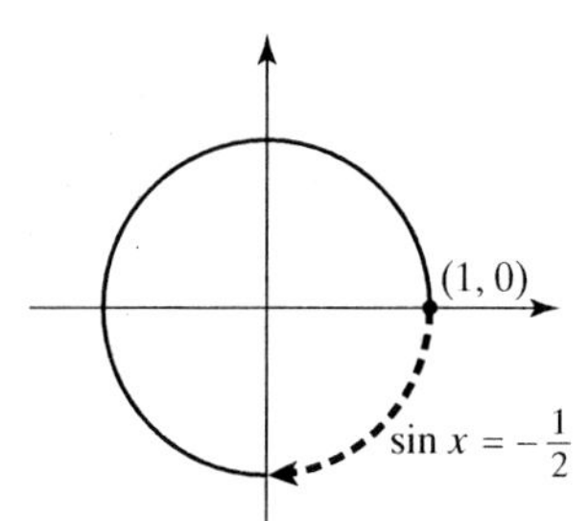

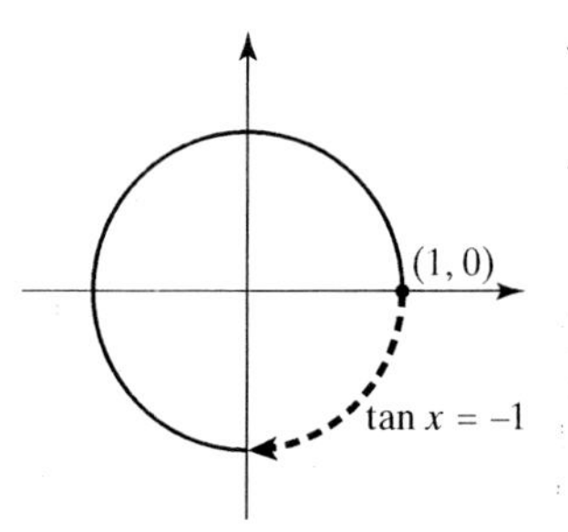

26. Use the definitions of the circular functions to rewrite the right side of the following equations in terms of $\cos x$ and $\sin x$. Then simplify to determine whether the resulting equation is an identity (a true statement for all values in which both sides are defined).

a. $\tan x = \tan^2 x \cdot \csc x \cdot \cos x$

b. $1 = \tan^2 x \cdot \cos^2 x + \cot^2 x \cdot \sin^2 x$

c. $\sec x = \cot x \cdot \sin x \cdot \csc x \cdot \sec x$

d. $\cot x = \sec^2 x \cdot \cos x \cdot \sin x$

e. $1 = 2 \sin^2 x + \cot^2 x \cdot \sin^2 x - \sin^2 x$

1.6

27. Are the following statements true or false? If a statement is false, explain why or give an example that shows why it is false.

a. $\cos(x + 2\pi) = \cos x$

b. $\sin(x + \pi) = \sin x$

c. $\tan(x + \pi) = \tan x$

d. $\csc(-x) = \csc x$

e. $\cos(-x) = \cos x$

f. $-\sin(-x) = \sin x$

28. Find all values of x within one wrap of the circle $(0 \le x < 2\pi)$ that make the statement true. Then use the periodic property to find all values that make it true.

a. $\sin x = \frac{\sqrt{3}}{2}$ **b.** $\sec x = \sqrt{2}$ **c.** $\tan x = -\frac{\sqrt{3}}{3}$

29. The horizontal displacement of a point on an oscillating pendulum with initial displacement of 12 cm is given by the equation $d(t) = 12\cos(2\pi t)$, where t is in seconds and $d(t)$ is in centimeters. Approximate the displacement to four decimal places when $t = \frac{1}{5}$ second.

30. The displacement of a spring is given by $d(t) = 1.3\cos(5.2t)$, where $d(t)$ is in inches and t is in seconds. Approximate the displacement to four decimal places when $t = 3\frac{1}{2}$ seconds.

31. Elka and Tom are both studying for a Chapter 1 test by doing some sample problems. They have both used their calculators (in radian mode) to find an approximate value to four decimal places for sec $4\pi/11$. Much to their dismay, they got different answers. Elka's answer was 2.4072 and Tom's was 1.0993.

    a. Why did they have to use their calculators for this problem?

    b. Which student is correct?

    c. The student with the wrong answer used an incorrect definition. What did the student input into the calculator to get the wrong answer?

Chapter 1 Test

1. Fill in the blanks to make a true statement.

   The sine t is defined to be the **(a)**_____ of the terminal point of arc t whose initial point is $(1, 0)$ on the **(b)**_____ circle, where t represents the length and **(c)**_____ of the arc on the unit circle. The Pythagorean identity is **(d)**_____ = 1. For the sine function, the domain is **(e)**_____, the range is **(f)**_____, and the period is **(g)**_____. For the tangent function, the domain is **(h)**_____, the range is **(i)**_____, and the period is **(j)**_____.

2. Copy the following table, and fill in the *exact* functional values for the common arcs. State where the values are undefined.

x	$\sin x$	$\tan x$	$\sec x$
0			
$\frac{\pi}{6}$			
$\frac{\pi}{4}$			
$\frac{\pi}{3}$			
$\frac{\pi}{2}$			
π			
$\frac{3\pi}{2}$			

3. Name the quadrant that contains the terminal point of arc x.

   a. $\cos x < 0$ and $\sin x > 0$ **b.** $\csc x < 0$ and $\tan x < 0$

4. For each expression sketch the arc and state the reference arc. Then find the *exact* functional value.

a. $\sin \frac{7\pi}{4}$ **b.** $\sec\left(-\frac{5\pi}{6}\right)$

5. Find the *exact* value for each expression.

a. $\cos \frac{5\pi}{4}$ **b.** $\sin^2 \frac{7\pi}{3}$ **c.** $\tan \frac{2\pi}{3}$ **d.** $\csc \frac{5\pi}{6}$

e. $\cos\left(-\frac{2\pi}{3}\right)$ **f.** $\left(\csc \frac{3\pi}{4}\right)^2 + \sec \pi$ **g.** $\cot \frac{11\pi}{6} + \cos\left(-\frac{7\pi}{6}\right)$

6. If $\cos x = -\frac{\sqrt{2}}{2}$ and $\pi \le x \le \frac{3\pi}{2}$, find *exact* values for each expression.

a. $\sin x$ **b.** $\tan x$ **c.** $\sec x$

7. If $\sin t = \frac{3}{4}$ and $\cos t < 0$, find *exact* values for:

a. the quadrant that contains the terminal point of t.

b. $\cos t$ **c.** $\tan t$

8. Suppose $\cos x = \frac{1}{2}$ and $\sin x = -\frac{\sqrt{3}}{2}$, where $0 \le x \le 2\pi$.

a. What quadrant contains the terminal point of the arc x? **b.** Find x.

9. Use a calculator to find an approximate value rounded to four decimal places for each expression.

a. $\cos^2 35.08$ **b.** $\tan(-4.76\pi)$ **c.** $\csc 3\left(\frac{\pi}{7} + \frac{\pi}{9}\right)$

d. $\cot 5 + \sec 2$ **e.** $\frac{5 \cos 8.9 - \tan 4.5}{\sin 9}$

10. Are the following statements true or false? If a statement is false, explain why or give an example that shows why it is false.

a. $\cot x = \csc x \cos x$ **b.** If $\cos x = 0$, then $\sin x = 1$.

c. $\cos^2 x = 1 - \sin^2 x$ **d.** $\tan^2 x = \tan x^2$

e. The period of the cosecant function is π. **f.** $\cos(-x) = -\cos x$

g. $\sin(x + 4\pi) = \sin x$ **h.** $1 = 2\cos^2 x + \sin^2 x - \cos^2 x$

i. $1 = \cos x + \sin x$ **j.** $\cos 2\pi = 2 \cos \pi$

11. Find all values within one wrap of the circle $(0 \le x \le 2\pi)$ that make the statement true. Then use the periodic property of the function to find all values that make it true.

a. $\cos x = \frac{\sqrt{3}}{2}$ **b.** $\tan x = -\sqrt{3}$

12. The horizontal displacement d of a bob on an oscillating pendulum is given by the equation $d(t) = 7\cos(2\pi t)$, where d is expressed in centimeters and t is expressed in seconds. Approximate the displacement d, rounded to four decimal places, when $t = 0.2$ seconds.

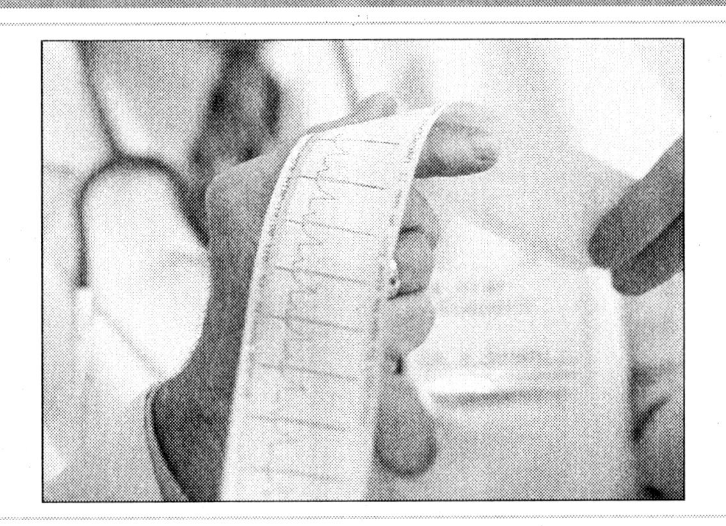

Chapter 2

Graphs of the Circular Functions

The electrical signals that flow through the heart can be picked up from the skin, amplified, and printed on paper for your doctor to examine. The recording of the periodic heartbeat is called an electrocardiogram, and certain changes in the normal EKG alert physicians to heart problems. The various and complex set of events that occur with each heartbeat are referred to as the cardiac cycle.

In this chapter we examine the graphs of the circular functions and observe what factors can change the characteristics of their cycles. As a result, we learn how they can model certain real-life periodic phenomena.

Important Concepts

- graphs of the circular functions
- sinusoidal waves
- amplitude and period (cycle)
- horizontal and vertical translations
- frequency
- periodic phenomena/models
- inverse circular functions

2.1 Graphs of the Sine and Cosine Functions

In Chapter 1 we defined six circular functions, which are periodic, and found many of their functional values. Now we will use what we have learned about these functions to graph them. We start with graphing the functions by plotting points. After we identify specific characteristics of these graphs, we can sketch the graphs quickly. This may be your first experience with graphs of periodic functions. You will see how the graphs exhibit the periodic property.

Graph of the Sine Function

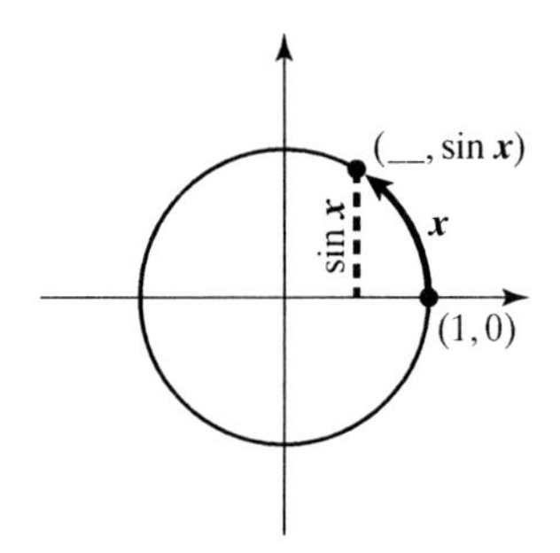

We begin with the sine function: $\{(x, \sin x)\}$, or $y = \sin x$.

Domain: $x \in \mathbb{R}$

Range: $-1 \leq \sin x \leq 1$

To graph *one period*, we set up a table with common arcs for x between and including 0 and 2π and calculator approximations for $y = \sin x$, when necessary: $\frac{1}{\sqrt{2}} \approx 0.7$, $\frac{\sqrt{3}}{2} \approx 0.87$. Since the common arcs are in terms of π, it is easier to mark off the x-axis in multiples of π. Because the range of y is -1 (minimum value) to 1 (maximum value), we mark off the y-axis in integers.

As we plot the points, we notice that the graph looks as if we are unwrapping our circular function.

x	0	$\pi/6$	$\pi/4$	$\pi/3$	$\pi/2$	$3\pi/4$	π	$5\pi/4$	$3\pi/2$	$7\pi/4$	2π
$\sin x$	0	$1/2$	$1/\sqrt{2}$	$\sqrt{3}/2$	1	$1/\sqrt{2}$	0	$-1/\sqrt{2}$	-1	$-1/\sqrt{2}$	0

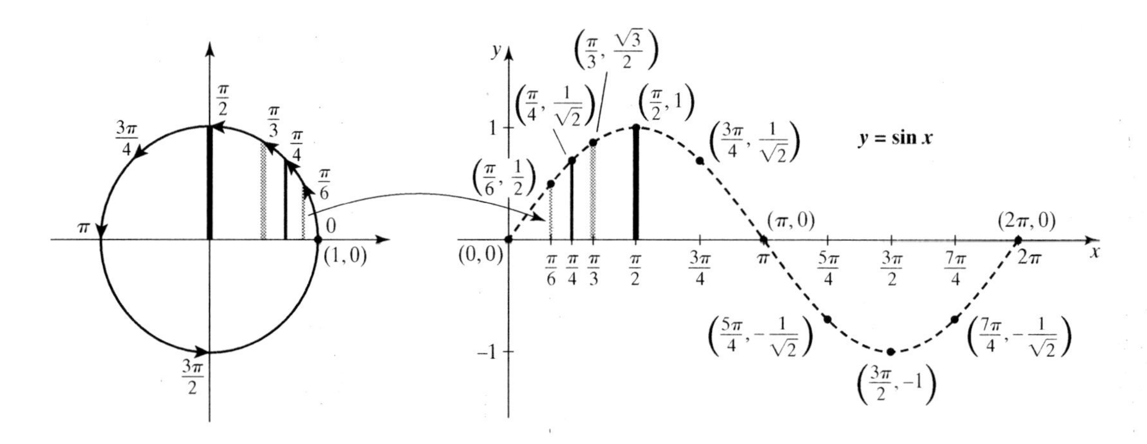

We could show by using calculus that the circular functions are continuous on their domains (that is, continuous wherever they are defined). A graph is *continuous* if you can draw it without lifting your pencil off the paper. Since $\sin x$ is defined for all $x \in \mathbb{R}$, the graph of $y = \sin x$ is continuous—it has no breaks—for all x. Therefore, we connect the points shown above with a smooth continuous curve to produce the graph of the sine function, which is called a **wave**. More precisely, the mathematical name is **sinusoidal wave**, which means "like the sine." Since the period of the sine function is 2π, the graph will repeat every 2π units. The graph over one period is called a **cycle**. The sine graph is also said to oscillate between -1 and 1.

If we look at one cycle of the sine graph, we notice there are *four identically shaped arc sections** in different positions. Each arc represents one-fourth the period, and each arc occurs over an interval along the x-axis that is one-fourth the period, or $\frac{1}{4}(2\pi) = \pi/2$ units, in length. There is either a maximum y-coordinate or minimum y-coordinate at one end of each arc, with an x-intercept at the other end. We call the endpoints of these arcs **critical points**. For $0 \le x \le 2\pi$, the first arc of the sine graph is increasing, the next two arcs are decreasing, and the last arc is increasing, forming a "cup-down, cup-up" wave.

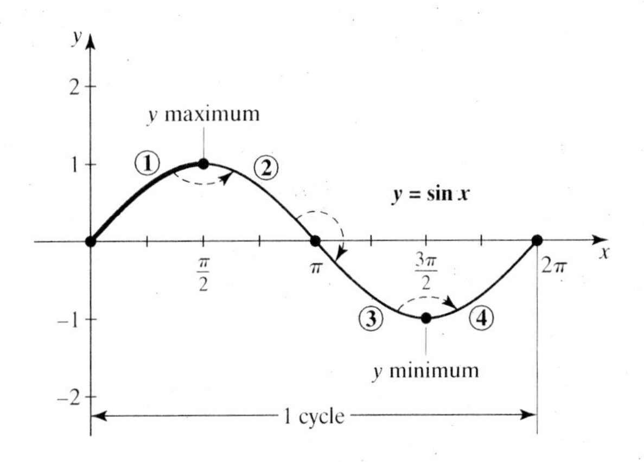

To graph $y = \sin x$ over the interval $-2\pi \le x \le 4\pi$, we repeat one cycle on the right from 2π to 4π and one on the left from -2π to 0. We can "copy and paste" the graph of one cycle to another cycle because the sine graph is continuous and periodic. Notice the repetition of the four arcs for each cycle.

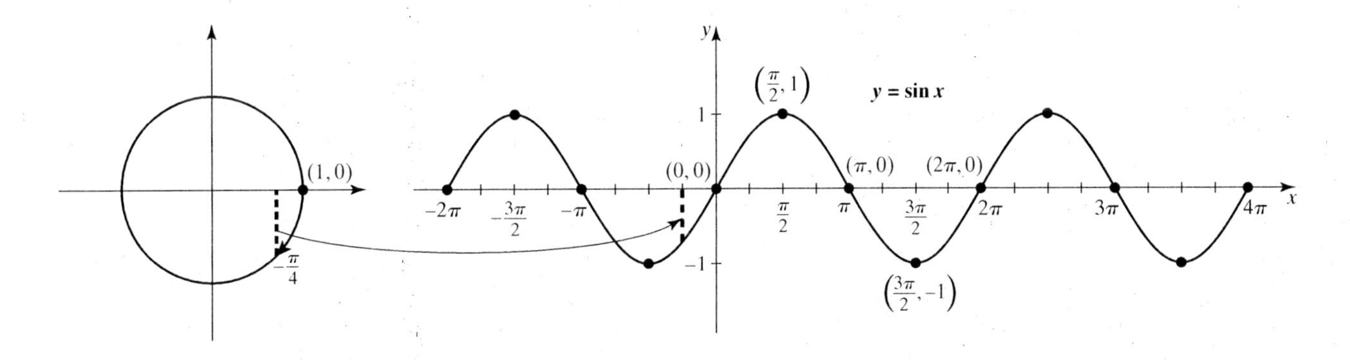

With the additional cycles we notice the graph of the sine is symmetric with respect to the origin. The x-intercepts (where $y = \sin x = 0$) are at $x = \ldots, -\pi, 0, \pi, 2\pi, 3\pi, \ldots$, which can be written as $x = k\pi$, where k is any integer. The x-intercepts are also referred to as the **zeros** or **roots** of the function. The y-intercept is at the origin.

*These arc sections are noncircular arc sections.

If you did the graphing calculator exploration Exercise 47 in Section 1.3, you may have already discovered some of these characteristics.

We define another characteristic of the graph of $y = \sin x$, which is called *amplitude.*

Amplitude of a Sinusoidal Function

The **amplitude** of a sinusoidal wave is one–half the distance between the maximum and minimum functional values.

$$\text{Amplitude} = \tfrac{1}{2}|\text{max } y\text{-coordinate} - \text{min } y\text{-coordinate}|$$

EXAMPLE 1 **a.** What is the amplitude of $y = \sin x$?

b. Label the critical points on one cycle of $y = \sin x$.

SOLUTION

a. Amplitude $= \frac{1}{2}|1 - (-1)| = 1$

b.

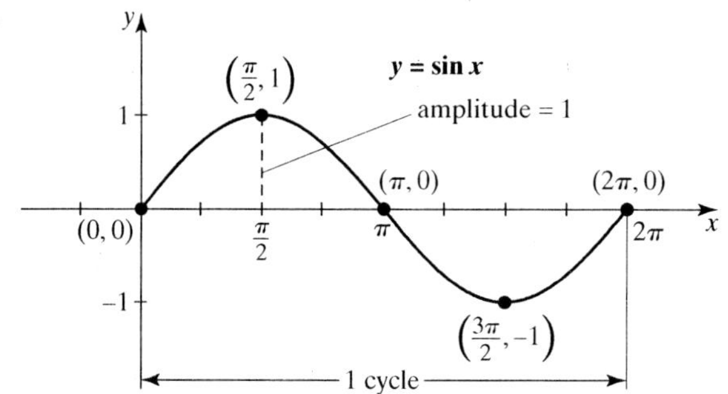

EXAMPLE 2 Below are four *incorrect* attempts to sketch one cycle of $y = \sin x$, $0 \le x \le 2\pi$. Find at least one error for each attempt.

a.

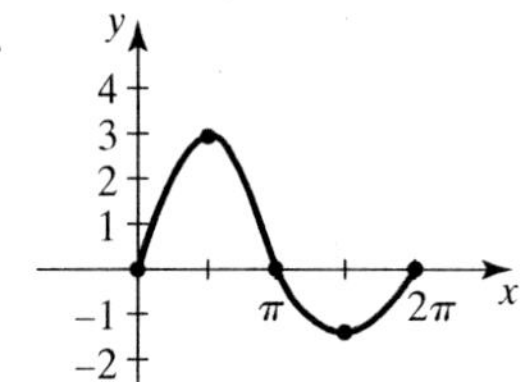

b.

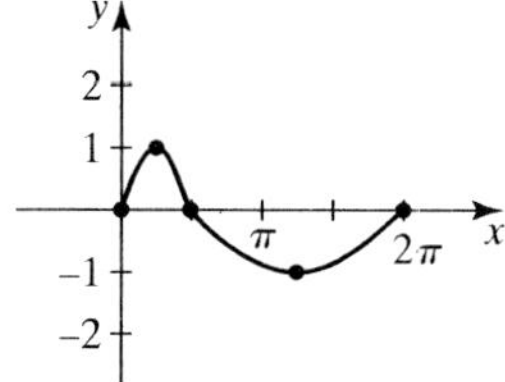

c.

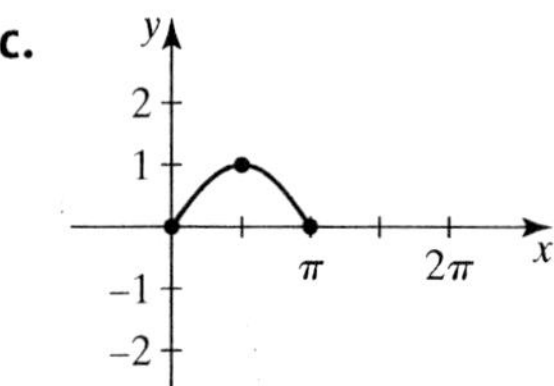

d.

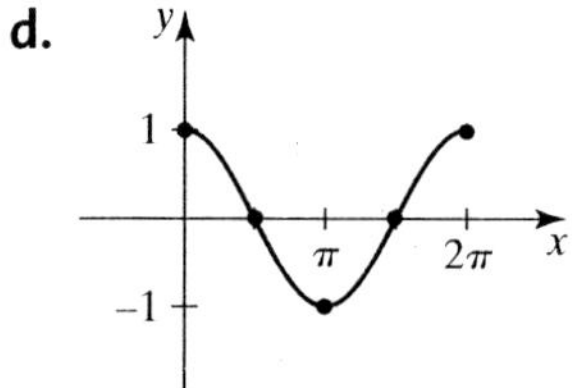

SOLUTION

a. The maximum and minimum values are incorrect. There are not four identical arc sections.

b. There are not four identical arc sections to the graph. Each arc does not correspond to an interval of length $\pi/2$ units along the x-axis.

c. The cycle is incomplete: It should extend to 2π. There are not four identical arc sections.

d. The x-intercepts are not at $x = k\pi$. The first arc section should increase, not decrease. ■

Graph of the Cosine Function

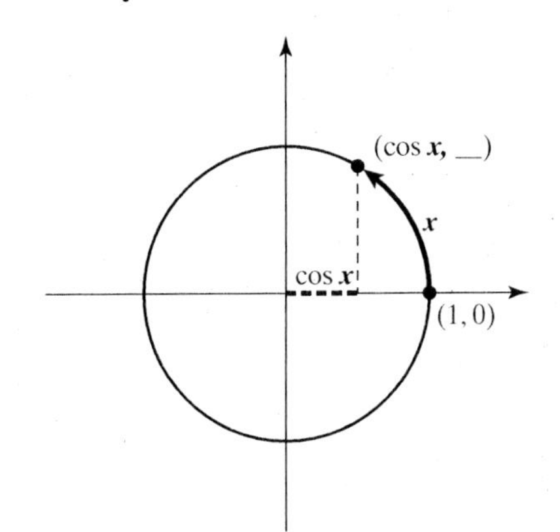

Next we graph the cosine function: $\{(x, \cos x)\}$ or $y = \cos x$.

$$\text{Domain: } x \in \mathbb{R}$$
$$\text{Range: } -1 \le \cos x \le 1$$

Following the same procedure used to graph one period of the sine function, we set up a table using common arcs for x between and including 0 and 2π and calculator approximations for $y = \cos x$, when necessary. Again we mark off the x-axis in multiples of π and plot the points.

x	0	$\pi/6$	$\pi/4$	$\pi/3$	$\pi/2$	$3\pi/4$	π	$5\pi/4$	$3\pi/2$	$7\pi/4$	2π
$\cos x$	1	$\sqrt{3}/2$	$1/\sqrt{2}$	$1/2$	0	$-1/\sqrt{2}$	-1	$-1/\sqrt{2}$	0	$1/\sqrt{2}$	1

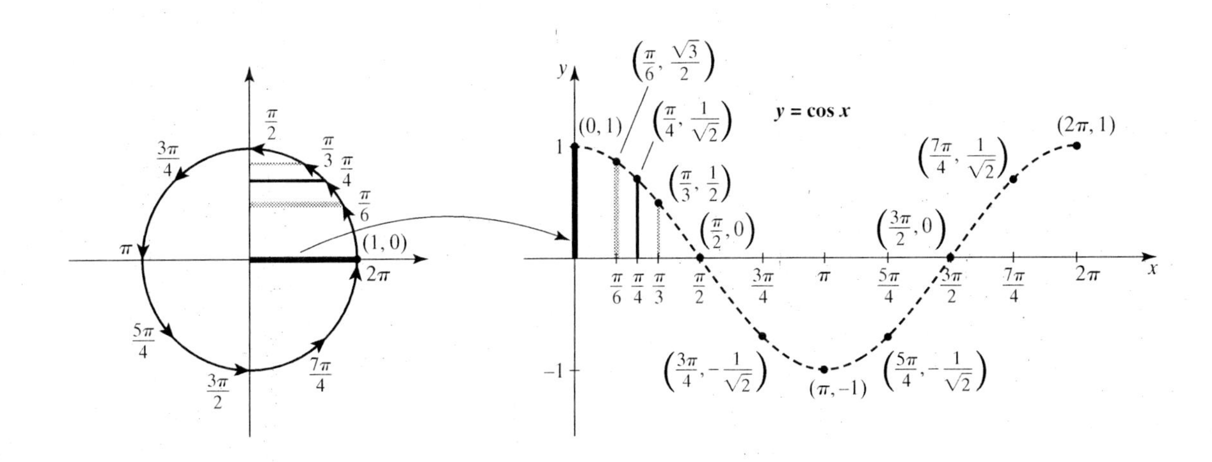

Since the cosine function is defined for all $x \in \mathbb{R}$ and periodic, the *graph of the cosine function is also a continuous wave, and the cycle, which also has four identical*

arc sections, repeats every 2π units. Both the sine and the cosine graphs are sinusoidal waves.

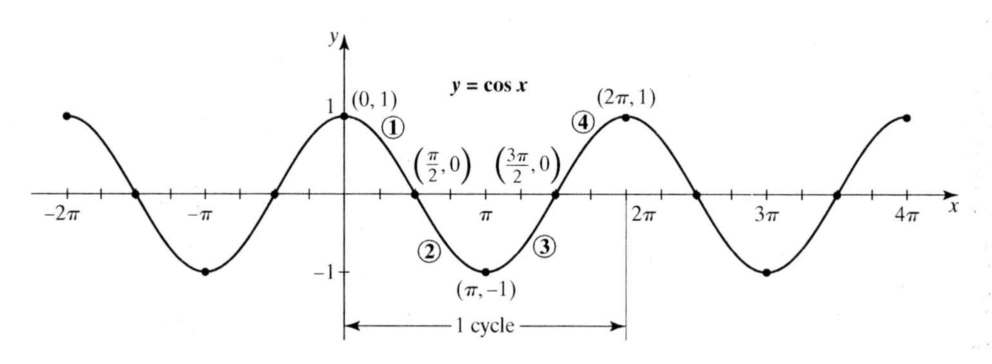

Important Characteristics of the Graphs of $y = \sin x$ and $y = \cos x$

Common Characteristics

- Sinusoidal waves with period 2π
- Amplitude $= \frac{1}{2}|1 - (-1)| = 1$
- Each cycle has four identically shaped arc sections. Each arc corresponds to an interval along the x-axis that is one-fourth the period, or $\frac{1}{4}(2\pi) = \pi/2$ units in length.
- The endpoints of the four arcs are critical points.

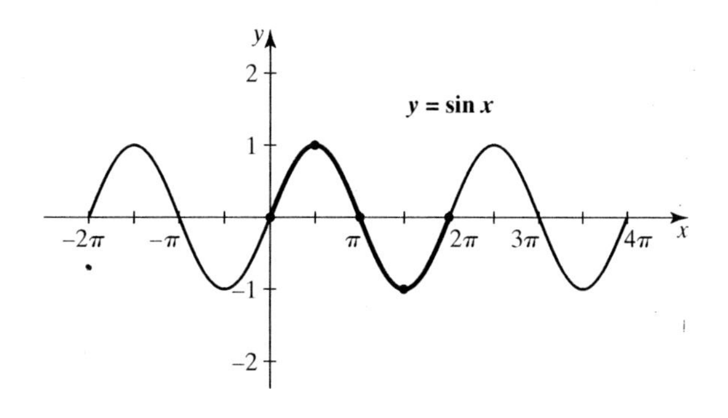

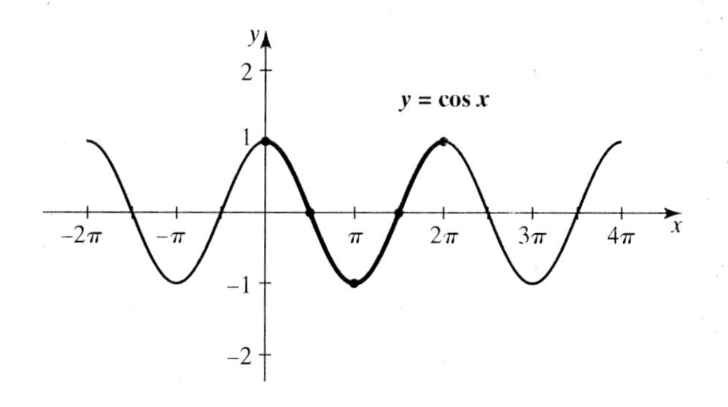

Different Characteristics

$y = \sin x$	$y = \cos x$
For $0 \le x \le 2\pi$, the first arc is increasing, the next two arcs are decreasing, and the last arc is increasing.	For $0 \le x \le 2\pi$, the first two arcs are decreasing and the next two arcs are increasing.
y-intercept: $(0, 0)$	y-intercept: $(0, 1)$
x-intercepts: $x = k\pi$	x-intercepts: $x = \dfrac{\pi}{2} + k\pi$
Symmetric with respect to the origin	Symmetric with respect to the y-axis

NOTE: You will need to study the common characteristics of these sinusoidal waves as well as their differences. The graphs of $y = \sin x$ and $y = \cos x$ can be used to generate graphs of other sinusoidal waves that are variations. Thus, the graphs of $y = \sin x$ and $y = \cos x$ are used as reference graphs, and so we call each graph the ***pure form***.

EXAMPLE 3 The following graphs are three *incorrect* attempts to sketch one cycle of $y = \cos x$. Find at least one error in each attempt.

a.

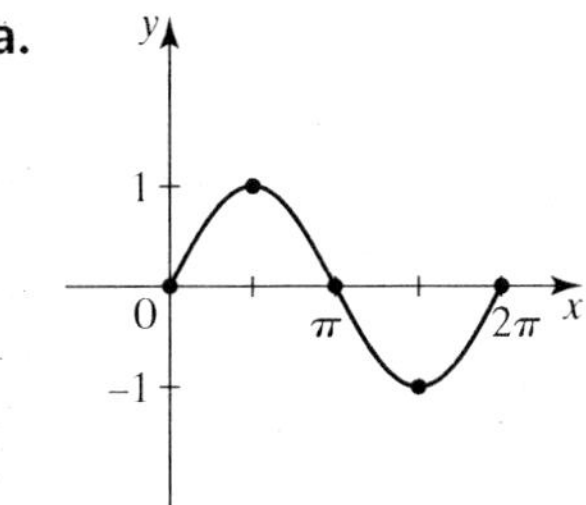

b.

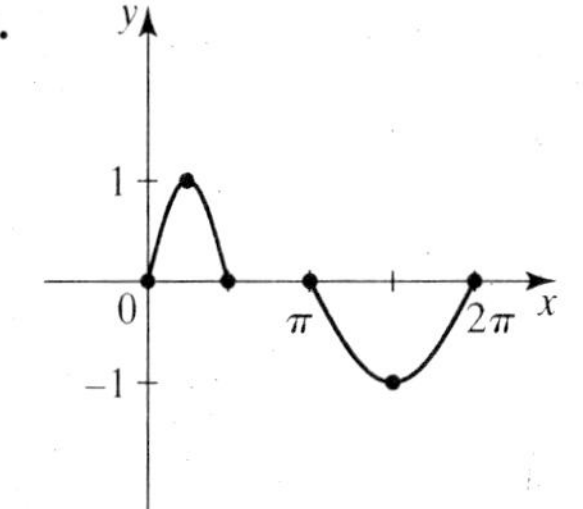

c.

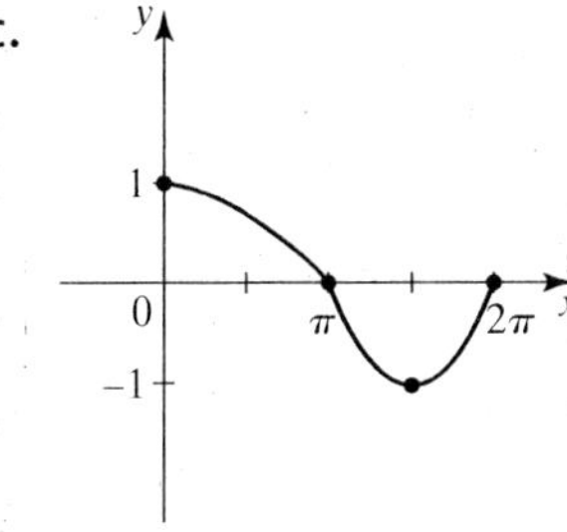

SOLUTION

a. The graph shown is the graph of $y = \sin x$, not $y = \cos x$.

b. The graph is not continuous. There are not four identical arc sections. The x-intercepts and y-intercept are incorrect.

c. There are not four identical arc sections. The x-intercepts are incorrect. ■

Unlike the graphing attempts in Examples 2 and 3, it is possible to have sinusoidal graphs that have modified characteristics of the pure forms. We begin with one of these modifications in the next example in which we investigate the effect on the pure form when we multiply the function by a nonzero real number.

EXAMPLE 4 Compare the graph of $y = \sin x$ with the graphs of $y = 3 \sin x$ and $y = -\frac{1}{2} \sin x$.

SOLUTION As a reference for each of the graphs we are to compare, we sketch one cycle of $y = \sin x$ using a dotted curve and label the critical points.

(continued)

Using a table or graphing utility, we obtain one cycle for

$$y = 3 \sin x \text{ and } y = -\tfrac{1}{2} \sin x.$$

x	$\sin x$	$3 \sin x$
0	0	0
$\frac{\pi}{2}$	1	3
π	0	0
$\frac{3\pi}{2}$	-1	-3
2π	0	0

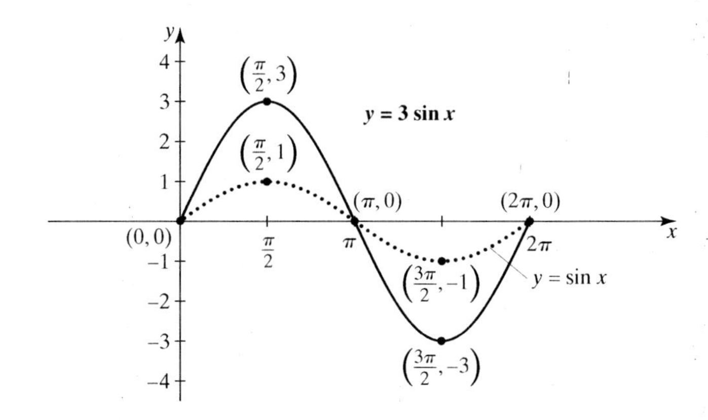

x	$\sin x$	$-\frac{1}{2} \sin x$
0	0	0
$\frac{\pi}{2}$	1	$-\frac{1}{2}$
π	0	0
$\frac{3\pi}{2}$	-1	$\frac{1}{2}$
2π	0	0

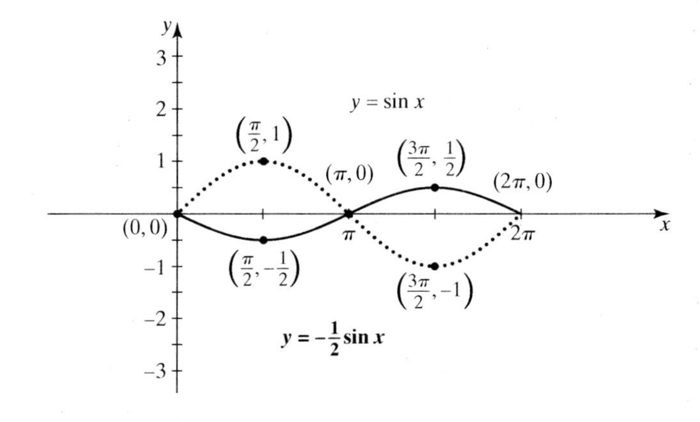

The graphs of $y = 3 \sin x$ and $y = -\frac{1}{2} \sin x$ are sinusoidal waves with the same period and same x-intercepts as $y = \sin x$.

One cycle of each graph has four identical arcs, but we notice the maximum and minimum y-coordinates of the critical points for $y = 3 \sin x$ and $y = -\frac{1}{2} \sin x$ are different from $y = \sin x$.

- Multiplying $\sin x$ by 3 changes the amplitude since it multiplies maximum and minimum y-coordinates of the pure form by 3. As a result, the amplitude of $y = 3 \sin x$ is $\frac{1}{2}|3 - (-3)| = 3$. Multiplying the function by a real number greater than 1 or less than -1 results in a **vertical stretch** of the pure form.
- Multiplying $\sin x$ by $-\frac{1}{2}$ not only changes the amplitude, but also as a result of the multiplier being negative, *reflects the graph across the x-axis.* The amplitude of $y = -\frac{1}{2} \sin x$ is $\frac{1}{2}\left|\frac{1}{2} - \left(-\frac{1}{2}\right)\right| = \frac{1}{2}$. Multiplying the function by a nonzero real number between -1 and 1 results in a **vertical shrink** of the pure form. ■

Generalizing the result of multiplying the function by a nonzero real number, we get the following description of the effect on the pure form.

Amplitude of a Sinusoidal Wave in the Form $y = A\cos x$ and $y = A\sin x$,

where A is any nonzero real number:

$|A|$ = amplitude

$|A| > 1$ produces a vertical stretch

$|A| < 1$ produces a vertical shrink

$A < 0$ produces a reflection across the x-axis

Using the information about amplitude and our knowledge of the pure form graphs, we should be able to quickly sketch any graph of the form $y = A\cos x$ and $y = A\sin x$.

EXAMPLE 5 Sketch $y = 4\cos x$ for $-2\pi \le x \le 2\pi$ and label the critical points on one cycle.

SOLUTION *Since $A = 4$, we have a vertical stretch.*
Sketch the dotted graph of $y = \cos x$ (pure form) on $0 \le x \le 2\pi$ for reference.
For $y = 4\cos x$, $A = 4$. So, amplitude $= |4| = 4$.
Plot the critical points for the graph of $y = 4\cos x$ on $0 \le x \le 2\pi$:

- The x-intercepts are the same as $y = \cos x$.
- The maximum and minimum y-coordinates of the critical points are 4 and -4, respectively.

Connect the critical points with four identical arc sections in the cosine form of a sinusoidal wave. Repeat the pattern for one more cycle (copy and paste) on the required interval.

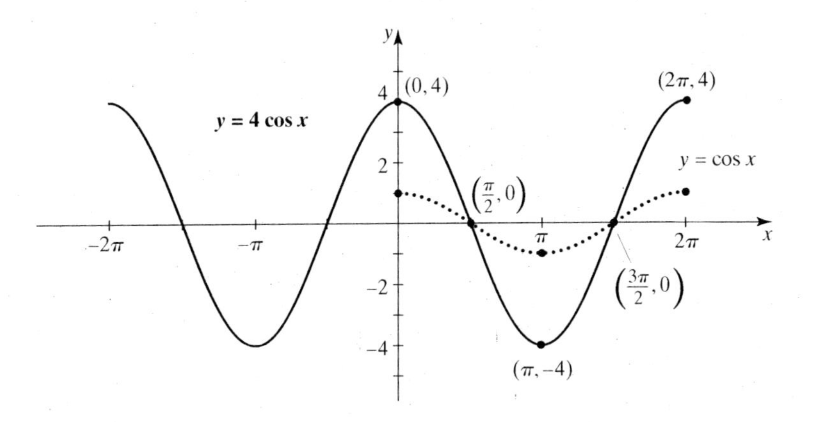

Notice the range is $|y| \le 4$ and the x-intercepts are $x = \ldots, -3\pi/2, -\pi/2, \pi/2, 3\pi/2, \ldots$, which can be written as $x = (\pi/2) + k\pi$.

EXAMPLE 6 Sketch $y = -5 \sin x$ for $0 \le x \le 4\pi$ and label the critical points on one cycle. State the range and the x-intercepts.

SOLUTION *Since $A = -5$, we have a vertical stretch and a reflection across the x-axis.*
Sketch the dotted graph of the pure form, $y = \sin x$ on $0 \le x \le 2\pi$ for reference.
For $y = -5 \sin x$, $A = -5$. So, amplitude $= |-5| = 5$.
Plot the critical points for the graph of $y = -5 \sin x$ on $0 \le x \le 2\pi$:

- The x-intercepts are the same as $y = \sin x$.
- The maximum and minimum values of the y-coordinates of the critical points are 5 and -5.
- Since A is negative, the critical points will reflect across the x-axis.

Connect the critical points with four identical arc sections in the sine form of a sinusoidal wave. Repeat the pattern for one more cycle.

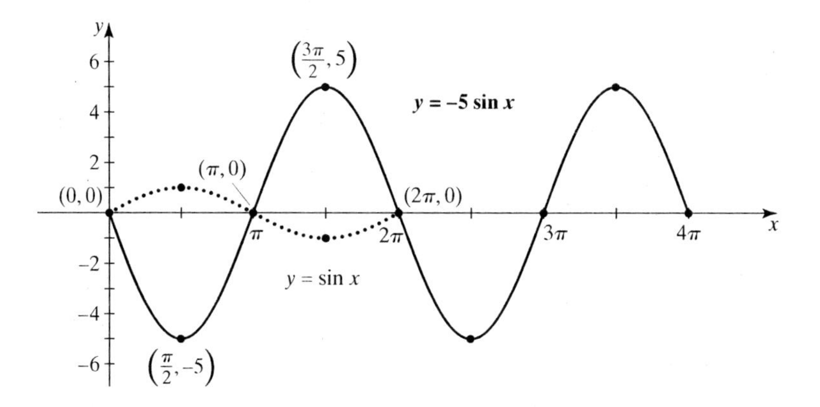

The range is $|y| \le 5$ and the x-intercepts are $x = \ldots, 0, \pi, 2\pi, 3\pi, 4\pi, \ldots$, which can be written as $x = k\pi$.

CONNECTIONS WITH TECHNOLOGY

Graphing Calculator/CAS

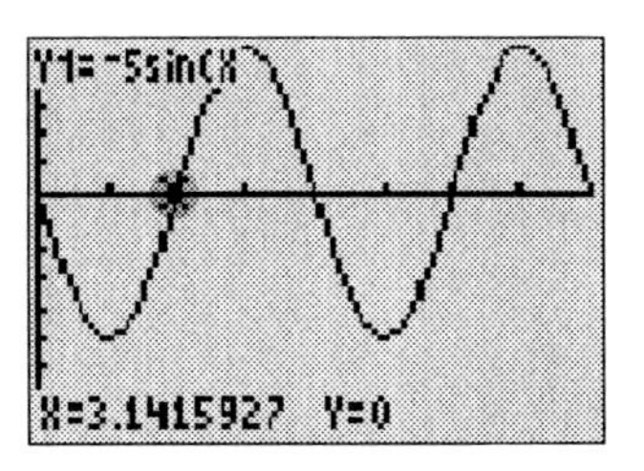

Each graph can be checked using a graphing utility.

To check Example 6, we use the following settings:

MODE	Radian
WINDOW	Xmin $= 0$ Xmax $= 4\pi$ Xscl $= \pi/2$
	Ymin $= -6$ Ymax $= 6$

and input $y = -5 \sin x$.

Trace along the graph (or use an EVALuation feature, if available) to verify critical points.

EXAMPLE 7 Examine the characteristics of the graphs of the following sinusoidal functions. Find an equation in the form $y = A \cos x$ or $y = A \sin x$ that represents each graph.

a.

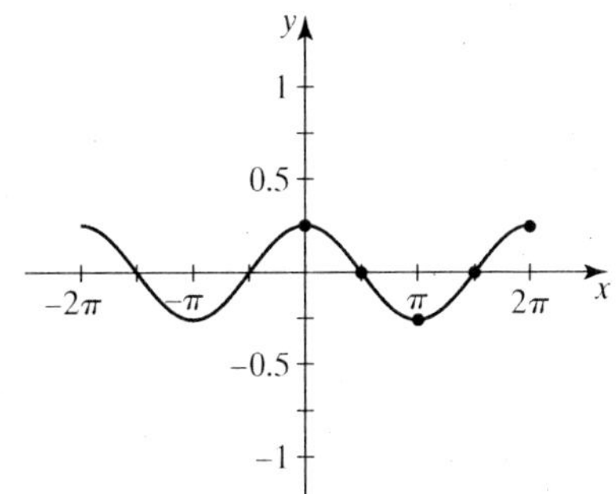

b.

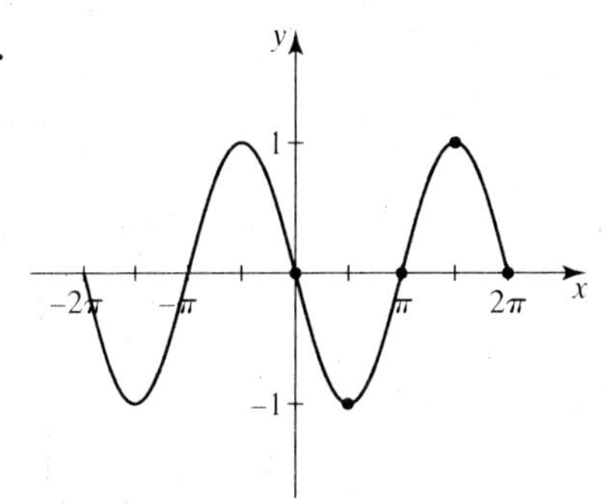

c.

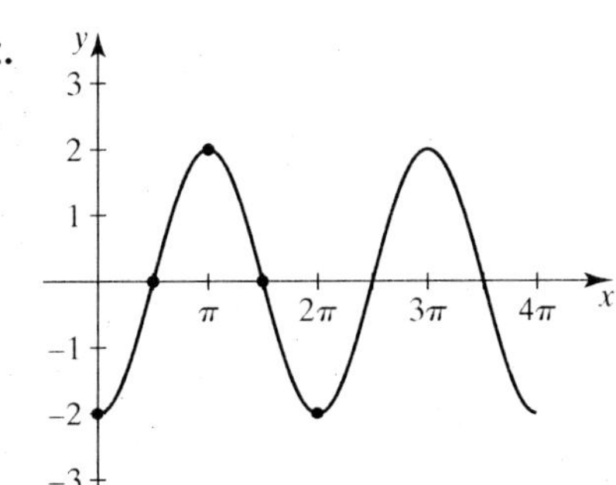

d.

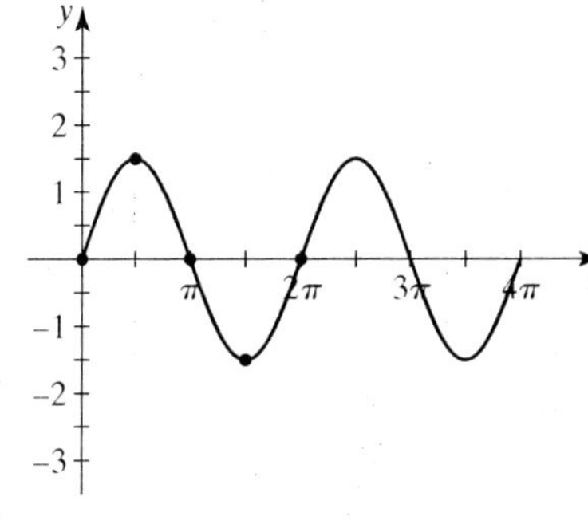

SOLUTION

a. The graph has cosine form with amplitude $\frac{1}{4}$. The equation is $y = \frac{1}{4} \cos x$.

b. The graph has sine form with amplitude 1, reflected across the x-axis. The equation is $y = -\sin x$.

c. $y = -2 \cos x$

d. $y = 1.5 \sin x$ ■

CONNECTIONS WITH TECHNOLOGY

Graphing Calculator/CAS

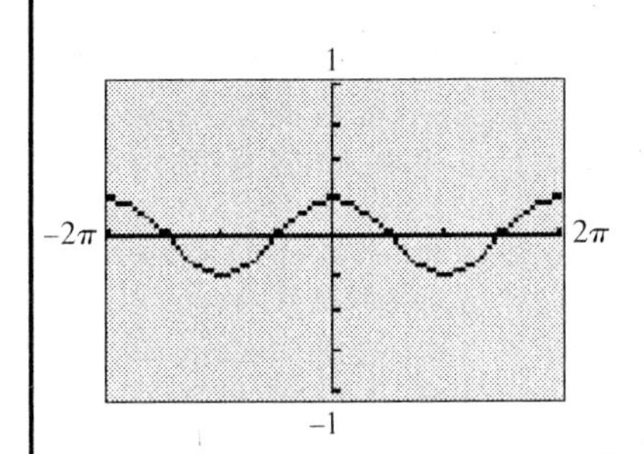

To check an equation to see if it represents the graph, input your equation using the same window setting as the given graph. If your equation is correct, the graphs should look the same. To check Example 7 (a), we use the following settings.

MODE	Radian		
WINDOW	Xmin $= -2\pi$	Xmax $= 2\pi$	Xscl $= \pi/2$
	Ymin $= -1$	Ymax $= 1$	Yscl $= 0.25$

Then we input $y = \frac{1}{4} \cos x$ and graph.

By knowing the pure form graphs $y = \sin x$ and $y = \cos x$, you should be able to identify or sketch graphs of functions in the form $y = A\cos x$ and $y = A\sin x$. Until you are more familiar with graphing, it is a good idea to first dot in the pure form as a reference. While graphing these sinusoidal functions, keep ***all these characteristics*** in mind:

- continuous periodic waves
- four identically shaped arcs for each cycle with critical points at the ends
- the x-intercepts and symmetry
- amplitude or reflection changes to the maximum and minimum functional values

It is recommended that you plot the critical points for one cycle first. Then connecting them with a smooth curve having four identically shaped arcs will produce a nice looking wave. When you are sketching your graphs, try to make the axes have a reasonable scale when possible. You can check any of your graphs using a graphing calculator or other graphing utility.

The important characteristics of the pure forms of the sinusoidal functions are summarized in the following chart.

Sinusoidal Functions

Function	Domain Amp. Period	Range	Graph (pure form)	x-intercepts Symmetry
$y = \sin x$	$x \in \mathbb{R}$ $A = 1$ $P = 2\pi$	$\lvert\sin x\rvert \le 1$ or $\lvert y\rvert \le 1$	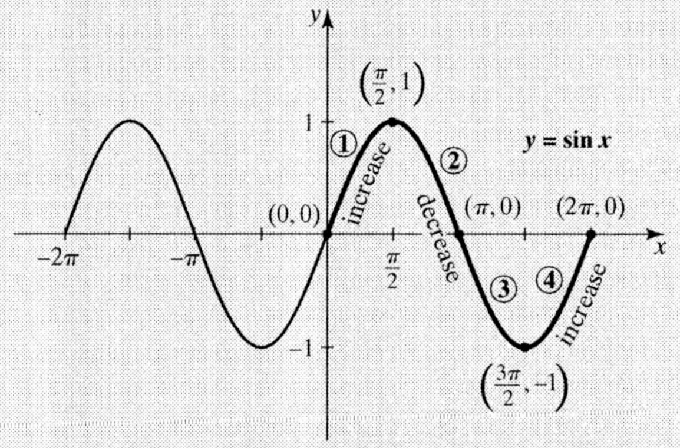	$x = k\pi$ origin
$y = \cos x$	$x \in \mathbb{R}$ $A = 1$ $P = 2\pi$	$\lvert\cos x\rvert \le 1$ or $\lvert y\rvert \le 1$	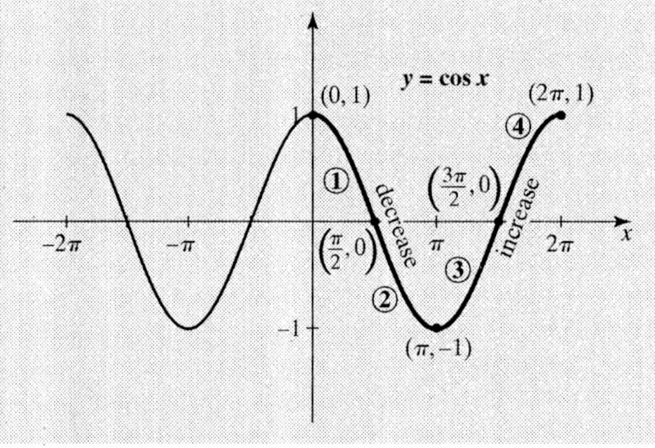	$x = \dfrac{\pi}{2} + k\pi$ y-axis

Modifications to characteristics in addition to the amplitude will be discussed in the next section after you have practiced amplitude and reflection changes to the pure forms.

Exercise Set 2.1

1–2. *Sketch the following sinusoidal functions on the indicated interval. Label the critical points for one cycle* $(0 \le x \le 2\pi)$.

1. a. $y = \cos x, 0 \le x \le 4\pi$ **b.** $y = \sin x, -2\pi \le x \le 2\pi$

2. a. $y = -\sin x, 0 \le x \le 4\pi$ **b.** $y = -\cos x, -2\pi \le x \le 2\pi$

3–12. **a.** *Without using a table or graphing utility, sketch the graph on* $-2\pi \le x \le 2\pi$.
b. *State the range and the x-intercepts.*
c. *Check your graph with a graphing utility.*

EXAMPLE $y = \frac{4}{5}\cos x$

SOLUTION *Since* $A = \frac{4}{5}$, *we have a vertical shrink.*

$$\text{Amplitude} = |A| = \left|\frac{4}{5}\right| = \frac{4}{5}.$$

Plot the critical points for $y = \frac{4}{5}\cos x$ on $0 \le x \le 2\pi$:

- The x-intercepts are the same as $y = \cos x$.
- The maximum and minimum y-coordinates of the critical points are $\frac{4}{5}$ and $-\frac{4}{5}$, respectively.

Connect the critical points with four identical arc sections in the cosine form of a sinusoidal wave. Repeat the pattern for one more cycle (copy and paste) on the required interval.

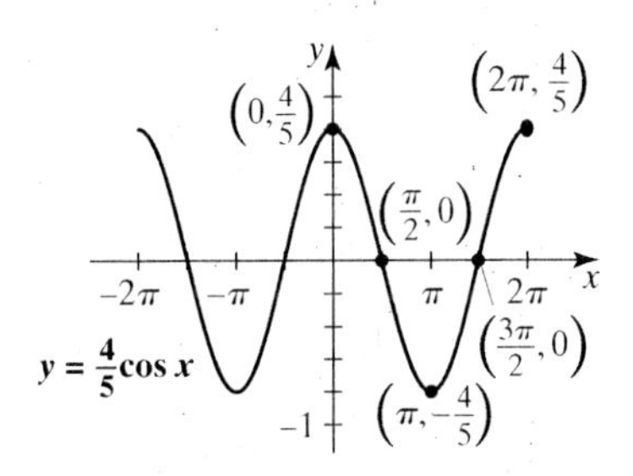

$$\text{Range: } |y| \le \frac{4}{5}$$

$$x\text{-intercepts: } x = -\frac{3\pi}{2}, -\frac{\pi}{2}, \frac{\pi}{2}, \frac{3\pi}{2} \left(\text{or } \frac{\pi}{2} + k\pi\right)$$

3. $y = 6\cos x$ **4.** $y = 4\sin x$

5. $y = -\pi\sin x$ **6.** $y = -\frac{2}{3}\cos x$

7. $y = 7\sin x$ **8.** $y = 2.5\sin x$

9. $y = -0.4\cos x$ **10.** $y = -\frac{7}{5}\cos x$

11. $y = \frac{3}{2}\cos x$ **12.** $y = -0.9\sin x$

13–18. *Find an equation in the form $y = A \cos x$ or $y = A \sin x$ that represents each graph. Check your answer using a graphing utility.*

13.

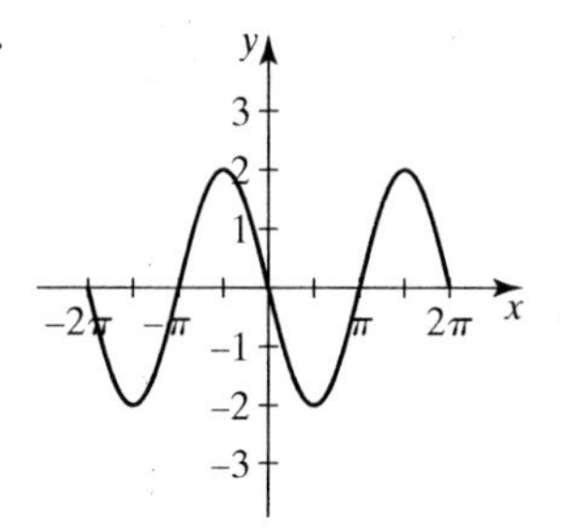

14.

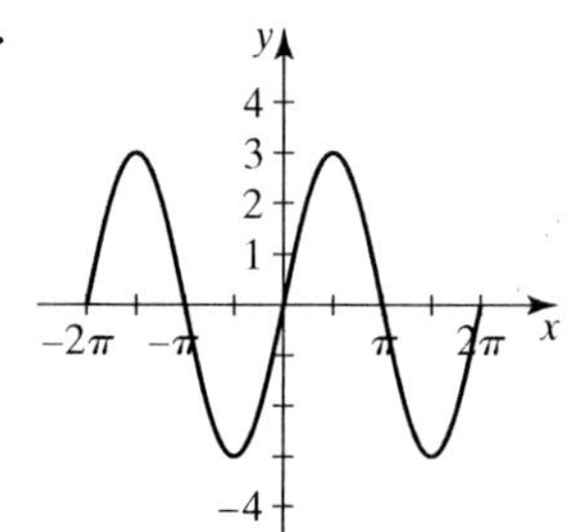

15.

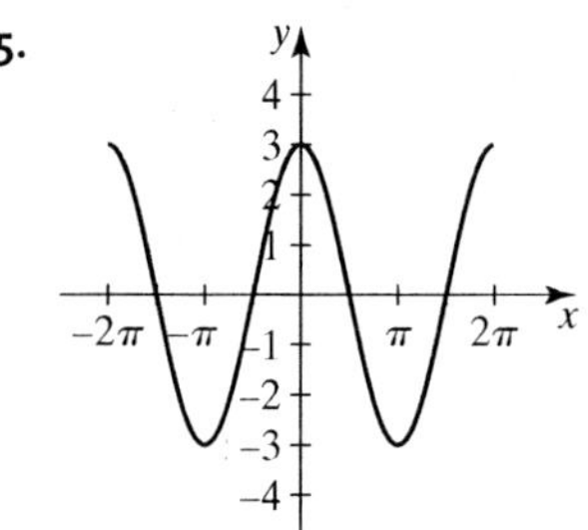

16.

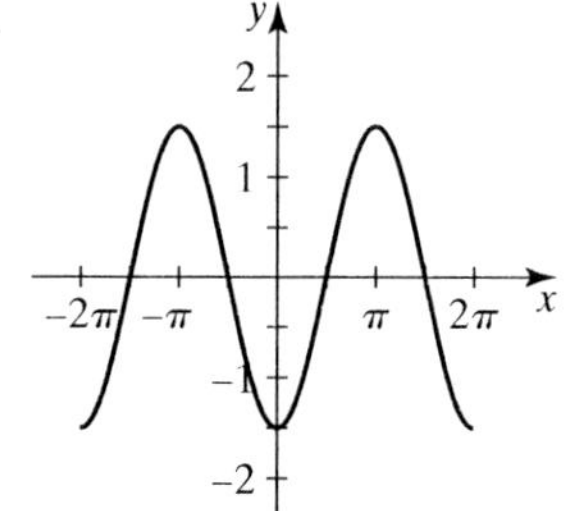

17.

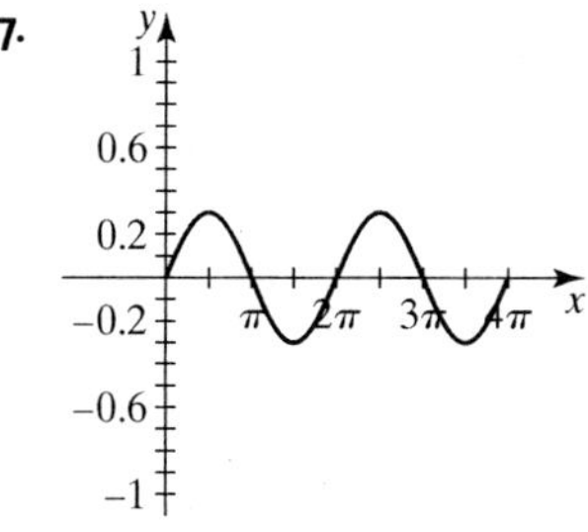

18.

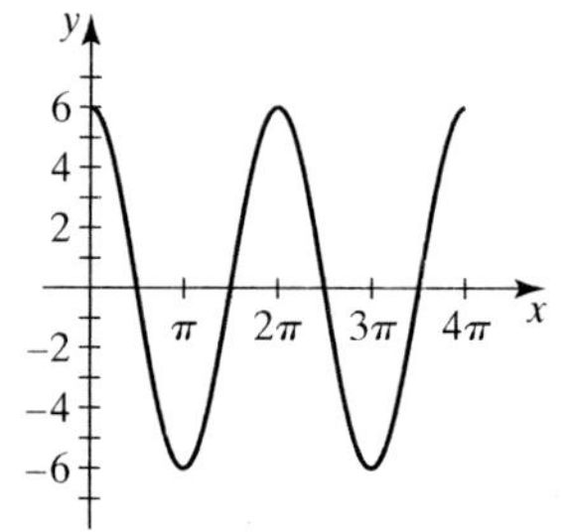

19–24. *Indicate whether the given graph represents one cycle of either $y = A \sin x$ or $y = A \cos x$. If the graph is not a representation of either function, explain why.*

19.

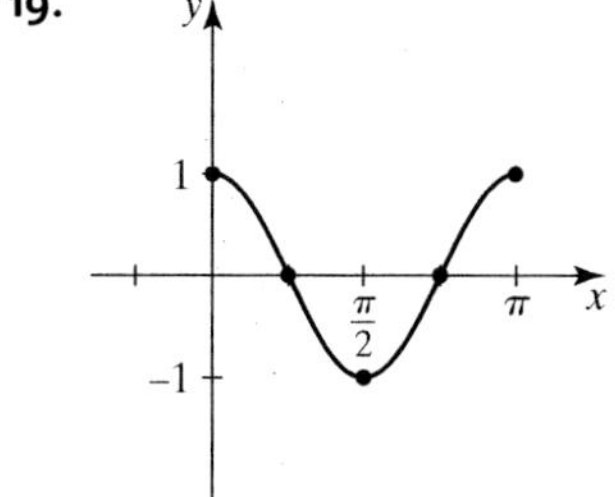

20.

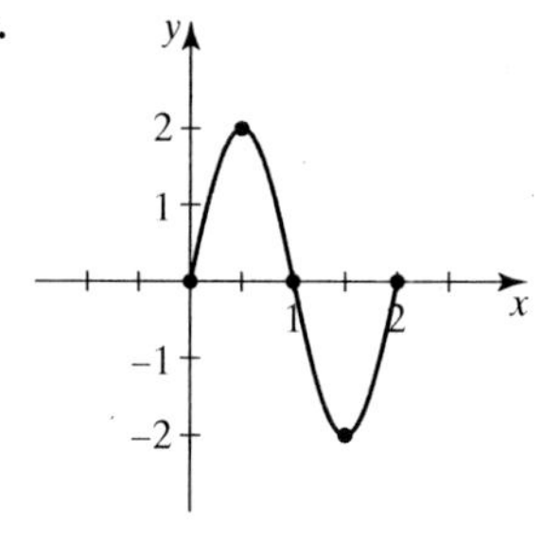

21.

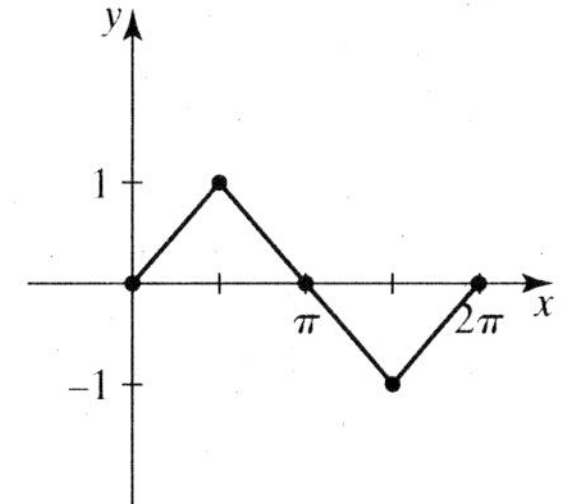

22.

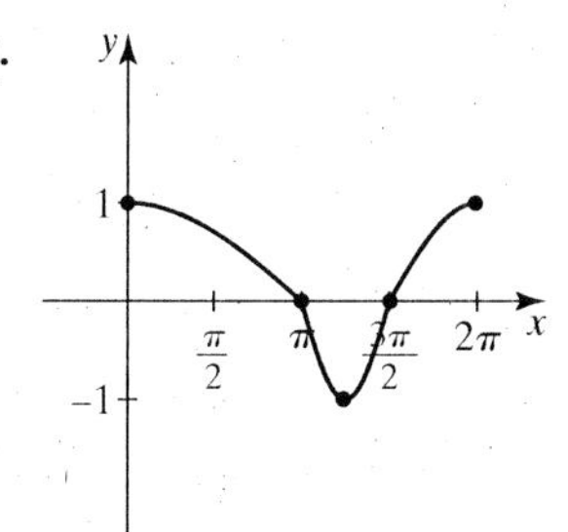

23.

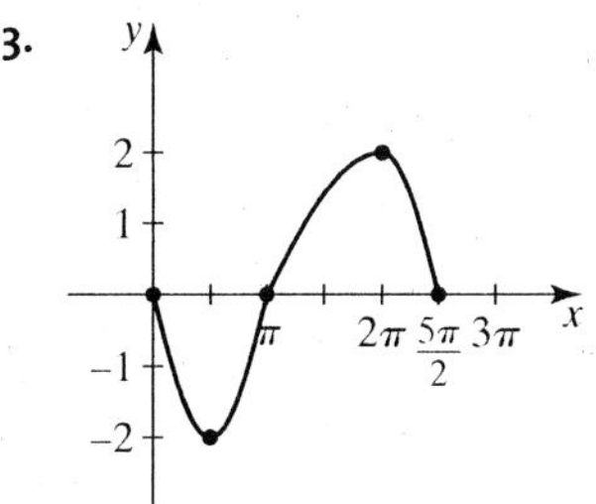

24.

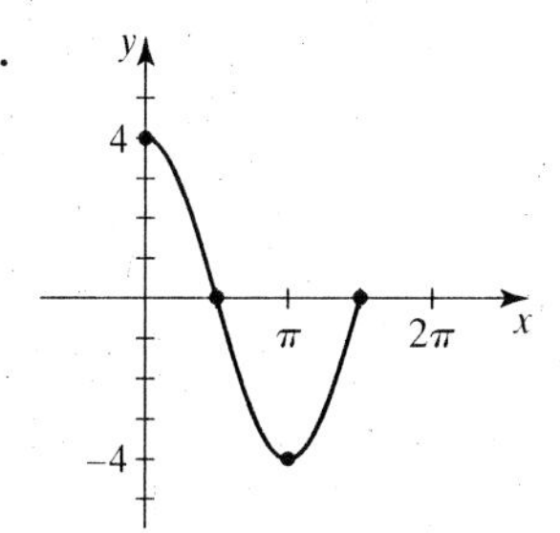

25. Specify the maximum number of complete cycles for $y = 4 \sin x$ on the given interval.

a. $-4\pi \leq x \leq 6\pi$ **b.** $-3\pi \leq x \leq 2\pi$

26. Specify the maximum number of complete cycles for $y = -9 \cos x$ on the given interval.

a. $-8\pi \leq x \leq -6\pi$ **b.** $\pi \leq x \leq 10\pi$

27. A function $f(x)$ is defined to be **even** if $f(-x) = f(x)$ for all x in the domain of f. The graph of an even function is *symmetric with respect to the y-axis.* Using the graph of $y = \cos x$, we can verify that the cosine function is an even function by checking the symmetry. Which identity from Chapter 1 can be used to show the cosine is an even function if we want to use the definition $f(-x) = f(x)$?

28. A function $f(x)$ is defined to be **odd** if $f(-x) = -f(x)$ for all x in the domain of f. The graph of an odd function is *symmetric with respect to the origin.* Using the graph of $y = \sin x$, we can verify that the sine function is an odd function by checking the symmetry. Which identity from Chapter 1 can be used to show the sine is an odd function if we want to use the definition $f(-x) = -f(x)$?

29. The horizontal displacement of a point on an oscillating pendulum is given by the equation $d(t) = 3 \sin t$, where t is in seconds and d is in inches.

a. Graph $d(t) = 3 \sin t$ on $0 \leq t \leq 4\pi$.

b. We consider the rest position to be when the displacement is zero $(d(t) = 0)$. Find the number of times the point is in the rest position.

c. What is the maximum displacement?

d. What is the relationship between the maximum displacement and the amplitude?

e. What is the displacement to the nearest tenth of an inch at $t = 2.3$ seconds? Plot the point on the graph that corresponds to this question.

30. The horizontal displacement of a weight attached to one end of a spring is given by $d(t) = -5 \cos t$, where t is in seconds and d is in centimeters.

a. Graph $d(t) = -5 \cos t$ on $0 \le t \le 4\pi$.

b. We consider the rest position to be when the displacement is zero $(d(t) = 0)$. Find the number of times the weight is in the rest position.

c. What is the minimum displacement?

d. What is the relationship between the minimum displacement and the amplitude?

e. What is the displacement to the nearest tenth of a centimeter at $t = 6$ seconds? Plot the point on the graph that corresponds to this question.

31–32. *Consider the given point graphed on the unit circle. Match the point that corresponds to it on both graphs of* $y = \cos x$ *and* $y = \sin x$.

31.

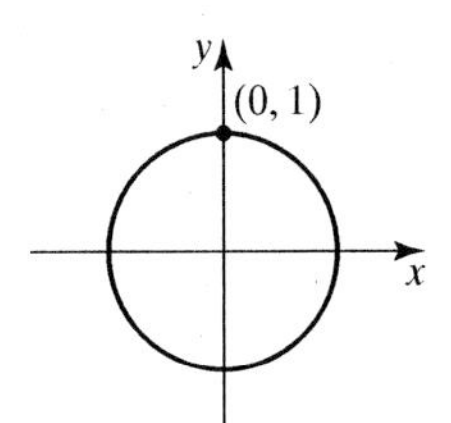

a. $y = \cos x$

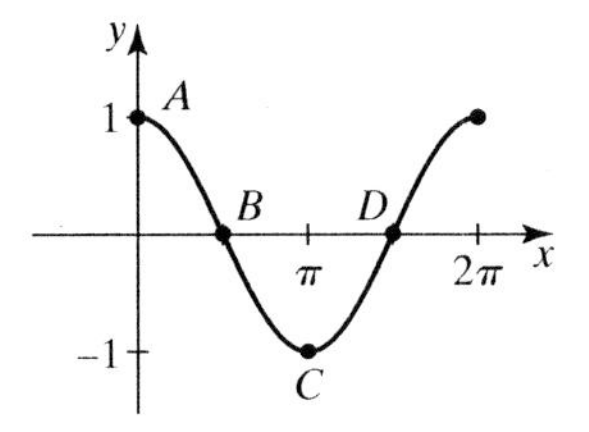

b. $y = \sin x$

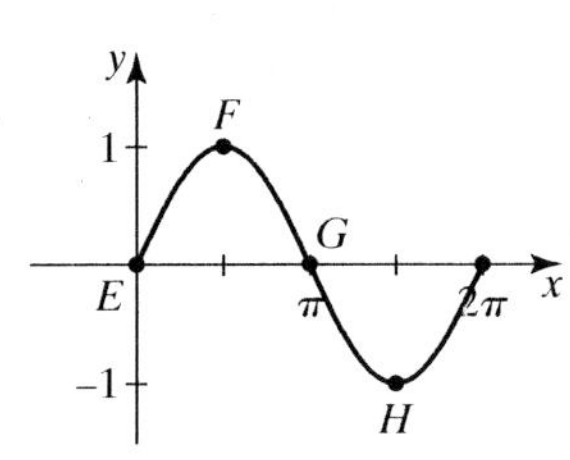

32.

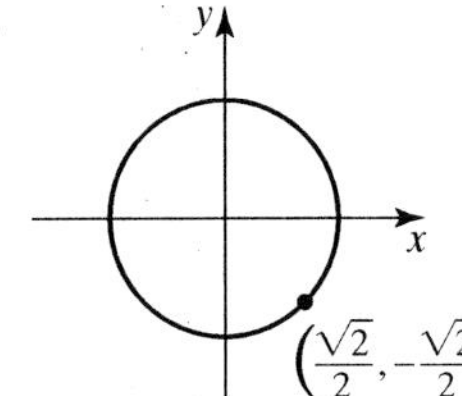

a. $y = \cos x$

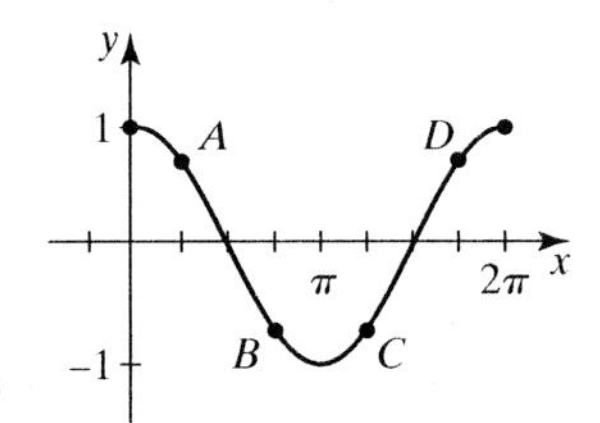

b. $y = \sin x$

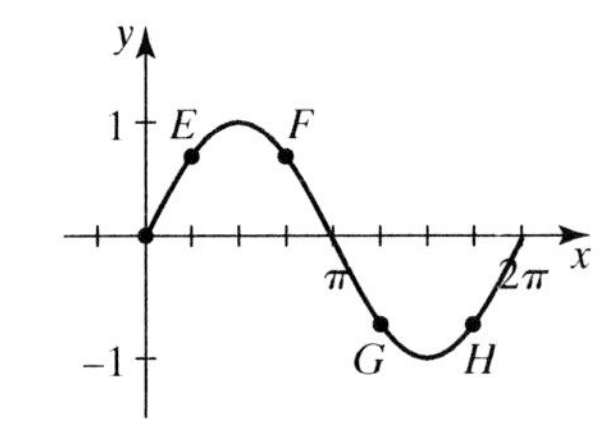

33–40. *Are the following statements true or false? If a statement is false, explain why or give an example that shows why it is false.*

33. The graphs of the sine and cosine function are sinusoidal.

34. Each cycle of a sinusoidal wave has four identically shaped arc sections.

35. The y-intercept of $y = A \cos x$ is $(0, A)$.
36. The y-intercept of $y = A \sin x$ is $(0, A)$.
37. One period of $y = A \sin x$ or $y = A \cos x$ is 2π.
38. The amplitude of $y = A \sin x$ or $y = A \cos x$ is A.
39. The graph of $y = A \sin x$ is symmetric with respect to the y-axis.
40. The graph of $y = A \cos x$ is symmetric with respect to the y-axis.

Graphing Calculator/CAS Explorations

41. **a.** Graph the pair of functions

$$y = \sin(x)$$
$$y = \sin(2x)$$

using the following settings:

MODE	Radian
WINDOW	Xmin $= 0$ Xmax $= 4\pi$ Xscl $= \pi/2$ Ymin $= -2$ Ymax $= 2$ Yscl $= 1$

b. What are the differences between the pure form $y = \sin(x)$ and $y = \sin(2x)$?

c. Use the same settings to graph the pair of functions

$$y = \sin(x)$$
$$y = \sin\left(\frac{1}{2}x\right).$$

d. What are the differences between the pure form $y = \sin(x)$ and $y = \sin\left(\frac{1}{2}x\right)$?

e. Predict the period of $y = \sin\left(\frac{1}{4}x\right)$ and verify your guess by using a graphing utility with an appropriate Xmin and Xmax.

42. **a.** Graph the pair of functions

$$y = \cos(x)$$
$$y = \cos(2x)$$

using the following settings:

MODE	Radian
WINDOW	Xmin $= 0$ Xmax $= 4\pi$ Xscl $= \pi/2$ Ymin $= -2$ Ymax $= 2$ Yscl $= 1$

b. What are the differences between the pure form $y = \cos(x)$ and $y = \cos(2x)$?

c. Use the same settings to graph the pair of functions

$$y = \cos(x)$$
$$y = \cos\left(\frac{1}{2}x\right).$$

d. What are the differences between the pure form $y = \cos(x)$ and $y = \cos\left(\frac{1}{2}x\right)$?

e. Predict the period of $y = \cos\left(\frac{1}{4}x\right)$ and verify your guess by using a graphing utility with an appropriate Xmin and Xmax.

43. **a.** Graph the pair of functions

$$y = \sin(x)$$
$$y = \sin\left(x + \frac{\pi}{4}\right)$$

using the following settings:

MODE	Radian
WINDOW	Xmin $= -2\pi$ Xmax $= 2\pi$ Xscl $= \pi/4$ Ymin $= -2$ Ymax $= 2$ Yscl $= 1$

b. What are the differences between the pure form $y = \sin(x)$ and $y = \sin(x + \frac{\pi}{4})$?

c. Use the same settings to graph the pair of functions

$$y = \cos(x)$$
$$y = \cos\left(x - \frac{\pi}{4}\right).$$

d. What are the differences between the pure form $y = \cos(x)$ and $y = \cos(x - \frac{\pi}{4})$?

e. Predict how much and in which direction the graph of $y = \sin(x + \pi)$ shifts the pure form of $y = \sin(x)$. Verify your guess by graphing $y = \sin(x)$ and $y = \sin(x + \pi)$.

44. **a.** Graph the pair of functions

$$y = \sin(x)$$
$$y = \sin(x) + 2$$

using the following settings:

MODE	Radian
WINDOW	Xmin $= -2\pi$ Xmax $= 2\pi$ Xscl $= \pi/2$ Ymin $= -4$ Ymax $= 4$ Yscl $= 1$

b. What are the differences between the pure form $y = \sin(x)$ and $y = \sin(x) + 2$?

c. Use the same settings to graph the pair of functions

$$y = \cos(x)$$
$$y = \cos(x) - 1.$$

d. What are the differences between the pure form $y = \cos(x)$ and $y = \cos(x) - 1$?

e. Predict how much and in which direction the graph of $y = \sin(x) - 2$ shifts the pure form of $y = \sin(x)$. Verify your guess by graphing $y = \sin(x)$ and $y = \sin(x) - 2$.

45–46. *Let's investigate what happens to the graph of $y = A \cos x$ or $y = A \sin x$ when A is a variable instead of a nonzero real number.*

45. Graph the pair of functions

$$y = x\sin(x)$$
$$y = 3\sin(x)$$

using the following settings:

MODE	Radian
WINDOW	Xmin = 0 Xmax = 4π Ymin = -13 Ymax = 13

a. Are the x-intercepts the same for both graphs?

b. What is the amplitude for $y = 3\sin(x)$?

c. What do you notice about the amplitude of $y = x\sin(x)$?

d. What effect does A being a variable have on the graph of $y = A\sin x$?

Use the same settings to graph the functions $y = x\sin(x)$, $y = x$, *and* $y = -x$.

e. What do you notice about the two linear graphs?

f. What do you know about the sine function that would allow you to anticipate that the two lines $y = x$ and $y = -x$ would be boundary graphs for $y = x\sin(x)$?

46. (Do Exercise 45 first.) Graph $y = \dfrac{1}{x+1}\cos(x)$ using the following settings:

MODE	Radian
WINDOW	Xmin = 0 Xmax = 6π Ymin = -1 Ymax = 1

a. What functions do you anticipate would be the boundary functions for $y = \dfrac{1}{x+1}\cos(x)$? Explain. Check your answer by graphing $y = \dfrac{1}{x+1}\cos(x)$ along with the two boundary functions.

b. Explain why this graph could represent the motion of a diving board.

47. In Exercise 45 of Section 1.3, we presented the following infinite power series formulas for the cosine and sine functions:

$$\cos x = 1 - \frac{x^2}{2!} + \frac{x^4}{4!} - \frac{x^6}{6!} + \frac{x^8}{8!} - \cdots + (-1)^{n-1}\frac{x^{2n-2}}{(2n-2)!} - \cdots$$

and

$$\sin x = x - \frac{x^3}{3!} + \frac{x^5}{5!} - \frac{x^7}{7!} + \cdots + (-1)^{n-1}\frac{x^{2n-1}}{(2n-1)!} - \cdots$$

a. Graph $y = \cos x$ using the following settings:

MODE	Radian
WINDOW	Xmin = $-\pi$ Xmax = π Ymin = -2 Ymax = 2

b. On the same screen, graph the polynomial

$$y = 1 - \frac{x^2}{2!} + \frac{x^4}{4!} - \frac{x^6}{6!},$$

which represents the first four terms of the infinite series for $\cos x$. Recall that $2! = 2 \cdot 1 = 2$ and $4! = 4 \cdot 3 \cdot 2 \cdot 1 = 24$. (Rather than determining the numerical values of the denominators, it would be much easier to use the "!" factorial key on your calculator.)

(continued)

c. On what interval along the x-axis do the cosine function and this polynomial appear to be converging to the same graph?

d. In the window settings, change Xmin to -2π and Xmax to 2π and explain what happens to these functions when they are graphed on this larger interval, away from the interval in which they seem to converge.

e. Using the same window settings given in part (d), input the next two terms in the formula for the infinite series:

$$y = 1 - \frac{x^2}{2!} + \frac{x^4}{4!} - \frac{x^6}{6!} + \frac{x^8}{8!} - \frac{x^{10}}{10!}.$$

Graph this along with $y = \cos x$. What is happening to the interval of convergence?

f. Change the window settings for Xmin to -4π and Xmax to 4π and experiment to see how many additional terms of the infinite series you need to include in order to have the graphs appear to converge on $[-4\pi, 4\pi]$.

Explore the Pattern

48. Without looking at the formula for the terms of the infinite series for $\sin x$, try to determine the pattern based on the first five terms:

$$x - \frac{x^3}{3!} + \frac{x^5}{5!} - \frac{x^7}{7!} + \frac{x^9}{9!}.$$

49. We know that $y = \sin x$ is an odd function (Section 2.1, Exercise 28) and oscillates between 1 and -1. How could someone use these characteristics of the sine function to remember the terms in its infinite series:

$$\sin x = x - \frac{x^3}{3!} + \frac{x^5}{5!} - \frac{x^7}{7!} + \frac{x^9}{9!} - \ldots?$$

2.2 Period, Phase Shift, and Other Translations

Many real-life occurrences have periodic phenomena. We mentioned a few in Chapter 1 and many have graphs that look sinusoidal. For example, if we plot the time of sunrise as a function of the day of the year, we could get a graph that looks like the following sinusoidal wave, whose period appears to be one year. In Section 2.1 all of the sine waves have a period of 2π, but most real-life occurrences won't have this same period. We also notice the entire sunrise graph is above the x-axis and has no x-intercepts.

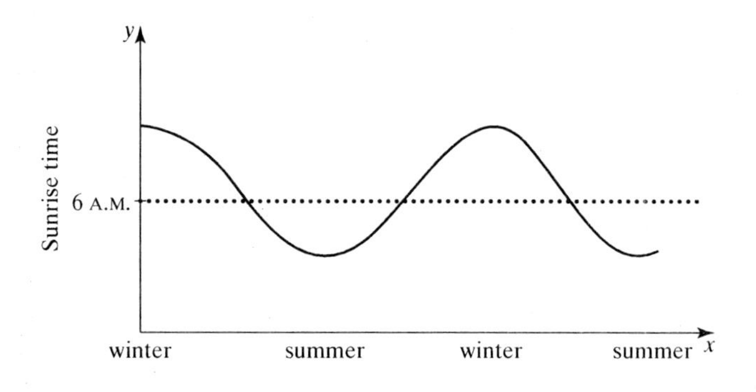

Since Earth's motion around the sun is periodic, it seems reasonable that sinusoidal waves can be used to model this phenomenon. But we cannot use the sinusoidal functions of the form $y = A\cos x$ or $y = A\sin x$ to model either this or certain other periodic phenomena because different values of A change only the amplitude and/or reflect the graph across the x-axis. We may also need to change the period (which produces a horizontal shrink or stretch) as well as the location of the maximum and minimum values (which shifts the graph up, down, to the right, or to the left). We now investigate changes to the pure forms that produce these additional modifications by first doing examples using a point-plotting method or graphing utility. Then, after you learn what characteristics of the equation cause these changes, you should be able to sketch the graphs quickly.

Period Change (Horizontal Shrink or Stretch)

In Example 1 we investigate the effect of multiplying the arc of a sinusoidal function by a nonzero real number.

EXAMPLE 1 Compare the graphs of the pure form $y = \sin x$ and $y = \sin(2x)$.

SOLUTION For reference, we sketch the dotted graph of $y = \sin x$ on $0 \le x \le 2\pi$. Using a table (or graphing utility), we graph $y = \sin(2x)$ on $0 \le x \le 2\pi$.

x	$2x$	$\sin(2x)$
0	0	0
$\frac{\pi}{4}$	$\frac{\pi}{2}$	1
$\frac{\pi}{2}$	π	0
$\frac{3\pi}{4}$	$\frac{3\pi}{2}$	-1
π	2π	0
$\frac{5\pi}{4}$	$\frac{5\pi}{2}$	1
$\frac{3\pi}{2}$	3π	0
$\frac{7\pi}{4}$	$\frac{7\pi}{2}$	-1
2π	4π	0

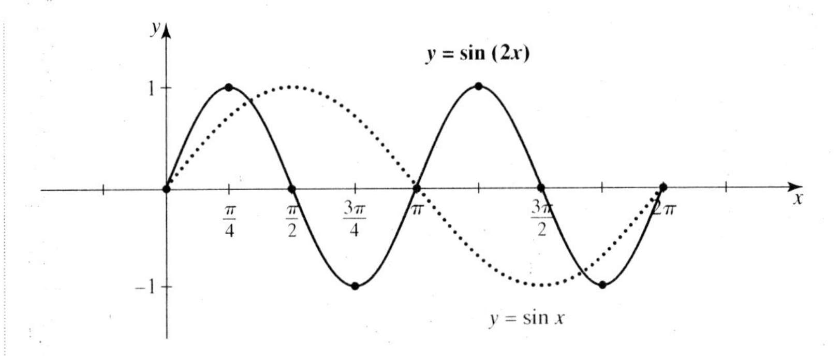

First we notice a *different period* for $y = \sin(2x)$. Multiplying the arc x by 2 results in two cycles of the graph within 2π units along the x-axis. Therefore, the period of $y = \sin(2x)$ is *half* the pure form's period. Also notice that as we change the period, we also change the x-intercepts. The x-intercepts of $y = \sin(2x)$ are found by dividing the pure form's x-intercepts by 2, which results in positions that are *half* as far as the pure form's positions. Comparing the graphs of $y = \sin x$ and $y = \sin(2x)$, we find that the period of $y = \sin(2x)$ is

$$P = \frac{\text{pure period}}{2} = \frac{2\pi}{2} = \pi,$$

and the x-intercepts are at

$$x = \frac{\text{pure } x\text{-intercepts}}{2} = \frac{k\pi}{2}, \; k \text{ any integer.}$$

We see that multiplying x by 2 shrinks the period, so we say it produces a **horizontal shrink** of the graph of $y = \sin x$. ■

Now let's investigate what happens if we multiply the arc by $\frac{1}{2}$.

EXAMPLE 2 Compare the graphs of the pure form $y = \cos x$ and $y = \cos\left(\frac{1}{2}x\right)$ on the interval $0 \le x \le 4\pi$.

SOLUTION For reference we sketch the dotted graph of $y = \cos x$. Using a table (or graphing utility), we graph $y = \cos\left(\frac{1}{2}x\right)$ on $0 \le x \le 4\pi$.

x	$\frac{1}{2}x$	$\cos\left(\frac{1}{2}x\right)$
0	0	1
π	$\frac{\pi}{2}$	0
2π	π	-1
3π	$\frac{3\pi}{2}$	0
4π	2π	1

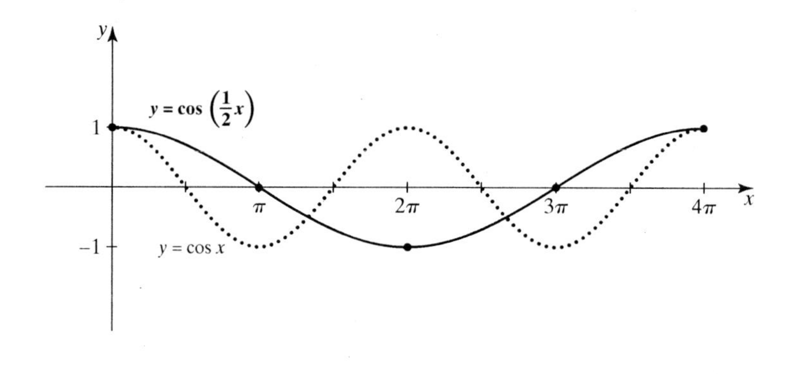

Multiplying the arc x by $\frac{1}{2}$ results in one-half of a cycle within 2π units along the x-axis. Therefore the period of $y = \cos\left(\frac{1}{2}x\right)$ is *twice* the pure form's period. Also, the x-intercepts of $y = \cos\left(\frac{1}{2}x\right)$ are found by dividing the pure form's x-intercepts by $\frac{1}{2}$, which results in positions that are *twice* as far as the pure form's positions. Comparing the graphs of $y = \cos x$ and $y = \cos\left(\frac{1}{2}x\right)$, we find that the period of $y = \cos\left(\frac{1}{2}x\right)$ is

$$P = \frac{\text{pure period}}{\frac{1}{2}} = \frac{2\pi}{\frac{1}{2}} = 2\pi \cdot 2 = 4\pi,$$

and the x-intercepts are at

$$x = \frac{\text{pure } x\text{-intercepts}}{\frac{1}{2}} = \frac{\frac{\pi}{2} + k\pi}{\frac{1}{2}} = 2\left(\frac{\pi}{2} + k\pi\right) = \pi + 2k\pi.$$

We see that multiplying x by $\frac{1}{2}$ produces a **horizontal stretch** of the graph of $y = \cos x$. ■

The changes shown in Examples 1 and 2 can be generalized for sinusoidal functions as follows.

Period and x-intercepts of a Sinusoidal Wave

For $y = \cos(Bx)$ and $y = \sin(Bx)$, where B is any nonzero real number:

$$\textbf{Period} = P = \frac{\text{pure period}}{|B|} = \frac{2\pi}{|B|}$$

$$x\textbf{-intercepts} = x = \frac{\text{pure } x\text{-intercepts}}{|B|} = \begin{cases} \dfrac{k\pi}{|B|} & \text{for } y = \sin(Bx) \\[2ex] \dfrac{\frac{\pi}{2} + k\pi}{|B|} & \text{for } y = \cos(Bx) \end{cases}$$

$|B| > 1$ produces a horizontal shrink.

$|B| < 1$ produces a horizontal stretch.

Combining Period and Amplitude Changes

If we are given an equation in the form $y = A\cos(Bx)$ or $y = A\sin(Bx)$, we should be able to sketch the graph quickly. Our next example gives a function in the form $y = A\sin(Bx)$, which involves two changes to the pure form. Because A affects the amplitude (a vertical stretch or shrink and possibly a reflection) and B affects the period (a horizontal stretch or shrink) of sinusoidal functions, we know that the range and x-intercepts will also change.

EXAMPLE 3 Graph $y = 2\sin(4x)$, $0 \le x \le 2\pi$. State the amplitude, period, x-intercepts, and range.

SOLUTION *Since $A = 2$ and $B = 4$, we have a vertical stretch and horizontal shrink.*
$A = 2$, so amplitude $= 2$. The maximum and minimum y-coordinates are 2 and -2, respectively.
$B = 4$, so the period $= P = 2\pi/4 = \pi/2$.

To plot the critical points for the graph of $y = 2\sin(4x)$ on $0 \le x \le \pi/2$ (one period), do the following:

- Start at the origin and mark off one period (an interval of length $\pi/2$) to the right along the x-axis. Divide this interval into four equal subintervals. Each subinterval will be

$$\frac{1}{4}(P) = \frac{1}{4}\left(\frac{\pi}{2}\right) = \frac{\pi}{8}$$

in length and will correspond to the four equal arc sections for one cycle. The critical points will be at the ends of each arc section. The x-coordinates of the critical points will correspond to the x-coordinate of these intervals.

(continued)

- Using the sine form, plot the x-intercepts and the maximum and minimum y-coordinates, 2 and -2. (See Graph (a).) Notice the x-intercepts are the same as those determined by the formula for $y = \sin(Bx)$:

$$x = \frac{\text{pure } x\text{-intercepts}}{|B|} = \frac{k\pi}{4}.$$

Letting $k = 0$, 1, and 2, we obtain the x-intercepts 0, $\pi/4$, and $\pi/2$, respectively.

Connect the critical points with four identically shaped arc sections to form a sinusoidal wave. (See Graph (b).) Repeat the pattern on $0 \leq x \leq 2\pi$. (See Graph (c).)

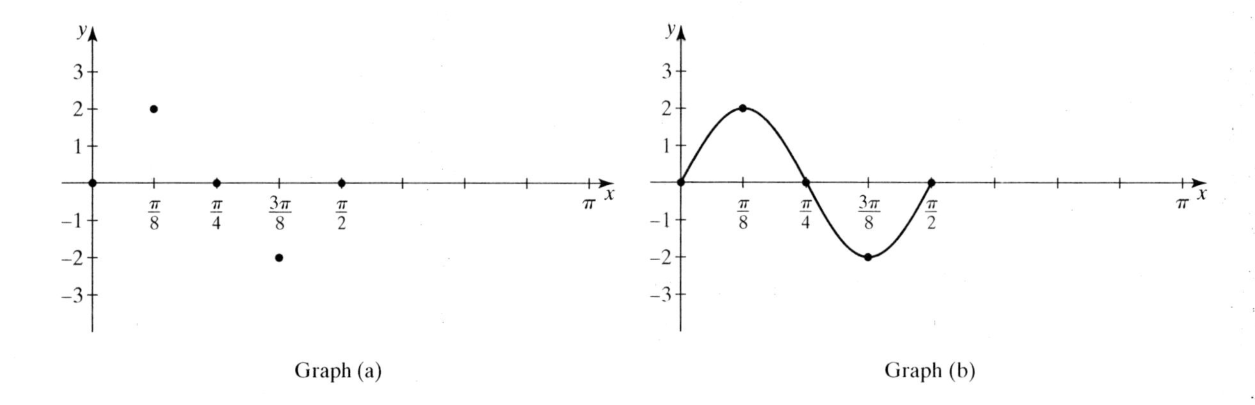

Graph (a) Graph (b)

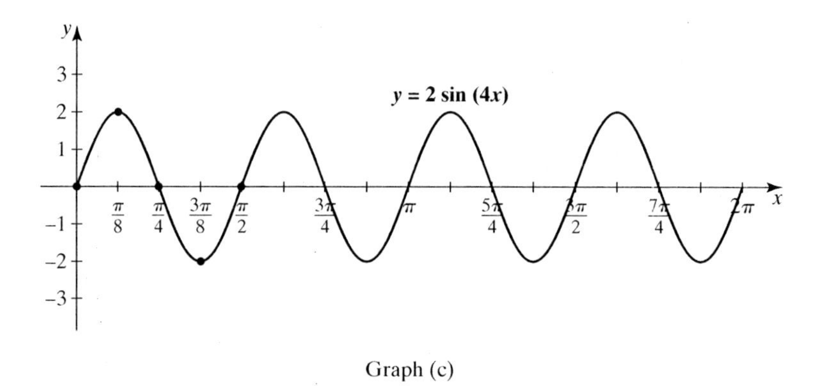

Graph (c)

The range is $|y| \leq 2$. The graph of $y = 2\sin(4x)$ has four cycles within the interval $0 \leq x \leq 2\pi$, which is what we would expect. ■

Translations

We still need to know how to shift the pure form graph up, down, to the right, or to the left. These movements are called **translations**. In the next example we investigate the effect of adding a nonzero real number to a sinusoidal function.

EXAMPLE 4 Compare the graphs of $y = \sin x$ and $y = \sin x + 3$.

SOLUTION We sketch the dotted graph of $y = \sin x$ on $0 \le x \le 2\pi$ for reference and label the critical points. Using a table (or graphing utility), we graph $y = \sin x + 3$ on $0 \le x \le 2\pi$.

x	$\sin x$	$\sin x + 3$
0	0	$0 + 3 = 3$
$\frac{\pi}{2}$	1	4
π	0	3
$\frac{3\pi}{2}$	-1	2
2π	0	3

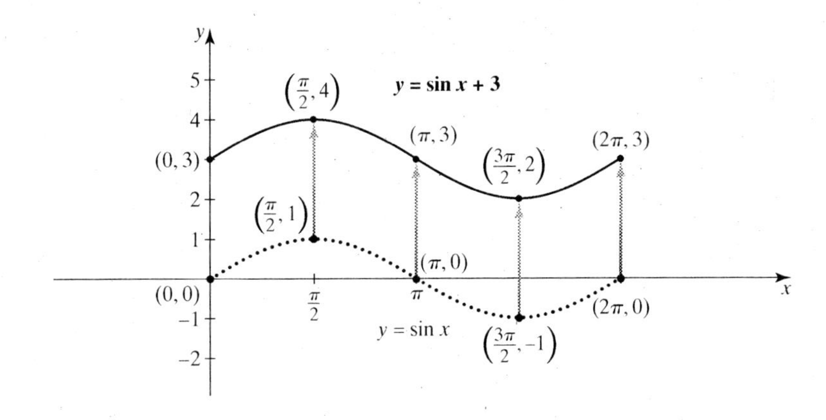

Notice that each point on the graph of $y = \sin x + 3$ is 3 units higher than the corresponding point on the graph of $y = \sin x$. The result is called a **vertical shift**, and in this case, the graph of $y = \sin x$ is shifted *up* 3 units. The range is $2 \le \sin x + 3 \le 4$, and there are no x-intercepts. Similarly, the graph of $y = \sin x - 3$ is the graph of $y = \sin x$ shifted *down* 3 units. ■

We generalize these two changes in the next statement.

Vertical Translation of a Sinusoidal Wave

For $y = \cos x + D$ and $y = \sin x + D$, where D is any nonzero real number:

$|D|$ is the vertical shift.
$D > 0$ produces a vertical shift up.
$D < 0$ produces a vertical shift down.

What about shifting the graphs horizontally to the right or to the left? In the next example, we investigate the effect of subtracting a nonzero real number from the arc of a sinusoidal function.

EXAMPLE 5 Compare the graphs of the pure form $y = \cos x$ and $y = \cos\left(x - \frac{\pi}{2}\right)$.

SOLUTION We sketch the dotted graph of $y = \cos x$ for reference and label the critical points. Using a table (or graphing utility), we graph $y = \cos\left(x - \frac{\pi}{2}\right)$ on $0 \le x \le 2\pi$.

(continued)

x	$\left(x - \frac{\pi}{2}\right)$	$\cos\left(x - \frac{\pi}{2}\right)$
0	$0 - \frac{\pi}{2} = -\frac{\pi}{2}$	$\cos\left(-\frac{\pi}{2}\right) = 0$
$\frac{\pi}{2}$	0	1
π	$\frac{\pi}{2}$	0
$\frac{3\pi}{2}$	π	-1
2π	$\frac{3\pi}{2}$	0

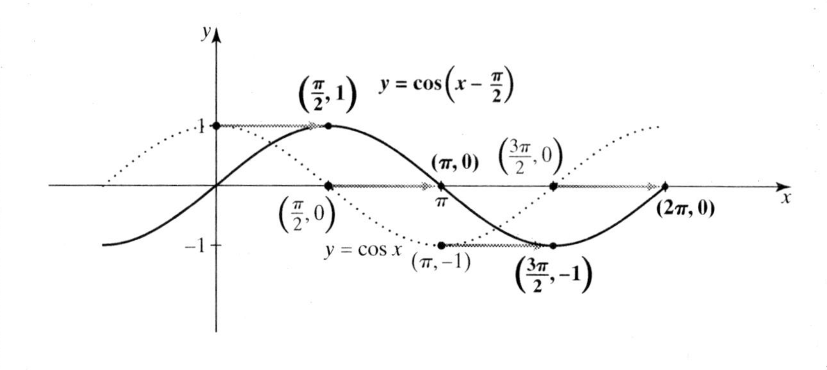

The effect of subtracting $\pi/2$ from the arc causes the graph of $y = \cos x$ to shift $\pi/2$ units to the *right*. Similarly, the graph of

$$y = \cos(x - (-\pi/2)) = \cos(x + \pi/2)$$

causes the graph of $y = \cos x$ to shift $\pi/2$ units to the *left*. A horizontal shift is called a **phase shift**. ■

Horizontal Translation (Phase Shift) of a Sinusoidal Wave

For $y = \cos(x - C)$ and $y = \sin(x - C)$, where C is any nonzero real number:

$|C|$ is the phase shift (horizontal shift).
$C > 0$ produces a shift to the right.
$C < 0$ produces a shift to the left.

NOTE: The graph of $y = \cos(x - \pi/2)$ in Example 5 should look familiar. It is the same as the graph of $y = \sin x$. We see that we can have two equations that represent the same graph.

Combining Translations with Period and Amplitude Changes

Sinusoidal functions of the form

$$y = A\cos[B(x - C)] + D \qquad \text{or} \qquad y = A\sin[B(x - C)] + D$$

have several changes to the characteristics of the pure forms. Graphing such functions can be a challenge. To determine the specific changes, make sure that your equation is in the indicated form. For graphs that include shifts, it may be best to first determine the A (amplitude) and B (period) values and dot in this graph with clearly labeled critical points. Then if there is a translation indicated by C or D, shift the critical points of the dotted graph in the direction(s) determined by these values to obtain the final graph. (Each graph can be checked using a graphing utility.) We summarize these steps at the end of this section.

EXAMPLE 6 Graph $y = 4\cos\left(\frac{1}{2}x + \frac{\pi}{2}\right)$ and label the critical points for one cycle.

SOLUTION In order to determine *A*, *B*, *C*, and *D*, we put the equation in the form $y = A\cos[B(x - C)] + D$ and get $y = 4\cos[\frac{1}{2}(x - (-\pi))]$, where $D = 0$. *Since* $A = 4$, $B = \frac{1}{2}$, *and* $C = -\pi$, *we have a vertical stretch, a horizontal stretch, and a phase (horizontal) shift.* We first sketch the graph determined by just the *A* and *B* values.

$$\text{Amplitude} = |A| = 4.$$

$$\text{Since } B = \frac{1}{2}, \text{ Period} = P = \frac{2\pi}{\frac{1}{2}} = 4\pi.$$

To plot the critical points for the graph of $y = 4\cos(\frac{1}{2}x)$ on $0 \le x \le 4\pi$ (one period), do the following:

- Start at the origin and mark off one period to the right along the *x*-axis. Divide $0 \le x \le 4\pi$ into four equal subintervals. Each subinterval will be $\frac{1}{4}(4\pi) = \pi$ in length and will correspond to the four equal arc sections for one cycle. The *x*-coordinates of the critical points will correspond to the *x*-coordinate of these intervals.
- Using the cosine form, plot the *x*-intercepts and the maximum and minimum *y*-coordinates, 4 and -4.

Connect the critical points with four identically shaped arc sections in a dotted sinusoidal wave.

Since $C = -\pi$, the dotted graph shifts to the left π units. Shift each critical point of the dotted graph π units to the left. Connect the shifted points with four identically shaped arc sections in a sinusoidal wave. Repeat the pattern.

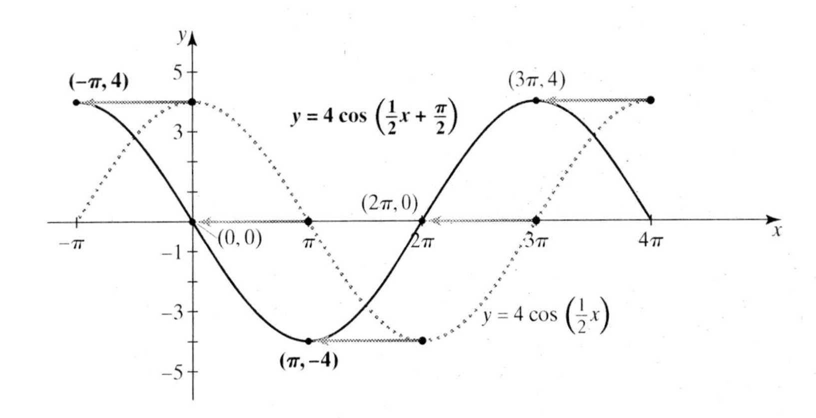

Finding Equations of Sinusoidal Graphs

Now that we know the effects of different A, B, C, and D values on the pure forms of sinusoidal graphs, we are able to find equations of graphs by determining the specific changes. Let's practice finding equations in the form of $y = A\cos[B(x - C)] + D$ or $y = A\sin[B(x - C)] + D$. Experience with this should help us find equations that model periodic phenomena in the next section.

EXAMPLE 7 Find an equation for a sinusoidal function that is represented by the following graph.

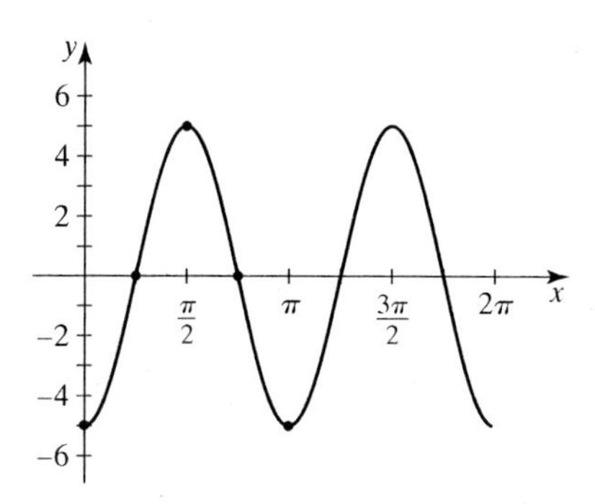

SOLUTION We identify the endpoints of four identically shaped arc sections.

Amplitude $= \frac{1}{2}|5 - (-5)| = 5$. Since it appears that the form of the cosine function has been reflected across the x-axis, $A = -5$.

By inspection the period is π. We know that

$$P = \pi = \frac{2\pi}{|B|},$$

so $B = 2$. Therefore, $y = -5\cos(2x)$.

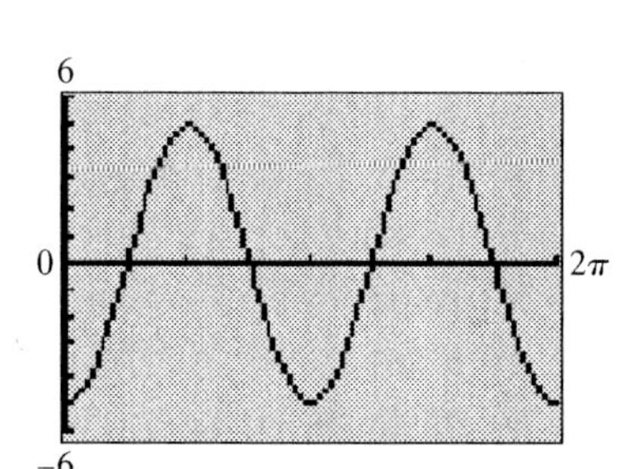

There are many other equations for this graph. For example, if we consider the graph of the sine function with the *same* amplitude (5) and period (π), that has been shifted $\pi/4$ units to the right, we obtain the equation

$$y = 5\sin\left[2\left(x - \frac{\pi}{4}\right)\right].$$

Either answer would be correct which can be verified using a graphing utility. ■

Examples 5, 7 and 8 illustrate that we can find different equations to represent the same graph. It is often the case in trigonometry that there are several ways to say something. For example, $\tan x = \sin x/\cos x$, $\sec x = 1/\cos x$, or $\cos^2 x + \sin^2 x = 1$, to name a few from Chapter 1.

EXAMPLE 8 Find two equations that describe the graph on the left.

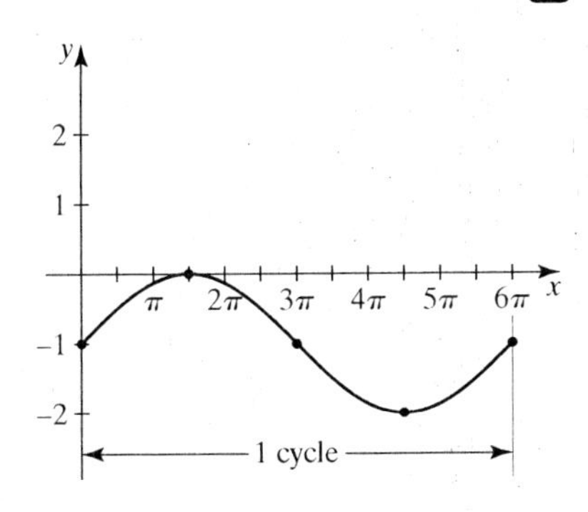

SOLUTION 1 We identify the endpoints of four identically shaped arc sections. Amplitude $= \frac{1}{2}|0 - (-2)| = 1$, so $A = 1$. It appears the sine graph has been shifted down 1 unit since 0 is the maximum y-coordinate, and -2 is the minimum y-coordinate. Therefore, $D = -1$. By inspection $P = 6\pi = 2\pi/|B|$. Solving $6\pi = 2\pi/B$ for $B > 0$, we get $B = \frac{1}{3}$.

One possible equation is $y = \sin\left(\frac{1}{3}x\right) - 1$.

SOLUTION 2 We can think of this sinusoidal wave as a cosine graph with the *same* period, amplitude, and vertical shift, but with a phase shift of $3\pi/2$ to the right. So $y = \cos\left[\frac{1}{3}(x - 3\pi/2)\right] - 1$ would also be correct. There are *many* other possibilities. ■

The ability to make several changes to the pure forms of the sinusoidal waves allows us to quickly sketch the graph and find equations of graphs that model many real-world periodic phenomena. The types of changes to the pure forms are summarized as follows.

Effects of A, B, C and D on the Graphs of $y = \cos x$ and $y = \sin x$

Let $y = A\cos[B(x - C)] + D$ and $y = A\sin[B(x - C)] + D$, where A, B, C, and D are nonzero real numbers:

Amplitude $= |A|$

- $|A| > 1$ produces a vertical stretch.
- $|A| < 1$ produces a vertical shrink.
- $A < 0$ produces a reflection across the x-axis.

Period $= \dfrac{\text{pure period}}{|B|} = \dfrac{2\pi}{|B|}$

- $|B| > 1$ produces a horizontal shrink.
- $|B| < 1$ produces a horizontal stretch.

There are four identically shaped arc sections per cycle. Each arc occurs over an interval along the x-axis that is one-fourth the period in length.

Phase shift (horizontal shift) $= |C|$

- If $C > 0$, shift right.
- If $C < 0$, shift left.

Vertical shift $= |D|$

- If $D > 0$, shift up.
- If $D < 0$, shift down.

NOTE: The preceding summary includes the effects of both positive and negative values of A, C, or D. *But what happens when B is negative?* We already know how to deal with this if we use the negative identities from Section 1.6. Recall $\cos(-x) = \cos x$, and $\sin(-x) = -\sin x$. This means the graph of $y = \cos(-6x)$ would be the same as $y = \cos(6x)$. Likewise, the graph of $y = \sin(-5x)$ would be the same as $y = -\sin(5x)$.

Guidelines for Graphing Sinusoidal Functions

- If necessary, write the equation in the form $y = A\cos[B(x - C)] + D$ or $y = A\sin[B(x - C)] + D$. If B is negative, use the negative identities so the coefficient of $(x - C)$ is positive. Observe the values of A, B, C, and D and anticipate the characteristics of the graph.
- Determine the amplitude $(|A|)$ and the period $(2\pi/|B|)$.
- Starting at the origin, mark off one period to the right along the x-axis. Divide this interval into four equal subintervals. The length of each subinterval will be $\frac{1}{4} \cdot$ (period) and will correspond to the four equal arc sections for one cycle.
- Using the appropriate cosine or sine form, plot the critical points of each of the four arc sections (watch amplitude). Remember, if A is negative, you will need to reflect these points across the x-axis.
- If $C = D = 0$, connect the critical points with a sinusoidal wave. Otherwise, connect the critical points with a dotted graph.
- If $C > 0$, shift the critical points C units to the right. If $C < 0$, shift the critical points $|C|$ units to the left.
- If $D > 0$, shift the critical points D units up. If $D < 0$, shift the critical points $|D|$ units down.
- Connect the shifted critical points with a sinusoidal wave.
- Repeat the pattern over the required interval.

Although all the graphs could be done more accurately and efficiently with the use of a graphing utility, graphing a few by hand gives you the necessary experience with the effects of A, B, C, and D on the characteristics of the pure forms. This becomes important when these values produce amplitudes, periods, and shifts that are too difficult to graph by hand, in which case a graphing utility is advisable. Your experience will provide you with the knowledge necessary to find an appropriate window setting when using a graphing utility, as demonstrated in the example on the next page.

CONNECTIONS WITH TECHNOLOGY

Graphing Calculator/CAS

EXAMPLE Using a graphing utility, graph two cycles of

$$y = 10\cos(30x - 1) + 2.$$

SOLUTION First we determine the amplitude, period, vertical shift, and the range of the function. Then we use these values to find appropriate window settings.

If we rewrite the equation in the form $y = 10\cos\left[30\left(x - \frac{1}{30}\right)\right] + 2$, we see that

$$A = 10,\ B = 30,\ C = \tfrac{1}{30},\ \text{and } D = 2.$$

$$\text{Amplitude} = |10| = 10$$

$$\text{Period} = \frac{2\pi}{|30|} = \frac{\pi}{15}$$

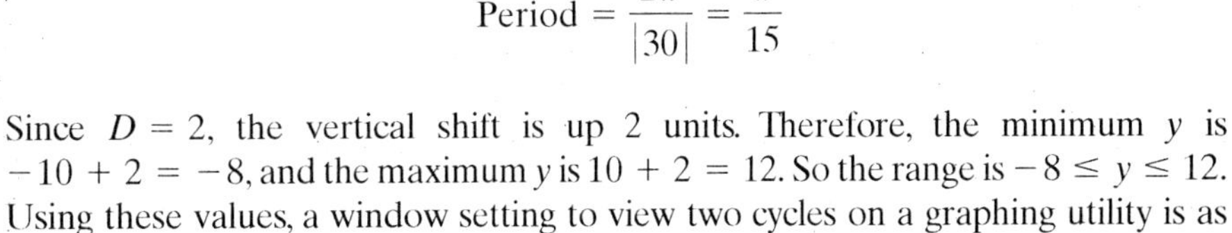

Since $D = 2$, the vertical shift is up 2 units. Therefore, the minimum y is $-10 + 2 = -8$, and the maximum y is $10 + 2 = 12$. So the range is $-8 \le y \le 12$. Using these values, a window setting to view two cycles on a graphing utility is as follows.

$$\text{Xmin} = -\pi/15 \qquad \text{Ymin} = -8$$
$$\text{Xmax} = \pi/15 \qquad \text{Ymax} = 12$$

The window we used to graph two cycles of $y = 10\cos(30x - 1) + 2$ is not in scale. Equal scaling along each axis is not always convenient.

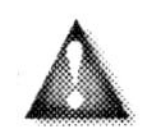

CAUTION: Most graphing calculators have a variety of zoom options that produce different window views of a graph. Graph (a), which was produced by using the "zoom trigonometry" option on a TI-83 Plus is not only difficult to use to see the period, range, and so on, but also quite misleading because $y = 10\cos(30x - 1) + 2$ does not even appear to be sinusoidal! Moreover, if you arbitrarily select the viewing window where Xmin $= -5$, Xmax $= 5$, Ymin $= -8$, and Ymax $= 12$, you get Graph (b), which also is not an accurate depiction of the graph. Your ability to set an appropriate window is important in order to avoid such pitfalls of technology.

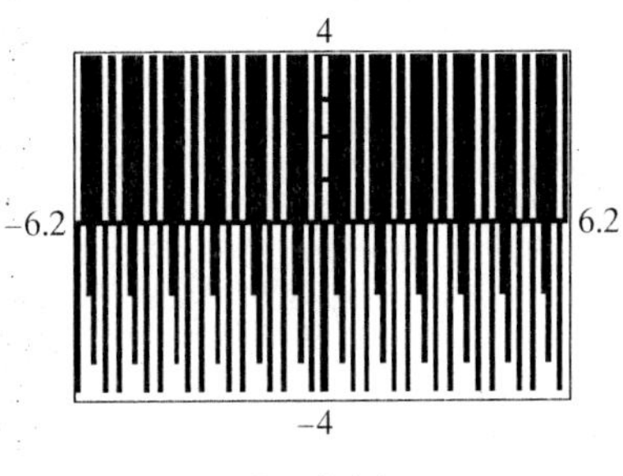

Graph (a)

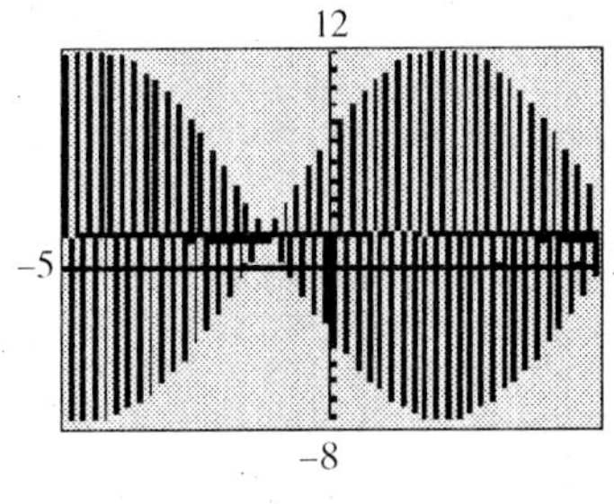

Graph (b)

Exercise Set 2.2

1–14. *Sketch the graph of each of the following functions for a minimum of two periods. Indicate the amplitude, period, and x-intercepts.*

EXAMPLE $y = -2\cos(-3x)$

SOLUTION Using negative identities, $-2\cos(-3x) = -2\cos(3x)$, so $y = -2\cos(3x)$. *Since $A = -2$, $B = 3$, we have a vertical stretch and a horizontal shrink. Because $A < 0$, the graph is a reflection across the x-axis.*

$$\text{Amplitude} = |-2| = 2; \qquad \text{period} = P = \frac{2\pi}{|B|} = \frac{2\pi}{3}$$

Starting at the origin, mark off one period, and divide it into four equal subintervals. Each subinterval will be $\frac{1}{4}\left(\frac{2\pi}{3}\right) = \frac{\pi}{6}$ along the x-axis. Using the cosine form reflected across the x-axis (recall $A < 0$), plot the critical points and connect them with a sinusoidal wave. (Since $D = C = 0$, there is no phase shift or vertical shift.) Repeat the pattern.

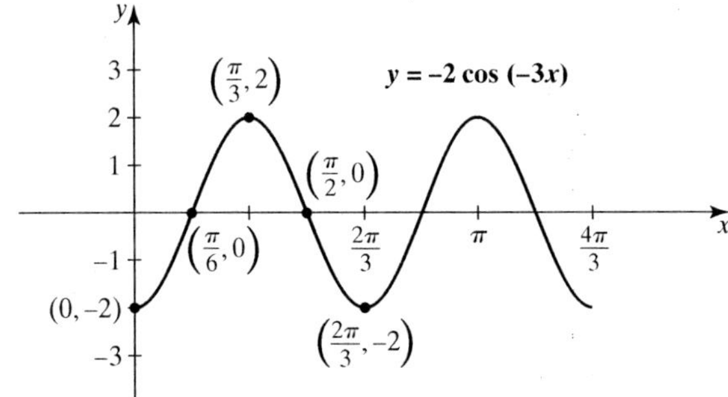

By inspection the x-intercepts are at $x = \frac{\pi}{6}, \frac{\pi}{2}, \frac{5\pi}{6}, \frac{7\pi}{6}$, or

$$\frac{\text{pure } x\text{-intercepts}}{|B|} = \frac{\frac{\pi}{2} + k\pi}{3} = \frac{\pi}{6} + \frac{k\pi}{3}.$$ ■

1. $y = \sin\left(\frac{1}{2}x\right)$
2. $y = \sin(-2x)$
3. $y = \cos(-4x)$
4. $y = \cos\left(\frac{3}{4}x\right)$
5. $y = \sin\left(-\frac{1}{3}x\right)$
6. $y = -\sin(-\pi x)$
7. $y = 3\cos\left(\frac{1}{4}x\right)$
8. $y = \frac{3}{2}\cos(2x)$
9. $y = -4\sin(2x)$
10. $y = 3.5\sin\left(-\frac{1}{4}x\right)$
11. $y = 0.5\sin(\pi x)$
12. $y = 6\sin\left(\frac{\pi}{3}x\right)$
13. $y = -2\cos\left(\frac{\pi}{4}x\right)$
14. $y = -5\cos\left(-\frac{\pi}{2}x\right)$

15–22. *Sketch the graph of the following functions for at least one period. Indicate the amplitude, period, phase shift (if any), vertical shift (if any), and range.*

EXAMPLE $y = 3\sin\left(\frac{1}{2}x + \frac{\pi}{2}\right)$

SOLUTION Rewrite as $y = 3\sin\left[\frac{1}{2}(x - (-\pi))\right]$.

$$\text{Amplitude} = |A| = 3; \quad \text{period} = P = \frac{2\pi}{|B|} = \frac{2\pi}{\frac{1}{2}} = 4\pi.$$

Dot in the graph of $y = 3\sin\left(\frac{1}{2}x\right)$.

- Each of the four arc sections, which are $\frac{1}{4}\cdot$ period $= \frac{1}{4}(4\pi) = \pi$ in length along the x-axis, have critical values that reflect the amplitude and period with sine form.

Since $C = -\pi$, the phase shift is π to the left, so we shift the critical points of the dotted graph π to the left. Connect the shifted points with a sinusoidal wave. (Since $D = 0$, there is no vertical shift.) Repeat the pattern.

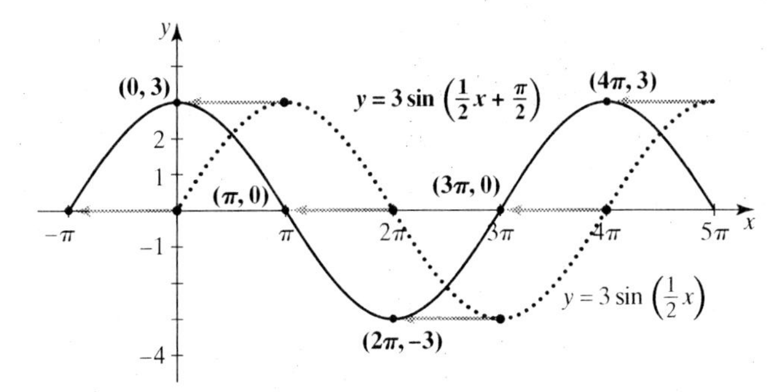

Range: $|y| \leq 3$

15. $y = 2\cos(x - \pi)$ **16.** $y = 5\sin(x + \pi)$ **17.** $y = 4\sin\left[\frac{1}{2}\left(x + \frac{\pi}{2}\right)\right]$

18. $y = 3\cos\left[2\left(x - \frac{\pi}{8}\right)\right]$ **19.** $y = -3\cos(4x + \pi)$ **20.** $y = -\sin\left(3x + \frac{3\pi}{2}\right)$

21. $y = 2\sin(2x) + 3$ **22.** $y = \cos(x + \pi) - 2$

23–26. *Without using a graphing utility, match each graph with two functions from the list below. Check your answers using a graphing utility.*

23.

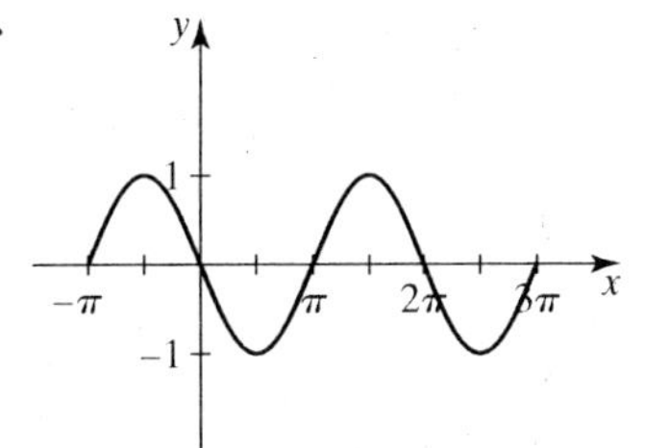

24.

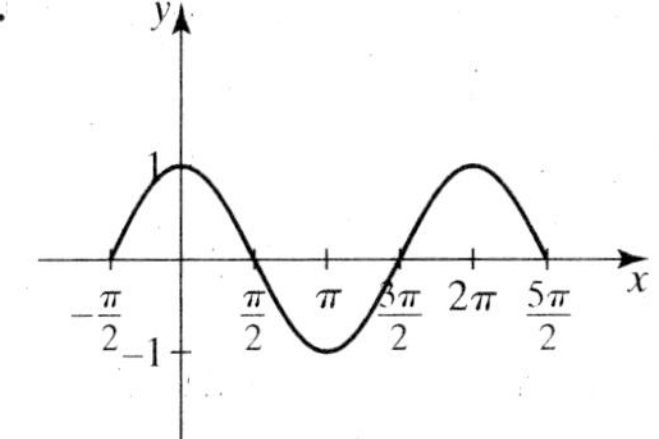

25.

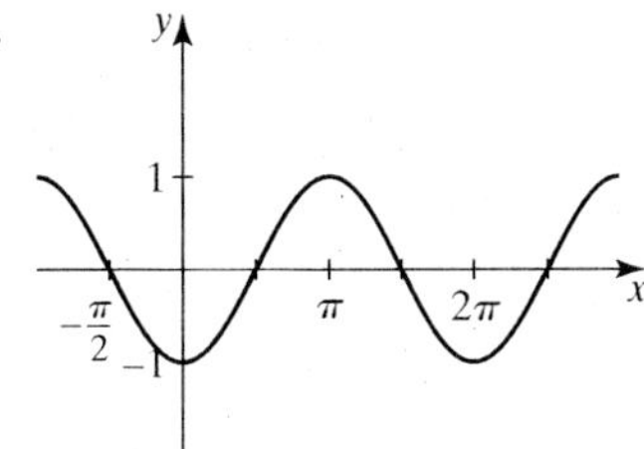

26.

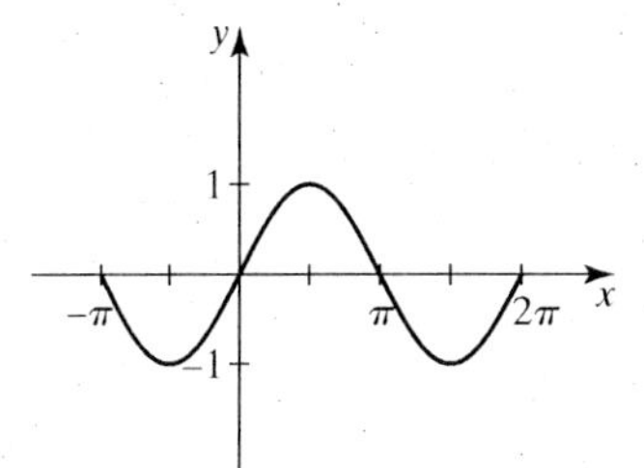

a. $y = \cos\left(x - \frac{\pi}{2}\right)$ **b.** $y = \cos\left(x + \frac{\pi}{2}\right)$ **c.** $y = \sin\left(x + \frac{\pi}{2}\right)$

d. $y = \sin\left(x - \frac{\pi}{2}\right)$ **e.** $y = \cos(-x)$ **f.** $y = -\sin(-x)$

g. $y = -\sin x$ **h.** $y = -\cos x$ **i.** $y = \cos\left(-x + \frac{\pi}{2}\right)$

27–32. *Find an equation of a function in the form*

$$y = A\cos(Bx) + D \quad \text{or} \quad y = A\sin(Bx) + D, \quad \text{for } B > 0,$$

that could represent the given periodic graph. Check your answer using a graphing utility.

27.

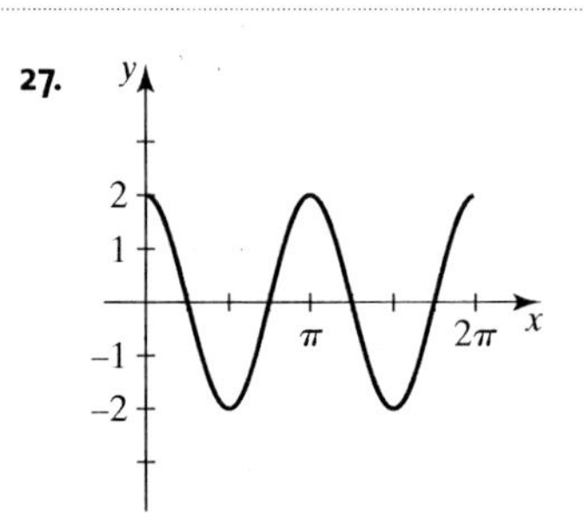

28.

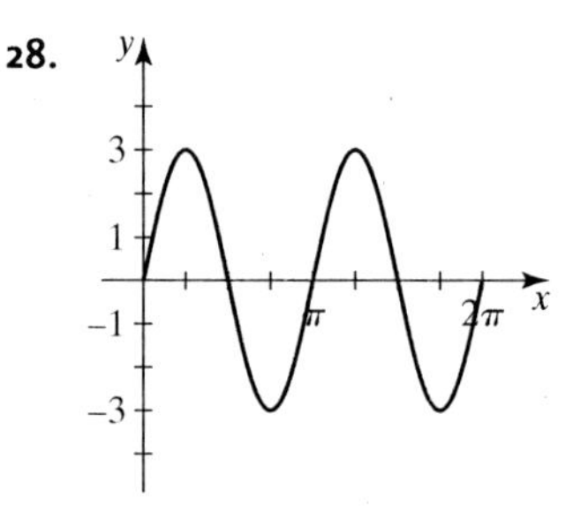

29.

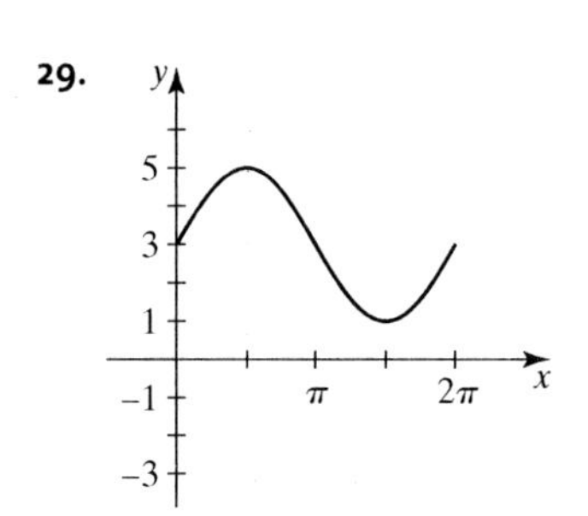

30.

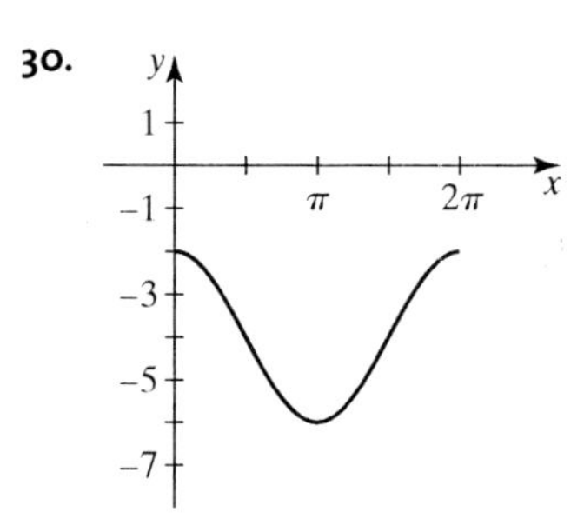

31.

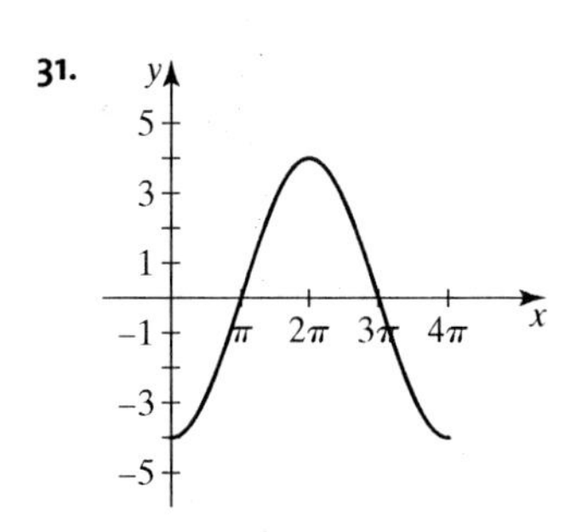

32.

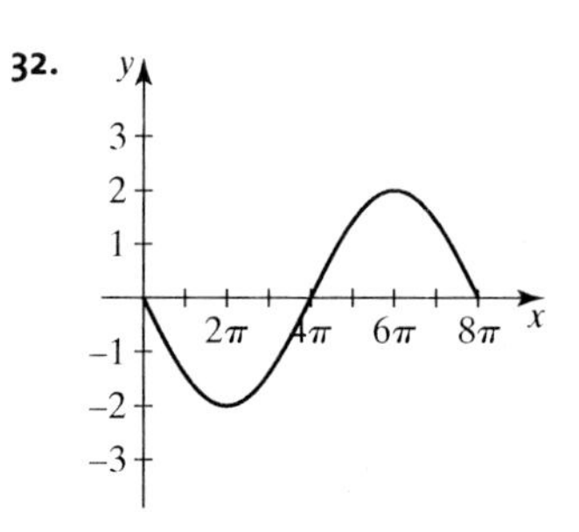

33–38. *Find an equation of a function in the form* $y = \cos[B(x - C)]$ *or* $y = \sin[B(x - C)]$ *for* $B > 0$, $-\pi < C < 0$, *and* $0 < C < \pi$, *that represents the given periodic graph. Check your answer using a graphing utility.*

33.

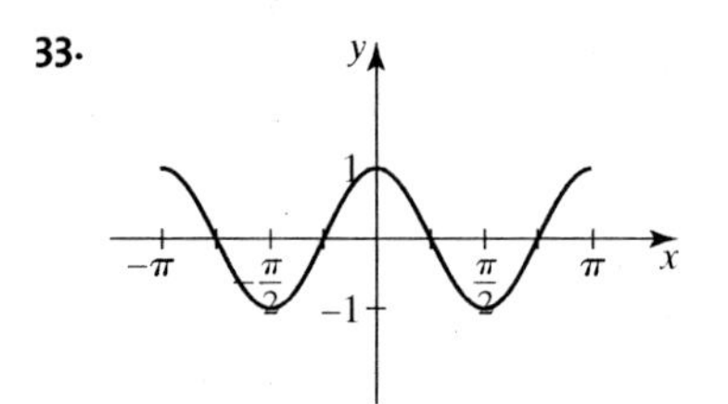

34.

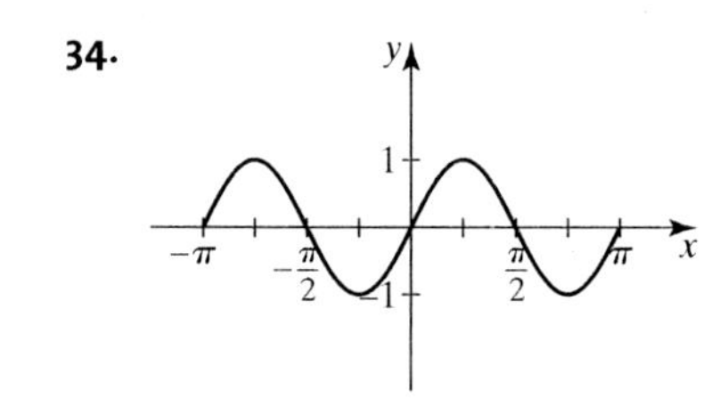

35.

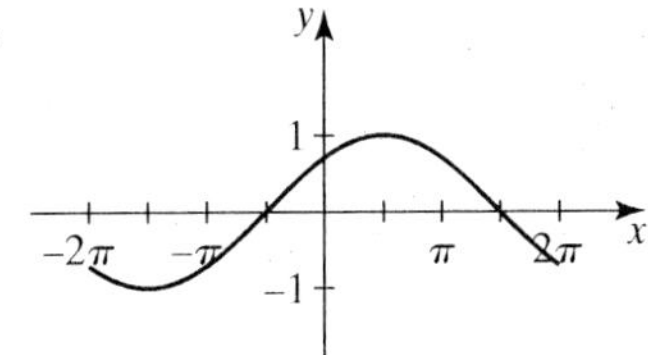

36.

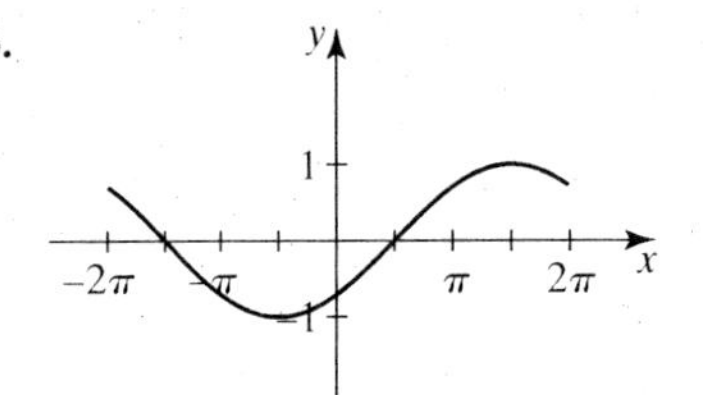

37.

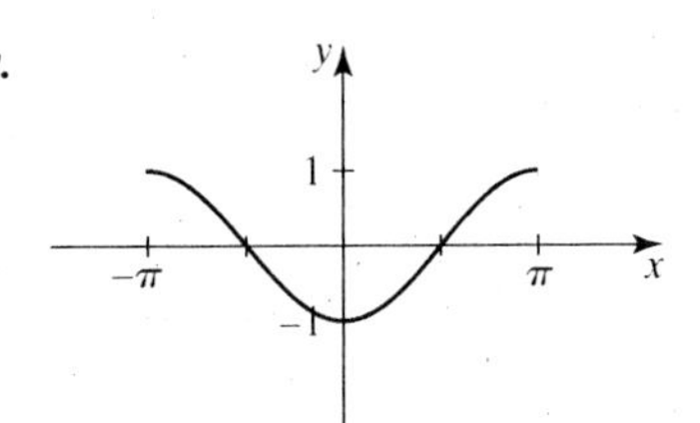

38.

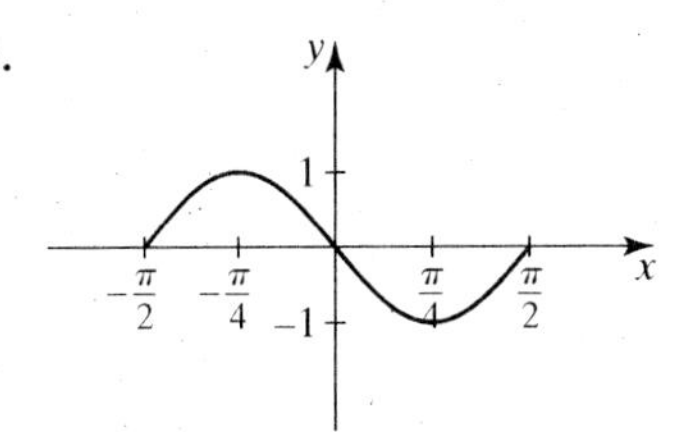

39–40. *Find two equations for each sinusoidal wave such that one equation is of the form* $y = A\sin(Bx)$ *or* $y = A\cos(Bx)$, $A,\ B \in \mathbb{R},\ B > 0$, *and the other is of the form* $y = A\sin[B(x - C)]$ *or* $y = A\cos[B(x - C)]$, *for* $A,\ B > 0,\ -\pi \le C < 0$ *and* $0 < C \le \pi$.

39.

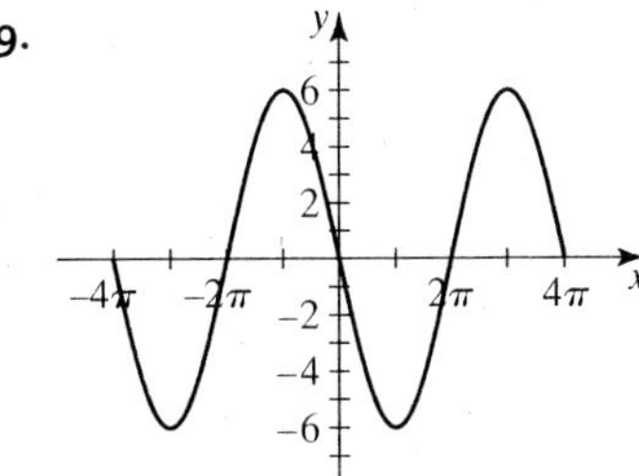

40.

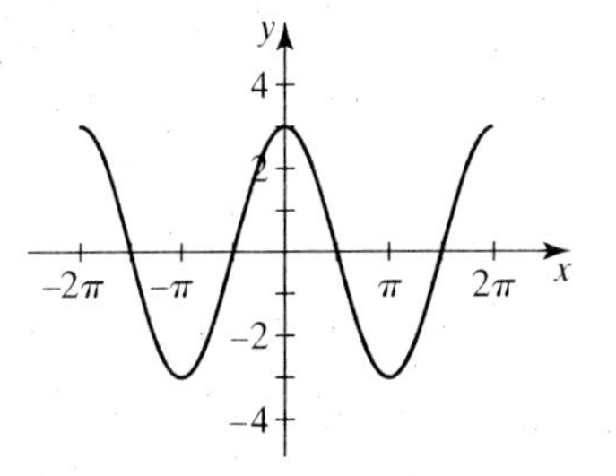

41–44. *State the amplitude, period, phase shift (if any), vertical shift (if any), and range of the given function. Use your answers along with a graphing utility to graph two cycles of the function in order to verify your answers.*

41. $y = \frac{1}{4}\sin(4x - 2)$

42. $y = 5.7\cos(3x + 12)$

43. $y = -2.3\cos\left(\frac{1}{6}x\right) + \pi$

44. $y = -\sqrt{2}\sin(0.5x) - 5$

45. Without graphing, explain the differences (if any) in the appropriate characteristics [period, amplitude, phase shift (horizontal shift), or vertical shift] for each pair of functions.

a. $y = \sin x + 2$ and $y = \sin(x + 2)$

b. $y = \sin(2x)$ and $y = 2\sin x$

c. $y = \frac{1}{2}\sin(2x)$ and $y = \sin x$

d. $y = \sin\left(x + \frac{\pi}{2}\right)$ and $y = \sin x + \sin\frac{\pi}{2}$

46. Without graphing, explain the differences (if any) in the appropriate characteristics [period, amplitude, phase shift (horizontal shift), or vertical shift] for each pair of functions.

a. $y = \cos\left(\frac{1}{2}x\right)$ and $y = \frac{1}{2}\cos x$

b. $y = \cos x + \cos\frac{\pi}{2}$ and $y = \cos\left(x + \frac{\pi}{2}\right)$

c. $y = \cos(-2x)$ and $y = \cos(2x)$

d. $y = \cos x - 5$ and $y = \cos(x - 5)$

Graphing Calculator/CAS Explorations

47. Graph $y = \cos^2 x$ using the following settings:

MODE	Radian		
WINDOW	Xmin $= 0$,	Xmax $= 2\pi$,	Xscl $= \frac{\pi}{2}$,
	Ymin $= -0.5$,	Ymax $= 1$,	Yscl $= 0.5$

a. What is the amplitude?

b. What is the period?

c. What is the vertical shift of the graph $y = A\cos(Bx)$?

d. Find an equation for this graph in the form $y = A\cos(Bx) + D$.

48. Graph $y = \sin^2 x$, using the same settings for Exercise 47.

a. What is the amplitude?

b. What is the period?

c. What is the vertical shift of the graph $y = A\cos(Bx)$?

d. Find an equation for this graph in the form $y = A\cos(Bx) + D$.

49. As a result of Exercises 47 and 48, are all sinusoidal waves the result of equations of the form $y = A\cos[B(x - C)] + D$ or $y = A\sin[B(x - C)] + D$?

50. As a result of Exercises 47 and 48, can sinusoidal waves have equations different from the form $y = A\cos[B(x - C)] + D$ or $y = A\sin[B(x - C)] + D$?

51. **a.** If you use a graphing utility to graph $y = \cos 2\left(x - \frac{\pi}{2}\right)$, will you need an additional set of parentheses? If so, explain where you would insert them and why.

b. Explain why no additional parentheses are needed when using a graphing utility to graph $y = \cos 2\left(x - \frac{\pi}{2}\right)$ if it is written as $y = \cos(2x - \pi)$.

52–54. *Noncircular functions can be periodic. Exercises 52–54 provide examples of periodic noncircular functions that involve the greatest integer function: $y = [x]$. The greatest integer function yields the greatest integer part of a real number x $(x \in \mathbb{R})$.*

EXAMPLES

a. If $x = 4.2$, then $y = [4.2] = 4$.

b. If $x = -4.2$, then $y = [-4.2] = -5$.

Locate the greatest integer key on your graphing utility (INT). *To ensure that the graphing calculator does not connect points and show segments that are not part of the graph, use dot mode.*

52. Graph the equation $y = x - [x]$, with settings

Xmin $= -5$,	Xmax $= 5$,	Xscl $= 1$,
Ymin $= -1$,	Ymax $= 2$,	Yscl $= 1$.

a. Is the graph periodic? If so, what is the period?

b. Approximate the amplitude to the nearest tenth.

c. What are the x-intercepts?

d. Is this graph continuous over the domain $-5 \le x \le 5$?

53. Graph the equation $y = \frac{1}{2}x - \left[\frac{1}{2}x\right]$. (Use the same settings as Exercise 52.)
    a. Is the graph periodic? If so, what is the period?
    b. Approximate the amplitude to the nearest tenth.
    c. What are the x-intercepts?
    d. Is this graph continuous over the domain $-5 \le x \le 5$?

54. Find an equation that involves the greatest integer function for these graphs.

a.

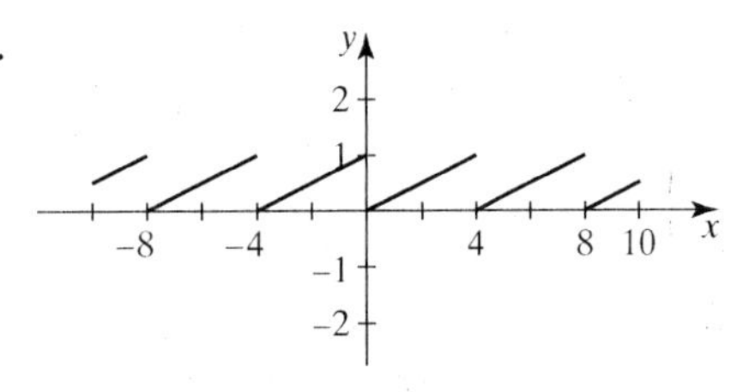

b.

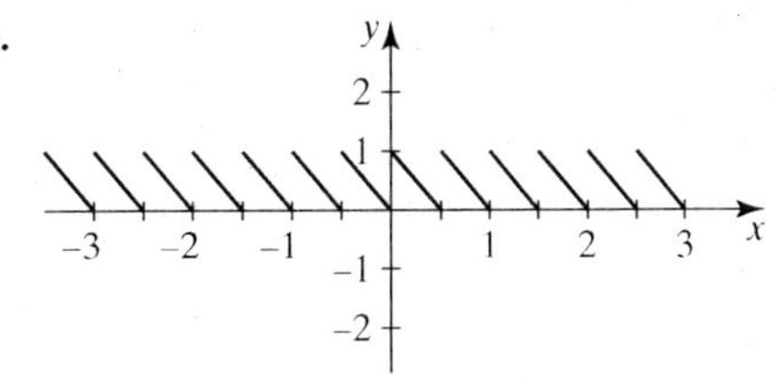

Discussion

55. Carefully explain the effects of A, B, C, and D on the pure forms of the sine and cosine function for

$$y = A\cos[B(x - C)] + D \text{ and } y = A\sin[B(x - C)] + D,$$

where A, B, C, and D are nonzero real numbers.

Web Activity

56. After some observation, Galileo claimed that the period of a pendulum was independent of the amplitude (*Two New Sciences*). Do you agree with this claim? Search the Internet (for example, Galileo + pendulum) for sites that provide information or show evidence that support or refute Galileo's claim. On the basis of the information you are able to find, is the claim valid? Write a paragraph or two that summarizes your findings and list the Web sites you used.

2.3 Applications and Modeling with Sinusoidal Functions

Simple Harmonic Motion

In Section 1.6 we presented simple harmonic motion as an application of the cosine and sine functions in the form $d(t) = A\cos(Bt)$ or $d(t) = A\sin(Bt)$. In the first examples of this section we examine the motion of an object to see how the values A and B are determined and which form to use. In doing so, we find the function that models the motion. We begin with a definition.

Frequency

The **frequency** of an object moving in simple harmonic motion is the number of periods (or cycles) of the motion per unit of time.

The **period** of the motion is the amount of time the object takes to complete one cycle. The period is given by $P = 2\pi/B$.

Since the frequency is the number of periods per unit of time, the frequency F and period P are related by

$$F = \frac{1}{P} = \frac{B}{2\pi}.$$

Since $F = B/2\pi$, it follows that $B = 2\pi \cdot F$.

EXAMPLE 1 Recall that Example 5 from Section 1.6 dealt with the harmonic motion of a spring with a weight attached to one end. The spring was stretched 5.2 inches below its rest position at time $t = 0$. The simple harmonic motion began as the weight on the spring bounced up and down 5.2 inches from its rest position. Since it took 2 seconds to go up and back down, or 2 seconds to complete one period, the spring has a frequency of $\frac{1}{2}$ period per second. We were given the equation $d(t) = -5.2\cos(\pi t)$.

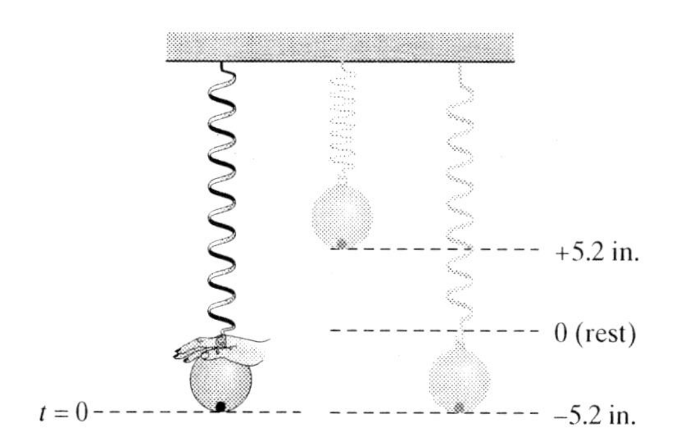

a. How did one arrive at the values of $A = -5.2$ and $B = \pi$?

b. Why use a sinusoidal function, in particular the cosine function, to model this motion?

SOLUTION

a. Since we are told the spring bounces 5.2 inches above and below the rest position, the amplitude $= 5.2 = |A|$. Thus, $A = -5.2$ or $A = 5.2$. And since $F = \frac{1}{2}$,

$$B = 2\pi F = 2\pi\left(\frac{1}{2}\right) = \pi.$$

So the choices are:

$$(1)\ d(t) = -5.2\cos(\pi t) \quad \text{or} \quad (2)\ d(t) = 5.2\cos(\pi t).$$

However, at time $t = 0$:

$$(1)\ d(0) = -5.2\cos(0) \quad \text{and} \quad (2)\ d(0) = 5.2\cos(0)$$
$$d(0) = -5.2 \text{ in.} \qquad\qquad d(0) = 5.2 \text{ in.}$$

We know the spring was below the rest position at $t = 0$, which indicates a negative value for the displacement. Therefore, we use

$$d(t) = -5.2\cos(\pi t),$$

which represents the displacement of the spring in inches from its rest position at time t in seconds.

b. To get a better look at why a sinusoidal function was used for this simple harmonic motion, we attach a pen to the weight at the end of the spring and record its motion on a strip of paper where the paper is moving at a constant pace. The resulting graph allows us to view the displacement of the weight from the rest position with respect to a particular time t. *Notice how the motion produces a sinusoidal wave*, in particular the graph of the cosine function.

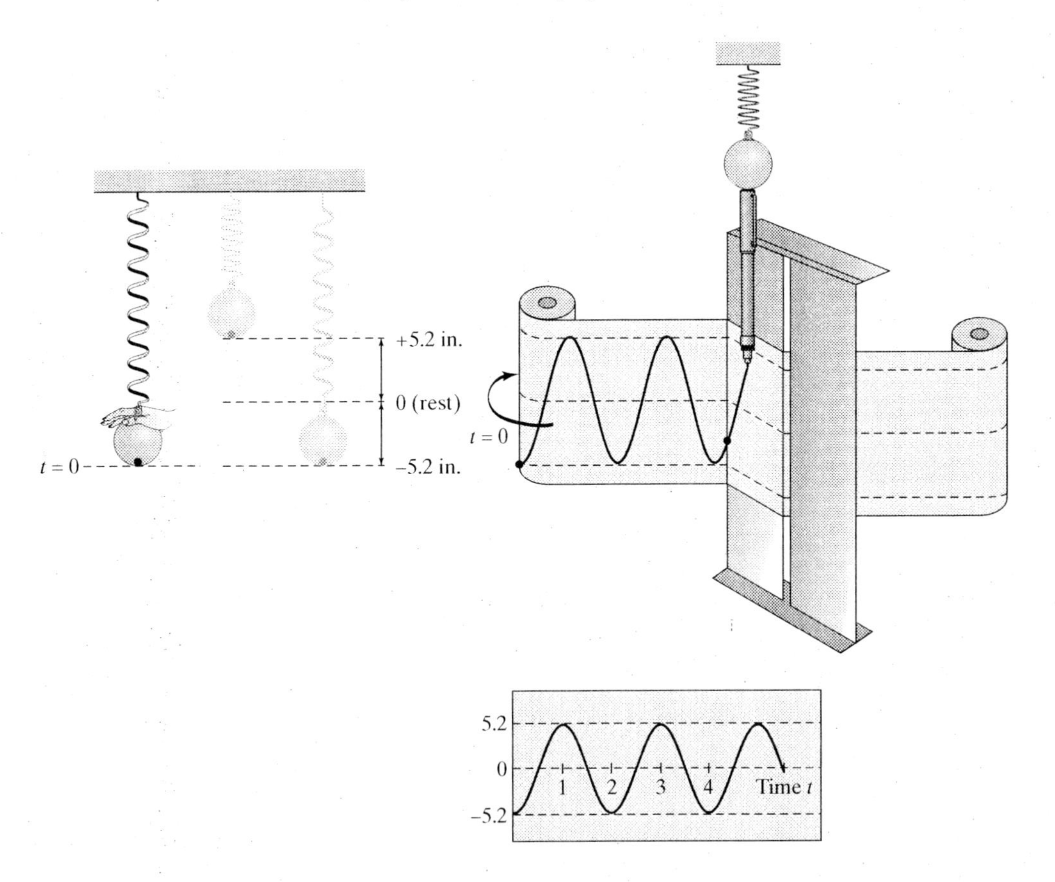

By inspection, the graph indicates that $A < 0$, which corresponds to the equation from part (a): $d(t) = -5.2\cos(\pi t)$. (Although the choice $d(t) = -5.2\sin(\pi t)$ has the correct A and B values, this equation would be incorrect because at $t = 0$ it indicates a wrong initial displacement of 0: $d(0) = -5.2\sin 0 = 0 \neq -5.2$.) ■

EXAMPLE 2 The diaphragm of a loudspeaker moves back and forth in simple harmonic motion to create sound. Usually one cycle per second is referred to as one hertz (Hz). This unit is named after Heinrich Hertz (1857–1894). One thousand cycles

(continued)

per second is called a kilohertz (kHz). The frequency of the motion of this speaker is $F = 1\text{ kHz} = 1000\text{ cycles}/1\text{ second}$, and the amplitude of the diaphragm movement is 0.2 mm. Assume that the diaphragm at $t = 0$ is at rest and moves out to the right first. Find an equation that represents this harmonic motion.

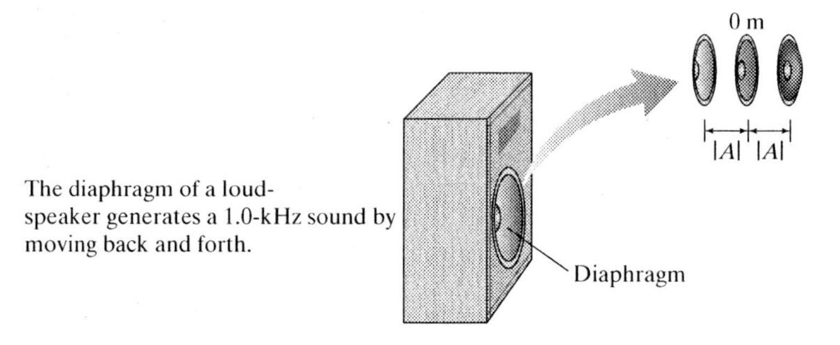

The diaphragm of a loudspeaker generates a 1.0-kHz sound by moving back and forth.

Solution At time $t = 0$, there is no displacement $(d = 0)$, so we use

$$d(t) = A\sin(Bt).$$

(Notice that if we used $d(t) = A\cos(Bt)$ at $t = 0$, we would get $d(0) = A\cos(B \cdot 0) = A\cos(0) = A(1) = A \neq 0$.) Since the amplitude is given as 0.2 mm, $A = 0.2$ or $A = -0.2$. We are told the diaphragm moves out to the right first (which we consider a positive direction), so $A = 0.2$. To find B, we know that $F = 1000$, so $B = 2\pi F = 2\pi(1000) = 2000\pi$. Therefore, the equation is

$$d(t) = 0.2\sin(2000\pi t).$$

We also could write this equation as $y = 0.2\sin(2000\pi x)$, where t or x represents time in seconds and $d(t)$, $d(x)$, or y represents the displacement in mm. ■

Now that we have the equation, we can ask questions about the motion.

Example 3 Use the equation $d(t) = 0.2\sin(2000\,\pi t)$, where t is in seconds and $d(t)$ in mm, for the motion of the diaphragm of the speaker described in Example 2.

a. Describe the movement of the diaphragm between the times $t = 0$ and $t = 1/4000$ second.

b. Explain the position of the diaphragm at times $t = 1$ and $t = 2$ seconds and then describe the movement of the diaphragm between these two times.

Solution

a. At $t = 0$, $d(0) = 0.2\sin(2000\,\pi(0)) = 0$, which means that the diaphragm is at rest. At $t = 1/4000$ sec,

$$d\left(\frac{1}{4000}\right) = 0.2\sin\left[(2000\pi)\left(\frac{1}{4000}\right)\right] = 0.2\sin\left(\frac{\pi}{2}\right) = 0.2(1) = 0.2,$$

which means that the diaphragm is 0.2 mm to the right.

Since the period $P = 1/F = 1/1000$ seconds per cycle, we know that from $t = 0$ sec to $t = \frac{1}{4000}$ sec, the diaphragm did not have time to go back and forth.

In fact, it had time to complete only $\frac{1}{4}$ of the period. Hence, it only went from its rest position to the right 0.2 mm in $\frac{1}{4}$ of a cycle.

b. At $t = 1$ sec, and $t = 2$ sec:

$$d(1) = 0.2 \sin[2000\,\pi(1)] = 0$$
$$d(2) = 0.2 \sin[2000\,\pi(2)] = 0.2 \sin(4000\,\pi) = 0.$$

Therefore, at both these times the diaphragm is in the rest position. Since the frequency is 1000 cycles per second, the diaphragm has moved back and forth 1000 times within this one second. ■

Models Involving Sinusoidal Functions

We would like to find equations of sinusoidal functions that model real-life data.

EXAMPLE 4 Consider the following data, which describe the monthly average number of daylight hours for Anchorage, Alaska. Plot the data and determine a sinusoidal function whose graph would closely fit (model) this data.

Jan	Feb	Mar	April	May	June	July	Aug	Sept	Oct	Nov	Dec
6.5	9	12	15	17.5	19	18	16	13.5	10	6.5	5.5

SOLUTION Since the data is periodic, we plot this data for two years by repeating the values, where January corresponds to $x = 1$ and $x = 13$. Notice the graph suggests that a sinusoidal function could model this data, which can be represented by using either a sine or cosine function.

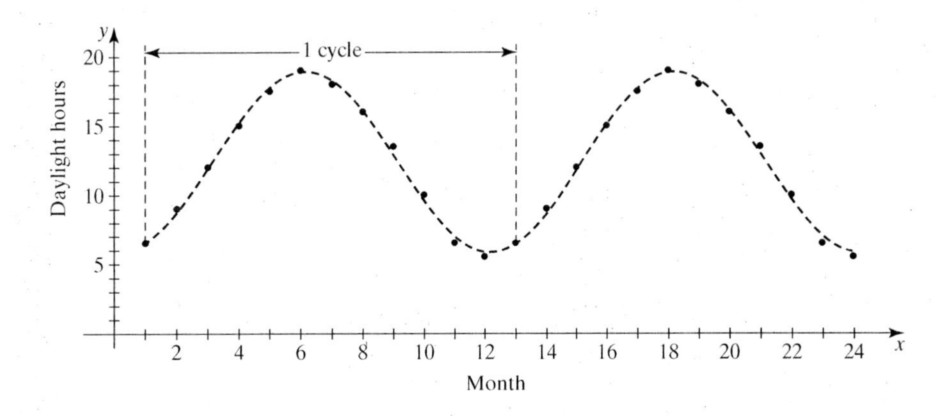

Let's choose to use the form $y = A \sin[B(x - C)] + D$, where x represents the month of the year. We need to determine A, B, C, and D.

- For the amplitude, the maximum number of hours is 19, and the minimum is 5.5. Therefore, amplitude $= \frac{1}{2}|19 - 5.5| = 6.75$. So $A = 6.75$ or $A = -6.75$
- Since the period is 12 months, $P = 2\pi/B = 12$. Solving this for B, we get $B = \pi/6$.

(continued)

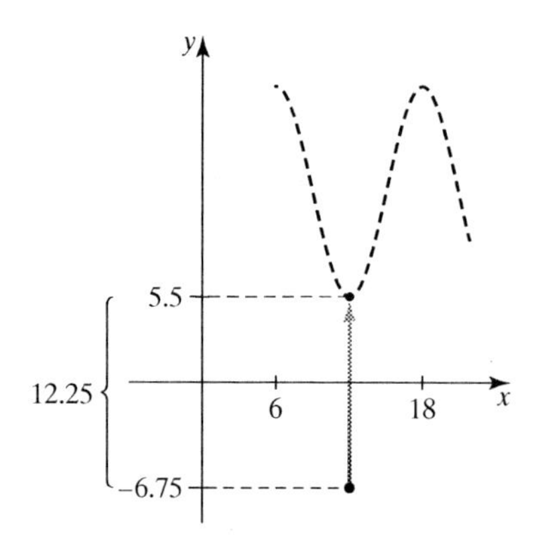

- The maximum sunrise time occurs when $x = 6$ or $x = 18$. The minimum sunrise time occurs when $x = 12$ or $x = 24$. These particular positions indicate that the sine has a phase shift of 3 to the right. Therefore, $A = 6.75$ and $C = 3$.
- Without a vertical shift, the minimum would be -6.75 (given the amplitude is 6.75). To move the minimum value of -6.75 to 5.5, we must shift up 12.25. Therefore, $D = 12.25$.

Putting together this information we obtain

$$S(x) = 6.75 \sin\left[\frac{\pi}{6}(x - 3)\right] + 12.25,$$

where $S(x)$ represents the monthly average number of daylight hours in Anchorage, Alaska, and x represents the month of the year. ■

CONNECTIONS WITH TECHNOLOGY

Graphing Calculator/CAS

Sine Regression

Many graphing calculators plot data (scatter plot) as well as find an equation of a sine function to model the data (SinReg: represents sine regression). The calculator usually produces an equation in the form $y = a\sin(bx + c) + d$. Notice that c does not reflect the phase shift, which is actually $-c/b$.

Below we show a scatter plot using the data in Example 4. Next to the graph is the sine regression equation generated by this data. The approximated values of a, b, c, and d are given below the screen.

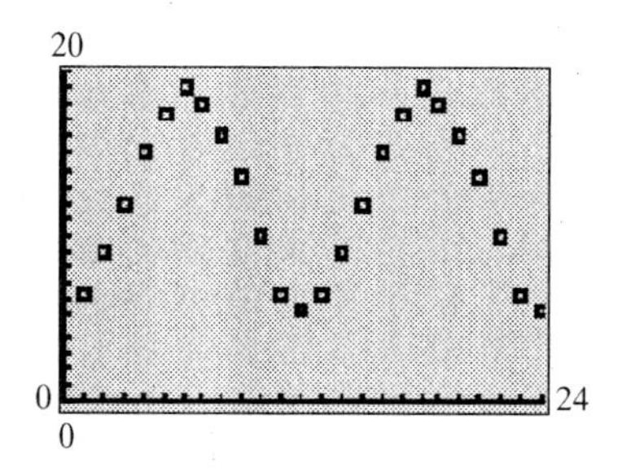

$a = 6.53$, $b = 0.522$, $c = -1.64$, $d = 12.35$

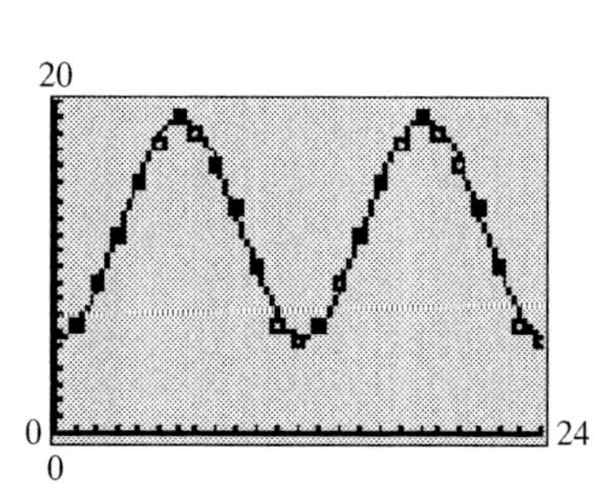

$y = 6.53\sin(0.522x - 1.64) + 12.35$

In order to see how well the equation generated by the sine regression models the data, we graph the equation with the a, b, c, and d values along with the original data.

The equation we determined in Example 4 has very similar values to those found using sine regression. You may find that using a different calculator/calculation model, or inputting a different number of periods of the data, will produce slightly different sine regression values. Although the calculator may provide us with a more efficient and accurate sinusoidal model for our data, doing a few by hand can improve your understanding of the process.

When given other real world situations that can be modeled by various sinusoidal functions, we can derive an equation (either by hand or by using technology), to describe the periodic occurrences. Once we have an equation, we can draw conclusions or make predictions about the situation. The exercises are designed to give you some experience using sinusoidal functions as mathematical models.

Exercise Set 2.3

1–4. *The graphs model the monthly average high or low Fahrenheit temperatures for the indicated city for one year, where x represents the month (January = 1, February = 2, and so on). For each graph, find:*

a. *the maximum and minimum values*

b. *the amplitude*

1. Average Monthly High Temperature (°F) in Honolulu, Hawaii

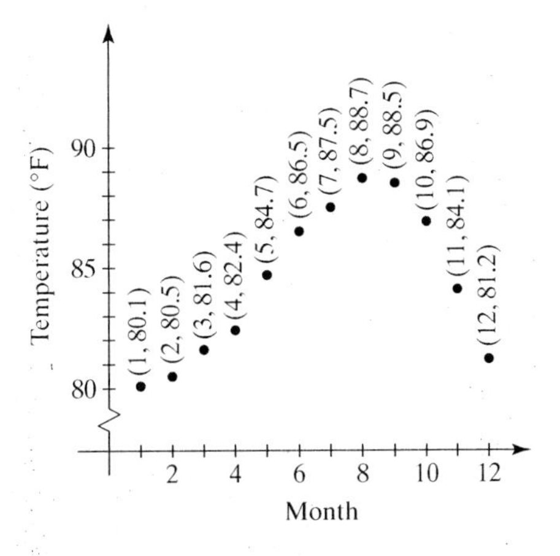

2. Average Monthly High Temperature (°F) in Chicago, Illinois

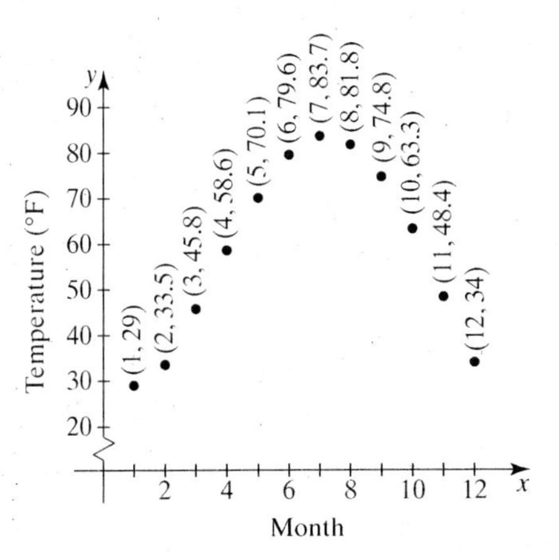

3. Average Monthly Low Temperature (°F) in Fort Lauderdale, Florida

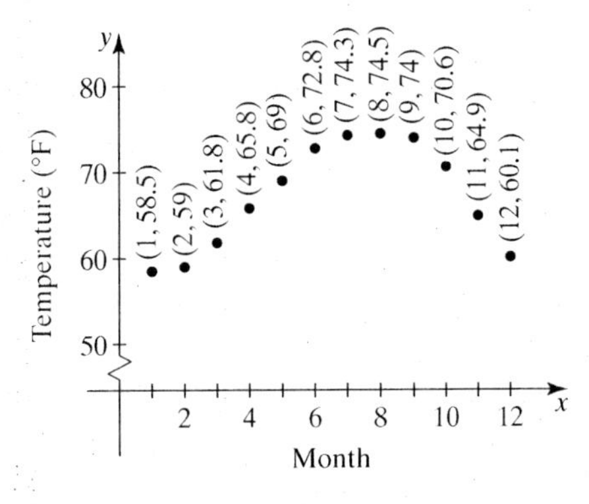

4. Average Monthly Low Temperature (°F) in New York, New York

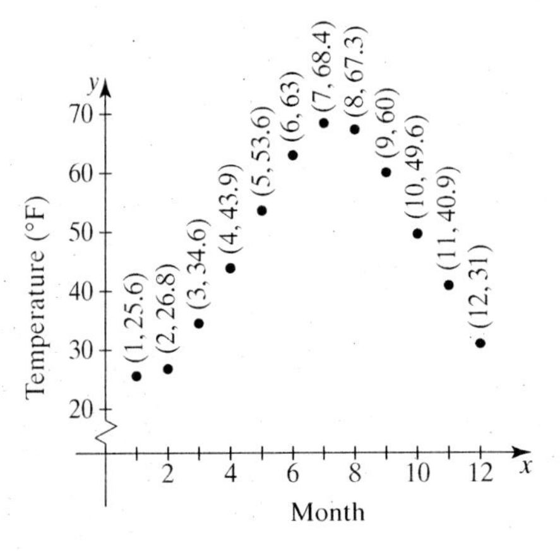

5–8. *Plot the data to determine whether a sinusoidal function can be used to model the given correspondence between the two quantities (x, y). Assume the data will be periodic.*

5.

x	1	2	3	4	5	6	7	8	9	10	11	12
y	7.5	7	6	5	3.5	4	4.5	5.5	5.5	6.5	7.5	8

6.

x	1	2	3	4	5	6	7	8	9	10
y	0	1	3.5	5	7	8.5	8	5.5	3	1

7.

x	1	2	3	4	5	6	7	8
y	0	2.5	3	7	6	15	10	25

8.

x	1	2	3	4	5	6	7	8
y	16	14	10	9	−9	5	0	−2

9.

x	1.2	5.9	10.6	15.3	20	24.7
y	2.5	−1.5	2.5	−1.5	2.5	−1.5

10.

x	1.9	2.8	3.7	4.6	5.5	6.4
y	−15	−28	−50	−59	−50	−28

11–14. *Find an equation in the form $y = A\sin[B(x - C)] + D$ that models the data in the following situations where $A > 0$.*

11.

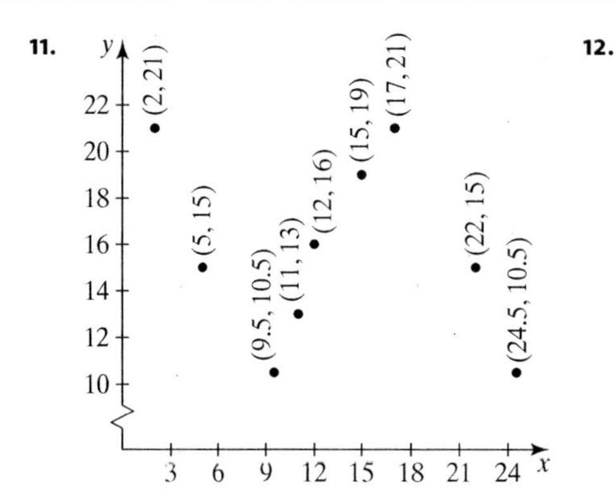

12.

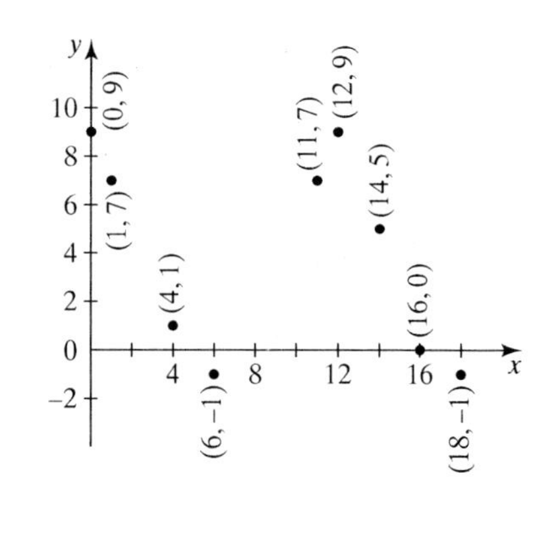

13.

x	1	2	3	4	5	6	7	8	9	10	11	12
y	73	71	74	79	82	85	87	90	87	83	79	74

14.

x	1	2	3	4	5	6	7	8	9	10	11	12
y	13	11	8	5	2.5	2	2.5	5	10.5	13	13.5	14.5

Graphing Calculator/CAS Explorations

15. Use the data from Exercise 11 to create a scatter plot. Then find the sine regression equation in the form $y = a\sin(bx + c) + d$ that models the data. Compare the regression equation with the one you found in Exercise 11.

16. Use the data from Exercise 12 to create a scatter plot. Then find the sine regression equation in the form $y = a\sin(bx + c) + d$ that models the data. Compare the regression equation with the one you found in Exercise 12.

17. The depth of water at the end of Frank's dock varies throughout the day, due to the tides. The following chart shows the depth measurements y, in feet, taken at specific times t.

t	12 A.M. (midnight)	2 A.M.	4 A.M.	6 A.M.	8 A.M.	10 A.M.	12 P.M. (noon)
y	4	4.5	5.7	7.2	5.5	4.7	3.8

a. Use sine regression to find a sinusoidal function to model this data.

b. Use this model to find the depth at 5 A.M. and 4 P.M.

c. Frank's boat needs a minimum depth of 5 feet of water to bring his boat in to dock. Will he be able to bring his boat in to dock at 9 P.M.?

18. The following table lists the U.S. unemployment rate for the month of July, 1992 through 2002. The unemployment rate represents the number unemployed as a percent of the labor force.

	7/92	7/93			. . .						7/02
t	0	1	2	3	4	5	6	7	8	9	10
r	7.7	6.9	6.1	5.7	5.5	4.9	4.6	4.3	4.1	4.6	5.9

a. Create a scatter plot of the data.

b. Do a sine regression to find an equation that models the data.

c. If we assume that the unemployment rate is cyclic, what is the length of this cycle?

d. What will the unemployment rate, based on this model, be in July, 2012?

19. A sound with a single frequency is called a pure tone. A healthy young person hears sounds with frequencies from around 20 to 20,000 Hz. Pure tones are used in push-button telephones. These phones simultaneously produce two pure tones when each button is pressed. Each number on the phone produces a difference pair of tones.

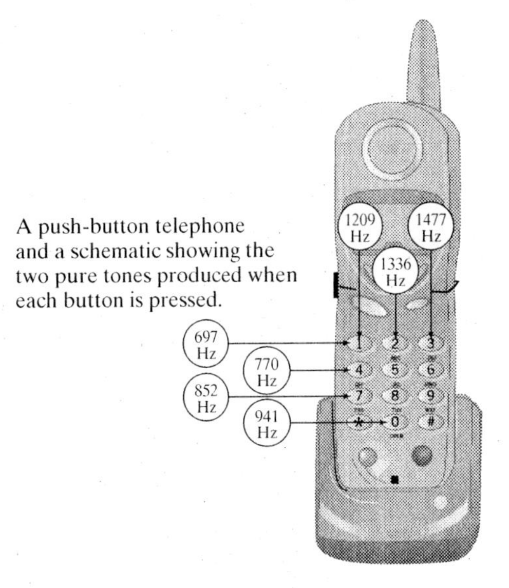

A push-button telephone and a schematic showing the two pure tones produced when each button is pressed.

(continued)

The following graphs display pressure patterns for pure tones of a 10-Hz wave and a 12-Hz wave for a time interval of one second:

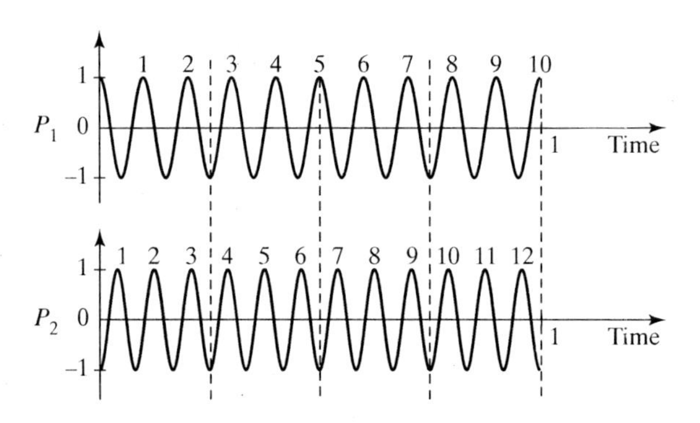

a. Find the equations to represent the pressure patterns for pure tones P_1 and P_2.

b. To see the resulting pressure when these two patterns overlap, use a graphing utility to graph $y = P_1 + P_2$. Determine whether this graph can be represented by a sinusoidal function of the form $y = A\sin[B(x - C)] + D$.

c. What is the period of $y = P_1 + P_2$?

20. Since we know $\sin(x + 2k\pi) = \sin x$, we said that $\sin(0 + 4000\pi) = \sin 0 = 0$ in Example 3. Determine how your calculator displays the functional value $\sin(4000\pi)$.

Discussion

21. You decide to take a ride on a ferris wheel, which rotates at 6 revolutions per minute. After you take your seat, the attendant continues to seat additional riders. As the last seat is filled, you notice that you are stopped about 6 feet off the ground. You start your stopwatch and notice that it takes you 4 seconds to reach the highest point, which is 48 feet above the ground. Your distance from the ground varies sinusoidally with time. One of the three following graphs is the best description of your ride. Select one and discuss why it is correct. Explain what is wrong with the other two.

a.

Height (in feet)
50
40
30
20
10
2 4 6 8 10
Time (in seconds)

b.

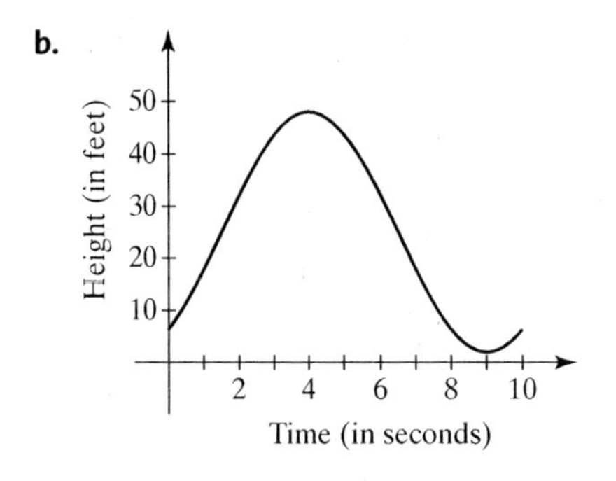

c.

Height (in feet)
50
40
30
20
10
−10
−20
−30
−40
−50
1 2 3 4 5 6 7 8 9 10
Time (in seconds)

Web Activity

22. There is a pseudoscience phenomenon known as *biorhythms*. It suggests that your body has an intellectual, emotional, and physical cycle. Search the Internet for biorhythms and find a site that will calculate your biorhythms once you input your date of birth. Many of these sites will also graph your cycles. If so, print your cycles. Determine the date of the current month that your biorhythm indicates you will reach your intellectual peak. (You may want to ask your instructor to plan a test on that date!) Conduct your own experiment to see if you agree with the theory.

2.4 Graphs of the Tangent, Cotangent, Secant, and Cosecant Functions

In this section we graph the four remaining circular functions: tangent, cotangent, secant, and cosecant functions. We begin each graph by recalling what we know about the function.

Graph of the Tangent Function

- Tangent function: $\{(x, \tan x)\}$ or $y = \tan x$
- Domain: $x \in \mathbb{R}$ such that $x \neq \dfrac{\pi}{2} + k\pi, k \in \{\ldots, -2, -1, 0, 1, 2, \ldots\}$.
- Range: $y \in \mathbb{R}$
- Period: π

Remember, $\tan x = (\sin x)/(\cos x)$ is defined when $\cos x \neq 0$, which means $x \neq (\pi/2) + k\pi$.

For $k = -1, 0,$ and 1, values such as $x = -\pi/2, \pi/2,$ and $3\pi/2$ are *not in the domain* of the tangent function. This means there will not be a point on the graph of the function corresponding to any of these x values. To indicate this, we draw dashed vertical lines called **vertical asymptotes** through $x = -\pi/2, \pi/2, 3\pi/2, \ldots,$ $x = (\pi/2) + k\pi$. The asymptotes are important to our graph. They serve as visual reminders of x values that are not in the domain. Therefore, the graph should never cross or touch them. To graph the tangent for one period, we set up a table of tangent functional values for common arcs, $-\pi/2 < x < \pi/2$, giving calculator approximations whenever necessary. We also use a calculator to find $\tan x$ for a few x-values close to the asymptotes. Further use of the calculator indicates that $\tan x$ becomes larger as x gets closer to the left side of an asymptote, and becomes smaller as x gets closer to the right side of an asymptote. Since we know that the circular functions are continuous on their domains, we connect the points on the graph with a smooth curve on the domain, which is between the asymptotes. We repeat the graph for two more cycles.

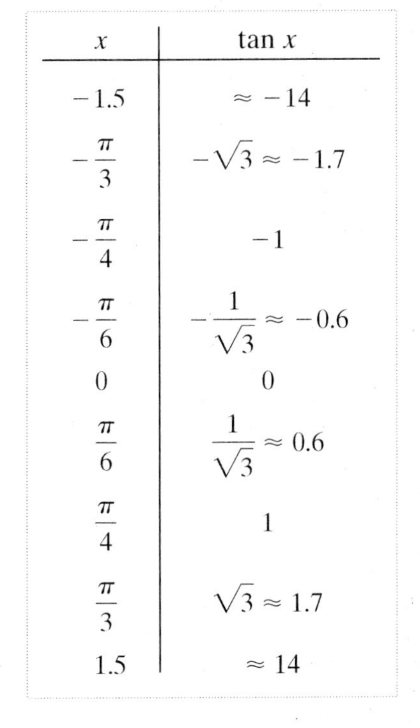

x	$\tan x$
-1.5	≈ -14
$-\dfrac{\pi}{3}$	$-\sqrt{3} \approx -1.7$
$-\dfrac{\pi}{4}$	-1
$-\dfrac{\pi}{6}$	$-\dfrac{1}{\sqrt{3}} \approx -0.6$
0	0
$\dfrac{\pi}{6}$	$\dfrac{1}{\sqrt{3}} \approx 0.6$
$\dfrac{\pi}{4}$	1
$\dfrac{\pi}{3}$	$\sqrt{3} \approx 1.7$
1.5	≈ 14

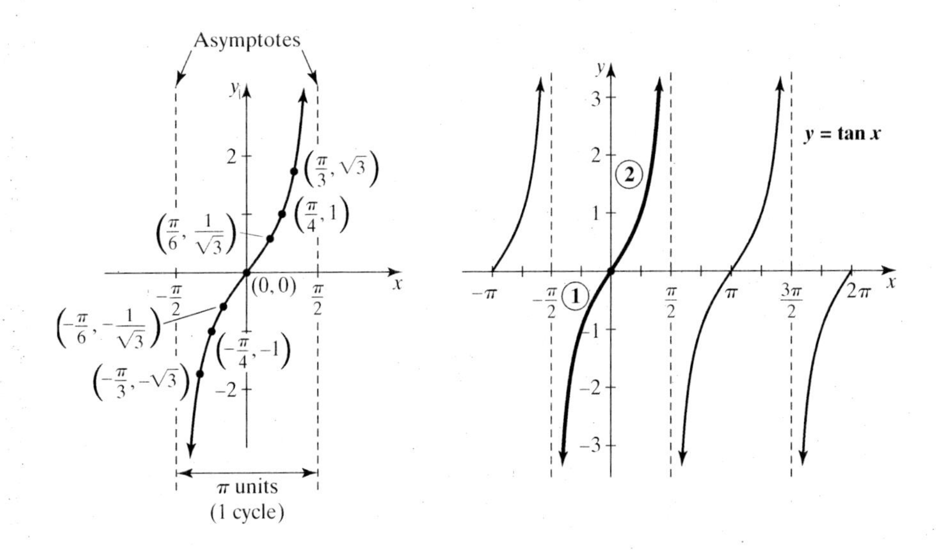

The graph of $y = \tan x$ is periodic, but not sinusoidal. The tangent function has no amplitude since there are no maximum and minimum functional values. The asymptotes at $x = (\pi/2) + k\pi$ are one period apart from each other. If we look at one cycle of the tangent graph on the previous page, we notice there are *two identically shaped arc sections.* Each arc occurs over an interval along the x-axis that is one-half of the period, or $(1/2)(\pi) = \pi/2$ units, in length. Each arc section is increasing and has an x-intercept on one end. The other end of the arc section is unbounded, which we indicate by drawing an arrow. The two arc sections are continuous between the asymptotes. The x-intercepts are at $x = k\pi$, and the y-intercept is at the origin, $(0, 0)$. The graph is symmetric with respect to the origin.

Graph of the Cotangent Function

- Cotangent function: $\{(x, \cot x)\}$ or $y = \cot x$
- Domain: $x \in \mathbb{R}$ such that $x \neq k\pi$, $k \in \{\ldots, -2, -1, 0, 1, 2, \ldots\}$.
- Range: $y \in \mathbb{R}$
- Period: π

Remember, $\cot x = \cos x/\sin x$ is defined when $\sin x \neq 0$, which means $x \neq k\pi$.

For $k = -1, 0$, and 1, values such as $x = -\pi, 0$, and π are not in the domain of the cotangent function. So, there will not be a point on the graph of the function corresponding to any of these x values. Therefore, we draw the asymptotes through $x = -\pi, 0, \pi, \ldots, x = k\pi$. Instead of using a table, we take advantage of the definition $\cot x = 1/\tan x$ by sketching the dotted graph of $y = \tan x$ for a reference. Next we plot corresponding points where the y-coordinates of $y = \cot x$ are the reciprocal values of $y = \tan x$, except for the indicated domain restrictions. As a result of their definitions, $\cot x = 0$ whenever $\tan x$ is undefined, and $\cot x$ is undefined whenever $\tan x = 0$. And as $\tan x$ increases, $\cot x$ decreases. We repeat the graph for two more cycles from $-\pi$ to 2π.

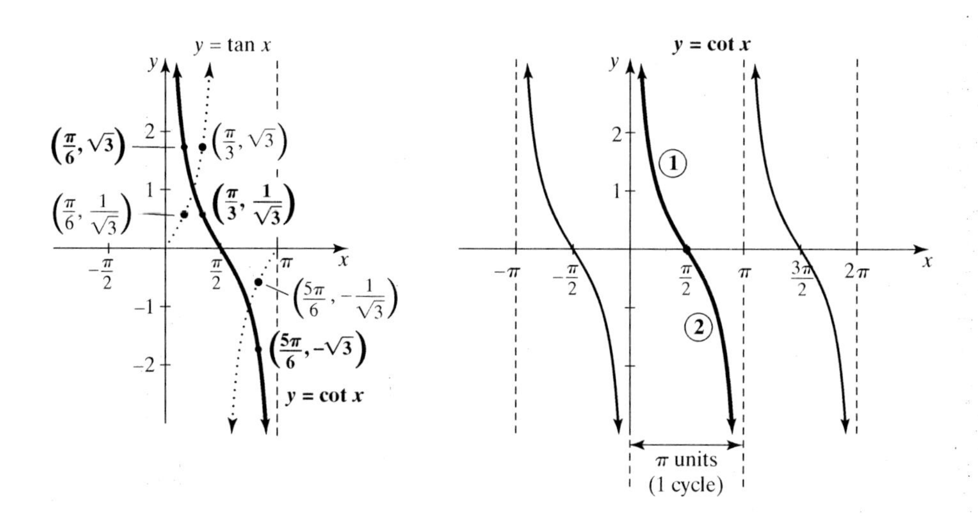

The graph of $y = \cot x$ is periodic but not sinusoidal. The cotangent function has no amplitude since there are no maximum and minimum functional values. The graph has no y-intercept.

Important Characteristics of the Graphs of $y = \tan x$ and $y = \cot x$

Common Characteristics

- Each cycle has length π.
- The asymptotes are one period apart from each other.
- If we look at one cycle, we notice there are two identically shaped arc sections. Each arc occurs over an interval along the x-axis that is one-half of the period in length. One endpoint of each arc is an x-intercept. The other end is unbounded. The two arc sections are continuous between the asymptotes.
- Both are symmetric with respect to the origin.

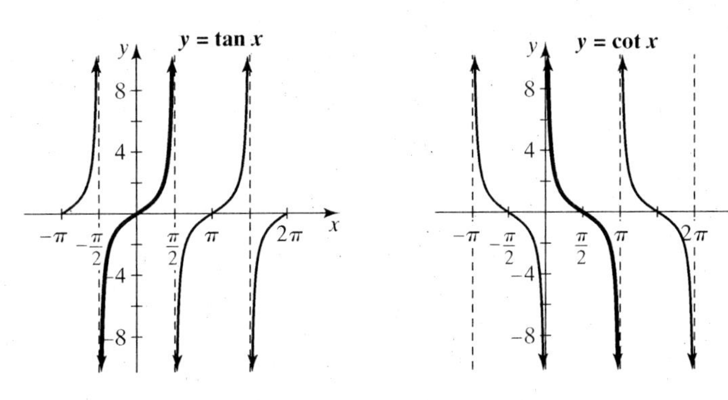

Different Characteristics

$y = \tan x$	$y = \cot x$
The arc sections increase.	The arc sections decrease.
asymptotes: $x = \dfrac{\pi}{2} + k\pi$	asymptotes: $x = k\pi$
x-intercepts: $x = k\pi$	x-intercepts: $x = \dfrac{\pi}{2} + k\pi$
y-intercept: $(0, 0)$	y-intercept: none

We refer to the graphs of $y = \tan x$ and $y = \cot x$ as the pure forms. You should study these forms and become familiar with the similarities and differences before you attempt to graph variations of these graphs having different periods and/or translations. We summarize the changes to these pure forms as follows.

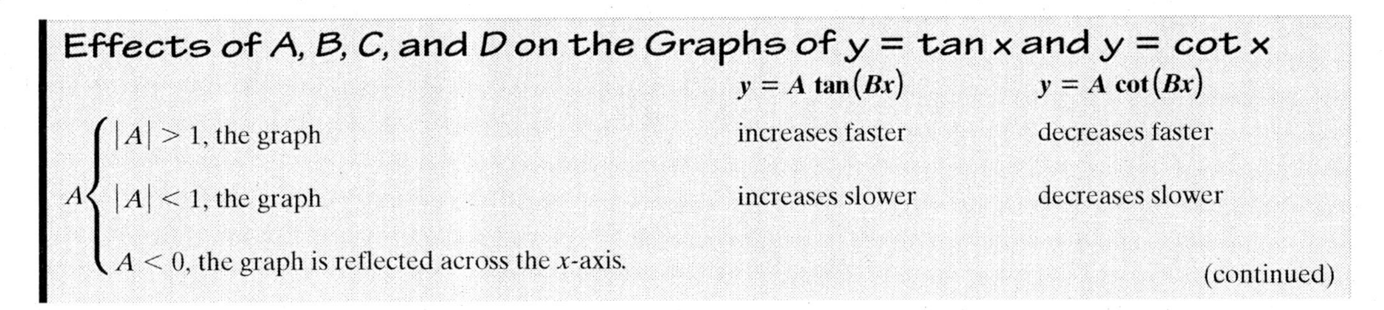

Effects of A, B, C, and D on the Graphs of $y = \tan x$ and $y = \cot x$

		$y = A\tan(Bx)$	$y = A\cot(Bx)$
A	$\lvert A\rvert > 1$, the graph	increases faster	decreases faster
	$\lvert A\rvert < 1$, the graph	increases slower	decreases slower
	$A < 0$, the graph is reflected across the x-axis.		

(continued)

Effects of A, B, C, and D on the Graphs of $y = \tan x$ and $y = \cot x$

		$y = A\tan(Bx)$	$y = A\cot(Bx)$
B	**Domain:** $x \in \mathbb{R}$ such that $x \neq \dfrac{\text{pure domain}}{\lvert B\rvert}$	$x \neq \dfrac{\frac{\pi}{2} + k\pi}{\lvert B\rvert}$	$x \neq \dfrac{k\pi}{\lvert B\rvert}$
	Asymptotes: $x = \dfrac{\text{pure asymptotes}}{\lvert B\rvert}$	$x = \dfrac{\frac{\pi}{2} + k\pi}{\lvert B\rvert}$	$x = \dfrac{k\pi}{\lvert B\rvert}$
	Period: $P = \dfrac{\text{pure period}}{\lvert B\rvert}$	$P = \dfrac{\pi}{\lvert B\rvert}$	$P = \dfrac{\pi}{\lvert B\rvert}$
	x-intercepts: $x = \dfrac{\text{pure } x\text{-intercepts}}{\lvert B\rvert}$	$x = \dfrac{k\pi}{\lvert B\rvert}$	$x = \dfrac{\frac{\pi}{2} + k\pi}{\lvert B\rvert}$

For $y = A\tan[B(x - C)] + D$ $\qquad$ $y = A\cot[B(x - C)] + D$

C indicates a phase (horizontal) shift.
D indicates a vertical shift.

Before doing an example, we outline steps for graphing equations of the form $y = A\tan(Bx)$ or $y = A\cot(Bx)$.

Graphing $y = A\tan(Bx)$ or $y = A\cot(Bx)$

Step 1. Determine the period.

Step 2. Find the formula for the asymptotes and plot two of the asymptotes. (Let $k = -1, 0$ for the tangent and $k = 0, 1$ for the cotangent.) *Your asymptotes should be one period apart.*

Step 3. Find the formula for the x-intercepts. Plot one x-intercept for $k = 0$, which should be one-half the distance between the asymptotes.

Step 4. Evaluate the function using two x values, each one-half the distance between the x-intercept and the asymptote. Plot these points.*

Step 5. Connect the points between the asymptotes, forming two identically shaped arc sections using the appropriate tangent or cotangent form.

Step 6. Repeat the pattern.

*Plotting these two additional points helps indicate how fast or slow the graph is increasing or decreasing. If you are interested only in a very rough sketch, you can omit Step 4 and just use the x-intercept with the appropriate form.

EXAMPLE 1 Graph $y = 3\tan(2x)$ on $-\pi \le x \le \pi$.

SOLUTION *Since $B = 2$, we have a horizontal shrink. For $A = 3$, the graph increases faster (vertical stretch).*

We follow the given preceding steps:

Step 1. $B = 2$, so $P = \dfrac{\text{pure period}}{|B|} = \dfrac{\pi}{2}$.

Step 2. Asymptotes: $x = \dfrac{\text{pure asymptotes}}{|B|} = \dfrac{\frac{\pi}{2} + k\pi}{2} = \dfrac{\pi}{4} + \dfrac{k\pi}{2}$

For $k = -1, 0$, we get $x = -\dfrac{\pi}{4}, \dfrac{\pi}{4}$, so we plot these two asymptotes.

Step 3. x-intercepts: $x = \dfrac{\text{pure } x\text{-intercepts}}{|B|} = \dfrac{k\pi}{2}$

For $k = 0$, $x = 0$ (halfway between the asymptotes).

Steps 4–5.

x	$(2x)$	$3\tan(2x)$
$-\dfrac{\pi}{8}$	$-\dfrac{\pi}{4}$	$3\tan\left(-\dfrac{\pi}{4}\right) = -3$
$\dfrac{\pi}{8}$	$\dfrac{\pi}{4}$	$3\tan\left(\dfrac{\pi}{4}\right) = 3$

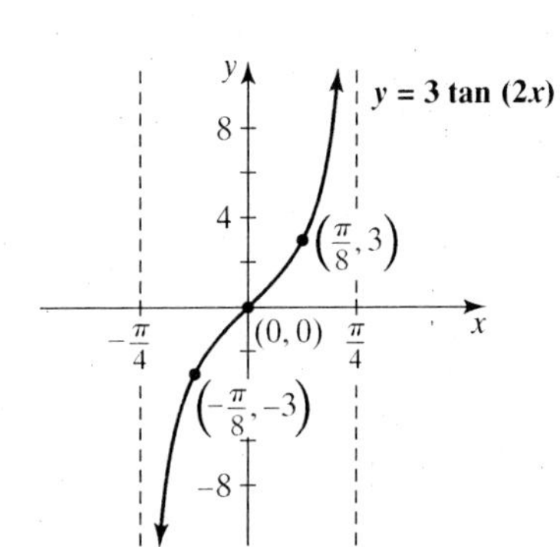

Step 6.

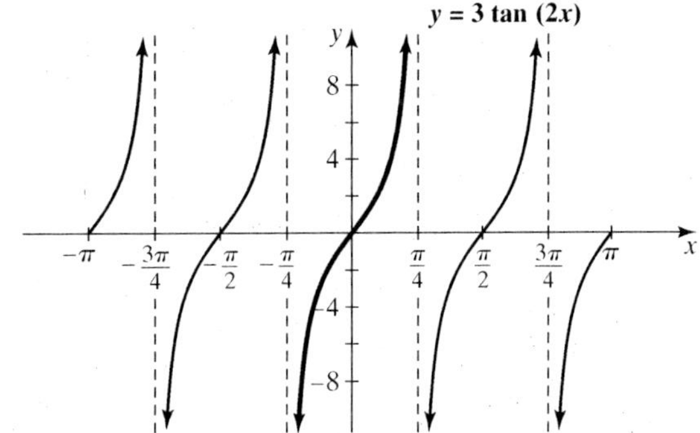

It is possible for the tangent and cotangent functions to also have a phase shift (C) and a vertical shift (D). To graph $y = A\tan[B(x - C)] + D$ or $y = A\cot[B(x - C)] + D$, where $C \ne 0$ or $D \ne 0$, it is advisable to follow a modification to the steps for $y = A\tan(Bx)$ or $y = A\cot(Bx)$ used in Example 1.

Graphing $y = A\tan[B(x - C)] + D$ or $y = A\cot[B(x - C)] + D$

- Find the period (see Step 1 for graphing $y = A\tan(Bx)$).
- The equation for the asymptotes can be found by finding two asymptotes for $y = A\tan(Bx)$ or $y = A\cot(Bx)$ (see Step 2) and then adding C to the value of each asymptote. Next, graph the two new asymptotes. These asymptotes should be one period apart from each other. If the graph has no vertical shift, the x-intercept will be one-half the distance between the asymptotes (see Step 3).
- If there is a vertical shift, the x-intercept should be shifted $|D|$ units in the appropriate direction.
- When you sketch the two equal arcs to form one period of the tangent or cotangent graph, let the value of A guide the shape or reflection (see Steps 4–6).

EXAMPLE 2 Graph one cycle of $y = \cot\left(x + \dfrac{\pi}{2}\right)$.

SOLUTION Rewriting the equation in the form $y = \cot(x - C)$, we get $y = \cot(x - (-\pi/2))$. *Since $C = -\pi/2$, we have a phase shift to the left.*

Step 1. $B = 1$, so $\quad P = \dfrac{\text{pure period}}{|B|} = \dfrac{\pi}{1} = \pi.$

Step 2. Asymptotes: First find two asymptotes for $y = \cot x$.

$$x = \frac{\text{pure asymptotes}}{|B|} = \frac{k\pi}{|B|} = \frac{k\pi}{1} = k\pi$$

When $k = 0, 1$, the asymptotes are $x = 0$ and $x = \pi$.
Since $C = -\pi/2$, the asymptotes for $y = \cot(x - (-\pi/2))$ are

$$x = 0 + \left(-\frac{\pi}{2}\right) = -\frac{\pi}{2}, \quad \text{and} \quad x = \pi + \left(-\frac{\pi}{2}\right) = \frac{\pi}{2}.$$

We graph these two asymptotes and notice they are one period (π) apart from each other.

Step 3. The x-intercept is at the origin (one-half the distance between the asymptotes).

Steps 4–6. Evaluate the function at two points, each one-half the distance between the x-intercept and the asymptote. We connect the points and repeat the pattern.

x	$\cot\left(x + \dfrac{\pi}{2}\right)$
$\dfrac{\pi}{4}$	-1
$-\dfrac{\pi}{4}$	1

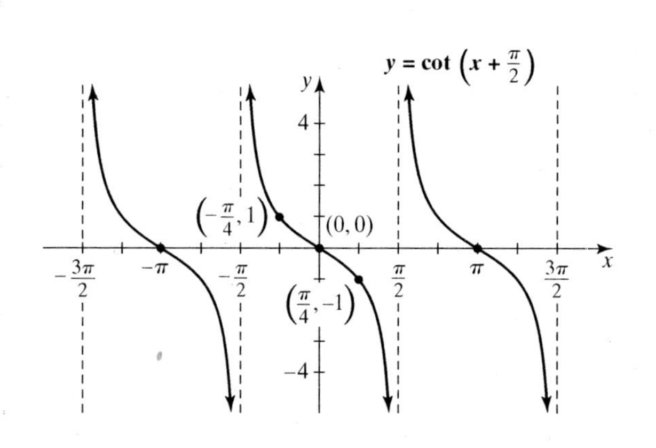

Does the final graph in Example 2 look familiar? The graph of $y = \cot(x + \pi/2)$ is also the graph of the tangent function reflected across the x-axis, or $y = -\tan x$. Again, we find there is more than one way to say the same thing in trigonometry.

Graphs of the Secant and Cosecant Functions

Function	$y = \sec x = \dfrac{1}{\cos x}$	$y = \csc x = \dfrac{1}{\sin x}$
Period	2π	2π
Domain	$x \neq (\pi/2) + k\pi$	$x \neq k\pi$
Range	$\lvert\sec x\rvert \geq 1$	$\lvert\csc x\rvert \geq 1$

To take advantage of the reciprocal definitions, we draw one dotted cycle of $y = \cos x$ on $0 \leq x \leq 2\pi$, as a reference for $y = \sec x = 1/\cos x$. Similarly, we draw one dotted cycle of $y = \sin x$ as a reference for $y = \csc x = 1/\sin x$ (See the graphs below). Next, we draw the asymptotes with dashed lines. For the secant function we draw the asymptotes through $x = (\pi/2) + k\pi$, that is, the x-intercepts of $y = \cos x$. For the cosecant function we draw the asymptotes through $x = k\pi$, that is, the x-intercepts of $y = \sin x$. Then, between the asymptotes, we plot corresponding points using reciprocal functional values to obtain the graphs of the secant and cosecant functions. If we repeat the pattern for more periods, we notice that the graph of $y = \sec x$ is symmetric with respect to the y-axis and the graph of $y = \csc x$ is symmetric with respect to the origin (See the graphs below). The graphs of $y = \sec x$ and $y = \csc x$ are the pure forms, which are periodic, but they are not sinusoidal.

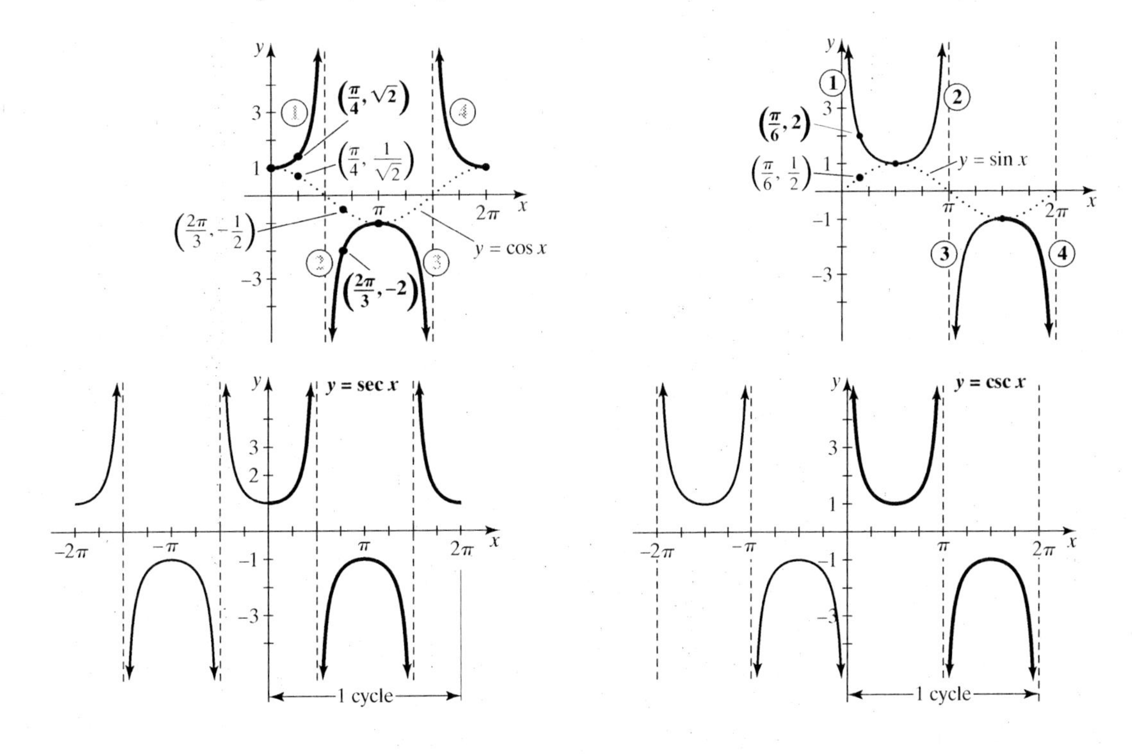

Like the graphs of the sine and cosine, each cycle of the secant and cosecant has four identically shaped arc sections. Each arc represents $\frac{1}{4}$ of the period and occurs over an interval along the x-axis that is one-fourth the period, or $(1/4)(2\pi) = \pi/2$ units, in length. These four arc sections for one period of the secant and cosecant appear in the form of two separate U-shaped branches between the asymptotes. The range of both the secant and cosecant function is $|y| \geq 1$, and the graphs have neither x-intercepts nor amplitude.

Our next example looks at a variation of the pure form of the secant.

EXAMPLE 3 Graph $y = \frac{1}{2} \sec x$ on $-2\pi \leq x \leq 2\pi$.

SOLUTION We need to determine the effect that $A = \frac{1}{2}$ has on the pure form $y = \sec x$ because there is no amplitude.

Since the range of $y = \sec x$ is $|\sec x| \geq 1$ $(y \geq 1$ or $y \leq -1)$, the y-values of $y = \frac{1}{2} \sec x$ are half as large or half as small as the y values of $y = \sec x$. Therefore, the range for $y = \frac{1}{2} \sec x$ will be $y \geq \frac{1}{2}$ or $y \leq -\frac{1}{2}$. (Since $B = 1$, and $C = D = 0$, there are no other changes to the pure form.)

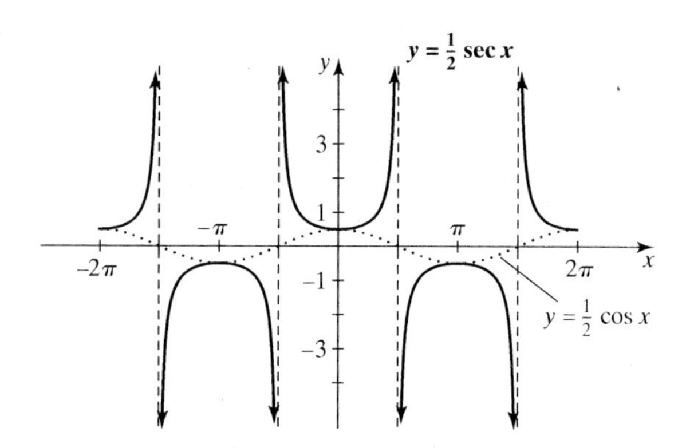

We notice that the dotted graph of $y = \frac{1}{2} \cos x$ can be used as a **guide function** to obtain the graph of $y = \frac{1}{2} \sec x$. We use this idea in Example 4. ■

CONNECTIONS WITH TECHNOLOGY

Graphing Calculator

We know we can check all of our circular function graphs with a graphing calculator. If we check graphs that have asymptotes, we can use different graph modes or styles. (Check your user manual.) One mode (dot) will not indicate the asymptotes. Another mode (connected) will incorrectly connect points on

(continued)

the graph so that it makes the asymptotes look like solid vertical lines. Each screen below is the graph of $y = \frac{1}{2}\sec x$.

Dot mode

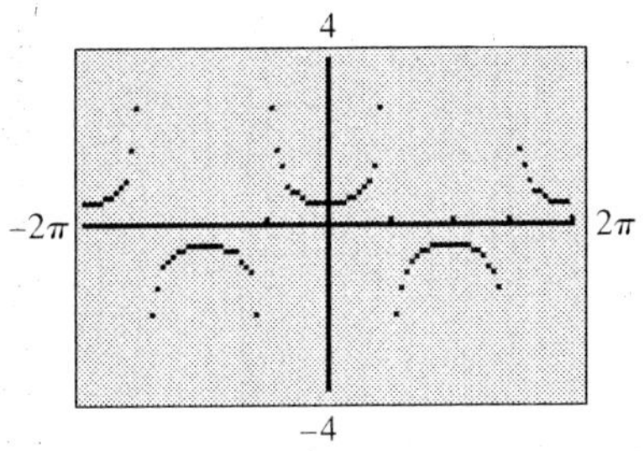

Connected graph mode

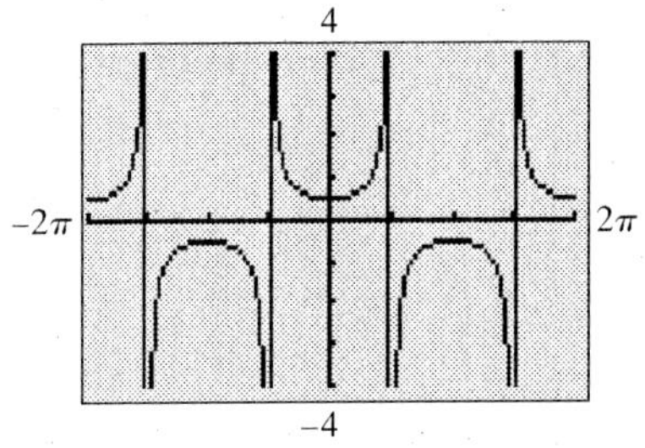

4

-2π 2π

-4

4

-2π 2π

-4

We now summarize the effects of the values of A, B, C, and D on the pure forms $y = \sec x$ and $y = \csc x$. Some of the effects are the same as the ones for the graphs of previous functions.

Effects of A, B, C, and D on the Graphs of $y = \sec x$ and $y = \csc x$

		$y = A\sec(Bx)$	$y = A\csc(Bx)$
A	**Range:**	$\lvert y\rvert \geq A$	$\lvert y\rvert \geq A$
		lowest y value of the $\cup$ (upper branch) is $\lvert A\rvert$	lowest y value of the upper branch is $\lvert A\rvert$
		highest y value of the $\cap$ (lower branch) is $-\lvert A\rvert$	highest y value of the lower branch is $-\lvert A\rvert$
		$A < 0$ the graph is reflected across the x-axis.	$A < 0$ the graph is reflected across the x-axis
B	**Domain:**		
	$x \neq \dfrac{\text{pure domain}}{\lvert B\rvert}$	$x \neq \dfrac{\frac{\pi}{2} + k\pi}{\lvert B\rvert}$	$x \neq \dfrac{k\pi}{\lvert B\rvert}$
	Asymptotes:		
	$x = \dfrac{\text{pure asymptotes}}{\lvert B\rvert}$	$x = \dfrac{\frac{\pi}{2} + k\pi}{\lvert B\rvert}$	$x = \dfrac{k\pi}{\lvert B\rvert}$
	Period:		
	$P = \dfrac{\text{pure period}}{\lvert B\rvert}$	$P = \dfrac{2\pi}{\lvert B\rvert}$	$P = \dfrac{2\pi}{\lvert B\rvert}$
	x-intercepts:	none	none

For $\quad y = A\sec[B(x - C)] + D \qquad y = A\csc[B(x - C)] + D$

C indicates a phase (horizontal) shift.
D indicates a vertical shift.

The following steps outline a procedure to help you graph the secant and cosecant functions. You will see that most of the difficulty in graphing complicated secant or cosecant functions is lessened by the ability to graph the *guide* function.

Graphing $y = A\sec[B(x - C)] + D$ or $y = A\csc[B(x - C)] + D$

Step 1. As a **guide** for $y = A\sec[B(x - C)] + D$, dot in one cycle of $y = A\cos[B(x - C)]$ and for $y = A\csc[B(x - C)] + D$, dot in one cycle of $y = A\sin[B(x - C)]$.
Make sure your guide function is only dotted in.

Step 2. Graph the asymptotes through the x-intercepts of the guide function graph.

Step 3. Draw the four arc sections in the form of two separate U-shaped branches between the asymptotes. One U-shaped branch is drawn upward from the maximum point of the guide function, and the second U-shaped branch is drawn downward from the minimum point of the guide function. (Remember, these four arc sections will form *two separate U-shaped branches per cycle.*)

Step 4. If $D \neq 0$, shift each branch the required number of units in the indicated direction.

EXAMPLE 4 Graph at least one cycle of $y = 3\csc\left(\frac{1}{2}x + \frac{\pi}{2}\right)$.

SOLUTION The guide function we use is

$$y = 3\sin\left(\frac{1}{2}x + \frac{\pi}{2}\right).$$

(Refer to Exercise Set 2.2 for the example that shows how to obtain this graph.) We dot in this graph. Graph the asymptotes through the x-intercepts. Form the two U-shaped branches: upward from maximum point of the guide function and downward from the minimum point of the guide function. (Since $D = 0$, there is no vertical shift.)

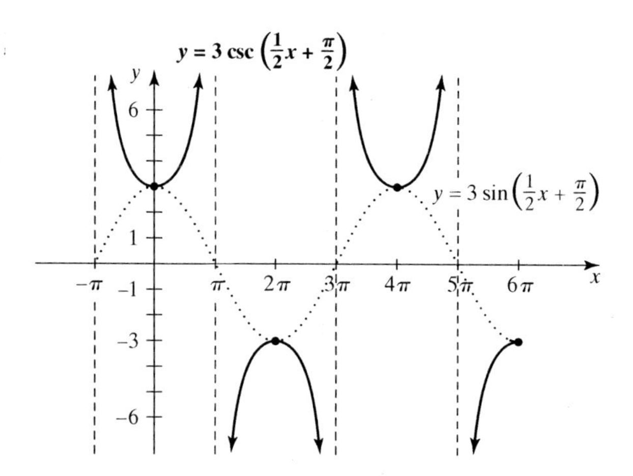

As we stated earlier, for more accurate and efficient graphs it is usually advisable to use a graphing utility, especially for challenging values of A, B, C, and D. In the exercises you will be asked to sketch some reasonable graphs by hand. For more difficult ones you will be asked to use a graphing utility. But even when you use a graphing utility, you know that you need to determine certain characteristics about the graph before the utility produces a good picture. Your graphing utility can graph, and you can think!

CONNECTIONS WITH TECHNOLOGY

Graphing Calculator/CAS

EXAMPLE Use a graphing utility to graph two cycles of $y = \pi \sec(0.2x)$. Determine the location of the asymptotes.

SOLUTION Be careful to correctly input the equation:

$$y = \pi \sec(0.2x) = \frac{\pi}{\cos(0.2x)}.$$

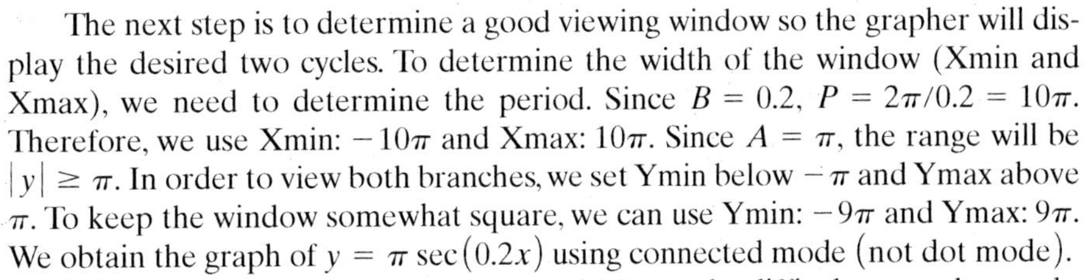

The next step is to determine a good viewing window so the grapher will display the desired two cycles. To determine the width of the window (Xmin and Xmax), we need to determine the period. Since $B = 0.2$, $P = 2\pi/0.2 = 10\pi$. Therefore, we use Xmin: -10π and Xmax: 10π. Since $A = \pi$, the range will be $|y| \geq \pi$. In order to view both branches, we set Ymin below $-\pi$ and Ymax above π. To keep the window somewhat square, we can use Ymin: -9π and Ymax: 9π. We obtain the graph of $y = \pi \sec(0.2x)$ using connected mode (not dot mode).

Where are the asymptotes for this graph? It may be difficult to use the graphing calculator to determine the exact location of these asymptotes, especially if we were in dot mode. Instead, we use the formula we developed to determine quickly and accurately that the asymptotes are at

$$x = \frac{(\pi/2) + k\pi}{|B|} = \frac{(\pi/2) + k\pi}{|0.2|} = \frac{5\pi}{2} + 5k\pi.$$

Exercise Set 2.4

1–6. *Copy the chart and fill in the requested information, when applicable, for each given circular function.*

		Domain	Range	Period	Asymptotes	x-intercepts	y-intercept	Graph
1.	$y = \sin x$							
2.	$y = \cos x$							
3.	$y = \tan x$							
4.	$y = \cot x$							
5.	$y = \sec x$							
6.	$y = \csc x$							

7–14. *Answer each question with as many of the functions that apply: $y = \cos x$, $y = \sin x$, $y = \tan x$, $y = \cot x$, $y = \sec x$, or $y = \csc x$.*

7. Which functions have graphs that go through the origin?
8. Which functions have graphs that are symmetric with respect to the y-axis?
9. Which functions have graphs that are symmetric with respect to the origin?
10. Which functions have graphs with disconnected U-shaped branches?
11. Which functions have graphs that are continuous on their domains?
12. Which functions have graphs that are not continuous on $0 < x < 2\pi$.
13. Which functions have graphs with asymptotes?
14. Which functions have amplitude?

15–22. *Sketch the graph of each function for at least one cycle. Indicate asymptotes and x-intercepts. Check your graph using a graphing utility.*

EXAMPLE $y = \cot\left(\frac{1}{2}x\right)$

SOLUTION We use the steps for graphing $y = A\cot(Bx)$.

Step 1. Since $B = \frac{1}{2}$, $P = \frac{\pi}{\frac{1}{2}} = 2\pi$.

Step 2. Asymptotes: $x = \frac{k\pi}{\frac{1}{2}} = 2k\pi$. For $k = 0, 1$, we get $x = 0, 2\pi$.

Step 3. x-intercepts: $x = \frac{\frac{\pi}{2} + k\pi}{\frac{1}{2}} = \pi + 2k\pi$. For $k = 0$, the x-intercept is at $x = \pi$ (which is halfway between the asymptotes).

Step 4.

x	$\cot\left(\frac{1}{2}x\right)$
$\frac{\pi}{2}$	1
$\frac{3\pi}{2}$	-1

Steps 5–6.

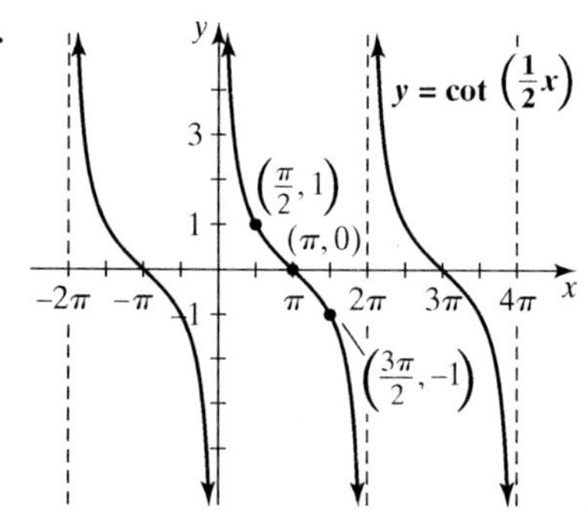

15. $y = \tan\left(\frac{1}{2}x\right)$
16. $y = \cot\left(\frac{1}{4}x\right)$
17. $y = \tan\left(x + \frac{\pi}{2}\right)$
18. $y = \cot\left(x - \frac{\pi}{4}\right)$
19. $y = -2\cot(2x)$
20. $y = \tan(2x - 2\pi)$
21. $y = -\cot x + 1$
22. $y = \tan x - 1$

23–28. *Sketch the graph of each function for at least one cycle. Indicate the period and the range. Check your graph using a graphing utility.*

EXAMPLE $y = 3\csc(2x)$

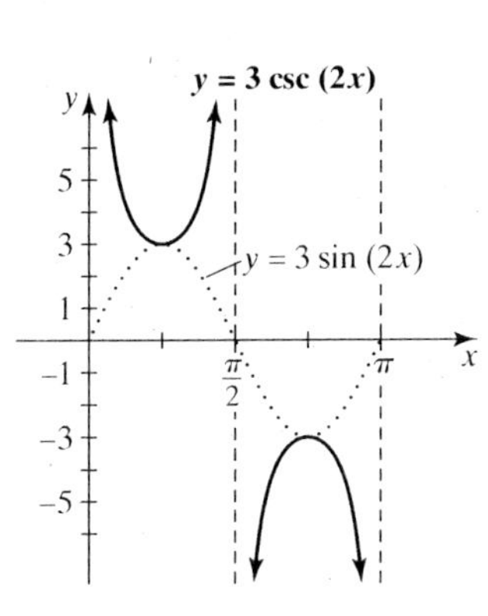

SOLUTION We use the steps for graphing $y = A\csc[B(x - C)] + D$, where $C = D = 0$.

Step 1. As a guide, we dot in the graph of $y = 3\sin(2x)$ that has amplitude 3 and period $2\pi/2 = \pi$.

Step 2. At each x-intercept, we plot the asymptotes.

Step 3. Draw the four arc sections in the form of two separate U-shaped branches between the asymptotes to obtain one cycle of $y = 3\csc(2x)$.

The period is π and the range is $|y| \geq 3$. ■

23. $y = -4\sec x$ **24.** $y = -\frac{1}{2}\csc x$ **25.** $y = \csc(-2x)$

26. $y = 3\sec\left(\frac{1}{2}x\right)$ **27.** $y = 4\sec\left(x + \dfrac{\pi}{2}\right)$ **28.** $y = \csc 2\left(x - \dfrac{\pi}{4}\right)$

29–32. *Without using a graphing calculator, match each function with its graph.*

29. $y = \sec\left(\frac{1}{2}x\right) + 2$ **30.** $y = -\csc x + 2$

31. $y = \cot(2x)$ **32.** $y = -\tan(2x)$

a.

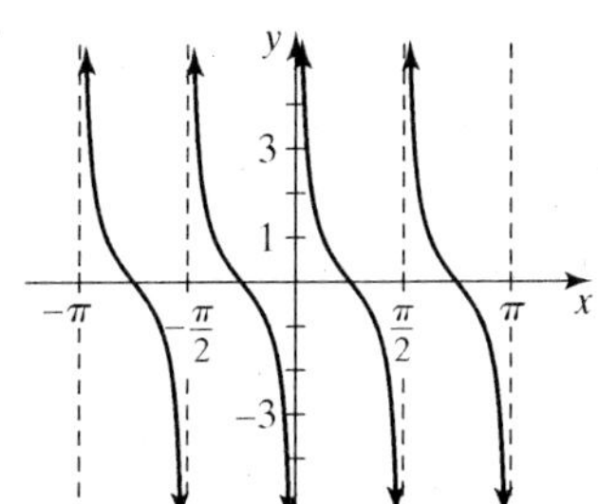

b.

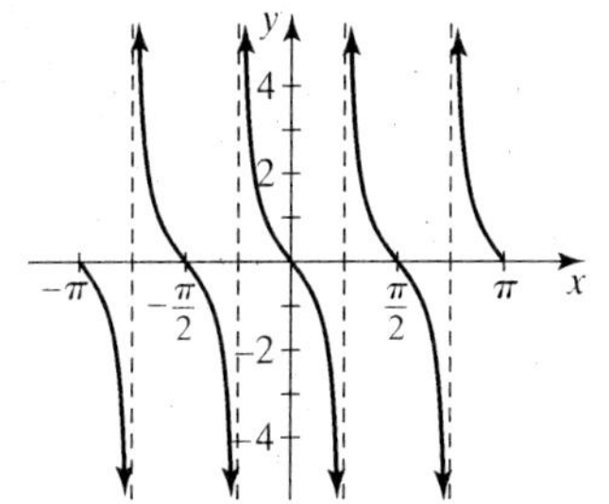

c.

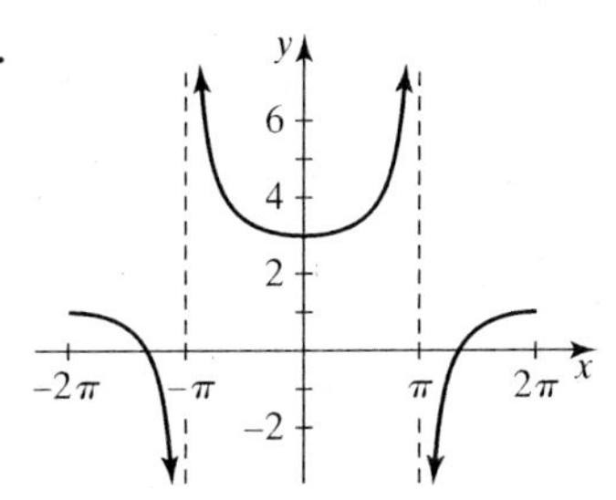

d.

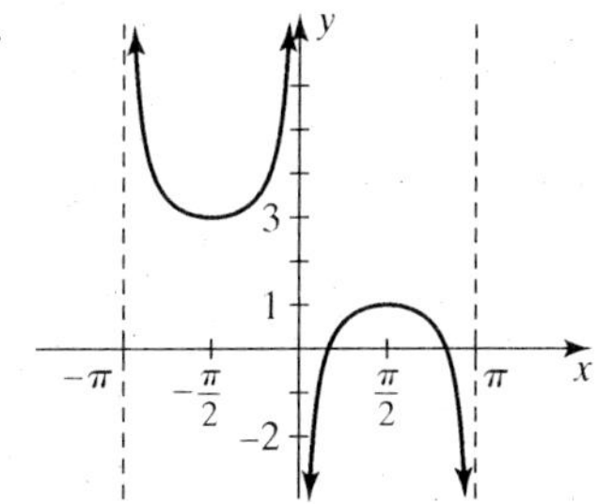

33–36. ***Incorrect*** *attempts to graph one cycle of the tangent, cotangent, secant, or cosecant functions are given. Find at least one mistake in each attempt.*

33. $y = \tan x$

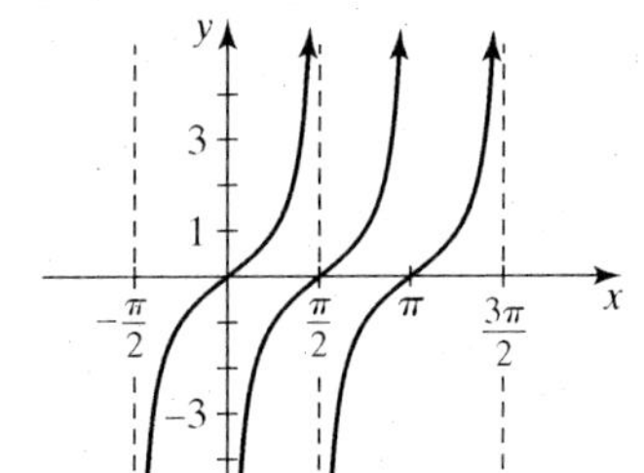

34. $y = \cot x$

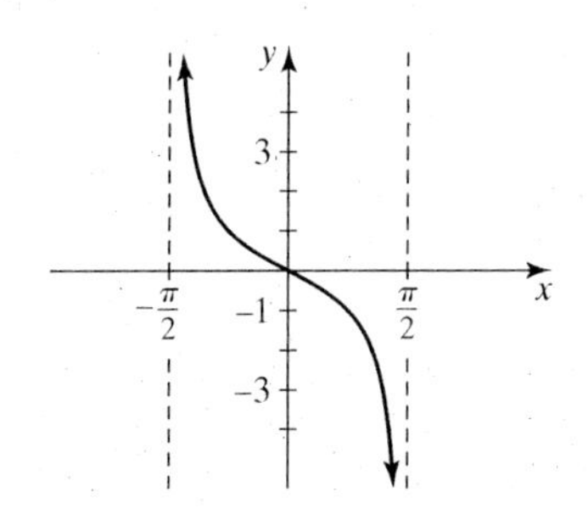

35. $y = \sec x$

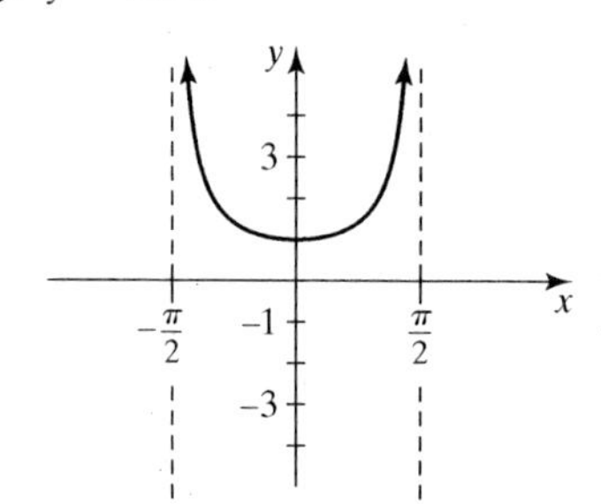

36. $y = \csc x$

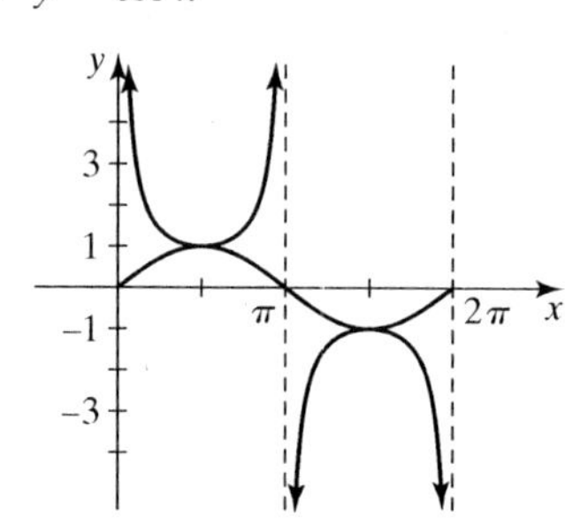

37. In Example 2, we graphed $y = \cot(x + \pi/2)$ and indicated that it is also the graph of $y = -\tan x$. Without a graphing calculator, use your knowledge of reflections and phase shifts to identify *two more* equations from the choices below that have the same graph as $y = -\tan x$. Check your answers with a graphing calculator.

a. $y = -\tan(x + \pi)$ **b.** $y = \cot(x - \pi)$

c. $y = \cot\left(x - \frac{\pi}{2}\right)$ **d.** $y = \tan\left(x + \frac{\pi}{2}\right)$

38. Without a graphing calculator, use your knowledge of reflections and phase shifts to identify *two* equations from the choices below that have the same graph as $y = -\sec x$. Check your answers with a graphing calculator.

a. $y = \csc\left(x - \frac{\pi}{2}\right)$ **b.** $y = -\csc\left(x + \frac{\pi}{2}\right)$

c. $y = -\csc(x + \pi)$ **d.** $y = \sec(x - \pi)$

Graphing Calculator/CAS Exploration

39–42. **a.** *Find the period and range for each function.*

b. *Use the information from part (a) to find an appropriate window and graph two cycles on a graphing calculator.*

39. $y = 9\tan(0.1\pi x)$

40. $y = \sqrt{2}\cot\frac{x}{5}$

41. $y = 5\csc[2(x - 1)]$

42. $y = \frac{2}{3}\sec(x - 3)$

43. Graph $y = \tan x + \cot x$ using connected (not dot) mode, Xmin $= -\pi$, Xmax $= \pi$, Ymin $= -8$, and Ymax $= 8$.

a. Where are the asymptotes?

b. What do you know about the two functions that could predict the location of the asymptotes?

c. Are there any x-intercepts?

d. What do you know about the functions that could predict whether or not there would be any x-intercepts?

Now graph $y = \csc x \sec x$ using the same settings.

e. What do you notice about the graphs of $y = \tan x + \cot x$ and $y = \csc x \sec x$?

f. If the graphs in (e) coincide, we should be able to verify the identity $\tan x + \cot x = \csc x \sec x$. We start by using the definitions of $\tan x$, $\cot x$, $\sec x$, and $\csc x$, and rewrite each of these functions in terms of $\sin x$ and/or $\cos x$ to get:

$$\tan x + \cot x = \csc x \sec x$$

$$\frac{\sin x}{\cos x} + \frac{\cos x}{\sin x} = \frac{1}{\sin x} \cdot \frac{1}{\cos x}$$

Use your knowledge of addition of fractions (find a common denominator) to prove that the left side of this equation can be made identical to the right side.

44. (Do Exercise 43 first.) If you want to determine whether the following statements are identities (equations that are true for all values in their domains), you have two choices as a result of your experience with Exercise 43. You could determine whether the following are identities by verifying the graphs of each side of the equation coincide, or you could rewrite one side of the statement in terms $\sin x$ and/or $\cos x$ and perform the indicated operations to see if you can make that side identical to the other. Use a graphing utility to determine whether the following are identities.

a. $\cot x + \csc x = \dfrac{\sin x}{1 - \cos x}$

b. $1 + \sin^2 x = \cos^2 x$

c. $\sec x - \sin x \tan x = \sin x$

45. (Do Exercise 44 first.) What identities did you discover from Exercise 37?

46. (Do Exercise 44 first.) What identities did you discover from Exercise 38?

2.5 Inverses of the Circular Functions

In Chapter 1 we are given an arc x, where x is a real number that represents the length and direction of the arc on the unit circle, and we are asked to find a circular functional value of x. The definition of a function states that for every element in the domain there corresponds only one element in the range, which is verifiable graphically by the vertical line test. As a result of the definitions for cosine, sine, tangent, secant, cotangent, and cosecant functions, for every x (arc), there is exactly one y (functional value). We use this fact to answer questions like the following.

given x → find y

$$\sin x = y$$

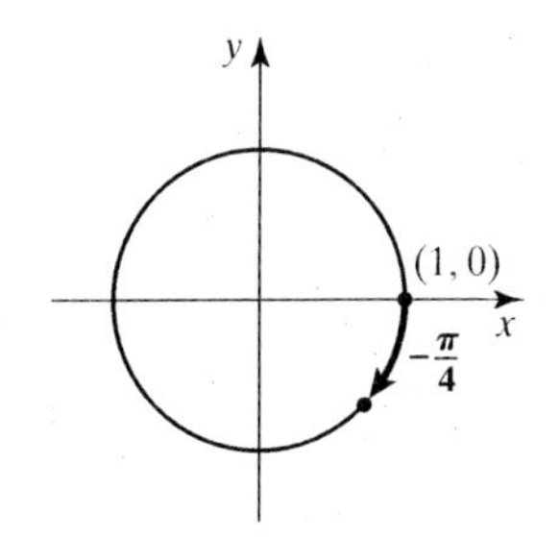

EXAMPLE Find: $\sin\left(-\dfrac{\pi}{4}\right) = y$

SOLUTION $y = -\dfrac{\sqrt{2}}{2}$

We refer to the problem "given x, find y" as the forward direction of the question. In Chapter 1 we are also asked this question in reverse: given y, find x. The reverse direction becomes the problem "given a circular functional value, find arc x" and is illustrated as follows.

$$\underset{\text{find } x}{\sin x} = \underset{\text{given } y}{y}$$

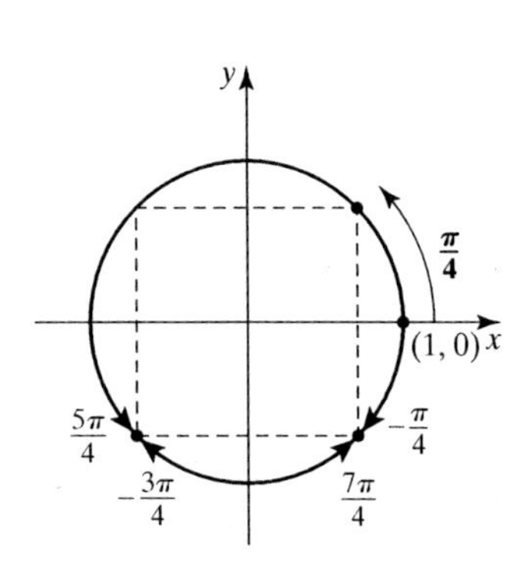

EXAMPLE Find x if $\sin x = -\dfrac{\sqrt{2}}{2}$, for $-2\pi \le x \le 2\pi$.

SOLUTION The sine functional value of $\sqrt{2}/2$ occurs at the reference arc $\hat{x} = \pi/4$.

Since $\sin x$ is negative in QIII and QIV, and $-2\pi \le x \le 2\pi$,

$$x = -\frac{3\pi}{4}, -\frac{\pi}{4}, \frac{5\pi}{4}, \text{ and } \frac{7\pi}{4}.$$

The reason we can answer the question in reverse is that every function has an inverse relation. However, we notice that even with the restriction on x, *there is more than one answer for this inverse relationship.* And without any restrictions on x, the periodic property would suggest many more solutions for x. If we want this problem to have only one answer for arc x, then the inverse relationship must be a function. To have both the forward and inverse relationships represent functions, we need a more suitable restriction on x so that for every arc x there is only one functional value y and for every functional value y there is only one arc x (one-to-one).

Referring to the previous example, once we restrict the domain of the sine function, the sine will represent a one-to-one function, and so the inverse relationship will also be a function. You may remember from algebra that the graph of a function that is one-to-one passes the horizontal line test.

In this section we focus on the inverses of three circular functions: sine, cosine, and tangent. The inverses of the remaining three circular functions are defined, but problems involving these inverses are better solved using the reciprocal identities.

If we look at the graphs of the sine, cosine, and tangent functions, we notice that none pass the horizontal line test—that is, none are one-to-one. This is not surprising since *each* function is periodic.

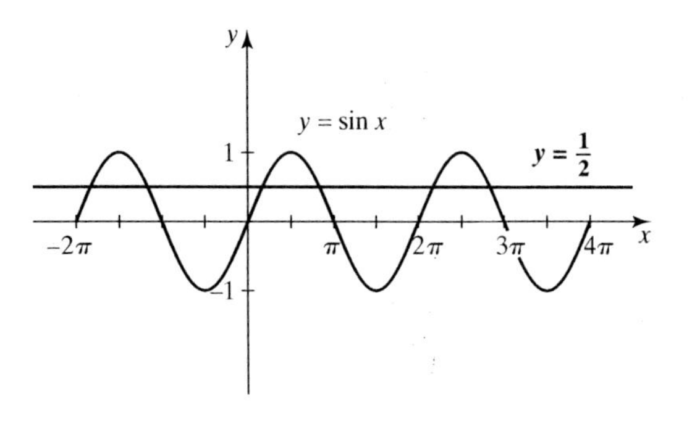

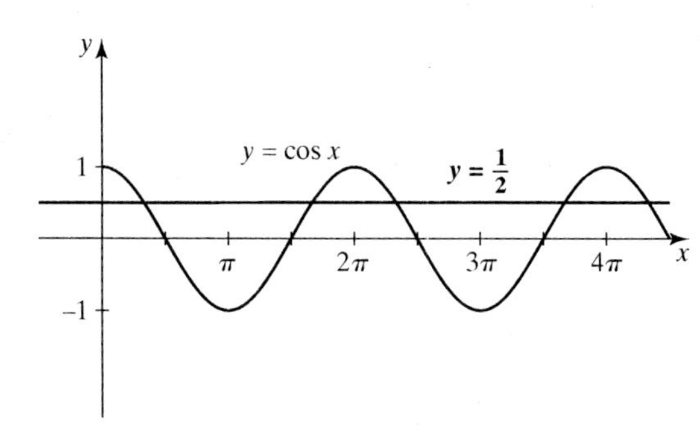

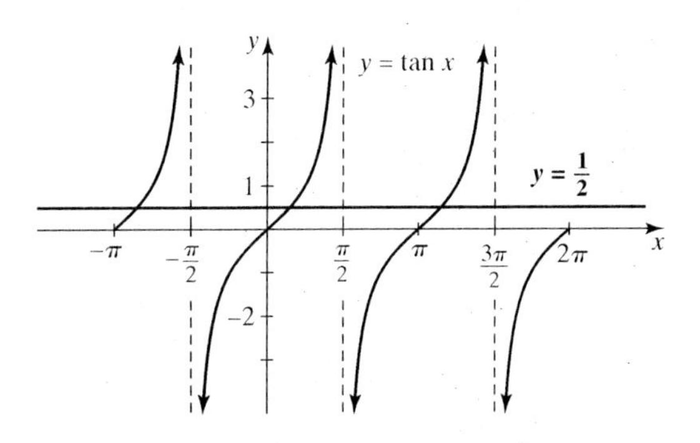

By suitably restricting the domain, each circular function becomes a one-to-one function on its restricted domain, and therefore, has an inverse function.

Inverse Sine

If we look at the graph of the sine function first and restrict its domain to $-\pi/2 \le x \le \pi/2$, we have only one x for any functional value y in the range, $-1 \le y \le 1$. Therefore, on this restricted domain the sine is a one-to-one function. (Notice on this interval, $y = \sin x$ passes the horizontal line test.) Thus, every arc has only one sine value and every sine value has only one arc.

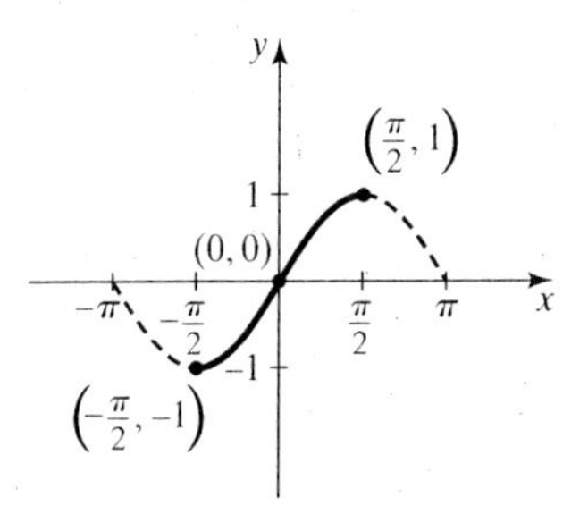

The forward direction of the sine function, defined in Section 1.2, is in the form $\{(x, y)\} = \{(x, \sin x)\}$.

Given the arc, find the functional value.

given x find y

$\sin x = y$

The reverse direction of the sine function (the inverse relation) interchanges the domain and range; that is, it interchanges x and y. So the inverse relation for the sine function is of the form $\{(x, y)\} = \{(\sin y, y)\}$.

Given the functional value, find the arc.

find y given x

$\sin y = x$

Since we are restricting the domain so that $y = \sin x$ is one-to-one, the inverse, $x = \sin y$, will be a function. And when we restrict the domain of a function to a specific interval, such as $-\pi/2 \le x \le \pi/2$, that interval is the range of the inverse function. So the inverse sine function is $x = \sin y$, where $-\pi/2 \leq y \leq \pi/2$. This equation is usually written with y in terms of x. To do this, we need to use the following notation.

Inverse Sine Function

The inverse sine function $x = \sin y$ is defined as

$$y = \arcsin(x) \quad \text{or} \quad y = \sin^{-1}(x)$$

$$\text{for } -1 \leq x \leq 1 \quad \text{and} \quad -\frac{\pi}{2} \leq y \leq \frac{\pi}{2}.$$

Notice that the *range* of the inverse function is included in the definition.

The graph of the inverse is found by reflecting the graph of the function about the line $y = x$.

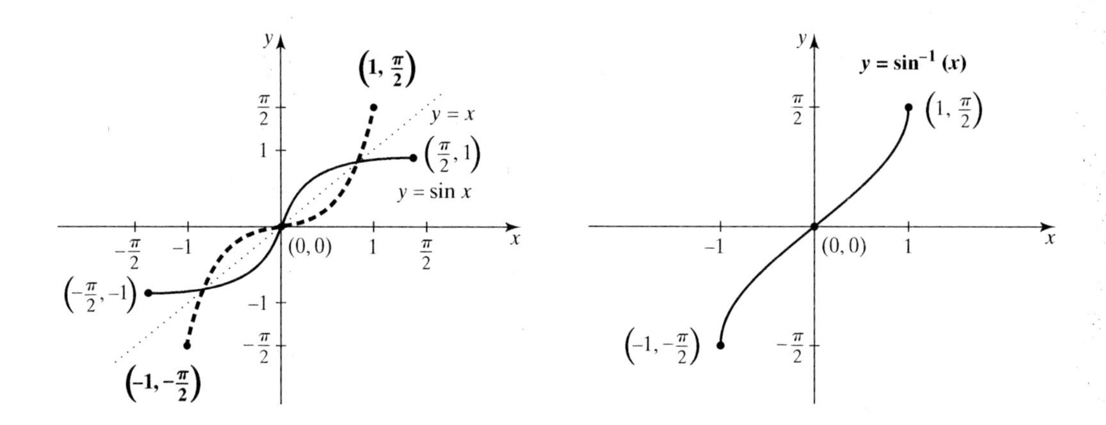

CAUTION: The inverse functional notation $\sin^{-1} x$ should not be confused with negative exponents:

$$\sin^{-1} x \neq \frac{1}{\sin x}$$

EXAMPLE 1 Find the exact value for y.

a. $y = \arcsin\left(-\frac{\sqrt{2}}{2}\right)$ **b.** $y = \arcsin\frac{1}{2}$ **c.** $y = \sin^{-1}\left(-\frac{\sqrt{3}}{2}\right)$

SOLUTION

a. The inverse functional notation $y = \arcsin(\sqrt{2}/2)$, which implies a restricted range, can be considered as an abbreviated form of the following question:

$$\text{Find the arc with sine value } \left(-\frac{\sqrt{2}}{2}\right).$$

The range for the arcsin x gives QI arcs in a positive direction and QIV arcs in a negative direction.

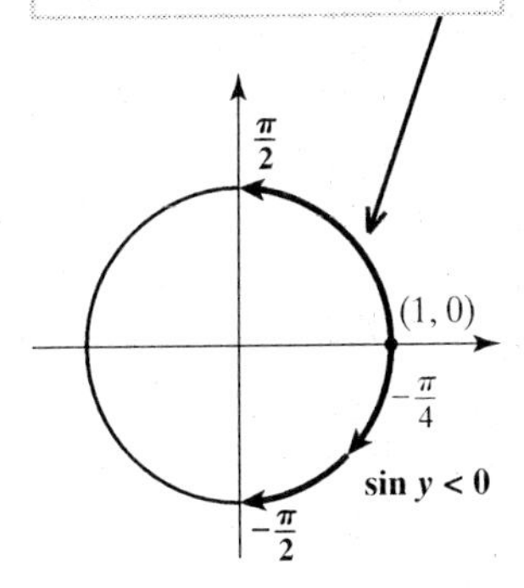

It may be easier to answer this question if it is first rewritten without the inverse notation:

$$y = \arcsin\left(-\frac{\sqrt{2}}{2}\right) \longleftrightarrow \sin y = -\frac{\sqrt{2}}{2}, \text{ for } -\frac{\pi}{2} \le y \le \frac{\pi}{2}$$

We recognize the sine functional value $-\sqrt{2}/2$ has the common reference arc $\hat{y} = \pi/4$. The only arc with this sine value in the restricted range is $y = -\pi/4$. Therefore, the

$$\text{arc with sine value } \left(-\frac{\sqrt{2}}{2}\right) = -\frac{\pi}{4}, \text{ or}$$

$$y = \arcsin\left(-\frac{\sqrt{2}}{2}\right) = -\frac{\pi}{4}.$$

You might want to provide answers like $y = 5\pi/4, 7\pi/4$, or $-3\pi/4$. Even though their sine value is $-\sqrt{2}/2$, these arcs are *incorrect* since they are not in the restricted interval for y: $-\pi/2 \le y \le \pi/2$. In other words, neither $5\pi/4$ nor $7\pi/4$ is a positive arc in the interval from 0 to $\pi/2$, and $-3\pi/4$ is not a negative arc in the interval from $-\pi/2$ to 0.

b. For $y = \arcsin \frac{1}{2}$, we are asked to find the

$$\text{arc with sine value } \frac{1}{2}.$$

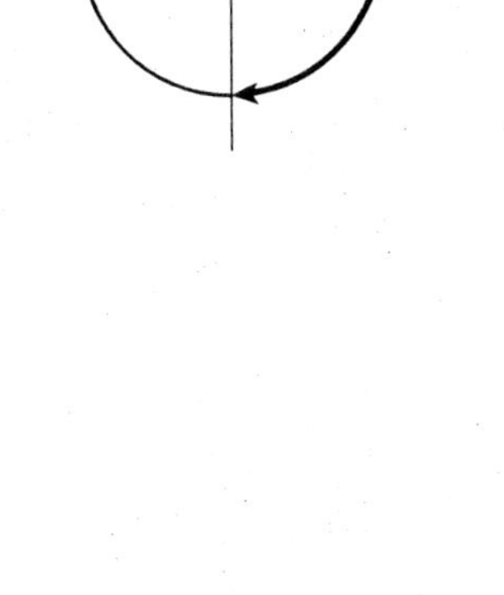

Rewriting without inverse notation, we get:

$$y = \arcsin \frac{1}{2} \longleftrightarrow \sin y = \frac{1}{2}, -\frac{\pi}{2} \le y \le \frac{\pi}{2}$$

Since $\hat{y} = \pi/6$, the only arc with this sine value in the restricted range is $y = \pi/6$.

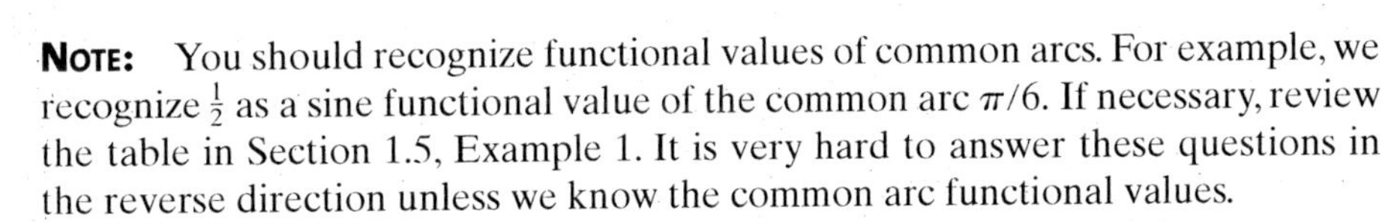

NOTE: You should recognize functional values of common arcs. For example, we recognize $\frac{1}{2}$ as a sine functional value of the common arc $\pi/6$. If necessary, review the table in Section 1.5, Example 1. It is very hard to answer these questions in the reverse direction unless we know the common arc functional values.

c. For $y = \sin^{-1}\left(-\frac{\sqrt{3}}{2}\right)$, we are asked to find the

$$\text{arc with sine value } \left(-\frac{\sqrt{3}}{2}\right).$$

Rewrite without inverse notation:

$$y = \sin^{-1}\left(-\frac{\sqrt{3}}{2}\right) \longleftrightarrow \sin y = \left(-\frac{\sqrt{3}}{2}\right), \text{ and } -\frac{\pi}{2} \le y \le \frac{\pi}{2}$$

Since $\hat{y} = \pi/3$, the only arc with this sine value in the restricted range is $y = -\pi/3$. ■

Inverse Cosine and Inverse Tangent

In a similar manner, we can define the cosine and tangent functions on suitable restricted domains so that their inverse relations are functions. The restrictions on the domains of the cosine and tangent functions are represented on the accompanying graphs. You can verify that the functions are one-to-one on these domains by using the horizontal line test. Mathematicians have generally agreed upon these restricted intervals, which coincide with most calculators.

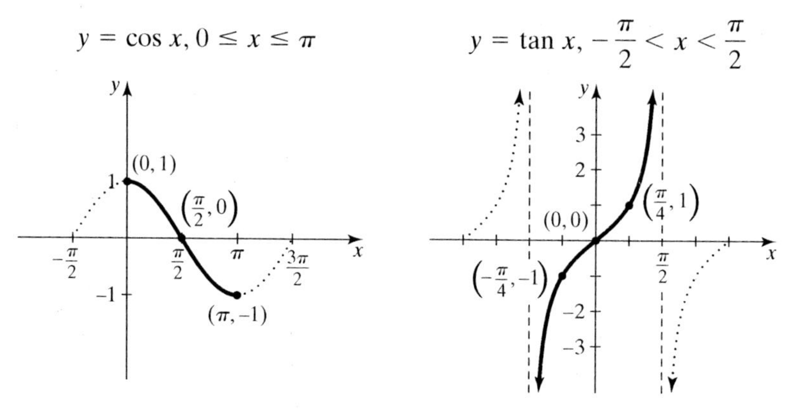

Using these domains, we now define and graph the inverse functions for both the cosine and the tangent.

Inverse Cosine and Inverse Tangent

The inverse cosine function $x = \cos y$ is defined as

$$y = \arccos x \quad \text{or} \quad y = \cos^{-1} x$$

for $-1 \le x \le 1$, and $0 \le y \le \pi$.

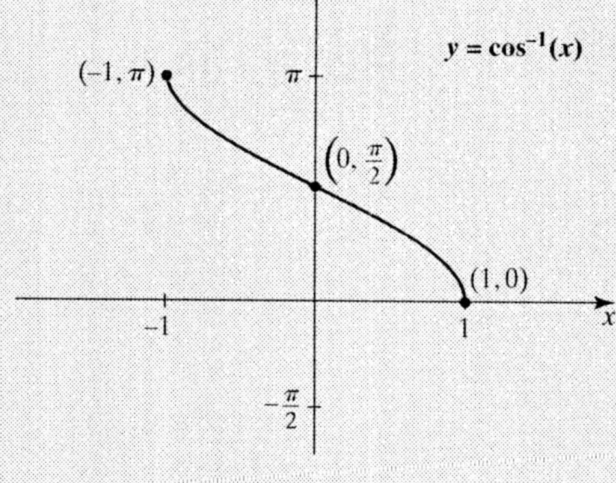

The inverse tangent function $x = \tan y$ is defined as

$$y = \arctan x \quad \text{or} \quad y = \tan^{-1} x$$

for $x \in \mathbb{R}$, and $-\frac{\pi}{2} < y < \frac{\pi}{2}$.

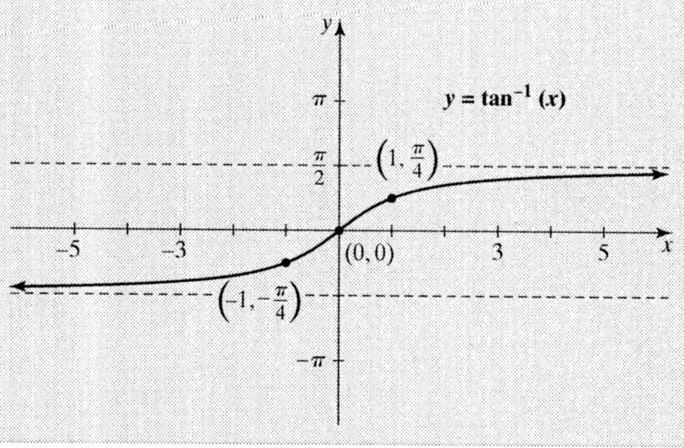

We see the graph of $y = \tan^{-1} x$ has horizontal asymptotes $y = \pi/2$, and $y = -\pi/2$.

The definitions for the inverse secant, inverse cosecant, and inverse cotangent are listed below along with the three inverse functions previously defined. There is not as much agreement on the ranges for these three new inverse functions, and they can be quite different in a calculus course or other texts. The inverse graphs are omitted and left as an exercise. The inverse functions are arranged to reveal the similarity of their ranges, which are demonstrated on the unit circle.

Inverses of the Circular Functions

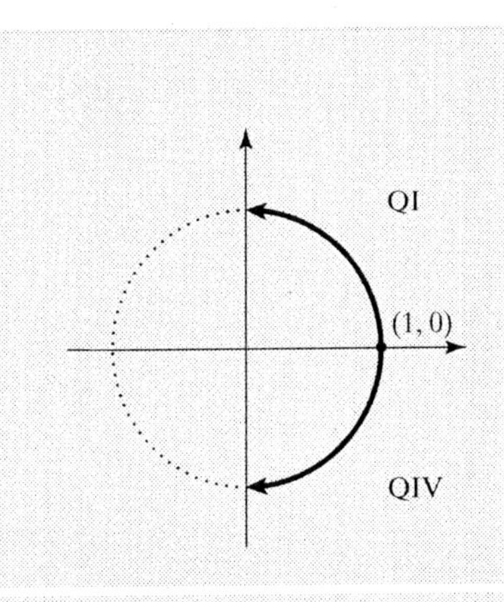

$y = \arcsin x \longleftrightarrow \sin y = x, -\dfrac{\pi}{2} \le y \le \dfrac{\pi}{2}$

$y = \sin^{-1} x$

$y = \operatorname{arccsc} x \longleftrightarrow \csc y = x, -\dfrac{\pi}{2} \le y < 0 \quad \text{or} \quad 0 < y \le \dfrac{\pi}{2}$

$y = \csc^{-1} x$

$y = \arctan x \longleftrightarrow \tan y = x, -\dfrac{\pi}{2} < y < \dfrac{\pi}{2}$

$y = \tan^{-1} x$

$y = \operatorname{arccot} x \longleftrightarrow \cot y = x, 0 < y < \pi$

$y = \cot^{-1} x$

$y = \arccos x \longleftrightarrow \cos y = x, 0 \le y \le \pi$

$y = \cos^{-1} x$

$y = \operatorname{arcsec} x \longleftrightarrow \sec y = x, 0 \le y < \dfrac{\pi}{2} \quad \text{or} \quad \dfrac{\pi}{2} < y \le \pi$

$y = \sec^{-1} x$

EXAMPLE 2 Find the exact inverse function values.

a. $\arctan \dfrac{\sqrt{3}}{3}$ **b.** $\cos^{-1}\left(-\dfrac{1}{2}\right)$ **c.** $\csc^{-1}(-\sqrt{2})$ **d.** $\operatorname{arccot}(-\sqrt{3})$

SOLUTION

a. $y = \arctan \dfrac{\sqrt{3}}{3} \longleftrightarrow \tan y = \dfrac{\sqrt{3}}{3}, -\dfrac{\pi}{2} < y < \dfrac{\pi}{2}$

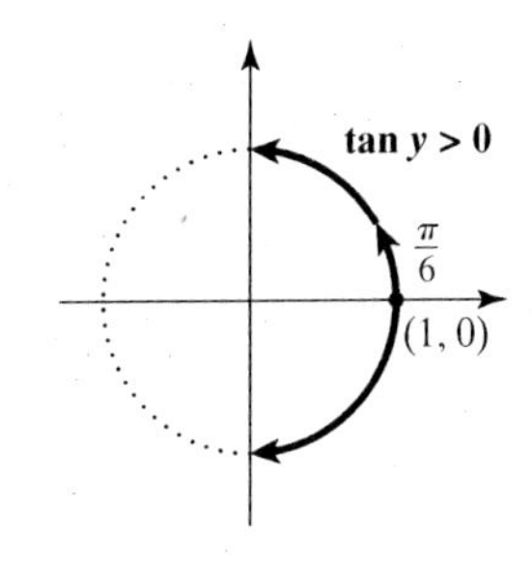

Since $\hat{y} = \pi/6$, the only arc with this tangent value in the restricted range is $y = \pi/6$. So,

$$\arctan \frac{\sqrt{3}}{3} = \frac{\pi}{6}.$$

b. $y = \cos^{-1}\left(-\frac{1}{2}\right) \longleftrightarrow \cos y = -\frac{1}{2}, 0 \le y \le \pi$
Since $\hat{y} = \pi/3$, the only arc with this cosine value in the restricted range is $y = \dfrac{2\pi}{3}$. So,

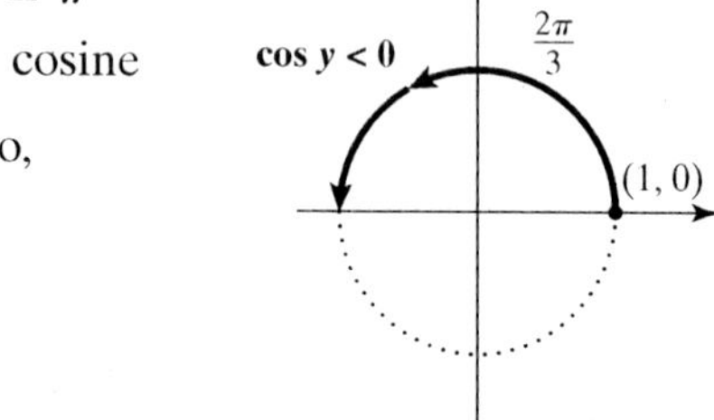

$$\cos^{-1}\left(-\frac{1}{2}\right) = \frac{2\pi}{3}.$$

c. $y = \csc^{-1}(-\sqrt{2}) \longleftrightarrow \csc y = (-\sqrt{2}), -\dfrac{\pi}{2} \le y < 0 \text{ or } 0 < y \le \dfrac{\pi}{2}$, so

$$\sin y = -\frac{1}{\sqrt{2}}$$

by using the reciprocal definitions.

Since $\hat{y} = \pi/4$, the only arc with this sine value in the restricted range is $y = -\pi/4$. So,

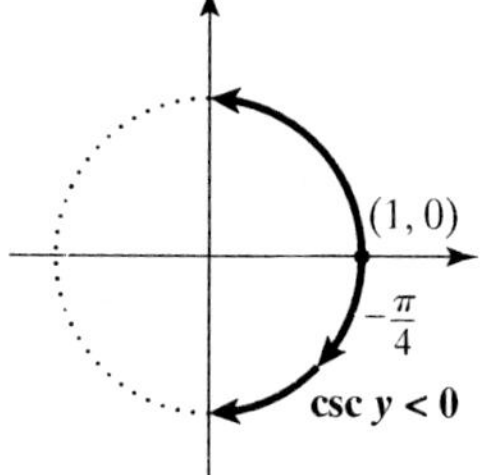

$$\csc^{-1}(-\sqrt{2}) = -\frac{\pi}{4}.$$

d. $y = \operatorname{arccot}(-\sqrt{3}) \longleftrightarrow \cot y = -\sqrt{3}, 0 < y < \pi$, or

$$\tan y = -\frac{1}{\sqrt{3}}.$$

Since $\hat{y} = \pi/6$, the only arc with this tangent value in the restricted range for the inverse cotangent is $y = 5\pi/6$. So,

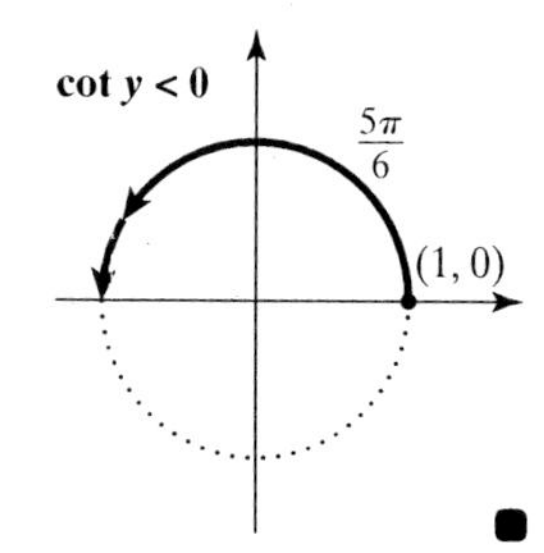

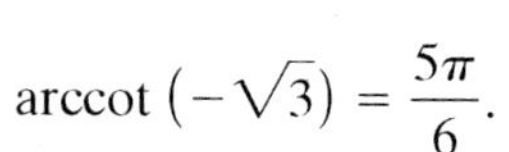

$$\operatorname{arccot}(-\sqrt{3}) = \frac{5\pi}{6}.$$

■

From Examples 2(c) and (d), we can see a particular type of relationship.

> The following relationships between the ***inverse of a function*** and the ***inverse of the related reciprocal function*** exist.
>
> $$\csc^{-1} x = \sin^{-1}\left(\frac{1}{x}\right) \qquad \sec^{-1}(x) = \cos^{-1}\left(\frac{1}{x}\right) \qquad \cot^{-1} x = \frac{\pi}{2} - \tan^{-1} x$$

The next examples give expressions that involve both the circular functions and the inverse circular functions. Let's see how to proceed.

EXAMPLE 3 Evaluate the following.

a. $\sin(\arctan(-1))$

b. $\cos^{-1}\left(\sin\dfrac{\pi}{3}\right)$

SOLUTION

a. For $\sin(\arctan(-1))$ we first evaluate the functional value in the parentheses. So we let $y = \arctan(-1)$:

$$y = \arctan(-1) \longleftrightarrow \tan y = -1, -\frac{\pi}{2} < y < \frac{\pi}{2}$$

Since $\hat{y} = \pi/4$, the only arc with this tangent value in the restricted interval is $y = -\pi/4$. Replacing the parentheses with its value, we have

$$\sin(\arctan(-1)) = \sin\left(-\frac{\pi}{4}\right) = -\frac{\sqrt{2}}{2}.$$

b. Evaluating the functional value inside the parentheses in $\cos^{-1}\left(\sin\frac{\pi}{3}\right)$, we get $\sin\frac{\pi}{3} = \frac{\sqrt{3}}{2}$.

Replacing the parentheses with its value, we get:

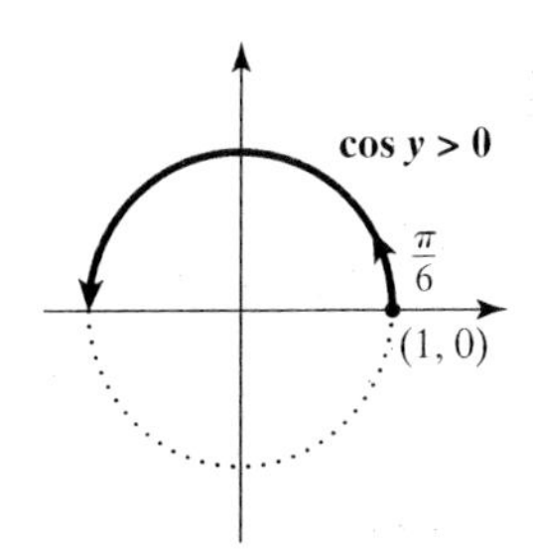

$$\cos^{-1}\left(\sin\frac{\pi}{3}\right) = \cos^{-1}\left(\frac{\sqrt{3}}{2}\right) \longleftrightarrow \cos y = \frac{\sqrt{3}}{2}, 0 \le y \le \pi$$

Since $\hat{y} = \pi/6$, the only arc with this cosine value in the restricted interval is $y = \pi/6$. Thus, we have

$$\cos^{-1}\left(\sin\frac{\pi}{3}\right) = \frac{\pi}{6}.$$

■

EXAMPLE 4 Evaluate $\sin(\cos^{-1}\frac{3}{5})$.

SOLUTION For $\sin(\cos^{-1}\frac{3}{5})$ we begin with evaluating the functional value inside the parentheses:

$$y = \cos^{-1}\tfrac{3}{5} \longleftrightarrow \cos y = \tfrac{3}{5}, 0 \le y \le \pi$$

We do not recognize the arc y that has the cosine value of $\frac{3}{5}$. Replacing y for $\cos^{-1}\frac{3}{5}$ in our original problem, we have

$$\sin(\cos^{-1}\tfrac{3}{5}) = \sin y.$$

Now we need to find $\sin y$, given that $\cos y = \frac{3}{5}, 0 \le y \le \pi$. We know from Chapter 1 that if we are given one of the functional values of an arc and the quadrant the arc is in, we can find any other functional value. *So in a much more abbreviated form, this is the same question*:

$$\sin(\cos^{-1}\tfrac{3}{5}) \longleftrightarrow \text{If } \cos y = \tfrac{3}{5}, 0 \le y \le \pi, \text{ find } \sin y.$$

$$\cos^2 y + \sin^2 y = 1 \qquad \text{Use the Pythagorean identity.}$$

$$\left(\frac{3}{5}\right)^2 + \sin^2 y = 1$$

$$\sin^2 y = \frac{16}{25}$$

$$\sin y = \pm\frac{4}{5}$$

(continued)

Since $0 \le y \le \pi$ indicates an arc in QI or QII, where $\sin y > 0$, we have $\sin y = \frac{4}{5}$.

So, $\sin\left(\cos^{-1}\frac{3}{5}\right) = \sin y = \frac{4}{5}$. ■

Using a Calculator for Inverse Functional Values

All the arcs involved in Examples 1–3 are either common arcs or their reference arcs are common. We can use a calculator to approximate these arcs and arcs that are not common. There are calculator keys for the inverse sine (SIN⁻¹), inverse cosine, and inverse tangent functions. Calculators are programmed with the same range restrictions for these inverse functions as indicated in this section. For the other inverses, you will need to apply the relationship between the inverse function and the inverse of the related reciprocal function.

EXAMPLE 5 Use a calculator (in radian mode) to approximate the inverse functional values to four decimal places.

a. $\cos^{-1} 0.65$ **b.** $\operatorname{arccsc}(-15.92)$ **c.** $\tan(\arcsin(-0.23))$

SOLUTION

A calculator gives these results.

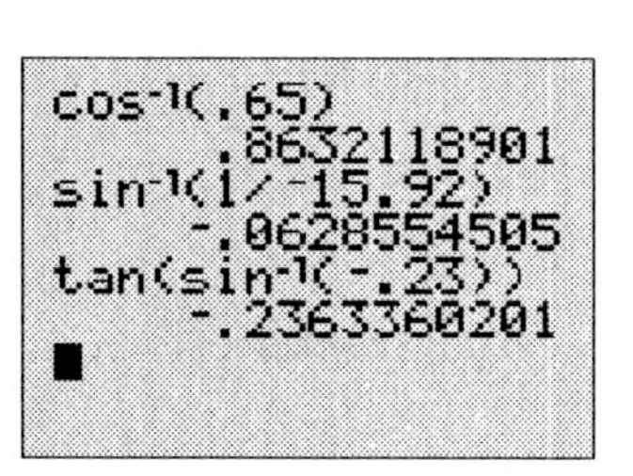

a. 0.8632 **b.** -0.0629 **c.** -0.2363 ■

By using inverse functions, we now have the ability to solve circular function equations. As an introduction to these equations, we assume that the specified arc restriction applies. We will solve trigonometric equations in greater detail in Chapter 5.

EXAMPLE 6 Solve the following equations for x.

a. $4\sin(3x) = 2$ **b.** $6\cos(5x) = 2$

SOLUTION

a. $4\sin(3x) = 2$

$$\sin(3x) = \frac{1}{2}$$ Multiply each side by $\frac{1}{4}$.

$$3x = \arcsin\left(\frac{1}{2}\right)$$ Use the definition of inverse functions to rewrite the equation.

$$3x = \frac{\pi}{6}$$ The **arc** whose **sine** value is $\frac{1}{2}$ is $\pi/6$.

$$x = \frac{\pi}{18}$$ Multiply each side by $\frac{1}{3}$.

SOLUTION

b. $6\cos(5x) = 2$

$$\cos(5x) = \frac{1}{3}$$ Multiply each side by $\frac{1}{6}$.

$$5x = \cos^{-1}\left(\frac{1}{3}\right)$$ Use the definition of inverse functions to rewrite the equation.

$$x = \frac{1}{5}\cos^{-1}\left(\frac{1}{3}\right)$$ Multiply each side by $\frac{1}{5}$.

$$x = \frac{1}{5}\cos^{-1}\left(\frac{1}{3}\right)$$ Leave your answer exact, or find a calculator approximation since $\frac{1}{3}$ is not a recognizable cosine value.

or $x \approx 0.2462$ ■

This section requires patience and a lot of memory recall. Until you are more familiar with inverse functions, you are advised to rewrite the problem without using inverse notation. Working through this section makes you realize how important it is to know the material from Chapter 1, as well as the graphs from the beginning of this chapter. Whenever you are having difficulty with the exercises, you are again advised to go back and review Chapter 1 along with rereading this section.

Exercise Set 2.5

1–6. *Rewrite each expression to eliminate the inverse notation. Then find the exact value for y in the restricted range without using a calculator.*

EXAMPLE $y = \tan^{-1} 0$

SOLUTION $y = \tan^{-1} 0 \longleftrightarrow \tan y = 0, \ -\frac{\pi}{2} < y < \frac{\pi}{2}$

$y = 0$ ■

1. $y = \arcsin\frac{\sqrt{3}}{2}$ **2.** $y = \arctan\sqrt{3}$ **3.** $y = \cos^{-1} 1$

4. $y = \sin^{-1}(-1)$ **5.** $y = \operatorname{arccsc}(-\sqrt{2})$ **6.** $y = \sec^{-1}\frac{2}{\sqrt{3}}$

7–20. *Find the exact value for each expression without using a calculator.*

7. $\tan^{-1} 1$ **8.** $\cos^{-1}\left(-\frac{\sqrt{2}}{2}\right)$ **9.** $\arccos\frac{1}{2}$

10. $\arcsin\left(-\frac{1}{2}\right)$ **11.** $\arcsin 0$ **12.** $\arccos 0$

13. $\sec^{-1}(-1)$ **14.** $\sin^{-1} 1$ **15.** $\cot^{-1} 1$

16. $\cot^{-1}\left(\frac{\sqrt{3}}{3}\right)$ **17.** $\arctan(-\sqrt{3})$ **18.** $\operatorname{arccsc} 2$

19. $\cos^{-1}\left(-\frac{\sqrt{3}}{2}\right)$ **20.** $\tan^{-1}\left(-\frac{\sqrt{3}}{3}\right)$

21–34. *Use a calculator to find an approximation rounded to four decimal places for each expression.*

21. $\arccos\left(-\frac{2}{3}\right)$ **22.** $\arcsin 0.6$ **23.** $\tan^{-1}\frac{4}{3}$

24. $\cos^{-1}\left(-\frac{3}{4}\right)$ **25.** $\sec^{-1}(-1.23)$ **26.** $\csc^{-1}(-22)$

27. $\operatorname{arccsc}\frac{13}{5}$ **28.** $\arctan\sqrt{7}$ **29.** $\arcsin 0.6445$

30. $\arccos(-0.2356)$ **31.** $\tan^{-1}(-5.967)$ **32.** $\sec^{-1} 55.6$

33. $\cos^{-1}\sqrt{\frac{2}{33}}$ **34.** $\cot^{-1} 0.71$

35–60. *Find the exact value for each expression without using a calculator.*

35. $\sin\left(\cos^{-1}\frac{1}{2}\right)$ **36.** $\cos\left(\sin^{-1}\frac{1}{2}\right)$ **37.** $\cos(\cos^{-1} 1)$

38. $\sin(\sin^{-1} 0)$ **39.** $\tan(\arctan\sqrt{3})$ **40.** $\sin(\tan^{-1}(-1))$

41. $\cos(\arcsin 1)$ **42.** $\cos(\arctan 0)$ **43.** $\csc\left(\cos^{-1}\left(-\frac{\sqrt{3}}{2}\right)\right)$

44. $\tan\left(\sin^{-1}\left(-\frac{\sqrt{2}}{2}\right)\right)$ **45.** $\cos\left(\arcsin\left(-\frac{\sqrt{2}}{2}\right)\right)$ **46.** $\sec\left(\cos^{-1}\left(-\frac{\sqrt{2}}{2}\right)\right)$

47. $\cot\left(\cos^{-1}\left(-\frac{1}{2}\right)\right)$ **48.** $\csc(\cos^{-1} 0)$ **49.** $\sin^{-1}\left(\cos\frac{7\pi}{6}\right)$

50. $\sin^{-1}\left(\tan\frac{3\pi}{4}\right)$ **51.** $\arctan\left(\cos\frac{\pi}{2}\right)$ **52.** $\operatorname{arccsc}(\cos\pi)$

53. $\sin^{-1}\left(\cos\left(-\frac{\pi}{3}\right)\right)$ **54.** $\tan^{-1}\left(\sin\frac{7\pi}{2}\right)$ **55.** $\operatorname{arcsec}\left(\tan\frac{3\pi}{4}\right)$

56. $\operatorname{arccot}\left(\sin\frac{3\pi}{2}\right)$ **57.** $\sin\left(\cos^{-1}\frac{12}{13}\right)$ **58.** $\cos\left(\sin^{-1}\frac{3}{5}\right)$

59. $\cos\left(\sin^{-1}\frac{8}{17}\right)$ **60.** $\sin\left(\arccos\frac{7}{25}\right)$

61–70. *Use a calculator to find an approximation to four decimal places for each expression.*

61. $\tan(\sin^{-1} 0.2657)$ **62.** $\sin(\tan^{-1} 55.67)$ **63.** $\sin(\cos^{-1} 0.4675)$

64. $\cos(\sin^{-1}(-0.7652))$ **65.** $\arctan(\sin 3.95)$ **66.** $\arccos(\tan 47.9)$

67. $\csc\left(\arcsin\left(-\frac{5}{9}\right)\right)$ **68.** $\sec\left(\arccos\left(-\frac{2}{15}\right)\right)$ **69.** $\sec^{-1}\left(\tan\frac{\sqrt{7}}{2}\right)$

70. $\operatorname{arccot}\left(\cos\left(-\frac{\sqrt{3}}{2}\right)\right)$

71. Identify the graph of the circular function on a restricted domain in parts (a), (b), and (c). Then match each graph with the graph of its corresponding inverse function in parts (i), (ii), and (iii).

a.

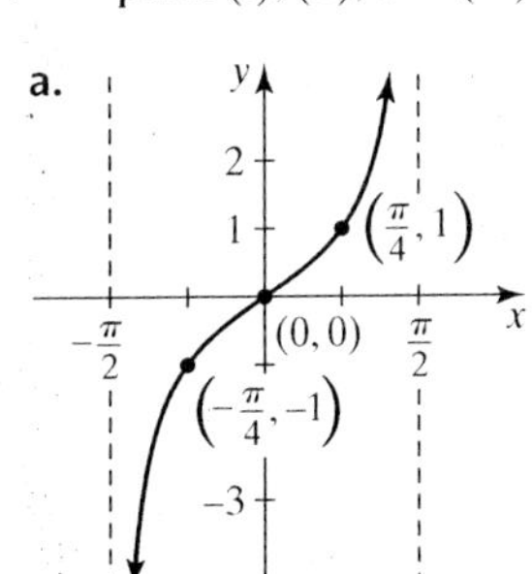

b.

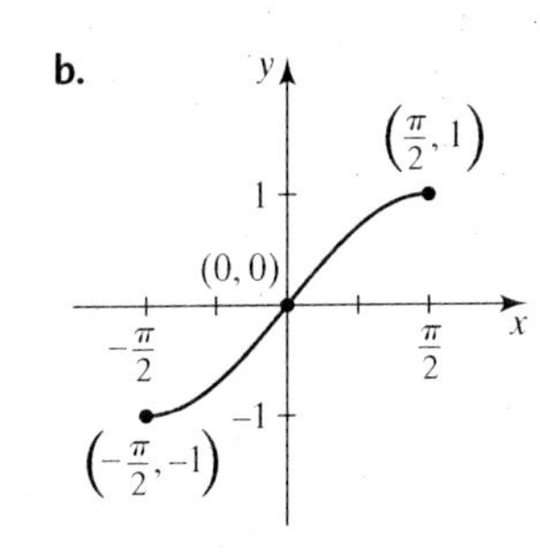

c.

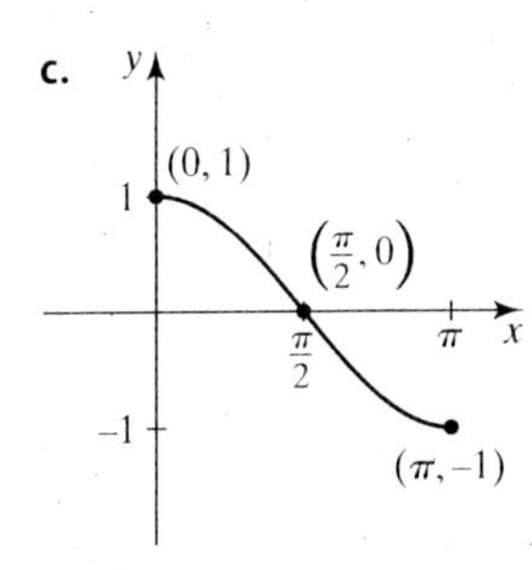

i.

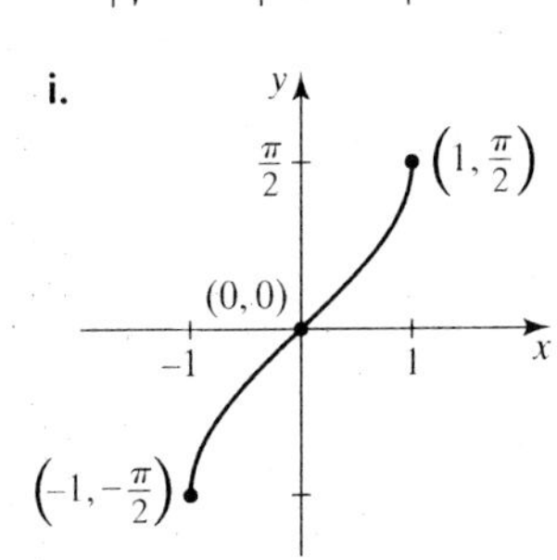

ii.

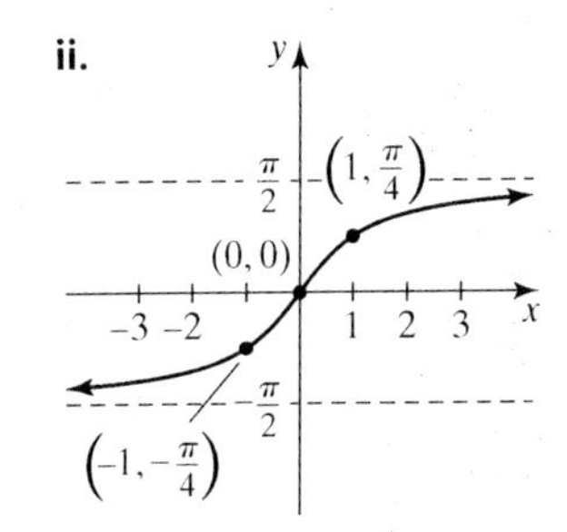

iii.

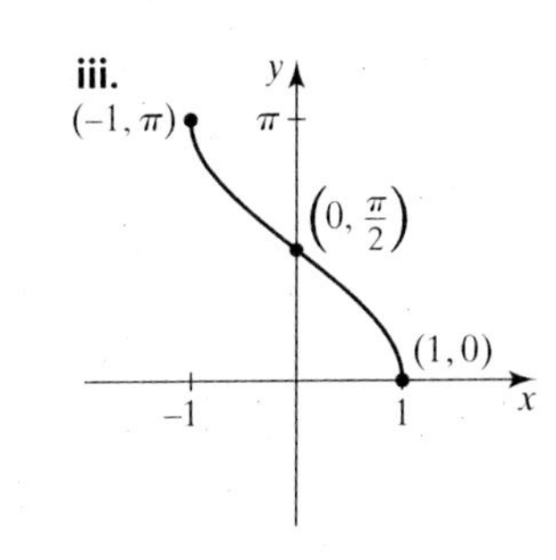

72. On the basis of the given graphs for the cotangent, secant, and cosecant functions on the specified domains, identify the graph of $y = \operatorname{arccot} x$, $y = \operatorname{arccsc} x$, and $y = \operatorname{arcsec} x$.

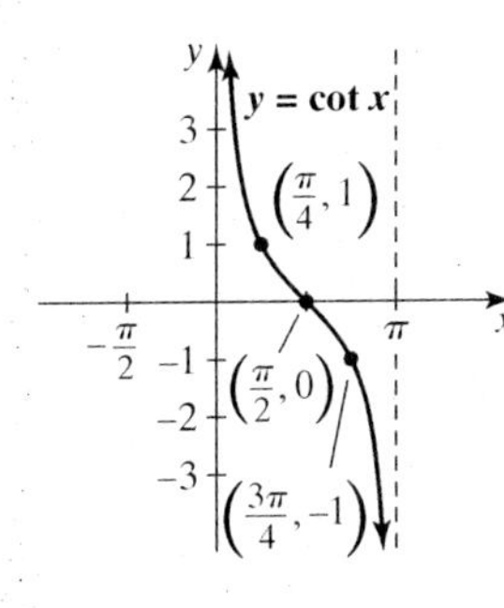

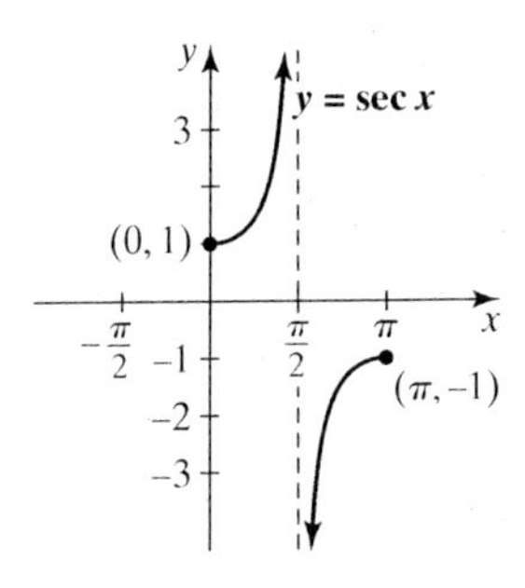

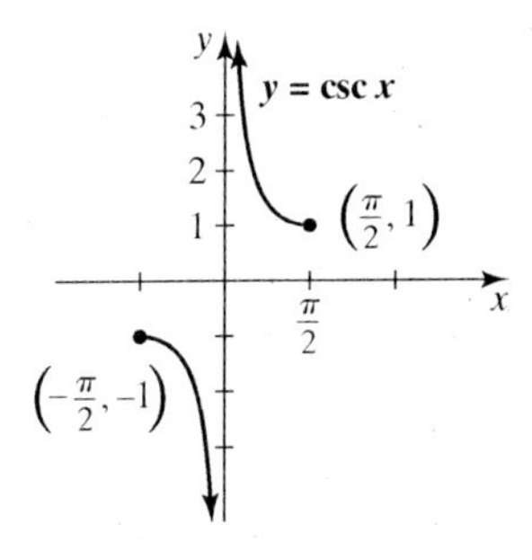

a.

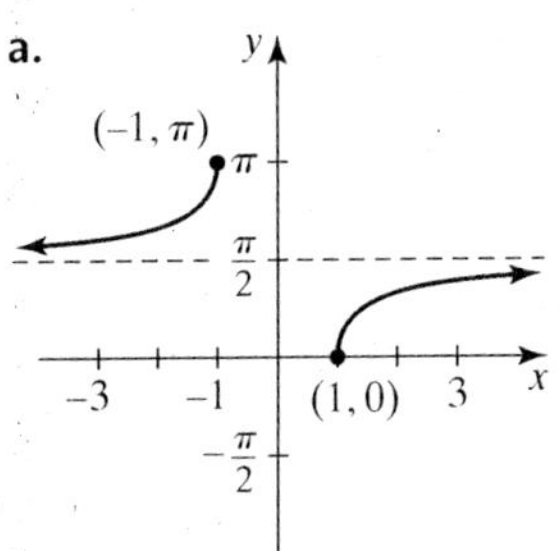

b.

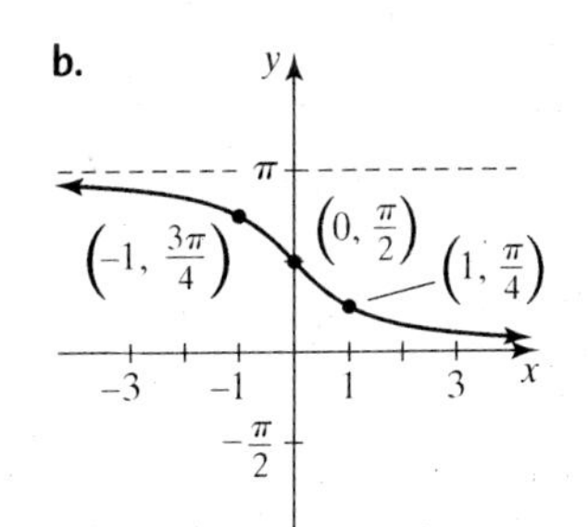

c.

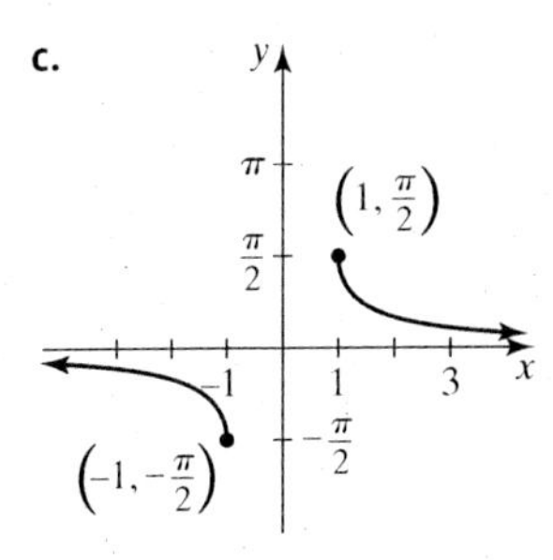

73–76. *Solve the following equations for x. Assume the specified restrictions on the arc, and approximate solutions (rounded to four decimal places) wherever necessary.*

73. $5\sin(6x) = -5$

74. $4\cos(3x) = 2\sqrt{3}$

75. $4\tan(9x) = 7$

76. $-2\tan\left(\frac{x}{2}\right) = 5$

77. Explain the difference between the functions $\sin^{-1} x$ and $(\sin x)^{-1}$, and give another name for both $\sin^{-1} x$ and $(\sin x)^{-1}$.

Discussion

78. Larry is unable to find an answer (a real number solution) to $\arcsin \frac{3}{2}$.
   a. What do you know about the domain of the inverse sine function that would explain his difficulty?
   b. Larry is still having trouble understanding why he cannot find an answer to this question. What are some other ways of helping Larry to understand why there is no arc whose sine value is $\frac{3}{2}$?
   c. Without the use of a calculator, determine which of these problems do not have real solutions.

$$\cos^{-1} 5 \qquad \sec^{-1}\frac{1}{2} \qquad \sin^{-1}(-2)$$

   d. Using a calculator, repeat part (c). Many calculators will display "error," "domain error," "input error," and so on, whereas others will display a complex (not real) answer. Make sure you understand the specific output of your calculator.

79. Tony is checking his homework and discovers that many of his answers do not agree with the solutions manual. Below is his work for Exercise 4.

$$\sin^{-1}(-1) = y \longleftrightarrow \sin y = -1$$
$$y = \frac{3\pi}{2}$$

The correct answer is $y = \pi/2$, but Tony says that his answer is also correct because he knows that $\sin \frac{3\pi}{2} = -1$.

   a. How many answers to Exercise 4 are possible?
   b. Is Tony correct when he says that $\sin 3\pi/2 = -1$? If so, what is wrong with Tony's answer?

Graphing Calculator Exploration

80. Following are the graphs of $y = \sec x$, $y = \csc x$, and $y = \cot x$ with the specified restricted domains.

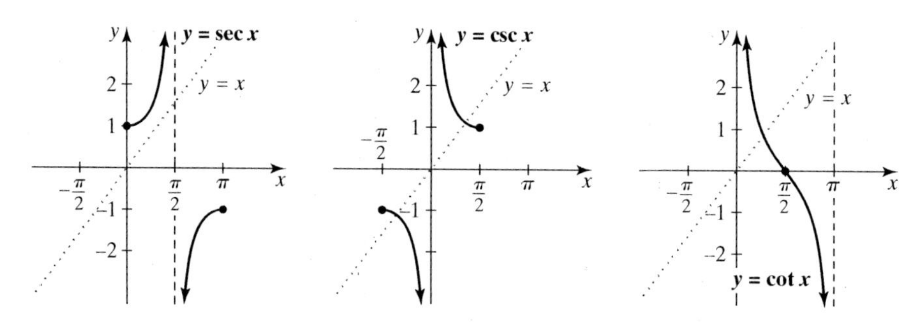

The functions $y = \cot^{-1} x$, $y = \sec^{-1} x$, and $y = \csc^{-1} x$ are not found on a graphing calculator, but they can be graphed using the relationships mentioned in this section and restated here in parts (a), (b) and (c). Obtain the graphs of these inverse functions using the following information. The graphs of the inverses should be the same as the ones *you* could find by reflecting the above graphs of the one-to-one functions about the line $y = x$.

a. For $y = \sec^{-1} x$, graph $y = \cos^{-1}(1/x)$ using Xmin $= -4$, Xmax $= 4$, Xscl $= 1$, Ymin $= -1$, Ymax $= 5$.

b. For $y = \csc^{-1} x$, graph $y = \sin^{-1}(1/x)$ using Xmin $= -4$, Xmax $= 4$, Xscl $= 1$, Ymin $= -2$, Ymax $= 2$.

c. For $y = \cot^{-1} x$, graph $y = \pi/2 - \tan^{-1} x$ using Xmin $= -4$, Xmax $= 4$, Xscl $= 1$, Ymin $= -1$, Ymax $= 5$.

Compare the graph of each function with the graph of its inverse function to determine whether the following statements are true or false:

d. If the function is increasing on its domain, then the inverse function is increasing on its domain.

e. If the function is decreasing on its domain, then the inverse function is decreasing on its domain.

Chapter 2 Summary

2.1 Graphs of the sine or cosine functions are called sinusoidal waves. There are several characteristics of the pure form of these graphs. The amplitude of the sinusoidal wave is found by taking one-half the absolute value of the difference between the maximum and minimum y-coordinates:

$$\frac{|y\max - y\min|}{2}.$$

The graph of one period is called a cycle. Each cycle of the sinusoidal functions has four identically shaped arc sections. Each arc represents one-fourth of the period, and each arc occurs over an interval along the x-axis that is one-fourth the period in length. The endpoints of these arc sections are called critical points. The graphs of the sine and cosine are continuous.

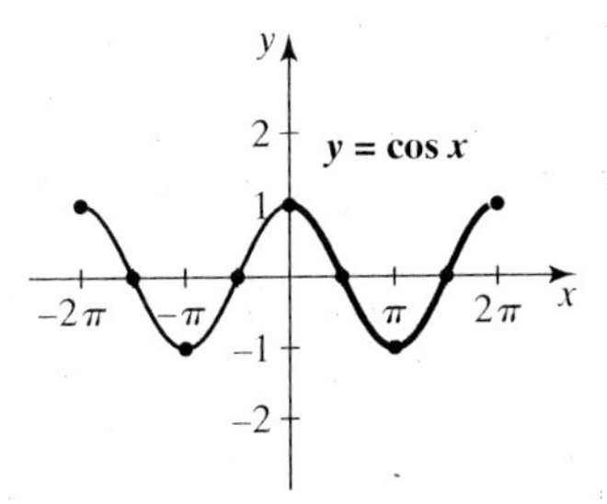

Domain: $x \in \mathbb{R}$
Range: $|y| \le 1$
Amplitude: 1
Period: 2π

x-intercepts: $x = \frac{1}{2}\pi + k\pi$

y-intercept: $(0, 1)$
Symmetry: y-axis

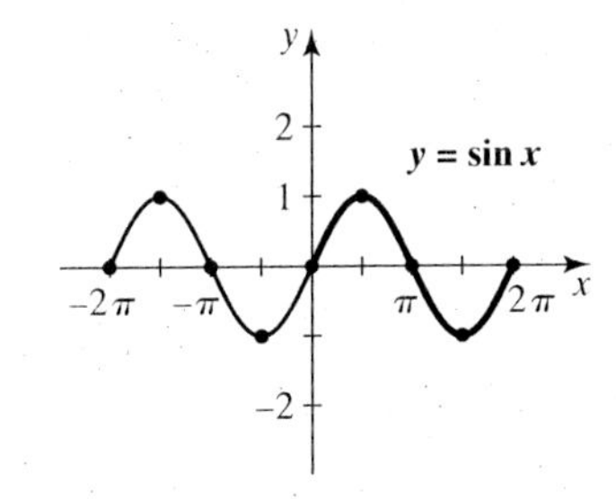

Domain: $x \in \mathbb{R}$
Range: $|y| \le 1$
Amplitude: 1
Period: 2π

x-intercepts: $x = k\pi$

y-intercept: $(0, 0)$
Symmetry: origin

Amplitude of a Sinusoidal Wave

For $y = A\cos x$ and $y = A\sin x$, amplitude $= |A|$.

$|A| > 1$ produces a vertical stretch.
$|A| < 1$ produces a vertical shrink.
$A < 0$ produces a reflection across the x-axis.

Equations, in the form $y = A\cos x$ or $y = A\sin x$, can be determined from their graphs.

2.2

Functions defined by equations of the form

$$y = A\cos[B(x - C)] + D \quad \text{or} \quad y = A\sin[B(x - C)] + D,$$

where $A, B, C, D, x, y \in \mathbb{R}$ $(A, B \neq 0)$, vary the pure form of the sine and cosine function graphs in that they produce sinusoidal waves with these changes.

- Amplitude: $|A|$
- Period: $\dfrac{\text{pure period}}{|B|} = \dfrac{2\pi}{|B|}$
- Phase Shift: $|C|$ units to the right if $C > 0$
  $|C|$ units to the left if $C < 0$
- Vertical Shift: $|D|$ units up if $D > 0$
  $|D|$ units down if $D < 0$

With the preceding variations, the sinusoidal functions continue to have four identically shaped arc sections for each period with each arc representing one-fourth the period, and occurring over an interval along the x-axis that is one-fourth the period in length, and critical points at the ends of each arc section. The range and x-intercepts of the function change with different values of A, B, C, and D.

All graphs can be done on a graphing calculator or other graphing utility. You need to know how to find the period and range in order to input this information to display the requested number of cycles.

Several equations can represent the same graph.

2.3

The frequency F of an object moving in simple harmonic motion is the number of periods of the motion per unit of time.

$$F = \frac{1}{P} = \frac{B}{2\pi},$$

where P is the period of the motion.

Many functions that are periodic can be modeled by equations of the form

$$y = A\cos[B(x - C)] + D \text{ or } y = A\sin[B(x - C)] + D,$$

where $A, B, C, D, x, y \in \mathbb{R}$ $(A, B \neq 0)$.

If data is periodic and sinusoidal in nature, one can also find a sinusoidal function that would model the data in either of these forms. An equation modeling sinusoidal data can be found using the sine regression feature of a graphing calculator. The equation will be in the form

$$y = a\sin(bx + c) + d.$$

2.4

The graphs of the circular functions defined by $y = \tan x$, $y = \cot x$, $y = \sec x$, and $y = \csc x$ also exhibit the periodic nature of the functions.

Function	Domain	Range	Period	x-intercepts	y-intercept	Symmetry
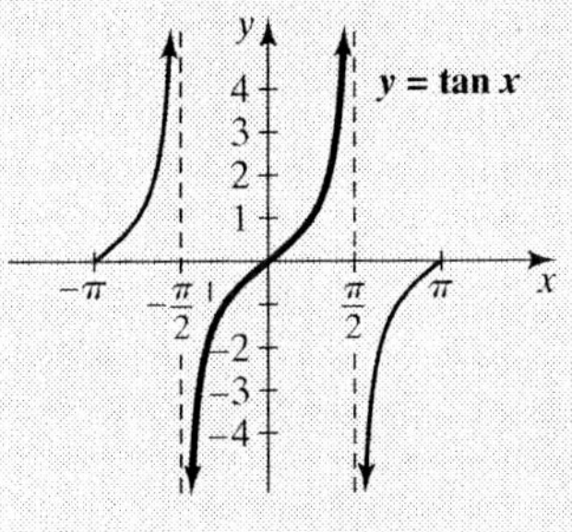 $y = \tan x$	$x \in \mathbb{R}$, $x \neq \frac{\pi}{2} + k\pi$	$y \in \mathbb{R}$	π	$x = k\pi$	$(0, 0)$	origin
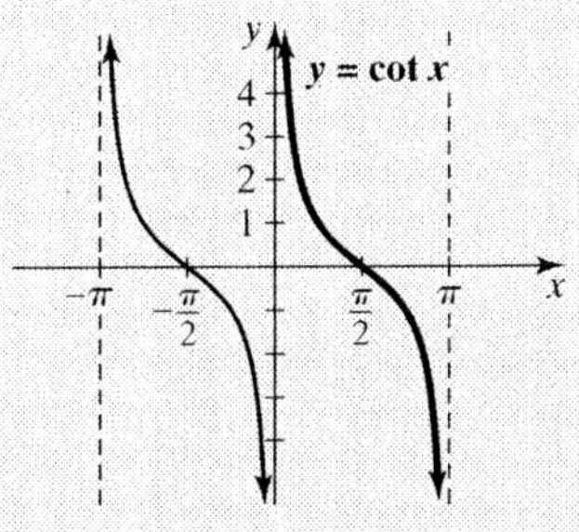 $y = \cot x$	$x \in \mathbb{R}$, $x \neq k\pi$	$y \in \mathbb{R}$	π	$x = \frac{\pi}{2} + k\pi$	none	origin
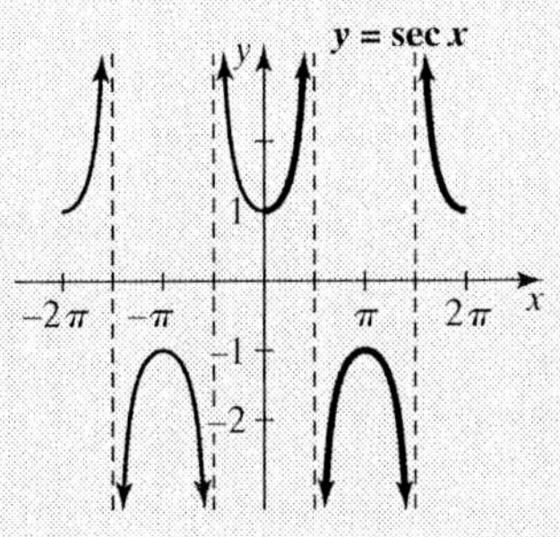 $y = \sec x$	$x \in \mathbb{R}$, $x \neq \frac{\pi}{2} + k\pi$	$\lvert y \rvert \geq 1$	2π	none	$(0, 1)$	y-axis
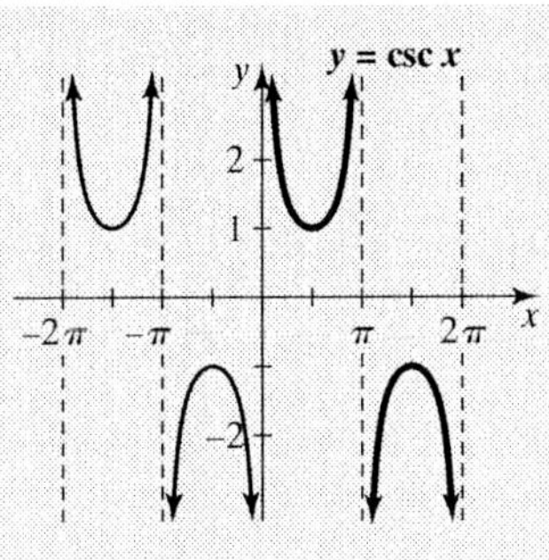 $y = \csc x$	$x \in \mathbb{R}$, $x \neq k\pi$	$\lvert y \rvert \geq 1$	2π	none	none	origin

The functions $y = \tan x$, $y = \cot x$, $y = \sec x$, and $y = \csc x$ are not sinusoidal and do not have amplitudes. The tangent and cotangent graphs have two equal arc sections, whereas the secant and cosecant have four equal arc sections for each period. Each graph has asymptotes at x values for which the function is not defined, and each function is continuous on its domain. The pure form of these functions also can be affected by A, B, C, and D.

2.5

We restrict the domain of each of the circular functions to an interval where the function is one-to-one so that the inverse relation is a function. The functional values of arccos x, arcsin x, arctan x, and so on, can be written in an equivalent form without the inverse notation. The ranges of the inverse functions are included in their definitions.

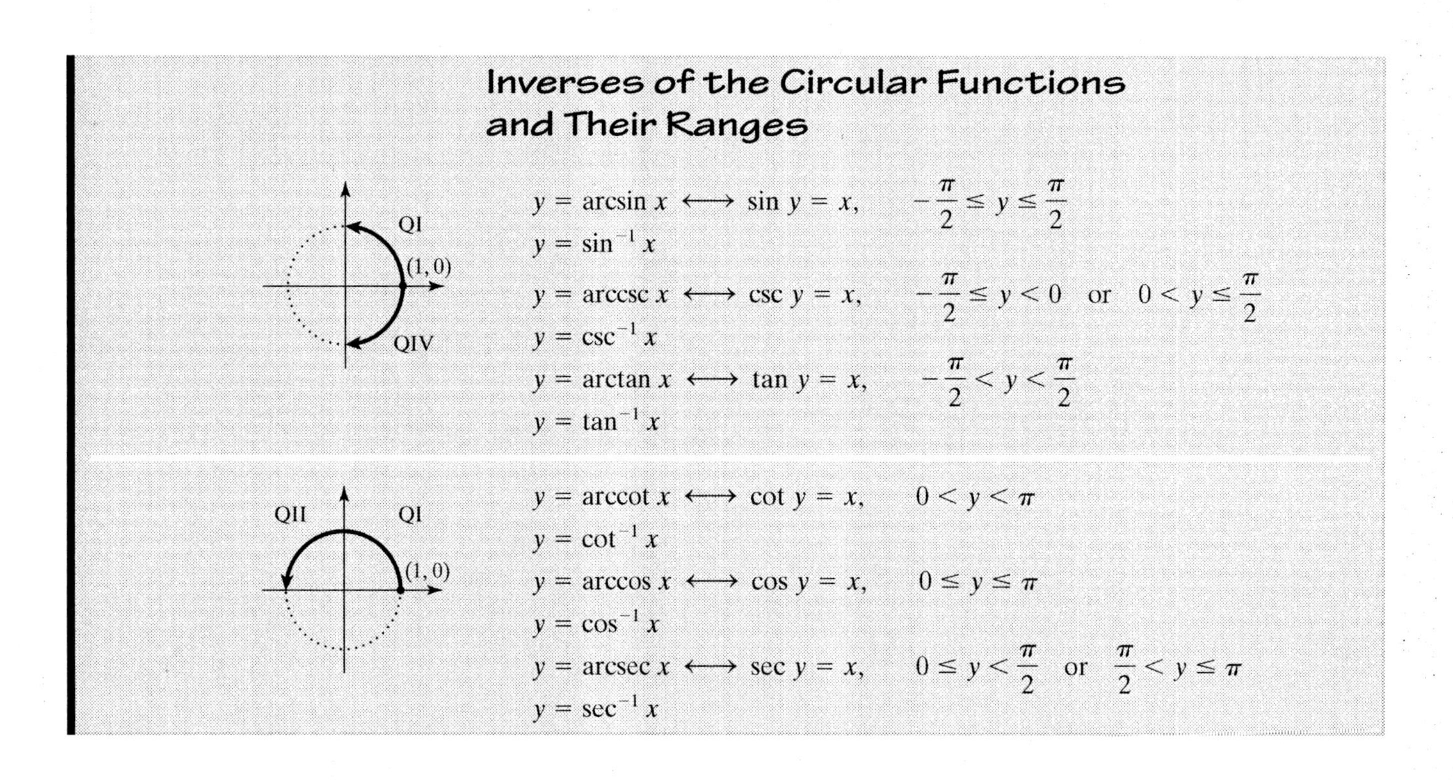

Inverses of the Circular Functions and Their Ranges

$y = \arcsin x \longleftrightarrow \sin y = x, \quad -\frac{\pi}{2} \le y \le \frac{\pi}{2}$

$y = \sin^{-1} x$

$y = \operatorname{arccsc} x \longleftrightarrow \csc y = x, \quad -\frac{\pi}{2} \le y < 0 \text{ or } 0 < y \le \frac{\pi}{2}$

$y = \csc^{-1} x$

$y = \arctan x \longleftrightarrow \tan y = x, \quad -\frac{\pi}{2} < y < \frac{\pi}{2}$

$y = \tan^{-1} x$

$y = \operatorname{arccot} x \longleftrightarrow \cot y = x, \quad 0 < y < \pi$

$y = \cot^{-1} x$

$y = \arccos x \longleftrightarrow \cos y = x, \quad 0 \le y \le \pi$

$y = \cos^{-1} x$

$y = \operatorname{arcsec} x \longleftrightarrow \sec y = x, \quad 0 \le y < \frac{\pi}{2} \text{ or } \frac{\pi}{2} < y \le \pi$

$y = \sec^{-1} x$

The following are the graphs of three inverse circular functions:

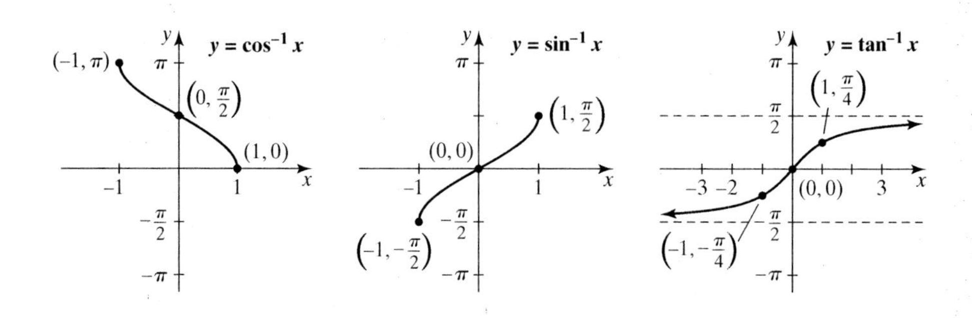

Chapter 2 Review Exercises

2.1 **1–6.** **a.** *Without using a graphing utility, graph each equation on* $-2\pi \le x \le 2\pi$.

b. *Indicate the amplitude.*

1. $y = 2 \sin x$ **2.** $y = -3 \cos x$ **3.** $y = -0.7 \cos x$

4. $y = \frac{1}{2} \sin x$ **5.** $y = -\frac{3}{2} \sin x$ **6.** $y = -2\pi \cos x$

2.2 **7–12.** **a.** *Without using a graphing utility, graph two cycles for the given function.*

b. *Label the critical points for one cycle.*

c. *State the period, range, and x-intercepts.*

7. $y = 4 \cos\left(\frac{1}{2}x\right)$ **8.** $y = 1.5 \sin(2x)$ **9.** $y = -3 \sin\left(-\frac{1}{4}x\right)$

10. $y = -2 \cos\left(-\frac{1}{2}x\right)$ **11.** $y = -\cos(-\pi x)$ **12.** $y = 5 \sin\left(-\frac{2\pi}{5}x\right)$

13–18. **a.** *Without using a graphing utility, graph two cycles for the given function.*

b. *State the period, phase shift, and vertical shift if applicable.*

13. $y = \cos\left(x - \frac{\pi}{2}\right)$ **14.** $y = -\sin\left(x + \frac{\pi}{2}\right)$

15. $y = -4 \sin\left[\frac{1}{2}(x + \pi)\right]$ **16.** $y = 2.5 \cos\left[2\left(x - \frac{\pi}{4}\right)\right]$

17. $y = 2 \cos\left(\frac{1}{3}x\right) + 4$ **18.** $y = \frac{1}{2}\sin(2x) - \frac{1}{2}$

19–22. *Identify each graph with an equation in the form* $y = A \sin x$ *or* $y = A \cos x$. *Check your answer with a graphing utility.*

19.

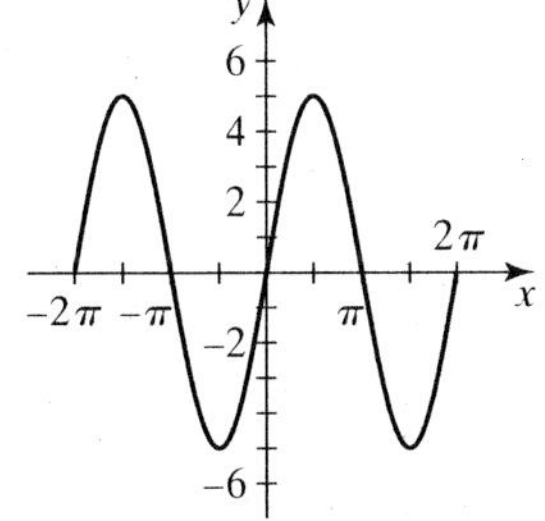

20.

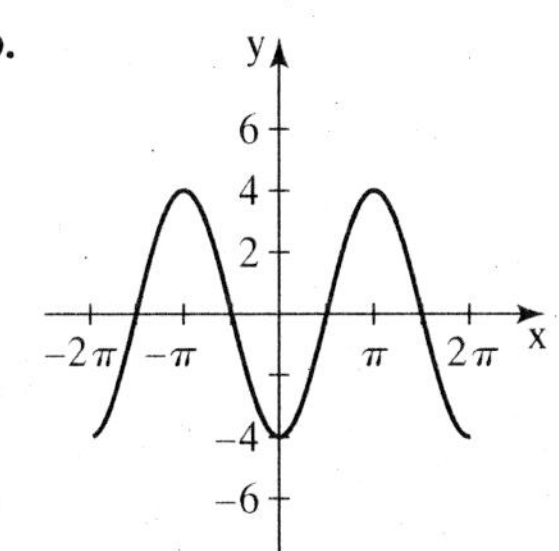

21.

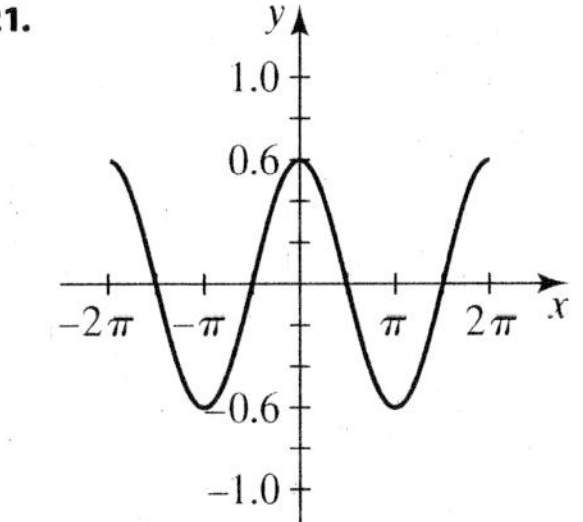

22.

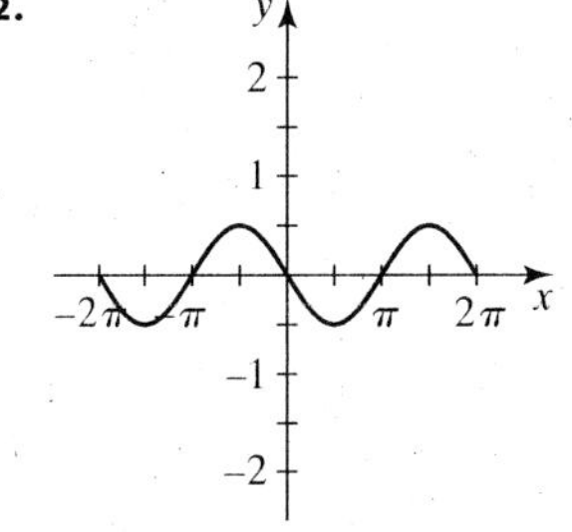

23–24. *Identify each graph with an equation in the form* $y = A \sin B(x - C)$ *or* $y = A \cos B(x - C)$, $B > 0$, $C \neq 0$. *Check your answer with a graphing calculator.*

23.

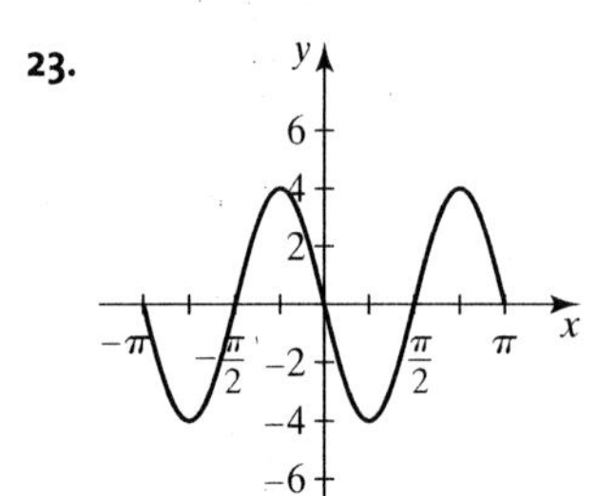

24.

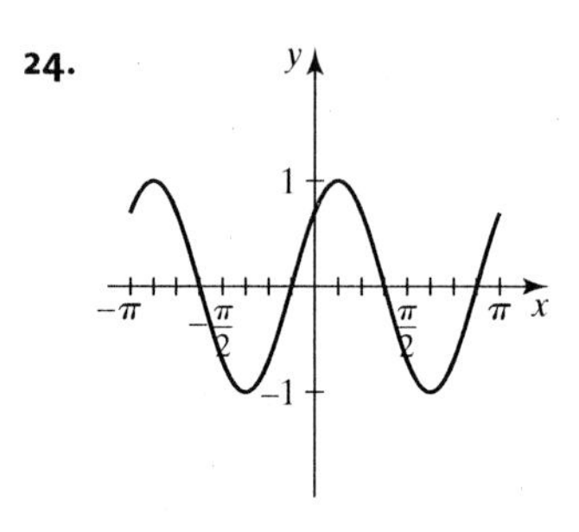

25. Find another equation for Exercise 23 . Check your answer with a graphing calculator.

26. Find another equation for Exercise 24. Check your answer with a graphing calculator.

27. The vertical displacement of a point on a weighted spring is given by the equation $d(t) = 4 \cos(\pi t)$, where t is in seconds and $d(t)$ is in inches.

a. Graph this equation for $0 \leq t \leq 4$.

b. If we consider the rest position to be when the displacement $d(t) = 0$, find the times the point is in the rest position.

c. Indicate on the graph the times when the displacement is 2 inches from the rest position.

28. The position of a pendulum, oscillating with maximum horizontal displacement of 4.5 inches where x is in seconds, is given by the formula $y = 4.5 \sin\left(\frac{1}{2}x\right)$.

a. Graph the equation for two cycles starting with $x = 0$.

b. Indicate on the graph each time the pendulum is displaced -3 inches from the rest position.

2.3

29. The table lists the number of daylight hours (to the nearest quarter hour) on the 21st day of the month (January 21 = 1) for San Francisco, California.

1	2	3	4	5	6	7	8	9	10	11	12
10	11	$12\frac{1}{4}$	$13\frac{1}{2}$	$14\frac{1}{2}$	$14\frac{3}{4}$	$14\frac{1}{2}$	$13\frac{1}{2}$	$12\frac{1}{4}$	11	10	$9\frac{1}{2}$

a. Find a sinusoidal function to model the data.

b. Use a graphing calculator to create a scatter plot.

c. Find the sine regression equation.

30. The table lists the number of daylight hours (to the nearest quarter hour) on the first of the month (January 1 = 1) for Edinburgh, Scotland.

1	2	3	4	5	6	7	8	9	10	11	12
7	$8\frac{1}{2}$	$10\frac{3}{4}$	$13\frac{1}{4}$	$15\frac{1}{4}$	$17\frac{1}{4}$	$17\frac{1}{2}$	16	$13\frac{3}{4}$	$11\frac{1}{2}$	$9\frac{1}{4}$	$7\frac{1}{2}$

a. Find a sinusoidal function to model the data.

b. Use a graphing calculator to create a scatter plot.

c. Find the sine regression equation.

2.4

31–38. **a.** *Without using a graphing utility, graph at least two cycles for the given function.*
b. *Indicate the range and period.*

31. $y = 3\cot\left(\frac{1}{3}x\right)$ **32.** $y = -\tan(2x)$ **33.** $y = \tan\left[\frac{1}{4}(x - \pi)\right]$

34. $y = \cot\left[\frac{1}{2}(x + \pi)\right]$ **35.** $y = -\frac{1}{2}\csc\left(\frac{1}{2}x\right)$ **36.** $y = 3\sec(2x + \pi)$

37. $y = \sec(2x) + 4$ **38.** $y = 2\csc x - 2$

39–42. *Find an equation for each circular function graph. Check your answer with a grapher.*

39.

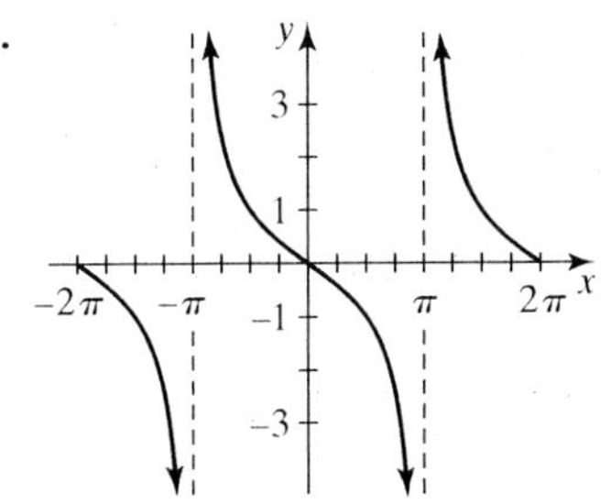

40.

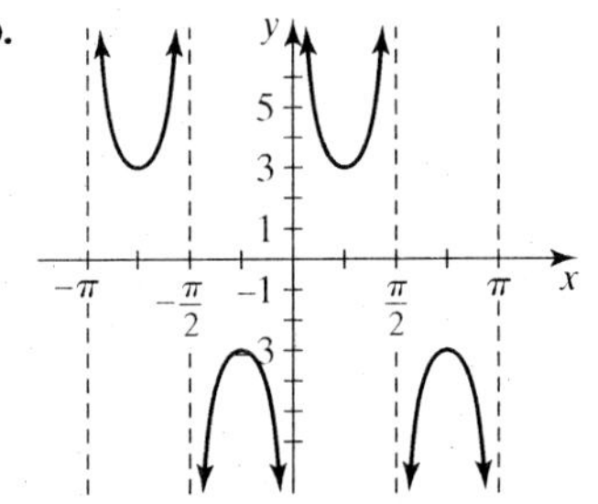

41.

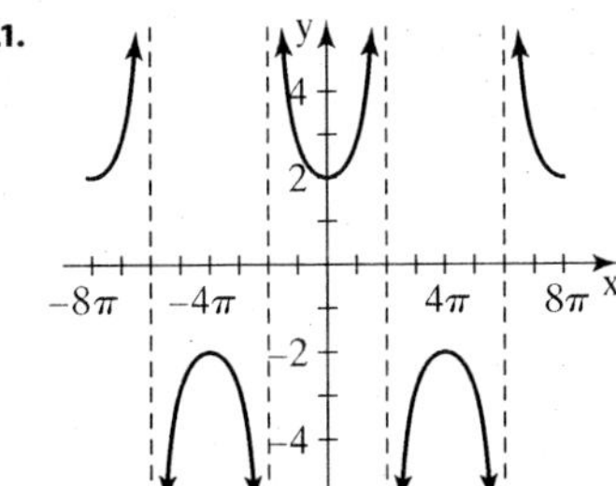

42.

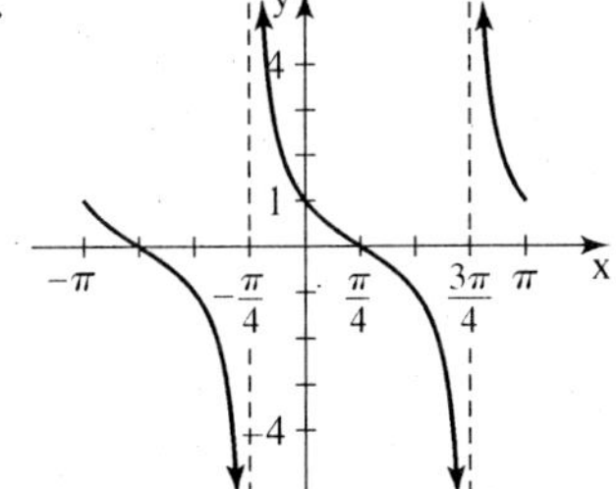

2.5

43–48. *Rewrite each equation without inverse notation. Then find the exact value of y.*

43. $y = \arcsin\dfrac{\sqrt{2}}{2}$ **44.** $y = \arccos(-1)$ **45.** $y = \tan^{-1}(-\sqrt{3})$

46. $y = \csc^{-1}\left(-\dfrac{2}{\sqrt{3}}\right)$ **47.** $y = \sec^{-1}(-\sqrt{2})$ **48.** $y = \cot^{-1}0$

49–56. *Without the use of a calculator, find the exact value for each expression.*

49. $\cos(\arctan 1)$

50. $\arcsin\left(\cos\dfrac{\pi}{3}\right)$

51. $\csc\left(\cos^{-1}\left(-\dfrac{\sqrt{2}}{2}\right)\right)$

52. $\cot(\sin^{-1}(-1))$

53. $\arcsin(\sec\pi)$

54. $\cos^{-1}\left(\sin\dfrac{11\pi}{6}\right)$

55. $\sin\left(\cos^{-1}\dfrac{5}{13}\right)$

56. $\tan\left(\sin^{-1}\dfrac{8}{17}\right)$

57–62. *Find an approximation to four decimal places for each expression.*

57. $\cos^{-1}(\sin 55)$

58. $\cos(\tan^{-1}(-3.76))$

59. $\cos(\csc^{-1} 7)$

60. $\sec^{-1}\left(\dfrac{22}{7}\right)$

61. $\tan^{-1}(\sin\sqrt{2.6})$

62. $\arcsin(\cot 2.4\pi)$

63–64. *Identify the graph of the inverse circular function.*

63.

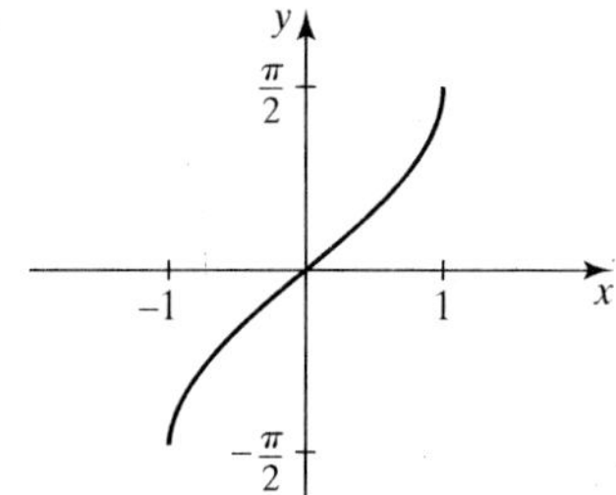

64.

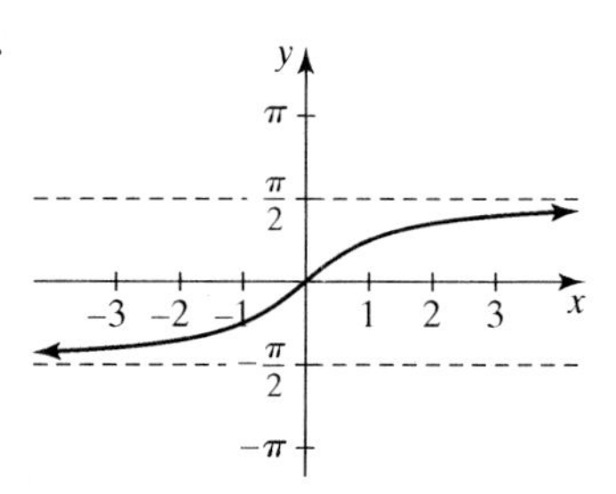

65–68. *Solve the following for x using inverse notation. Assume the arc restrictions are those specified in the respective inverse function definition.*

65. $0.5 = 4\sin(2x)$

66. $3.2\cos\left(\tfrac{1}{2}x\right) = 0.9$

67. $-6\tan(4x) = 2y$

68. $\pi\sin(-4x) = y$

69–84. *Are the following statements true or false? If a statement is false, explain why or give an example that shows why it is false.*

69. All circular functions have graphs that are sinusoidal.

70. The period of $y = \tan\left(\tfrac{1}{2}x\right)$ is π.

71. The graphs of $y = 3\sin x$ and $y = \sin 3x$ are the same.

72. The graph of $y = \sin x$ is symmetric about the origin.

73. There is no real value for $\arccos(-100)$.

74. Amplitude $= |y\text{ max} - y\text{ min}|$.

75. The graph of $y = \cos(-x)$ is the same as the graph of $y = -\cos x$.

76. The graph of $y = A\cos B(x - C) + D$ is a vertical shift of $y = A\cos B(x - C)$ up D units if $D > 0$, or down $|D|$ units if $D < 0$.

77. $\csc^{-1}\left(\tfrac{5}{3}\right) = \sin^{-1}\left(\tfrac{3}{5}\right)$

78. $\cos^{-1} x = \dfrac{1}{\cos x}$

79. The range for arcsin $x = y$ is $-\frac{\pi}{2} \le y \le \frac{\pi}{2}$.

80. $\arccos\left(-\frac{1}{2}\right) = -\frac{\pi}{3}$

81. $\sin^{-1}(-1) = \frac{3\pi}{2}$

82. The graph of any circular function is represented by only one equation.

83. The graph of $y = \tan^{-1} x$ has horizontal asymptotes at $y = -\pi/2$ and $y = \pi/2$ because the graph of $y = \tan x$ has vertical asymptotes at $x = -\pi/2$ and $x = \pi/2$.

84. The guide function to be used for $y = 4\csc(2x)$ is $y = \frac{1}{4}\sin(2x)$.

85. Jack and Jo are arguing about which guide function to use and the procedure to graph $y = \sec(2x) + 4$. Jack says, "Dot in the guide function $y = \cos(2x) + 4$, draw the asymptotes through its x-intercepts, and then draw the two U-shaped branches." Jo responds, "You should dot in the guide function $y = \cos(2x)$, draw the asymptotes through its x-intercepts, draw the two U-shaped branches, and then shift each branch up 4 units." Are they both correct, is only one correct, or are they both wrong? If one or both is wrong, explain why their procedure would not produce the correct graph.

Chapter 2 Test

1–6. *Without using a graphing utility, graph two cycles of the given function. Provide the requested information.*

1. $y = 5\sin\left(-\frac{1}{2}x\right)$

   Indicate the period, amplitude, and x-intercepts .

2. $y = \cos\left[2\left(x - \frac{\pi}{2}\right)\right]$

   Indicate the period, phase shift, and x-intercepts.

3. $y = \tan\left(\frac{1}{2}x\right)$

   Indicate the period, asymptotes, and range.

4. $y = 3\sec x$

   Indicate the period, asymptotes, and range.

5. $y = 2\sin(\pi x) + 2$

   Indicate the period, x-intercepts, and range.

6. $y = 5\csc\left(x - \frac{\pi}{2}\right)$

   Indicate the range and asymptotes.

7. Find an equation for each of the following circular functions.

a.

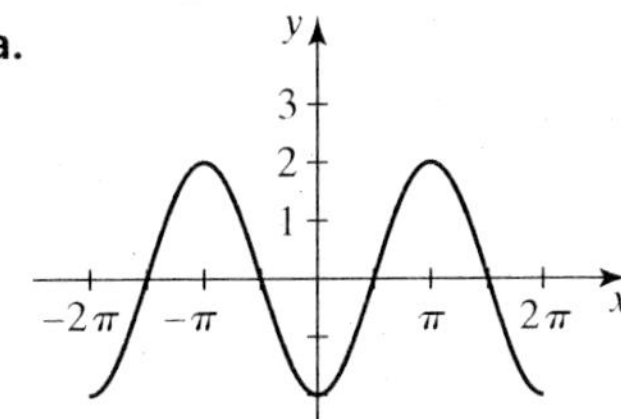

b.

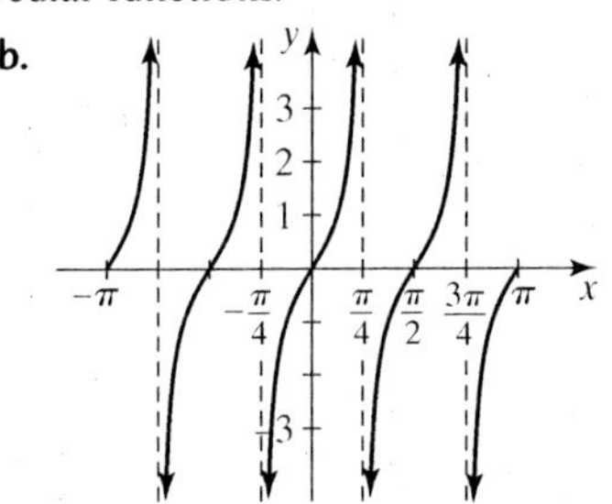

(continued)

c.

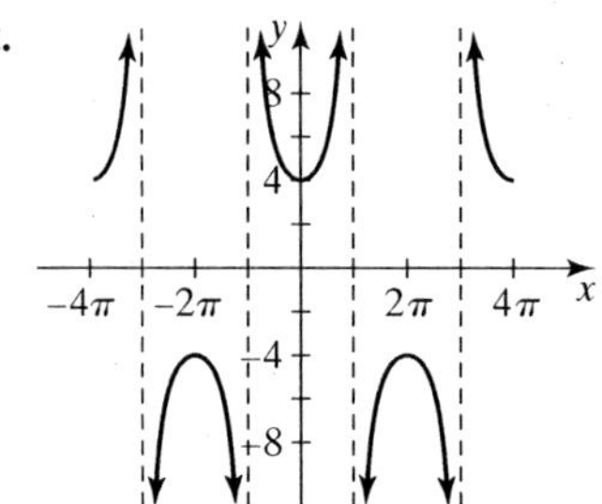

d.

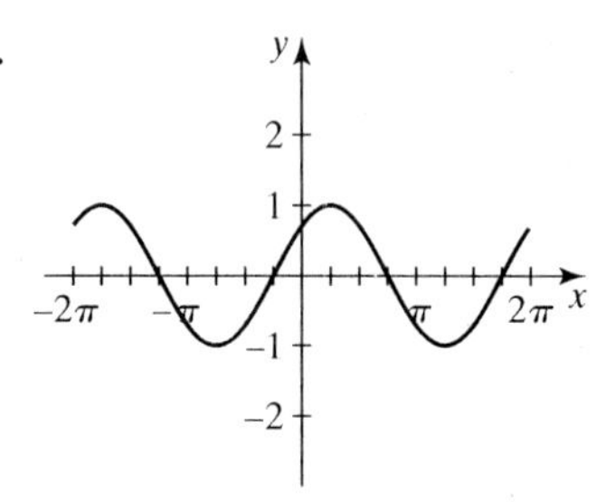

8. Find a second equation for the graph in Problem 7(a).

9. The simple harmonic motion of a spring that is initially displaced 5 cm from its rest position is given by the equation $y = 5\cos((\pi/2)x)$, where x is in seconds and y in centimeters.

   a. Graph this equation for time $x = 0$ to $x = 8$ seconds.

   b. What is the maximum displacement?

   c. Mark the graph with an asterisk (*) *each time* the spring is 4 cm below the rest position.

10. Find the exact value for each expression.

   a. $\arccos \frac{1}{2}$ b. $\arctan 0$ c. $\sin^{-1} \frac{\sqrt{2}}{2}$ d. $\operatorname{arccsc} 2$

   e. $\sec^{-1}(-\sqrt{2})$ f. $\arctan(-1)$ g. $\arccos(-1)$ h. $\cot^{-1} \frac{1}{\sqrt{3}}$

   i. $\sin\left(\arccos \frac{3}{5}\right)$ j. $\sec^{-1}(\cos \pi)$

11. Use a calculator to find an approximation of the following rounded to four decimal places.

   a. $\tan^{-1} 13.4$ b. $\arccos(-0.89)$ c. $\sec^{-1}\left(-\frac{13}{7}\right)$

   d. $\sin^{-1} 0.56$ e. $\tan(\arctan(-3.67))$ f. $\cos(\operatorname{arccot} 4.65)$

12. Solve the equation for x. Assume the arc restrictions are those specified in the definition for the respective inverse function.

   a. $6 \sin 8x = 3$ b. $4 \tan 2x = y$

13. Identify the inverse circular function graph.

a.

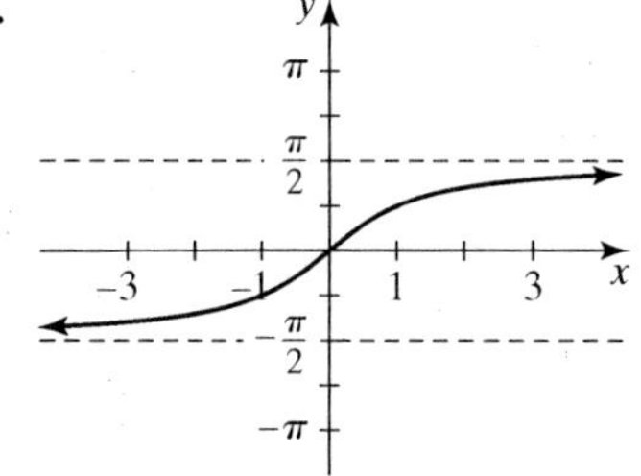

b.

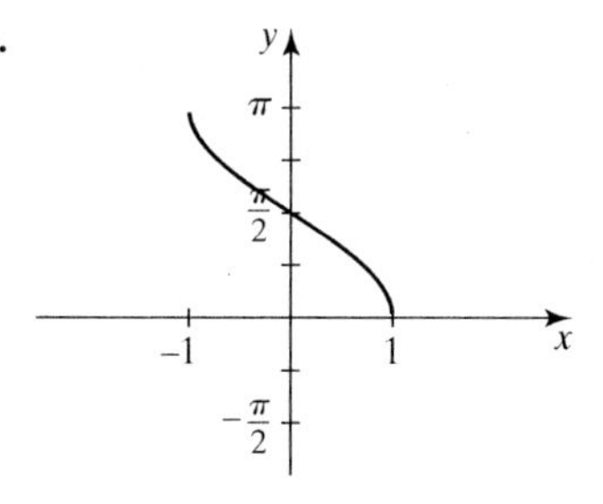

14. Are the following statements true or false? If a statement is false, explain why or give an example that shows why it is false.

a. All the circular function graphs are sine waves.

b. The graph of a circular function has asymptotes where it is not defined.

c. The graph of $y = \sec x$ is symmetric about the y-axis and has no x-intercepts.

d. The graph of $y = A \tan(-Bx)$ is the same as the graph of $y = A \tan(Bx)$.

e. The graph of $y = \sin 2x$ is the same as the graph of $y = 2 \sin x$.

f. All the circular functions are continuous on their domains.

g. Sinusoidal functions can be used to represent some periodic phenomena.

h. The graphs of $y = \sin\left(x + \frac{\pi}{2}\right)$ and $y = \sin x + \sin \frac{\pi}{2}$ are the same.

i. There is no real value for $\arcsin(25.6)$.

15. The following graph models the number of daylight hours per month (January corresponds to $x = 1$).

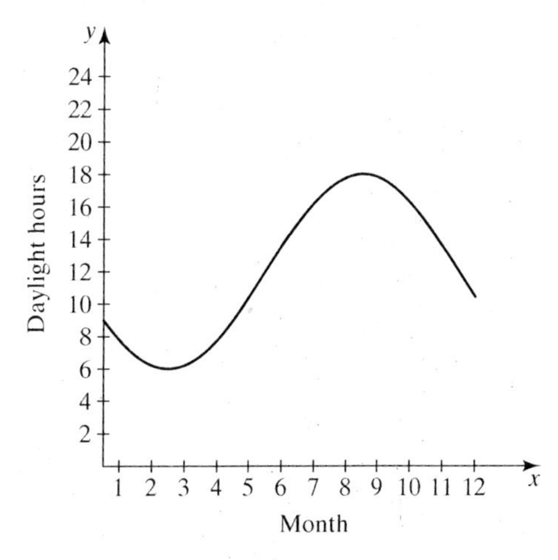

a. Approximate the minimum number of daylight hours. In what month(s) does this occur?

b. In what month(s) will there be at least 8 hours of daylight?

c. Approximate the maximum number of daylight hours. In what month(s) does this occur?

Chapter 3

The Trigonometric Functions

At 11:30 A.M. on May 29, 1953, Tenzing Norgay Sherpa and Edmund Hillary were the first mountain climbers to reach the summit of Mount Everest, the highest mountain in the world. In 1954, B.L. Gulatee, a surveyor, arrived at the height of 29,028 feet or approximately 8850 meters.

As a result of the work we do with triangles in this chapter, you will be able to find a good approximation of the height of a mountain—without the arduous task of mountain climbing!

Important Concepts

- Angle measure
  - —radians and degrees
- coterminal angles
- arc length
- linear and angular velocity
- trigonometric functions
- inverse trigonometric functions
- common angles
  - —trigonometric functional values of common angles
- solutions of triangles
- trigonometric ratios for right triangles
- law of sines
- law of cosines
- angles of elevation and depression
- bearing

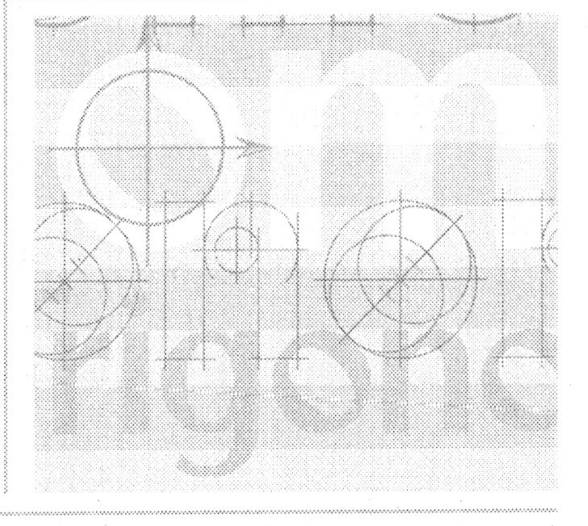

3.1 Angles and Their Measure

The word *trigonometry* stems from the Greek words for *triangle measurement.* Trigonometry arises from the early connection between mathematics and astronomy and was used to solve geometry problems involving triangles. In trigonometry, triangle problems are solved using functions that associate the measure of an angle with a ratio of lengths of certain sides of a right triangle containing the angle. The domain of these functions consists of angle measurements. In order to better understand these functions and their connection with the circular functions, we start with a review of some important geometric ideas and definitions. For a more detailed review see Appendix B.

Angles

In a plane, a line drawn through two points A and B can be referred to as **line** AB, denoted $\overleftrightarrow{AB}$. The portion of the line between and including points A and B is called **segment** AB, denoted $\overline{AB}$. The portion of $\overleftrightarrow{AB}$ that starts at point A and continues infinitely in the direction of B is called **ray** AB, and is denoted $\overrightarrow{AB}$, where A is the endpoint. An **angle** is formed by rotating a ray about its endpoint. The ray in its initial position is called the **initial side** of the angle, and the position of the ray after it has been rotated is called the **terminal side** of the angle. The endpoint of the ray is called the **vertex** of the angle.

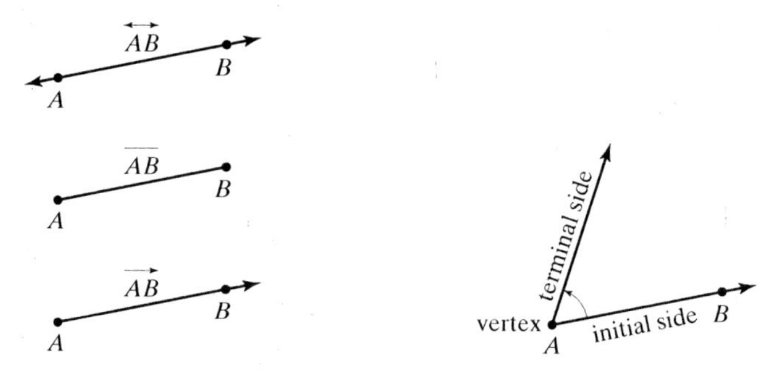

α alpha
β beta
γ gamma
θ theta
ϕ phi
ρ rho

We use a curved arrow to indicate the direction of the rotation. If the rotation of the ray is *counterclockwise,* the angle is called a **positive angle**. If the rotation is *clockwise,* the angle is called a **negative angle**. If no direction is indicated, we assume the angle is a positive angle. There are different ways we can name an angle. Greek letters, like α, β, γ, θ, ϕ, or ρ, are often used to name angles.

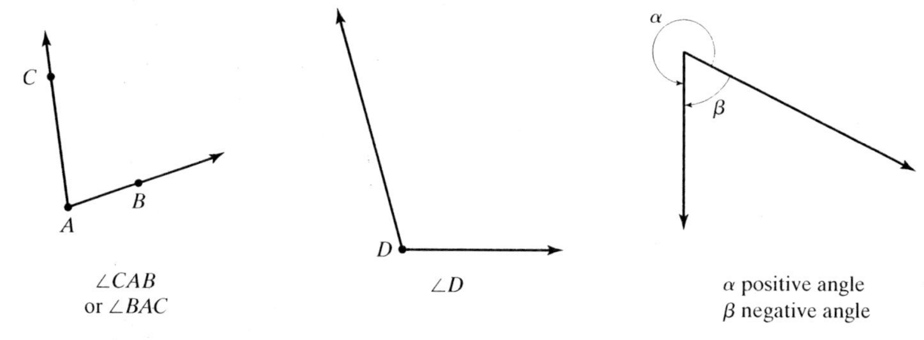

Notice that counterclockwise rotation refers to a positive angle just as a counterclockwise wrap referred to a positive arc on the unit circle.

In the xy-plane if the vertex of the angle is at the origin and the initial side is along the positive x-axis, the angle is said to be in **standard position**.

The terminal side of a standard position angle can lie in a quadrant or on an axis. The terminal side of the angle determines the quadrant designation for the angle. If the terminal side is on an axis, the angle is called a **quadrantal angle**.

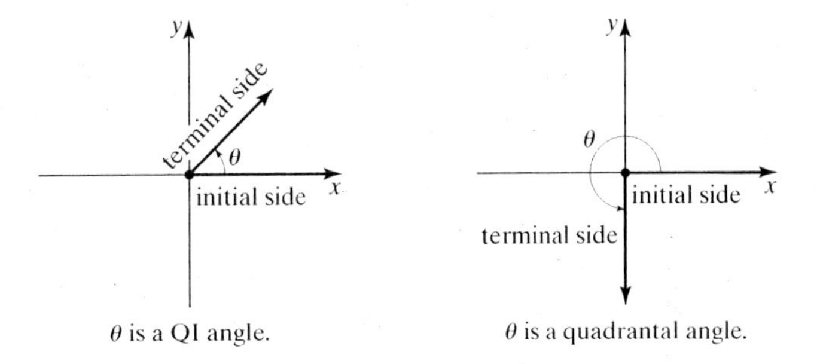

θ is a QI angle. θ is a quadrantal angle.

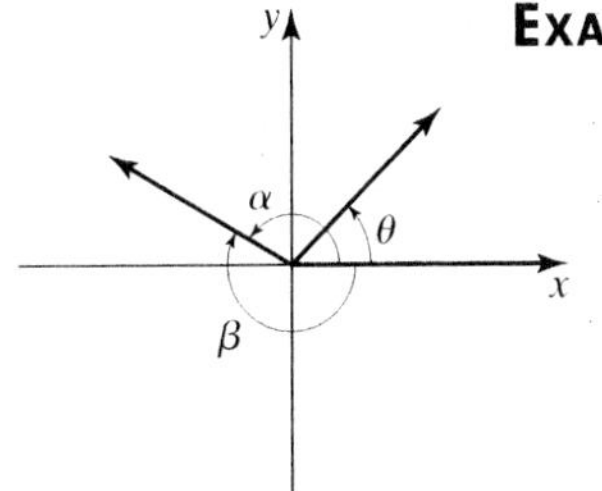

EXAMPLE 1 In the graph shown at the left, which are QII angles?

SOLUTION α and β are QII angles because their terminal sides lie in QII. ■

Measure of an Angle

The two common units of measurement for angles are radians and degrees. We define these units and show how to convert from radians to degrees and vice versa.

Radians

A circle with radius r has circumference $C = 2\pi r \approx 6.28r$. On this circle if we mark off arcs so that the length of each arc is equal to the radius r, there are $2\pi \approx 6.28$ of these arcs (or radius lengths) along the circumference of the circle. An angle whose vertex is at the center of the circle is a **central angle**. Any central angle intercepting an arc whose length is equal to the radius has a measure of *one* **radian**.

The number of these arcs intercepted by a central angle gives the radian measure of the angle.

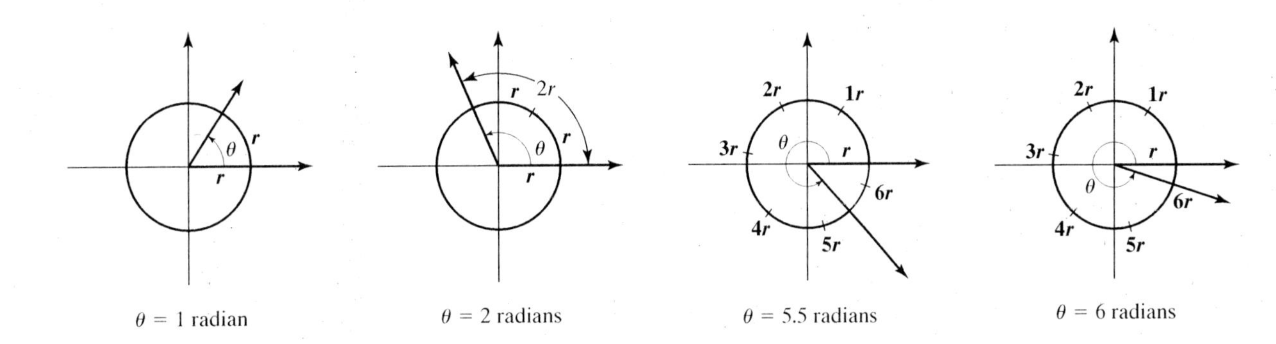

$\theta = 1$ radian $\theta = 2$ radians $\theta = 5.5$ radians $\theta = 6$ radians

NOTE: In this text we will not distinguish between the angle and its measure. For example, we will use $\theta = 1$ radian to mean *the measure of angle* $\theta = 1$ radian.

An angle that makes one complete counterclockwise rotation, or revolution, has a measure of 2π radians. Similarly, one-half of a rotation is π radians, one-fourth of a rotation is $\pi/2$ radians, and so on. A negative sign is used with the radian measure to indicate a clockwise rotation (negative angle).

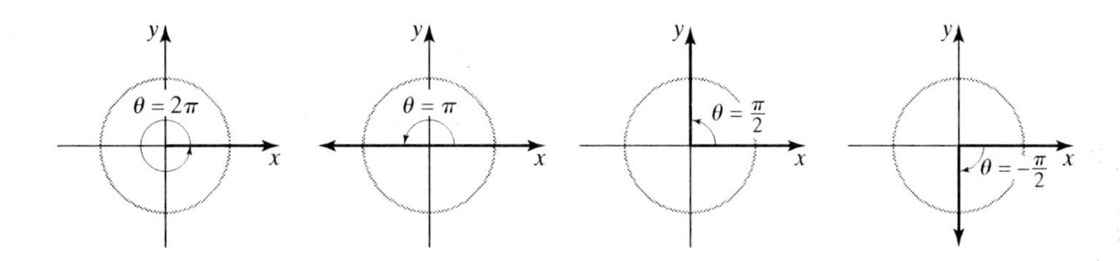

We often omit the word radians (abbreviated rad) when using radian measure. For example, $\theta = 2\pi$ radians or $\theta = 2\pi$ rad can be written as simply $\theta = 2\pi$.

Another useful and common way to measure an angle is in degree units.

Degrees

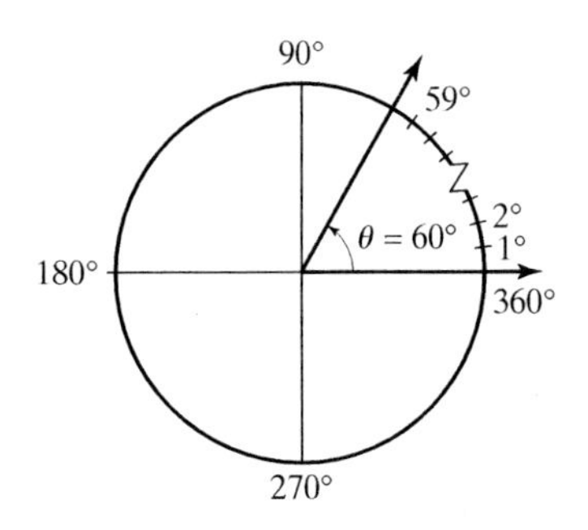

If the circle is divided into 360 arcs of equal length, the number of such arcs intercepted by a central angle is the measure of the angle in degrees, denoted by "°." Each degree is divided into 60 minutes $(1° = 60')$ and each minute is divided into 60 seconds $(1' = 60'')$.

An angle that makes one complete counterclockwise rotation, or revolution, has a measure of 360 degrees. Similarly, one-half of a rotation is 180 degrees, one-fourth of a rotation is 90 degrees, and so on. A negative sign is used with the degree measure to indicate a clockwise rotation (negative angle).

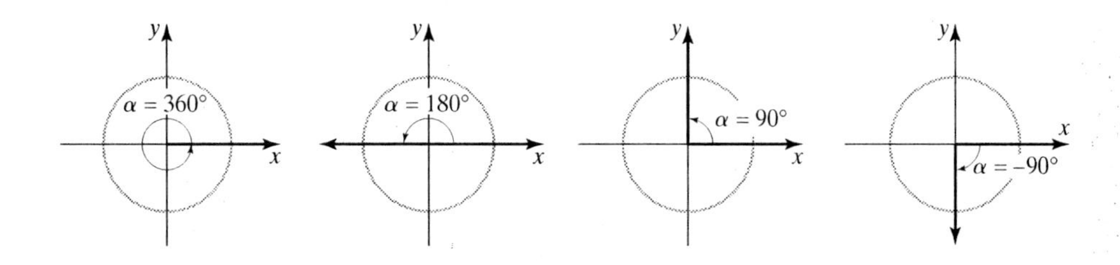

CAUTION: Be careful to always write the degree symbol ° when using degree measure. For example, $\alpha = 90°$ is *not* the same as $\alpha = 90$.

Since one-half of a rotation is either π radians or 180°, π radians = 180°. If we divide both sides of this equality by π or 180, we obtain two forms of the equality. We convert from radians to degrees or degrees to radians by using the appropriate form to obtain the desired unit of measure:

Conversions between Degrees and Radians

$$1 \text{ radian} = \frac{180}{\pi} \text{ degrees}$$

$$1 \text{ degree} = \frac{\pi}{180} \text{ radians}$$

How do we convert $\pi/3$ radians to degrees? Since each radian is $180/\pi$ degrees and we have $\pi/3$ radians, we get

$$\frac{\pi}{3} \text{ radians} = \frac{\pi}{3} \cdot \frac{180}{\pi} \text{ degrees} = 60°.$$

How do we convert 135° to radians? Since each degree is $\pi/180$ radians and we have 135°, we get

$$135° = 135 \cdot \frac{\pi}{180} \text{ radians} = \frac{3\pi}{4} \text{ rad} = \frac{3\pi}{4}.$$

Notice that if we approximate the conversion equations, we get:

$$1 \text{ radian} = \frac{180}{\pi} \text{ degrees} \approx 57.3°$$

$$1 \text{ degree} = \frac{\pi}{180} \text{ radians} \approx 0.0175 \text{ rad}$$

EXAMPLE 2 Convert each radian measure to degrees (or nearest degree), and then draw the angle in standard position. When no unit is given, we assume that the unit is radians.

a. $\alpha = \dfrac{7\pi}{6}$ **b.** $\beta = 2.6$ **c.** $\theta = -\dfrac{7\pi}{4}$

SOLUTION

a.

$$\alpha = \frac{7\pi}{6} \cdot \frac{180°}{\pi} = 210°$$

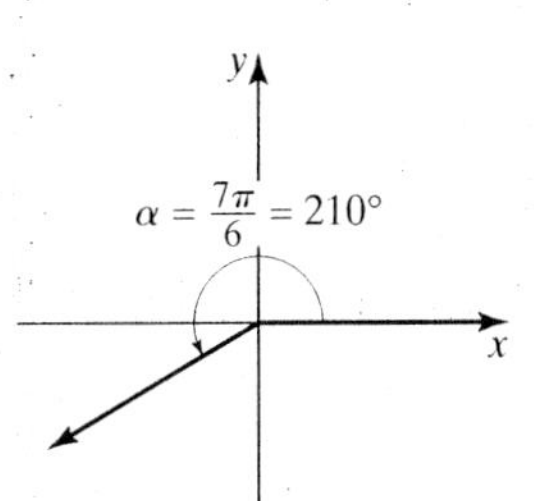

b.

$$\beta = 2.6 \cdot \frac{180°}{\pi} \approx 149°$$

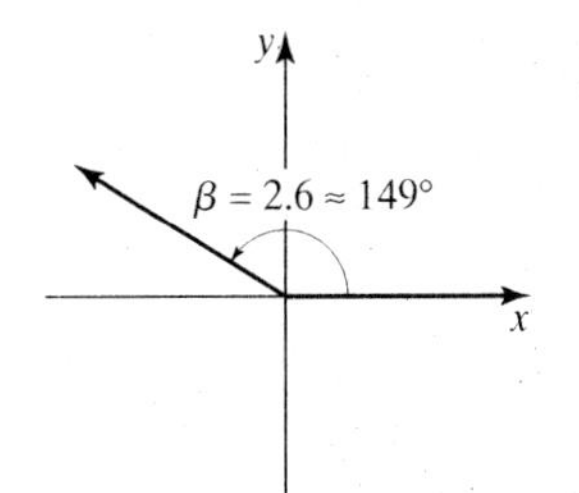

c.

$$\theta = -\frac{7\pi}{4} \cdot \frac{180°}{\pi} = -315°$$

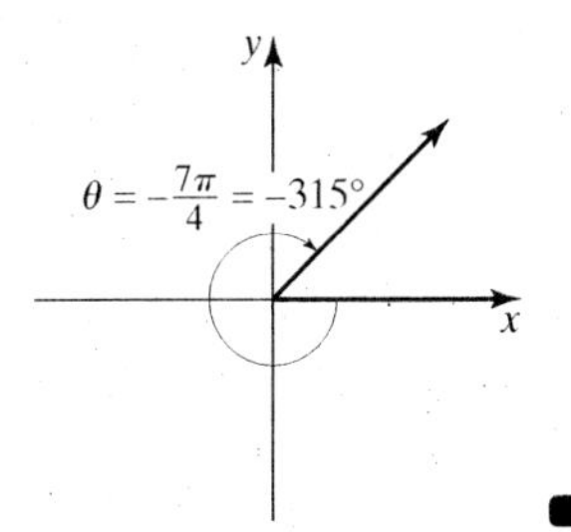

EXAMPLE 3 Draw the angle in standard position, and then convert each degree measure to radians. Leave your answers in terms of π.

a. $\alpha = 30°$ **b.** $\beta = -135°$ **c.** $\theta = 40°30'$

SOLUTION

a.

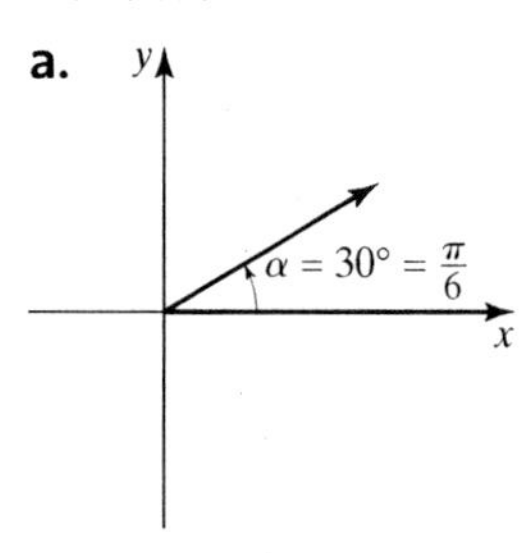

$$\alpha = 30 \cdot \frac{\pi}{180}$$
$$= \frac{\pi}{6}$$

b.

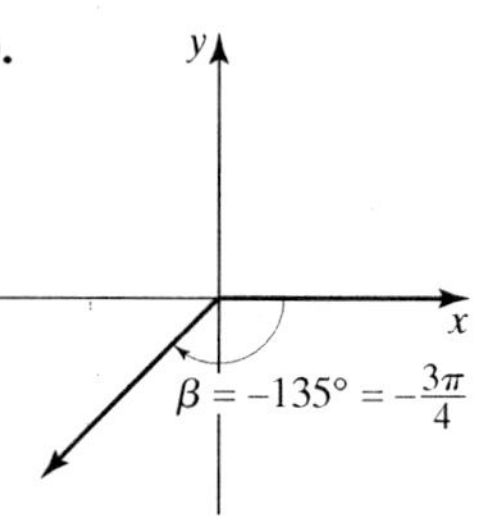

$$\beta = -135 \cdot \frac{\pi}{180}$$
$$= -\frac{3\pi}{4}$$

c.

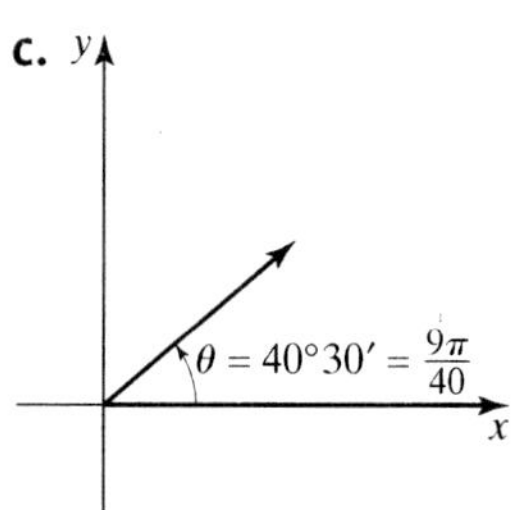

Note: $60' = 1° \rightarrow 1' = \frac{1}{60}°$.

So $30' = 30 \cdot \frac{1}{60}° = \frac{1}{2}°$.

$$\theta = 40°30' = 40\tfrac{1}{2}°$$
$$= (40\tfrac{1}{2}) \cdot \left(\frac{\pi}{180}\right)$$
$$= \frac{81}{2} \cdot \frac{\pi}{180} = \frac{9\pi}{40}$$

■

Special Angles Degree and Radian Measure

The following circles indicate degree and radian measures for some positive and negative angles. They represent angles whose importance will be established in this chapter. Positive angles that measure less than 90° are called **acute angles**. Those that measure 90° are called **right angles**, and angles that measure between 90° and 180° are called **obtuse angles**. Two angles are called **complementary** if the sum of their measures is 90°, and **supplementary** if the sum is 180°.

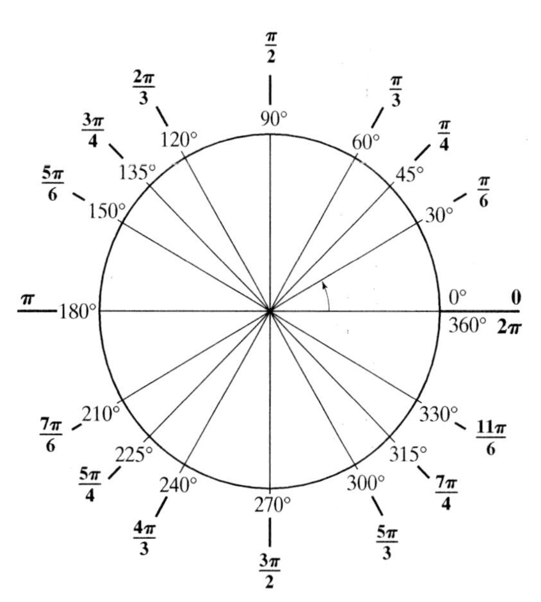

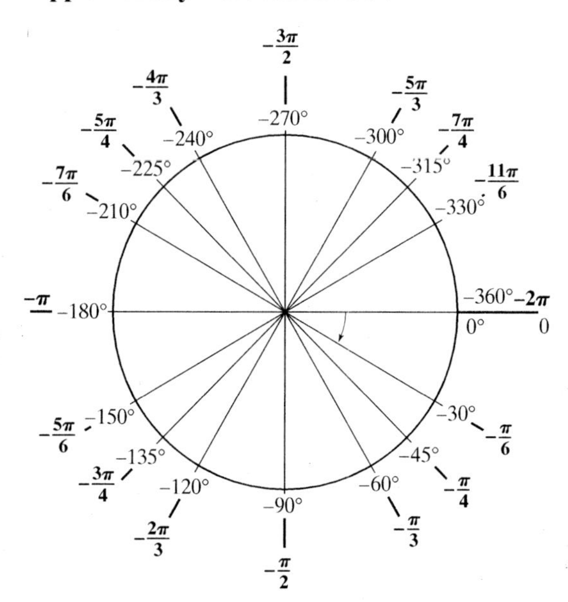

Many calculators convert radians to degrees and degrees to radians, but often only find approximations rather than exact values for the conversions.

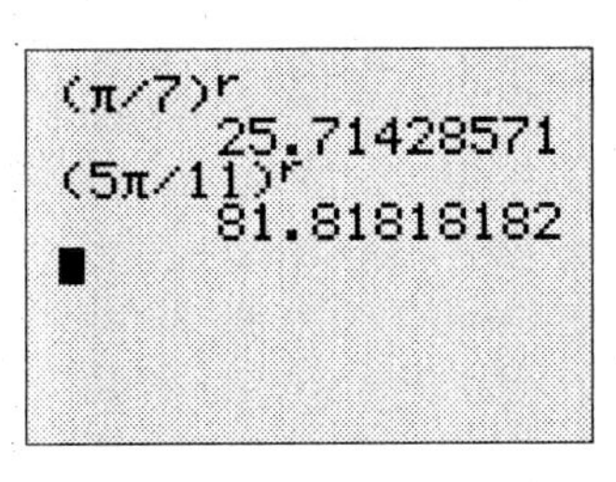

in degree mode

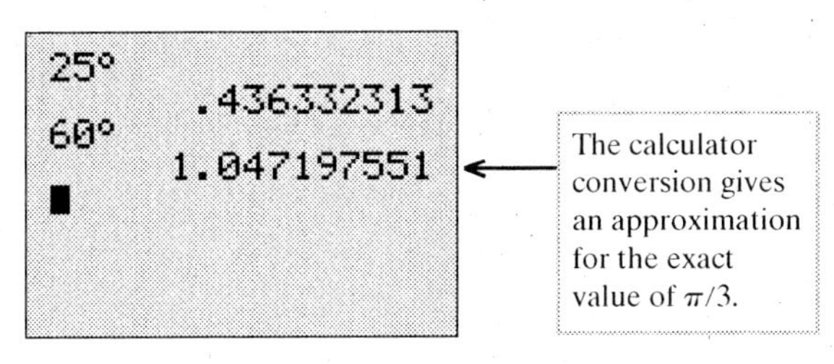

in radian mode

Coterminal Angles

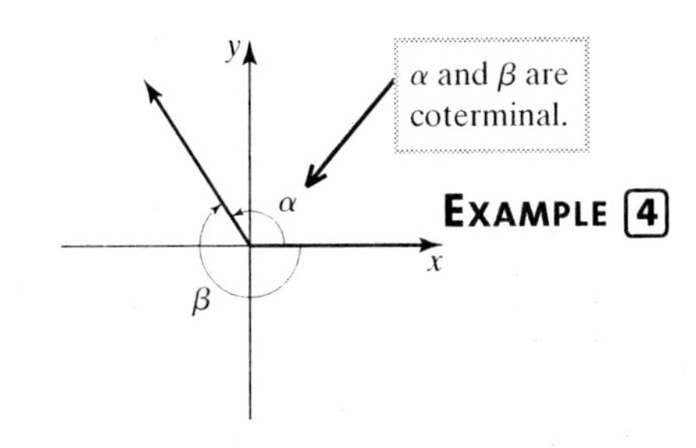

Angles with the same initial and terminal sides are called **coterminal angles**. Just as coterminal arcs are very useful, so are coterminal angles.

EXAMPLE 4 Draw the angle in standard position. Then find its smallest positive coterminal angle with the same unit of measure.

a. $\alpha = 485°$ **b.** $\beta = 6.85$ **c.** $\gamma = -150°$ **d.** $\theta = -\dfrac{13\pi}{6}$

SOLUTION

a.

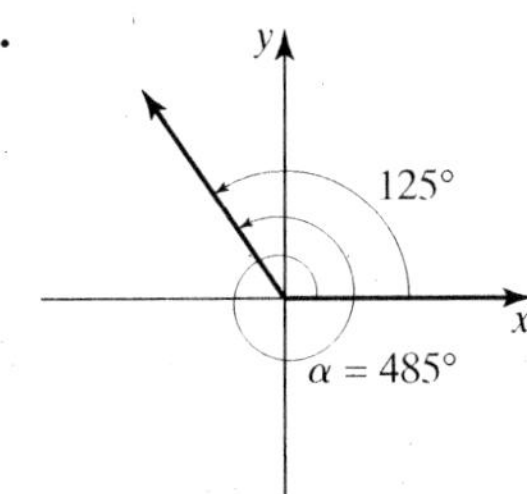

$$485° - 360° = 125°$$

The smallest positive angle coterminal with 485° is 125°.

b.

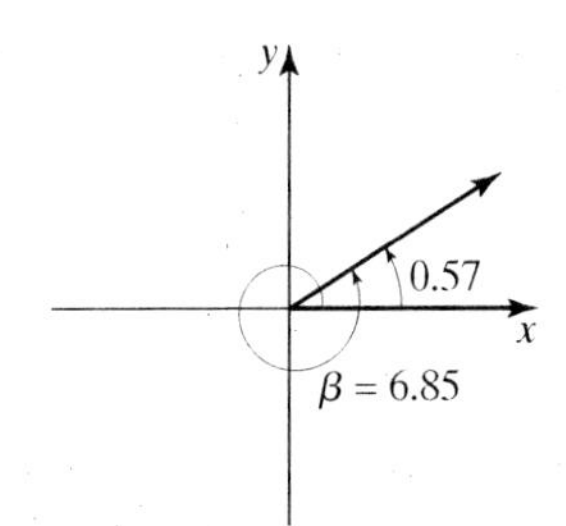

$$6.85 - 2\pi \approx 0.57$$

The smallest positive angle coterminal with 6.85 is approximately 0.57.

c.

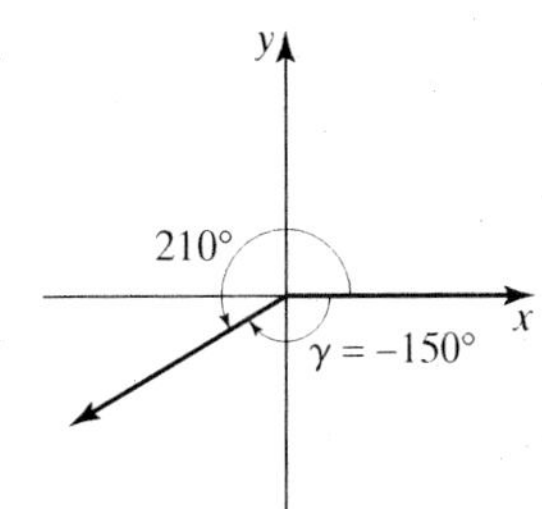

$$-150° + 360° = 210°$$

The smallest positive angle coterminal with −150° is 210°.

d.

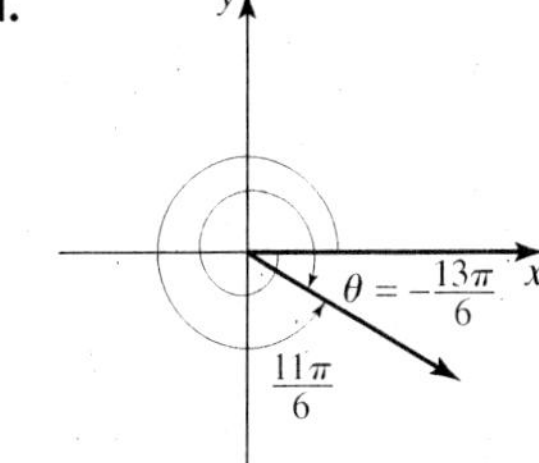

$$-\frac{13\pi}{6} + 4\pi = \frac{11\pi}{6}$$

The smallest positive angle coterminal with $-\dfrac{13\pi}{6}$ is $\dfrac{11\pi}{6}$. ■

To find a formula for all angles coterminal with a given angle θ, we add multiples of 2π or $360°$ to the angle (as we did with coterminal arcs in Chapter 1). For example, all angles coterminal with $45°$ can be expressed in the form

$$45° + k \cdot 360° \quad \text{or} \quad \frac{\pi}{4} + k \cdot 2\pi, \quad \text{where } k \in \{\ldots, -2, -1, 0, 1, 2, \ldots\}.$$

Applications

There are many applications found in physics, calculus, and engineering, to name a few, that involve working with angles (typically measured in radians). Let's look at a few of these applications. We begin with arc length.

Arc Length of a Circle

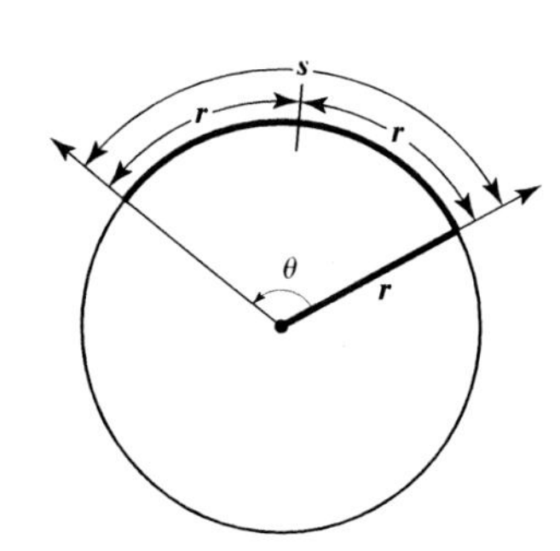

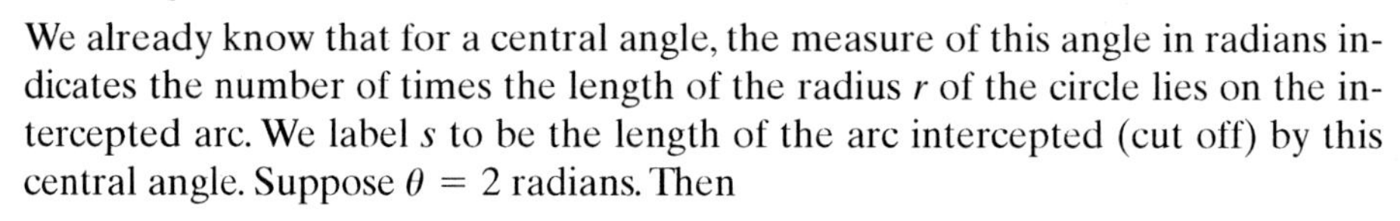

We already know that for a central angle, the measure of this angle in radians indicates the number of times the length of the radius r of the circle lies on the intercepted arc. We label s to be the length of the arc intercepted (cut off) by this central angle. Suppose $\theta = 2$ radians. Then

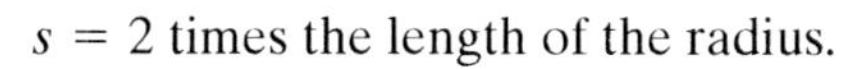

$$\begin{aligned} s &= 2 \text{ times the length of the radius.} \\ s &= 2 \cdot r \\ s &= \theta \cdot r \\ \text{or} \quad s &= r \cdot \theta \end{aligned}$$

> **Arc Length**
>
> On a circle with radius r, the arc length s intercepted by a central angle with radian measure θ is
>
> $$s = r\theta.$$

In order to use the arc length formula, the central angle θ *must be in radians.*

CAUTION: Arc length is always a nonnegative value. Recall from Chapter 1 that we assigned t to represent the length of the arc *and* its direction. Here s only represents the length.

EXAMPLE 5 On a circle with the given radius, find the length s of the arc, intercepted by the central angle.

a.

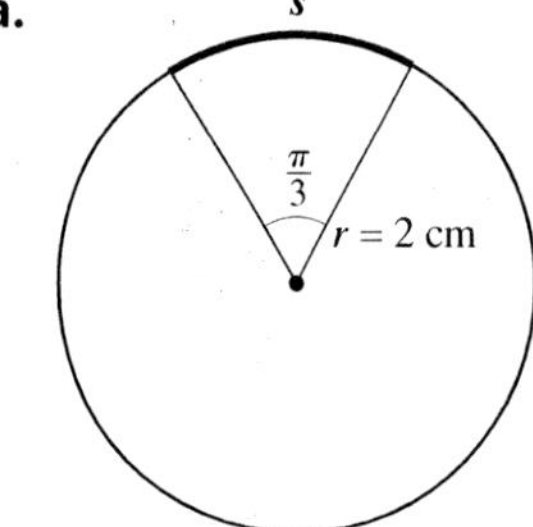

b.

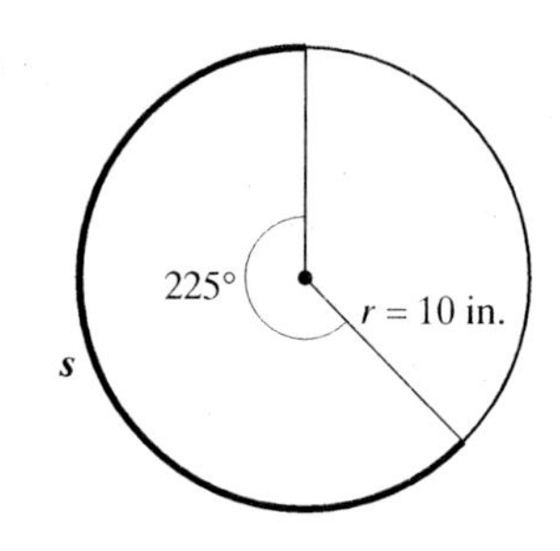

SOLUTION

a. $s = r\theta$

$$s = (2\text{ cm})\left(\frac{\pi}{3}\right)$$

$$= \frac{2\pi}{3}\text{ cm} \approx 2.09\text{ cm}$$

b. First we must convert θ to radians.

$$\theta = 225° = 225 \cdot \frac{\pi}{180} = \frac{5\pi}{4}$$

$$s = r\theta$$

$$s = (10\text{ in.})\left(\frac{5\pi}{4}\right) = 12.5\,\pi\text{ in.} \approx 39.27\text{ in.}$$ ■

NOTE: Radian measure has no units associated with it. As a result, when we multiply, for example, inches by radians, the result is measured in inches.

If an object moves at a constant speed on a circular path, we might ask, "How fast is it moving?" In other words, we ask about its velocity. We can answer this question with two different, yet closely related, velocities: *linear* and *angular velocity*.

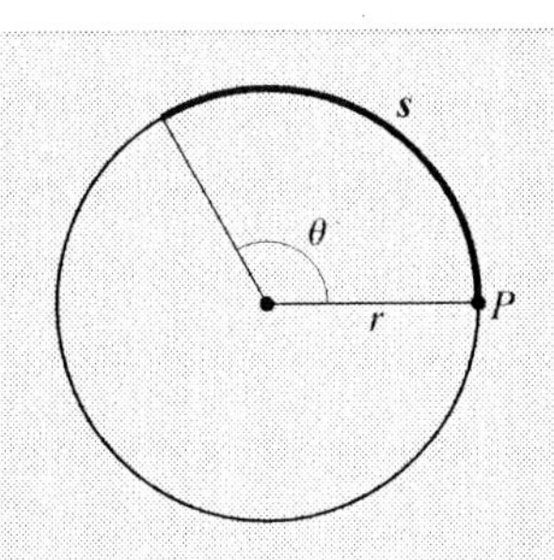

Linear Velocity

Consider a point P on a circle of radius r moving a distance of s in time t. The **linear velocity** v of the point P is the ratio of the length s of the arc to the time t it takes to travel this length.

$$v = \frac{\text{arc length}}{\text{time elapsed}} = \frac{s}{t} = \frac{r\theta}{t},$$

where θ is measured in radians.

You may have done problems in algebra that used a similar ratio when you used the formula for average rate $r = d/t$, which comes from the distance formula, distance equals rate (speed) times time, or $d = rt$. Since the distance traveled is the arc length, and rate is referred to as linear velocity, it follows that $v = s/t$. Linear velocity tells us how fast the position of the point is changing and can be expressed in miles per hour (mi/hr), centimeters per minute (cm/min), and so on.

EXAMPLE 6 A point on a tricycle wheel of radius 3 inches rotates $5\pi/6$ radians in 2 seconds. Find the linear velocity of the point.

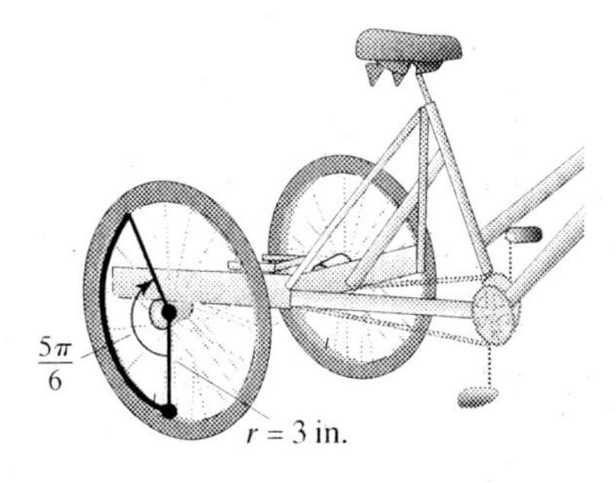

SOLUTION

$$v = \frac{s}{t} = \frac{r\theta}{t} = \frac{3\text{ in.} \times \frac{5\pi}{6}\text{ rad}}{2\text{ sec}} = \frac{\frac{5\pi}{2}\text{ in.}}{2\text{ sec}} = \frac{5\pi\text{ in.}}{4\text{ sec}} \approx 3.93\text{ in./sec}$$ ■

Another way to describe how fast an object is moving at a constant speed on a circular path is called *angular velocity.*

Angular Velocity

Consider that a point, moving with constant velocity (uniform circular motion) on a circle of radius r, travels on an arc that corresponds to a central angle of measure θ. The **angular velocity** ω of the point is the ratio of the radian measure of θ to the time it takes to sweep out this angle.

$$\omega = \frac{\theta}{t},$$

where θ is measured in radians.

Angular velocity tells us how fast the central angle is changing. That is, angular velocity measures the speed of the rotation. Angular velocity can be expressed in radians per second (rad/sec), radians per hour (rad/hr), and so on. With conversions these measurements can be expressed in degrees per minute, degrees per hour, and so on.

You probably have heard the phrase *revolutions per minute* (rev/min or rpm) which describes angular velocity. Since one revolution is 2π radians, the relationship between x revolutions per minute and the way we define angular velocity is:

$$\omega = x \text{ rev/min} = x\, 2\pi \text{ rad/min} = 2\pi x \text{ rad/min}$$

The same relationship applies for revolutions per second and radians per second, and similarly, with other units of time.

You will be asked in the exercises to find the relationship between linear velocity (v) and angular velocity (ω).

A helicopter blade revolves $3\frac{1}{4}$ times in $\frac{1}{2}$ second. Find the angular velocity of a point on the blade.

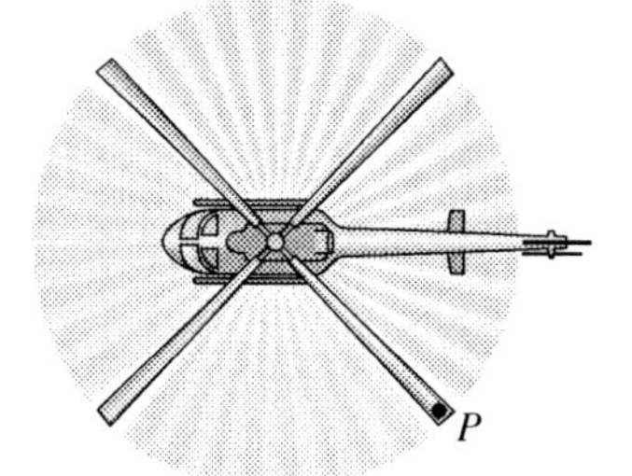

SOLUTION Since one revolution is 2π radians, $\theta = (3\frac{1}{4})(2\pi) = \frac{13\pi}{2}$ rad.

Therefore, the angular velocity is:

$$\omega = \frac{\theta}{t} = \frac{\frac{13\pi}{2}\text{ rad}}{\frac{1}{2}\text{ sec}} = \left(\frac{13\pi}{2} \cdot \frac{2}{1}\right) \text{rad/sec}$$

$$= 13\pi \text{ rad/sec} \approx 40.84 \text{ rad/sec}$$

Exercise Set 3.1

1–2. *Fill in the numerator and denominator with either 180 or π in the following conversion formulas.*

1. θ radians $= \theta \cdot \left(\frac{?}{?}\right)$ degrees

2. θ degrees $= \theta \cdot \left(\frac{?}{?}\right)$ radians

3. Convert to the missing radian or degree unit.

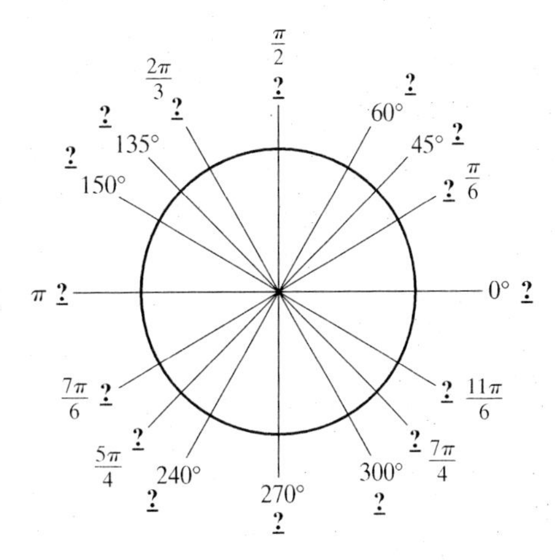

4. Convert to the missing radian or degree unit.

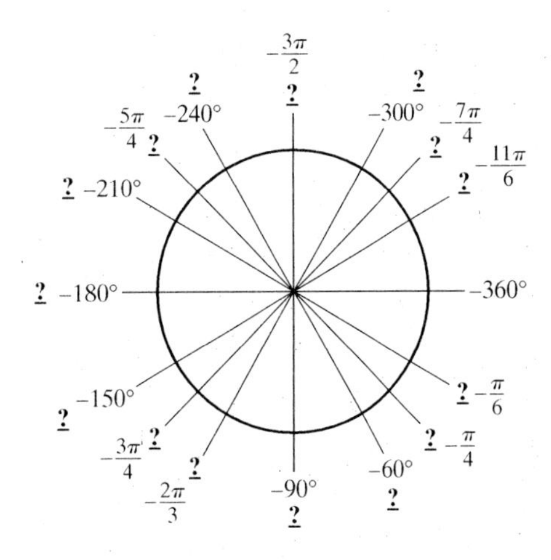

5–16. *Convert each radian measure to degrees (or nearest degree). Then draw the angle in standard position.*

5. $\frac{\pi}{4}$ **6.** $\frac{\pi}{6}$ **7.** $\frac{5\pi}{6}$ **8.** $\frac{2\pi}{3}$

9. $\frac{8\pi}{3}$ **10.** $\frac{7\pi}{4}$ **11.** $-\frac{7\pi}{6}$ **12.** $-\frac{\pi}{30}$

13. -3π **14.** $-\pi$ **15.** 1 **16.** 3

17–24. *Draw the angle in standard position. Then convert each degree measure to radians. Leave your answers in terms of π.*

17. $75°$ **18.** $160°$ **19.** $225°$ **20.** $315°$

21. $-12°$ **22.** $-56°$ **23.** $-120°$ **24.** $-270°$

25–28. *Convert each degree measure to the nearest hundredth of a radian.*

25. $16°$ **26.** $-157°$ **27.** $238°30'$ **28.** $310°55'$

29–36. *For each of the given angles, do the following:*

a. *Find the smallest positive coterminal angle with the same measurement unit. If necessary, approximate to the nearest degree or to the nearest hundredth of a radian.*

b. *Use this smallest positive coterminal angle to write a formula for all its coterminal angles.*

29. $\frac{8\pi}{3}$ **30.** $-\frac{13\pi}{4}$ **31.** $412°$ **32.** $513°$

33. -5.21 **34.** -3.98 **35.** $-710°30'$ **36.** $-600.3°$

37–42. *On a circle with the given radius r, find the length s (to the nearest hundredth of a unit when necessary) of the arc intercepted by the central angle θ.*

37. $r = 5$ in., $\theta = \frac{7\pi}{6}$

38. $r = 2.5$ m, $\theta = \frac{2\pi}{3}$

39. $r = 3$ cm, $\theta = 3.5$

40. $r = 6$ ft, $\theta = 5.4$

41. $r = 16$ m, $\theta = 120°$

42. $r = 0.6$ in., $\theta = 225°$

43. Find the radius r of the circle.

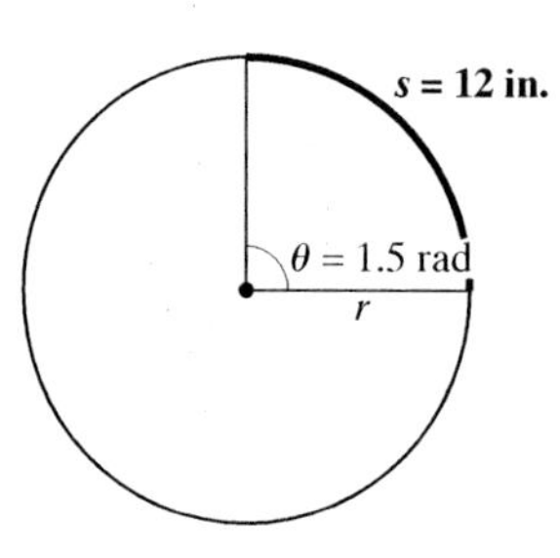

44. Find the central angle θ.

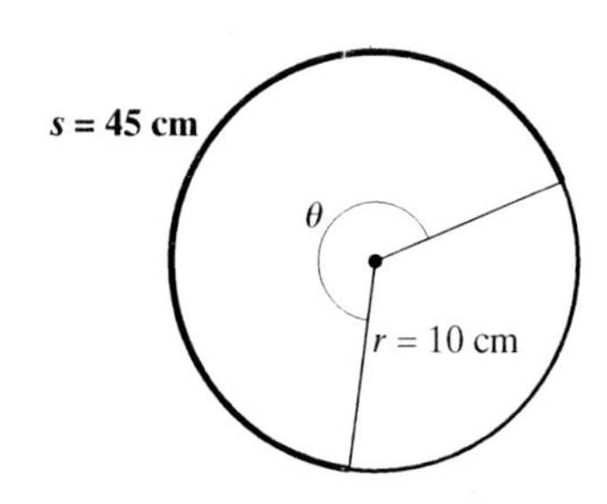

45. In Chapter 1, we discussed arcs on the unit circle. Find the measure in radians of the central angle θ that intercepts the arc with length s on the unit circle.

EXAMPLE $s = \frac{\pi}{6}$

SOLUTION

$$s = \theta \cdot r$$

$$\frac{\pi}{6} = \theta \cdot 1$$

$$\frac{\pi}{6} = \theta$$

■

a. $s = \frac{\pi}{4}$ **b.** $s = \frac{\pi}{3}$ **c.** $s = \pi$ **d.** $s = \frac{\pi}{2}$

46. What conclusion can you make from Exercise 45 about the radian measure of a central angle and the length of the arc it intercepts on the unit circle?

47. Find a formula that relates linear velocity to angular velocity.

48. If angular velocity is given as revolutions per unit of time, what should be done to this value before using the formula you found in Exercise 47 for linear velocity?

49–58. *Consider a point that is moving at a constant velocity on a circle of radius r. Approximate the requested value to the nearest hundredth of a unit.*

EXAMPLE Find v if $r = 34.7$ ft, $\theta = 45°$, and $t = 3$ min.

SOLUTION First we find the radian measure for θ:

$$\theta = 45 \cdot \frac{\pi}{180} \text{ rad} = \frac{\pi}{4}.$$

$$v = \frac{s}{t} = \frac{r\theta}{t} = \frac{34.7 \text{ ft} \cdot \frac{\pi}{4}}{3 \text{ min}} \approx 9.08 \text{ ft/min}.$$

49. Find v if $r = 3$ ft, $\theta = 30°$, and $t = 2$ hr.

50. Find v if $r = 10$ mi, $\theta = 180°$, and $t = 6$ min.

51. Find ω if $\theta = \frac{\pi}{2}$ and $t = 10$ sec.

52. Find ω if $\theta = \frac{\pi}{10}$ and $t = 5$ sec.

53. Find v if $r = 4$ ft and $\omega = 2$ rad/min.

54. Find v if $r = 24$ in. and $\omega = 5.6$ rad/sec.

55. Find s if $v = 35$ ft/sec and $t = 4$ sec.

56. Find s if $v = 2.6$ cm/min and $t = 15$ min.

57. Find s if $\omega = 6$ rad/sec, $r = 2.6$ ft, and $t = 4$ sec.

58. Find s if $\omega = 2.5$ rad/min, $r = 8$ mi, and $t = 10$ min.

59–64. *Are the following statements true or false? If a statement is false, explain why or give an example that shows why it is false.*

59. $\pi = 180$

60. $\pi = 180°$

61. $r\omega = v$

62. $\omega = \frac{v}{r}$

63. $s = r\theta$, where θ is measured in degrees.

64. On the unit circle, the length s of an arc has the same numerical value as the radian measure of the corresponding central angle.

65. A tachometer on the dashboard of a car indicates that the engine is "turning over" at an angular speed of 2500 rpm. Express this angular velocity in rad/min (to the nearest hundredth).

66. A string trimmer is a grass and weed cutting tool that utilizes a 0.21 m length of nylon string that rotates on a circular path with angular velocity of 650 rad/sec. Express this angular velocity in rev/sec (to the nearest hundredth).

67–70. *Assume that Earth travels around the sun in a circular orbit with constant velocity at a distance of 93,000,000 miles from the sun and makes one complete orbit per year (1 year = 365 days).*

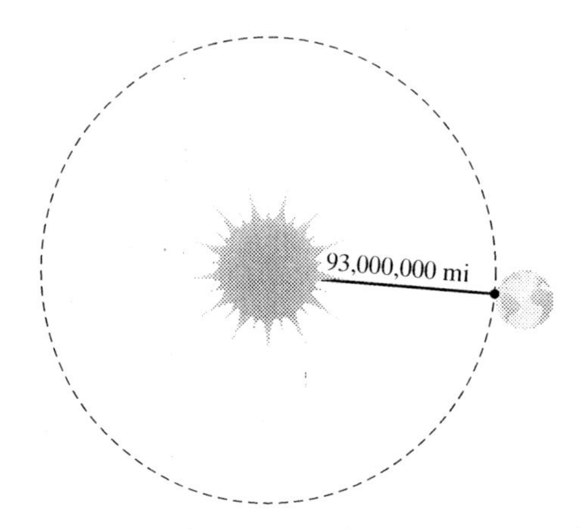

67. To the nearest mile, how far does Earth travel in its orbit in one day?

68. To the nearest mile, how far does Earth travel in its orbit in one week?

69. To the nearest mile per week, find the linear velocity of Earth's orbit.

70. To the nearest degree per day, find the angular velocity of Earth's orbit.

71. If an angle whose vertex is at the center of Earth has a measure of $1'$, which is $\frac{1}{60}°$, then the arc it intercepts on Earth's surface has a measure of 1 nautical mile. If Earth's radius is approximately 4000 miles (statute), find to the nearest hundredth the number of miles (statute) in 1 nautical mile.

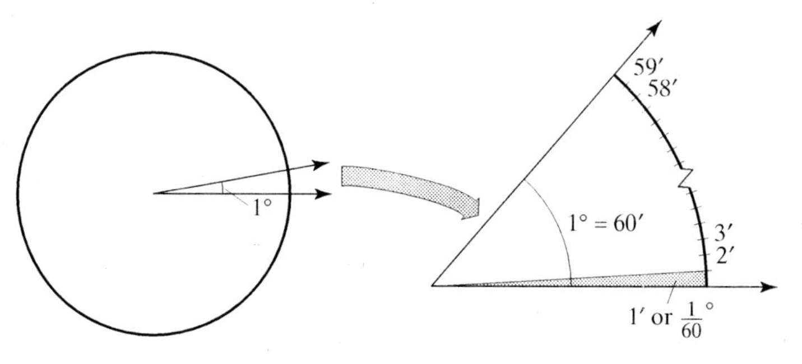

72. Two cruise ships are 33 nautical miles apart on the Atlantic Ocean. To the nearest mile, find the distance in miles (statute) between them. (Use the results from Exercise 71 to compute the distance.)

73. A $3\frac{1}{2}$-inch floppy diskette has an angular velocity of 300 rpm when placed in a computer's disk drive. Find the linear velocity to the nearest inch per minute of a point that is $1\frac{1}{4}$ inches from the center of the diskette.

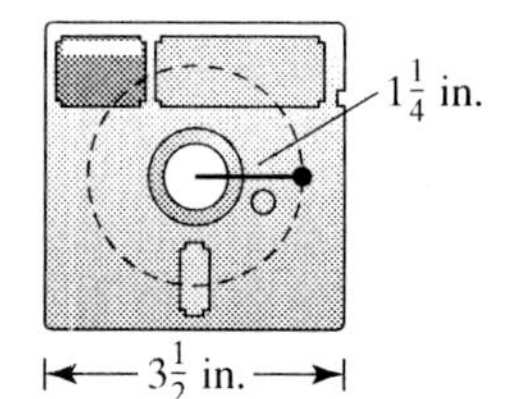

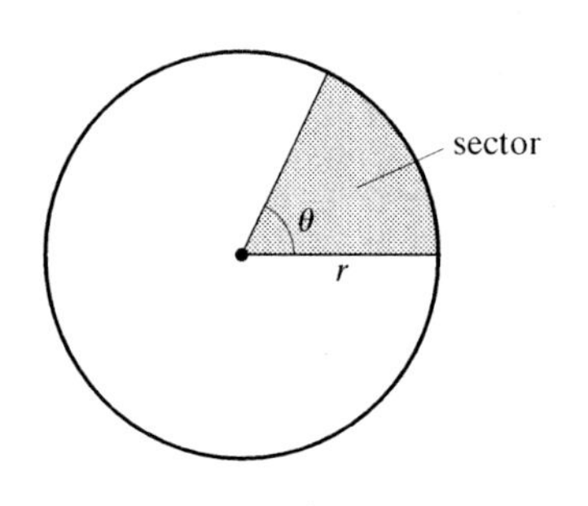

74. The area of a circle with radius r is $A = \pi r^2$. The central angle that sweeps out the area of the entire circle is 2π. Prove that the area of the sector swept out by the central angle θ is

$$A = \frac{\theta}{2} r^2,$$

where θ is measured in radians.

75. Use the formula in Exercise 74 to find the area of a sector with the given central angle θ in a circle of radius r.

a. $\theta = 2, r = 7$ in. **b.** $\theta = \frac{\pi}{2}, r = 4$ ft **c.** $\theta = 30°, r = 10$ cm

Discussion

76. A gymnast on a high bar swings through 2 revolutions in a time of 1.5 seconds.

a. Find the angular velocity and linear velocity of the point P on the gymnast.

b. Would there be a different angular velocity if the point P was moved from the elbow to a point on the shoulder of the gymnast? Explain why or why not.

c. Would there be a different linear velocity of the point P if P was moved from the elbow to a point on the shoulder of the gymnast? Explain why or why not.

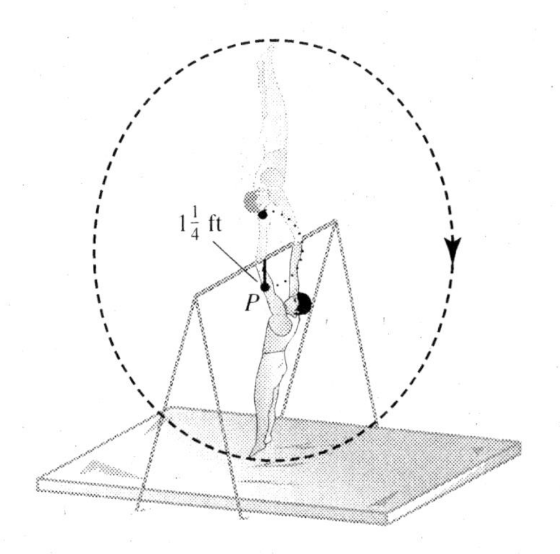

3.2 Trigonometric and Circular Functions

In Section 1.2 we defined the circular functions cos t and sin t as functions whose domains are real numbers.

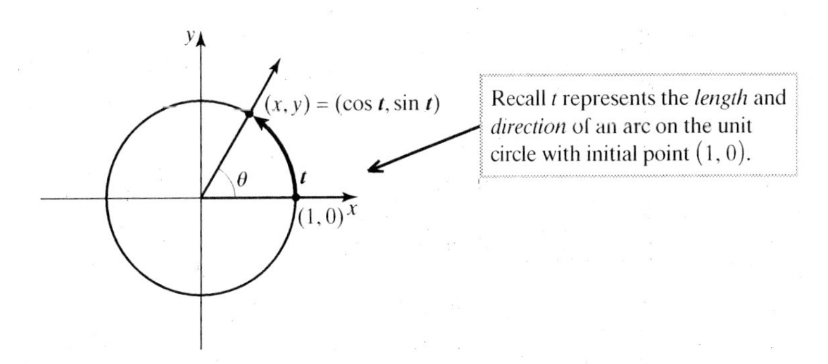

In this section we define new functions whose domains are angles based on the relationship between the arc t and its corresponding central angle, θ. To find this relationship, we begin with the fact that for each arc on a circle there corresponds a central angle θ. We know that the length of the arc can be found by using the formula $s = r\theta$, where θ is in radians. (See Section 3.1.) On the *unit* circle, where

$r = 1$, we have the special relationship between the length of an arc and its central angle: $s = r \cdot \theta = 1 \cdot \theta$, or $s = \theta$. If the measure of the central angle θ indicates direction of rotation, then the arc t has length and direction, or $t = \theta$. So if an angle θ in standard position intercepts arc t on the unit circle at the point $(x, y) = (1/2, \sqrt{3}/2)$, and since we know that

$$\cos t = \frac{1}{2} \quad \text{and} \quad \sin t = \frac{\sqrt{3}}{2},$$

it seems reasonable to say that

$$\cos \theta = \frac{1}{2} \quad \text{and} \quad \sin \theta = \frac{\sqrt{3}}{2}.$$

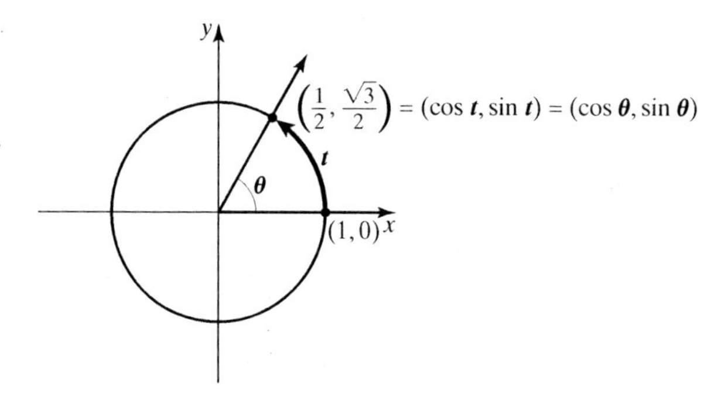

This new relationship requires us to define new functions because the domain is no longer a real number (corresponding to arc t), but the measure of an angle θ. The elements in the range of these new functions correspond to the elements in the range of the circular functions. The new functions are called the **trigonometric functions**. Before we define them, we need to look at some geometric ideas.

In the diagram shown P_1 is on the unit circle and P is any other point (other than $(0, 0)$) on the terminal side of angle θ. In geometry we can prove triangles are similar $(\sim)$ if two angles of one triangle have the same measure as two angles of the other triangle. We know $\triangle OA_1P_1 \sim \triangle OAP$ because each triangle has a right angle and they share the common angle θ. So if the triangles are similar, then their corresponding sides are proportional. Therefore, we get the following proportions:

$$\frac{x_1}{1} = \frac{x}{r}, \qquad \frac{y_1}{1} = \frac{y}{r}, \qquad \text{and} \qquad \frac{y_1}{x_1} = \frac{y}{x}.$$

Here, r is the distance from $P(x, y)$ to the origin. That is, $r = \sqrt{x^2 + y^2}$.

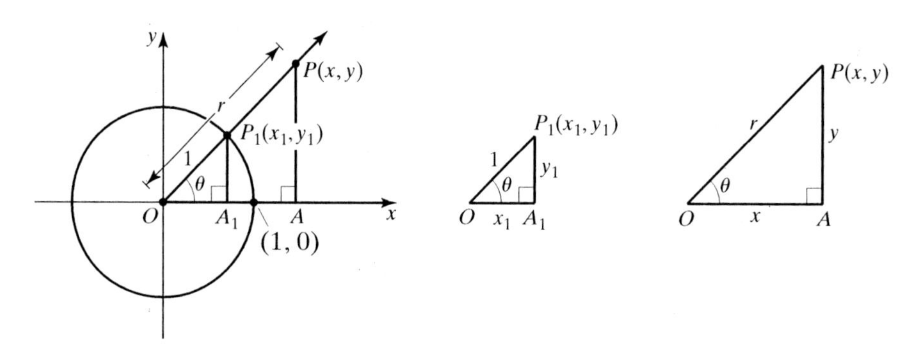

We are now ready to define these new trigonometric functions. *Notice in each definition, the relationship with the corresponding circular functions definition in which $r = 1$.*

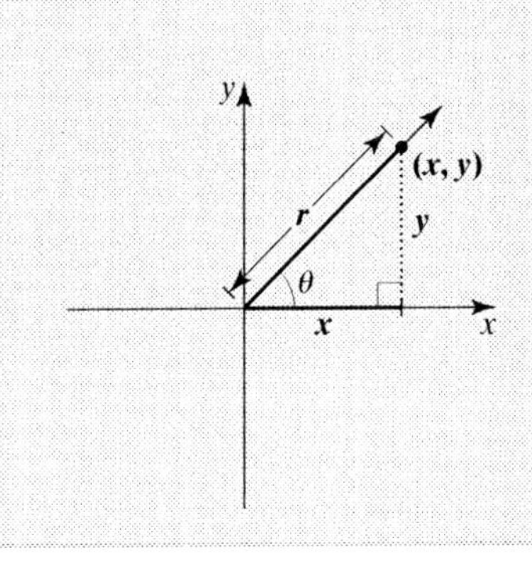

Trigonometric Functions

For any point (x, y) other than the origin on the terminal side of an angle θ in standard position, the trigonometric functions of θ are defined as:

$$\cos\theta = \frac{x}{r} \qquad \sin\theta = \frac{y}{r} \qquad \tan\theta = \frac{y}{x},\quad x \neq 0$$

$$\sec\theta = \frac{r}{x},\quad x \neq 0 \qquad \csc\theta = \frac{r}{y},\quad y \neq 0 \qquad \cot\theta = \frac{x}{y},\quad y \neq 0$$

where $r = \sqrt{x^2 + y^2}$ and $r > 0$ (since (x, y) is not the origin).

From this definition we make the following observations:

- The trigonometric ratios define functions since they assign to each angle in the domain a unique real number.
- The reciprocal relationships are consistent with the circular functions.
- Angles in the domain of the trigonometric functions correspond to the arcs in the domain of the corresponding circular functions. For example, the domain of the tangent (trigonometric function) is all angles θ such that $\theta \neq (\pi/2) + k\pi$ since points on the terminal sides of these angles have $x \neq 0$. This domain corresponds to the domain of the tangent (circular function), which is all $t \in \mathbb{R}$ such that $t \neq (\pi/2) + k\pi$.

The following is a summary for the domains of the trigonometric functions, where A represents the set of all angles and k is any integer. The ranges are omitted because they are identical to the ranges of the corresponding circular functions.

Trigonometric Function	Domain
$\cos\theta$, $\sin\theta$	$\theta \in A$
$\tan\theta$, $\sec\theta$	$\theta \neq \frac{\pi}{2} + k\pi$ or $\theta \neq 90° + k \cdot 180°$
$\csc\theta$, $\cot\theta$	$\theta \neq k\pi$ or $\theta \neq k \cdot 180°$

Later, we will see that these trigonometric functions *share* other properties and identities with the circular functions.

EXAMPLE 1 Find the six trigonometric functional values for an angle θ in standard position with the point $(3, 4)$ on its terminal side.

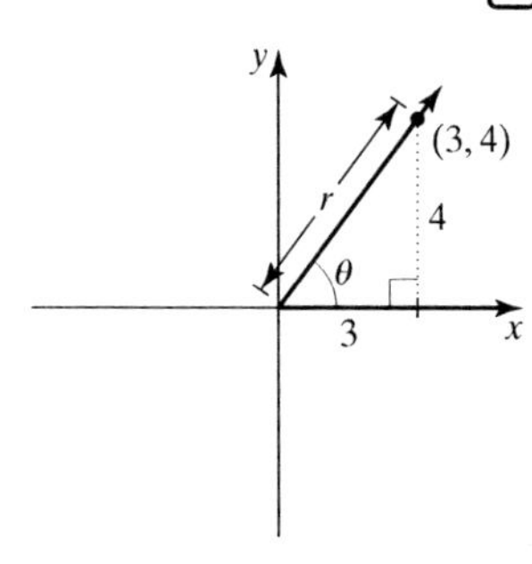

SOLUTION It is always a good idea to sketch the angle and label the point on the terminal side. In order to find the six trigonometric functional values, we need r. Since $(x, y) = (3, 4)$,

$$r = \sqrt{x^2 + y^2} = \sqrt{(3)^2 + (4)^2} = \sqrt{9 + 16} = \sqrt{25} = 5.$$

Using our definitions, we get:

$$\cos\theta = \frac{x}{r} = \frac{3}{5} \qquad \sin\theta = \frac{y}{r} = \frac{4}{5} \qquad \tan\theta = \frac{y}{x} = \frac{4}{3}$$

$$\sec\theta = \frac{5}{3} \qquad \csc\theta = \frac{5}{4} \qquad \cot\theta = \frac{3}{4}$$

When we labeled θ in the diagram for Example 1, we could have indicated any other angle coterminal with θ. Each coterminal angle would have the same point on the terminal side and therefore would have the same functional values. Since coterminal angles for θ are of the form $\theta + k(2\pi)$ or $\theta + k(360°)$, we have

$$\cos\theta = \cos[\theta + k(2\pi)] \quad \text{or} \quad \cos\theta = \cos[\theta + k(360°)].$$

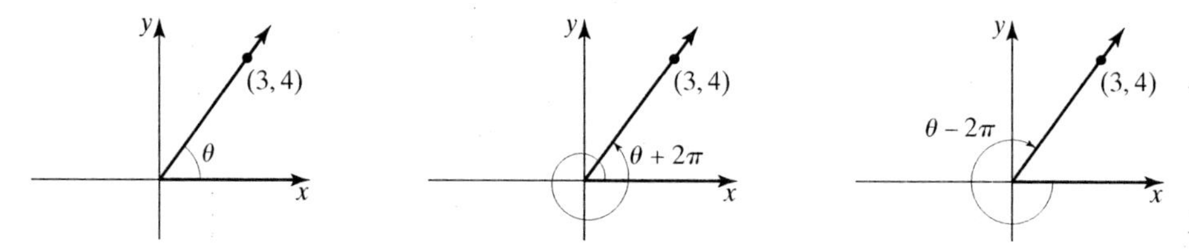

This would apply to $\sin\theta$, $\tan\theta$, or any of the other trigonometric functions. Therefore, the *trigonometric functions are periodic.*

Since r is always positive, the positive or negative nature of the trigonometric ratio is determined by the values of x and y, which depend on the quadrant in which the angle's terminal side lies. Therefore, the trigonometric values for an angle in a particular quadrant have the same positive and negative nature as the circular functions of arcs that terminate in the same quadrant.

EXAMPLE 2

CIRCULAR FUNCTION QUESTION Name the quadrant where $\sin t < 0$ and $\cos t > 0$.

SOLUTION In QIV, $(\cos t, \sin t) = (+, -)$.

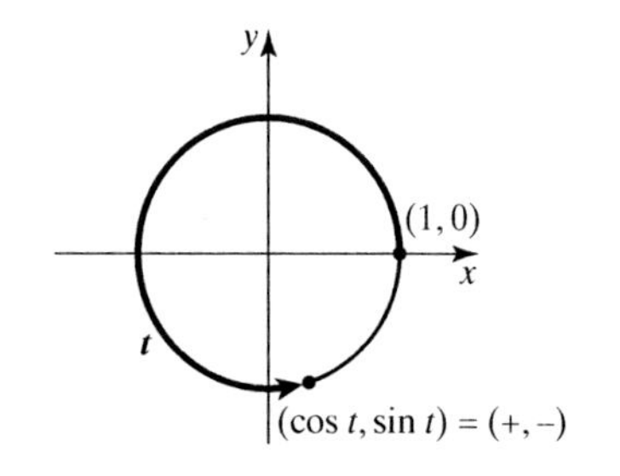

TRIGONOMETRIC FUNCTION QUESTION Name the quadrant where $\sin\theta < 0$ and $\cos\theta > 0$.

SOLUTION If $\sin\theta < 0$ and $\cos\theta > 0$, any point (x, y) on the terminal side of θ is of the form $(+, -)$, which is in QIV.

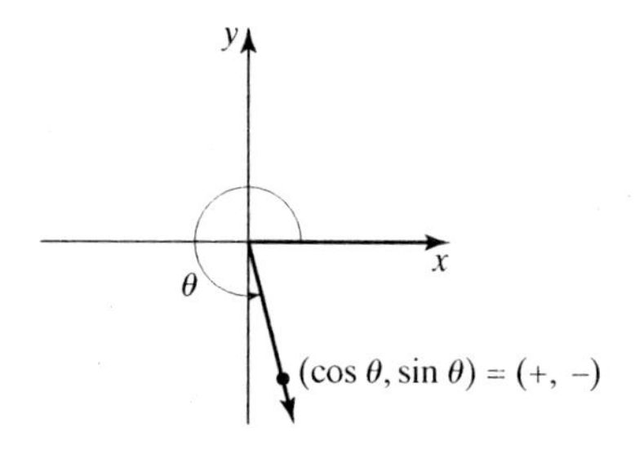

EXAMPLE 3 Does the Pythagorean identity hold for the trigonometric functions?

SOLUTION The Pythagorean identity for circular functions is

$$(\cos t)^2 + (\sin t)^2 = 1.$$

From the definitions of the trigonometric functions $\cos\theta$ and $\sin\theta$, we get:

$$\begin{aligned}(\cos\theta)^2 + (\sin\theta)^2 &= \left(\frac{x}{r}\right)^2 + \left(\frac{y}{r}\right)^2 \\ &= \frac{x^2 + y^2}{r^2} = \frac{r^2}{r^2} \qquad r = \sqrt{x^2 + y^2} \rightarrow r^2 = x^2 + y^2 \\ &= 1\end{aligned}$$

or $\quad \cos^2\theta + \sin^2\theta = 1$

Therefore, the Pythagorean identity also holds for the trigonometric functions. ■

As a result of Example 3, if we are given one of the trigonometric functional values and the quadrant of the angle, we can find all remaining functional values by using either the definitions of the trigonometric functions, or the Pythagorean identity.

EXAMPLE 4 If $\sec\theta = -\sqrt{26}/5$ and θ is in QIII, find all remaining functional values of θ.

SOLUTION 1 Use the definitions of the new trigonometric functions.

$$\sec\theta = -\frac{\sqrt{26}}{5} = \frac{r}{x}$$

So $x = -5$ and $r = \sqrt{26}$. To find the remaining functional values, we need y. Using the relationship $r = \sqrt{x^2 + y^2}$, we get:

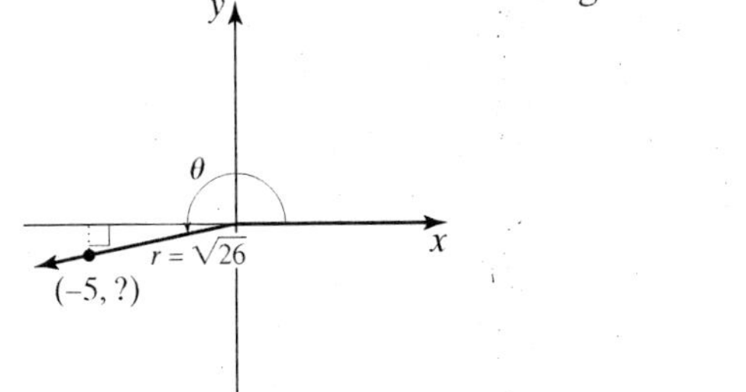

$$\begin{aligned} r &= \sqrt{x^2 + y^2} \\ \sqrt{26} &= \sqrt{(-5)^2 + y^2} && \text{Substitute } r = \sqrt{26},\ x = -5. \\ 26 &= 25 + y^2 && \text{Square both sides.} \\ y^2 &= 1 && \text{Subtract 25 from both sides.} \\ y &= \pm 1 && \text{Square root both sides.} \end{aligned}$$

Since we are given θ is in QIII, $y = -1$. We complete the functional values using the trigonometric definitions.

$$\cos\theta = \frac{x}{r} = \frac{-5}{\sqrt{26}} = -\frac{5\sqrt{26}}{26}$$

$$\sin\theta = \frac{y}{r} = \frac{-1}{\sqrt{26}} = -\frac{\sqrt{26}}{26} \qquad \csc\theta = \frac{r}{y} = -\sqrt{26}$$

$$\tan\theta = \frac{y}{x} = \frac{-1}{-5} = \frac{1}{5} \qquad \cot\theta = \frac{x}{y} = 5$$

(continued)

SOLUTION 2 Use the Pythagorean identity for the trigonometric functions. If $\sec\theta = -\sqrt{26}/5$, then $\cos\theta = -5/\sqrt{26}$.

$$\begin{aligned}(\cos\theta)^2 + (\sin\theta)^2 &= 1\\ (-5/\sqrt{26})^2 + (\sin\theta)^2 &= 1\\ 25/26 + (\sin\theta)^2 &= 1\\ (\sin\theta)^2 &= 1/26\\ \sin\theta &= \pm 1/\sqrt{26}\end{aligned}$$

Since θ is in QIII,

$$\sin\theta = -1/\sqrt{26} = -\sqrt{26}/26, \quad \csc\theta = -\sqrt{26} \qquad \text{reciprocal relationship,}$$

$$\tan\theta = \frac{\sin\theta}{\cos\theta} = \frac{-1/\sqrt{26}}{-5/\sqrt{26}} = \frac{1}{5}, \quad \text{and} \quad \cot\theta = 5 \qquad \text{reciprocal relationship.}$$

While Solution 2 may be a more familiar way to find functional values, you might want to use Solution 1 in order to become more familiar with the trigonometric functions definitions. ■

Functional Values for Common Angles

The correspondence between t and θ allows us to find exact trigonometric functional values of angles that correspond to common arcs. We refer to these particular values of θ as common angles, and we use these exact values as we did with circular functions. We can also use a calculator to find approximations for functional values of other angles.

θ(rad)	$\theta°$	$\cos\theta$	$\sin\theta$	$\tan\theta$
0	$0°$	1	0	0
$\frac{\pi}{6}$	$30°$	$\frac{\sqrt{3}}{2}$	$\frac{1}{2}$	$\frac{1}{\sqrt{3}}$ or $\frac{\sqrt{3}}{3}$
$\frac{\pi}{4}$	$45°$	$\frac{1}{\sqrt{2}}$ or $\frac{\sqrt{2}}{2}$	$\frac{1}{\sqrt{2}}$ or $\frac{\sqrt{2}}{2}$	1
$\frac{\pi}{3}$	$60°$	$\frac{1}{2}$	$\frac{\sqrt{3}}{2}$	$\sqrt{3}$
$\frac{\pi}{2}$	$90°$	0	1	undefined
π	$180°$	-1	0	0
$\frac{3\pi}{2}$	$270°$	0	-1	undefined

For any angle θ in Quadrants I, II, III, or IV, we use a reference angle $\hat{\theta}$ (whose sides are the terminal side of θ and the x-axis), where $0° < \hat{\theta} < 90°$, in the same way we used a reference arc ($\hat{t}$ or $\hat{x}$) to find functional values.

EXAMPLE 5 Find the exact trigonometric functional value.

a. $\sin 330°$ **b.** $\tan \frac{5\pi}{4}$

SOLUTION

a.

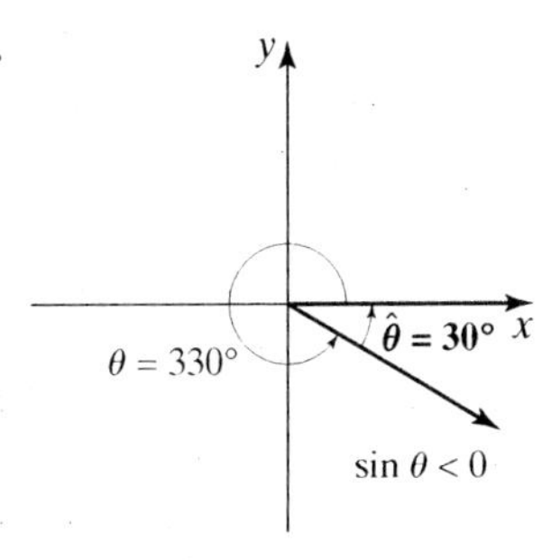

$$\hat{\theta} = 360° - 330° = 30°$$

$$\sin 330° = -\sin 30° = -\frac{1}{2}$$

b.

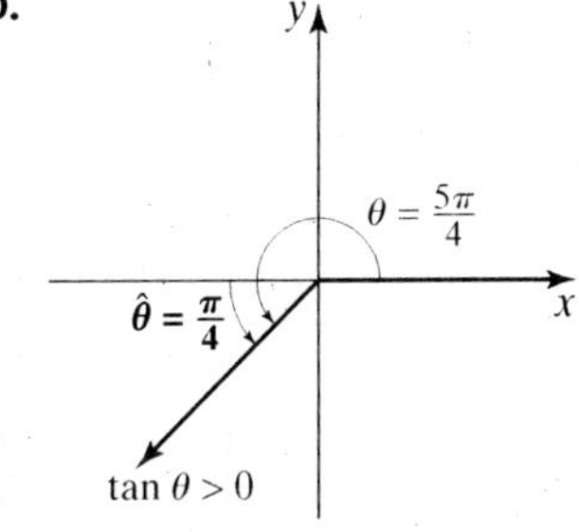

$$\hat{\theta} = \frac{5\pi}{4} - \pi = \frac{\pi}{4}$$

$$\tan \frac{5\pi}{4} = \tan \frac{\pi}{4} = 1$$

NOTE: If we are not given the specific domain (angles in radians or arcs as real numbers) in a problem like Example 5 (b), we could interpret the problem as either a trigonometric or circular function question. Would we get a different answer? (See Exercise 17 in Section 1.5.)

The periods of the trigonometric functions and the ranges for their inverse functions are analogous to those of the circular functions.

Trigonometric Function	Period
$\cos\theta, \sin\theta$ $\sec\theta, \csc\theta$	2π or $360°$
$\tan\theta, \cot\theta$	π or $180°$

Inverse Trigonometric Function	Range
$\theta = \cos^{-1} x$	$0 \le \theta \le \pi$ or $0° \le \theta \le 180°$
$\theta = \sin^{-1} x$	$-\frac{\pi}{2} \le \theta \le \frac{\pi}{2}$ or $-90° \le \theta \le 90°$
$\theta = \tan^{-1} x$	$-\frac{\pi}{2} < \theta < \frac{\pi}{2}$ or $-90° < \theta < 90°$

EXAMPLE 6 Find the exact radian measure of θ.

a. $\theta = \tan^{-1} 1$ **b.** $\theta = \sec^{-1}(-2)$ **c.** $\theta = \arcsin \frac{\sqrt{3}}{2}$

SOLUTION We rewrite without using inverse notation and indicate the range.

a. $\theta = \tan^{-1} 1 \longleftrightarrow \tan\theta = 1,\ -\frac{\pi}{2} < \theta < \frac{\pi}{2}$ (θ is in QI since $\tan\theta > 0$)

$\hat{\theta} = \frac{\pi}{4}$, so $\theta = \frac{\pi}{4}$

b. $\theta = \sec^{-1}(-2) \longleftrightarrow \sec\theta = -2,\ 0 \le \theta < \frac{\pi}{2}$ or $\frac{\pi}{2} < \theta \le \pi$

$\cos\theta = -\frac{1}{2}$ (θ is in QII since $\cos\theta < 0$)

$\hat{\theta} = \frac{\pi}{3}$, so $\theta = \frac{2\pi}{3}$

c. $\theta = \arcsin \frac{\sqrt{3}}{2} \longleftrightarrow \sin\theta = \frac{\sqrt{3}}{2},\ -\frac{\pi}{2} \le \theta \le \frac{\pi}{2}$ (θ is in QI since $\sin\theta > 0$)

$\hat{\theta} = \frac{\pi}{3}$, so $\theta = \frac{\pi}{3}$ ■

EXAMPLE 7 Find the degree measure of θ.

a. $\theta = \cos^{-1} \frac{1}{2}$ **b.** $\theta = \csc^{-1}(-\sqrt{2})$ **c.** $\theta = \arctan \frac{\sqrt{3}}{3}$

SOLUTION We rewrite without using inverse notation and indicate the range.

a. $\theta = \cos^{-1} \frac{1}{2} \longleftrightarrow \cos\theta = \frac{1}{2},\ 0 \le \theta \le 180°$ (θ is in QI since $\cos\theta > 0$)

$\hat{\theta} = 60°$, so $\theta = 60°$

b. $\theta = \csc^{-1}(-\sqrt{2}) \longleftrightarrow \csc\theta = -\sqrt{2},\ -90° \leq \theta < 0 \text{ or } 0 < \theta \leq 90°$

$$\sin\theta = -\frac{1}{\sqrt{2}} \quad (\theta \text{ is in QIV since } \sin\theta < 0)$$

$\hat{\theta} = 45°$, so $\qquad \theta = -45°$

c. $\theta = \arctan\frac{\sqrt{3}}{3} \longleftrightarrow \tan\theta = \frac{\sqrt{3}}{3},\ -90° < \theta < 90°$

$\hat{\theta} = 30°$, so $\qquad \theta = 30° \quad (\theta \text{ is in QI since} \tan\theta > 0)$ ■

Because of your experience with inverse functions in Section 2.5, you might find that with problems like those in Example 7 you will first answer inverse functions in radian measure. That's ok. If you want the answer in degrees, just convert!

EXAMPLE 8 Find the exact value for $\cos\left(\arctan\frac{2}{7}\right)$.

SOLUTION $\cos\left(\arctan\frac{2}{7}\right) = \cos\theta \qquad$ Substitute θ for $\arctan\left(\frac{2}{7}\right)$.

So,

$$\theta = \arctan\left(\frac{2}{7}\right), \text{ where } -\frac{\pi}{2} < \theta < \frac{\pi}{2}.$$

Rewriting, we get

$$\theta = \arctan\left(\frac{2}{7}\right) \longleftrightarrow \tan\theta = \frac{2}{7},$$

and θ is in QI because $\tan\theta > 0$.

Again, we have the old question where we are given one functional value $(\tan\theta)$ and asked to find another functional value $(\cos\theta)$. Now we use the definition for the trigonometric tangent function.

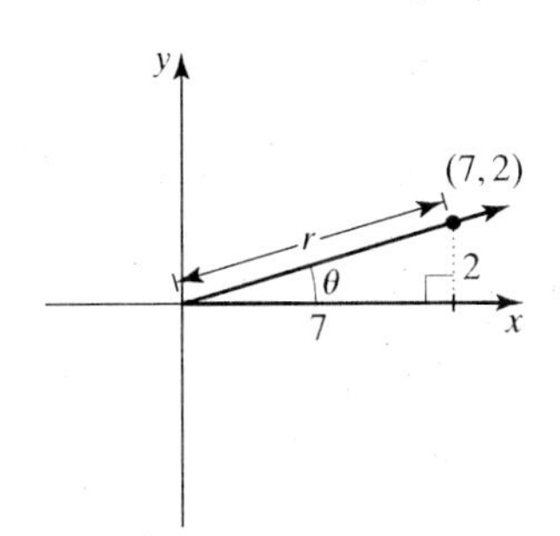

$$\tan\theta = \frac{y}{x} = \frac{2}{7} \rightarrow x = 7,\ y = 2$$

$$r = \sqrt{7^2 + 2^2}$$

$$r = \sqrt{53}$$

Our original substitution indicates that we want $\cos\theta$. Therefore, we use the trigonometric definition for $\cos\theta$.

$$\cos\theta = \frac{x}{r} = \frac{7}{\sqrt{53}}$$

So,

$$\cos\left(\arctan\frac{2}{7}\right) = \cos\theta = \frac{7}{\sqrt{53}} = \frac{7\sqrt{53}}{53}.$$ ■

EXAMPLE 9 Being careful to use the correct mode (degree or radian), use a calculator to find the following. Approximate the answer by rounding to four decimal places.

a. $\tan 265°$ **b.** $\sec(-2.1)$

c. arctan 53 in degrees **d.** $\sec^{-1} 5$ in radians

SOLUTION

a. $\tan 265° \approx 11.4301$

b. $\sec(-2.1) \approx -1.9808$

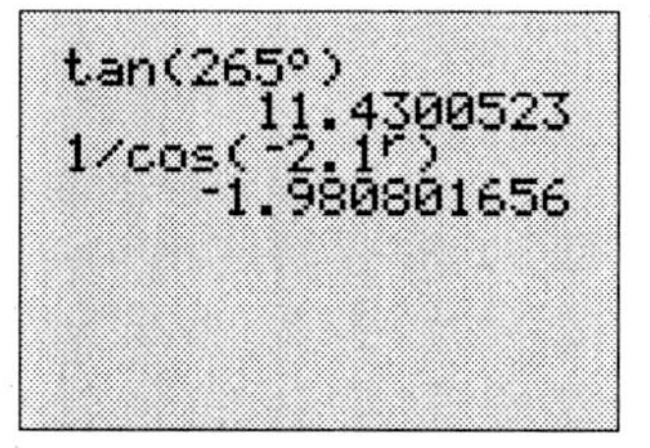

c. $\tan^{-1} 53 \approx 88.9191°$

d. $\sec^{-1} 5 \approx 1.3694$ radians

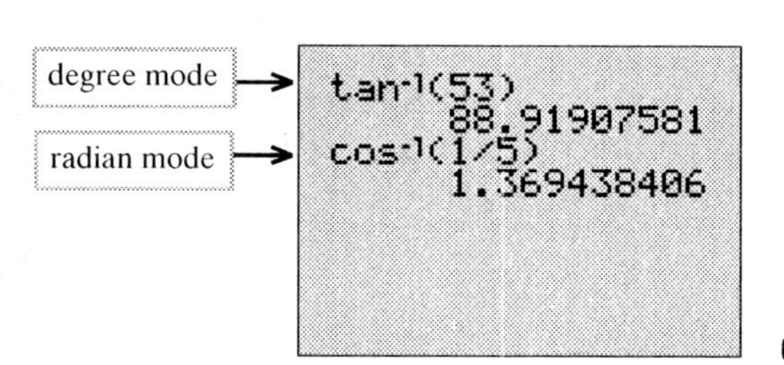

You will eventually encounter problems having two or more angles. So that you become more familiar with different angle names, such as α, β, or γ, we vary these symbols in the exercises. The exercises for this section are quite extensive because they include ideas from Chapters 1 and 2. Remember that the domain for the trigonometric functions is angles and the range for the inverse trigonometric functions is angles.

Exercise Set 3.2

1–10. *Find* $\cos\theta$, $\sin\theta$, *and* $\tan\theta$ *for an angle* θ *in standard position if the given point is on its terminal side. Sketch* θ *as the smallest positive angle and label the given point on the terminal side.*

EXAMPLE $(-8, 6)$

SOLUTION If $(x, y) = (-8, 6)$, then $x = -8$ and $y = 6$.

Using $r = \sqrt{x^2 + y^2} = \sqrt{(-8)^2 + 6^2} = 10$ and our definitions, we get:

$$\cos\theta = \frac{x}{r} = \frac{-8}{10} = -\frac{4}{5}$$

$$\sin\theta = \frac{y}{r} = \frac{6}{10} = \frac{3}{5}$$

$$\tan\theta = \frac{y}{x} = \frac{6}{-8} = -\frac{3}{4}$$

1. $(8, 15)$ **2.** $(5, 12)$ **3.** $(2, -2)$ **4.** $(-1, 5)$ **5.** $(-4, 3)$

6. $(-6, -8)$ **7.** $(-\sqrt{3}, -1)$ **8.** $(2, -2\sqrt{3})$ **9.** $(-5, 0)$ **10.** $(0, 4)$

11–14. *Find the quadrant that contains the terminal side of angle α.*

11. $\sin\alpha < 0$ and $\sec\alpha > 0$

12. $\tan\alpha > 0$ and $\sin\alpha < 0$

13. $\csc\alpha > 0$ and $\cos\alpha < 0$

14. $\cos\alpha > 0$ and $\tan\alpha > 0$

15–22. *Find the five other trigonometric values for each angle with the given information.*

EXAMPLE $\cot\beta = \frac{5}{2}$ and β is in QIII.

SOLUTION Since β is in QIII where $(x, y) = (-, -)$,

$$\cot\beta = \frac{x}{y} = \frac{5}{2} = \frac{-5}{-2}.$$

So $x = -5$ and $y = -2$.

$$r = \sqrt{x^2 + y^2} = \sqrt{(-5)^2 + (-2)^2} = \sqrt{29}$$

$$\cos\beta = \frac{x}{r} = \frac{-5}{\sqrt{29}} = -\frac{5\sqrt{29}}{29} \qquad \sec\beta = -\frac{\sqrt{29}}{5}$$

$$\sin\beta = \frac{y}{r} = \frac{-2}{\sqrt{29}} = -\frac{2\sqrt{29}}{29} \qquad \csc\beta = -\frac{\sqrt{29}}{2}$$

$$\tan\beta = \frac{2}{5}$$

15. $\cos\beta = \frac{1}{4}$ and β is in QIV.

16. $\tan\beta = -5$ and β is in the QII.

17. $\sec\beta = 1.5$ and $0° < \beta < 90°$.

18. $\sin\beta = -\frac{2}{3}$ and $180° < \beta < 270°$.

19. $\csc\beta = -4$ and $\cos\beta < 0$.

20. $\cot\beta = -2$ and $\cos\beta > 0$.

21. $\tan\beta = -\frac{1}{2}$ and β is in QII.

22. $\cos\beta = \frac{12}{13}$ and β is in QI.

23–36. *Find the exact functional value if it is defined.*

EXAMPLE $\tan(-300°)$

SOLUTION For $\theta = -300°$, $\hat{\theta} = -300° + 360° = 60°$

Since θ is in QI, $\tan\theta > 0$.

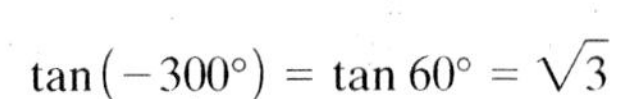

$$\tan(-300°) = \tan 60° = \sqrt{3}$$

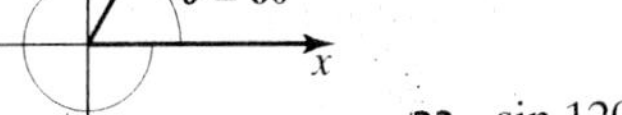

23. $\sin 120°$

24. $\cot 210°$

25. $\csc 330°$

26. $\tan 405°$

27. $\sec^2 900°$

28. $\sin^2 660°$

29. $\cot\frac{7\pi}{4}$

30. $\cos\frac{7\pi}{6}$

31. $\cos\left(-\frac{\pi}{2}\right)$

32. $\sec\frac{8\pi}{3}$

33. $\tan(-150°)$

34. $\sec(-495°)$

35. $\tan\left(-\frac{5\pi}{2}\right)$

36. $\csc(-3\pi)$

37–48. *Being careful to set your calculator in the correct mode, find an approximation rounded to four decimal places for each of the following expressions.*

```
cos(5.2r)
        .4685166713
1/tan(-158.5°)
        2.538647896
```

Example

a. $\cos 5.2$ **b.** $\cot(-158.5°)$

Solution

a. $\cos 5.2 \approx 0.4685$ **b.** $\cot(-158.5°) \approx 2.5386$ ■

37. $\tan 25°$ **38.** $\sec 317°$ **39.** $\sin \frac{5\pi}{9}$ **40.** $\cot \frac{\pi}{7}$

41. $\cos 53°24'$ **42.** $\tan 44°6'$ **43.** $\sec\left(-125\tfrac{3}{4}°\right)$ **44.** $\csc 27\tfrac{2}{5}°$

45. $\cot 9.25$ **46.** $\sin 13.5$ **47.** $\sin\left(\cos^{-1} 0.55\right)$ **48.** $\tan\left(\sin^{-1} 0.26\right)$

49–54. *Find the exact value of the angle α in degrees.*

Example $\alpha = \cos^{-1}\left(-\frac{\sqrt{2}}{2}\right)$

Solution Writing without inverse notation, we get:

$$\cos \alpha = -\frac{\sqrt{2}}{2} \quad \text{for} \quad 0 \le \alpha \le 180°.$$

We recognize $\hat{\alpha} = 45°$, and since $\cos \alpha < 0$, α is in QII.

So, $\alpha = 135°$. ■

49. $\alpha = \arccos \frac{\sqrt{3}}{2}$ **50.** $\alpha = \sin^{-1}(-1)$ **51.** $\alpha = \sin^{-1}\left(-\frac{1}{2}\right)$

52. $\alpha = \arctan\left(-\sqrt{3}\right)$ **53.** $\alpha = \cot^{-1} 0$ **54.** $\alpha = \cos^{-1} \frac{\sqrt{2}}{2}$

55–60. *Find an approximate value to the nearest tenth of a degree of the angle θ.*

55. $\theta = \cos^{-1} 0.225$ **56.** $\theta = \tan^{-1}(-2.56)$ **57.** $\theta = \sin^{-1}(-0.456)$

58. $\theta = \operatorname{arcsec} 5.1$ **59.** $\theta = \arctan(-0.4577)$ **60.** $\theta = \cos^{-1}(-0.25)$

61–68. *Find the exact functional value. Compare your answer with one found using a calculator when possible.*

Example $\sin\left(\cos^{-1} \frac{2}{3}\right)$

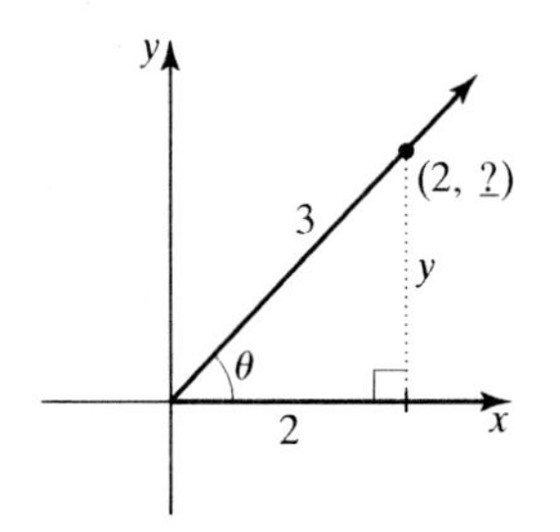

Solution $\sin\left(\cos^{-1} \frac{2}{3}\right) = \sin \theta$ Substitute θ for $\cos^{-1} \frac{2}{3}$.

$$\theta = \cos^{-1} \frac{2}{3}, \quad 0 \le \theta \le \pi$$

$$\theta = \cos^{-1} \frac{2}{3} \longleftrightarrow \cos \theta = \frac{2}{3}$$ θ is in QI because $\cos \theta > 0$.

```
sin(cos⁻¹(2/3))
      .7453559925
√(5)/3
      .7453559925
```

$$\cos\theta = \frac{x}{r} = \frac{2}{3} \longrightarrow x = 2,\ r = 3$$

$$r^2 = x^2 + y^2 \qquad r = \sqrt{x^2 + y^2} \longrightarrow r^2 = x^2 + y^2$$

$$3^2 = 2^2 + y^2$$

$$y = \sqrt{5} \qquad \text{Since } \theta \text{ is in QI, } y > 0.$$

$$\sin\theta = \frac{y}{r} = \frac{\sqrt{5}}{3}$$

$$\sin\left(\cos^{-1}\frac{2}{3}\right) = \sin\theta = \frac{\sqrt{5}}{3}$$

61. $\cos\left(\arcsin\frac{1}{4}\right)$ **62.** $\sec\left(\tan^{-1}\frac{8}{3}\right)$ **63.** $\tan\left(\cos^{-1}\frac{3}{5}\right)$

64. $\sin\left(\arccos\frac{5}{13}\right)$ **65.** $\csc\left(\tan^{-1}\frac{5}{12}\right)$ **66.** $\csc\left(\cot^{-1}\frac{4}{3}\right)$

67. $\cos\left(\arcsin\frac{x}{2}\right),\ -2 \le x \le 2$ **68.** $\sin\left(\cos^{-1}\frac{x}{4}\right),\ -4 \le x \le 4$

3.3 Solving Right Triangles and Applications

When we use information given about one, two, or three sides of a triangle and one or two angles to find all of the remaining sides and angles, we are solving a triangle. If you take a closer look at some examples from Section 3.2, in which you are asked to find trigonometric values of an angle in standard position given a point on its terminal side, you will notice that you were beginning to solve a right triangle. To better understand this, let's recall Example 1 from Section 3.2.

EXAMPLE: Find the six trigonometric functional values for an angle θ in standard position with the point $(3, 4)$ on its terminal side.

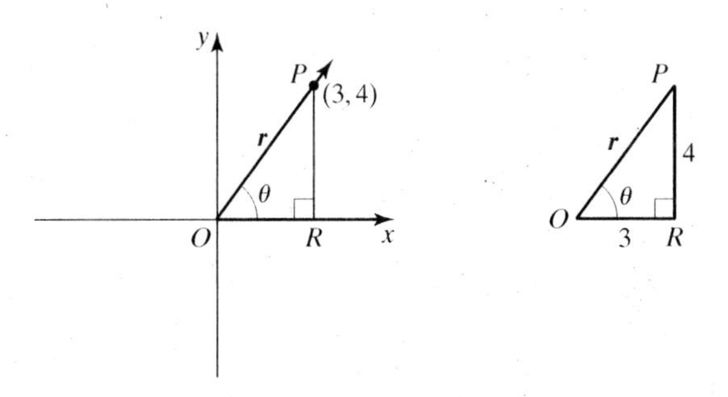

If we look at the right triangle formed by O, P, and R, we have the right angle at R, and we know the lengths of two sides. This example asked for the trigonometric functional values, so we began by finding the length of r by using $r = \sqrt{x^2 + y^2}$. If we look at r in the context of $\triangle ORP$, r is the hypotenuse of this right triangle. So, when we found the length of this unknown side, we solved one part of a right triangle problem. At this point we know the lengths of all three sides of $\triangle ORP$ and the measure of one angle. In order to finish the solution of this triangle, we need to find the measures of the two remaining angles, which we learn how to do in this section.

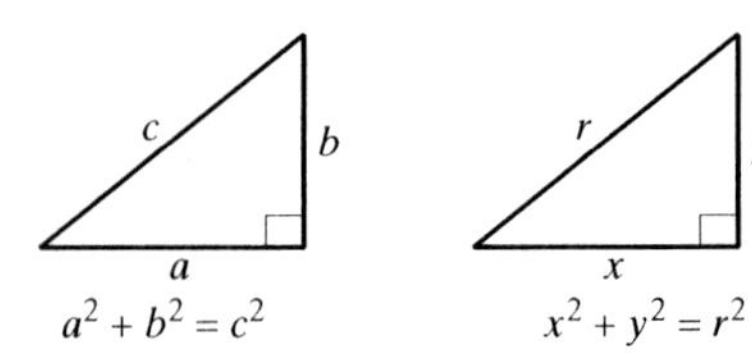

You may be familiar with a theorem from geometry called the *Pythagorean Theorem,* which says that the square of the length of the hypotenuse (longest side) of a right triangle is equal to the sum of the squares of the lengths of the legs (remaining sides), or $a^2 + b^2 = c^2$. Viewing the previous example as a right triangle problem, we are applying the Pythagorean Theorem in the form $x^2 + y^2 = r^2$, where x and y are the sides and r is the hypotenuse of a right triangle.

Right triangles can be solved by using the trigonometric functions ratios of the acute angles of the triangle. To see how this can be done, let any point (x, y) on the terminal side of an acute angle θ in standard position determine a right triangle with sides of length x and y and the hypotenuse of length $r = \sqrt{x^2 + y^2}$. The hypotenuse is the side opposite the right angle. If we rename sides x and y in terms of their positions with respect to θ we have x as the side adjacent to θ and y as the side opposite θ. Replacing these names in the trigonometric ratios, we get the trigonometric ratios for right triangles.

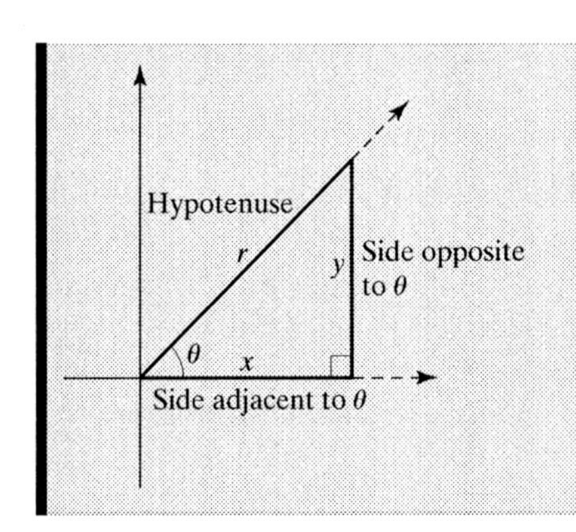

Trigonometric Ratios for Right Triangles

For any acute angle θ of a right triangle,

$$\sin\theta = \frac{\text{length of side opposite } \theta}{\text{length of hypotenuse}} \qquad \csc\theta = \frac{\text{length of hypotenuse}}{\text{length of side opposite } \theta}$$

$$\cos\theta = \frac{\text{length of side adjacent } \theta}{\text{length of hypotenuse}} \qquad \sec\theta = \frac{\text{length of hypotenuse}}{\text{length of side adjacent } \theta}$$

$$\tan\theta = \frac{\text{length of side opposite } \theta}{\text{length of side adjacent } \theta} \qquad \cot\theta = \frac{\text{length of side adjacent } \theta}{\text{length of side opposite } \theta}$$

NOTE: Exercise 66 gives a mnemonic to help you memorize these ratios.

Once we have the trigonometric functions ratios for a right triangle, it is not necessary that the acute angle θ be in standard position. We can apply these ratios to a right triangle in any position. The lengths of the sides of a triangle are labeled a, b, and c. *The angles across* from a, b, and c are labeled α, β, and γ, respectively. Sometimes the angles will be labeled using names for the vertices, such as A, B, or C. In a right triangle, γ or C is customarily used to name the right angle. Angles of triangles typically are measured in degrees.

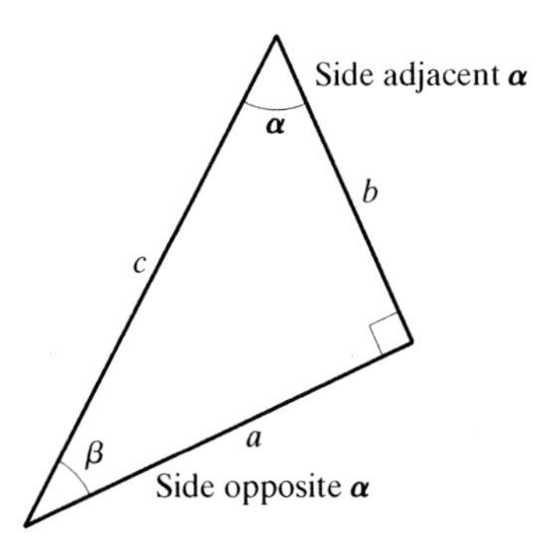

$$\sin\alpha = \frac{a}{c} \qquad \csc\alpha = \frac{c}{a}$$

$$\cos\alpha = \frac{b}{c} \qquad \sec\alpha = \frac{c}{b}$$

$$\tan\alpha = \frac{a}{b} \qquad \cot\alpha = \frac{b}{a}$$

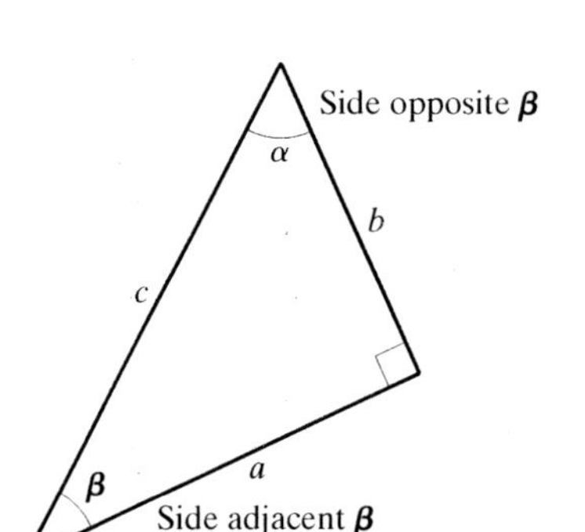

$$\sin\beta = \frac{b}{c} \qquad \csc\beta = \frac{c}{b}$$

$$\cos\beta = \frac{a}{c} \qquad \sec\beta = \frac{c}{a}$$

$$\tan\beta = \frac{b}{a} \qquad \cot\beta = \frac{a}{b}$$

CAUTION: Notice that side a, the side opposite angle α, is adjacent to angle β. Don't incorrectly refer to side c as the side adjacent to β because c represents the hypotenuse. When you are setting up these ratios, be very careful to identify the sides correctly.

EXAMPLE 1 Find the values of the trigonometric functions for angles α and β.

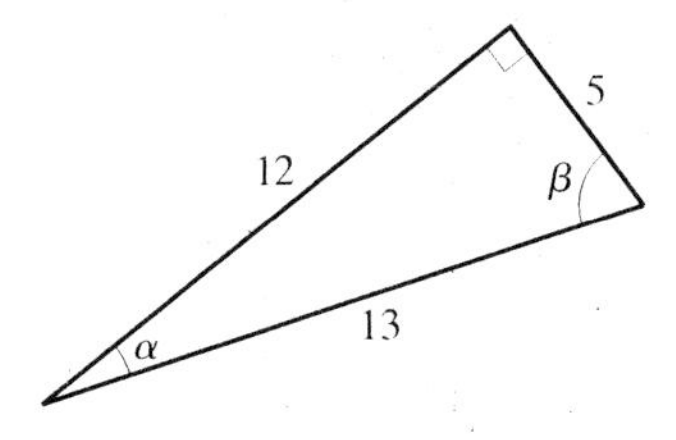

SOLUTION The length of the side opposite α is 5, the length of the side adjacent to α is 12, and the length of the hypotenuse is 13. The length of the side opposite β is 12, the length of the side adjacent to β is 5. Using the trigonometric ratios for right triangles, we get:

$$\sin\alpha = \frac{\text{opp}}{\text{hyp}} = \frac{5}{13} \qquad \csc\alpha = \frac{13}{5} \qquad \sin\beta = \frac{\text{opp}}{\text{hyp}} = \frac{12}{13} \qquad \csc\beta = \frac{13}{12}$$

$$\cos\alpha = \frac{\text{adj}}{\text{hyp}} = \frac{12}{13} \qquad \sec\alpha = \frac{13}{12} \qquad \cos\beta = \frac{\text{adj}}{\text{hyp}} = \frac{5}{13} \qquad \sec\beta = \frac{13}{5}$$

$$\tan\alpha = \frac{\text{opp}}{\text{adj}} = \frac{5}{12} \qquad \cot\alpha = \frac{12}{5} \qquad \tan\beta = \frac{\text{opp}}{\text{adj}} = \frac{12}{5} \qquad \cot\beta = \frac{5}{12}$$

In any triangle the sum of the measures of the three angles is 180°. Since one angle in a right triangle is 90°, the measures of the remaining two acute angles add to 90°, $\alpha + \beta = 90°$ (complementary).

EXAMPLE 2 In Example 1, what is the relationship between the values of $\sin\alpha$ and $\cos\beta$?

SOLUTION $\sin\alpha = \frac{5}{13}$ and $\cos\beta = \frac{5}{13}$. Therefore,

$$\sin\alpha = \cos\beta, \text{ where } \alpha + \beta = 90°.$$

Example 2 demonstrates the following relationship between the sine and cosine.

Cofunctions

The sine and cosine functions are called **cofunctions**, as are the tangent and the cotangent, the secant and the cosecant.

For two angles α and β such that $\alpha + \beta = 90°$, we have the following relationship between cofunctions:

$$\sin\alpha = \cos(90° - \alpha) = \cos\beta$$

$$\tan\alpha = \cot(90° - \alpha) = \cot\beta$$

$$\sec\alpha = \csc(90° - \alpha) = \csc\beta$$

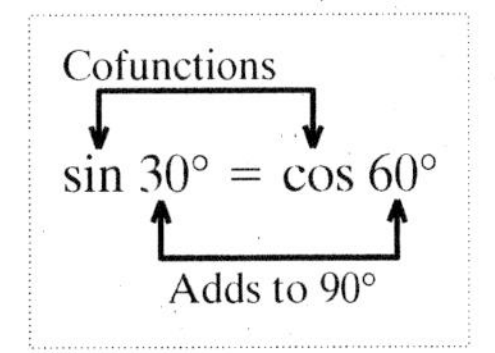

EXAMPLE 3 Complete the statement with a function or an acute angle that makes the statement true. Use a calculator in degree mode to verify the statement.

a. $\sin 56° = \cos$ ______ **b.** $\cot 13.5° = \tan$ ______

c. $\sec 23° =$ ______ $67°$

SOLUTION

a. $\sin 56° = \cos(90° - 56°)$
$= \cos 34°$

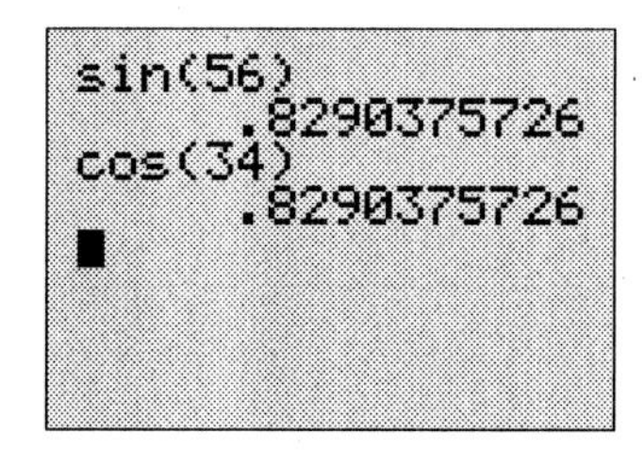

b. $\cot 13.5° = \tan(90° - 13.5°)$
$= \tan 76.5°$

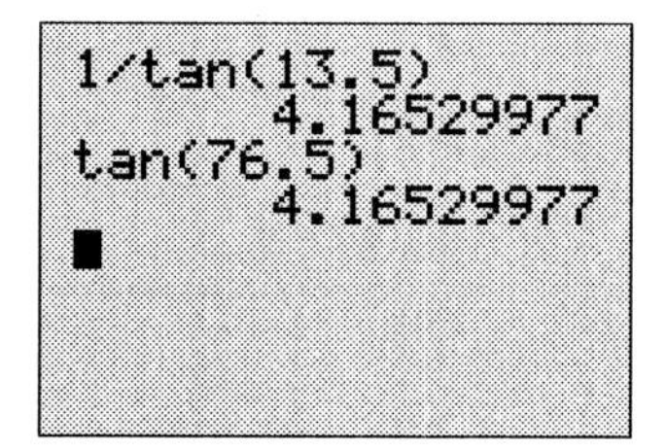

c. $\sec 23° = \csc(90° - 23°)$
$= \csc 67°$

Solving Right Triangles

In the following examples we solve right triangles by using the trigonometric ratios and the fact that the sum of the measures of the angles of any triangle is 180°. *When we solve triangles, we must be given the measures of at least three parts, and at least one of those parts must be the length of a side.* If we are given three angles, we know only the shape of the triangle, not the lengths of the sides, because similar triangles have equal angles.

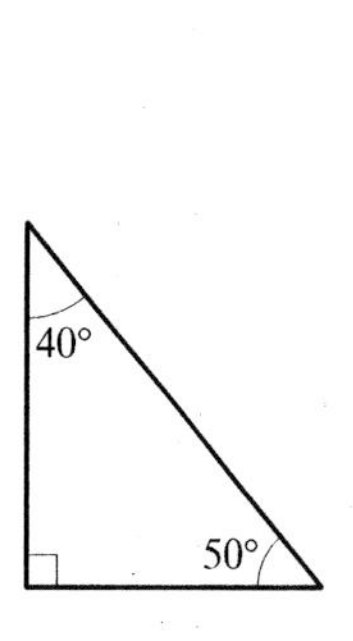

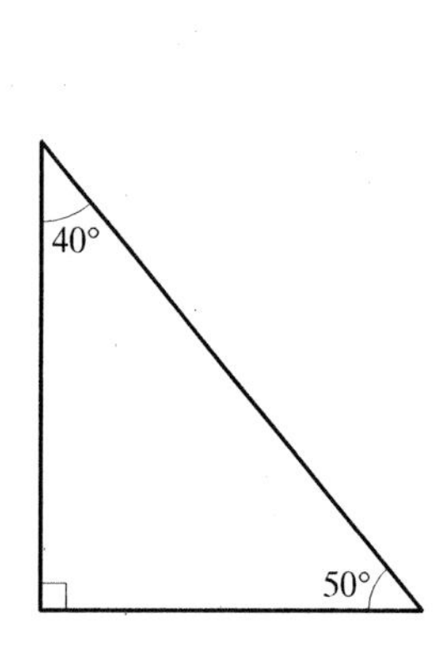

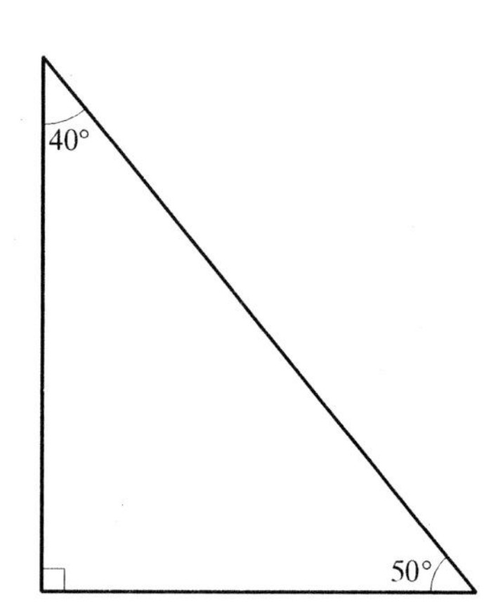

For the most part, all measurements are approximations. As you have seen previously, many of our functional values are approximations. In solving triangles we will approximate the answers when necessary in the following manner, unless otherwise indicated.

- lengths of sides rounded to two decimal places (the nearest hundredth of a unit)
- angles in degrees rounded to one decimal place (the nearest tenth of a degree)

In order to simplify the presentation of our examples, the values in the intermediate steps will display only four decimal places. This is not meant to indicate any rounding during the problem. *You should not round until the final answer.* In addition, in this text we will not always distinguish between the side and its length.

EXAMPLE 4 In a right triangle $\beta = 30°$ and side $c = 10$. Find the remaining angle and sides of the triangle, that is, *solve the triangle.*

SOLUTION We begin by drawing a right triangle. This is critical for visualizing the relationship between the information we are given and what we need to find. We label the triangle with the given information and enter these values in a chart like the one shown. As we solve the triangle, we continue to label the triangle and enter values in the chart. Doing so keeps our solution organized.

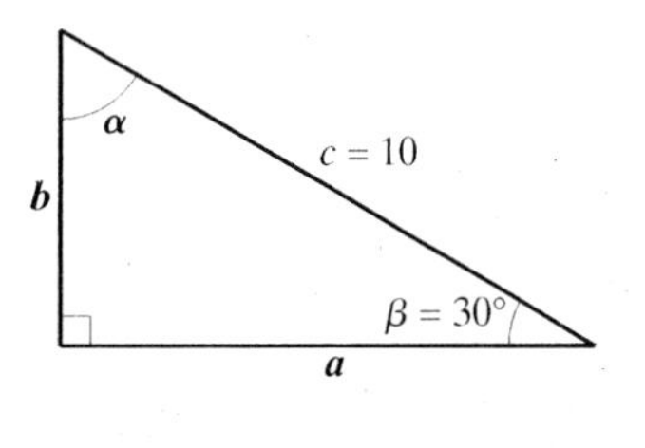

$a = ?$	$\alpha = ?$
$b = ?$	$\beta = 30°$
$c = 10$	$\gamma = 90°$

The sum of the angles is 180°.

$$\alpha + \beta + \gamma = 180°, \quad \text{where } \beta = 30° \quad \text{and} \quad \gamma = 90°.$$

$$\alpha + 30° + 90° = 180° \rightarrow \alpha = 60°$$

To find a and b, we can use the trigonometric ratios for α or β. We arbitrarily select β:

$$\cos 30° = \frac{\text{adj}}{\text{hyp}} = \frac{a}{10}$$

$$\frac{\sqrt{3}}{2} = \frac{a}{10} \quad \text{Solve for } a.$$

$$a = 10\left(\frac{\sqrt{3}}{2}\right) = 5\sqrt{3} \approx 8.66$$

$$\sin 30° = \frac{\text{opp}}{\text{hyp}} = \frac{b}{10}$$

$$\frac{1}{2} = \frac{b}{10} \quad \text{Solve for } b.$$

$$b = 10\left(\frac{1}{2}\right) = 5$$

$a = 5\sqrt{3}$ ≈ 8.66	$\alpha = 60°$
$b = 5$	$\beta = 30°$
$c = 10$	$\gamma = 90°$

Notice in Example 4 that after finding the *exact value* for a, we would get the same answer for b if we use the Pythagorean Theorem instead of using a trigonometric ratio.

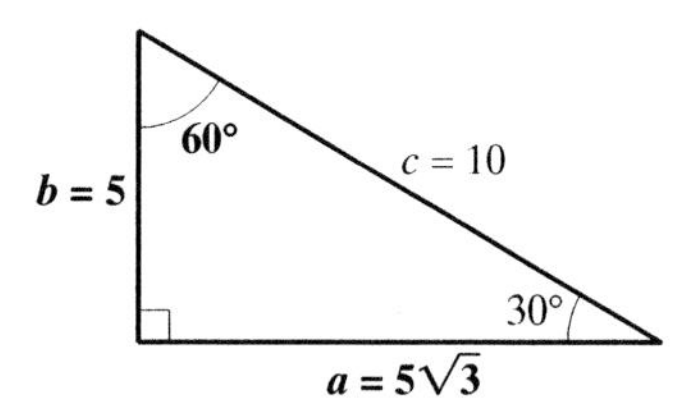

$$
\begin{aligned}
a^2 + b^2 &= c^2 \\
(5\sqrt{3})^2 + b^2 &= 10^2 && \text{Substitute } a = 5\sqrt{3} \text{ and } c = 10. \\
75 + b^2 &= 100 \\
b^2 &= 25 \\
b &= 5
\end{aligned}
$$

However, if either of the two sides used in the Pythagorean Theorem are the result of approximations, it is highly recommended you use the information given in the problem with a trigonometric ratio to find a more precise answer for the third side.

NOTE: As a general rule, if you are given a choice, it is best when solving triangles to use the given information and exact values rather than approximations that are the result of rounding.

The triangle in Example 4 has angles with measures 30°, 60° and 90°. See Exercise 68 to discover the pattern for the sides of these special triangles.

EXAMPLE 5

a. Solve the right triangle with one side 3.4 feet and the hypotenuse 7.35 feet.

b. Find the area of the triangle.

SOLUTION

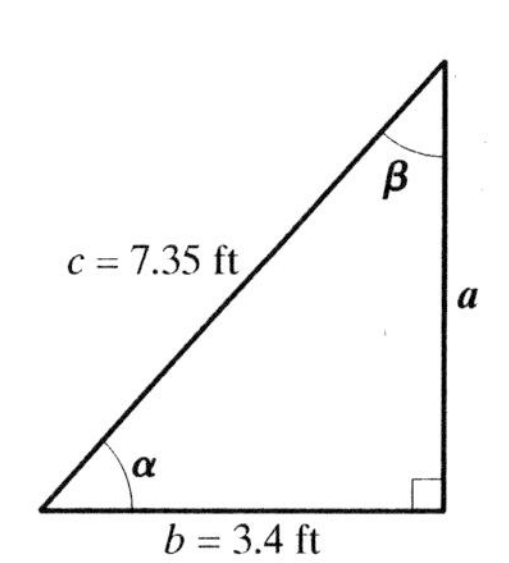

a. Draw and label the triangle. The hypotenuse $c = 7.35$ ft. Since one side of the two remaining sides is given, we can label this side either a or b. We arbitrarily let $b = 3.4$ ft and then begin a chart.

$a = ?$	$\alpha = ?$
$b = 3.4$ ft	$\beta = ?$
$c = 7.35$ ft	$\gamma = 90°$

From the diagram we could start by finding one of the angles. Let's choose α. Since we are given the adjacent side to α and the hypotenuse, we use the cosine trigonometric ratio:

$$\cos\alpha = \frac{\text{adj}}{\text{hyp}}$$

$$\cos\alpha = \frac{3.4 \cancel{\text{ft}}}{7.35 \cancel{\text{ft}}} \qquad \text{Substitute.}$$

$$\alpha = \cos^{-1}\left(\frac{3.4}{7.35}\right) \qquad \text{Use the inverse cosine function.}$$

$$\alpha \approx 62.4° \qquad \text{Find the } \cos^{-1} \text{ value on a calculator in degree mode (to the nearest tenth of a degree).}$$

Since $\alpha + \beta = 90°$, $\beta \approx 90° - 62.4°$ or $\beta \approx 27.6°$.

As in Example 4, we can find the length of side a in different ways. But, if possible, it is better to use the information we are originally given. Since we were given b and c, and we need a, we use the Pythagorean Theorem.

$$
\begin{aligned}
(a)^2 + (3.4\text{ ft})^2 &= (7.35\text{ ft})^2 \\
a &= \sqrt{42.4625\text{ ft}^2} \\
a &\approx 6.52\text{ ft}
\end{aligned}
$$

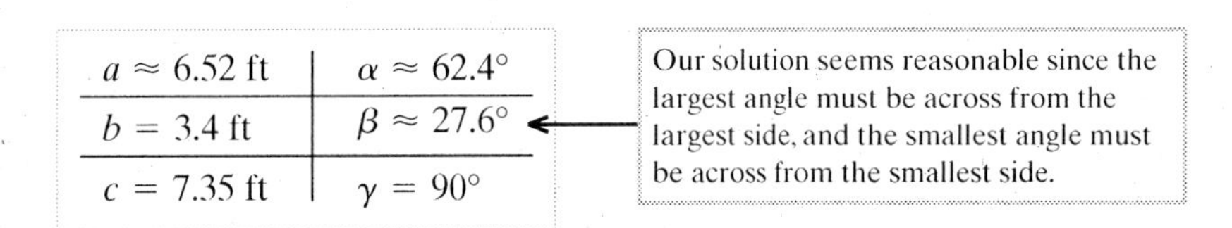

$a \approx 6.52$ ft	$\alpha \approx 62.4°$
$b = 3.4$ ft	$\beta \approx 27.6°$
$c = 7.35$ ft	$\gamma = 90°$

Our solution seems reasonable since the largest angle must be across from the largest side, and the smallest angle must be across from the smallest side.

b. The area of the triangle is

$$A = \frac{1}{2}(\text{base})(\text{height})$$

Since side a is perpendicular to side b, a acts as the height, and b the base.

$$A \approx \frac{1}{2}(3.4 \text{ ft})(6.5163 \text{ ft})$$

$$\approx 11.08 \text{ ft}^2$$

■

Applications of Right Triangles

As you work application problems, make sure you do the following first:

- Draw a diagram of the problem and label all given information.
- Label the side or angle you are asked to find with an appropriate letter.
- In order to determine the correct trigonometric ratio, look carefully at the relationship between the given information and what you are trying to find.

To begin a discussion of the applications of right triangles, we look at two special types of angles.

Angles of Depression or Elevation

Sometimes we look up to see an object, and sometimes we look down. An angle whose initial side is horizontal and has a rotation such that the terminal side's position is above the horizontal is called an **angle of elevation**. If the rotation is such that the terminal side position is below the horizontal, the angle is called an **angle of depression**. As we will see, there are applications of right triangles that use angles of elevation and depression.

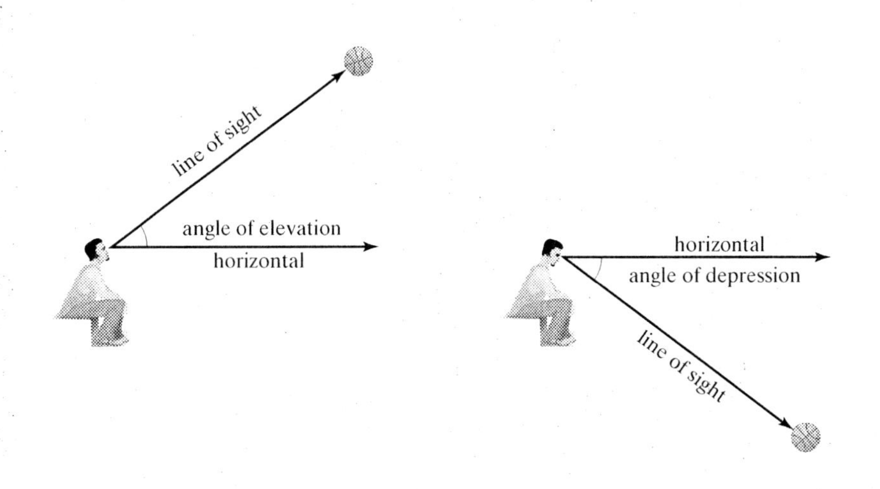

EXAMPLE 6 Looking out the window of her New York loft, Diana can see the Empire State Building. She measures the angle of elevation from the base of the window to the top of the Empire State Building to be 48°. If the base of the window is 80 feet above ground and if her apartment building and the Empire State Building are 1237 feet apart, approximate to the nearest foot the height of the Empire State Building.

SOLUTION Our diagram indicates that we should use the tangent trigonometric ratio because we know $\alpha = 48°$ and the adjacent side $b = 1237$ ft, and we want the opposite side, a. Therefore,

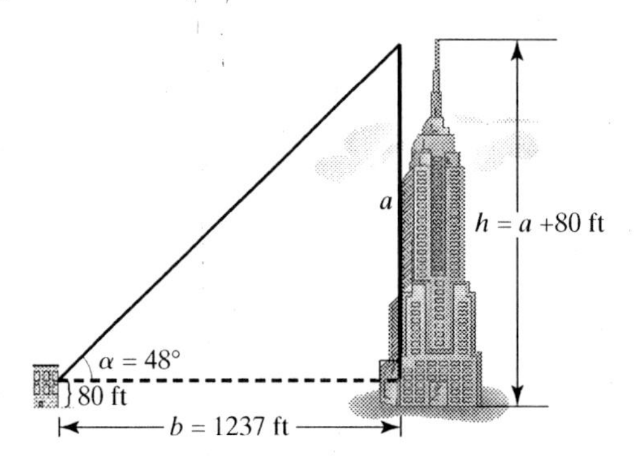

$$\tan \alpha = \frac{\text{opp}}{\text{adj}} = \frac{a}{b}$$

$$\tan 48° = \frac{a}{1237 \text{ ft}}$$

$$a = 1237 \text{ ft} \cdot \tan 48°$$

$$a = 1373.8276808\ldots \text{ ft} \approx 1374 \text{ ft}$$

The height of the Empire State Building, approximated to the nearest foot, is

$$80 \text{ ft} + 1374 \text{ ft} = 1454 \text{ ft}.$$

Bearing

In navigation and surveying, bearing is a way to give a direction. One way to describe bearing is the direction of one point from another point. In this case, we begin at the north–south axis (N or S) and move through an angle in degrees toward the east–west axis to arrive at the ray containing the point in question. The following illustrations are examples of the bearings of a point A from O.

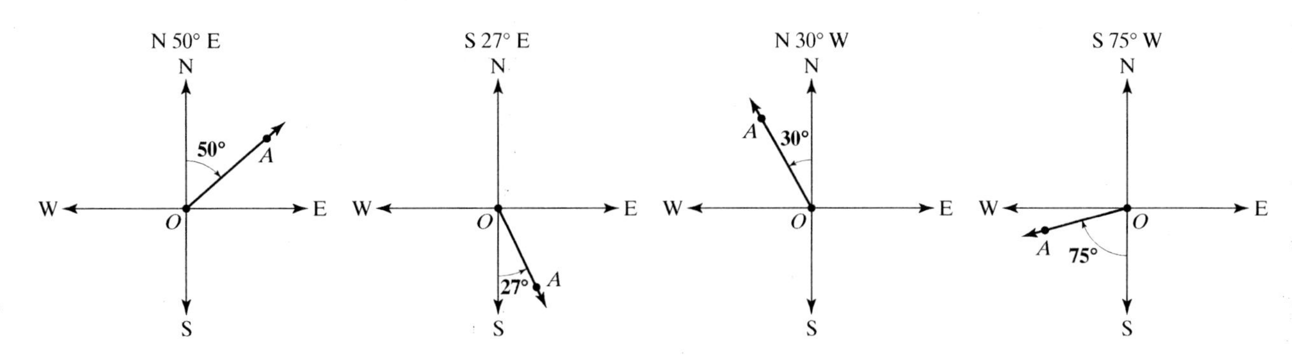

A different way to describe bearing will be discussed in Section 3.5.

EXAMPLE 7 A cruise ship travels from Rome to Palermo on a course bearing S 15° E for 150 miles. Approximate, to the nearest mile, how many miles south and how many miles east the ship has traveled.

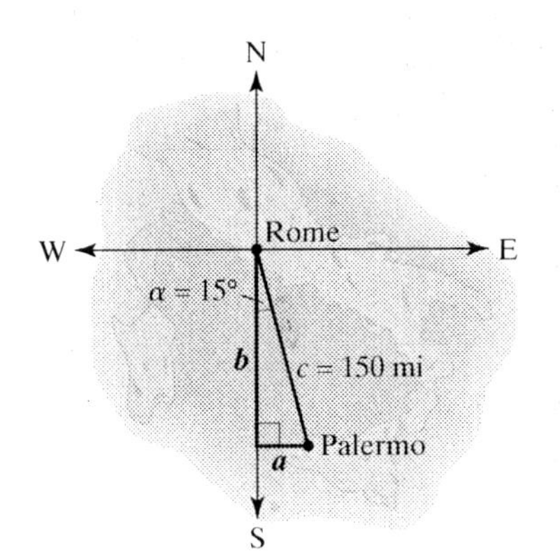

SOLUTION Our diagram indicates we should use the sine ratio because we know the measure of an acute angle $(\alpha = 15°)$ and the hypotenuse $(c = 150 \text{ mi})$ and we want the length of the opposite side, a. In addition, we use the cosine trigonometric ratio to find the adjacent side, b.

$$\sin \alpha = \frac{\text{opp}}{\text{hyp}} \qquad \cos \alpha = \frac{\text{adj}}{\text{hyp}}$$

$$\sin 15° = \frac{a}{150 \text{ mi}} \qquad \cos 15° = \frac{b}{150 \text{ mi}}$$

$$a = 150 \text{ mi} \cdot \sin 15° \qquad b = 150 \text{ mi} \cdot \cos 15°$$

$$a \approx 38.8229 \text{ mi} \qquad b \approx 144.8889 \text{ mi}$$

To the nearest mile, the ship has traveled 39 miles east and 145 miles south. ■

EXAMPLE 8 A marathon runner is on a new course bearing N 39° W. The old course went west and then 15 miles north to get to the same finish point. To the nearest mile, how many fewer miles is the new course?

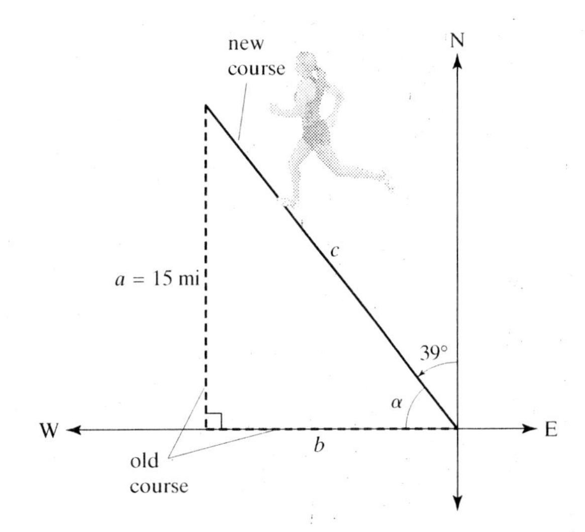

SOLUTION We first find α, and then b and c.

$$\alpha = 90° - 39° = 51°$$

$$\tan 51° = \frac{\text{opp}}{\text{adj}} = \frac{15 \text{ mi}}{b} \qquad \sin 51° = \frac{\text{opp}}{\text{hyp}} = \frac{15 \text{ mi}}{c}$$

$$b = \frac{15 \text{ mi}}{\tan 51°} \qquad c = \frac{15 \text{ mi}}{\sin 51°}$$

$$b \approx 12.1468 \text{ mi} \qquad c \approx 19.3014 \text{ mi}$$

The old course $(a + b) \approx 12.1468 \text{ mi} + 15 \text{ mi} = 27.1468 \text{ mi}$; the new course $c \approx 19.3014$ mi. To the nearest mile, there are 8 fewer miles on the new course. ■

Summary of Information to Solve Right Triangles

Draw a triangle and label any information and part(s) you want to find.

Use the chart for organization and to determine if the solution is reasonable.

$\gamma = 90°, \alpha + \beta = 90°$

$$\sin \alpha = \frac{\text{opp}}{\text{hyp}} = \frac{a}{c}$$

$$\cos \alpha = \frac{\text{adj}}{\text{hyp}} = \frac{b}{c}$$

$$\tan \alpha = \frac{\text{opp}}{\text{adj}} = \frac{a}{b}$$

$$a^2 + b^2 = c^2$$

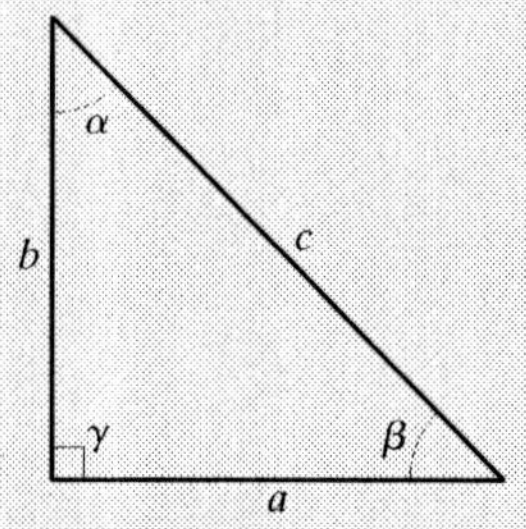

$a =$	$\alpha =$
$b =$	$\beta =$
$c =$	$\gamma = 90°$
	total $= 180°$

When finding the measure of an angle:

- If you know the other two angles, use the fact that the sum of all three is 180°.
- Identify two known sides in relation to their position to the angle. If one of the sides is opposite the angle and the other the hypotenuse, use the sine ratio; if one of the sides is adjacent to the angle and the other the hypotenuse, use the cosine ratio; if one side is opposite the angle and the other adjacent to the angle, use the tangent ratio.

  Then use the appropriate inverse function to find the angle.

When finding the measure of a side:

- If you know the other two sides, use the Pythagorean Theorem.
- Identify one known side and one known angle. Determine how these three parts are related and use the appropriate sin, cos, or tan ratio.

Whenever possible in solving a triangle, select the ratio or formula that uses given information or exact values as opposed to approximate values.

In this section we have learned how to solve right triangles. The trigonometric ratios and the Pythagorean Theorem apply only to right triangles, but not all triangles are right triangles. In the next section we will deal with triangles that do not have a right angle or triangles in which it is not known if there is a right angle.

We should keep the following geometric properties in mind when we are solving *any* triangle.

- The sum of the measures of the angles of a triangle is 180°.
- The sum of the lengths of any two sides of a triangle must be greater than the length of the third side.
- The largest side is opposite the largest angle, the smallest side is opposite the smallest angle of the triangle.

Exercise Set 3.3

1–6. *Write each of the following functions of α in terms of its cofunction of an acute angle β.*

Example

a. $\sin 36°15'$

b. $\csc 45.2°$

Solution

a. $\sin 36°15' = \cos(90° - 36°15')$
$= \cos(89°60' - 36°15')$
$= \cos 53°45'$

b. $\csc 45.2° = \sec(90° - 45.2°)$
$= \sec 44.8°$

1. $\cos 30°$

2. $\sin 60°$

3. $\sec 83.7°$

4. $\csc 48°$

5. $\tan 13°17'$

6. $\cot 35°58'$

7–8. *For $\triangle RST$ shown, find the following.*

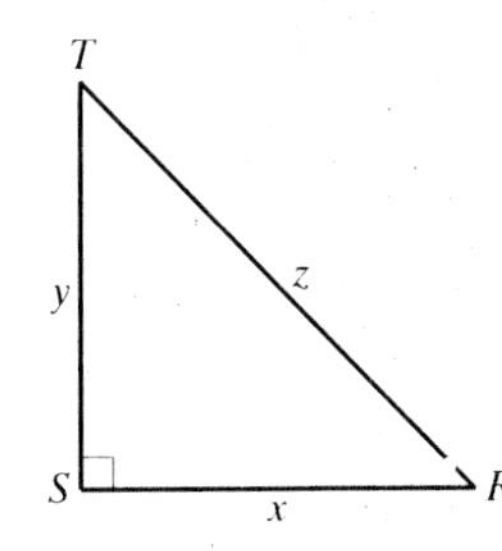

7. a. The side opposite angle R

b. The side adjacent to angle R

c. $\cos R$

d. $\tan T$

e. The angle across from side x

8. a. The side adjacent to angle T

b. The hypotenuse

c. $\sin R$

d. $\cos T$

e. area of $\triangle RST$

9–14. *Consider the information you are given in each triangle. If possible, set up the trigonometric ratio for right triangles that you would use to find x. You are not asked to find x.*

Example

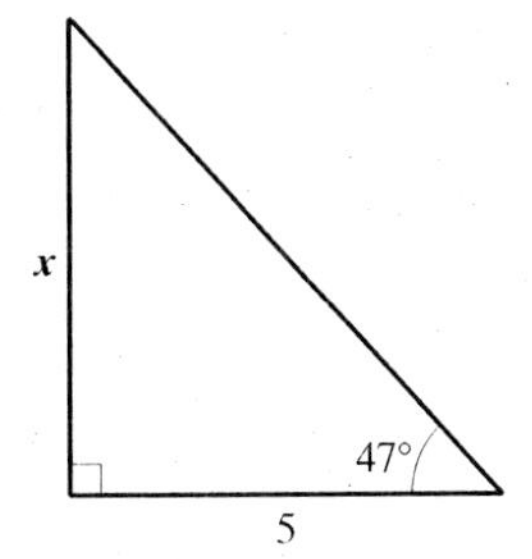

Solution $\tan 47° = \dfrac{x}{5}$

9.

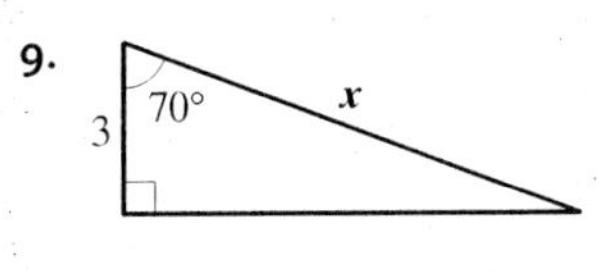

10.

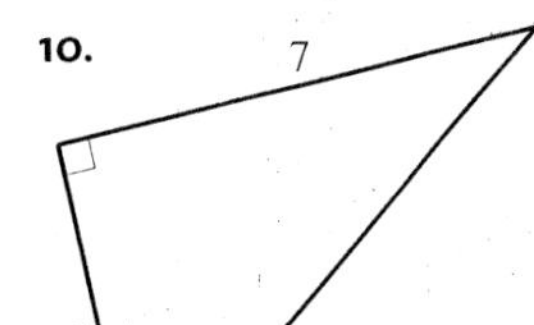

11.

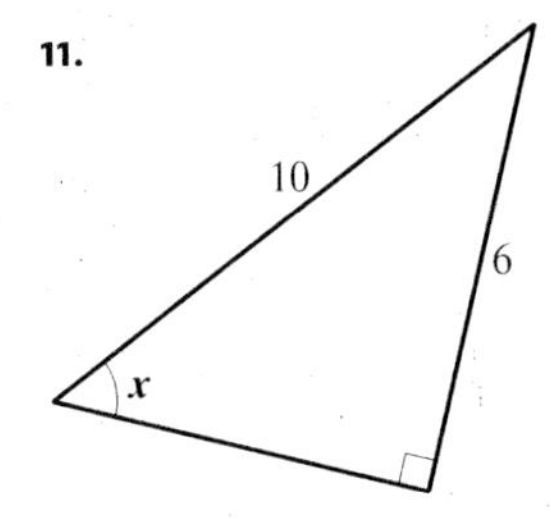

12.

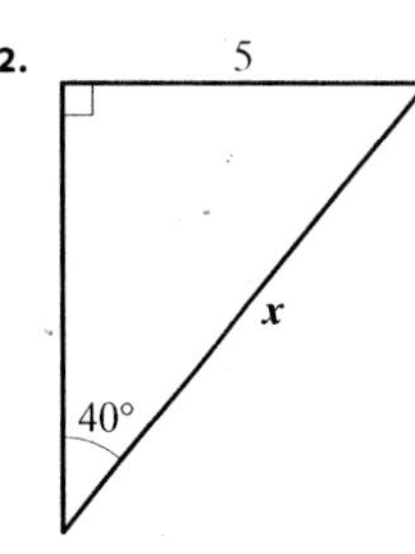

13.

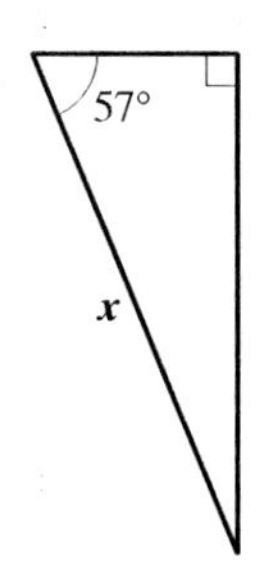

14.

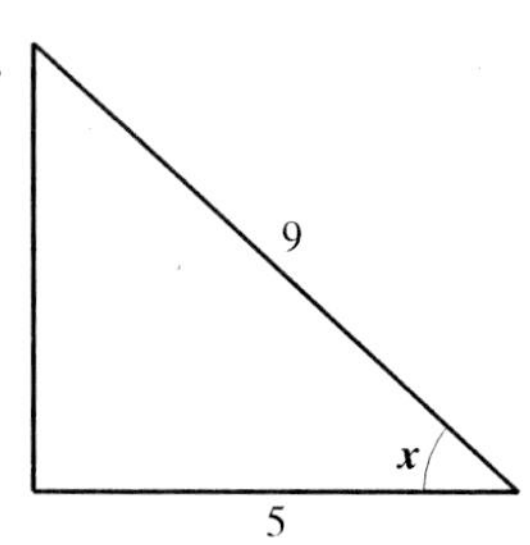

15–18. *Find* sin α, cos α, *and* tan α.

15.

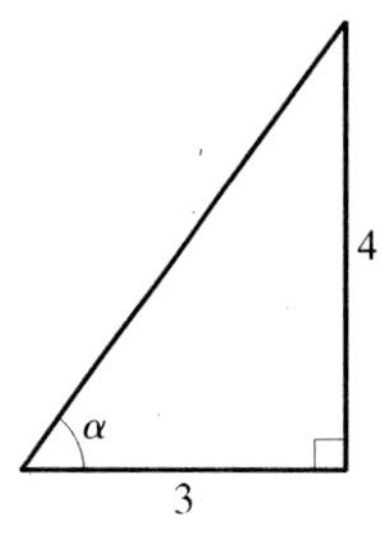

16.

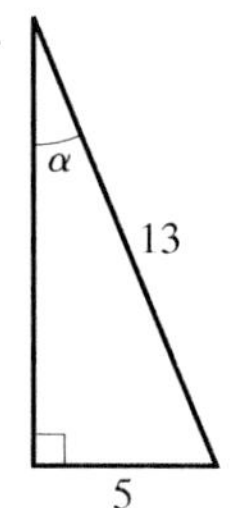

17.

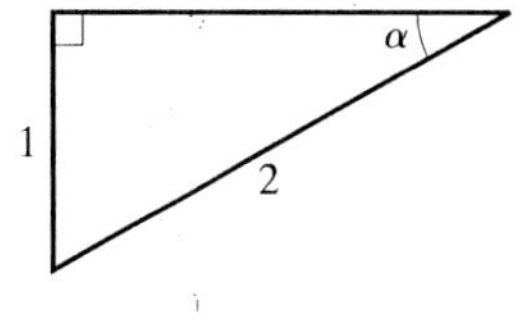

18.

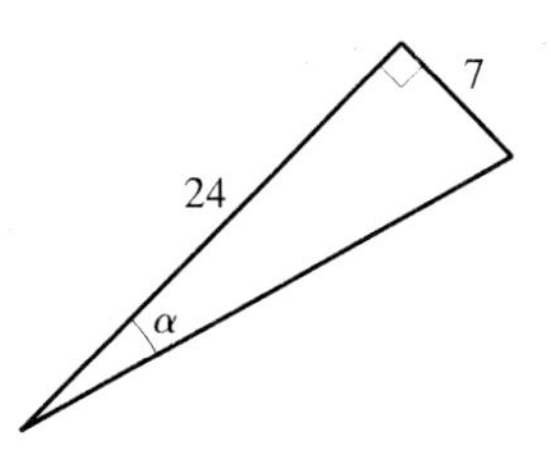

19–28. *Find the length of the side or measure of the angle in each right triangle* $(\gamma = 90°)$.

EXAMPLE

a. $\alpha = 32.1°$, find β. **b.** $a = 5, c = 7$, find α. **c.** $a = 3, b = 2.4$, find c.

SOLUTION Sketch a triangle for each case.

a.

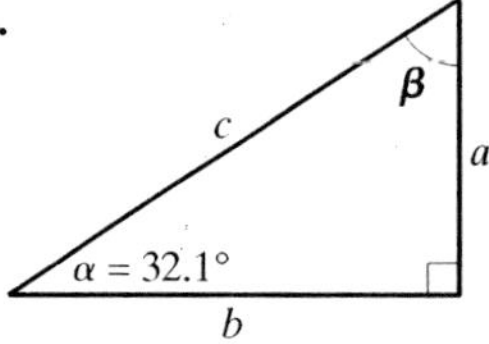

$$\alpha + \beta = 90°$$
$$\beta = 90° - 32.1°$$
$$= 57.9°$$

b.

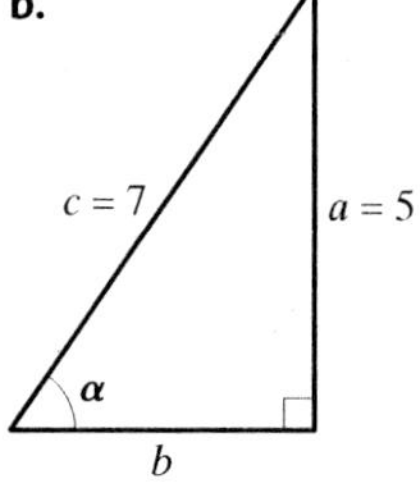

$$\sin \alpha = \frac{5}{7}$$
$$\alpha = \sin^{-1}\left(\frac{5}{7}\right)$$
$$\alpha \approx 45.6°$$

c.

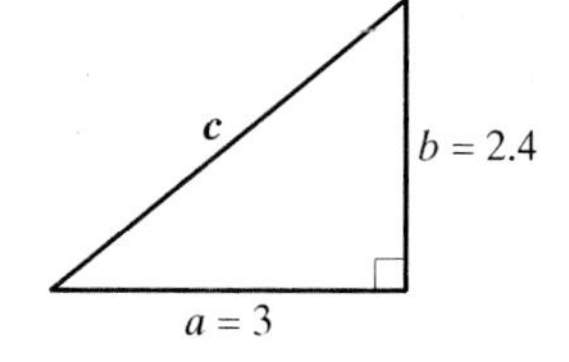

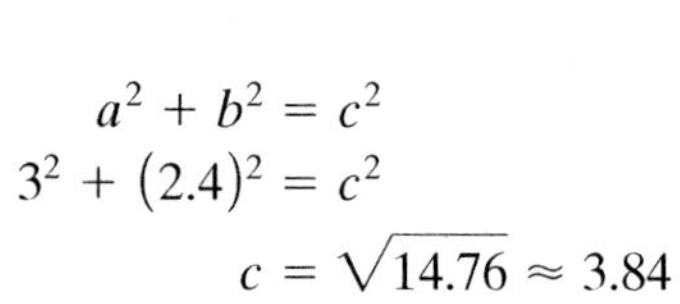

$$a^2 + b^2 = c^2$$
$$3^2 + (2.4)^2 = c^2$$
$$c = \sqrt{14.76} \approx 3.84$$

19. $\beta = 43.6°$, find α.

20. $\alpha = 30°16'$, find β.

21. $\alpha = 31°6'$ and $b = 8$, find c.

22. $\beta = 26°24'$ and $c = 9$, find a.

23. $a = 3$ and $b = 9$, find c.

24. $\alpha = 70.6°$ and $b = 7$, find a.

25. $\alpha = 63°30'$ and $a = 6.2$, find b.

26. $a = 5$ and $c = 9$, find b.

27. $b = 2$ and $c = 6$, find β.

28. $a = 6$ and $c = 9$, find α.

29–38. *Solve each right triangle ($\gamma = 90°$) with the given angle measures and/or lengths of sides.*

EXAMPLE $\alpha = 35°, c = 17$

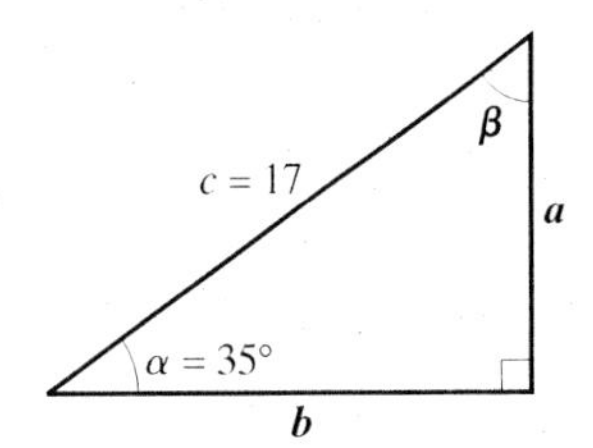

SOLUTION

$$\beta = 90° - 35° = 55°$$

$$\sin\alpha = \frac{a}{c} \qquad \sin\beta = \frac{b}{c}$$

$$\sin 35° = \frac{a}{17} \qquad \sin 55° = \frac{b}{17}$$

$$a \approx 9.75 \qquad b \approx 13.93$$

$a \approx 9.75$	$\alpha = 35°$
$b \approx 13.93$	$\beta = 55°$
$c = 17$	$\gamma = 90°$

29. $\beta = 15.2°, a = 7$

30. $b = 6, \alpha = 56°$

31. $a = 3, b = 7$

32. $b = 10.7, c = 15.2$

33. $\alpha = 80°, a = 21$

34. $\beta = 68°, b = 34$

35. $c = 10, b = 8.5$

36. $c = 15, a = 9.12$

37. $c = 23.8, \beta = 78°36'$

38. $c = 38.1, \alpha = 39°48'$

39. If you only know the measures of the angles of the right triangle, explain why you are not able to use any trigonometric ratios to solve the triangle.

40. If you are given one side and an acute angle of a right triangle, what unknown part of the triangle would be the easiest to find?

41. Our trigonometric ratios were defined for only the two acute angles of a right triangle. It may help you understand why this is so if you try to apply these ratios to the right angle ($\gamma = 90°$) of the triangle.

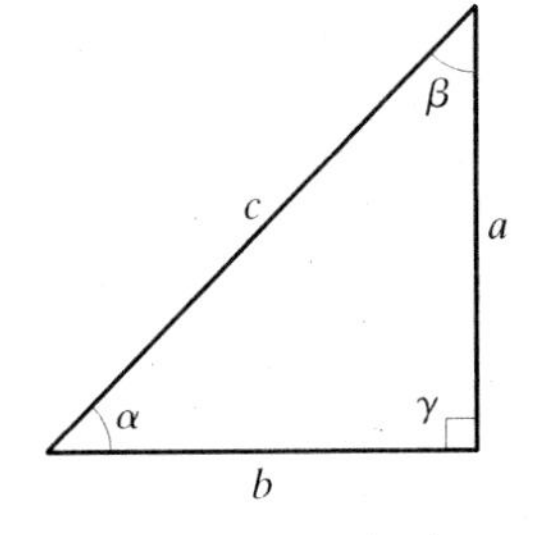

   a. Would there be a clearly defined adjacent side to γ?

   b. What difficulty would you have using $\tan\gamma$?

   c. What difficulty would you have using $\cos\gamma$?

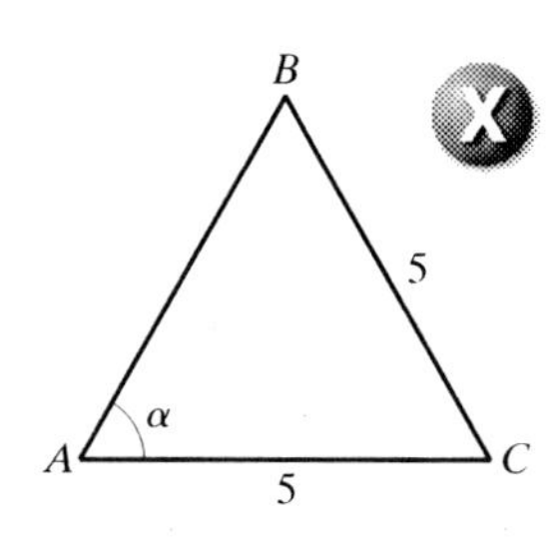

42. What is wrong with the procedure shown to find the measure of α in $\triangle ABC$?

$$\tan \alpha = \frac{5}{5} = 1$$

$$\alpha = 45^\circ$$

43. **a.** Find the angles of elevation.

b. Find the angles of depression.

c. Explain why $\angle 13$ is not an angle of elevation or depression.

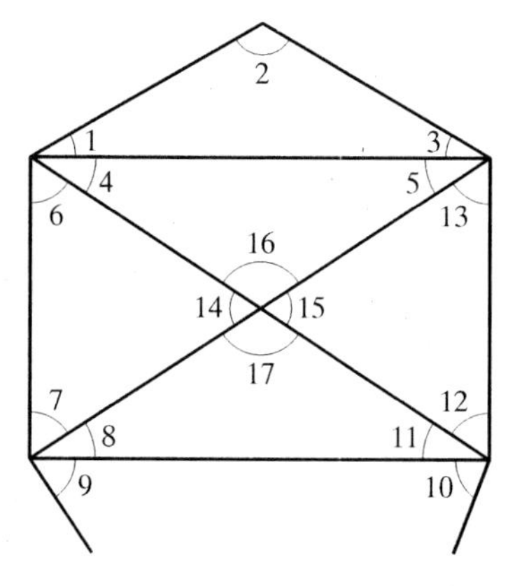

44. What must be true about one side of either an angle of elevation or depression?

45. When the angle of elevation of the sun is 53°, a tree casts a 25-foot shadow along the ground. How tall is the tree (to the nearest foot)?

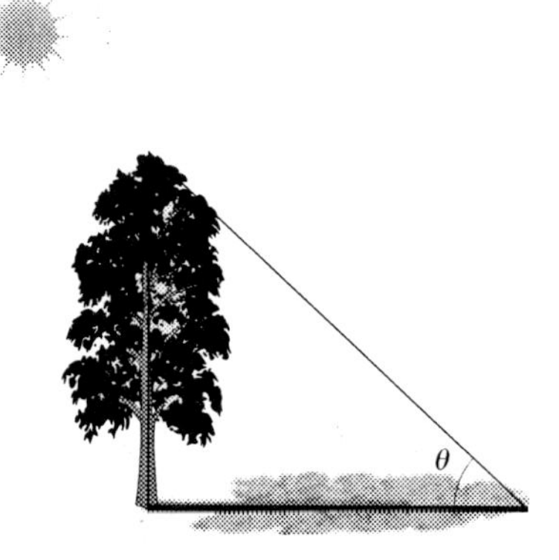

46. A television camera in a blimp is focused on a football field with an angle of depression of 20.3°. If the field is 450 meters away from a point on the ground directly below the blimp, find to the nearest tenth of a meter the altitude of the blimp.

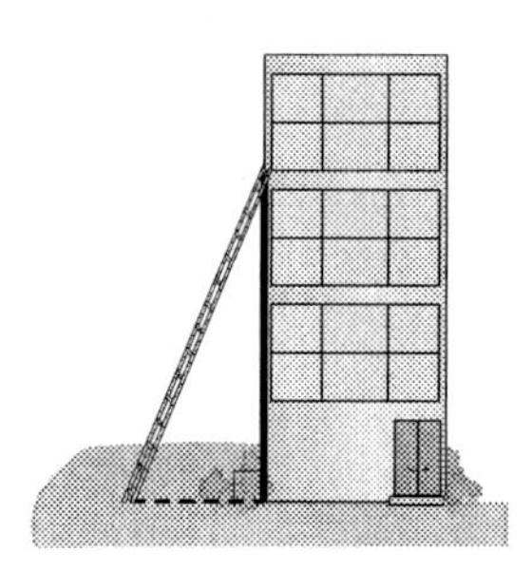

47. A submarine on the surface of the ocean makes a dive of 10° with the surface of the ocean. If it continues 180 meters on the same path, what will be the depth (to the nearest tenth of a meter) below sea level?

48. Kirby Clark Cleaning Service uses a 30-foot ladder to clean the windows of a building. For safety reasons, the ladder should never make an angle of more than 50° with the ground (horizontal). What is the *highest* window the ladder can reach (to the nearest foot)?

49. An observer has a video camera located at ground level, 1000 feet from the launch pad of a rocket. The rocket will ascend vertically for 500 feet at which time it is supposed to release its booster. What angle of elevation should the observer use so the camera will record the rocket booster release?

50. Thomas, standing on a bridge, is directly over the lowest point of a gorge which is 200 feet below. Thomas is 18 feet from the edge of the bridge where Nicole is standing. Find the angle of depression from the edge of the bridge that Nicole should use to see this lowest point of the gorge.

51. If the sides of an equilateral triangle (shown at the left below) are 4 inches, find the length of the altitude and the area of the triangle.

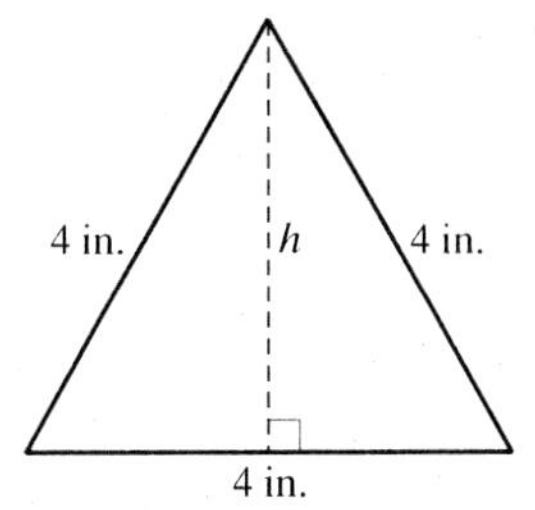

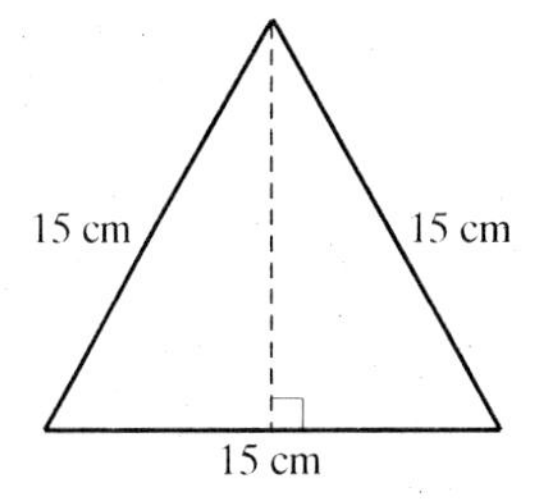

52. If the sides of an equilateral triangle (shown at the right above) are 15 centimeters, find the length of the altitude and the area of the triangle.

53. Debby is standing 14 feet in front of a statue and notices that the angle of elevation from eye level to the top of the statue is 35°, whereas the angle of depression to the bottom of the statue if 22°. To the nearest foot, find the height of the statue.

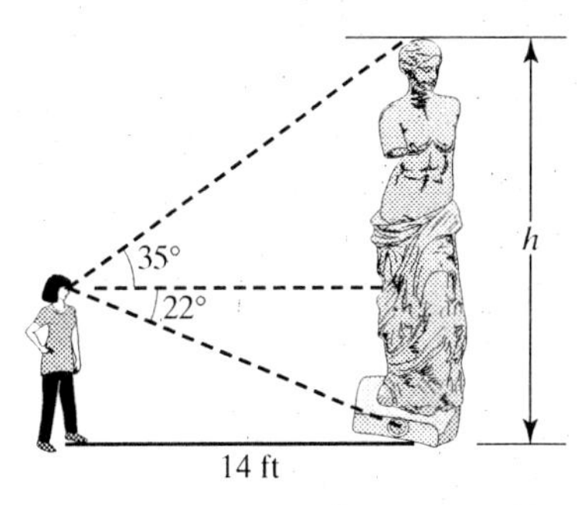

54. Two buildings are 120 feet apart. From a point on top of the taller building the angle of depression to the top of the shorter building is 17°. From a point on the ground directly below the point on the top of the taller building, the angle of elevation to the top of the shorter building is 47°. To the nearest foot, how tall are the buildings?

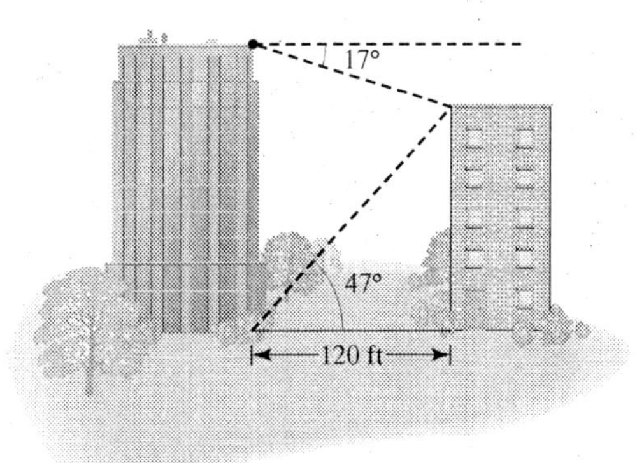

55. **a.** Which angle has bearing N 22° E?

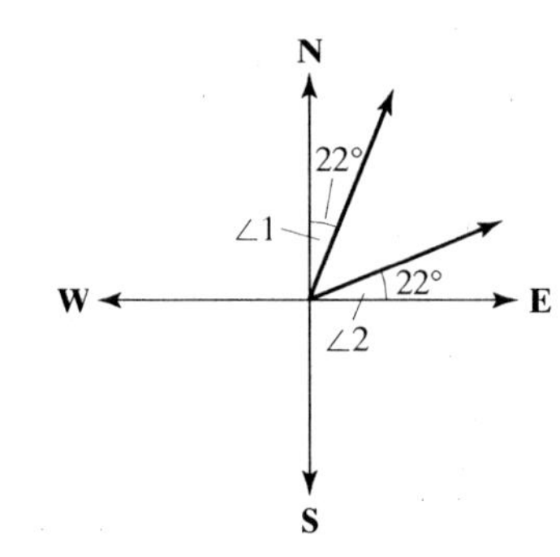

b. Which angle has bearing S 30° W?

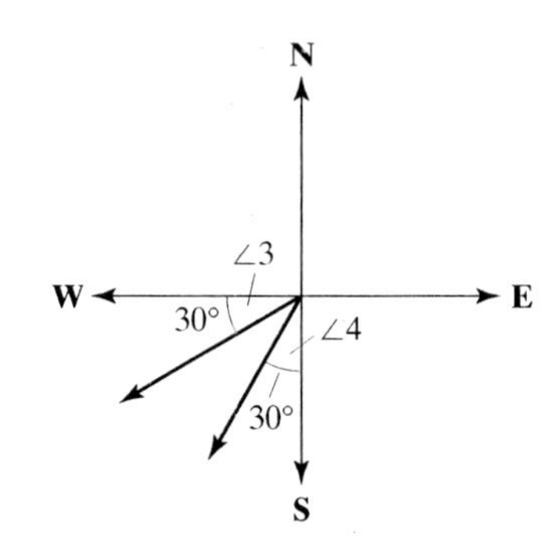

56. Daryl is trying to help Ryan understand bearing. Which of the following statements could Daryl use to help describe bearing? If the statement is incorrect, explain why.

 a. No matter what the bearing is, you always begin with an initial side along the north or south axis, whichever is stated. Then you rotate the given number of degrees toward the indicated direction (east or west).

 b. No matter what the bearing is, one side will always be in an east or west direction.

 c. For example, S 13° E means you should start with the initial side of the angle along the east axis, then rotate 13° toward the south axis.

 d. S 13° E means you should start with the initial side of the angle along the south axis, then rotate 13° toward the east axis.

57. Taking a stroll in the park, Karen walks in the direction N 42° E for 3 miles. Gary leaves from the same place but walks directly north for 6 miles to his favorite park bench, where he decides to sit and read his trigonometry book.

 a. How many miles east of Gary is Karen?

 b. What is the bearing and the number of miles Karen should walk if she wants to find Gary?

58. A boat travels 83.6 miles on a course of bearing N 47° W. How many miles north and how many miles west has it traveled?

59. Find h.

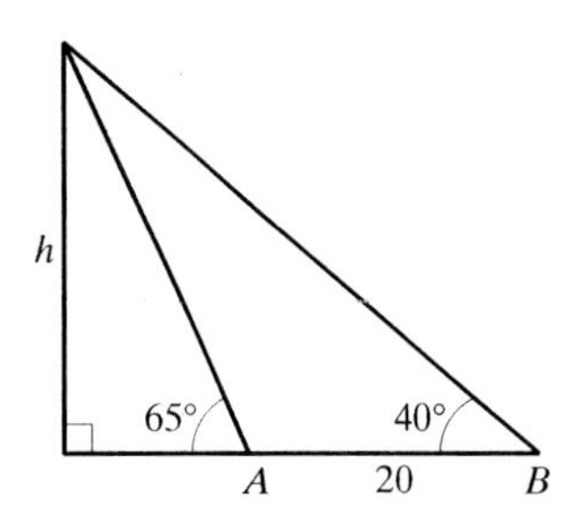

60. Find h in terms of b.

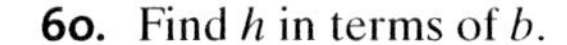

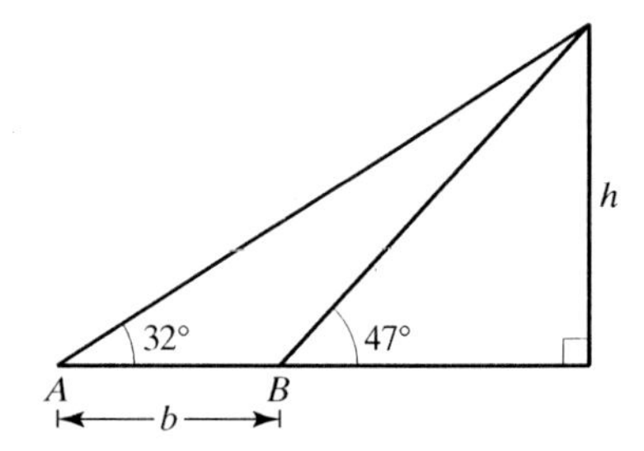

61. Terry determines the angles of elevation from ground level to the top of a pyramid from two points A and B, are 72.8° and 60.5°, respectively. If A and B are 120 feet apart, how high is the pyramid?

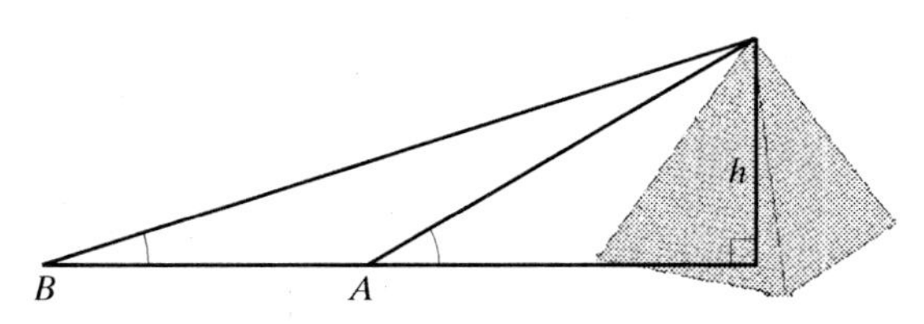

62. A vertical pendulum whose length is 27 inches has a maximum displacement of 6 inches from its rest position

a. Find the angle θ created by the pendulum as it swings back and forth.

b. At the point of its maximum displacement, determine how high the pendulum is above its rest position (lowest point).

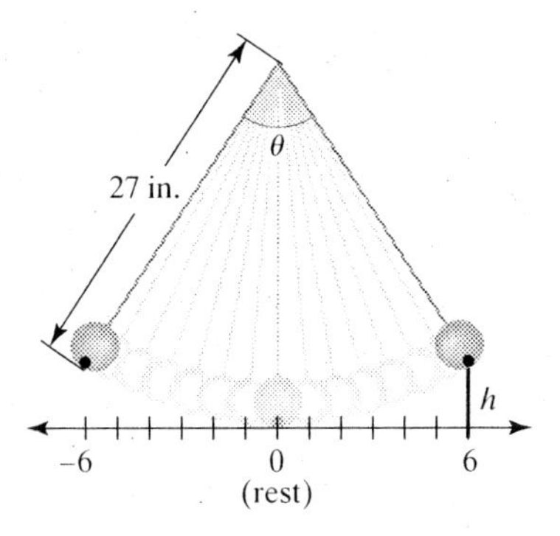

63. John's Market wants to place a security camera at one corner of the roof on the top of the parking lot side of the building so that it will focus on the opposite corner of the parking lot. If the building is 15 feet tall and the rectangular parking lot is 40 feet wide and 120 feet long, find the angle of depression of the camera.

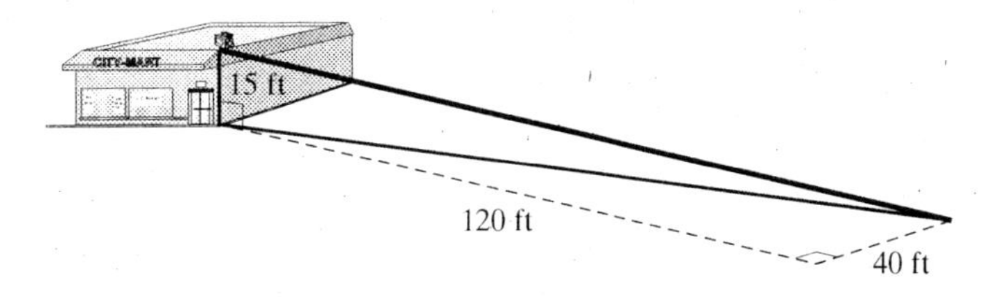

64. Astronomers use a method called **parallax** to measure distances to some stars. To understand this method, hold your thumb up at arm's length, close one eye, and place your thumb in front of a distant object. Then, without moving your thumb, switch eyes, and you will see your finger seems to shift with respect to the distant object behind it. The effect is called parallax. Astronomers measure parallax by very carefully measuring the position of a nearby star with respect to more distant stars behind it. Six months later when Earth is on the opposite side of its orbit, they measure those positions again. If the star is close enough, a measurable parallax will be seen. The position of the star relative to the more distant background stars will have shifted. Even

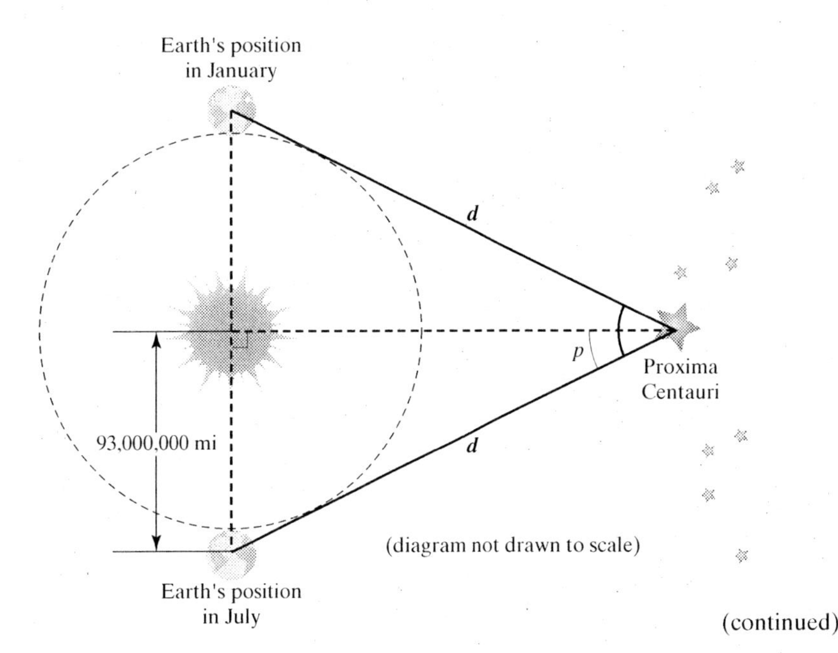

(continued)

for the nearest star, the shift is very small, less than an arcsecond, or $(\frac{1}{3600})°$. In the accompanying illustration, the line of sight to the star in January is different than that in July, when Earth is on the other side of its orbit. As seen from Earth, the position of the nearby star appears to sweep through the angle shown. Half of this angle is the parallax, p. If $p \approx 0.00021111°$ to the nearest star Proxima Centauri, find the distance from Earth to this star if the distance from the sun to Earth is 93,000,000 miles.

65. Romeo is longing to see Juliet. He stands 20 feet from the castle and looks up 65° (angle of elevation) to see Juliet. He calls to Juliet, but she can't find him.

a. What angle of depression does Juliet need in order to see Romeo?

b. In general, what is true about the relationship between the measure of the angle of depression and the measure of the angle of elevation between two points?

66. Some people use the mnemonic SOHCAHTOA to help them remember the trigonometric ratios. Determine the meaning of this mnemonic as it relates to the trigonometric right triangle ratios. (See Exercise 70 for assistance.)

Discussion

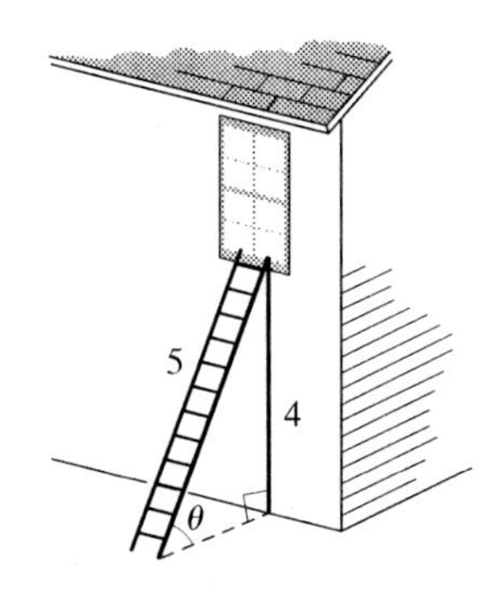

67. Discuss how the following three problems are related.

From Section 1.2: If $\sin t = \frac{4}{5}$ and t is in QI, find $\cos t$.

From Section 2.5: Find $\cos\left(\arcsin \frac{4}{5}\right)$.

From Section 3.3: A 5-ft ladder is leaning against a building and meets the building at a height of 4 ft. What is the cosine of the angle of elevation the ladder makes with the ground?

Explore the Pattern

68. Right triangles that contain common angles are triangles whose angles are 30°, 60°, and 90° or triangles whose angles are 45°, 45°, and 90°. For each 30–60–90 triangle shown, you are given a, the side opposite the 30° angle.

a. Find the exact lengths of b, the side opposite the 60° angle, and c, the side opposite the 90° angle, for each triangle. Then fill in the table with these values.

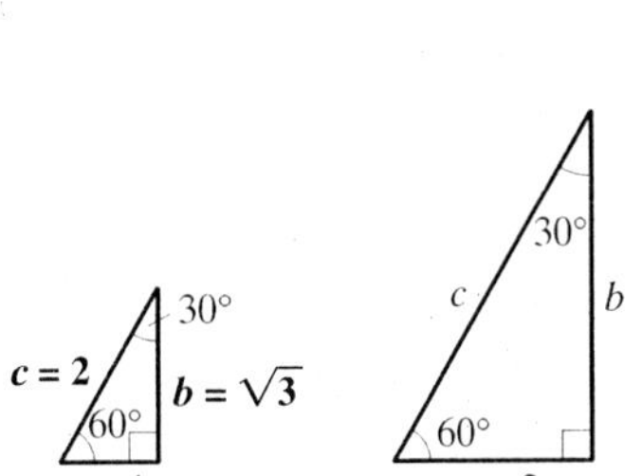

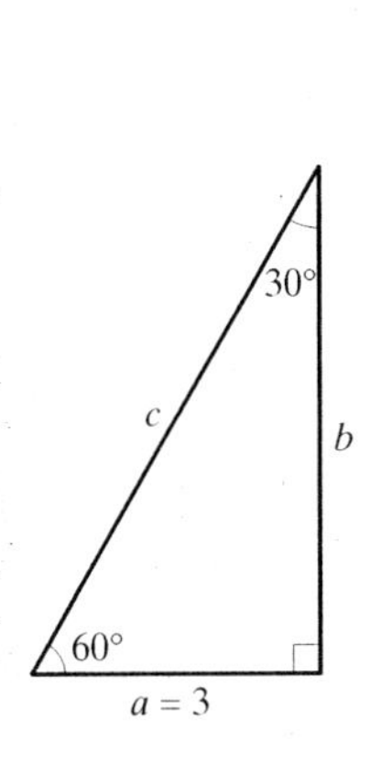

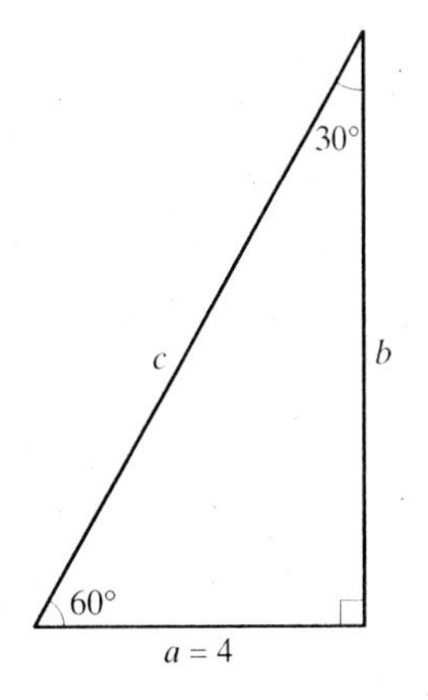

30°	60°	90°
a	b	c
1	$\sqrt{3}$	2
2		
3		
4		
a		

EXAMPLE $a = 1$

SOLUTION

$$\tan 60° = \frac{b}{1} \qquad \cos 60° = \frac{1}{c}$$

$$\sqrt{3} = \frac{b}{1} = b \qquad \frac{1}{2} = \frac{1}{c}$$

$$2 = c$$

■

b. If we let a represent the side opposite 30°, use the completed portion of the table to determine the pattern for the sides opposite the other angles in terms of a. Describe the pattern by completing the last row of the table for sides b and c in terms of a.

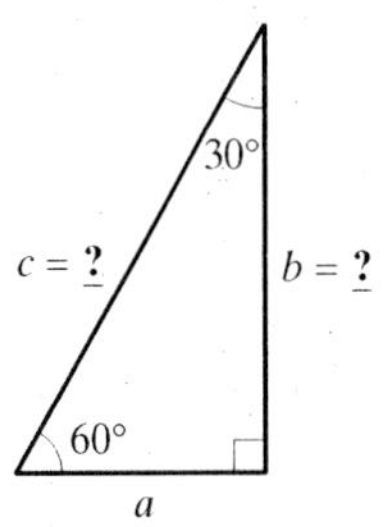

Then use your pattern to quickly find the remaining sides of the following 30–60–90 triangle.

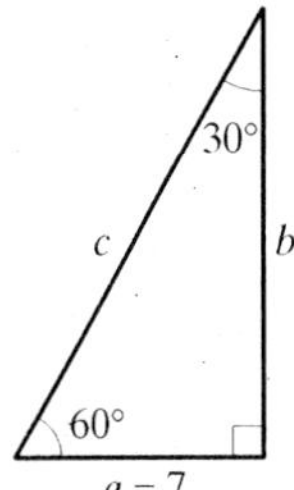

45°	45°	90°
a	b	c
1	1	$\sqrt{2}$
2		
3		
4		
a		

69. For each 45–45–90 triangle, you are given one of the sides opposite a 45° angle.

a. Find the exact lengths of the remaining sides and fill in the table.

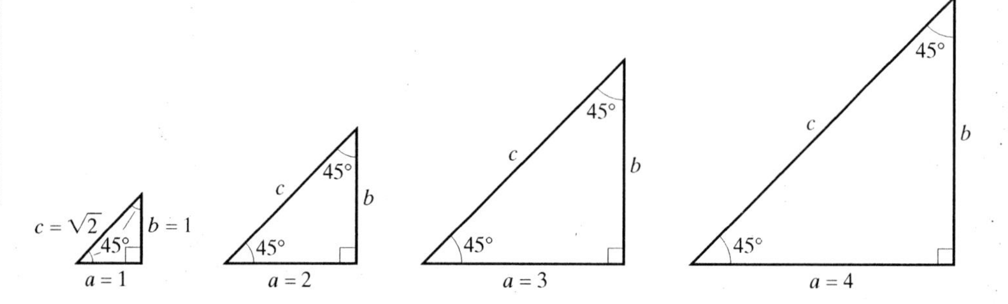

EXAMPLE $a = 1$

SOLUTION

$$\tan 45° = \frac{b}{1} \qquad \cos 45° = \frac{1}{c}$$

$$1 = b \qquad \frac{1}{\sqrt{2}} = \frac{1}{c}$$

$$c = \sqrt{2}$$

b. If we let a represent the side opposite one of the 45° angles, use the completed portion of the table to determine the pattern for the sides opposite the other angles in terms of a. Describe the pattern by completing the last row of the table for sides b and c in terms of a.

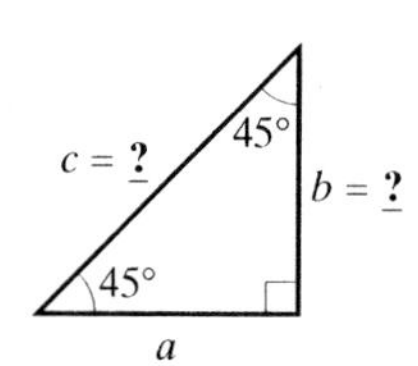

Then use your pattern to quickly find the missing sides of the following 45–45–90 triangle.

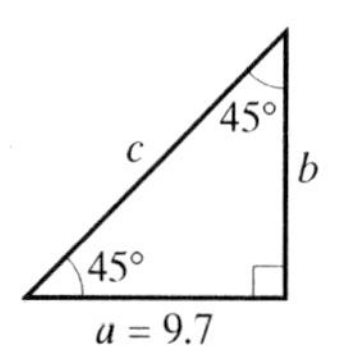

Web Activity

70. Search SOHCAHTOA for additional information on this mnemonic for trigonometric ratios.

3.4 Solution of Triangles Using Law of Sines

In Section 3.3 we solved right triangles using the trigonometric ratios for right triangles. We also want to be able to solve triangles that do not have a right angle, which are called **oblique triangles**. In this and the next section we find in order to solve *any triangle,* in particular any oblique triangle, we can use two important laws: *Law of Sines* and *Law of Cosines.*

As we stated in Section 3.3, when we are solving triangles, we must be given the measures of at least three parts, and at least one of those parts must be the length of a side. But sometimes this given information does not always determine a triangle, and sometimes it determines more than one triangle. Let's look at the possible situations, along with their abbreviations.

Solving Oblique Triangles

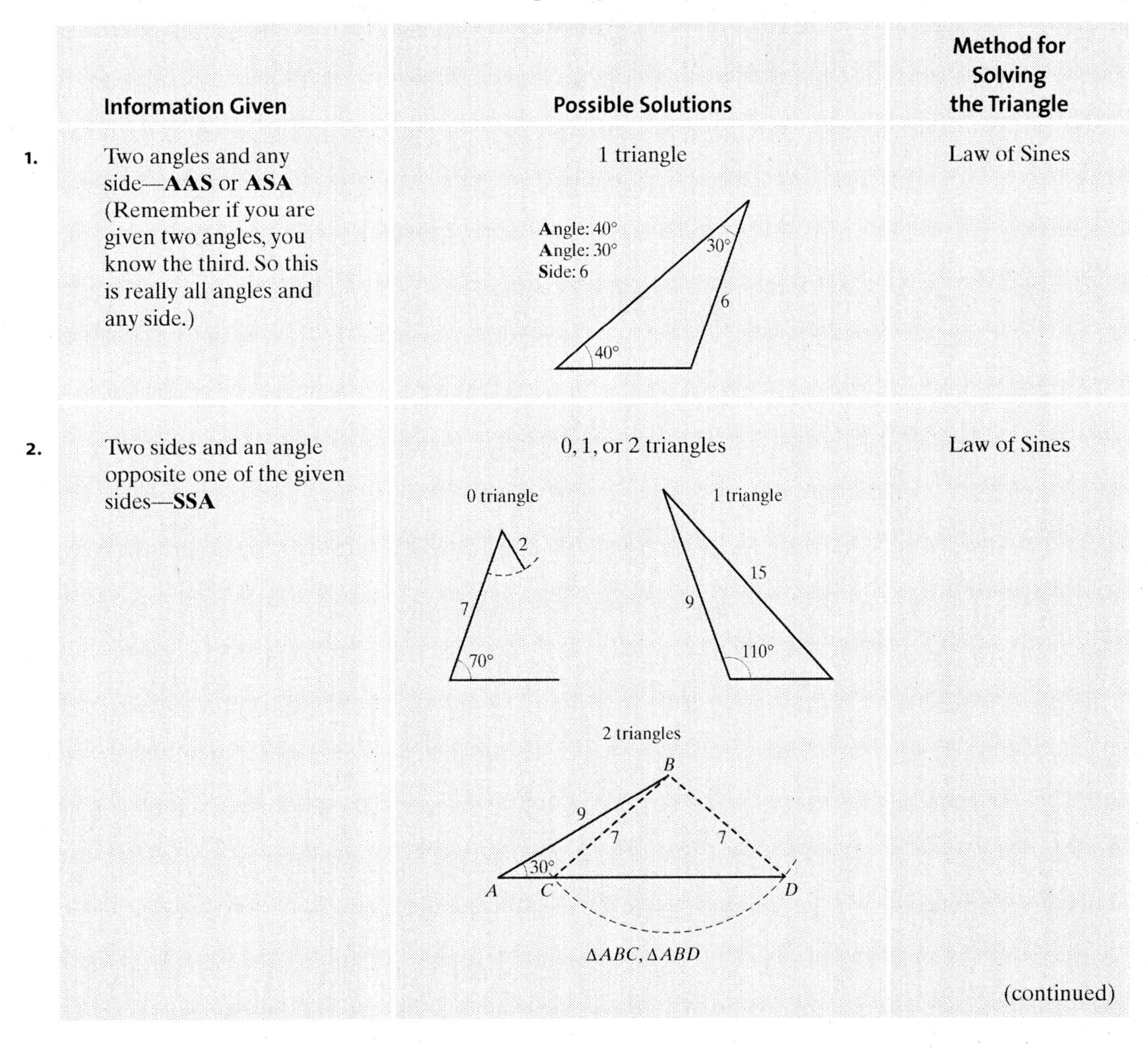

	Information Given	Possible Solutions	Method for Solving the Triangle
1.	Two angles and any side—**AAS** or **ASA** (Remember if you are given two angles, you know the third. So this is really all angles and any side.)	1 triangle	Law of Sines
2.	Two sides and an angle opposite one of the given sides—**SSA**	0, 1, or 2 triangles	Law of Sines

(continued)

NOTE: SSA is called the **ambiguous case** since this situation could produce any one of the three possible solutions. When we are given this situation, we first determine the number of solutions, which depends on the specific values of the two sides and opposite angle.

	Information Given	Possible Solutions	Method for Solving the Triangle
3.	All three sides—**SSS**	0 or 1 triangle 0 triangle: $a = 3$, $b = 2$, $c = 9$; $3 + 2 \not> 9$ The sum of any two sides must be greater than the third. 1 triangle: 7, 8, 5	Law of Cosines (Next section)
4.	Two sides and their included angle—**SAS**	1 triangle 5, 80°, 4	Law of Cosines (Next section)

We begin with the development of the law of sines, which we use to solve triangles with given information in the form of AAS and SSA. The law of cosines is discussed in Section 3.5, and it will address solving triangles with given information in the form SSS and SAS.

Law of Sines

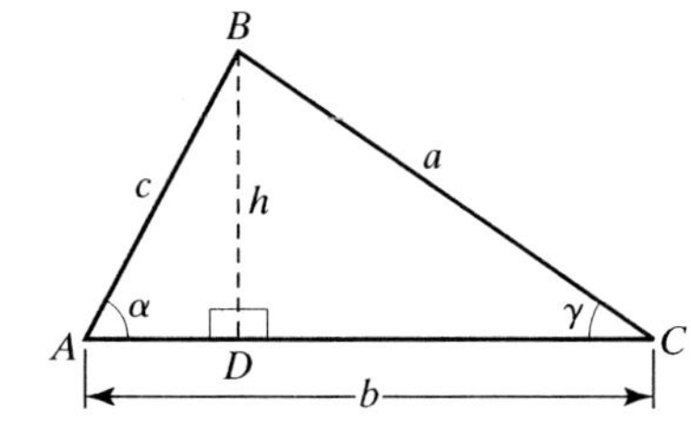

In an oblique triangle, we can draw an altitude to get two right triangles.

For $\triangle ABD$, $\sin\alpha = h/c$, and for $\triangle CBD$, $\sin\gamma = h/a$. Solving each equation for h, we get:

$$h = c\sin\alpha \qquad \text{and} \qquad h = a\sin\gamma$$

Therefore $c\sin\alpha = a\sin\gamma$. If we divide both sides of this equation by ac, we get

$$\frac{\sin\alpha}{a} = \frac{\sin\gamma}{c}.$$

Drawing a different altitude in $\triangle ABC$ and working in a similar manner, we get

$$\frac{\sin\alpha}{a} = \frac{\sin\beta}{b}.$$

Combining these two results tells us the following relationship between sides and angles of a triangle. It states that in a triangle the ratio of the sine of an angle to the length of the opposite side is constant, which is known as the law of sines for obvious reasons.

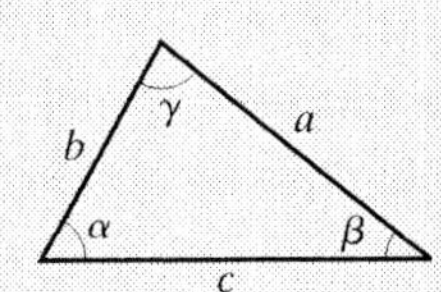

Law of Sines

In triangle ABC, if a, b, and c are the sides opposite angles α, β and γ, respectively, then

$$\frac{\sin\alpha}{a} = \frac{\sin\beta}{b} = \frac{\sin\gamma}{c}, \quad \text{or equivalently} \quad \frac{a}{\sin\alpha} = \frac{b}{\sin\beta} = \frac{c}{\sin\gamma}.$$

When we solve a triangle using the law of sines, we use two equivalent ratios at a time. Typically, we select the two ratios that contain three given parts of the triangle along with a part we need to find. And, if possible, we try (as we did with the trigonometric right triangle ratios) to use the ratios that involve the given information or exact values, which should improve the accuracy of the answer.

EXAMPLE 1 (AAS) In the triangle shown $\alpha = 30°$, $\beta = 45.25°$, and $a = 8$. Solve the triangle.

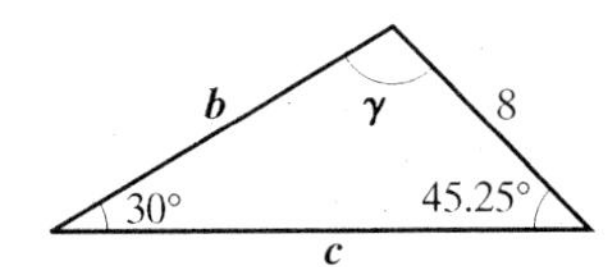

SOLUTION Since we know two angles, we can quickly find the third.

$$\gamma = 180° - \alpha - \beta = 180° - 30° - 45.25° = 104.75°$$

We select two ratios of the law of sines that contain the given values a, α, and β. Then we solve for b.

$$\frac{a}{\sin\alpha} = \frac{b}{\sin\beta}$$

$$\frac{8}{\sin 30°} = \frac{b}{\sin 45.25°} \qquad \text{Substitute the known values.}$$

$$b = \frac{8\sin 45.25°}{\sin 30°} \approx 11.36 \qquad \text{To solve for } b\text{, multiple each side by } \sin 45.25°.$$

To find c, we use two ratios of the law of sines that involve a, α, and γ.

$$\frac{c}{\sin\gamma} = \frac{a}{\sin\alpha}$$

$$\frac{c}{\sin 104.75°} = \frac{8}{\sin 30°} \qquad \text{Substitute the known values.}$$

$$c = \frac{8\sin 104.75°}{\sin 30°} \approx 15.47 \qquad \text{To solve for } c\text{, multiply each side by } \sin 104.75°.$$

We show the solution to this triangle by completing the chart.

$a = 8$	$\alpha = 30°$
$b \approx 11.36$	$\beta = 45.25°$
$c \approx 15.47$	$\gamma = 104.75°$

■

CAUTION: In Example 1, once we have a and b you might *incorrectly* try to use $c^2 = a^2 + b^2$ to find c. *Since this is not a right triangle, you cannot use the Pythagorean Theorem.*

Ambiguous Case SSA of the Law of Sines

As stated at the beginning of this section, when we are given the ambiguous case (two sides and an opposite angle), we could get 0, 1, or 2 triangle solutions. We can determine the number of triangle solutions by using the law of sines, as well as the law we will discuss in the next section. Examples 3 and 4 will be given again in the next section to demonstrate two ways to deal with the ambiguous case. For now, we use the Law of Sines.

EXAMPLE 2 (SSA) Solve the triangle(s) if $b = 3$, $c = 5$ and $\beta = 58°$.

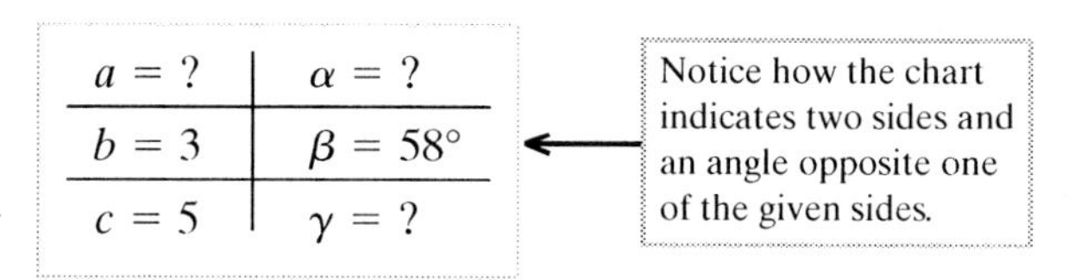

SOLUTION Since this is the ambiguous case, we need to determine the number of triangles determined by the given information. Since we know b, c, and β, we begin by finding γ:

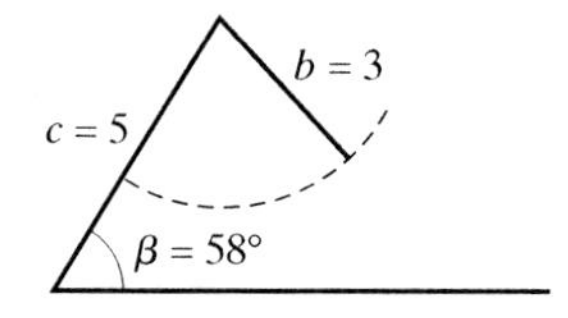

$$\frac{\sin \gamma}{c} = \frac{\sin \beta}{b}$$

$$\frac{\sin \gamma}{5} = \frac{\sin 58°}{3}$$

$$\sin \gamma = \frac{5 \sin 58°}{3} \approx 1.41$$

Since the sine function is never greater than 1, there is no solution for γ, which means there is no triangle solution. ■

CAUTION: If you were to continue without recognizing that the value of the sine is too large, and use a calculator for $\sin^{-1}(1.41)$, some will display "error" or "domain error," or an answer in the complex numbers. The first display should alert you to the problem with the sine value, but do not misinterpret this last display. This last display does not mean there are two solutions, rather the real and imaginary parts of a complex number.

sin⁻¹ (1.41)
(1.57079632679, -.877...

Example 3 (SSA) Solve the triangle if $a = 5$, $b = 7$, and $\beta = 60°$.

Solution Since this is the ambiguous case, we need to determine the number of triangles as a result of the given information. Since we know a, b, and β, we begin by finding α.

$$\frac{\sin \alpha}{a} = \frac{\sin \beta}{b}$$

$$\frac{\sin \alpha}{5} = \frac{\sin 60°}{7}$$

$$\sin \alpha = \frac{5 \sin 60°}{7} \approx 0.6186$$

There are two possible angles in a triangle, one acute and one obtuse, that have this sine value. So,

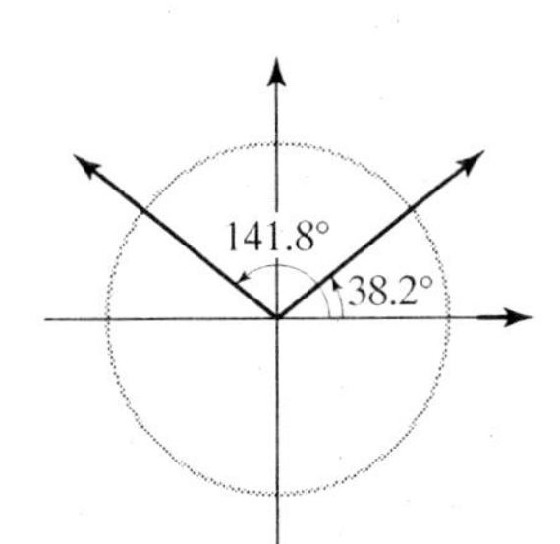

$$\alpha \approx \sin^{-1}(0.6186) \approx 38.2° \quad \text{or} \quad \alpha' \approx (180° - 38.2°) = 141.8°.$$

Each angle is possible in a triangle, but we must see if the larger of the two, $\alpha' \approx 141.8°$, is possible in *this* triangle in which $\beta = 60°$. Since $\alpha' + \beta \approx 141.8° + 60° = 201.8°$ is greater than 180°, we do not have a second triangle solution. We only have one triangle solution for $\alpha \approx 38.2°$ and $\beta = 60°$.

Solution 1

$a = 5$	$\alpha \approx 38.2°$
$b = 7$	$\beta = 60°$
$c \approx 8.0$	$\gamma \approx 81.8°$

~~**Solution 2**~~

$a = 5$	$\alpha' = 141.8°$
$b = 7$	$\beta = 60°$
c'	γ'

} more than 180°

To finish this solution, we first find $\gamma \approx 180° - 38.2° - 60° = 81.8°$. Then we find c.

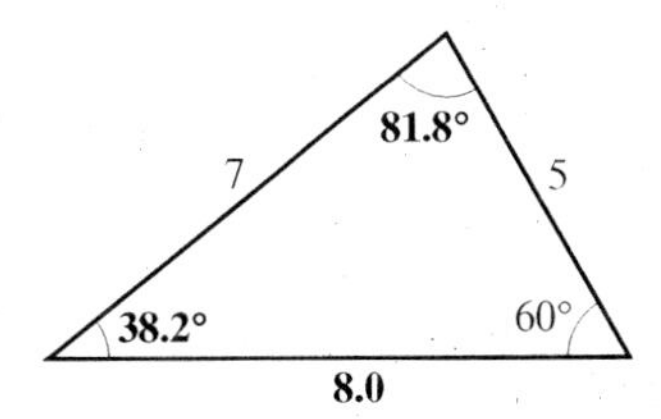

$$\frac{c}{\sin \gamma} = \frac{b}{\sin \beta}$$

$$\frac{c}{\sin 81.8°} = \frac{7}{\sin 60°}$$

$$c = \frac{7 \sin 81.8°}{\sin 60°} \approx 8.0$$

■

Example 4 (SSA) Solve the triangle if $b = 3$, $c = 4$, and $\beta = 28°$.

Solution Since this is the ambiguous case, we need to determine the number of triangles as a result of the given information. Since we know b, c, and β, we begin by finding γ.

$$\frac{\sin \gamma}{c} = \frac{\sin \beta}{b}$$

$$\frac{\sin \gamma}{4} = \frac{\sin 28°}{3}$$

$$\sin \gamma = \frac{4 \sin 28°}{3} \approx 0.6260$$

(continued)

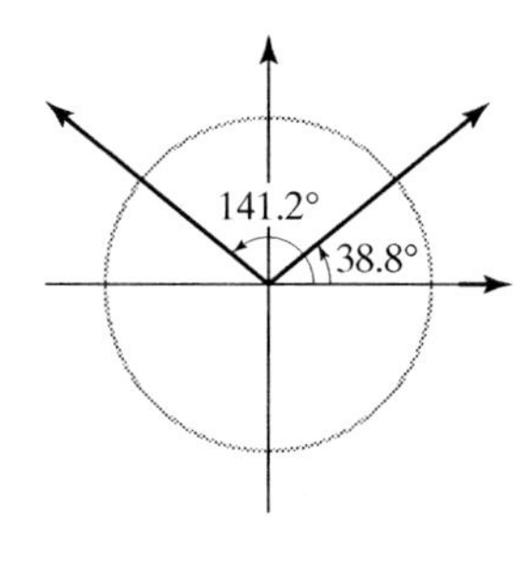

There are two possible angles in a triangle that have this sine value, namely

$$\gamma \approx \sin^{-1}(0.6260) \approx 38.8° \quad \text{or} \quad \gamma' \approx (180° - 38.8°) = 141.2°.$$

Each angle is possible in a triangle, but we must see if the larger of the two, $\gamma' \approx 141.2°$, is possible in *this* triangle in which $\beta = 28°$. Because $\gamma' + \beta \approx 169.2°$ is less than $180°$, we have a second triangle solution.

To finish Solution 1:

We first find α:

$$\alpha \approx 180° - 28° - 38.8° = 113.2°$$

To find a, we use the law of sines.

$$\frac{a}{\sin\alpha} = \frac{b}{\sin\beta}$$

$$\frac{a}{\sin 113.2°} = \frac{3}{\sin 28°}$$

$$a = \frac{3\sin 113.2°}{\sin 28°} \approx 5.87$$

To finish Solution 2:

$$\alpha' \approx 180° - 28° - 141.2° = 10.8°$$

To find a', we use the law of sines.

$$\frac{a'}{\sin\alpha'} = \frac{b}{\sin\beta}$$

$$\frac{a'}{\sin 10.8°} = \frac{3}{\sin 28°}$$

$$a' = \frac{3\sin 10.8°}{\sin 28°} \approx 1.20$$

SOLUTION 1

$a \approx 5.87$	$\alpha \approx 113.2°$
$b = 3$	$\beta = 28°$
$c = 4$	$\gamma \approx 38.8°$

SOLUTION 2

$a' \approx 1.20$	$\alpha' \approx 10.8°$
$b = 3$	$\beta = 28°$
$c = 4$	$\gamma' \approx 141.2°$

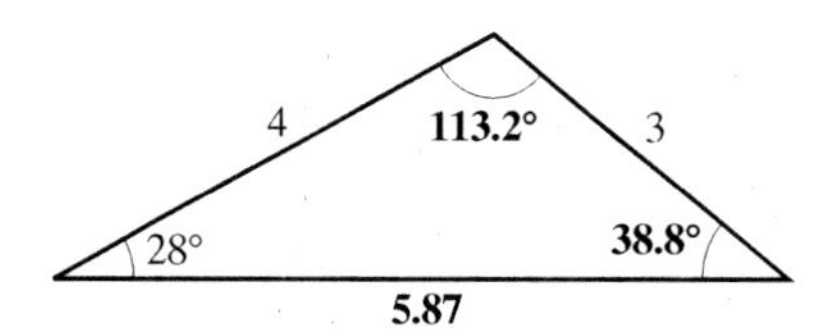

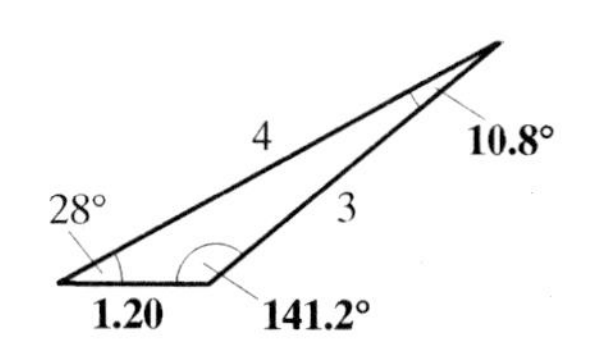

Solving the Ambiguous Case

When we use the law of sines with the ambiguous SSA situation, we always begin by solving for an angle in the triangle, such as α.

- If $|\sin\alpha| > 1$ then no solution is possible (Example 2).
- If $|\sin\alpha| < 1$, there are *two angles* to consider: α and $\alpha' = (180° - \alpha)$. We know we have at least one triangle solution that contains α. We immediately check to see if a *second triangle is possible* by adding α' to the angle given in the problem. If their sum is greater than $180°$, only one triangle solution is possible (Example 3); if their sum is less than $180°$, a second triangle solution is possible (Example 4).

What if $|\sin\alpha| = 1$? (See Exercise 39.)

Application of the Law of Sines

EXAMPLE 5 Coast Guard Station A and Station B are 3 miles apart. Station B is directly east of Station A. Station A receives a distress message from a cruise ship and determines that the bearing of the ship from Station A is N 35° E. Station B also receives the same distress message from the ship and determines that the bearing of the ship from Station B is N 70° W. We know Station A is closer. (Why?) Determine how much closer (to the nearest half mile) Station A is to the cruise ship.

SOLUTION Draw a diagram and label the information. Since Station B is east of Station A, $c = 3$, $\alpha = 55°$, $\beta = 20°$, and $\gamma = 105°$. We need to find a and b. (We know that b should be smaller than a, since it is across from the smallest angle in the triangle.)

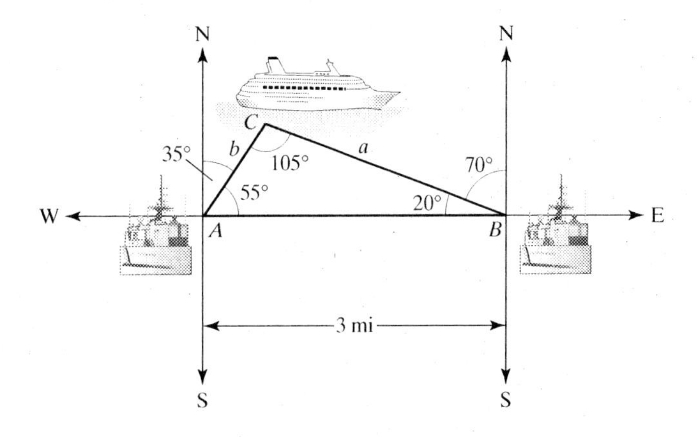

Using the law of sines since we have AAS, we get:

$$\frac{a}{\sin\alpha} = \frac{c}{\sin\gamma}$$

$$\frac{a}{\sin 55°} = \frac{3}{\sin 105°}$$

$$a = \frac{3\sin 55°}{\sin 105°} \approx 2.54$$

$a \approx 2.54$	$\alpha = 55°$
$b \approx 1.06$	$\beta = 20°$
$c = 3$	$\gamma = 105°$

$$\frac{b}{\sin\beta} = \frac{c}{\sin\gamma}$$

$$\frac{b}{\sin 20°} = \frac{3}{\sin 105°}$$

$$b = \frac{3\sin 20°}{\sin 105°} \approx 1.06$$

Finally, we get $a - b \approx 2.54 - 1.06 = 1.48$. Thus to the nearest half mile, Station A is $1\frac{1}{2}$ miles closer to the cruise ship. ■

EXAMPLE 6 A helicopter that has been hired by a local radio station to do traffic reporting is hovering above an accident scene which is causing a traffic jam ahead of you on the freeway. Your friend, who is directly ahead of you on the freeway, calls you on your cell phone to tell you that she too is stuck in traffic and has not passed the accident scene. You see the helicopter at a 45° angle of elevation. If the distance between you and the helicopter is 80 feet and the distance between your friend and the helicopter is 60 feet, what is the distance between you and your friend?

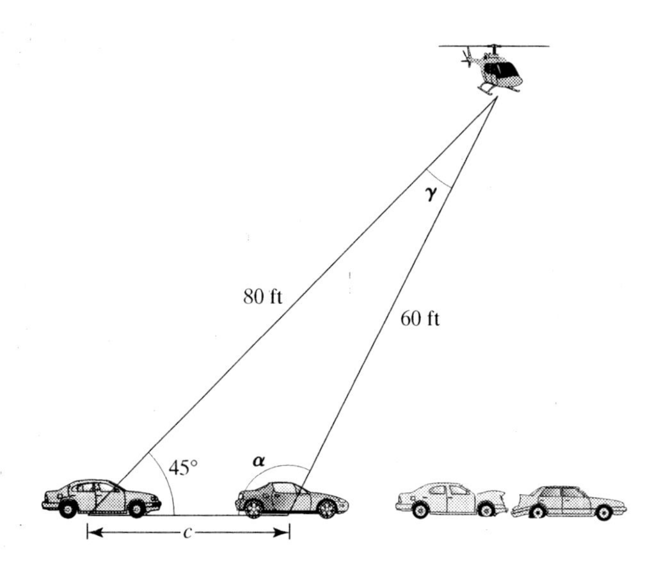

SOLUTION Since we have SSA, the ambiguous case, we may have two triangles. Since we know $a = 80$, $b = 60$, and $\beta = 45°$, we start by finding α.

$$\frac{\sin\alpha}{a} = \frac{\sin\beta}{b} \longrightarrow \frac{\sin\alpha}{80} = \frac{\sin 45°}{60}$$

$$\sin\alpha = \frac{80\sin 45°}{60} \approx 0.9428$$

$$\alpha \approx \sin^{-1}(0.9428) \approx 70.5°, \quad \text{or} \quad \alpha' \approx 180° - 70.5° = 109.5°$$

Since $\alpha' + \beta \approx 154.5°$, we have a second triangle solution.

To finish Solution 1:

We first find $\gamma \approx 180° - 45° - 70.5° = 64.5°$.
To find c, we use the law of sines.

$$\frac{c}{\sin\gamma} = \frac{b}{\sin\beta}$$

$$\frac{c}{\sin 64.5°} = \frac{60}{\sin 45°}$$

$$c = \frac{60\sin 64.5°}{\sin 45°} \approx 76.5869$$

To finish Solution 2:

$$\gamma' \approx 180° - 45° - 109.5° = 25.5°$$

$$\frac{c'}{\sin\gamma'} = \frac{b}{\sin\beta}$$

$$\frac{c'}{\sin 25.5°} = \frac{60}{\sin 45°}$$

$$c' = \frac{60\sin 25.5°}{\sin 45°} \approx 36.5301$$

Solution 1

$a = 80$	$\alpha \approx 70.5°$
$b = 60$	$\beta = 45°$
$c \approx 76.59$	$\gamma \approx 64.5°$

Solution 2

$a = 80$	$\alpha' \approx 109.5°$
$b = 60$	$\beta = 45°$
$c' \approx 36.53$	$\gamma' \approx 25.5°$

For the answer c, the distance between you and your friend, we have two choices. Obviously, your friend can't be in two places. So which answer is correct?

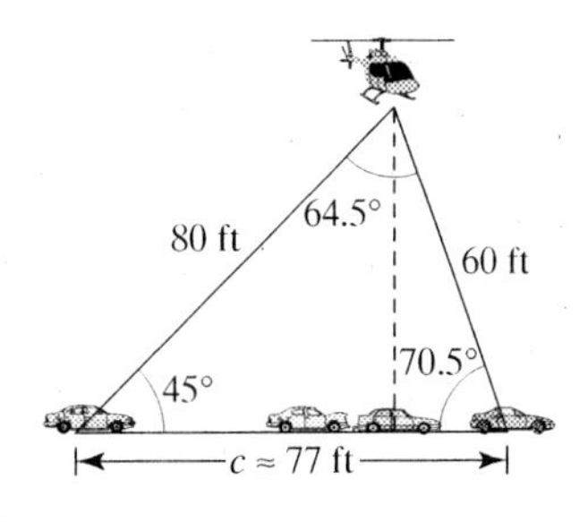

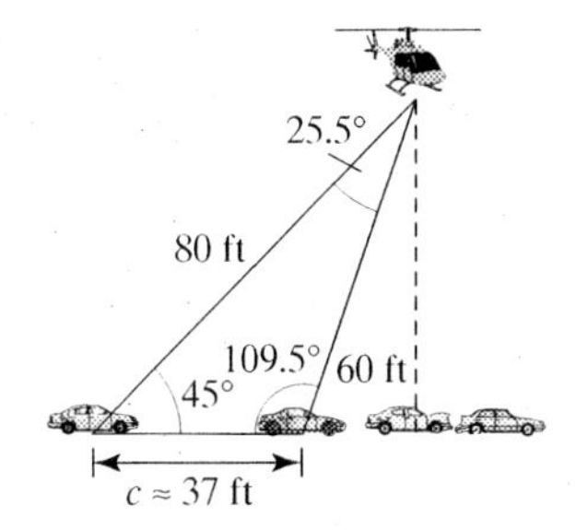

If we look at the triangle with Solution 1, it indicates your friend would be past the accident scene. On the other hand, Solution 2 correctly indicates she has not passed the scene and, like you, she is still stuck in traffic. So your friend is approximately 37 feet ahead of you. ■

As you begin the exercises in this section, remember to draw a triangle, label the given information, label what you are asked to find, and fill in a chart with this information. The diagram or the chart should indicate what situation you have.

AAS or ASA (two angles and any side)

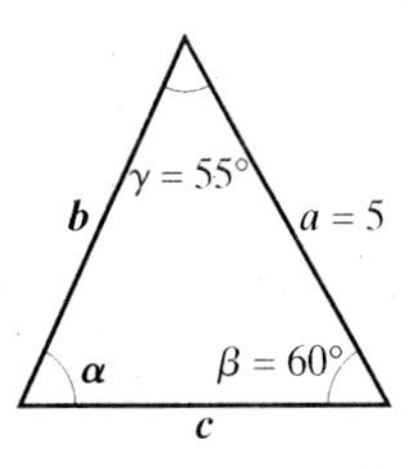

SSA (two sides and an opposite angle)—the ambiguous case

Solution 1

$a = 12$	α
$b = 13$	$\beta = 50°$
c	γ

Solution 2 (?)

$a = 12$	α'
$b = 13$	$\beta = 50°$
c'	γ'

SSS or SAS

$a = 5$	$\alpha = ?$
$b = 7$	$\beta = ?$
$c = 8$	$\gamma = ?$

What if we are given the case SSS (three sides)? Can we use the law of sines to solve this triangle? Consider the information at the left. If we select any two of the ratios, we are not able to solve any of these equations individually because each equation then has two different unknown values.

$$\frac{\sin\alpha}{5} = \frac{\sin\beta}{7} \quad \text{or} \quad \frac{\sin\beta}{7} = \frac{\sin\gamma}{8} \quad \text{or} \quad \frac{\sin\alpha}{5} = \frac{\sin\gamma}{8}$$

Therefore, we cannot use the law of sines for the SSS situation.

$a = 4$	$\alpha = ?$
$b = ?$	$\beta = 80°$
$c = 5$	$\gamma = ?$

Likewise, if we are given two sides and the included angle (SAS), we cannot use the law of sines to solve this triangle for the same reason. To see this, consider the information given at the left which gives:

$$\frac{\sin\alpha}{4} = \frac{\sin 80°}{b} \quad \text{or} \quad \frac{\sin 80°}{b} = \frac{\sin\gamma}{5} \quad \text{or} \quad \frac{\sin\alpha}{4} = \frac{\sin\gamma}{5}$$

These two situations require the law of cosines, which we will discuss in the next section.

Exercise Set 3.4

1–8. *Use the given information for each $\triangle ABC$, to determine the situation: AAS, SSA, SSS, or SAS. Also, indicate if the situation applies to the law of sines and if so, whether it is the ambiguous case.*

1.

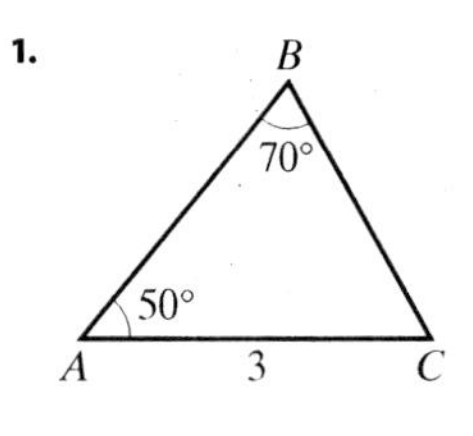

2.

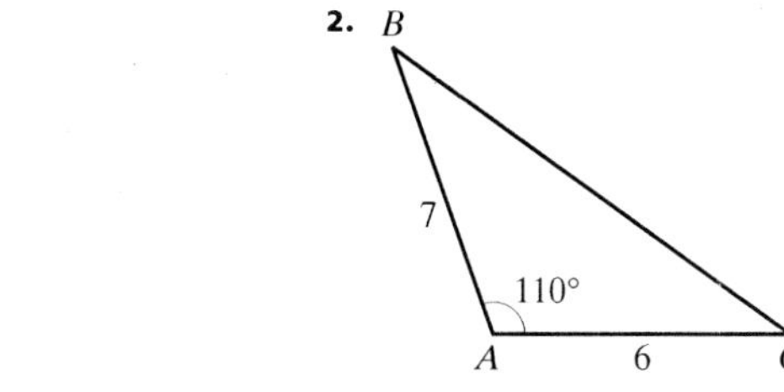

3.

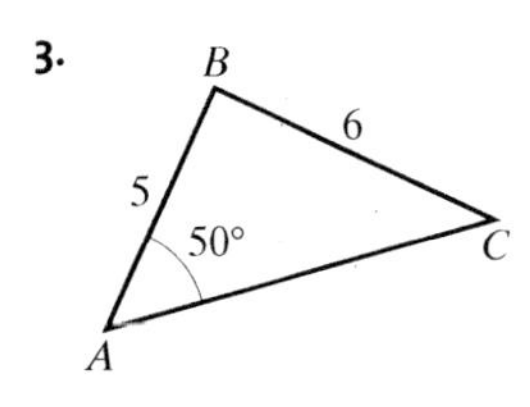

4.

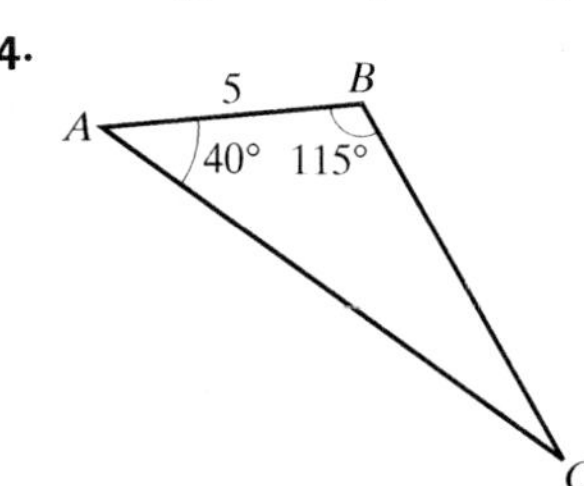

5.

$a = 6$	α
b	$\beta = 20°$
$c = 5$	γ

6.

a	$\alpha = 20°$
$b = 5$	β
$c = 2$	γ

7.

$a = 5$	α
$b = 6$	β
$c = 3$	γ

8.

$a = 7$	$\alpha = 32°$
$b = 9$	β
c	γ

9–22. *Solve each triangle.*

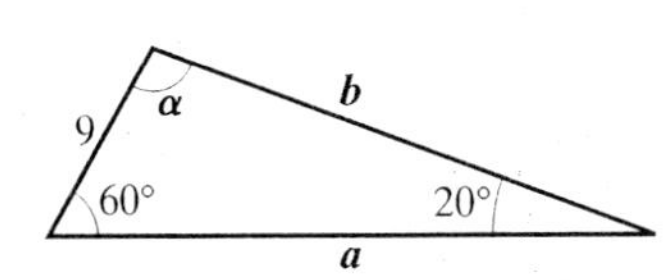

EXAMPLE $c = 9, \beta = 60°, \gamma = 20°$

SOLUTION Draw and label the triangle, and fill in a chart.
We are given two angles and a side (AAS), so we quickly find α and apply the law of sines.

$$\alpha = 180° - 60° - 20° = 100°.$$

$$\frac{b}{\sin 60°} = \frac{9}{\sin 20°} \rightarrow b = \frac{9 \sin 60°}{\sin 20°} \approx 22.79$$

$$\frac{a}{\sin 100°} = \frac{9}{\sin 20°} \rightarrow a = \frac{9 \sin 100°}{\sin 20°} \approx 25.91$$

$a \approx 25.91$	$\alpha = 100°$
$b \approx 22.79$	$\beta = 60°$
$c = 9$	$\gamma = 20°$

9. $a = 7, \beta = 14°, \gamma = 62°$

10. $\alpha = 78°, c = 3, \beta = 15°$

11. $\beta = 100°, \alpha = 15.6°, a = 6$

12. $\alpha = 30°54', c = 15.3, \gamma = 18°6'$

13. $\beta = 30°, b = 9, c = 7$

14. $a = 10, b = 5, \alpha = 48°$

15. $b = 16.8, \gamma = 110°, c = 19$

16. $a = 3.2, b = 1.6, \alpha = 72°$

17. $a = 10, b = 5.9, \beta = 45°$

18. $c = 6, \alpha = 68°, a = 4.2$

19. $b = 7, c = 5, \gamma = 37°$

20. $c = 15, \beta = 42°, b = 10.7$

21. $a = 1.6, c = 3.2, \alpha = 30°$

22. $c = 15, a = 30, \gamma = 30°$

23. Find the length of the two equal sides of an isosceles triangle if the base angles each measure 54° and the length of the base is 24.8 inches.

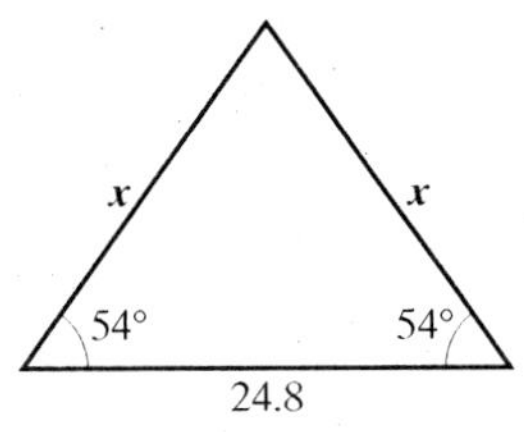

24. The diagonal of a parallelogram is 10 inches long. The diagonal makes angles of 33° and 25° with the sides of the parallelogram. Find the lengths of the sides of the parallelogram.

25. If $a = 32$, $\beta = 48°$, and $c = 25$, find the altitude to side a and then find the area of the triangle to the nearest square unit.

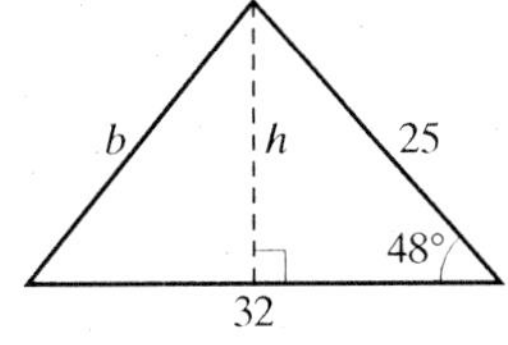

26. If $a = 19$, $\alpha = 55°$, and $c = 18$, find the altitude to the side b and then find the area of the triangle to the nearest square unit.

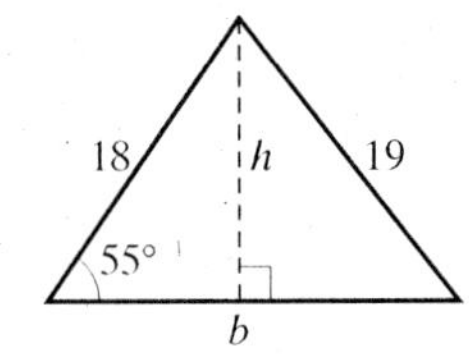

27. Prove that the area of a triangle is $\frac{1}{2}$ the product of the lengths of two consecutive sides and the sine of the included angle. (Area of $\triangle ABC = \frac{1}{2}ab \sin \gamma = \frac{1}{2}bc \sin \alpha = \frac{1}{2}ac \sin \beta$)

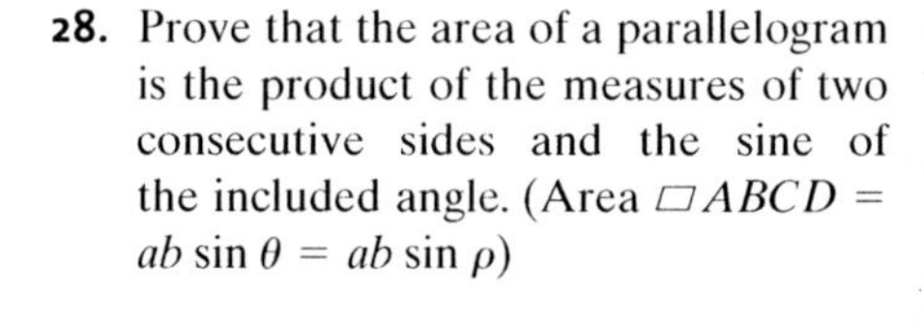
28. Prove that the area of a parallelogram is the product of the measures of two consecutive sides and the sine of the included angle. (Area $\square ABCD = ab \sin \theta = ab \sin \rho$)

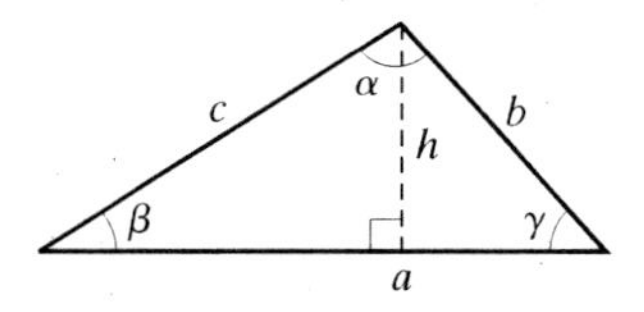

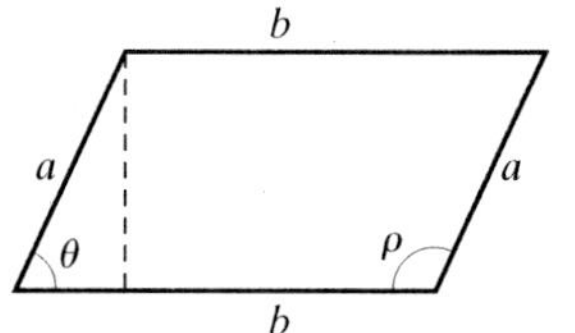

29. You are scheduled for a one-way direct flight from City A to City B, which is a 270-mile trip, but your flight gets cancelled. You are then rescheduled on the next available flight, which will leave City A, stop in City C, and then go on to City B, which is directly south of City A. If the bearing of City C from City A is S 42° E and the bearing of City C from City B is N 34° E, find the total miles (to the nearest mile) of your rescheduled trip.

30. The path of a satellite orbiting Earth causes it to pass directly over two tracking stations A and B that are 100 kilometers apart. When the satellite is west of the two stations, the angles of elevation of the satellite from Station A and from Station B are 75° and 60°, respectively. How far (to the nearest kilometer) is the satellite from Station A?

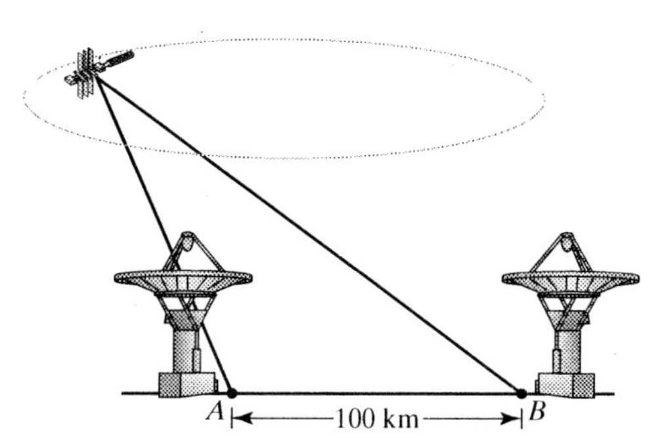

31. A baseball diamond is a square that is 90 feet on each side. The pitcher's mound is directly between home plate and second base, but it is *closer* to home plate. If the pitcher's mound is 63 feet 9 inches from first base, how far (to the nearest half-foot) is the pitcher from home plate? (Be careful to consider both situations.)

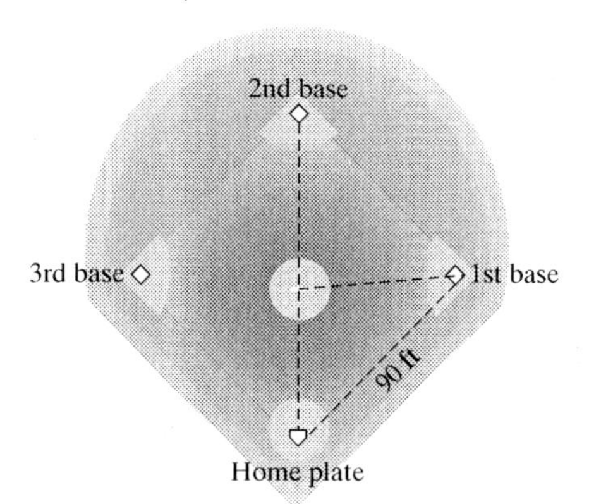

32. Anthony is standing near a radio station and observes that the angle of elevation to the top of the radio station's antenna from ground level is 65°. He walks straight back until he is 100 feet further away and measures the angle of elevation from ground level to the top of the antenna to be 46°. Find the height (to the nearest foot) of the antenna.

33. A man is flying in a hot-air balloon on a straight path of constant altitude at a rate of 5 feet per second. As he approaches the parking lot of a market, he notices that the angle of depression to a friend's car is 35°. After $1\frac{1}{2}$ minutes and after flying directly over his friend's car, he looks back to see the car at a 36° angle of depression. At that time, what is the distance between the man and his friend's car?

34. Tim, whose eye level is 6 feet above the floor, is standing in a museum and notices that the bottom of a mural is at floor level. He looks down 13° to see the bottom of the mural and upward 17° to see the top.

a. How far is Tim from the mural?

b. How tall is the mural?

35. A live TV talk show is broadcast with the use of two identical TV transmission cameras. One camera is 18 feet away from the host of the show, while the other camera is only 14 feet away from him. The cameraman who is farthest away notices that the angle formed by his line of sight to the other camera and his line of sight to the talk show host is 45°. The two cameras share a 15-foot electrical cord, which means they can never be more than 15 feet apart. Find the distance to the nearest foot between the two cameras. (Be careful to consider both situations.)

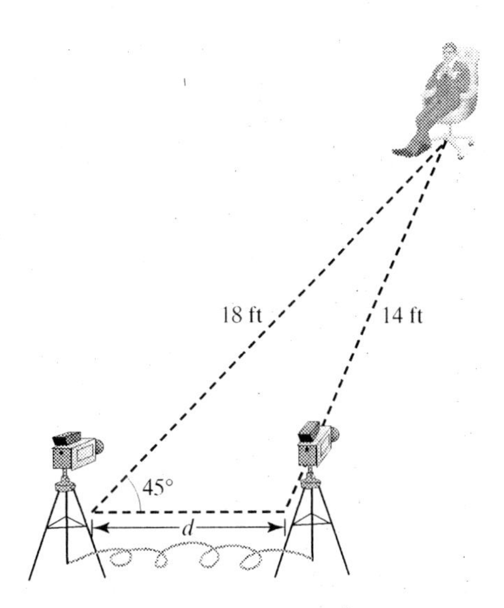

36. A newly planted tree is secured with several wires. Each wire is attached to the tree at the same height, then anchored in level ground. One of the wires is 20 feet in length and makes an angle of 40° with the ground. A second wire is 15 feet in length and is placed directly between the trunk of the tree and the first wire. To the nearest half-foot, how far apart are these two wires? (Be careful to consider both situations.)

37–38. *Use the law of sines to rewrite the right side of the equation in order to show that in any triangle ABC, the following identities are true.*

37. $\dfrac{a+b}{b} = \dfrac{\sin\alpha + \sin\beta}{\sin\beta}$

38. $\dfrac{a-c}{c} = \dfrac{\sin\alpha - \sin\gamma}{\sin\gamma}$

Discussion

39. Suppose you are solving a triangle with the law of sines and, as a result of substituting values for a, b, and β, the following appears:

$$\sin \alpha = \frac{a \sin \beta}{b}$$

$$\sin \alpha = 1$$

What would you know about the triangle? How would you continue to solve the triangle?

40. We have said that the situation in which we are given two sides and an opposite angle is called the ambiguous case. What opposite angle could be given that would immediately determine the number of solutions?

41. You would like to determine the height of a mountain. You do not have any climbing gear, no altimeter, nor any earth moving equipment. But you do have a pencil, paper, calculator, an angle finder, and a 16-foot measuring tape. Write a detailed set of steps that outline how you would use the law of sines to find the height of the mountain.

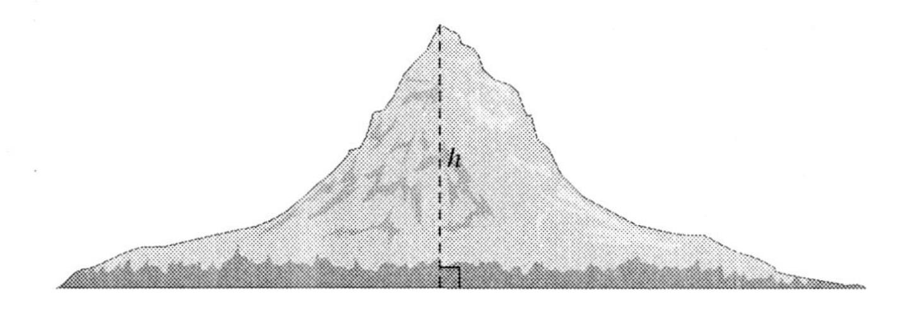

42. You are on one side of a river and you would like to determine the width of the river. On the bank of the other side of the river, you spot a beautiful tree. You know the two banks of a river are parallel, and all you have is a pencil, paper, calculator, angle finder, and a 16-foot measuring tape. Explain how you can find the width of the river.

3.5 Solution of Triangles Using Law of Cosines

In this section we discuss the second law for solving triangles, the law of cosines, which is a generalization of the Pythagorean Theorem. This law is used to solve triangles when we are given three sides (SSS) or two sides and the included angle (SAS). We can also apply this law to solve the ambiguous case, which is two sides and an opposite angle (SSA).

Law of Cosines

Suppose we are given in $\triangle ABC$ the lengths of two sides $(a$ and $b)$ and the measure of the included angle (γ). We place the triangle on the xy-plane such that angle (γ) is in standard position. If $A(x, y)$ is a point on the terminal side of angle γ and we use the definition of the trigonometric functions (Section 3.2), then we have

$$\cos(\gamma) = \frac{x}{b} \quad \text{and} \quad \sin(\gamma) = \frac{y}{b}.$$

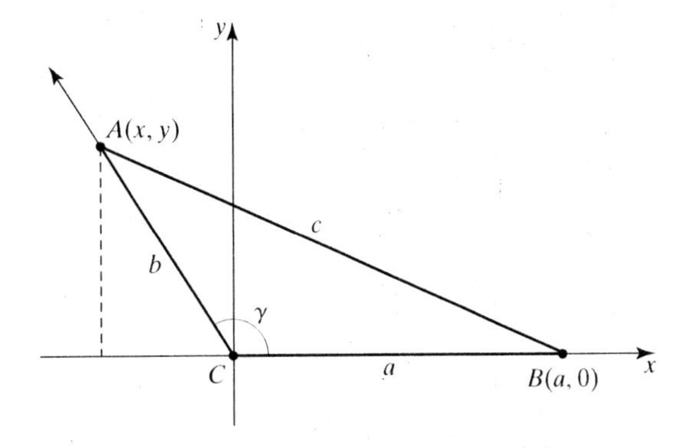

Solving for x and y, the coordinates of A are

$$x = b\cos(\gamma) \quad \text{and} \quad y = b\sin(\gamma).$$

Using the distance formula from $A(x, y)$ to $B(a, 0)$, we get:

$$c = \sqrt{(x-a)^2 + (y-0)^2}$$

$$c^2 = (x-a)^2 + (y-0)^2 \qquad \text{Square both sides.}$$

$$= (b\cos\gamma - a)^2 + (b\sin\gamma - 0)^2 \qquad \text{Substitute for } x \text{ and } y.$$

$$= b^2\cos^2\gamma - 2ab\cos\gamma + a^2 + b^2\sin^2\gamma \qquad \text{Multiply.}$$

$$= b^2(\cos^2\gamma + \sin^2\gamma) - 2ab\cos\gamma + a^2 \qquad \text{Factor.}$$

$$= b^2(1) - 2ab\cos\gamma + a^2 \qquad \cos^2\gamma + \sin^2\gamma = 1$$

Thus,

$$c^2 = a^2 + b^2 - 2ab\cos\gamma.$$

This proves one part of the following, known as the *law of cosines.* Proofs of the other two parts would follow in a similar manner by placing each of the other angles in standard position.

Law of Cosines

In any triangle ABC, if a, b, and c are the sides opposite angles α, β, and γ, respectively, then

$$c^2 = a^2 + b^2 - 2ab\cos\gamma$$
$$b^2 = a^2 + c^2 - 2ac\cos\beta$$
$$a^2 = b^2 + c^2 - 2bc\cos\alpha$$

EXAMPLE 1 What happens when you apply the law of cosines to a right triangle?

SOLUTION In a right triangle we have $\gamma = 90°$, so $c^2 = a^2 + b^2 - 2ab \cos 90°$. Since $\cos 90° = 0$, the law of cosines reduces to $c^2 = a^2 + b^2$, which is the Pythagorean Theorem.

This example does not indicate that the law of cosines proves the Pythagorean Theorem. In fact the opposite is true since the Pythagorean Theorem was used (in the distance formula) to derive the law of cosines. ■

We now solve general triangles with given information that allows us to use the law of cosines: three sides (SSS) and two sides and the included angle (SAS).

EXAMPLE 2 (SSS) Find γ if $a = 9$, $b = 5$, and $c = 7$.

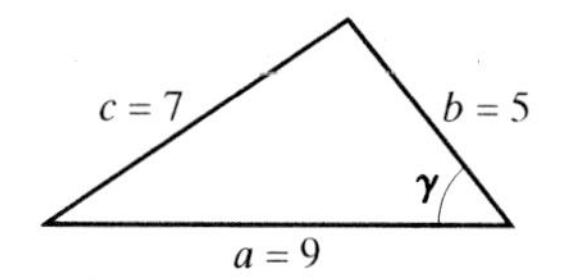

SOLUTION Since we know a, b, and c and want to find γ, we use the part of the law of cosines that contains these values.

$$c^2 = a^2 + b^2 - 2ab \cos \gamma$$
$$7^2 = 9^2 + 5^2 - 2(9)(5)\cos \gamma$$
$$49 = 81 + 25 - 2(9)(5)\cos \gamma$$
$$\frac{49 - 81 - 25}{-2(9)(5)} = \cos \gamma$$
$$0.6333 \approx \cos \gamma$$
$$\gamma \approx \cos^{-1}(0.6333)$$
$$\gamma \approx 50.7°$$ ■

We were asked to find only γ in Example 2. If we were asked to solve the triangle, then we could continue solving by using either the law of sines or another part of the law of cosines.

- To continue with the law of sines (using SSA), we have:

$$\frac{\sin \beta}{5} = \frac{\sin 50.7°}{7} \rightarrow \sin \beta = \frac{5 \sin 50.7°}{7} \rightarrow \beta \approx 33.6°, \quad \text{or} \quad \beta' \approx 146.4°.$$

$a = 9$	$\alpha \approx 95.7°$
$b = 5$	$\beta \approx 33.6°$
$c = 7$	$\gamma \approx 50.7°$

Here we select $\beta \approx 33.6°$ because it is opposite the smallest side b. (Also $\beta' \approx 146.4°$ is too large since $\gamma = 50.7°$.)

- If instead we use the law of cosines (SSA), even though it may be a little more work, we immediately get the same solution for β.

$$5^2 = 7^2 + 9^2 - 2(7)(9)\cos \beta \rightarrow \cos \beta = \frac{25 - 49 - 81}{-2(7)(9)} \rightarrow \beta \approx 33.6°.$$

Notice that the law of cosines immediately produced the correct value for β, whereas with the law of sines we had to determine the correct answer from two choices. Answers from the law of sines may differ slightly than those obtained using the law of cosines, due to whether or not you are using approximations.

EXAMPLE 3 (SSS) If $a = 5$, $c = 10$, and $b = 3$, find α.

SOLUTION To find α, we use the part of the law of cosines that includes α along with the known values a, b, and c.

$$a^2 = b^2 + c^2 - 2bc\cos\alpha$$
$$5^2 = 3^2 + 10^2 - 2(3)(10)\cos\alpha$$
$$\frac{25 - 9 - 100}{-60} = \cos\alpha$$
$$\cos\alpha = 1.4$$

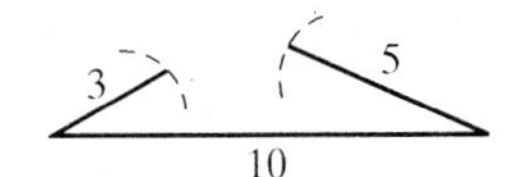

Since the cosine function is never larger than 1, there is no solution. ■

NOTE: It might be easier to determine that Example 3 has no solution by noticing that the sides do not satisfy the condition that the sum of the lengths of any two sides of a triangle must be greater than the length of the third side: $3 + 5 \not> 10$. Before you begin to solve the triangle in which you are given three sides (SSS), you might want to check to make sure the sides of the triangle satisfy this inequality. If you forget to check, you can still apply the law of cosines.

EXAMPLE 4 (SAS) In $\triangle ABC$ $a = 9$, $c = 7$, and $\beta = 105°$. Find the length of b.

SOLUTION Sketch a triangle with the given information. Use the law of cosines that contains the known values and what we are asked to find.

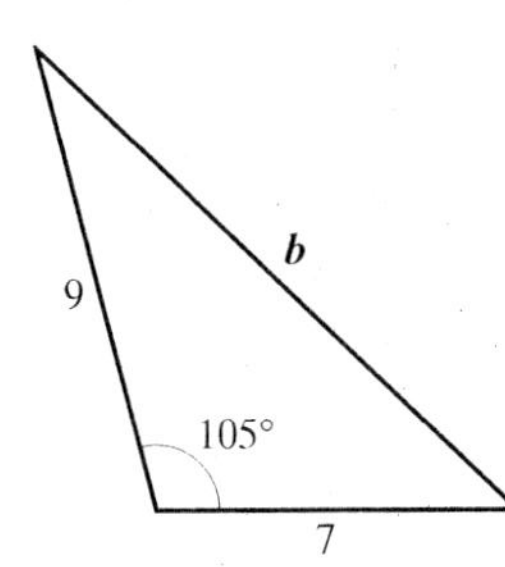

$$b^2 = a^2 + c^2 - 2ac\cos\beta$$
$$b^2 = 9^2 + 7^2 - \underline{2(9)(7)\cos 105°}$$
$$b^2 \approx 81 + 49 - \underline{126(-0.2588)}$$
$$b \approx \sqrt{162.6112} \approx 12.75$$

Make sure to do all the multiplication before you begin to add or subtract. (Order of Operations) ■

Using the Law of Cosines to Solve the Ambiguous Case (Optional)

The following two examples are intended to demonstrate how to apply the law of cosines when we are given two sides and an opposite angle (SSA), or the ambiguous case. The first is the same as Example 3 of Section 3.4, in which we solved the triangle using the law of sines.

EXAMPLE 5 (SSA) Solve the triangle if $a = 5$, $b = 7$, and $\beta = 60°$.

SOLUTION We need to determine the number of possible triangles, because SSA could have 0, 1, or 2 triangle solutions. We use the part of the law of cosines that contains our known values a, b, and β.

$$b^2 = a^2 + c^2 - 2ac\cos\beta$$
$$7^2 = 5^2 + c^2 - 2(5)c\cos 60°$$
$$49 = 25 + c^2 - 2(5)c\left(\tfrac{1}{2}\right)$$
$$0 = c^2 - 5c - 24$$
$$0 = (c - 8)(c + 3)$$
$$c = 8 \quad \text{or} \quad c = -3$$

(continued)

Since c is the length of a side of the triangle and cannot be negative, we have only one solution for c. (If both values of c were negative, we would have no solution. If both values of c were positive, we would have two triangle solutions.) Continuing to solve the triangle, we can find α by using:

$a = 5$	$\alpha \approx 38.2°$
$b = 7$	$\beta = 60°$
$c = 8$	$\gamma \approx 81.8°$

$$a^2 = b^2 + c^2 - 2bc\cos\alpha$$
$$5^2 = 7^2 + 8^2 - 2(7)(8)\cos\alpha$$
$$\frac{25 - 49 - 64}{-112} = \cos\alpha$$
$$0.7857 \approx \cos\alpha$$
$$\alpha \approx \cos^{-1}(0.7857) \approx 38.2°$$

To find γ:

$$\gamma = 180° - \alpha - \beta \approx 180° - 38.2° - 60°$$
$$\gamma \approx 81.8°$$

The first part of the solution in Example 5 produced a quadratic equation that we solved by factoring. Should we choose to apply the law of cosines for SSA, there will be times when we need to use the quadratic formula (see Appendix A) to solve the resulting equation. Our second example demonstrates the use of the quadratic formula as well as two triangle solutions using the law of cosines while solving Example 4 of Section 3.4.

EXAMPLE 6 (SSA) Solve the triangle if $b = 3$, $c = 4$, and $\beta = 28°$.

SOLUTION Again, we need to determine the number of solutions for SSA. Using the part of the law of cosines that contains our known values, we get:

$$b^2 = a^2 + c^2 - 2ac\cos\beta$$
$$3^2 = a^2 + 4^2 - 2a(4)\cos 28°$$
$$9 \approx a^2 + 16 - 8a(0.8829)$$
$$0 \approx a^2 - 7.0636a + 7$$

So,

$$a \approx \frac{7.0636 \pm \sqrt{(-7.0636)^2 - 4(1)(7)}}{2(1)} \approx \frac{7.0636 \pm 4.6792}{2}$$

$a \approx 5.87$ or $a' \approx 1.19$ We have two solutions.

SOLUTION 1

$a \approx 5.87$	α
$b = 3$	$\beta = 28°$
$c = 4$	γ

SOLUTION 2

$a' \approx 1.19$	α'
$b = 3$	$\beta = 28°$
$c = 4$	γ'

We set up two charts and leave it for you to complete the solutions either as we did in Example 4 of Section 3.4 using the law of sines, or by using the law of cosines. You will notice that you can't avoid using approximate values in the solution of this example with either the law of cosines or law of sines. As a result, you may have slightly different answers in the second triangle solution for a'. (Refer to Example 4 in Section 3.4 for the complete solution.) ■

Bearing

In Section 3.3 we used bearing to indicate the direction of one point from another. There is a second way to describe bearing used in aerial navigation. When a single angle is given, its direction is understood to be the number of degrees measured *clockwise from due north.* These illustrations are a few examples of aerial directions, or bearings.

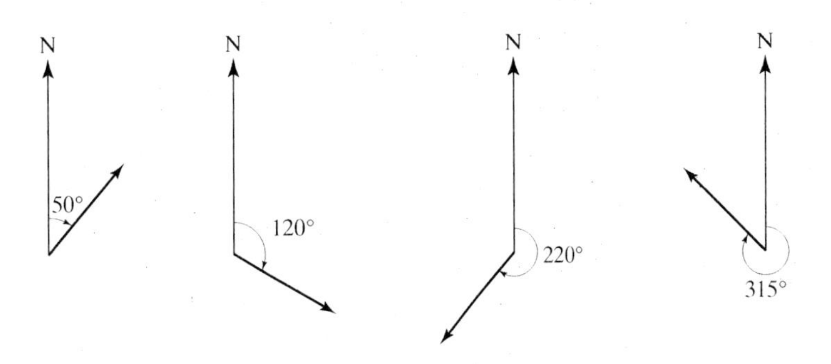

These situations can be stated as "in the direction 50°," "on a course of 120°," "at a bearing of 220°," and "heading 315°," respectively.

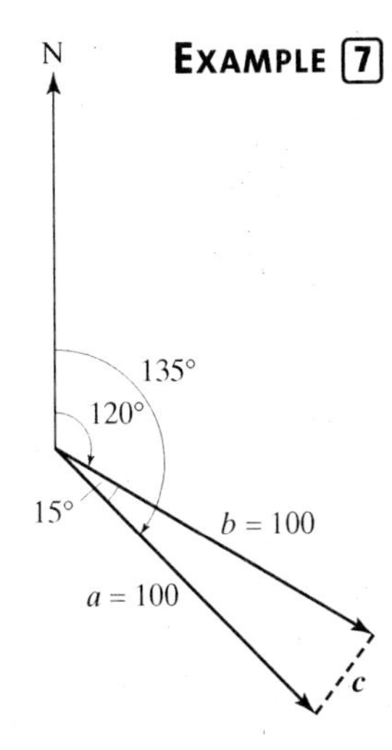

EXAMPLE 7 Plane A has traveled 100 miles at a bearing of 120° when the pilot realizes he should have been traveling the 100 miles at a bearing of 135°. Determine to the nearest mile the distance between the current and intended positions.

SOLUTION Our diagram indicates that we have SAS: $a = 100$, $b = 100$, and $\gamma = 135° - 120° = 15°$. Using the law of cosines, we get:

$$c^2 = a^2 + b^2 - 2ab\cos\gamma$$
$$c^2 = 100^2 + 100^2 - 2(100)(100)\cos 15°$$
$$c^2 \approx 10{,}000 + 10{,}000 - 19318.5165$$
$$c^2 \approx 681.4835$$
$$c \approx \sqrt{681.4835} \approx 26.11$$

So the airplane is approximately 26 miles from its intended position. ■

We are now able to solve any triangle, and in some situations we can use more than one law or trigonometric ratio. On the basis of the given information, it is important to first determine which triangle situation is presented: right triangle or oblique. If oblique, determine if the situation is SSS, SAS, AAS, or SSA. Then, select and apply the correct law or trigonometric ratio to solve the triangle. Practice makes perfect, so let's get started.

Exercise Set 3.5

1–8. *Indicate what situation applies to the given information for $\triangle ABC$: SSS, SAS, SSA, or AAS. Then state the method(s) you could use to find the indicated part. Do not solve.*

1.

a
5
50°
7

2.

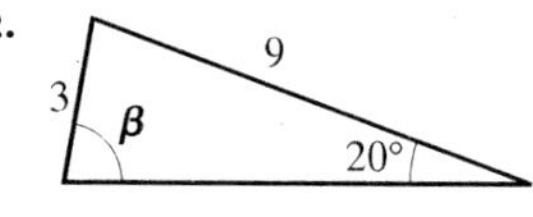

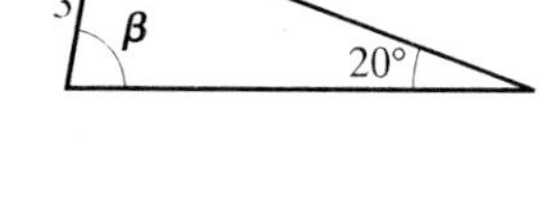

3.

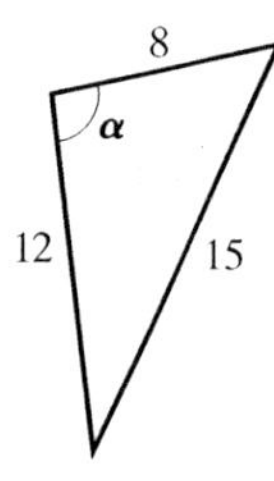

4.

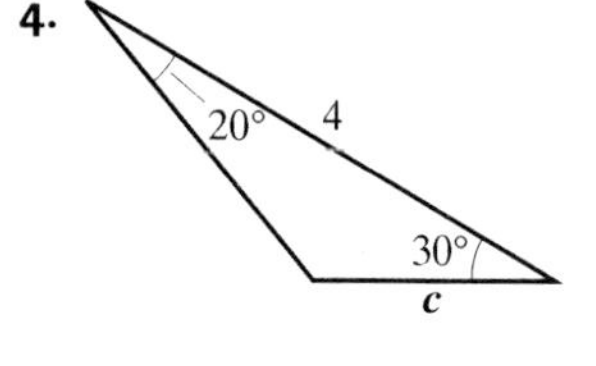

5. $a = 6, c = 4$, and $\gamma = 60°$, find b.

6. $\beta = 42.6°, a = 10$, and $c = 9$, find b.

7. $\alpha = 26°, \beta = 64°$, and $b = 8$, find c.

8. $a = 5, b = 6$, and $c = 8$, find α.

9–24. *Solve each triangle if possible.*

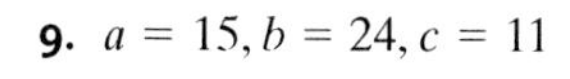

9. $a = 15, b = 24, c = 11$

10. $a = 4, b = 12, c = 9$

11. $a = 6, b = 9, \gamma = 73°$

12. $c = 9.1, a = 5.6, \beta = 19°$

13. $c = 9$ ft, $a = 16$ ft, $\beta = 94° 24'$

14. $b = 6.2$ in., $c = 4$ in., $\alpha = 95°6'$

15. $\gamma = 28.7°, a = 7$ cm, $b = 5$ cm

16. $\beta = 55.6°, a = 5$ mi, $c = 3.2$ mi

17. $\alpha = 56°, b = 9$

18. $\alpha = 22°, \beta = 170°, a = 9$

19. $a = 5, b = 9, c = 3$

20. $c = 22.1, b = 12.9, a = 8.1$

21. $c = 4, a = 5, b = 6$

22. $a = 7, b = 5, c = 8$

23. $\alpha = 82°, \beta = 34°, c = 9$

24. $\gamma = 120°, c = 14, \beta = 30°$

25. If the sides of a triangle are 9, 12, and 15, find the measure of its largest angle.

26. If the sides of a triangle are 24, 6, and 28, find the measure of its smallest angle.

27. Two airplanes head out from the same airport at the same time. The first plane is flying at 300 mph on a course of 42°, whereas the second is flying 245 mph in the direction of 97°. If each continues on course how far apart (to the nearest mile) are the airplanes after 2 hours?

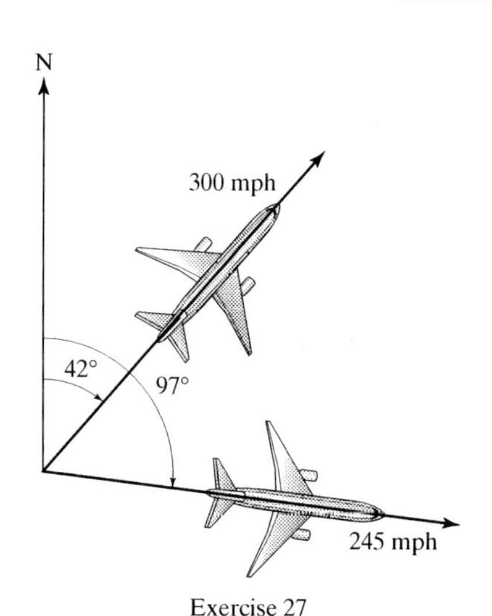

Exercise 27

28. At noon an airplane is directly over Las Vegas flying at 270 mph on a course of 265°. It continues on the same course for 3 hours, then changes to a course of 250° and continues flying at 300 mph for $1\frac{1}{2}$ hours. How many miles is the plane from its noontime position over Las Vegas? (Assume that it is flying at a constant altitude.)

29. A pilot traveling 220 mph at a constant altitude is passing over Madison, Wisconsin, heading towards Detroit, Michigan, a distance of 330 miles. After $\frac{1}{2}$ hour the pilot detects an error indicating he has been 10° off course since Madison.

  a. How many miles is the plane from Detroit when the error is detected?

  b. What angle should the pilot turn to head toward Detroit?

30. A ship sailing due east in the North Atlantic has been warned to change course to avoid a group of icebergs. The captain turns the ship in a northeasterly direction and sails for 27 miles, and then changes course again turning southeast and sailing for 36 miles until the ship reaches the original course at a point 50 miles due east of the point where he turned northeast. What was the bearing of the first turn?

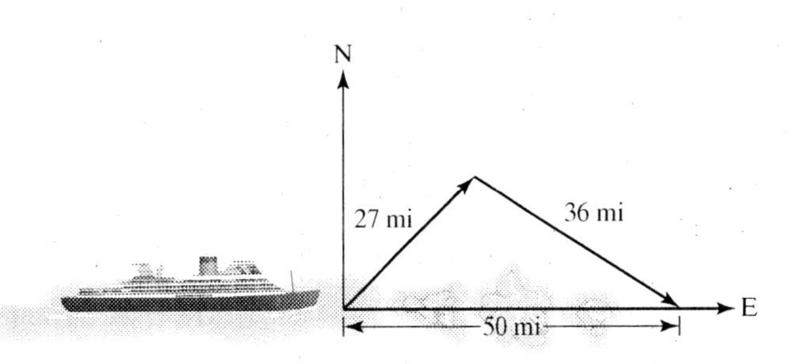

31. Three sides of a triangular lot are 50′ by 70′ by 60′.

a. What is the angle between the 50′ and 70′ sides?

b. What is the area of the triangle, to the nearest square foot? (See Exercise 27 of Section 3.4.)

32. A 50-ft tower sits on a hill. An 80-ft wire from the top of the tower is anchored to the ground 40 feet from the base of the tower.

a. Determine the angle between the tower and the ground.

b. Determine the angle the ground makes with the horizontal.

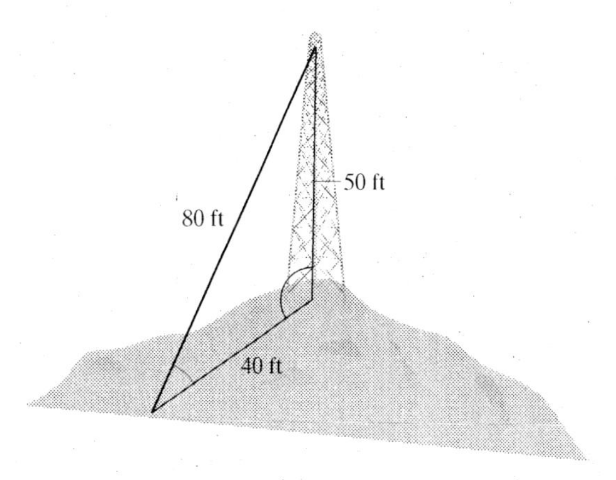

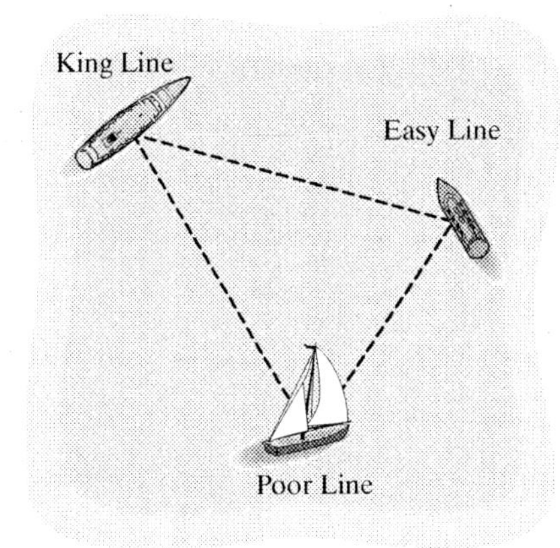

33. King Line, Poor Line, and Easy Line, three cruise ships, are on the ocean. King Line is traveling in the direction of the Easy Line when he receives a radio message to help Poor Line. The distance from the King Line to the Poor Line is 56 miles, the distance from the King Line to the Easy Line is 46 miles, while the distance between the Poor Line and the Easy Line is 30 miles. What is the angle that King Line will use to change its course to turn in the direction of Poor Line ?

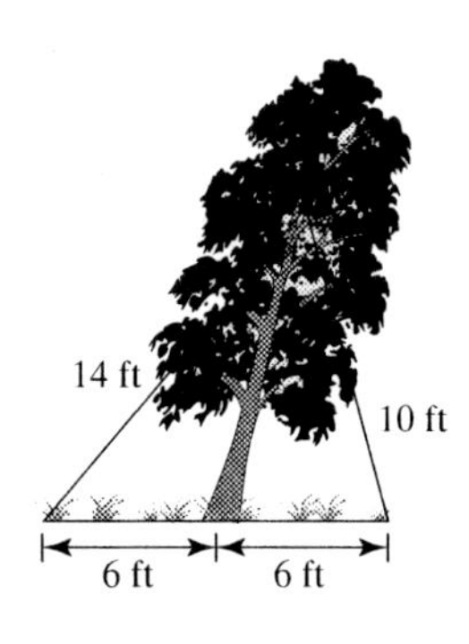

34. Two wires, one 10 feet long and the other 14 feet long, secure a tree. Each wire is attached to the slanted tree at the same height and secured on the ground on opposite sides of the trunk of the tree. If each wire is secured on the ground 6 feet from the trunk of the tree, what is the angle that the tree makes with the ground as it leans towards the 10-ft wire?

35. Comerica Park, home of the Detroit Tigers baseball team, features a main scoreboard larger than any other scoreboard in existence. To find its width, Mark uses his 120-ft tape measure and selects a spot in center field. From this point he measures the distances to points directly below the lower corners of the scoreboard to be 86 ft 6 in. and 120 ft. He also measures the angle at center field between these two lengths to be 120°. Using these measurements, determine (to the nearest foot) the width of the scoreboard.

36. Two aircraft carriers, one cruising at 25 knots and the other at 20 knots, left the same base at the same time. Three hours later they were 100 nautical miles apart. What was the measure of the angle between their courses? (1 knot = 1 nautical mile per hour.)

37. Two sides of a parallelogram measure 7 and 9 meters, and a diagonal is 15 meters.

a. Find the measures of the angles of the parallelogram.

b. Is the 15-meter diagonal the shorter or longer diagonal of the parallelogram?

38. The measures of two sides of a parallelogram are 50 and 80 inches, while one diagonal is 90 inches long. Find the length of the other diagonal.

39–42. *Use the law of cosines to solve the ambiguous case.*

39. $a = 7, b = 3, \alpha = 60°$

40. $a = 8, b = 7, \beta = 60°$

41. $b = 3, c = 4, \beta = 40°$

42. $a = 9, c = 6, \alpha = 80°$

43–47. *For a triangle whose sides are a, b, and c and if* $s = \frac{1}{2}(a + b + c)$, *use the law of cosines to rewrite the right side of the equation in order to verify the identity.*

43. $1 + \cos\alpha = \dfrac{(b + c + a)(b + c - a)}{2bc}$

44. $1 - \cos\alpha = \dfrac{(a - b + c)(a + b - c)}{2bc}$

45. $\dfrac{1 - \cos\alpha}{2} = \dfrac{(s - b)(s - c)}{bc}$

46. $\dfrac{1 + \cos\alpha}{2} = \dfrac{s(s - a)}{bc}$

47–48. *Hero's Formula for the area of a triangle with sides a, b, and c, is* $A = \sqrt{s(s - a)(s - b)(s - c)}$, *where* $s = \frac{1}{2}(a + b + c)$. *This formula says that the area of a triangle can be found from the lengths of the three sides.*

47. Use the results from Exercises 45 and 46 and Exercise 27 of Section 3.4 to prove Hero's Formula.

48. Use Hero's Formula to find the area of triangle ABC.

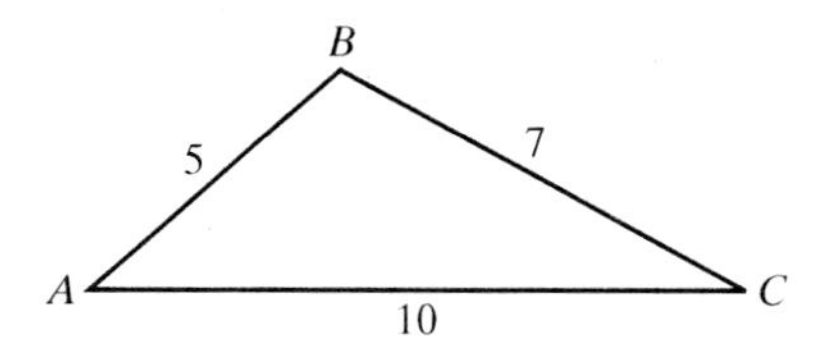

Chapter 3 Summary

3.1 An angle whose vertex is at the origin and whose initial side is on the positive side of the x-axis is said to be in standard position. The angle is said to be in the quadrant that contains its terminal side. If the terminal side is on an axis, the angle is said to be quadrantal. Angles with the same initial and terminal sides are said to be coterminal. An angle whose vertex is at the center of a circle is called a central angle.

Converting Angle Measurement

The two angle measurement units most commonly used are radians and degrees. Since π radians $= 180°$, converting between radians and degrees is accomplished by one of the following.

Radians to degrees: $\theta \cdot \dfrac{180°}{\pi}$ Degrees to radians: $\theta \cdot \dfrac{\pi}{180°}$

Arc Length

If θ (measured in radians) is a central angle of a circle of radius r, then the formula for length of the arc s intercepted by θ is

$$s = r\theta.$$

Linear Velocity

If s is the distance an object traveled in time t around a circle, then the linear velocity is defined as

$$v = \frac{s}{t} = \frac{r\theta}{t}.$$

Angular Velocity

If θ is the angle swept out in time t, then the angular velocity is defined as

$$\omega = \frac{\theta}{t}.$$

3.2 The values of the circular functions are equal to the corresponding values of the trigonometric functions. The domain of the trigonometric functions is the measure of an angle.

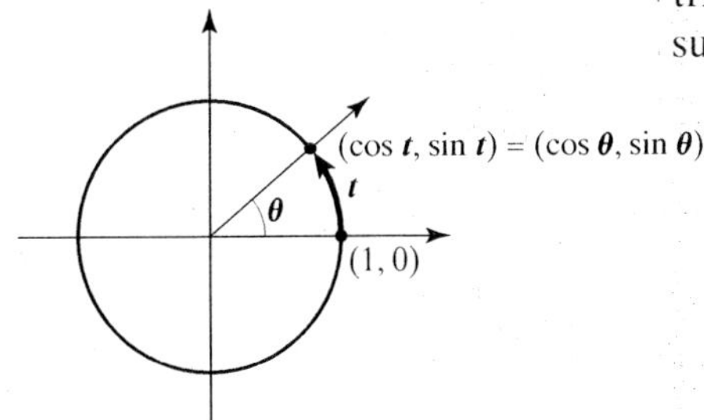

Trigonometric Functions

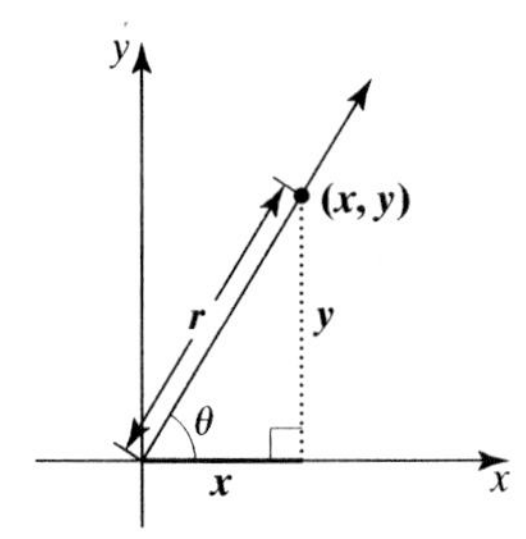

If (x, y) is a point other than the origin on the terminal side of angle θ in standard position, and $r = \sqrt{x^2 + y^2}$, then the trigonometric functions of θ are defined as follows:

$$\cos\theta = \frac{x}{r} \qquad \sin\theta = \frac{y}{r} \qquad \tan\theta = \frac{y}{x},\ x \neq 0$$

$$\sec\theta = \frac{r}{x},\ x \neq 0 \qquad \csc\theta = \frac{r}{y},\ y \neq 0 \qquad \cot\theta = \frac{x}{y},\ y \neq 0$$

The angles in the domain of the trigonometric functions correspond to the arcs in the domain of the circular functions. The ranges, definitions of the inverse functions, identities, and other properties for the trigonometric functions are analogous to the circular functions.

3.3–3.5 Solving Triangles

In solving any triangle,

- You must be given the measures of at least three parts of the triangle and one of these parts must be a side.
- Draw the triangle and mark given side(s) and angle(s) in order to determine a method to use. Use a chart when solving for *all* unknown parts of the triangle.
- $\alpha + \beta + \gamma = 180°$.
- The sum of any two sides of the triangle must be greater than the third.
- The largest angle is across from the largest side, and the smallest angle is across from the smallest side of the triangle.

3.3 Right Triangles

For angles $\alpha, \beta, \gamma = 90°$ of a right triangle, sides a, b, and c are opposite the angles α, β, and γ, respectively. The trigonometric ratios and the Pythagorean Theorem are used to find parts of a right triangle.

$$\sin\alpha = \frac{\text{side opposite } \alpha}{\text{hypotenuse}}$$

$$\cos\alpha = \frac{\text{side adjacent } \alpha}{\text{hypotenuse}}$$

$$\tan\alpha = \frac{\text{side opposite } \alpha}{\text{side adjacent } \alpha}$$

$$\alpha + \beta = 90°$$

$$a^2 + b^2 = c^2$$

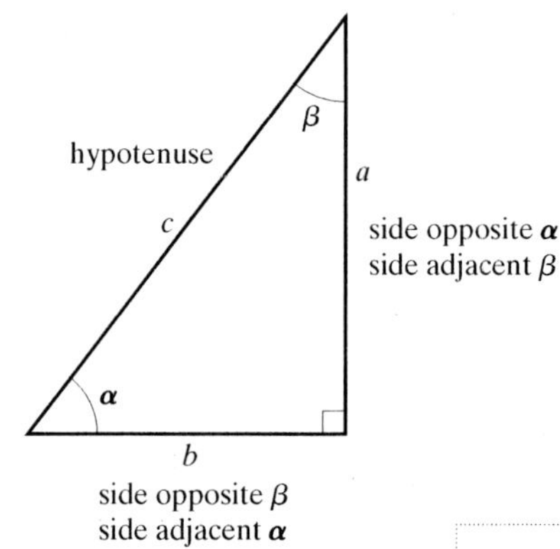

$a =$	$\alpha =$
$b =$	$\beta =$
$c =$	$\gamma = 90°$

180°

Angles of Elevation or Depression

An angle whose initial side is horizontal and whose terminal side is above the horizontal (up) is called an angle of elevation. If the terminal side is below the horizontal (down), it is called an angle of depression.

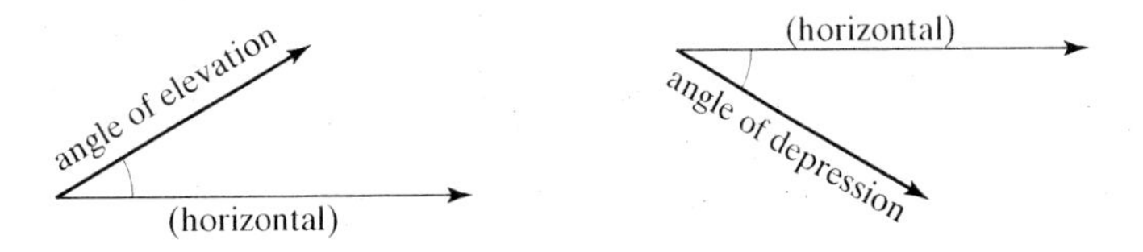

Cofunctions

$$\text{If } \alpha + \beta = 90^\circ, \quad \sin\alpha = \cos(90^\circ - \alpha) = \cos\beta$$
$$\tan\alpha = \cot(90^\circ - \alpha) = \cot\beta$$
$$\sec\alpha = \csc(90^\circ - \alpha) = \csc\beta$$

These pairs of functions are called cofunctions.

3.4 Law of Sines

For angles α, β, and γ of a triangle with sides a, b, and c opposite the angles α, β, and γ, respectively, then

$$\frac{\sin\alpha}{a} = \frac{\sin\beta}{b} = \frac{\sin\gamma}{c}.$$

The law of sines is used when you are given the measure of two angles and the length of one side (AAS) or the lengths of two sides and the measure of an angle opposite one of these two sides (SSA). The SSA case causes certain ambiguities since the information can produce two, one, or no triangles.

3.5 Law of Cosines

For angles α, β, and γ of a triangle with sides a, b, and c opposite the angles α, β, and γ, respectively:

$$c^2 = a^2 + b^2 - 2ab\cos\gamma$$
$$b^2 = a^2 + c^2 - 2ac\cos\beta$$
$$a^2 = b^2 + c^2 - 2bc\cos\alpha.$$

The law of cosines is used when you are given the lengths of three sides (SSS) or the lengths of two sides and the included angle (SAS). This law can also be used when you are given two sides and an opposite angle (SSA), the ambiguous case.

3.3–3.5 Bearing

There are two ways to specify direction or bearing.

- One way to give the direction to point B from point A is to use the notation for bearing that begins with N or S, followed by an angle measured in degrees, and ends with E or W.

The bearing from A to B is S 48° E.

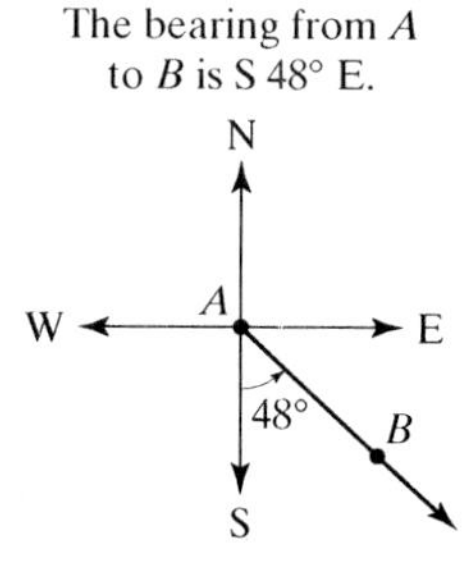

- The second way is to use a single angle measurement that represents the direction from the north in a clockwise direction.

The bearing from A to B is 132°.

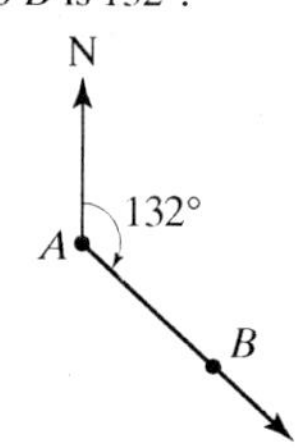

Chapter 3 Review Exercises

3.1

1–2. *Convert each radian measure to degrees (or nearest degree). Draw the angle in standard position and determine the quadrant that the angle is in.*

1. $\dfrac{5\pi}{6}$ **2.** 4

3–4. *Convert each degree measure to radians. Leave your answer in terms of π.*

3. 55° **4.** $-140°$

5–6. **a.** *Find the smallest positive coterminal angle in the same measurement unit.*

b. *Use this smallest positive coterminal angle to write a formula for all its coterminal angles.*

5. 993° **6.** $\dfrac{25\pi}{4}$

7. Find the length of an arc (to the nearest tenth of a foot) intercepted by a central angle of 120° on a circle of radius 6 feet.

8. A speed skater at the 2002 Olympics did five laps around a circular rink of radius 10 feet in 15.6 seconds. What is the skater's linear velocity (to the nearest tenth of a foot per second)?

9. A snowboarder at the 2002 Olympics did a 900° flip (which is two and one-half rotations) in 2.7 seconds. What is his angular velocity (to the nearest tenth of a degree per second)?

10. Find the measure of the central angle in radians that intercepts an arc of length 18 m on a circle of radius 12 m.

3.2

11–14. *Find* $\cos\theta$, $\sin\theta$, *and* $\tan\theta$, *for an angle* θ *in standard position if the given point is on the terminal side.*

11. $(-15, 8)$ **12.** $(5, 12)$ **13.** $(2\sqrt{3}, -2)$ **14.** $(-5, -2)$

15. Find the quadrant that contains the terminal side of θ.

a. $\sec\theta > 0, \cot\theta < 0$ **b.** $\sin\theta > 0, \tan\theta < 0$

16. Copy the diagram and complete the missing information.

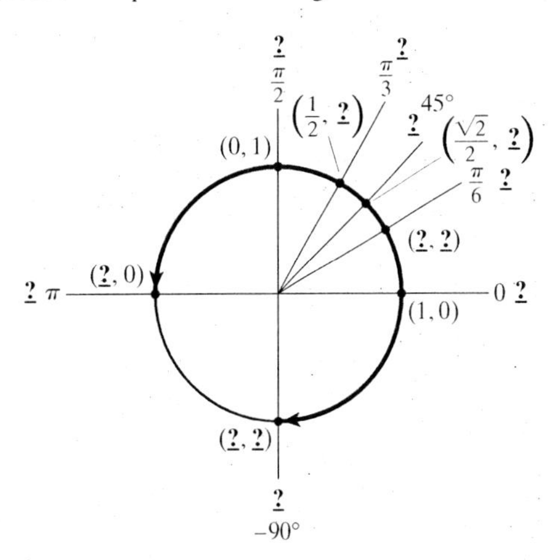

17–28. *Find the exact value of the following, if defined.*

17. $\sin\dfrac{7\pi}{6}$ **18.** $\cos\left(-\dfrac{5\pi}{3}\right)$ **19.** $\tan\left(-\dfrac{\pi}{2}\right)$ **20.** $\cos^2 120°$

21. $\sec 3\pi$ **22.** $\cot(-60°)$ **23.** $\sin(-390°)$ **24.** $\cos 135°$

25. $\tan\dfrac{9\pi}{4}$ **26.** $\csc(-270°)$ **27.** $\sec^2(-45°)$ **28.** $\sin 180°$

29–40. *Find the exact value of the angle in degrees, if possible.*

29. $\arccos\left(-\dfrac{1}{\sqrt{2}}\right)$ **30.** $\sin^{-1}(-1)$ **31.** $\tan^{-1}\sqrt{3}$

32. $\operatorname{arcsec} 2$ **33.** $\operatorname{arccsc}\dfrac{1}{2}$ **34.** $\tan^{-1} 0$

35. $\cot^{-1}(-\sqrt{3})$ **36.** $\cos^{-1}\dfrac{\sqrt{3}}{2}$ **37.** $\sin^{-1}\dfrac{1}{2}$

38. $\tan^{-1}(-1)$ **39.** $\cos^{-1} 3$ **40.** $\operatorname{arccsc}\left(-\dfrac{1}{\sqrt{2}}\right)$

41–42. *Find the exact value.*

41. $\tan\left(\arccos\dfrac{1}{5}\right)$ **42.** $\sin\left(\tan^{-1}\dfrac{1}{4}\right)$

3.3

43–46. *Write the following functions of* α *in terms of its cofunction of an acute angle* β.

43. $\sin 45° =$ ______ **44.** $\tan 60° =$ ______

45. $\sec 15°35' =$ ______ **46.** $\cos 72° =$ ______

47–50. *Solve each right triangle* $(\gamma = 90°)$.

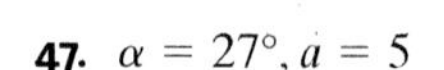

47. $\alpha = 27°, a = 5$

48. $c = 14, \beta = 52°$

49. $a = 5, c = 8$

50. $a = 93.1, b = 49$

51. An aircraft is traveling on a course of bearing N 30° E and passes over a control tower and continues on this course for 3 miles. At that point the pilot turns the aircraft 90° and heads toward the northwest. If the plane then travels for 2 miles in this direction, what bearing should the control tower use to locate the aircraft?

3.4

52–55. *Solve each triangle.*

52. $a = 10, \beta = 12°, \gamma = 100°$

53. $a = 6, b = 5, \beta = 110°$

54. $a = 3, b = 6, \alpha = 10°$

55. $b = 35, a = 50, \alpha = 122°30'$

56. The tallest freestanding structure in the world is the CN Tower in Toronto, Canada. From ground level the angle of elevation from point A to an object on the observation deck of the CN Tower is 75°, and 111 feet away at point B, the angle of elevation is 70°. Find the height to the nearest foot of the object on the observation deck of the CN Tower.

57. The Leaning Tower of Pisa was built to be 185 feet tall. An observer noticed that from a point at ground level 120 feet away from the base of the tower, the angle of elevation to the top of the tower is 62°. What is the measure of the acute angle that the tower makes with the horizontal?

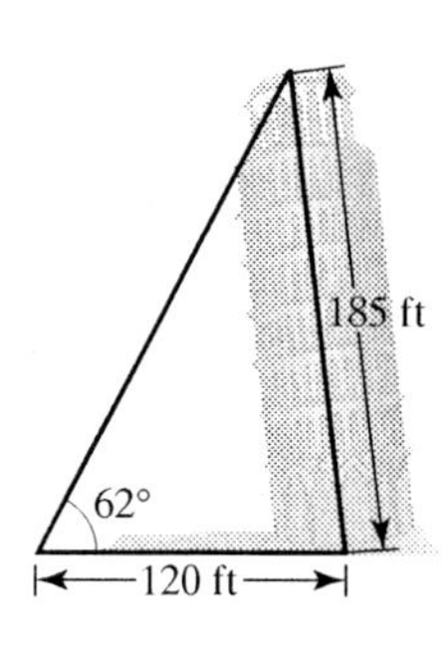

3.5

58–61. *Solve each triangle.*

58. $a = 3, c = 2, \beta = 100°$

59. $a = 3, b = 4, c = 1$

60. $a = 5, b = 2, c = 6$

61. $b = 1.2, a = 2.3, \gamma = 58.2°$

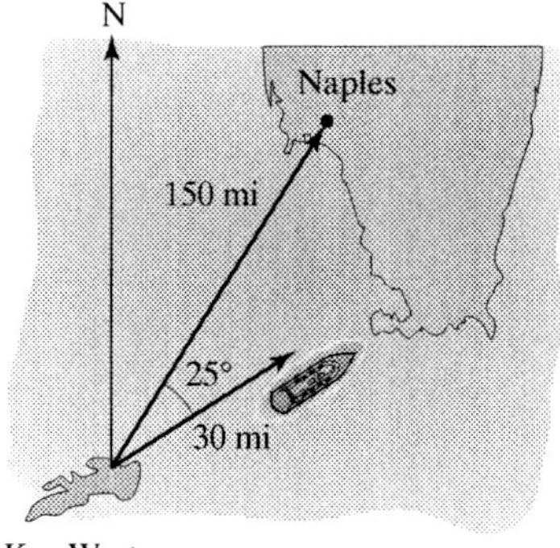

62. A tourist boat is traveling from Key West to Naples, Florida, which is approximately 150 miles away. After traveling for 30 miles, the captain notices that he is 25° off course due to heavy winds. At that point, determine how far the tourist boat is from Naples and the angle the boat should turn to correct its course.

63. Indicate a method you would use (right triangle, law of sines, or law of cosines) to solve, if possible, the triangle if the following information were given.

a. three sides

b. two sides and the included angle

c. three angles

d. two sides and the opposite angle

e. one right angle and two sides

f. two sides

g. one side and two angles

64. Are the following statements true or false? If a statement is false, explain why or give an example that shows why it is false.

   a. The sum of two sides of a triangle is always greater than the third side.
   b. You can solve a triangle for its missing parts with any three pieces of information.
   c. You may have two triangle solutions when you are given two sides and the included angle.
   d. You may have two triangle solutions when you are given two sides and the opposite angle.
   e. The sum of the measures of the angles of a triangle is 180.
   f. The largest angle in the triangle is always across from the largest side.
   g. An isosceles triangle has two equal angles.
   h. SOHCAHTOA is what you can use when you are solving any triangle.
   i. SOHCAHTOA is what you can use when you are solving only right triangles.
   j. It is possible to begin to solve a triangle with the law of cosines, then use another method to finish solving the triangle.
   k. $a^2 + b^2 = c^2$ is true for any triangle with sides a, b, and c.

65. You only have to look around you to see the many uses of trigonometry. Compose a real-world problem that involves using a triangle. The solution to your problem must include one or more methods used in this chapter to solve triangles. The measurements you give should be attainable using simple tools. Make sure all necessary information is given in the problem and that you are using correct terminology. Supply any necessary diagrams and make sure your information is not contradictory. You should complete the solution to your problem and your answer should be reasonable and believable.

Chapter 3 Test

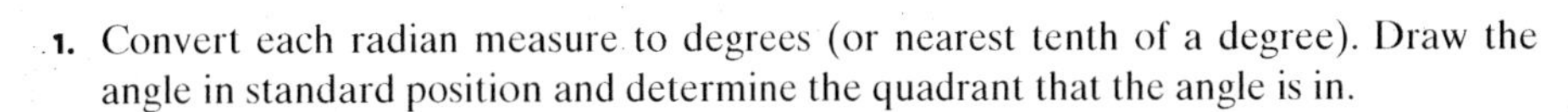

1. Convert each radian measure to degrees (or nearest tenth of a degree). Draw the angle in standard position and determine the quadrant that the angle is in.

   a. $\frac{7\pi}{8}$ b. 1

2. Convert each degree measure to radians. Leave your answer in terms of π.

   a. $-15°$ b. $210°$

3. A particle with constant speed is moving around the circle from point P to R in 6 seconds. The radius of the circle is 100 feet, and the measure of the central angle α is 330°.

   a. Find the length of the arc s (to nearest foot).
   b. Find the linear velocity v (to nearest foot per second).
   c. Find the angular velocity ω (to nearest radian per second).

4. Consider an angle θ in standard position with $(3, -2)$ on the terminal side. Draw the angle and find the exact functional values for this angle.

   a. $\cos\theta$ b. $\sin\theta$ c. $\cot\theta$

5. If $\tan\alpha = -\frac{4}{3}$ and α is in QII, draw α, then find:

   a. $\cos\alpha$ b. $\csc\alpha$

6. Find the exact value for each of the following expressions.

a. $\cos(-120°)$ b. $\cot 150°$ c. $\tan \frac{7\pi}{3}$

d. $\csc(-45°)$ e. $\sec(-\pi)$ f. $\sin(-855°)$

7. Find the exact measure in degrees if possible. If not possible, tell why.

a. $\cos^{-1}(-1)$ b. $\csc^{-1} 2$ c. $\arctan \sqrt{3}$ d. $\arcsin 5.6$

8. Find the approximate value rounded to four decimal places.

a. $\cos 0.56$ b. $\tan 123°$ c. $\sec 20°$ d. $\cot(-2.89)$

e. $\arcsin 0.9$ (to the nearest tenth of a degree)

f. $\operatorname{arcsec}(-3.5)$ (in radians to four decimal places)

9. Consider the following right triangles. Solve for x by using a trigonometric ratio.

a.

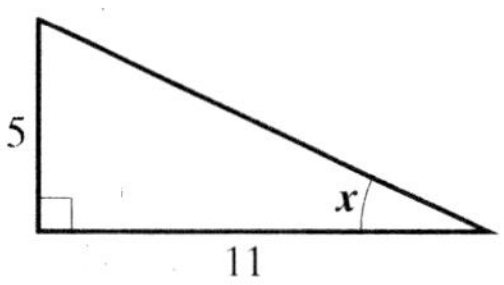

b.

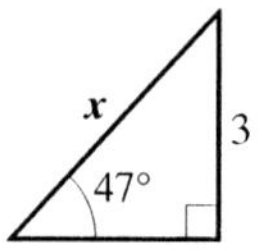

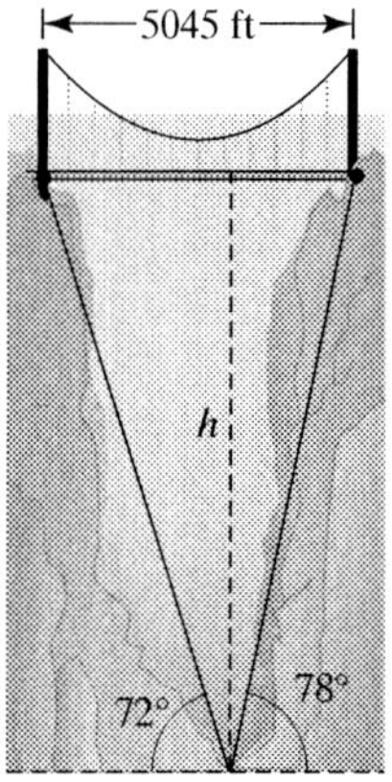

10. State the method (law of cosines, law of sines, or right triangle) that you would use to solve the triangle on the basis of the information that is given. If there is no method you could use, state why.

a. $a = 7, b = 5, c = 10$ b. $c = 5, \beta = 42°, \gamma = 67°$

c. $\alpha = 27°, \beta = 63°$ d.

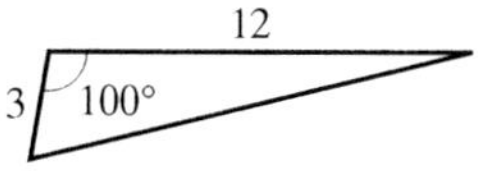

e. $\gamma = 90°, a = 5, \beta = 40°$

11. Solve the triangle for the indicated value.

a. If $c = 12, \beta = 42°, \gamma = 90°$, find a. b. If $\alpha = 75°, a = 15, b = 4$, find β.

12. A bridge is being built across a canyon. The length of the bridge is 5045 feet. From the deepest point in the canyon, the angles of elevation to the ends of the bridge are 78° and 72°. Find the height (to nearest tenth) of the bridge over the deepest point in the canyon. (See diagram on the left.)

13. Anthony and Rosalie went boating. Anthony headed out from the dock at a bearing of S 42° E going 36 mph. Rosalie headed out from the dock due south going 40 mph. If each continues on their course at their same speed, how far apart (to the nearest mile) are they after $1\frac{1}{2}$ hours?

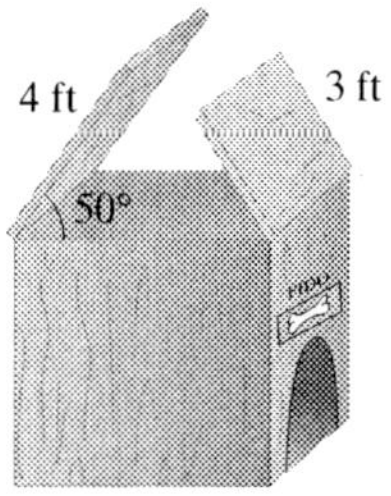

14. David is trying to build a doghouse. The roof of the house needs to have a pitch (an angle of elevation) of 50°. To frame the roof, he finds one piece of wood 4 feet long and uses it for one side of the roof and nails it in place. His brother Mike finds another similar piece of wood 3 feet long to be used for the side across from the 50° angle, for the other side of the roof. Is the 3-foot piece of wood long enough? Explain why or why not.

15. Use the right triangle to answer the following questions.

a. $\alpha + \beta =$ ______

b. Why does $\cos \alpha = \sin \beta$?

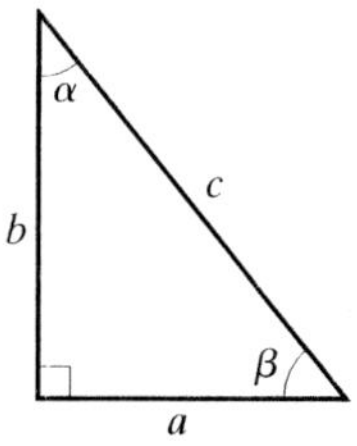

Chapter 4

Identities

The trigonometric functions that you have become familiar with in the previous chapters are interrelated in a number of very useful and interesting ways.

For example, if we look at the graphs of the functions

$$y = \cos^2 x - \sin^2 x \qquad \text{and} \qquad y = \cos(2x),$$

we notice they are identical.

Although these functions look quite different from each other, they are the same. This means that for all values of x, $\cos^2 x - \sin^2 x = \cos(2x)$. *In this chapter you will become aware of many of these relationships that are powerful tools in more advanced mathematics and physics. These tools allow us to change trigonometric functions into equivalent forms that may involve other trigonometric functions.*

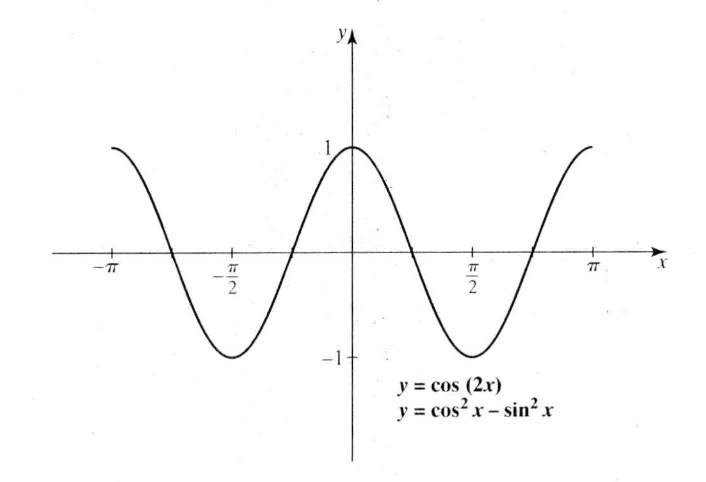

Important Concepts

- identities
- Pythagorean identities
- proofs of identities
- sum and difference identities —cosine, sine, tangent
- double-angle identities
- half-angle identities
- sum-to-product identities
- product-to-sum identities

4.1 Proving Identities

In Chapter 1 we defined an identity as a statement that is true for all values for which it is defined. The definitions of the ratios and reciprocals of the trigonometric (or circular) functions $\cos x$ and $\sin x$ are identities because they are true for all possible replacement values of the variable x. Identities, such as the important ones listed below that were introduced in previous chapters, will continue to be used to simplify expressions, solve problems or equations, verify other identities, and prove new theorems and formulas. You should be quite familiar with these basic identities and most likely have memorized them. (*If not, memorize them now!*)

Basic Identities

$$\cos^2 x + \sin^2 x = 1$$

$$\tan x = \frac{\sin x}{\cos x}$$

$$\sec x = \frac{1}{\cos x}, \quad \cos x = \frac{1}{\sec x}$$

$$\csc x = \frac{1}{\sin x}, \quad \sin x = \frac{1}{\csc x}$$

$$\cot x = \frac{1}{\tan x} = \frac{\cos x}{\sin x}, \quad \tan x = \frac{1}{\cot x}$$

$$\cos(-x) = \cos x$$

$$\sin(-x) = -\sin x$$

$$\tan(-x) = -\tan x$$

As we have often said, there are several ways to express the same idea in the language of trigonometry. And the Pythagorean identity is no exception since it has several important equivalent identities that are the result of rewriting $\cos^2 x + \sin^2 x = 1$.

Pythagorean Identities

$$\cos^2 x + \sin^2 x = 1 \quad (1)$$

$$\cos^2 x + \sin^2 x - \cos^2 x = 1 - \cos^2 x \quad \rightarrow \quad \sin^2 x = 1 - \cos^2 x \quad (2)$$

$$\cos^2 x + \sin^2 x - \sin^2 x = 1 - \sin^2 x \quad \rightarrow \quad \cos^2 x = 1 - \sin^2 x \quad (3)$$

$$\frac{\cos^2 x}{\cos^2 x} + \frac{\sin^2 x}{\cos^2 x} = \frac{1}{\cos^2 x} \quad (\cos x \neq 0) \quad \rightarrow \quad 1 + \tan^2 x = \sec^2 x \quad (4)$$

$$\frac{\cos^2 x}{\sin^2 x} + \frac{\sin^2 x}{\sin^2 x} = \frac{1}{\sin^2 x} \quad (\sin x \neq 0) \quad \rightarrow \quad \cot^2 x + 1 = \csc^2 x \quad (5)$$

It seems that there is no end to finding equivalent forms of identities. For example, to name a few, we have:

$$\sin^2 x = 1 - \cos^2 x \quad \rightarrow \quad \sin x = \pm\sqrt{1 - \cos^2 x}$$
$$\cos^2 x = 1 - \sin^2 x \quad \rightarrow \quad \cos x = \pm\sqrt{1 - \sin^2 x}$$
$$1 + \tan^2 x = \sec^2 x \quad \rightarrow \quad \tan^2 x = \sec^2 x - 1$$
$$\cot^2 x + 1 = \csc^2 x \quad \rightarrow \quad \cot^2 x = \csc^2 x - 1$$

It is not necessary that you commit all equivalent forms of the Pythagorean identity to memory. But you should be able to recognize basic identities in their equivalent forms or be able to quickly derive them from the basic identity. You may want to memorize those that you feel are not readily apparent.

Simplify Expressions

Using basic identities to transform complicated trigonometric expressions to simpler ones is similar to rewriting algebraic expressions. For example, you have had practice rewriting

$$x + 5(x + 10) - 3x \quad \text{as} \quad 3x + 50 \qquad \text{and} \qquad \frac{x+x^3}{1+x^2} \quad \text{as} \quad x.$$

Due mainly to identities, you may become more aware of different ways a problem can be simplified. The next example demonstrates this situation and is followed by steps to help simplify trigonometric expressions.

EXAMPLE 1 Simplify each expression.

a. $\cot x \cdot \sin x$

b. $\tan x\,(\sin x + \cot x \cdot \cos(-x))$

c. $\cos^2 x\,(1 + \tan^2 x)$

SOLUTION

a. The expression $\cot x$ can be rewritten in terms of $\sin x$ and $\cos x$. We simplify the expression for values for which it is defined.

$$\cot x \cdot \sin x$$
$$= \frac{\cos x}{\sin x} \cdot \sin x$$
$$= \frac{\cos x}{\cancel{\sin x}} \cdot \frac{\cancel{\sin x}}{1}$$
$$= \cos x,\ x \neq k\pi$$

Recall $\cot x$ is defined when $x \neq k\pi$ because $\sin x \neq 0$.

NOTE: Even though we will not continue to explicitly state the specific values in which the original and simplified expressions are defined, it should be understood that the simplified form is true only for these specific values.

b. The expression contains $\tan x$ and $\cot x$, which can be rewritten in terms of $\sin x$ and $\cos x$, and $\cos(-x) = \cos x$. Once we make the substitutions, we can either use the distributive property (Solution 1) or perform the operations inside parentheses first (Solution 2). Both methods arrive at the same simplified expression.

SOLUTION 1

$$\tan x\left(\sin x+\cot x\cdot\cos(-x)\right)$$

$$=\frac{\sin x}{\cos x}\left(\frac{\sin x}{1}+\frac{\cos x}{\sin x}\cdot\frac{\cos x}{1}\right)$$

$$=\frac{\sin^2 x}{\cos x}+\frac{\sin x}{\cos x}\cdot\frac{\cos x}{\sin x}\cdot\frac{\cos x}{1}$$ Use the distributive property.

$$=\frac{\sin^2 x}{\cos x}+\frac{\cos x}{1}$$ Simplify.

$$=\frac{\sin^2 x}{\cos x}+\frac{\cos^2 x}{\cos x}$$ Find a common denominator.

$$=\frac{\sin^2 x+\cos^2 x}{\cos x}$$ Add the fractions.

$$=\frac{1}{\cos x}$$ $\sin^2 x+\cos^2 x=1$

$$=\sec x$$

SOLUTION 2

$$\tan x\left(\sin x+\cot x\cdot\cos(-x)\right)$$

$$=\frac{\sin x}{\cos x}\left(\frac{\sin x}{1}+\frac{\cos x}{\sin x}\cdot\frac{\cos x}{1}\right)$$

$$=\frac{\sin x}{\cos x}\left(\frac{\sin x}{1}+\frac{\cos^2 x}{\sin x}\right)$$ Multiply the fractions.

$$=\frac{\sin x}{\cos x}\left(\frac{\sin^2 x}{\sin x}+\frac{\cos^2 x}{\sin x}\right)$$ Find a common denominator.

$$=\frac{\sin x}{\cos x}\left(\frac{\sin^2 x+\cos^2 x}{\sin x}\right)$$ Add the fractions.

$$=\frac{\sin x}{\cos x}\left(\frac{1}{\sin x}\right)$$ $\sin^2 x+\cos^2 x=1$

$$=\frac{1}{\cos x}$$ Simplify.

$$=\sec x$$

Since $1/\cos x$ and $\sec x$ are equivalent, it may be debatable which form is simpler. Nevertheless, we will agree to use $\sec x$ as the simplified expression.

c. Since this expression contains $1+\tan^2 x$ and we know the identity $1+\tan^2 x=\sec^2 x$, we rewrite $1+\tan^2 x$ as $\sec^2 x$, which we then write in terms of $\cos x$ (Solution 1). If you fail to recognize $1+\tan^2 x=\sec^2 x$, you can use the basic identity $\tan x=\sin x/\cos x$ (Solution 2).

SOLUTION 1

$$\cos^2 x\left(1+\tan^2 x\right)$$

$$=\cos^2 x\left(\sec^2 x\right)$$ $1+\tan^2 x=\sec^2 x$

$$=\cos^2 x\left(\frac{1}{\cos^2 x}\right)$$ $\sec x=\dfrac{1}{\cos x}$

$$=1$$ Simplify.

SOLUTION 2

$$\cos^2 x\left(1+\tan^2 x\right)$$

$$=\cos^2 x\left(1+\frac{\sin^2 x}{\cos^2 x}\right)$$ $\tan x=\dfrac{\sin x}{\cos x}$

$$=\cos^2 x\cdot 1+\cos^2 x\left(\frac{\sin^2 x}{\cos^2 x}\right)$$ Distribute.

$$=\cos^2 x+\sin^2 x$$ Simplify.

$$=1$$ $\cos^2 x+\sin^2 x=1$ ■

Notice that different solutions depend on which identities and which algebraic or arithmetic procedures are selected. Even though there are different ways to simplify an expression, we use the same general steps.

Steps to Simplify an Expression

Step 1. Try to recognize basic identities in the expression. Use identities to rewrite the expression either in an equivalent form or in terms of the sine and cosine.

Step 2. Perform the indicated operations. Continue, as needed, to use trigonometric identities along with arithmetic or algebraic skills (such as finding common denominators) to simplify.

Proving Identities

Many times we are given two expressions that are proposed to be equal (or equivalent) for all values of the variable for which they are defined. If they are equivalent, the resulting equation is an identity. Thus to **prove an identity**, we must verify that one side of the equation is a simplified or equivalent form of the expression on the other side.

You already have had some experience with proving identities in previous chapters. Although proving identities may look familiar in this section, we outline steps to follow for proving more challenging identities. Many of the identities you are asked to prove are not ones you will want to memorize. What is important here is to develop the ability to verify identities using the basic identities and their equivalent forms.

Remember, in order to prove an identity, we want to think of it as rewriting the expression on one side of the equation so as to obtain the form on the other side of the equation. If we adopt this strategy, we agree to work on only one side of the equation.

EXAMPLE 2 Prove $\sec x\left(\sec x - \cos x\right) = \tan^2 x$ is an identity.

SOLUTION We select the left side of this equation as the side to be simplified. We can begin by rewriting $\sec x$ in terms of $\cos x$.

$$\sec x\left(\sec x - \cos x\right) = \tan^2 x$$

$$\frac{1}{\cos x}\left(\frac{1}{\cos x} - \cos x\right) = \qquad \sec x = \frac{1}{\cos x}$$

$$\frac{1}{\cos x}\cdot\frac{1}{\cos x} - \frac{1}{\cos x}\cdot\frac{\cos x}{1} = \qquad \text{Use the distributive property.}$$

$$\frac{1}{\cos^2 x} - \frac{\cos x}{\cos x}\cdot\frac{\cos x}{\cos x} = \qquad \text{Find a common denominator.}$$

$$\frac{1 - \cos^2 x}{\cos^2 x} = \qquad \text{Add the fractions.}$$

$$\frac{\sin^2 x}{\cos^2 x} = \qquad 1 - \cos^2 x = \sin^2 x$$

$$\tan^2 x = \tan^2 x \qquad \frac{\sin x}{\cos x} = \tan x$$

■

Notice the steps we used in Example 2 to simplify the left side are identical to those used to simplify an expression. Also, we decided to simplify the left side because it is more complicated than the right side. Sometimes it may be debatable which side is more complicated. If the proposed equation is an identity, you should be able to select either side and rewrite it in the form on the other side.

CAUTION: It is tempting to use algebraic techniques in which you do the same operation to both sides of a given equation that is a true statement, such as adding the same terms to both sides or multiplying both sides by the same quantity. Such rules are not valid steps for proving an identity since we are trying to prove that the given statement is true. In other words, we can't use rules that apply to a true statement before we know it is a true statement.

EXAMPLE 3 Prove that the following equations are identities.

a. $\tan x + \cot x = \csc x \sec x$ **b.** $\csc \theta = \csc \theta \cos^2 \theta + \sin \theta$

SOLUTION

a. Selecting the left side, we rewrite tan x and cot x in terms of sin x and cos x, and then perform the indicated operations. Looking at the desired form on the other side of the equation, we arrive at the requested form using basic identities.

$$\tan x + \cot x = \csc x \sec x$$

$$\frac{\sin x}{\cos x} + \frac{\cos x}{\sin x} = \qquad \text{Rewrite in terms of } \sin x \text{ and } \cos x.$$

$$\frac{\sin x}{\cos x} \cdot \frac{\sin x}{\sin x} + \frac{\cos x}{\sin x} \cdot \frac{\cos x}{\cos x} = \qquad \text{Find a common denominator.}$$

$$\frac{\sin^2 x + \cos^2 x}{\sin x \cos x} = \qquad \text{Add the fractions.}$$

$$\frac{1}{\sin x \cos x} = \qquad \sin^2 x + \cos^2 x = 1$$

$$\frac{1}{\sin x} \cdot \frac{1}{\cos x} = \qquad \text{Factor.}$$

$$\csc x \sec x = \csc x \sec x \qquad \frac{1}{\sin x} = \csc x, \frac{1}{\cos x} = \sec x$$

b. Selecting the right side of the equation to work on, we look at two ways to prove this identity. (The more approaches you see, the more you become aware of different strategies.)

SOLUTION 1 If we look at the left side of the equation, csc θ, we notice that it is also a part of the right side. Since we need csc θ as the simplified form of the right side, we first factor out csc θ.

$$\csc \theta = \csc \theta \cos^2 \theta + \sin \theta$$

$$= \csc \theta \left(\cos^2 \theta + \frac{\sin \theta}{\csc \theta} \right) \qquad \text{Factor out } \csc \theta.$$

$$= \csc \theta \left(\cos^2 \theta + \sin \theta \cdot \sin \theta \right) \qquad \frac{\sin \theta}{\csc \theta} = \sin \theta \cdot \frac{1}{\csc \theta} = \sin \theta \cdot \sin \theta$$

$$= \csc \theta \left(\cos^2 \theta + \sin^2 \theta \right) \qquad \text{Simplify.}$$

$$\csc \theta = \csc \theta \qquad \cos^2 \theta + \sin^2 \theta = 1$$

SOLUTION 2 What if we didn't think to factor $\csc\theta$ out of the right side of the equation? We could have instead rewritten each function in terms of $\sin\theta$ and $\cos\theta$ and performed the indicated operations:

$$\csc\theta = \csc\theta\cos^2\theta + \sin\theta$$

$$= \frac{1}{\sin\theta}\cos^2\theta + \sin\theta \qquad \csc\theta = \frac{1}{\sin\theta}$$

$$= \frac{\cos^2\theta}{\sin\theta} + \frac{\sin\theta}{1}\cdot\frac{\sin\theta}{\sin\theta} \qquad \text{Find a common denominator.}$$

$$= \frac{\cos^2\theta + \sin^2\theta}{\sin\theta} \qquad \text{Add the fractions.}$$

$$= \frac{1}{\sin\theta} \qquad \cos^2\theta + \sin^2\theta = 1$$

$$\csc\theta = \csc\theta \qquad \frac{1}{\sin\theta} = \csc\theta$$

■

It is sometimes unsettling when it is possible to arrive at the answer in different ways. But if you are not aware of the fact that these problems can be solved in different correct ways, you may think that because you are not doing the problem in the same manner as your instructor, the author, or solutions manual, you must be doing it wrong. Not the case. As long as you are using identities and good mathematics, your approach should be valid. Of course, you must know the basic identities.

Before we proceed with more examples, let's list steps that can be used to prove identities.

Steps to Prove Identities

Step 1. Select *one* side of the equation to work on. It is usually easier to start with the more complicated side of the equation. Take a good look at the other side of the equation as it may give you insight as to what steps to take. *Always look where you are going!*

Step 2. Use the basic identities to rewrite functions or expressions in an equivalent form or in terms of the sine and cosine. Continue to look at the other side for clues as to how to proceed.

Step 3. Perform the indicated operations. Use trigonometry and arithmetic or algebraic skills that may simplify the next step. For example:

- If there are two or three terms and you want a single term, add terms by finding common denominators, factor the expression, or use a basic identity (or an equivalent form).
- If there are no denominators or different denominators than those on the other side, multiply by the appropriate fractional form of 1 to obtain the necessary denominators.
- Look for identities that replace functions with ones needed or ones on the other side.

NOTE: Another procedure to prove identities allows you to simplify each side of the equation, as long as you work the two sides independently. In this text we will continue to follow the procedure outlined above, working with only one side.

EXAMPLE 4 Prove that $\dfrac{\sin x}{1 + \cos x} = \dfrac{1 - \cos x}{\sin x}$ is an identity.

SOLUTION Since either side arguably could be the more complicated one, we arbitrarily select the left side. The expression is already in terms of $\sin x$ and $\cos x$ with no basic identity. If we look at the right side for clues, we notice the form we need to obtain has $\sin x$ in the denominator. Since the denominator on the left side contains $\cos x$, we want to use the identity that rewrites $\cos x$ in terms of $\sin x$: $\sin^2 x = 1 - \cos^2 x = (1 + \cos x)(1 - \cos x)$. To get the denominator of the left side to be $(1 + \cos x)(1 - \cos x)$, we need to multiply the fraction by 1 in the form $(1 - \cos x)/(1 - \cos x)$.

$$\frac{\sin x}{1 + \cos x} = \frac{1 - \cos x}{\sin x}$$

$$\frac{\sin x}{1 + \cos x} \cdot \frac{1 - \cos x}{1 - \cos x} = \qquad \text{Multiply by 1 in the form } \frac{1 - \cos x}{1 - \cos x}.$$

$$\frac{\sin x\,(1 - \cos x)}{1 - \cos^2 x} = \qquad \text{Multiply the fractions.}$$

$$\frac{\sin x\,(1 - \cos x)}{\sin^2 x} = \qquad 1 - \cos^2 x = \sin^2 x$$

$$\frac{1 - \cos x}{\sin x} = \qquad \text{Reduce to lowest terms.}$$

■

Observe in Example 4 the importance of focusing on where we were going (that is, to see how the side we selected needed to be rewritten).

EXAMPLE 5 Prove the following identity.

$$\frac{1 + \cot^2 x}{\sec^2 x} = \cot^2 x$$

SOLUTION

$$\frac{1 + \cot^2 x}{\sec^2 x} = \cot^2 x$$

$$\frac{\csc^2 x}{\sec^2 x} = \qquad 1 + \cot^2 x = \csc^2 x$$

$$\frac{\left(\dfrac{1}{\sin^2 x}\right)}{\left(\dfrac{1}{\cos^2 x}\right)} = \qquad \text{Use reciprocal identities.}$$

$$\frac{1}{\sin^2 x} \cdot \frac{\cos^2 x}{1} = \qquad \text{Divide the fractions: } \frac{\frac{a}{b}}{\frac{c}{d}} = \frac{a}{b} \cdot \frac{d}{c}$$

$$\frac{\cos^2 x}{\sin^2 x} = \qquad \text{Simplify.}$$

$$\cot^2 x = \qquad \frac{\cos x}{\sin x} = \cot x$$

■

Identities and Graphing

In Chapter 2 we determined that the graph of $y = \cot\left(x + (\pi/2)\right)$ has the same graph as $y = -\tan x$. Hence, $\cot\left(x + (\pi/2)\right) = -\tan x$ is an identity. This suggests that one way to verify an identity is to use a graphing utility to see if the graphs of the expressions on both sides of the equation are the same. Depending on the graphing utility, we may only be able to determine that the graphs on the limited viewing window appear to coincide. Unless we are knowledgeable about the entire graph, the graphs we are able to view on a graphing utility that appear to coincide suggest only the likelihood of an identity. The analytic approach, which transforms the expression on one side of the equation into the equivalent form on the other side, proves the identity. Of course, if the graphs do not coincide, we conclude that we do not have an identity.

CONNECTIONS WITH TECHNOLOGY

Graphing Calculator/CAS

EXAMPLE Use a graphing utility to determine whether the statement could be an identity by graphing each side of the equation on the same screen.

a. $\cos x = \sec x - \tan x \sin x$

b. $1 + \tan x = \dfrac{\sin x + \cos x}{\sin x}$

SOLUTION

a. $y_1 = \cos x,\ y_2 = \sec x - \tan x \sin x$

Yes, it could be an identity since the graphs appear to coincide.

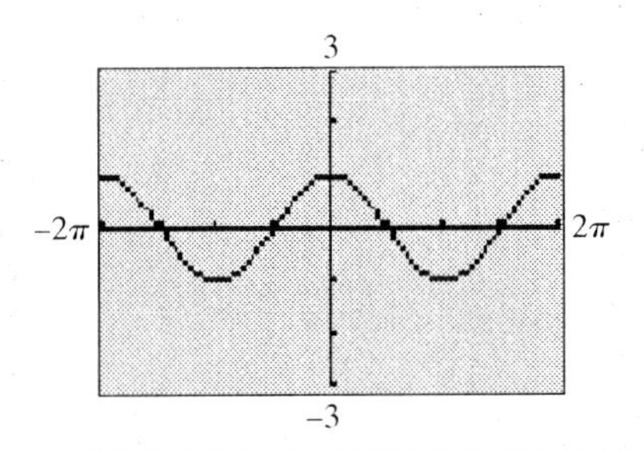

b. $y_1 = 1 + \tan x,$
$y_2 = \left(\sin x + \cos x\right)/\sin x$

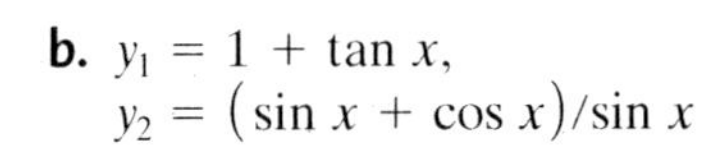

No, it is not an identity.

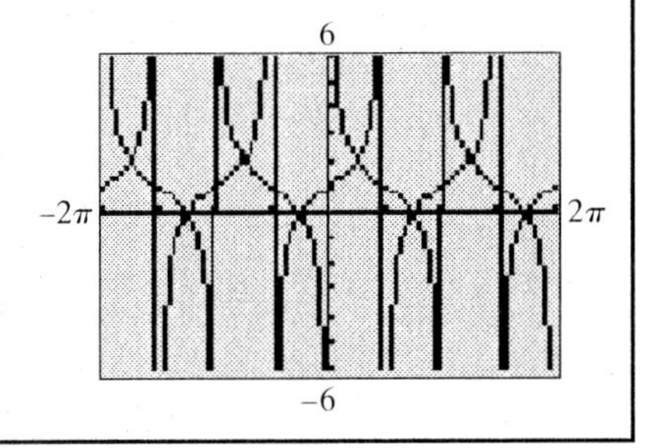

One thing we notice about identities is that, at first glance, it may not look possible that both sides are equivalent. That's the case with many trigonometric identities. Don't let your eyes fool you. Identities are not proven by appearance. They are verified on the basis of known identities, arithmetic, algebra skills, hard work, and a lot of experience. For instance, suppose a calculus student has the expression $\csc^2 x - \csc x \cot x$ as an answer to a homework problem, while the answer key has $\dfrac{1 - \cos x}{\sin^2 x}$. The calculus student having lots of practice with trigonometric identities would know that although these expressions may not look alike, the two may be equivalent. It's not a matter of redoing the calculus problem, but using trigonometry to see if the student's answer is an equivalent form of the correct answer, thus proving an identity.

As we have indicated, one quick way to determine that an equation is not an identity is to demonstrate that the graphs of each side of the equation do not coincide. The analytical way to prove an equation is not an identity is to find a specific value of the variable for which both sides are defined that makes the statement false. This value is called a **counterexample**. Common arcs or angles are typically used as counterexamples.

EXAMPLE 6 Use a graphing utility to determine if the given equation could be an identity. If so, prove it. If it cannot be an identity, find a counterexample by showing that the equation is not true for one possible replacement value of x.

$$\sin x = \tan x \cdot \cos^2 x \cdot \csc x$$

SOLUTION Graph each side.

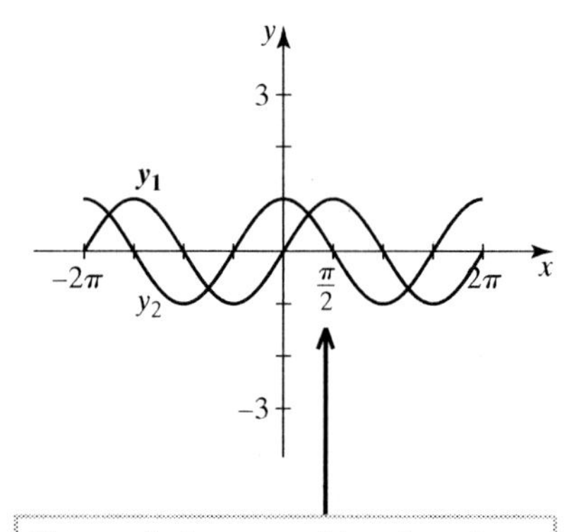

The graph shows possible values that can be used as counterexamples since we can determine an x value where the y values are not equal. But, be careful to select values where each graph is defined. While it may appear $\pi/2$ would be a choice, y_2 is not defined at this value since it contains $\tan x$.

$$y_1 = \sin x$$
$$y_2 = \tan x \cdot \cos^2 x \cdot \csc x$$

Since the graphs do not coincide, the equation is not an identity.

We are also asked to find a counterexample. One such possible value is $x = \pi/3$ because each side of the equation is defined for $x = \pi/3$, and it yields a false statement.

$$\sin x = \tan x \cdot \cos^2 x \cdot \csc x$$

$$\sin\frac{\pi}{3} \stackrel{?}{=} \tan\frac{\pi}{3}\left(\cos\frac{\pi}{3}\right)^2 \csc\frac{\pi}{3} \qquad \text{Let } x = \frac{\pi}{3}.$$

$$\frac{\sqrt{3}}{2} \stackrel{?}{=} (\sqrt{3})\left(\frac{1}{2}\right)^2\left(\frac{2}{\sqrt{3}}\right)$$

$$\frac{\sqrt{3}}{2} \neq \frac{1}{2}$$

The counterexample, $x = \pi/3$, proves that the given equation is not an identity. ■

CAUTION: Suppose we had selected $x = \pi/4$ for our counterexample in Example 6.

$$\sin x = \tan x \cdot \cos^2 x \cdot \csc x$$

$$\sin\frac{\pi}{4} \stackrel{?}{=} \tan\frac{\pi}{4} \cdot \left(\cos\frac{\pi}{4}\right)^2 \cdot \frac{1}{\sin \pi/4} \qquad \text{Let } x = \frac{\pi}{4}.$$

$$\frac{\sqrt{2}}{2} \stackrel{?}{=} 1 \cdot \left(\frac{1}{\sqrt{2}}\right)^2 \cdot \frac{\sqrt{2}}{1}$$

$$\frac{\sqrt{2}}{2} = \frac{\sqrt{2}}{2}$$

Both sides are equal when $x = \pi/4$. However, finding one value that makes both sides equal does not prove an identity because it must be true for *all* possible replacement values. And yet we need only *one* value that makes the statement false to prove the equation is not an identity. When the first attempt to find a counterexample does not produce a false statement, try a different possible replacement value.

Your work with identities will be very important to your understanding of many of the remaining sections of this text. Practice will make you much more confident and proficient, so let's get started simplifying expressions and proving identities. Since there is typically more than one way to prove an identity, don't be afraid to try a different approach if your first attempt is unsuccessful. When you feel that you have mastered proving identities, try finding a second solution to the problem.

NOTE: Most of the examples in this section used the variable x to represent the arc or angle in the identity. The problems in the exercises use different variables to accommodate flexibility.

Exercise Set 4.1

1–12. *Select the correct expression from Column A to complete the basic identity.*

1. $1 - \cos^2 x =$ ______

2. $1 + \tan^2 x =$ ______

3. $\csc^2 x = 1 +$ ______

4. $\sin(-x) =$ ______

5. $(1 - \sin x)(1 + \sin x) = 1 - \sin^2 x =$ ______

6. $-\tan x =$ ______

7. $\sec^2 x - 1 =$ ______

8. $\sin^2 x =$ ______

9. $\dfrac{1}{\sec^2 x} =$ ______

10. $1 + \cot^2 x =$ ______

11. $\cos x =$ ______

12. $\dfrac{1}{\csc x} =$ ______

COLUMN A

- $\sin x$
- $-\sin x$
- $\sin^2 x$
- $\cos^2 x$
- $\pm\sqrt{1 - \sin^2 x}$
- $\tan^2 x$
- $\dfrac{\sin(-x)}{\cos(-x)}$
- $1 - \cos^2 x$
- $\csc^2 x$
- $\sec^2 x$
- $\cot^2 x$

13–28. *Simplify each expression.*

EXAMPLE $\cos x \csc x$

SOLUTION $\cos x \csc x$

$$= \cos x \cdot \frac{1}{\sin x}$$

$$= \frac{\cos x}{\sin x}$$

$$= \cot x$$

■

13. $\sin x \sec(-x) \cot x$

14. $\cos x \tan(-x) \csc x$

15. $\cos x \tan x + \sin x \csc x + \sin(-x)$

16. $\sin x \tan x + \cos(-x)$

17. $(1 - \cos\alpha)(1 + \cos\alpha)$

18. $(1 + \sin\alpha)(1 - \sin\alpha)$

19. $(\cos A + \sin A)^2$

20. $(\sin A - \cos A)^2$

21. $\dfrac{\sec\beta}{\tan\beta}$

22. $\dfrac{\csc\beta}{\cot\beta}$

23. $\sin^3 x + \sin x \cos^2 x$

24. $\sec x + \sec x \tan^2 x$

25. $\cos^4 B - \sin^4 B$

26. $\csc^2 B + \csc^2 B \cot^2 B$

27. $\dfrac{1 - \sin^2 x}{1 - \sin x}$

28. $\dfrac{\tan^2 x + 1}{\sec^2 x}$

29–56. *Prove that each equation is an identity.*

EXAMPLE $\sin^2 t\,(\cot^2 t + 1) = 1$

SOLUTION 1

$$\sin^2 t\,(\cot^2 t + 1) = 1$$
$$\sin^2 t\left(\frac{\cos^2 t}{\sin^2 t} + 1\right) =$$
$$\sin^2 t \cdot \frac{\cos^2 t}{\sin^2 t} + \sin^2 t =$$
$$\cos^2 t + \sin^2 t =$$
$$1 = 1$$

or

SOLUTION 2

$$\sin^2 t\,(\cot^2 t + 1) = 1$$
$$\sin^2 t\,(\csc^2 t) =$$
$$\sin^2 t\left(\frac{1}{\sin^2 t}\right) =$$
$$1 = 1$$

■

29. $\cos A \tan A = \sin A$

30. $\csc A \tan A = \sec A$

31. $\sin^2 x \sec x \csc x = \tan x$

32. $\cos^2 x \sec x \csc x = \cot x$

33. $\sec t - \cos t = \tan t \sin t$

34. $\sec t - \tan t \sin t = \cos t$

35. $1 + \cot x = \dfrac{\cos x + \sin x}{\sin x}$

36. $\cot x - \tan x = \dfrac{1 - \tan^2 x}{\tan x}$

37. $\cos^2 x - \sin^2 x = 2\cos^2 x - 1$

38. $\cos^2 x - \sin^2 x = 1 - 2\sin^2 x$

39. $\tan^2\beta - \sin^2\beta = \tan^2\beta \sin^2\beta$

40. $2\sec^2\beta = \dfrac{1}{1 + \sin\beta} + \dfrac{1}{1 - \sin\beta}$

41. $\dfrac{\sin^4 x - \cos^4 x}{\sin^2 x - \cos^2 x} = 1$

42. $\dfrac{\cos^4 x - \sin^4 x}{2\cos^2 x - 1} = 1$

43. $\dfrac{1}{1 + \cos A} = \csc^2 A - \csc A \cot A$

44. $\dfrac{1}{1 - \sin A} = \sec A \tan A + \sec^2 A$

45. $\dfrac{1 - \cos\alpha}{\sin\alpha} + \dfrac{\sin\alpha}{1 - \cos\alpha} = 2\csc\alpha$

46. $\dfrac{\cos\alpha}{\sec\alpha - 1} - \dfrac{\cos\alpha}{\tan^2\alpha} = \cot^2\alpha$

47. $\sec^2\theta \csc^2\theta = \sec^2\theta + \csc^2\theta$

48. $\dfrac{1}{\sec\theta - \tan\theta} = \sec\theta + \tan\theta$

49. $(\sin x + \cos x)^2 - 1 = 2\sin x \cos x$

50. $(\tan x + \cot x)^2 = \sec^2 x + \csc^2 x$

51. $\dfrac{\csc^2 x - \cot^2 x}{\sec^2 x} = \cos^2 x$

52. $\dfrac{\csc x - \cot x}{\sec x - 1} = \cot x$

53. $2\cos^2 B - 1 = 1 - 2\sin^2 B$

54. $\cot^2 B - \cos^2 B = \cot^2 B \cos^2 B$

55. $\sec(-A)\cot(-A)\sin(-A) = 1$

56. $\tan(-A)\csc A \cos(-A) = -1$

57–60. *An identity is an equation that is true for all values for which it is defined. State the restrictions on the variable for the following identities.*

EXAMPLE $\sec x \cos x = 1$

SOLUTION The left side contains $\sec x$ that has a restriction on the variable. Since $\sec x = 1/\cos x$ is only defined where $\cos x \neq 0$, the identity is true for $x \neq (\pi/2) + k\pi$, (k is any integer). ■

57. $\tan x \cdot \cos x = \sin x$

58. $\cot x \cdot \sin x = \cos x$

59. $\dfrac{1 - \cos^2 x}{\sin^2 x} = \sin^2 x + \cos^2 x$

60. $\dfrac{\cos^2 x}{1 - \sin^2 x} = \cos^2 x + \sin^2 x$

61. Krystn, a calculus student, has the answer $-\tan x + \dfrac{\sin x}{\cos x - 1}$ to a homework question. The answer key has $\tan x\left(\dfrac{1}{\cos x - 1}\right)$ as the answer to the question. Does Krystn have an equivalent form of the answer, or should the calculus homework problem be redone? Show the work to justify your answer.

62. Jameson, a calculus student, has the answer $2 \sin x\,(\sec x + \tan x \sec^2 x)$ to a homework question. The answer key has $2 \tan x \sec^2 x$ as the answer to the question. Does Jameson have an equivalent form of the answer, or should the homework problem be redone? Show the work to justify your answer.

Graphing Calculator/CAS Exercises

63–72. *Use a graphing utility to determine whether the given equation could be an identity. If the equation appears to be an identity, prove it. Otherwise, find a counterexample.*

EXAMPLE $\cos x + \sin x = \sec x$

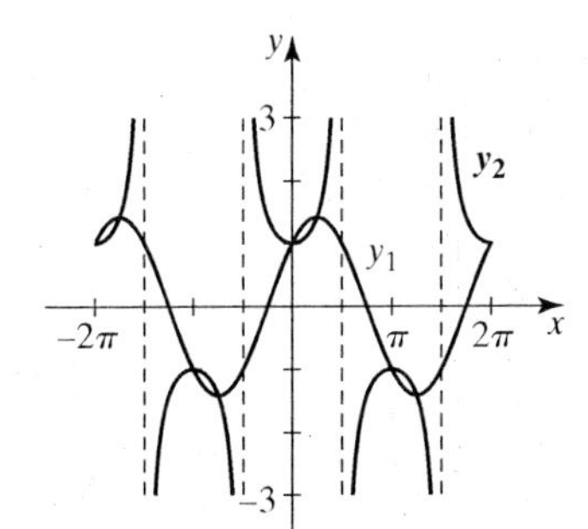

SOLUTION Since the graphs $y_1 = \cos x + \sin x$ and $y_2 = \sec x$ do not coincide, we select $x = \pi/3$ for a counterexample.

$$\cos x + \sin x = \sec x$$

$$\cos\frac{\pi}{3} + \sin\frac{\pi}{3} \stackrel{?}{=} \sec\frac{\pi}{3} \qquad \text{Let } x = \pi/3.$$

$$\frac{1}{2} + \frac{\sqrt{3}}{2} \neq 2$$

■

63. $\sin x \cot x \sec x = 1$

64. $\cos^2 x + \sec^2 x = 1$

65. $(\tan x + \cot x)^2 = \sec^2 x$

66. $\sec^4 x - \tan^4 x = 2\tan^2 x$

67. $\dfrac{\tan x \sin x}{\tan x + \sin x} = \dfrac{\tan x - \sin x}{\tan x \sin x}$

68. $\dfrac{\cot x \cos x}{\cot x + \cos x} = \dfrac{\cot x - \cos x}{\cot x \cos x}$

69. $2\cos x = \cos 2x$

70. $\frac{1}{4}\sin 4x = \sin x$

71. $\dfrac{1 - \cos^4 x}{\sin x} + \sin^3 x = 2\sin x$

72. $\sec^4 x - \tan^4 x = 1 + 2\tan^2 x$

73. a. Nicole uses a graphing utility (Xmin $= -2\pi$, Xmax $= 2\pi$, Ymin $= -2$, Ymax $= 2$) and says that $(\sin x + 0.01)^2 = (\sin x)^2 + (0.01)^2$ is an identity since the graphs of each side of the equation appears to coincide. Is Nicole correct when she says the graphs appear to coincide?

(continued)

b. Robert says that the equation is not an identity since it is false if $x = \pi/6$. Is Robert correct that the statement is false when $x = \pi/6$?

c. When the graphs appear to coincide, can we be sure that it is an identity? Explain why or why not.

74. Graph

$$y_1 = \frac{\sin x}{1 - \cos x} - \frac{1 - \cos x}{\sin x}.$$

Find a simpler equation, y_2, to represent this graph. Then, prove the identity $y_1 = y_2$.

Discussion

75. We know that we can prove that $\sin x = -\sin x$ is not an identity by using $x = \pi/2$ as a counterexample:

$$\sin\frac{\pi}{2} \neq -\sin\frac{\pi}{2}$$
$$1 \neq -1$$

Yet Roger insists that it *is* an identity based on his proof:

$\sin x = -\sin x$

$\sin^2 x = (-\sin x)^2$ Square both sides.

$\sin^2 x = \sin^2 x$ Simplify.

Therefore, $\sin x = -\sin x$ is an identity.

Discuss the incorrect reasoning Roger used in his proof.

76. Discuss why it is permissible to work the following problem by subtracting values from both sides of the equation.

$$1 + \tan^2 x = \sec^2 x$$
$$1 + \tan^2 x - \tan^2 x = \sec^2 x - \tan^2 x$$
$$1 = \sec^2 x - \tan^2 x$$

Therefore, $1 = \sec^2 x - \tan^2 x$ is an identity.

77. Marc says that $\sin\theta + \cos\theta = 1$ is an identity since the equation is true when $\theta - \pi/2$ and when $\theta = 0$. What is wrong with this reasoning?

78. Referring to Exercise 77, Gina corrected Marc by saying that $\sin^2 + \cos^2 = 1$ is the correct form of the identity. What is wrong with this statement?

4.2 Sum and Difference Identities for Cosine

In the previous section we simplified expressions and proved identities using basic identities. In this section there are new important identities that are a little more difficult to prove, and therefore they should be memorized. These identities involve trigonometric functional values of the sum or difference of two angles. We begin with the proof of the identity called the **cosine difference identity**.

Cosine Difference Identity

On the unit circle, we have two arcs, A and B, whose initial points are at $(1, 0)$. The terminal points of the arcs have coordinates $(\cos A, \sin A)$ and $(\cos B, \sin B)$, respectively. The arc between the terminal points of these two arcs is $(A - B)$. If we draw an arc equal to $(A - B)$ so that its initial point is also $(1, 0)$, the coordinates of its terminal point will be $(\cos(A - B), \sin(A - B))$. Since we have two arcs equal to $(A - B)$ in the same circle, the distance between the endpoints of each arc are equal.

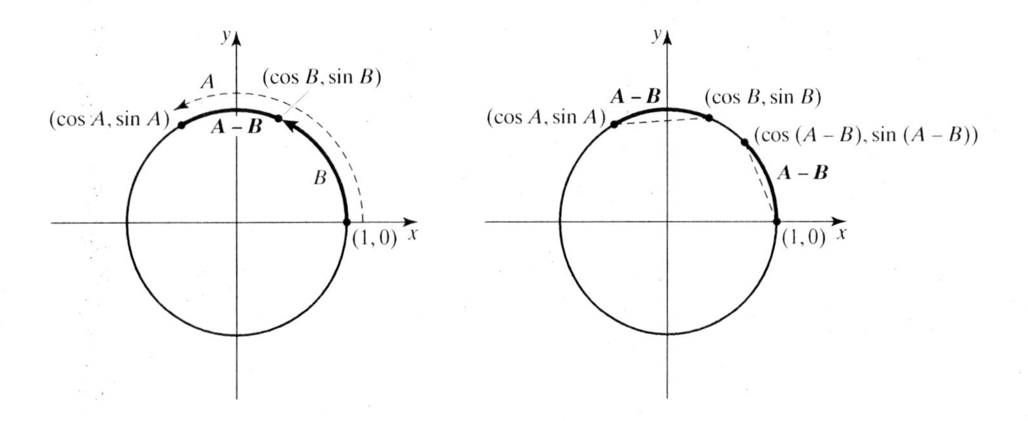

We set the distance between points $(\cos(A - B), \sin(A - B))$ and $(1, 0)$ equal to the distance between points $(\cos A, \sin A)$ and $(\cos B, \sin B)$, square both sides of the equation, and then simplify. In simplifying, we use the basic identity $\cos^2 x + \sin^2 x = 1$ several times.

$$\sqrt{(\cos(A - B) - 1)^2 + (\sin(A - B) - 0)^2} = \sqrt{(\cos A - \cos B)^2 + (\sin A - \sin B)^2}$$
$$(\cos(A - B) - 1)^2 + (\sin(A - B))^2 = (\cos A - \cos B)^2 + (\sin A - \sin B)^2$$
$$\cos^2(A - B) - 2\cos(A - B) + 1 + \sin^2(A - B) = \cos^2 A - 2\cos A\cos B + \cos^2 B + \sin^2 A - 2\sin A\sin B + \sin^2 B$$
$$-2\cos(A - B) + 2 = -2\cos A\cos B - 2\sin A\sin B + 2$$
$$\cos(A - B) = \cos A\cos B + \sin A\sin B$$

The result is called the *cosine difference identity*.

Cosine Sum Identity

We now can derive the identity for $\cos(A + B)$, called the **cosine sum identity**, by using the fact that $\cos(A + B) = \cos(A - (-B))$.

$$\begin{aligned}\cos(A + B) &= \cos(A - (-B)) \\ &= \cos(A)\cos(-B) + \sin(A)\sin(-B) \\ &= \cos A\cos B - \sin A\sin B\end{aligned}$$

$\cos(-B) = \cos B$,
$\sin(-B) = -\sin B$

Recall our work in Chapter 3 that relates an arc on the unit circle to the radian measure of the corresponding central angle. As a result, these identities can be applied to two angles (in radians or degrees).

> The cosine *sum* formula *subtracts* the second term, and the cosine *difference* formula *adds* the second term. The first term only involves the cosine and the second term only involves the sine.

Cosine Sum and Difference Identities (Formulas)

$$\cos(A+B) = \cos A \cos B - \sin A \sin B$$
$$\cos(A-B) = \cos A \cos B + \sin A \sin B$$

CAUTION: Notice that $\cos(A+B) \neq \cos A + \cos B$ and $\cos(A-B) \neq \cos A - \cos B$.

The cosine sum and difference identities should be memorized. Like the basic identities these are important in trigonometry as well as in other areas of mathematics.

EXAMPLE 1 Prove the identity $\cos\left(\frac{\pi}{2} - x\right) = \sin x$.

SOLUTION Using the cosine difference identity $\cos(A-B) = \cos A \cos B + \sin A \sin B$ and letting $A = \frac{\pi}{2}$ and $B = x$, we get:

$$\begin{aligned} \cos\left(\frac{\pi}{2} - x\right) &= \sin x \\ \cos\frac{\pi}{2}\cos x + \sin\frac{\pi}{2}\sin x &= \\ 0 \cdot \cos x + 1 \cdot \sin x &= \\ \sin x &= \end{aligned}$$

■

The identity in Example 1 should look familiar because it was the definition of cofunctions in Chapter 3. (Recall that $\pi/2 = 90°$.) We now refer to this as a **cofunction identity**, which is applicable to any angle. We develop similar cofunction identities for the sine and tangent in the next section, and a summary of all three is included in the chapter review.

Using Cosine Sum or Difference Identities to Find Exact Functional Values

In the next example we look at an important use of the cosine sum and difference identities—finding exact cosine values of angles that can be rewritten as the sum or difference of angles whose exact functional values are already known.

EXAMPLE 2 Find the exact functional value.

a. $\cos\frac{\pi}{12}$ **b.** $\cos 105°$

SOLUTION

a. Prior to this section, we needed a calculator to find an approximation for this functional value since $\pi/12$ is not a common angle. But if $\pi/12$ can be expressed as the difference or sum of common angles, whose functional values are known, we now can find the exact functional value for $\cos \pi/12$. Since

$$\frac{\pi}{12} = \frac{4\pi}{12} - \frac{3\pi}{12} = \frac{\pi}{3} - \frac{\pi}{4},$$

we use the cosine difference identity with $A = \pi/3$ and $B = \pi/4$.

$$\begin{aligned}
\cos(A - B) &= \cos A \cos B + \sin A \sin B \\
\cos\frac{\pi}{12} &= \cos\left(\frac{\pi}{3} - \frac{\pi}{4}\right) \\
&= \cos\frac{\pi}{3}\cos\frac{\pi}{4} + \sin\frac{\pi}{3}\sin\frac{\pi}{4} \\
&= \frac{1}{2}\cdot\frac{\sqrt{2}}{2} + \frac{\sqrt{3}}{2}\cdot\frac{\sqrt{2}}{2} \\
&= \frac{\sqrt{2} + \sqrt{6}}{4}
\end{aligned}$$

NOTE: We could have said $\pi/12 = \pi/4 - \pi/6$, or we could have converted $\pi/12$ to degrees, getting $\pi/12 = 15° = 45° - 30°$.

b. Since $105° = 60° + 45°$, we use the cosine sum identity with $A = 60°$ and $B = 45°$ to find the exact value.

$$\begin{aligned}
\cos(A + B) &= \cos A \cos B - \sin A \sin B \\
\cos 105° &= \cos(60° + 45°) \\
&= \cos 60° \cos 45° - \sin 60° \sin 45° \\
&= \frac{1}{2}\cdot\frac{\sqrt{2}}{2} - \frac{\sqrt{3}}{2}\cdot\frac{\sqrt{2}}{2} \\
&= \frac{\sqrt{2} - \sqrt{6}}{4}
\end{aligned}$$

■

EXAMPLE 3 Find the exact value of $\cos 57° \cos 12° + \sin 57° \sin 12°$

SOLUTION At first glance, you might think that we need to use a calculator for an approximate answer to this problem since 57° and 12° are neither common angles nor the sum or difference of common angles. A closer look indicates that the expression *is the expanded side of the cosine difference identity.* We are able to take advantage of this identity and rewrite this as the cosine of a single angle that has an exact functional value.

$$\begin{aligned}
\cos A \cos B + \sin A \sin B &= \cos(A - B) \\
\cos 57° \cos 12° + \sin 57° \sin 12° &= \cos(57° - 12°) \\
&= \cos(45°) \\
&= \frac{\sqrt{2}}{2}
\end{aligned}$$

■

CONNECTIONS WITH TECHNOLOGY

Graphing Calculator

same

- We already know that many answers can be verified or approximated using a calculator. Let's verify the result found in Example 3 with a calculator in degree mode: $\cos 57^\circ \cos 12^\circ + \sin 57^\circ \sin 12^\circ = \dfrac{\sqrt{2}}{2}$
- We can also use a graphing utility to find (or verify) a simpler form of the expanded side of the cosine sum or cosine difference identity. This simpler form rewrites the expanded side as the cosine of an angle, which we demonstrate in the next example.

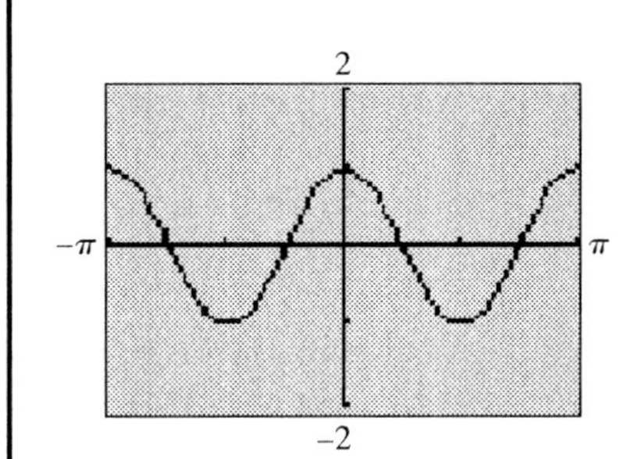

EXAMPLE Simplify $\cos(5x)\cos(3x) + \sin(5x)\sin(3x)$.

SOLUTION Graph: $y_1 = \cos(5x)\cos(3x) + \sin(5x)\sin(3x)$

WINDOW: $\text{Xmin} = -\pi$, $\text{Xmax} = \pi$,
$\text{Ymin} = -2$, $\text{Ymax} = 2$

By inspection, a simpler equation to describe this graph is $y = \cos 2x$ $(A = 1$, and $P = \pi$, so $B = 2)$.

The simplified form is consistent with the cosine difference identity since

$$\cos(5x)\cos(3x) + \sin(5x)\sin(3x) = \cos(5x - 3x) = \cos 2x.$$

In the next example, for each of two angles we are given a functional value of the angle and the quadrant that contains the angle, and we are asked to find the exact value of the cosine of the sum of these two angles. Let's see how the cosine sum or difference identity can help.

EXAMPLE 4 If $\cos A = 1/2$ and $\sin B = -5/\sqrt{26}$ with angles A and B in QIV, find $\cos(A + B)$.

SOLUTION Using the identity $\cos(A + B) = \cos A \cos B - \sin A \sin B$, we need to find $\sin A$ and $\cos B$. We first find $\sin A$ by recognizing that when $\cos A = 1/2$, $\sin A$ will be $\pm\sqrt{3}/2$. Since A is in QIV,

$$\sin A = -\frac{\sqrt{3}}{2}.$$

(If you fail to recognize this, you can always proceed with the Pythagorean identity.)

Next, to find $\cos B$, we use the Pythagorean identity (or we can use trigonometric ratios, as demonstrated in Example 5).

$$\cos^2 B + \sin^2 B = 1$$

$$\cos^2 B + \left(\frac{5}{\sqrt{26}}\right)^2 = 1$$

$$\cos^2 B + \frac{25}{26} = 1$$

$$\cos B = \pm\sqrt{\frac{1}{26}}$$

Since B is in QIV, $\cos B = 1/\sqrt{26}$.

Using this box format can help you reduce errors when you substitute values into the cosine sum identity.

QIV	QIV
$\cos A = \dfrac{1}{2}$	$\cos B = \dfrac{1}{\sqrt{26}}$
$\sin A = -\dfrac{\sqrt{3}}{2}$	$\sin B = -\dfrac{5}{\sqrt{26}}$

We now have all the functional values we need for our identity. So, we substitute these values in the cosine sum identity. (Observe that the products in each term produce common denominators.)

$$\begin{aligned}\cos(A + B) &= \cos A \cos B - \sin A \sin B \\ &= \frac{1}{2} \cdot \frac{1}{\sqrt{26}} - \left(-\frac{\sqrt{3}}{2}\right) \cdot \left(-\frac{5}{\sqrt{26}}\right) \\ &= \frac{1 - 5\sqrt{3}}{2\sqrt{26}} && \text{Add the fractions.} \\ &= \frac{\sqrt{26} - 5\sqrt{78}}{52} && \text{Rationalize the denominator.}\end{aligned}$$

Therefore, $\cos(A + B) = \dfrac{\sqrt{26} - 5\sqrt{78}}{52}$. ■

CAUTION: Problems like Example 4 require you to be very careful. It is a common error to substitute $\cos B$ when you should be substituting $\cos A$, and so on. (Substitution errors usually produce products with uncommon denominators.)

EXAMPLE 5 Find the exact value for $\cos(A - B)$ if $\tan A = \frac{1}{2}$ and $\sin B = \frac{3}{5}$, where $-90° < A < 90°$ and $-90° \leq B \leq 90°$.

SOLUTION Since $\tan A > 0$ and we know that $-90° < A < 90°$, A is in QI. Likewise, B is also in QI since $\sin B > 0$ and $-90° \leq B \leq 90°$. Once again we have a functional value of the angle with its quadrant, and we need to find other functional values. This time we find the other functional values using trigonometric ratios.

If $\tan A = \dfrac{1}{2} = \dfrac{y\ (\text{opp})}{x\ (\text{adj})}$ and A is in QI, then $y = 1$ and $x = 2$.

$$\begin{aligned} r^2 &= x^2 + y^2 \\ r^2 &= 2^2 + 1^2 = 5 \\ r &= \sqrt{5} \end{aligned}$$

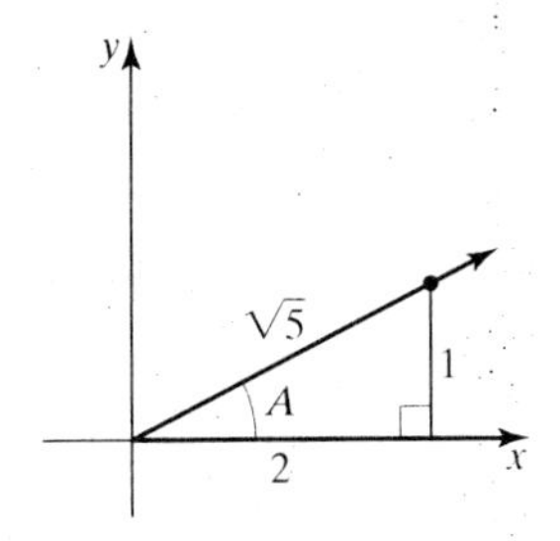

If $\sin B = \dfrac{3}{5} = \dfrac{y\ (\text{opp})}{r\ (\text{hyp})}$ and B is in QI, then $y = 3$, and $r = 5$.

$$\begin{aligned} x^2 + y^2 &= r^2 \\ x^2 + 3^2 &= 5^2 \\ x^2 &= 16 \\ x &= 4 \end{aligned}$$

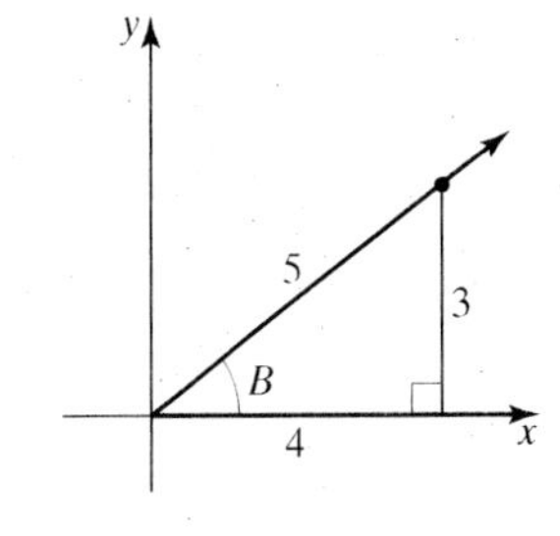

(continued)

Therefore,

QI	QI
$\cos A = \dfrac{2}{\sqrt{5}}$	$\cos B = \dfrac{4}{5}$
$\sin A = \dfrac{1}{\sqrt{5}}$	$\sin B = \dfrac{3}{5}$

Now we substitute these values into the cosine difference identity.

$$\begin{aligned}\cos(A - B) &= \cos A \cos B + \sin A \sin B \\ &= \frac{2}{\sqrt{5}} \cdot \frac{4}{5} + \frac{1}{\sqrt{5}} \cdot \frac{3}{5} \\ &= \frac{8+3}{5\sqrt{5}} = \frac{11}{5\sqrt{5}} \\ &= \frac{11\sqrt{5}}{25}\end{aligned}$$

Therefore, $\cos(A - B) = \dfrac{11\sqrt{5}}{25}$. ■

As you can see, we are able to find exact functional values as a result of using the cosine sum and difference identities. Although it would be much quicker to use a calculator to find approximate values for many problems in the exercise set, finding the exact functional values gives you experience with these identities and, as a result, will make them easier to remember.

Exercise Set 4.2

1–8. *Find the exact functional value using the cosine sum or difference identity. Verify your result with a calculator.*

1. $\cos 15°$ **2.** $\cos 75°$ **3.** $\cos 195°$ **4.** $\cos(-195°)$

5. $\cos\left(\dfrac{7\pi}{12}\right)$ **6.** $\cos\left(\dfrac{5\pi}{12}\right)$ **7.** $\sec\left(-\dfrac{\pi}{12}\right)$ **8.** $\sec\left(\dfrac{13\pi}{12}\right)$

9–16. *Use the appropriate cosine sum or difference identity to rewrite the expanded side of the cosine sum or difference identity as the cosine of an angle. Then find the exact functional value. Verify your answer with a calculator. Be careful to use the appropriate radian or degree setting.*

EXAMPLE $\cos\dfrac{\pi}{7}\cos\dfrac{5\pi}{14} - \sin\dfrac{\pi}{7}\sin\dfrac{5\pi}{14}$

SOLUTION This expression is the expanded form of the cosine sum formula.

```
cos(π/7)cos(5π/1
4)-sin(π/7)sin(5
π/14)
                0
```

$$\begin{aligned}\cos\frac{\pi}{7}\cos\frac{5\pi}{14} - \sin\frac{\pi}{7}\sin\frac{5\pi}{14} &= \cos\left(\frac{\pi}{7} + \frac{5\pi}{14}\right) \\ &= \cos\left(\frac{2\pi}{14} + \frac{5\pi}{14}\right) \\ &= \cos\frac{\pi}{2} \\ &= 0\end{aligned}$$

■

9. $\cos 190° \cos 10° + \sin 190° \sin 10°$

10. $\cos 210° \cos 180° + \sin 210° \sin 180°$

11. $\cos 27° \cos 63° - \sin 27° \sin 63°$

12. $\cos 279° \cos 36° - \sin 279° \sin 36°$

13. $\cos \frac{5\pi}{6} \cos \frac{\pi}{12} + \sin \frac{5\pi}{6} \sin \frac{\pi}{12}$

14. $\cos \frac{10\pi}{9} \cos \frac{5\pi}{36} - \sin \frac{10\pi}{9} \sin \frac{5\pi}{36}$

15. $\cos 25\frac{1}{2}° \cos 4\frac{1}{2}° - \sin 25\frac{1}{2}° \sin 4\frac{1}{2}°$

16. $\cos 1° \cos 271° + \sin 1° \sin 271°$

17–26. *Without using a calculator, determine if each statement is true or false.*

17. $\cos 80° \cos 35° + \sin 80° \sin 35° = \frac{\sqrt{2}}{2}$

18. $\cos 160° \cos 20° - \sin 160° \sin 20° = \frac{1}{2}$

19. $\cos \frac{5\pi}{8} \cos \frac{3\pi}{8} - \sin \frac{5\pi}{8} \sin \frac{3\pi}{8} = -1$

20. $\cos \frac{4\pi}{10} \cos \frac{\pi}{10} - \sin \frac{4\pi}{10} \sin \frac{\pi}{10} = 1$

21. $\cos 112° \cos 8° + \sin 112° \sin 8° = -\frac{\sqrt{3}}{2}$

22. $\cos 3° \cos 3° + \sin 3° \sin 3° = 1$

23. $\cos 9x \cos 8x - \sin 9x \sin 8x = \cos x$

24. $\cos 5x \cos 7x + \sin 5x \sin 7x = \cos 2x$

25. $\dfrac{1}{\cos \frac{x}{2} \cos 3x + \sin \frac{x}{2} \sin 3x} = \sec\left(\frac{5x}{2}\right)$

26. $\dfrac{1}{\cos x \cos 7x - \sin x \sin 7x} = \sec 8x$

27–30. *Without using a graphing utility, select an equation from Column A that represents the graph. More than one answer may be possible.*

27.

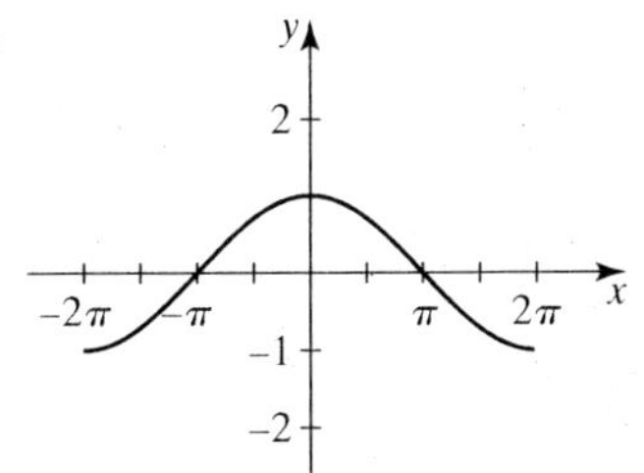

28.

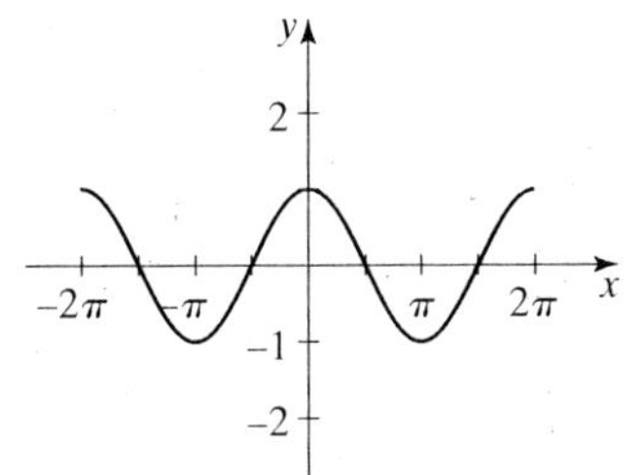

29.

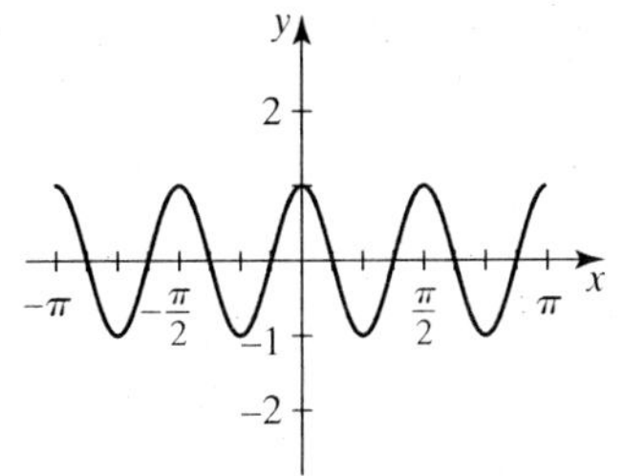

30.

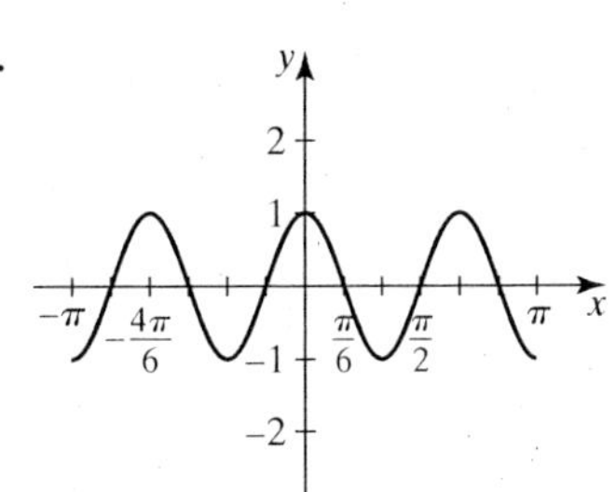

COLUMN A

a. $y = \cos 2x \cos x + \sin 2x \sin x$

b. $y = \cos 2x \cos x - \sin 2x \sin x$

c. $y = \cos \frac{1}{2}x \cos \frac{1}{2}x - \sin \frac{1}{2}x \sin \frac{1}{2}x$

d. $y = \cos x \cos \frac{1}{2}x + \sin x \sin \frac{1}{2}x$

e. $y = \cos 5x \cos x + \sin 5x \sin x$

f. $y = \cos 10x \cos 6x + \sin 10x \sin 6x$

g. $y = \cos 11\frac{1}{2}x \cos 11x + \sin 11\frac{1}{2}x \sin 11x$

31–38. *Use the cosine sum or difference identity to write each expression as a single function of x.*

EXAMPLE $\cos(360° - x)$

SOLUTION

$$\begin{aligned}\cos(360° - x) &= \cos 360° \cos x + \sin 360° \sin x \\ &= 1 \cdot \cos x + 0 \cdot \sin x \\ &= \cos x\end{aligned}$$

31. $\cos(x - 270°)$

32. $\cos(90° + x)$

33. $\cos\left(x - \frac{\pi}{2}\right)$

34. $\cos\left(\frac{\pi}{2} - x\right)$

35. $\cos(180° - x)$

36. $\cos(x + 180°)$

37. $\cos\left(\frac{3\pi}{2} + x\right)$

38. $\cos\left(\frac{3\pi}{2} - x\right)$

39–46. *Find* $\cos(A + B)$ *and* $\cos(A - B)$.

EXAMPLE $\sin A = 1$ and $\cos B = -\frac{1}{3}$, where $\pi/2 < B < \pi$

SOLUTION In order to find either $\cos(A + B)$ or $\cos(A - B)$, we need $\cos A$ and $\sin B$. Since $\sin A = 1$, we know that $\cos A = 0$. To find $\sin B$, we use the Pythagorean identity and note that B is in QII.

$$\cos^2 B + \sin^2 B = 1$$

$$\left(-\frac{1}{3}\right)^2 + \sin^2 B = 1$$

$$\sin^2 B = 1 - \frac{1}{9} = \frac{8}{9}$$

$$\sin B = \sqrt{\frac{8}{9}} = \frac{2\sqrt{2}}{3} \qquad \sin B > 0 \text{ in QII}$$

	QII
$\cos A = 0$	$\cos B = -\frac{1}{3}$
$\sin A = 1$	$\sin B = \frac{2\sqrt{2}}{3}$

$$\cos(A + B) = \cos A \cos B - \sin A \sin B = 0 \cdot \left(-\frac{1}{3}\right) - 1 \cdot \frac{2\sqrt{2}}{3} = -\frac{2\sqrt{2}}{3}$$

$$\cos(A - B) = \cos A \cos B + \sin A \sin B = 0 \cdot \left(-\frac{1}{3}\right) + 1 \cdot \frac{2\sqrt{2}}{3} = \frac{2\sqrt{2}}{3}$$

39. $\cos A = -\frac{1}{2}$ and $\sin B = -\frac{1}{2}$, with A and B in QIII.

40. $\sin A = \frac{1}{2}$ and $\cos B = -\frac{1}{2}$, with A and B in QII.

41. $\cos A = -\frac{4}{5}$ and $\cos B = \frac{12}{13}$, where $90° < A < 180°$ and $270° < B < 360°$.

42. $\sin A = \frac{3}{5}$ and $\cos B = -\frac{\sqrt{3}}{2}$, with A and B in QII.

43. $\sin A = \frac{1}{\sqrt{10}}$ and $\cos B = -1$, where $0 \le A \le \frac{\pi}{2}$.

44. $\cos A = -\frac{8}{17}$ and $\sin B = 1$, where $180° < A < 270°$.

45. $\sin A = -\frac{3}{5}$ and $\tan B = \frac{4}{3}$, with A in QIV and B in QI.

46. $\tan A = -\frac{5}{12}$ and $\cos B = -\frac{3}{5}$, with A in QII and B in QIII.

47–52. *Find the exact value for each expression.*

EXAMPLE $\cos\left(\arctan(-1) + \arccos\left(\frac{\sqrt{3}}{2}\right)\right)$

SOLUTION Let $A = \arctan(-1)$, and $B = \arccos(\sqrt{3}/2)$. Then

$$\cos\left(\arctan(-1) + \arccos\left(\frac{\sqrt{3}}{2}\right)\right) = \cos(A + B).$$

$$A = \arctan(-1) \longleftrightarrow \tan A = -1, -90° < A < 90°$$

$$B = \arccos\left(\frac{\sqrt{3}}{2}\right) \longleftrightarrow \cos B = \frac{\sqrt{3}}{2}, 0° \le B \le 180°$$

So, $A = -45°$ and $B = 30°$.

$$\begin{aligned}\cos(A + B) &= \cos(-45° + 30°)\\ &= \cos(-45°)\cos 30° - \sin(-45°)\sin 30°\\ &= \frac{\sqrt{2}}{2} \cdot \frac{\sqrt{3}}{2} - \left(-\frac{\sqrt{2}}{2}\right)\left(\frac{1}{2}\right)\\ &= \frac{\sqrt{6} + \sqrt{2}}{4}\end{aligned}$$

47. $\cos\left(\operatorname{arcsec}\sqrt{2} + \arcsin\frac{1}{2}\right)$

48. $\cos\left(\arccos\frac{1}{2} - \arctan 1\right)$

49. $\cos\left(\tan^{-1} 1 + \arccos\frac{1}{2}\right)$

50. $\cos\left(\sin^{-1}\left(-\frac{1}{2}\right) - \cos^{-1}\frac{\sqrt{2}}{2}\right)$

51. $\cos\left[\cos^{-1}\left(-\frac{1}{2}\right) - \arcsin\left(-\frac{\sqrt{2}}{2}\right)\right]$

52. $\cos\left[\operatorname{arcsec} 2 + \sin^{-1}\left(-\frac{\sqrt{2}}{2}\right)\right]$

53. In Chapter 3 we defined *cofunctions* with respect to the cosine and sine functions as $\cos(90° - \alpha) = \sin\alpha$. Use the cosine difference identity to prove this cofunction identity.

54. Verify the identity $\cos\left(\frac{\pi}{2} + x\right) = -\sin x$.

55. Verify the identity $\cos(2\pi - A) = \cos A$.

56. Verify the identity $\cos(360° + A) = \cos A$.

57. Use a counterexample to verify that $\cos(A + B) = \cos A + \cos B$ is *not* an identity.

58. Use a counterexample to verify that $\cos(A - B) = \cos A - \cos B$ is *not* an identity.

59. Verify the identity $\cos(2x) = \cos^2 x - \sin^2 x$. (*Hint*: $\cos 2x = \cos(x + x)$)

60. Verify the identity $\cos(2x) = 2\cos^2 x - 1$. (*Hint*: $\cos 2x = \cos(x + x)$)

Graphing Calculator/CAS Exercises

61. Graph $y = \cos 2x \cos 3x + \sin 2x \sin 3x$.

MODE: Radian

WINDOW: Xmin $= -2\pi$ Xmax $= 2\pi$

Ymin $= -2$ Ymax $= 2$

a. What is a simpler equation to describe your graph? Verify your answer using the cosine difference identity.

b. Without graphing $y = \cos 3x \cos x + \sin 3x \sin x$, use the cosine difference identity to find a simpler equation. Use a graphing utility to support your answer by graphing both equations.

62. Graph $y = \cos 3x \cos x - \sin 3x \sin x$.

WINDOW: Xmin $= -\pi/2$ Xmax $= \pi/2$

Ymin $= -2$ Ymax $= 2$

a. What is a simpler equation to describe your graph? Verify your answer using the cosine sum identity.

b. Without graphing $y = \cos x \cos(-3x) + \sin x \sin(-3x)$, use the cosine sum identity to find a simpler equation. Use a graphing utility to support your answer by graphing both equations.

4.3 Sum and Difference Identities for Sine and Tangent

Sum and difference identities for the sine function can be obtained using the cosine sum and difference identities from Section 4.2 along with the cofunction definitions from Section 3.3.

Sine Sum and Difference Identities

Using the cofunction identity $\sin x = \cos((\pi/2) - x)$ where $x = A + B$, we have:

$$\sin(A + B) = \cos\left(\frac{\pi}{2} - (A + B)\right)$$
Apply the cofunction definition.

$$= \cos\left(\left(\frac{\pi}{2} - A\right) - B\right)$$
Use the distributive then the associative property.

$$= \cos\left(\frac{\pi}{2} - A\right)\cos B + \sin\left(\frac{\pi}{2} - A\right)\sin B$$
Apply the cosine difference identity.

$$= \sin A \cos B + \cos A \sin B$$
Use the cofunction definitions.

Therefore, $\sin(A + B) = \sin A \cos B + \cos A \sin B$. This is called the **sine sum identity**. Similar reasoning produces the **sine difference identity**,

$$\sin(A - B) = \sin A \cos B - \cos A \sin B.$$

These identities are also referred to as the sine sum and difference formulas.

The sine *sum* formula *adds* the terms, and the sine *difference* formula *subtracts* the terms. And each term involves both the sine and the cosine.

Sine Sum and Difference Identities (Formulas)

$$\sin(A + B) = \sin A \cos B + \cos A \sin B$$
$$\sin(A - B) = \sin A \cos B - \cos A \sin B$$

Tangent Sum and Difference Identities

Using the identity $\tan x = \dfrac{\sin x}{\cos x}$ along with the sum and difference identities for the sine and cosine, we can find the tangent sum and difference identities.

$$\tan(A + B) = \frac{\sin(A + B)}{\cos(A + B)} = \frac{\sin A \cos B + \cos A \sin B}{\cos A \cos B - \sin A \sin B}$$

Since we would like the result to involve the tangent function, we multiply by 1 in a form that produces the tangent in both the numerator and denominator.

$$\tan(A + B) = \frac{\dfrac{\sin A \cos B + \cos A \sin B}{1} \cdot \dfrac{1}{\cos A \cos B}}{\dfrac{\cos A \cos B - \sin A \sin B}{1} \cdot \dfrac{1}{\cos A \cos B}}$$

$$= \frac{\dfrac{\sin A \cancel{\cos B}}{\cos A \cancel{\cos B}} + \dfrac{\cancel{\cos A} \sin B}{\cancel{\cos A} \cos B}}{\dfrac{\cancel{\cos A \cos B}}{\cancel{\cos A \cos B}} - \dfrac{\sin A \sin B}{\cos A \cos B}}$$

$$= \frac{\tan A + \tan B}{1 - \tan A \tan B}$$

Similar reasoning produces the identity

$$\tan(A - B) = \frac{\tan A - \tan B}{1 + \tan A \tan B}.$$

Tangent Sum and Difference Identities (Formulas)

$$\tan(A + B) = \frac{\tan A + \tan B}{1 - \tan A \tan B}$$
$$\tan(A - B) = \frac{\tan A - \tan B}{1 + \tan A \tan B}$$

We use the sum and difference identities for the sine and tangent the same way we used the sum and difference identities for the cosine. Example 1 in Section 4.2 verified the first cofunction identity, $\cos((\pi/2) - x) = \sin x$. In the exercises you are asked to verify the remaining two cofunction identities using the sine and tangent difference identity.

Cofunction Identities

$$\cos\left(\frac{\pi}{2} - x\right) = \sin x \qquad \sin\left(\frac{\pi}{2} - x\right) = \cos x \qquad \tan\left(\frac{\pi}{2} - x\right) = \cot x$$

Using Sine and Tangent Sum or Difference Identities to Find Exact Functional Values

EXAMPLE 1 Find the exact functional value.

a. $\sin 15°$ **b.** $\sin \frac{5\pi}{12}$ **c.** $\tan 105°$

SOLUTION

a.

$$\sin 15° = \sin(45° - 30°)$$

$$= \sin 45° \cos 30° - \cos 45° \sin 30° \quad \text{Apply the sine difference formula.}$$

$$= \frac{\sqrt{2}}{2} \cdot \frac{\sqrt{3}}{2} - \frac{\sqrt{2}}{2} \cdot \frac{1}{2} \quad \text{Substitute exact functional values.}$$

$$= \frac{\sqrt{6} - \sqrt{2}}{4} \quad \text{Simplify.}$$

b.

$$\sin \frac{5\pi}{12} = \sin\left(\frac{3\pi}{12} + \frac{2\pi}{12}\right) = \sin\left(\frac{\pi}{4} + \frac{\pi}{6}\right)$$

$$= \sin \frac{\pi}{4} \cos \frac{\pi}{6} + \cos \frac{\pi}{4} \sin \frac{\pi}{6} \quad \text{Apply the sine sum formula.}$$

$$= \frac{\sqrt{2}}{2} \cdot \frac{\sqrt{3}}{2} + \frac{\sqrt{2}}{2} \cdot \frac{1}{2} \quad \text{Substitute exact functional values.}$$

$$= \frac{\sqrt{6} + \sqrt{2}}{4} \quad \text{Simplify.}$$

c.

$$\tan 105° = \tan(45° + 60°)$$

$$= \frac{\tan 45° + \tan 60°}{1 - \tan 45° \tan 60°} \quad \text{Apply the tangent sum formula.}$$

$$= \frac{1 + \sqrt{3}}{1 - 1 \cdot \sqrt{3}} = \frac{1 + \sqrt{3}}{1 - \sqrt{3}} \quad \text{Substitute exact functional values.}$$

$$= \frac{(1 + \sqrt{3})}{(1 - \sqrt{3})} \cdot \frac{(1 + \sqrt{3})}{(1 + \sqrt{3})} \quad \text{Rationalize the denominator.}$$

$$= \frac{1 + \sqrt{3} + \sqrt{3} + 3}{1 - 3}$$ Multiply the fractions.

$$= \frac{4 + 2\sqrt{3}}{-2}$$ Simplify.

$$= \frac{2(2 + \sqrt{3})}{-2}$$ Factor 2 out of the numerator.

$$= -2 - \sqrt{3}$$ Reduce to lowest terms and simplify. ■

EXAMPLE 2 Use either the sine sum or difference or the tangent sum or difference identities to write each expression as a single function of x.

a. $\sin(90° + x)$ **b.** $\tan(180° - x)$ **c.** $\sin(\pi + x)$

SOLUTION

a.
$$\begin{aligned}\sin(90° + x) &= \sin 90° \cos x + \cos 90° \sin x \\ &= 1 \cdot \cos x + 0 \cdot \sin x \\ &= \cos x\end{aligned}$$

b. $$\tan(180° - x) = \frac{\tan 180° - \tan x}{1 + \tan 180° \tan x} = \frac{0 - \tan x}{1 + 0 \cdot \tan x} = -\tan x$$

c.
$$\begin{aligned}\sin(\pi + x) &= \sin \pi \cos x + \cos \pi \sin x \\ &= 0 \cdot \cos x + (-1) \cdot \sin x = -\sin x\end{aligned}$$ ■

EXAMPLE 3 Find the exact value.

a. $\sin 122° \cos 88° + \cos 122° \sin 88°$ **b.** $\dfrac{\tan 7° + \tan 38°}{1 - \tan 7° \tan 38°}$

SOLUTION

a. $\sin 122° \cos 88° + \cos 122° \sin 88° = \sin(122° + 88°) = \sin(210°) = -\frac{1}{2}$

b. $$\frac{\tan 7° + \tan 38°}{1 - \tan 7° \tan 38°} = \tan(7° + 38°) = \tan 45° = 1$$ ■

Having the sum and difference identities for sine, cosine, and tangent again expands our ability to find exact functional values. As we saw in Chapters 1 and 3, by knowing one of the six exact functional values of an angle (or arc) and the specific quadrant, we were able to find the remaining five exact functional values. Now if we know an exact functional value for two angles and their respective quadrants, we can find any of the six exact functional values of the sum or difference of these angles. We demonstrate this in the next example.

Example 4 If $\sin A = \frac{3}{5}$ with A in QII and $\cos B = \frac{2}{3}$ with B in QIV, find the exact value. Then approximate the value rounded to four decimal places.

a. $\sin(A - B)$ **b.** $\tan(A - B)$

Solution

a. To find $\cos A$ and $\sin B$, we use the Pythagorean identity.

$$\cos^2 A + \sin^2 A = 1$$
$$\cos^2 A + \left(\frac{3}{5}\right)^2 = 1 \qquad \sin A = \tfrac{3}{5}$$
$$\cos^2 A = \frac{16}{25}$$
$$\cos A = \pm\sqrt{\frac{16}{25}} = \pm\frac{4}{5}$$

Since A is in QII, $\cos A = -\frac{4}{5}$.

$$\cos^2 B + \sin^2 B = 1$$
$$\left(\frac{2}{3}\right)^2 + \sin^2 B = 1 \qquad \cos B = \tfrac{2}{3}$$
$$\sin^2 B = 1 - \frac{4}{9} = \frac{5}{9}$$
$$\sin B = \pm\sqrt{\frac{5}{9}} = \pm\frac{\sqrt{5}}{3}$$

Since B is in QIV, $\sin B = -\frac{\sqrt{5}}{3}$.

QII	QIV
$\cos A = -\frac{4}{5}$	$\cos B = \frac{2}{3}$
$\sin A = \frac{3}{5}$	$\sin B = -\frac{\sqrt{5}}{3}$

We now substitute these functional values into the sine difference identity.

$$\begin{aligned}\sin(A - B) &= \sin A \cos B - \cos A \sin B \\ &= \left(\frac{3}{5}\right)\left(\frac{2}{3}\right) - \left(-\frac{4}{5}\right)\left(-\frac{\sqrt{5}}{3}\right) \\ &= \frac{6-4\sqrt{5}}{15} \approx -0.1963\end{aligned}$$

b. From part (a), since $\sin A = \frac{3}{5}$ and $\cos A = -\frac{4}{5}$, $\tan A = -\frac{3}{4}$. Likewise, since $\sin B = -\frac{\sqrt{5}}{3}$ and $\cos B = \frac{2}{3}$, $\tan B = -\frac{\sqrt{5}}{2}$. Therefore,

$$\begin{aligned}\tan(A - B) &= \frac{\tan A - \tan B}{1 + \tan A \tan B} \\ &= \frac{-\frac{3}{4} - \left(-\frac{\sqrt{5}}{2}\right)}{1 + \left(-\frac{3}{4}\right)\left(-\frac{\sqrt{5}}{2}\right)} = \frac{\frac{-3 + 2\sqrt{5}}{4}}{\frac{8 + 3\sqrt{5}}{8}} \\ &= \frac{-6 + 4\sqrt{5}}{8 + 3\sqrt{5}} = \frac{-108 + 50\sqrt{5}}{19} \approx 0.2002.\end{aligned}$$

By determining the functional values of the sum or difference of two angles, $(A + B)$ or $(A - B)$, we also can determine the quadrants for $(A + B)$ or $(A - B)$, as demonstrated in the next example.

EXAMPLE 5 Using the information in Example 4, determine the quadrant that contains $(A - B)$.

SOLUTION Since $\tan(A - B) > 0$ and $\sin(A - B) < 0$, $(A - B)$ is in QIII. ■

Since this section presents the sum and difference identities for sine and tangent, you will find the exercises similar in format to those in the previous section which involved the cosine sum and difference identities.

Exercise Set 4.3

1–8. *Find the exact functional value using either the sine sum or difference or the tangent sum or difference identities. Verify your result with a calculator.*

1. $\sin 75°$
2. $\sin 165°$
3. $\tan 195°$
4. $\tan(-15°)$
5. $\sin\left(-\frac{\pi}{12}\right)$
6. $\sin\frac{7\pi}{12}$
7. $\tan\frac{5\pi}{12}$
8. $\tan\frac{13\pi}{12}$

9–16. *Use the appropriate sum or difference identity to rewrite the expanded side as the sine or tangent of an angle. Then find the exact functional value. Verify your answer with a calculator. Be aware of the appropriate radian or degree setting.*

9. $\sin 140° \cos 40° + \cos 140° \sin 40°$
10. $\sin 107° \cos 17° - \cos 107° \sin 17°$
11. $\sin\frac{7\pi}{8}\cos\frac{\pi}{8} - \cos\frac{7\pi}{8}\sin\frac{\pi}{8}$
12. $\sin\frac{\pi}{12}\cos\frac{13\pi}{12} + \cos\frac{\pi}{12}\sin\frac{13\pi}{12}$
13. $\frac{\tan 13° + \tan 32°}{1 - \tan 13° \tan 32°}$
14. $\frac{\tan 120° + \tan 60°}{1 - \tan 120° \tan 60°}$
15. $\frac{\tan 233° - \tan 113°}{1 + \tan 233° \tan 113°}$
16. $\frac{\tan 68° - \tan 98°}{1 + \tan 68° \tan 98°}$

17–24. *Without using a calculator, determine whether each statement is true or false.*

17. $\sin 102° \cos 72° - \cos 102° \sin 72° = \frac{\sqrt{3}}{2}$
18. $\sin 67.5° \cos 22.5° + \cos 67.5° \sin 22.5° = \frac{\sqrt{2}}{2}$
19. $\frac{\tan 15° - \tan 45°}{1 + \tan 15° \tan 45°} = -\frac{\sqrt{3}}{3}$
20. $\frac{\tan 13° + \tan 47°}{1 - \tan 13° \tan 47°} = \frac{\sqrt{3}}{3}$
21. $\sin 3x \cos 2x - \cos 3x \sin 2x = \sin x$
22. $\sin 5x \cos 3x + \sin 5x \cos 3x = \sin 8x$
23. $\frac{\tan 3x + \tan 2x}{1 - \tan 3x \tan 2x} = \tan x$
24. $\frac{\tan x - \tan 2x}{1 + \tan x \tan 2x} = -\tan x$

25–28. *Without using a graphing utility, select an equation from Column A that represents each graph. More than one answer may be possible. Check your answers with a graphing utility.*

25.

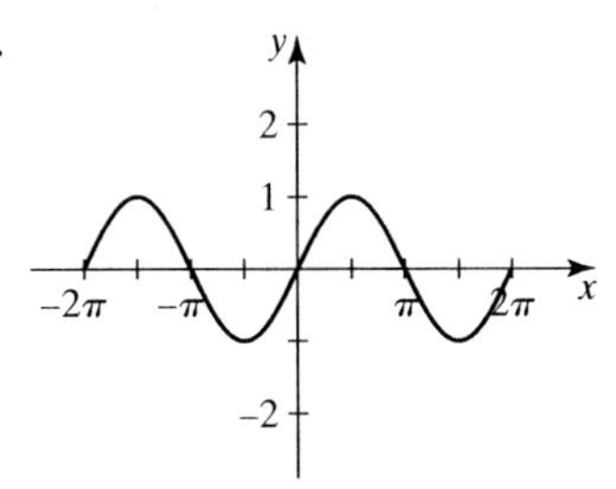

26.

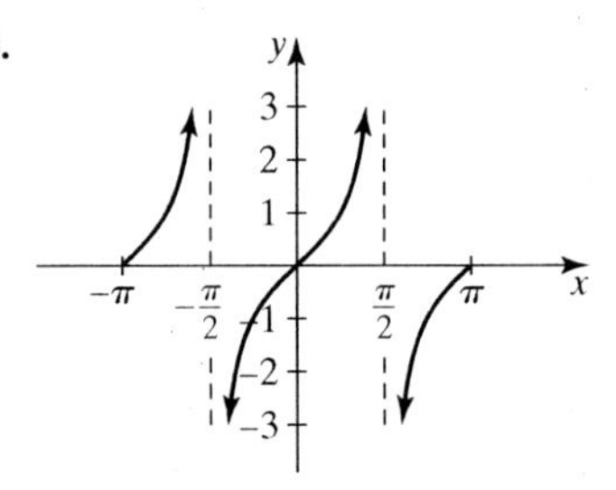

27.

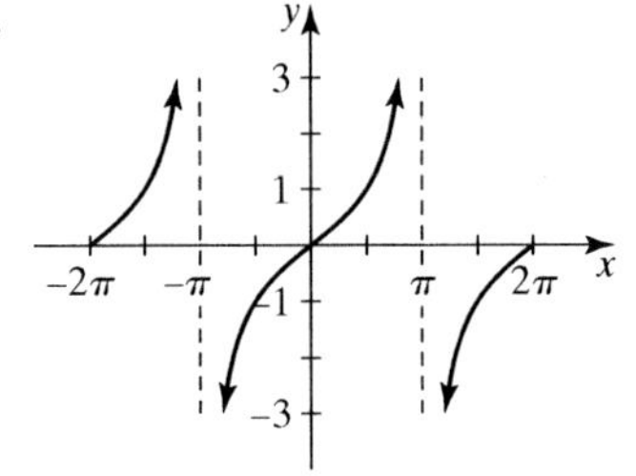

28.

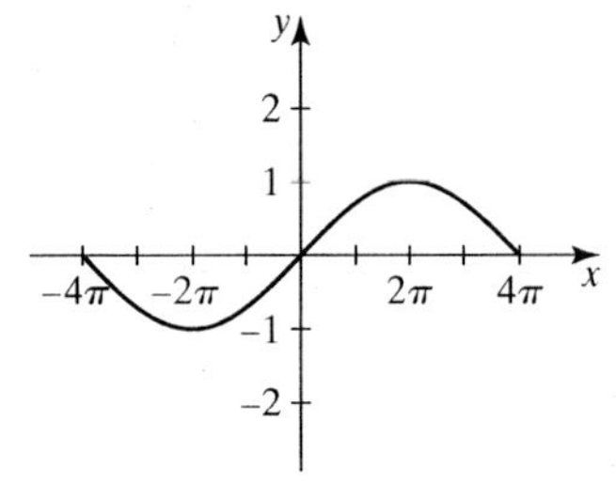

Column A

a. $y = \sin\left(\frac{1}{8}x\right)\cos\left(\frac{1}{8}x\right) + \cos\left(\frac{1}{8}x\right)\sin\left(\frac{1}{8}x\right)$

b. $y = \dfrac{\tan 2x - \tan x}{1 + \tan 2x \tan x}$

c. $y = \sin 13x \cos 11x - \cos 13x \sin 11x$

d. $y = \dfrac{\tan\frac{1}{4}x + \tan\frac{1}{4}x}{1 - \tan\frac{1}{4}x \tan\frac{1}{4}x}$

e. $y = \sin(1.25x)\cos x - \cos(1.25x)\sin x$

f. $y = \sin 2x \cos(-x) + \cos 2x \sin(-x)$

29–34. *Use either the sine sum or difference or the tangent sum or difference identities to write each expression as a single function of x.*

29. $\sin(\pi - x)$

30. $\sin\left(\dfrac{3\pi}{2} + x\right)$

31. $\sin(x - 90°)$

32. $\sin(360° + x)$

33. $\tan(x + 180°)$

34. $\tan(360° - x)$

35–38. *Find* $\sin(A + B)$.

35. $\cos A = -\dfrac{4}{5}$ and $\sin B = \dfrac{5}{13}$, with A and B in QII.

36. $\sin A = \dfrac{3}{5}$ and $\sin B = -\dfrac{12}{13}$, where $\dfrac{\pi}{2} \le A \le \pi$ and $\pi \le B \le \dfrac{3\pi}{2}$.

37. $\sin A = -\dfrac{7}{25}$ and $\cos B = \dfrac{3}{5}$, with A and B in QIV.

38. $\cos A = \dfrac{24}{25}$ and $\cos B = \dfrac{1}{2}$, where $\dfrac{3\pi}{2} \le A \le 2\pi$ and $0 \le B \le \dfrac{\pi}{2}$.

39. If $\sin A = -\dfrac{5}{13}$ and $\cos B = \dfrac{3}{5}$, with A in QIII and B in QIV, find the following.

a. $\sin(A - B)$ **b.** $\tan(A + B)$ **c.** $\cos(A + B)$
d. the quadrant for $(A + B)$

40. If $\cos A = \dfrac{12}{13}$ and $\sin B = \dfrac{3}{5}$, with A in QIV and B in QII, find the following.

a. $\sin(A - B)$ **b.** $\tan(A + B)$ **c.** $\cos(A + B)$
d. the quadrant for $(A + B)$

41–46. *Evaluate each expression using exact functional values. Check your answer with a calculator.*

EXAMPLE $\sin\left(\cos^{-1}\left(-\frac{1}{\sqrt{2}}\right) + \arccos\frac{\sqrt{3}}{2}\right)$

SOLUTION Evaluate what is inside the parentheses first.

$$A = \cos^{-1}\left(-\frac{1}{\sqrt{2}}\right) = 135° \quad \text{and} \quad B = \arccos\frac{\sqrt{3}}{2} = 30°$$

Therefore,

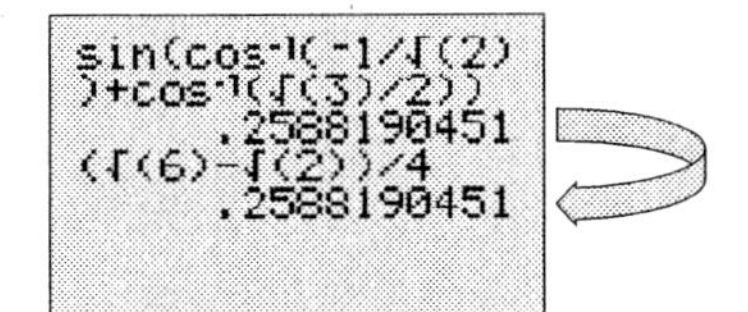

$$\sin\left(\cos^{-1}\left(-\frac{1}{\sqrt{2}}\right) + \arccos\frac{\sqrt{3}}{2}\right) =$$

$$\sin(A + B) =$$

$$\sin(135° + 30°) = \sin(135°)\cos(30°) + \cos(135°)\sin(30°)$$

$$= \frac{\sqrt{2}}{2} \cdot \frac{\sqrt{3}}{2} + \left(-\frac{\sqrt{2}}{2}\right)\frac{1}{2}$$

$$= \frac{\sqrt{6} - \sqrt{2}}{4}$$

41. $\sin\left(\sin^{-1}\frac{\sqrt{2}}{2} - \cos^{-1}\frac{1}{2}\right)$

42. $\sin(\sec^{-1}\sqrt{2} - \csc^{-1}1)$

43. $\tan(\tan^{-1}\sqrt{3} + \csc^{-1}\sqrt{2})$

44. $\tan\left(\arctan\frac{\sqrt{3}}{3} + \sec^{-1}2\right)$

45. $\tan\left(\sec^{-1}\frac{3}{2} - \cos^{-1}\frac{2}{3}\right)$

46. $\tan(\tan^{-1}1 - \cot^{-1}1)$

47. If $\tan(A + B) = 3$ and $\tan A = \frac{1}{2}$, find $\tan B$.

48. If $\tan(A - B) = -3$ and $\tan B = 1$, find $\tan A$.

49–56. *Prove the following identities.*

49. $\dfrac{\sin(A - B)}{\cos A \cos B} = \tan A - \tan B$

50. $\dfrac{\cos(A - B)}{\cos A \sin B} = \tan A + \cot B$

51. $\sin 2x = 2\sin x\cos x$ [*Hint*: $\sin 2x = \sin(x + x)$]

52. The cofunction identity: $\sin\left(\frac{\pi}{2} - x\right) = \cos x$

53. The cofunction identity: $\tan\left(\frac{\pi}{2} - x\right) = \cot x$

54. $\tan 2x = \dfrac{2\tan x}{1 - \tan^2 x}$ [*Hint:* $\tan 2x = \tan(x + x)$]

55. $\sin x + \cos x = \sqrt{2}\sin\left(x + \frac{\pi}{4}\right)$

56. $\sqrt{3}\sin x + \cos x = 2\sin\left(x + \frac{\pi}{6}\right)$

57. Use a counterexample to verify that $\sin(A + B) = \sin A + \sin B$ is *not* an identity.

58. Use a counterexample to verify that $\tan(A + B) = \tan A + \tan B$ is *not* an identity.

59. Donald completed all Section 4.2 exercises including Exercise 1, which asked for the exact value of cos 15°. As he began the exercises in this section, he noticed that Exercise 1 asks for the exact value of sin 75°. Donald said he immediately knew at a glance the answer to sin 75°. What identity did Donald use that allowed him to know the exact value of sin 75° without doing any additional work?

60. Explain why you cannot use the tangent difference identity to prove the identity $\cot(90° - x) = \tan x$. Prove this is an identity by first rewriting $\cot(90° - x)$ in terms of the sine and cosine functions.

Graphing Calculator/CAS Exercises

61. Demonstrate that Exercise 51 is an identity by graphing $y_1 = \sin 2x$ and $y_2 = 2 \sin x \cos x$.

62. Demonstrate that Exercise 54 is an identity by graphing $y_1 = \tan 2x$ and $y_2 = \dfrac{2 \tan x}{1 - \tan^2 x}$.

4.4 Double-Angle Identities

We know that $\sin 2x = 2 \sin x$ is *not* an identity because the graphs of $y = \sin 2x$ and $y = 2 \sin x$ do not coincide due to their different periods and amplitudes. We can also find counterexamples. In this section we find an equivalent expression for $\sin 2x$, $\cos 2x$, and $\tan 2x$ that are referred to as the **double-angle identities** (formulas), which are developed using the respective sum identities.

Double-Angle Identities for Sine and Cosine

To find the double-angle identity for the sine function, we use the sine sum identity.

$$\begin{aligned}
\sin 2A &= \sin(A + A) && \text{Let } 2A = A + A. \\
&= \sin A \cos A + \cos A \sin A && \text{Use the sine sum identity.} \\
&= \sin A \cos A + \sin A \cos A && \text{Use the commutative property of multiplication.} \\
&= 2 \sin A \cos A && \text{Add like terms.}
\end{aligned}$$

So, we find that $\sin 2A = 2 \sin A \cos A$, which is called the **sine double-angle identity**.

We can derive the double-angle identity for $\cos 2A$ in a similar manner using the cosine sum identity.

$$\begin{aligned}
\cos 2A = \cos(A + A) &= \cos A \cos A - \sin A \sin A \\
&= \cos^2 A - \sin^2 A
\end{aligned}$$

So, we find that $\cos 2A = \cos^2 A - \sin^2 A$, which is called the **cosine double-angle identity**.

We can rewrite this identity two different ways.

$$\begin{aligned}
\cos 2A &= \cos^2 A - \sin^2 A \\
&= (1 - \sin^2 A) - \sin^2 A && \cos^2 A = 1 - \sin^2 A \\
&= 1 - 2 \sin^2 A
\end{aligned}$$

$$\begin{aligned}
\cos 2A &= \cos^2 A - \sin^2 A \\
&= \cos^2 A - (1 - \cos^2 A) && \sin^2 A = 1 - \cos^2 A \\
&= 2 \cos^2 A - 1
\end{aligned}$$

Therefore, the cosine double-angle identity has *three* forms. The derivation of the tangent double-angle identity, which is included in the following summary, is left as an exercise.

Double-Angle Identities (Formulas)

$$\sin 2A = 2\sin A\cos A$$

$$\begin{aligned}\cos 2A &= \cos^2 A - \sin^2 A\\ &= 1 - 2\sin^2 A\\ &= 2\cos^2 A - 1\end{aligned}$$

$$\tan 2A = \frac{2\tan A}{1 - \tan^2 A}$$

Double-angle identities for $\csc 2A$, $\sec 2A$, and $\cot 2A$ are usually defined in terms of the respective reciprocal of the above identities.

Using Double-Angle Identities to Find Exact Functional Values

If we know one functional value of an angle and the quadrant that contains the angle, we now can use the double-angle identities to find exact functional values of double the angle and the quadrant that contains this double-angle.

EXAMPLE 1 If $\sin x = -\frac{4}{5}$ with x in QIII, find the following.

a. $\sin 2x$ **b.** $\cos 2x$ **c.** $\tan 2x$ **d.** the quadrant that contains $2x$

SOLUTION

a. Since $\sin 2x = 2\sin x\cos x$, we need both $\sin x$ and $\cos x$. We use the Pythagorean identity to find $\cos x$.

$$\cos^2 x + \sin^2 x = 1$$

$$\cos^2 x + \left(-\frac{4}{5}\right)^2 = 1 \qquad \sin x = -\tfrac{4}{5}$$

$$\cos^2 x = \frac{9}{25}$$

$$\cos x = \pm\frac{3}{5}$$

Since x is in QIII, $\cos x = -\frac{3}{5}$. We substitute these values into the double-angle identity.

$$\sin 2x = 2\sin x\cos x$$

$$= 2\left(-\frac{4}{5}\right)\left(-\frac{3}{5}\right) = \frac{24}{25} \qquad \text{Substitute } \sin x = -\tfrac{4}{5} \text{ and } \cos x = -\tfrac{3}{5}.$$

Thus, $\sin 2x = 24/25$.

b. We can use any of the three equivalent forms of identities for $\cos 2x$. Here, we select $\cos 2x = \cos^2 x - \sin^2 x$.

$$\begin{aligned} \cos 2x &= \cos^2 x - \sin^2 x \\ &= \left(-\frac{3}{5}\right)^2 - \left(-\frac{4}{5}\right)^2 && \text{Substitute } \sin x = -\tfrac{4}{5}, \text{ and from part (a), } \cos x = -\tfrac{3}{5}. \\ &= \frac{9}{25} - \frac{16}{25} \\ &= -\frac{7}{25} \end{aligned}$$

So, $\cos 2x = -7/25$.

c. There are two ways we can find $\tan 2x$, but since we already know $\sin 2x$ and $\cos 2x$, we use the following.

$$\tan 2x = \frac{\sin 2x}{\cos 2x} = \frac{\frac{24}{25}}{-\frac{7}{25}} = -\frac{24}{7}$$

You will be asked in an exercise to verify this value using the identity $\tan 2x = \dfrac{2\tan x}{1 - \tan^2 x}$.

d. Since $\sin 2x > 0$, $\cos 2x < 0$, $2x$ is in QII. ■

In the next example, we reverse the question asked in Example 1. That is, if we know the functional value and the quadrant of a double-angle, we can find exact functional values of the single-angle.

EXAMPLE 2 If $\cos 2x = -\frac{12}{13}$ and $2x$ is in QIII (that is, $180° < 2x < 270°$), find $\cos x$.

SOLUTION We use the double-angle identity for the $\cos 2x$ that involves only $\cos x$.

$$\begin{aligned} 2\cos^2 x - 1 &= \cos 2x \\ 2\cos^2 x - 1 &= -\frac{12}{13} && \cos 2x = -\tfrac{12}{13} \\ 2\cos^2 x &= -\frac{12}{13} + 1 && \text{Add 1 to both sides.} \\ 2\cos^2 x &= \frac{1}{13} && \text{Simplify.} \\ \cos^2 x &= \frac{1}{26} && \text{Multiply both sides by } \tfrac{1}{2}. \\ \cos x &= \pm\sqrt{\frac{1}{26}} = \pm\frac{\sqrt{26}}{26} && \text{Take the square root both sides.} \end{aligned}$$

Next, we find the quadrant in which x lies so as to determine the positive or negative nature of $\cos x$. Since $2x$ is in QIII we get:

$$\begin{aligned} 180° &< 2x < 270° \\ \frac{180°}{2} &< \frac{2x}{2} < \frac{270°}{2} && \text{Divide each term by 2 and simplify.} \\ 90° &< x < 135° \end{aligned}$$

So x is in QII, where $\cos x < 0$. Therefore, $\cos x = -\dfrac{\sqrt{26}}{26}$. ■

Using Double-Angle Identities to Rewrite Expressions

EXAMPLE 3 Use the double-angle identities to rewrite each expression as a single function of a multiple angle.

a. $2 \sin 3x \cos 3x$ **b.** $\cos^2 \frac{x}{2} - \sin^2 \frac{x}{2}$ **c.** $4 \cos^2 5x - 2$

SOLUTION

a. If $A = 3x$, then $2 \sin 3x \cos 3x$ is the left side of the sine double-angle identity.

$$\begin{aligned} 2 \sin A \cos A &= \sin 2A \\ 2 \sin(3x)\cos(3x) &= \sin 2(3x) \qquad A = 3x \\ &= \sin 6x \end{aligned}$$

Therefore, $2 \sin 3x \cos 3x = \sin 6x$.

b. If $A = \frac{x}{2}$, then $\cos^2 \frac{x}{2} - \sin^2 \frac{x}{2}$ is the left side of the cosine double-angle identity.

$$\begin{aligned} \cos^2 A - \sin^2 A &= \cos 2A \\ \cos^2\left(\frac{x}{2}\right) - \sin^2\left(\frac{x}{2}\right) &= \cos 2\left(\frac{x}{2}\right) \qquad A = \frac{x}{2} \\ &= \cos x \end{aligned}$$

Therefore, $\cos^2 \frac{x}{2} - \sin^2 \frac{x}{2} = \cos x$.

c. If we factor $4 \cos^2 5x - 2 = 2(2 \cos^2 5x - 1)$ and let $A = 5x$, we have twice the cosine double-angle identity.

$$\begin{aligned} 2 \cos^2 A - 1 &= \cos 2A \\ 2 \cos^2(5x) - 1 &= \cos 2(5x) \qquad A = 5x \\ &= \cos 10x \end{aligned}$$

Therefore, $4 \cos^2 5x - 2 = 2(2 \cos^2 5x - 1) = 2(\cos 10x) = 2 \cos 10x$. ■

Double-angle identities not only allow us to rewrite some expressions as a single function of a multiple angle, they also allow us to rewrite multiple angles in terms of single angles.

EXAMPLE 4 Rewrite each expression as a trigonometric function of x.

a. $\frac{1 - \cos 2x}{\sin 2x}$ **b.** $\cos 3x$

SOLUTION

a. $\dfrac{1-\cos 2x}{\sin 2x} = \dfrac{1-(1-2\sin^2 x)}{2\sin x\cos x}$ Write $\cos 2x$ and $\sin 2x$ as functions of x.

$= \dfrac{2\sin^2 x}{2\cos x\sin x}$ Simplify the numerator.

$= \dfrac{\sin x}{\cos x}$ Reduce to lowest terms.

$= \tan x$

b. $\cos 3x = \cos(2x+x) = \cos 2x\cos x - \sin 2x\sin x$ Use the cosine sum formula.

$= (2\cos^2 x - 1)\cos x - (2\sin x\cos x)\sin x$ Use the double-angle identities.

$= 2\cos^3 x - \cos x - 2\sin^2 x\cos x$ Simplify.

$= 2\cos^3 x - \cos x - 2(1-\cos^2 x)\cos x$ $\sin^2 x = 1 - \cos^2 x$

$= 4\cos^3 x - 3\cos x$ Simplify. ■

EXAMPLE 5 Prove the identity $(\cos x - \sin x)^2 + \sin 2x = 1$.

SOLUTION

$$(\cos x - \sin x)^2 + \sin 2x = 1$$
$$\cos^2 x - 2\cos x\sin x + \sin^2 x + 2\sin x\cos x =$$
$$\cos^2 x + \sin^2 x =$$
$$1 = \;\downarrow$$ ■

It will require a lot of practice for you to become familiar with the forms and uses of the double-angle identities. Are you ready?

Exercise Set 4.4

1–10. *Match each expression with the equivalent form from Column A. You may use an expression from Column A more than once.*

1. $\cos^2 x - \sin^2 x$
2. $2\cos^2 x - 1$
3. $2 - 4\sin^2 x$
4. $2\sin 5x\cos 5x$
5. $6\sin 3x\cos 3x$
6. $3 - 6\sin^2 x$
7. $2\cos^2 8x - 1$
8. $\cos^2 6x - \sin^2 6x$
9. $\dfrac{2\tan x}{1-\tan^2 x}$
10. $\dfrac{2\tan 2x}{1-\tan^2 2x}$

COLUMN A

$2\cos 2x$

$3\cos 2x$

$\tan 2x$

$\cos 16x$

$\sin 10x$

$3\sin 6x$

$\cos 12x$

$\tan 4x$

$\cos 2x$

11–20. *Use a double-angle identity to find the exact value. Check your answer with a calculator. Be aware of the radian or degree setting.*

EXAMPLE $2\cos^2 22.5° - 1$

```
2(cos(22.5))²-1
 .7071067812
√(2)/2
 .7071067812
```

SOLUTION Using $2\cos^2 A - 1 = \cos 2A$, we let $A = 22.5°$.

$$2\cos^2 22.5° - 1 = \cos 2(22.5°) = \cos 45° = \frac{\sqrt{2}}{2}$$

11. $2\sin 15° \cos 15°$

12. $\cos^2 22\tfrac{1}{2}° - \sin^2 22\tfrac{1}{2}°$

13. $1 - 2\sin^2 105°$

14. $2\cos^2 75° - 1$

15. $\dfrac{2\tan 75°}{1 - \tan^2 75°}$

16. $4\sin\dfrac{\pi}{8}\cos\dfrac{\pi}{8}$

17. $6\cos^2\dfrac{\pi}{12} - 3$

18. $5 - 10\sin^2\dfrac{3\pi}{8}$

19. $16\cos^2\dfrac{\pi}{8} - 16\sin^2\dfrac{\pi}{8}$

20. $\dfrac{\tan 22.5°}{2 - 2\tan^2 22.5°}$

21–30. *Find the exact functional value.*

EXAMPLE If $\sin A = \frac{5}{13}$ with $90° < A < 180°$, find $\cos 2A$.

SOLUTION

$$\cos 2A = 1 - 2\sin^2 A$$

$$= 1 - 2\left(\frac{5}{13}\right)^2 = 1 - 2\left(\frac{25}{169}\right) = 1 - \frac{50}{169}$$

$$= \frac{119}{169}$$

The other choices for $\cos 2A$ would require us to first find $\cos A$. Using this form of the identity saves us some work.

21. If $\cos A = \frac{3}{5}$ with A in QIV, find $\sin 2A$.

22. If $\sin B = -\frac{12}{13}$ with B in QIII, find $\cos 2B$.

23. If $\sin x = \frac{24}{25}$ and $\cot x < 0$, find $\cos 2x$.

24. If $\cos x = -\frac{7}{25}$ and $\tan x > 0$, find $\sin 2x$.

25. If $\tan A = \frac{4}{3}$, find $\tan 2A$.

26. If $\tan A = -\frac{3}{4}$, find $\tan 2A$.

27. If $\cos 2\beta = \frac{5}{13}$ and $270° < 2\beta < 360°$, find $\sin \beta$.

28. If $\sin 2\alpha = -\frac{8}{17}$ and $180° < 2\alpha < 270°$, find $\cos \alpha$.

29. If $\tan\theta = \frac{12}{5}$ with θ in QIII, find $\sec 2\theta$.

30. If $\csc\theta = 3$ with θ in QII, find $\cot 2\theta$.

31–42. *Prove the following identities.*

31. $\sin 2x = \dfrac{2\tan x}{1 + \tan^2 x}$

32. $\cos 2x = \dfrac{2\cot^2 x - \csc^2 x}{1 + \cot^2 x}$

33. $\cot\theta = \dfrac{\sin 2\theta}{1 - \cos 2\theta}$

34. $\cos 2\alpha = \dfrac{1 - \tan^2\alpha}{1 + \tan^2\alpha}$

35. $\cos^2 A = \dfrac{1 + \cos 2A}{2}$

36. $\sin^2 A = \dfrac{1 - \cos 2A}{2}$

37. $1 = 2\sin^2 x + \cos 2x$

38. $\csc x \sin 2x = 2\cos x$

39. $\sin 3x = 3\sin x - 4\sin^3 x$
(*Hint:* $3x = 2x + x$)

40. $\cos 4x = 1 - 8\sin^2 x \cos^2 x$
(*Hint:* $4x = 2x + 2x$)

41. $(\sin x - \cos x)^2 = 1 - \sin 2x$

42. $\tan x = \dfrac{\sin 2x}{1 + \cos 2x}$

43–50. *Are the following statements true or false? If a statement is false, explain why or give an example that shows why it is false.*

43. $\tan(x + y) = \tan x + \tan y$

44. $\sin(a + b) = \sin a \sin b + \cos a \cos b$

45. $\cos(x - y) = \cos x \cos y - \sin x \sin y$

46. $(\cos x - \sin x)(\cos x + \sin x) = \cos 2x$

47. $\cos^2 3y - \sin^2 3y = \cos 6y$

48. $\cos 2x = 1 - 2\cos^2 x$

49. $\dfrac{\sin 2x}{2} = \sin x$

50. $-\tan(2x) = \tan(-2x)$

51. Derive the tangent double-angle identity: $\tan 2x = \dfrac{2\tan x}{1 - \tan^2 x}$.

52. In this section, Example 1(c) finds the value for $\tan 2x = -\dfrac{24}{7}$ by using $\tan 2x = \dfrac{\sin 2x}{\cos 2x}$. Verify that you obtain the same result if you use the tangent-double angle identity, $\tan 2x = \dfrac{2\tan x}{1 - \tan^2 x}$.

53. A security device located on the floor 6 feet away from a gallery wall is angled so that it projects a beam that reaches a height of 4 feet along the wall. What will be the height along the wall that the beam will reach if the angle is doubled?

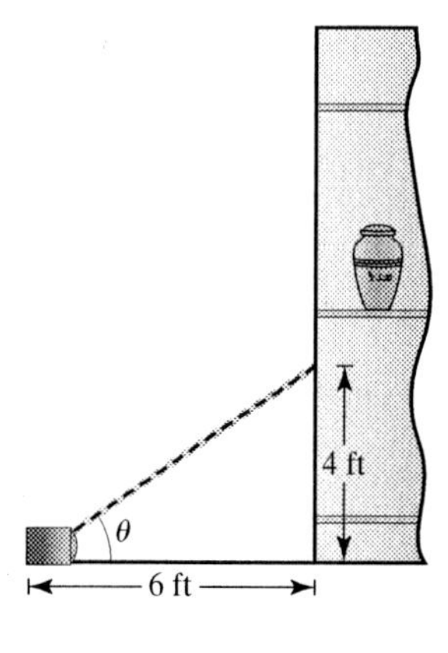

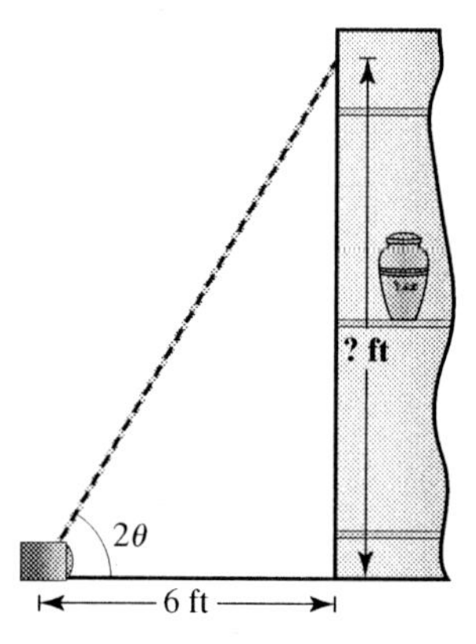

54. A cargo boat, anchored at a horizontal distance of 8 feet to the seawall alongside the dock, needs a 10-foot ramp to reach the top of the dock so it can unload its freight. If

the water level drops so that the angle of elevation from the boat to the top of the dock is doubled, what is the length of the ramp (to the nearest foot) necessary to reach the top of the dock?

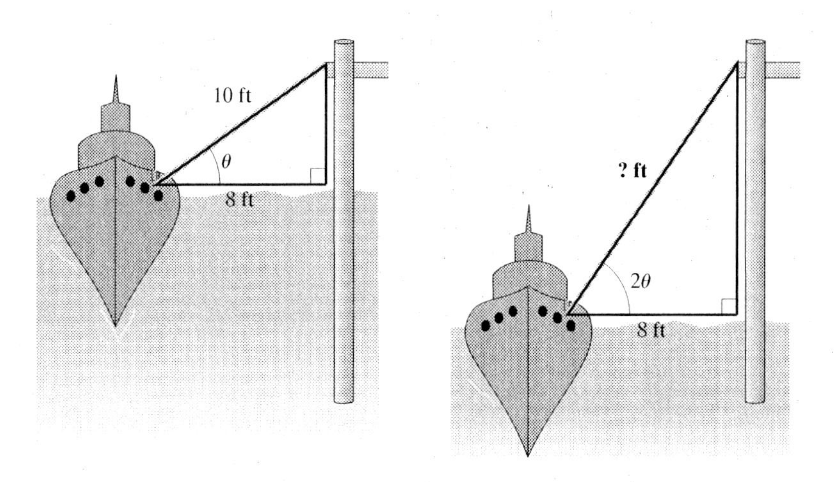

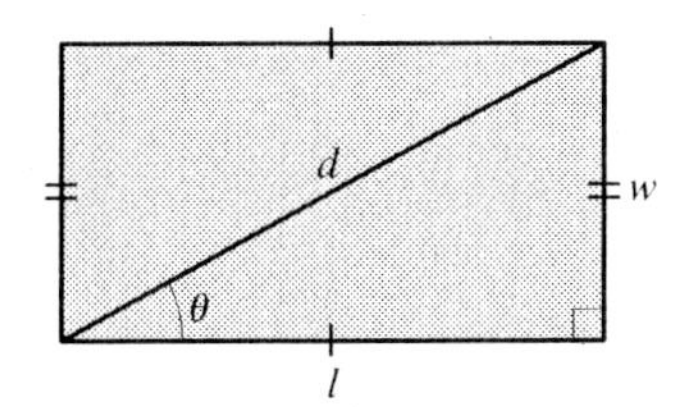

55. Show that the area of a rectangle can be expressed as $A(\theta) = \frac{d^2}{2}\sin(2\theta)$, where d is the diagonal and θ is the angle the diagonal makes with one of the sides.

56. Use the formula for the area of a rectangle given Exercise 55 to find the angle θ that would yield the greatest area A.

Graphing Calculator/CAS Exercises

57–60. *Graph the equation y_1. (MODE: radian; WINDOW: Xmin $= -\pi$, Xmax $= \pi$, Ymin $= -2$, Ymax $= 2$) Then find a simpler equation y_2 for the graph in the form of a single function of a multiple angle. Graph y_2 to see if the graphs coincide. If so, prove the identity $y_1 = y_2$.*

57. $y_1 = \cos^4 x - \sin^4 x$

58. $y_1 = \dfrac{\tan x}{1 - \sec^2 x}$

59. $y_1 = \dfrac{1 - \tan^2 x}{1 + \tan^2 x}$

60. $y_1 = \dfrac{1 + \cos 2x}{\sin 2x}$

4.5 Half-Angle and Additional Identities

We know one result of identities from previous sections of this chapter is that we are able to find many more exact functional values: We have found exact functional values of the sum of two angles (or arcs), the difference of two angles, or double an angle. In this section we find exact functional values of half the angle.

Half-Angle Identities (Formulas)

We consider two forms of the cosine double-angle identity and solve it for the indicated function of angle A.

Solve for $\sin A$.	Solve for $\cos A$.
$\cos 2A = 1 - 2\sin^2 A$	$\cos 2A = 2\cos^2 A - 1$
$2\sin^2 A = 1 - \cos 2A$	$2\cos^2 A = \cos 2A + 1$
$\sin^2 A = \dfrac{1 - \cos 2A}{2}$	$\cos^2 A = \dfrac{1 + \cos 2A}{2}$
or	or
$\sin A = \pm\sqrt{\dfrac{1 - \cos 2A}{2}}$	$\cos A = \pm\sqrt{\dfrac{1 + \cos 2A}{2}}$

Notice these equivalent forms enable us to write second-degree terms, $\sin^2 A$ and $\cos^2 A$, as first-degree terms involving $\cos 2A$.

If we let $2A = x$, then $A = \frac{x}{2}$. Using this substitution we obtain the half-angle identities (formulas) for the sine and cosine. Half-angle identities for the tangent can be derived using sine and cosine half-angle identities and are left as an exercise. The positive or negative value is determined by the quadrant of the half-angle.

Half-Angle Identities (Formulas)

$$\sin^2\frac{x}{2} = \frac{1 - \cos x}{2} \quad \longleftrightarrow \quad \sin\frac{x}{2} = \pm\sqrt{\frac{1 - \cos x}{2}}$$

$$\cos^2\frac{x}{2} = \frac{1 + \cos x}{2} \quad \longleftrightarrow \quad \cos\frac{x}{2} = \pm\sqrt{\frac{1 + \cos x}{2}}$$

$$\tan\frac{x}{2} = \pm\sqrt{\frac{1 - \cos x}{1 + \cos x}} = \frac{\sin x}{1 + \cos x} = \frac{1 - \cos x}{\sin x}$$

Using Half-Angle Identities to Find Exact Functional Values

We can use the half-angle identities to find exact functional values of angles that are half of given angles with known exact functional values. For example, we can find exact functional values of a $22\frac{1}{2}°$ angle because it is half of a $45°$ angle.

EXAMPLE 1 Use a half-angle identity to find the exact value.

a. $\cos\dfrac{\pi}{12}$ **b.** $\sin\dfrac{9\pi}{8}$ **c.** $\tan 22\frac{1}{2}°$

SOLUTION

a. For $\cos\dfrac{\pi}{12}$, let $\dfrac{x}{2} = \dfrac{\pi}{12}$. So, $x = \dfrac{\pi}{12} \cdot 2 = \dfrac{\pi}{6}$.

Because $\pi/12$ is in QI and $\cos \pi/12 > 0$, we choose the positive value.

$$\cos\frac{x}{2} = \sqrt{\frac{1+\cos x}{2}}$$

$$\cos\frac{\pi}{12} = \cos\frac{\left(\frac{\pi}{6}\right)}{2} = \sqrt{\frac{1+\cos\frac{\pi}{6}}{2}} = \sqrt{\frac{1+\frac{\sqrt{3}}{2}}{2}} \qquad \cos\frac{\pi}{6} = \frac{\sqrt{3}}{2}$$

$$= \sqrt{\frac{\left(1+\frac{\sqrt{3}}{2}\right)\cdot 2}{2\cdot 2}} = \sqrt{\frac{2+\sqrt{3}}{4}}$$

$$= \frac{\sqrt{2+\sqrt{3}}}{2}$$

NOTE: This is the same problem as Example 2 (a) in Section 4.2 in which we used the cosine difference formula to find $\cos \pi/12$. Use your calculator to verify that although the answers look different, they are equal.

b. For $\sin\frac{9\pi}{8}$, let $\frac{x}{2} = \frac{9\pi}{8}$. So, $x = \frac{9\pi}{4}$.

Because $9\pi/8$ is in QIII and $\sin 9\pi/8 < 0$, we choose the negative value.

$$\sin\frac{x}{2} = -\sqrt{\frac{1-\cos x}{2}}$$

$$\sin\frac{9\pi}{8} = \sin\frac{\left(\frac{9\pi}{4}\right)}{2} = -\sqrt{\frac{1-\cos\frac{9\pi}{4}}{2}} \qquad \cos\frac{9\pi}{4} = \frac{\sqrt{2}}{2}$$

$$= -\sqrt{\frac{1-\frac{\sqrt{2}}{2}}{2}} = -\sqrt{\frac{\left(1-\frac{\sqrt{2}}{2}\right)2}{2\cdot 2}}$$

$$= -\frac{\sqrt{2-\sqrt{2}}}{2}$$

c. For $\tan 22\frac{1}{2}^\circ$, $\frac{x}{2} = 22\frac{1}{2}^\circ$. Therefore, $x = 45^\circ$. Any one of the three choices of the tangent half-angle identities will work, but the two forms that do not involve a radical may be easier to use and often produce less complicated forms of the answer. So we select $\tan\frac{x}{2} = \frac{\sin x}{1+\cos x}$.

$$\tan 22\tfrac{1}{2}^\circ = \tan\frac{45^\circ}{2} = \frac{\sin 45^\circ}{1+\cos 45^\circ}$$

$$= \frac{\frac{\sqrt{2}}{2}}{1+\frac{\sqrt{2}}{2}} = \frac{\left(\frac{\sqrt{2}}{2}\right)\cdot 2}{\left(1+\frac{\sqrt{2}}{2}\right)\cdot 2} = \frac{\sqrt{2}}{2+\sqrt{2}} \qquad \text{Substitute exact functional values and simplify.}$$

(continued)

$$= \frac{\sqrt{2}}{2+\sqrt{2}} \cdot \frac{2-\sqrt{2}}{2-\sqrt{2}}$$ Rationalize the denominator.

$$= \frac{2\sqrt{2}-2}{2} = \frac{2(\sqrt{2}-1)}{2}$$ Factor the numerator.

$$= \sqrt{2}-1$$ Reduce to lowest terms. ■

In the next example, we are given the exact functional value of an angle and the quadrant and asked to find all six trigonometric functional values of the half-angle.

EXAMPLE 2 If $\cos x = \frac{3}{5}$ and x is in QIV, find all trigonometric functional values of $x/2$.

SOLUTION We start by finding $\cos \frac{x}{2}$ and $\sec \frac{x}{2}$.

$$\cos \frac{x}{2} = \pm\sqrt{\frac{1+\cos x}{2}}$$

$$= \pm\sqrt{\frac{1+\frac{3}{5}}{2}} = \pm\sqrt{\frac{\left(1+\frac{3}{5}\right)\cdot 5}{2\cdot 5}}$$ $\cos x = \frac{3}{5}$

$$= \pm\sqrt{\frac{8}{10}} = \pm\frac{2}{\sqrt{5}}$$

$$= \pm\frac{2\sqrt{5}}{5}$$

Next we find the quadrant in which $x/2$ lies.

Since x is in QIV, we get:

$$270° < x < 360°$$

$$\frac{270°}{2} < \frac{x}{2} < \frac{360°}{2}$$ Divide by 2.

$$135° < \frac{x}{2} < 180° \rightarrow \frac{x}{2} \text{ is in QII, so } \cos\frac{x}{2} < 0.$$

Therefore,

$$\cos\frac{x}{2} = -\frac{2\sqrt{5}}{5} \quad \text{and} \quad \sec\frac{x}{2} = -\frac{5}{2\sqrt{5}} = -\frac{\sqrt{5}}{2}.$$

Since we just found $\cos x/2$, we have several ways to find the $\sin x/2$. We can use either the Pythagorean identity, the trigonometric ratios, or the sine half-angle identity. We use the sine half-angle identity here since you have had less practice with it. Because $x/2$ is in QII where $\sin x/2 > 0$, we choose the positive value.

$$\sin\frac{x}{2} = \sqrt{\frac{1-\cos x}{2}}$$

$$= \sqrt{\frac{1-\frac{3}{5}}{2}} = \sqrt{\frac{\left(1-\frac{3}{5}\right)\cdot 5}{2\cdot 5}} \qquad \cos x = \frac{3}{5}$$

$$= \sqrt{\frac{2}{10}} = \sqrt{\frac{1}{5}}$$

$$= \frac{1}{\sqrt{5}} = \frac{\sqrt{5}}{5}$$

Therefore,

$$\sin\frac{x}{2} = \frac{\sqrt{5}}{5} \quad \text{and} \quad \csc\frac{x}{2} = \sqrt{5}.$$

Next, we find $\tan x/2$ and $\cot x/2$ by using the basic identities.

$$\tan\frac{x}{2} = \frac{\sin\frac{x}{2}}{\cos\frac{x}{2}} = \frac{\frac{\sqrt{5}}{5}}{-\frac{2\sqrt{5}}{5}} = \frac{\left(\frac{\sqrt{5}}{5}\right)\cdot 5}{\left(-\frac{2\sqrt{5}}{5}\right)\cdot 5} = \frac{\sqrt{5}}{-2\sqrt{5}} = -\frac{1}{2}$$

$$\cot\frac{x}{2} = -2$$

■

NOTE: To get more practice with the half-angle identities in the previous example, we can also use a half-angle identity to find $\tan x/2$.

$$\tan\frac{x}{2} = \frac{1-\cos x}{\sin x} \qquad \cos x = \frac{3}{5} \text{ and } x \text{ in QIV} \rightarrow \sin x = -\frac{4}{5}$$

$$= \frac{1-\cos x}{\sin x}$$

$$= \frac{1-\frac{3}{5}}{-\frac{4}{5}} = \frac{\left(1-\frac{3}{5}\right)\cdot 5}{\left(-\frac{4}{5}\right)\cdot 5} = -\frac{1}{2}$$

Using Half-Angle Identities to Prove Identities and Simplify Expressions

EXAMPLE 3 Prove the identity $\tan\frac{x}{2} + \cot\frac{x}{2} = 2\csc x$.

SOLUTION We choose to work the left side for two reasons. First, it is the most complicated side. Second and most importantly, we have identities for rewriting

$\tan x/2$ in terms of functions of x. Since we have $\tan x/2$ and $\cot x/2 = \dfrac{1}{\tan x/2}$, we select two different identities for $\tan x/2$ in order to obtain common denominators. (Again, we are *looking where we are going:* we want a denominator of $\sin x$.)

$$\tan\frac{x}{2} + \cot\frac{x}{2} = 2\csc x$$

$$\frac{1-\cos x}{\sin x} + \frac{1}{\dfrac{\sin x}{1+\cos x}} =$$

$$\frac{1-\cos x}{\sin x} + \frac{1+\cos x}{\sin x} =$$

$$\frac{2}{\sin x} =$$

$$2\csc x =$$

■

EXAMPLE 4 Prove the identity $2\sin^2 x + \cos 2x = 1$.

SOLUTION We choose the left side for obvious reasons and use the half-angle identity that rewrites a second-degree function, $\sin^2 x$, in terms of a first-degree function, $\cos 2x$ (called power reducing).

$$2\sin^2 x + \cos 2x = 1$$

$$2\left(\frac{1-\cos 2x}{2}\right) + \cos 2x =$$

$$1 - \cos 2x + \cos 2x =$$

$$1 =$$

■

In addition to proving identities, the half-angle identities help to simplify expressions.

EXAMPLE 5 Simplify the following expressions by rewriting each as a single trigonometric function and using the half-angle identities.

a. $\sqrt{\dfrac{1-\cos 12°}{2}}$ **b.** $\pm\sqrt{\dfrac{1+\cos 6\theta}{2}}$ **c.** $\sqrt{\dfrac{1-\cos\dfrac{\pi}{14}}{1+\cos\dfrac{\pi}{14}}}$

SOLUTION

a. We use the sine half-angle identity with $x = 12°$. Thus, $x/2 = 6°$.

$$\pm\sqrt{\frac{1-\cos x}{2}} = \sin\frac{x}{2}$$

$$\sqrt{\frac{1-\cos 12°}{2}} = \sin 6° \qquad \text{Since } 6° \text{ is in QI, } \sin 6° > 0.$$

b. We use the cosine half-angle identity with $x = 6\theta$ and $x/2 = 3\theta$.

$$\pm\sqrt{\frac{1+\cos x}{2}} = \cos\frac{x}{2}$$

$$\pm\sqrt{\frac{1+\cos 6\theta}{2}} = \cos 3\theta$$

NOTE: Since the quadrant for 3θ is unknown, we retain the $\pm$ notation.

c. We use the tangent half-angle identity with $x = \frac{\pi}{14}$ and $\frac{x}{2} = \frac{\pi}{14}\cdot\frac{1}{2} = \frac{\pi}{28}$.

$$\pm\sqrt{\frac{1-\cos x}{1+\cos x}} = \tan\frac{x}{2}$$

$$\sqrt{\frac{1-\cos\frac{\pi}{14}}{1+\cos\frac{\pi}{14}}} = \tan\frac{\pi}{28}$$

Since $\frac{\pi}{28}$ is in QI, $\tan\frac{\pi}{28} > 0$.

Product-to-Sum and Sum-to-Product Identities

The following identities can be derived from the sum and difference formulas for the sine and cosine functions, and the proofs are left as exercises. Products of cosine values and/or sine values can be rewritten as the sums or differences of cosine or sine values as a result of these identities. Thus, these identities are usually called **product-to-sum identities**.

Product-to-Sum Identities (Formulas)

$$\cos A\cos B = \tfrac{1}{2}[\cos(A+B) + \cos(A-B)]$$
$$\sin A\sin B = \tfrac{1}{2}[\cos(A-B) - \cos(A+B)]$$
$$\sin A\cos B = \tfrac{1}{2}[\sin(A+B) + \sin(A-B)]$$
$$\cos A\sin B = \tfrac{1}{2}[\sin(A+B) - \sin(A-B)]$$

EXAMPLE 6 Write $\cos 2x \sin 8x$ as a sum or difference.

SOLUTION Use $\cos A\sin B = \frac{1}{2}[\sin(A+B) - \sin(A-B)]$ with $A = 2x$ and $B = 8x$.

$$\begin{aligned}\cos 2x\sin 8x &= \tfrac{1}{2}[\sin(2x+8x) - \sin(2x-8x)]\\ &= \tfrac{1}{2}[\sin 10x - \sin(-6x)]\\ &= \tfrac{1}{2}[\sin 10x + \sin 6x]\end{aligned}$$

$\sin(-x) = -\sin x$

We also have identities that allow us to do just the opposite: rewrite sums or differences as products. Again, the proofs of these identities, which are usually called **sum-to-product identities**, are left as exercises.

Sum-to-Product Identities (Formulas)

$$\cos A + \cos B = 2\cos\left(\frac{A+B}{2}\right)\cos\left(\frac{A-B}{2}\right) \qquad \sin A + \sin B = 2\sin\left(\frac{A+B}{2}\right)\cos\left(\frac{A-B}{2}\right)$$

$$\cos A - \cos B = -2\sin\left(\frac{A+B}{2}\right)\sin\left(\frac{A-B}{2}\right) \qquad \sin A - \sin B = 2\sin\left(\frac{A-B}{2}\right)\cos\left(\frac{A+B}{2}\right)$$

In this section we discussed identities that can be used to write products as sums or sums as products. We also discussed half-angle identities, some of which can be used to reduce the power of a function. You need to look carefully at the identities to see what is accomplished by replacing one side of the identity with an equivalent form (the other side). Although you may not need to memorize all of the identities from this section, you should memorize the half-angle identities and be aware of the other identities and how they work.

Exercise Set 4.5

1–10. *Use the half-angle identities to find the exact value of each expression.*

1. $\cos 22\frac{1}{2}°$ **2.** $\sin 22\frac{1}{2}°$ **3.** $\sin 195°$ **4.** $\cos 202.5°$

5. $\tan 165°$ **6.** $\tan 67.5°$ **7.** $\cos\frac{\pi}{8}$ **8.** $\sin\frac{\pi}{12}$

9. $\sin^2\frac{5\pi}{12}$ **10.** $\tan^2\frac{9\pi}{8}$

11–20. *Match the expression with the equivalent form from Column A using half-angle identities.*

11. $\sqrt{\dfrac{1+\cos\frac{\pi}{7}}{2}}$ **12.** $\sqrt{\dfrac{1-\cos\frac{\pi}{7}}{2}}$

13. $-\sqrt{\dfrac{1-\cos 488°}{2}}$ **14.** $-\sqrt{\dfrac{1+\cos 330°}{2}}$

15. $\pm\sqrt{\dfrac{1+\cos x}{2}}$ **16.** $\pm\sqrt{\dfrac{1-\cos x}{2}}$

17. $\sin^2 x$ **18.** $\cos^2 x$

19. $\dfrac{\sin 20°}{1+\cos 20°}$ **20.** $\sqrt{\dfrac{1-\cos 18°}{1+\cos 18°}}$

Column A

$\cos\frac{x}{2}$

$\sin\frac{\pi}{14}$

$\cos\frac{\pi}{14}$

$\sin\frac{x}{2}$

$\tan 10°$

$\dfrac{1+\cos 2x}{2}$

$\tan 9°$

$\dfrac{1-\cos 2x}{2}$

$\cos 165°$

$\sin 244°$

21–28. *Find the exact functional value.*

21. If $\cos x = -\frac{4}{5}$ and x is in QII, find the following.

a. $\cos\frac{x}{2}$ **b.** $\sin\frac{x}{2}$ **c.** $\tan\frac{x}{2}$

22. If $\cos x = \frac{12}{13}$ and x is in QIV, find the following.

a. $\cos\frac{x}{2}$ **b.** $\sin\frac{x}{2}$ **c.** $\tan\frac{x}{2}$

23. If $\sin x = -\frac{1}{2}$ and x is in QIII, find the following.

a. $\cos\frac{x}{2}$ **b.** $\sin^2\frac{x}{2}$ **c.** $\tan\frac{x}{2}$

24. If $\sin x = -\frac{3}{5}$ and x is in QIV, find the following.

a. $\cos\frac{x}{2}$ **b.** $\sin^2\frac{x}{2}$ **c.** $\tan\frac{x}{2}$

25. If $\tan A = 2$ and $\sin A > 0$, find $\cos\frac{A}{2}$.

26. If $\tan B = -2$ and $\cos B < 0$, find $\sin\frac{B}{2}$.

27. If $\sin\frac{B}{2} = \frac{\sqrt{3}}{3}$, find $\cos B$.

28. If $\cos\frac{x}{2} = \frac{3}{7}$, find $\cos x$.

29–44. *If* $\sin A = \frac{3}{5}$ *with* $\frac{\pi}{2} < A < \pi$ *and* $\cos B = \frac{1}{2}$ *with* $\sin B > 0$, *find the exact value of each expression.*

29. $\cos A$

30. $\sin B$

31. $\sin\frac{A}{2}$

32. $\cos\frac{A}{2}$

33. $\tan\frac{A}{2}$

34. $\csc\frac{A}{2}$

35. $\cos 2A$

36. $\sin 2A$

37. $\cos^2\frac{B}{2}$

38. $\sin^2\frac{B}{2}$

39. $\sin(A - B)$

40. $\cos(A - B)$

41. $\cos(A + B)$

42. $\sin(A + B)$

43. $\tan(A + B)$

44. $\tan(A - B)$

45–54. *Prove the following identities.*

45. $\sin^2\frac{\theta}{2} = \frac{\csc\theta - \cot\theta}{2\csc\theta}$

46. $\csc 2B = \frac{\csc B}{2\cos B}$

47. $\left(\cos\frac{\beta}{2} - \sin\frac{\beta}{2}\right)^2 = 1 - \sin\beta$

48. $\tan\frac{x}{2}\tan x = \sec x - 1$

49. $\tan\dfrac{x}{2} = \csc x - \cot x$

50. $\sin y - \cos y \tan\dfrac{y}{2} = \tan\dfrac{y}{2}$

51. $1 - \tan^2\dfrac{A}{2} = \dfrac{2\cos A}{1 + \cos A}$

52. $\sec^2 x = \dfrac{2}{1 + \cos 2x}$

53. $\sin^2\dfrac{x}{2} = \dfrac{\sec x - 1}{2\sec x}$

54. $\tan\dfrac{A}{2}\sin A = \dfrac{\tan A - \sin A}{\sin A \sec A}$

55–60. *Write each expression as a sum or difference.*

55. $\sin 16^\circ \cos 44^\circ$

56. $\cos 15^\circ \cos 99^\circ$

57. $\sin 2x \sin 10x$

58. $\sin 19x \cos x$

59. $10\cos 5^\circ \cos(-3^\circ)$

60. $4\sin(-18^\circ)\sin 10^\circ$

61–66. *Write each expression as a product. Simplify if possible.*

61. $\cos 75^\circ + \cos 15^\circ$

62. $\sin 15^\circ - \sin 105^\circ$

63. $\sin 2y + \sin 4y$

64. $\cos 8x - \cos 9x$

65. $\cos 5x - \cos x$

66. $\sin 18y + \sin 20y$

67. Derive each half-angle identity for the tangent.

a. $\tan\dfrac{x}{2} = \sqrt{\dfrac{1 - \cos x}{1 + \cos x}}$

b. $\tan\dfrac{x}{2} = \dfrac{\sin x}{1 + \cos x}$ (*Hint:* Let $\dfrac{x}{2} = A$ and $x = 2A$.)

c. $\tan\dfrac{x}{2} = \dfrac{1 - \cos x}{\sin x}$ (*Hint:* Use the result of part (b).)

68–70. *Prove the following product-to-sum identities.*

EXAMPLE $\cos A \cos B = \frac{1}{2}[\cos(A + B) + \cos(A - B)]$

SOLUTION We select the right side of the identity and use the cosine sum and difference identities.

$$\begin{aligned}\cos A \cos B &= \tfrac{1}{2}[\cos(A + B) + \cos(A - B)] \\ &= \tfrac{1}{2}[(\cos A \cos B - \sin A \sin B) + (\cos A \cos B + \sin A \sin B)] \\ &= \tfrac{1}{2}[2\cos A \cos B] \\ &= \cos A \cos B\end{aligned}$$

68. $\sin A \sin B = \frac{1}{2}[\cos(A - B) - \cos(A + B)]$

69. $\sin A \cos B = \frac{1}{2}[\sin(A + B) + \sin(A - B)]$

70. $\cos A \sin B = \frac{1}{2}[\sin(A + B) - \sin(A - B)]$

71–74. *Prove the following sum-to-product identities.*

71. $\cos A + \cos B = 2\cos\left(\dfrac{A + B}{2}\right)\cos\left(\dfrac{A - B}{2}\right)$

72. $\sin A + \sin B = 2\sin\left(\dfrac{A + B}{2}\right)\cos\left(\dfrac{A - B}{2}\right)$

73. $\cos A - \cos B = -2\sin\left(\dfrac{A + B}{2}\right)\sin\left(\dfrac{A - B}{2}\right)$

74. $\sin A - \sin B = 2\cos\left(\dfrac{A + B}{2}\right)\sin\left(\dfrac{A - B}{2}\right)$

Graphing Calculator/CAS Exercises

75–76. *Match each graph with two equations from Column A. Then prove the expressions on the right side of the equations are equivalent.*

75.

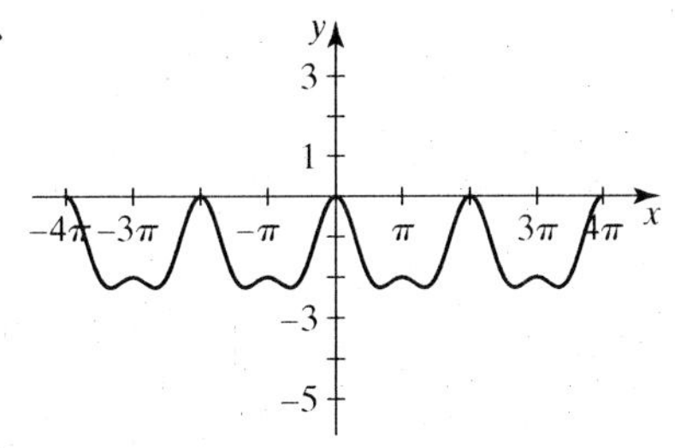

76.

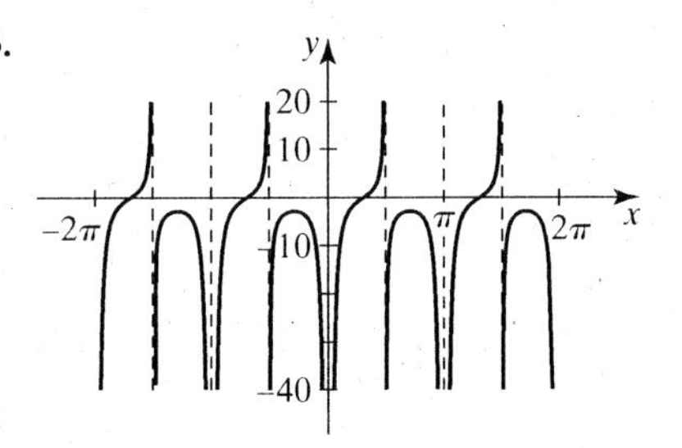

COLUMN A

$y = (\cos x - 1)(\cos x + 2)$

$y = \tan x + \sin 2x$

$y = \cos x + \dfrac{1}{2}\cos 2x - \dfrac{3}{2}$

$y = \tan x - \csc^2 x$

$y = \cot x - \sin^2 \dfrac{x}{2}$

$y = \dfrac{\sin 2x}{2\cos^2 x} - 1 - \cot^2 x$

77–80. *Graph the equation y_1. (MODE: radian; WINDOW: $Xmin = -2\pi$, $Xmax = 2\pi$, $Ymin = -2$, $Ymax = 2$) Then find a simpler equation for the graph in the form of a single function of a multiple angle, y_2. Graph y_2 to see if the graphs coincide. If so, prove the identity $y_1 = y_2$.*

77. $y_1 = 2\cot x + \tan \dfrac{x}{2}$

78. $y_1 = \dfrac{2\tan \frac{x}{2}}{1 + \tan^2 \frac{x}{2}}$

79. $y_1 = \cos^4 \dfrac{x}{2} - \sin^4 \dfrac{x}{2}$

80. $y_1 = \dfrac{1 - \cos 2x}{\tan x}$

Discussion

81. The following identities are sometimes referred to as power-reducing identities.

$$\sin^2 x = \frac{1 - \cos 2x}{2} \qquad \cos^2 x = \frac{1 + \cos 2x}{2} \qquad \tan^2 x = \frac{1 - \cos 2x}{1 + \cos 2x}$$

a. Explain why this reference is justified.

b. In using these power-reducing identities to reduce powers, explain what increases.

82. As stated earlier in this section, we can use the half-angle identities to find exact functional values of angles that are half of angles with known exact functional values. We can use the half-angle identities to find the exact value of cos 15° because 15° is half of 30°, which has known exact functional values. We now can find $\cos 7\frac{1}{2}°$, and so on. Discuss the implications of knowing these functional values if we continue to use sum, difference, and half-angle identities.

Chapter 4 Summary

4.1–4.5 Identities are statements that are true for all values for which they are defined.

Basic Identities (Formulas)

Pythagorean Identity

$$\cos^2 x + \sin^2 x = 1$$

$$\cos^2 x = 1 - \sin^2 x, \qquad \sin^2 x = 1 - \cos^2 x$$

$$1 + \tan^2 x = \sec^2 x, \qquad \cot^2 x + 1 = \csc^2 x$$

Ratio and Reciprocal Identities

$$\tan x = \frac{\sin x}{\cos x}$$

$$\sec x = \frac{1}{\cos x}, \quad \cos x = \frac{1}{\sec x}$$

$$\csc x = \frac{1}{\sin x}, \quad \sin x = \frac{1}{\csc x}$$

$$\cot x = \frac{1}{\tan x} = \frac{\cos x}{\sin x}, \quad \tan x = \frac{1}{\cot x}$$

Negative Identities

$$\cos(-x) = \cos x$$

$$\sin(-x) = -\sin x$$

$$\tan(-x) = -\tan x$$

Sum and Difference Identities

$$\cos(A + B) = \cos A \cos B - \sin A \sin B, \qquad \cos(A - B) = \cos A \cos B + \sin A \sin B$$

$$\sin(A + B) = \sin A \cos B + \cos A \sin B, \qquad \sin(A - B) = \sin A \cos B - \cos A \sin B$$

$$\tan(A + B) = \frac{\tan A + \tan B}{1 - \tan A \tan B}, \qquad \tan(A - B) = \frac{\tan A - \tan B}{1 + \tan A \tan B}$$

Double-Angle Identities

$$\cos 2x = \cos^2 x - \sin^2 x$$
$$= 2\cos^2 x - 1$$
$$= 1 - 2\sin^2 x$$

$$\sin 2x = 2 \sin x \cos x$$

$$\tan 2x = \frac{2 \tan x}{1 - \tan^2 x}$$

Cofunction Identities

$$\cos\left(\frac{\pi}{2} - x\right) = \sin x$$

$$\sin\left(\frac{\pi}{2} - x\right) = \cos x$$

$$\tan\left(\frac{\pi}{2} - x\right) = \cot x$$

Half-Angle Identities

$$\cos^2 x = \frac{1 + \cos 2x}{2} \qquad \cos\frac{x}{2} = \pm\sqrt{\frac{1 + \cos x}{2}}$$

$$\sin^2 x = \frac{1 - \cos 2x}{2} \qquad \sin\frac{x}{2} = \pm\sqrt{\frac{1 - \cos x}{2}}$$

$$\tan\frac{x}{2} = \pm\sqrt{\frac{1 - \cos x}{1 + \cos x}} = \frac{\sin x}{1 + \cos x} = \frac{1 - \cos x}{\sin x}$$

The identities summarized are important and *should be memorized* because they are used to prove other identities.

Steps to Prove Identities

Step 1. Select one side of the equation to work on.

Step 2. Use any of the identities previously listed to rewrite the identity in its equivalent form.

Step 3. Perform the indicated operations. While looking at the side you are not working on for clues as to how to proceed, use trigonometry along with any arithmetic or algebraic skills that may simplify the next step. For example,

- If there are two or three terms and you want a single term, add terms by finding common denominators, factor the expression, and use known identities (or an equivalent form).
- If there are no denominators or different denominators on the side you are working than those on the other side, multiply by the appropriate fractional form of 1 to obtain the necessary denominators.
- Look for identities that replace functions with ones needed or ones in the requested form.

Always look where you're going!

To Prove That an Equation Is Not an Identity

To prove that the statement is not an identity, find a counterexample—replace the variable with a defined value that demonstrates the statement is false. You can also use a graphing utility to show that the graphs of each side of the statement do not coincide.

Using Identities to Find Exact Values

Basic identities are also used to find exact functional values. If we know a functional value of an angle and the quadrant, we can use identities to find the functional values of double the angle or half the angle. In addition, if we know the functional values of a second angle and its quadrant, we can find all functional values of the sum or difference of these two angles. Identities can also express double angles in terms of single angles or half-angles with full angles and vice-versa. Furthermore, we can reduce powers on trigonometric functions and rewrite products to sums and vice versa.

Product to Sum/Difference Identities (Formulas)

$$\cos A \cos B = \tfrac{1}{2}[\cos(A+B) + \cos(A-B)]$$
$$\sin A \sin B = \tfrac{1}{2}[\cos(A-B) - \cos(A+B)]$$
$$\sin A \cos B = \tfrac{1}{2}[\sin(A+B) + \sin(A-B)]$$
$$\cos A \sin B = \tfrac{1}{2}[\sin(A+B) - \sin(A-B)]$$

Sum/Difference to Product Identities (Formulas)

$$\cos A + \cos B = 2\cos\left(\frac{A+B}{2}\right)\cos\left(\frac{A-B}{2}\right) \qquad \sin A + \sin B = 2\sin\left(\frac{A+B}{2}\right)\cos\left(\frac{A-B}{2}\right)$$
$$\cos A - \cos B = -2\sin\left(\frac{A+B}{2}\right)\sin\left(\frac{A-B}{2}\right) \qquad \sin A - \sin B = 2\cos\left(\frac{A+B}{2}\right)\sin\left(\frac{A-B}{2}\right)$$

Chapter 4 Review Exercises

4.1–4.5

1–16. *Select the correct expression from Column A to complete the identity. Some expressions may be used more than once or not at all.*

1. $1 - \sin^2 x =$ ______
2. $1 - 2\sin^2 x =$ ______
3. $1 + \tan^2 x =$ ______
4. $2\sin x \cos x =$ ______
5. $\sec(-x) =$ ______
6. $\dfrac{1}{\tan x} =$ ______
7. $\cos^2 x - \sin^2 x =$ ______
8. $\cos^2 x + \sin^2 x =$ ______
9. $\pm\sqrt{\dfrac{1 + \cos 2x}{2}} =$ ______
10. $\pm\sqrt{\dfrac{1 - \cos 2x}{2}} =$ ______
11. $\cos(x - y) =$ ______
12. $\sin x \csc x =$ ______
13. $\sin(x + y) =$ ______
14. $1 + \cot^2 x =$ ______
15. $\cos(\pi + x) =$ ______
16. $\sin(90° - x) =$ ______

Column A

1

$\cos x$

$\sin x$

$\csc^2 x$

$\sin^2 x$

$\cos^2 x$

$\cos x$

$\cot x$

$-\sin x$

$\dfrac{1}{\cos x}$

$\sin 2x$

$\cos 2x$

$\sec^2 x$

$\cos x \cos y + \sin x \sin y$

$\sin x \cos y - \cos x \sin y$

$\cos x \cos y - \sin x \sin y$

$\sin x \cos y + \cos x \sin y$

17–26. *Use identities to simplify the expressions to a single function.*

17. $2 \csc x \sin 2x$

18. $2 \sec x \sin 2x$

19. $\sin^2 x + \cos 2x$

20. $\sec(-x)\cot(-x)$

21. $2 \sin 5x \cos 5x$

22. $1 - 2\sin^2 10x$

23. $\cos 7x \cos 6x + \sin 7x \sin 6x$

24. $\sin 5x \cos 3x + \cos 5x \sin 3x$

25. $\sqrt{\dfrac{1 - \cos 8x}{2}}$

26. $\dfrac{\sin 4x}{1 + \cos 4x}$

27–40. *Prove the following identities.*

27. $\cos^2 x(\sec^2 x - 1) = \sin^2 x$

28. $\tan^2 A \csc^2 A - \tan^2 A = 1$

29. $\dfrac{\cos y \tan y}{\sin y} = 1$

30. $\cot \alpha - \tan \alpha = \dfrac{\cos 2\alpha}{\sin \alpha \cos \alpha}$

31. $\dfrac{1 + \sin \theta}{\cos \theta} = \dfrac{\cos \theta}{1 - \sin \theta}$

32. $\csc t = 2 \cos t \csc 2t$

33. $\sin 2x = 2\sin^2 x \cos x \csc x$

34. $\cos^4 \beta - \sin^4 \beta = \cos 2\beta$

35. $\tan y + \sec y = \dfrac{1}{\sec y - \tan y}$

36. $\dfrac{\cos x}{1 - \tan x} + \dfrac{\sin x}{1 - \cot x} = \cos x + \sin x$

37. $\cos(90° + \theta) = -\sin \theta$

38. $\sin(90° + x) = \cos x$

39. $\tan\left(\alpha - \dfrac{\pi}{4}\right) = \dfrac{\tan \alpha - 1}{1 + \tan \alpha}$

40. $\tan \dfrac{A}{2} = \dfrac{\sec A \sin A}{\sec A + 1}$

41–42. *Find a counterexample to prove each statement is not an identity.*

41. $\tan A \cos A \sin A = \sec A$

42. $\cos 2x = 1 + 2\cos^2 x$

43–44. *Use a graphing calculator to determine whether the statement could be an identity. If it appears to be an identity, prove it. If not, find a counterexample.*

43. $\sin^2 \dfrac{x}{2} = \dfrac{\cos^2 x}{2 - 2\cos x}$

44. $\cot^2 x = \dfrac{1 - \cos^2 2x}{\sin x}$

45–46. *Find the exact value.*

45. $\cos\left(-\dfrac{5\pi}{12}\right)$

46. $\sin(105°)$

47–54. *Determine whether each statement is true or false. If it is false, rewrite the right side of the equation to make the statement true.*

47. $\cos 77° \cos 32° - \sin 77° \sin 32° = \dfrac{\sqrt{2}}{2}$

48. $\sin 30° \cos 60° + \cos 30° \sin 60° = 1$

49. $\cos 105° = \cos(60° + 45°) = \dfrac{\sqrt{6} - \sqrt{2}}{4}$

50. $\tan \dfrac{\pi}{12} = \tan\left(\dfrac{\pi}{3} - \dfrac{\pi}{4}\right) = -\sqrt{3} + 2$

51. $\sin 22.5° = \sqrt{\dfrac{1 - \cos 45°}{2}}$

52. $4 \sin 8x \cos 8x = 2 \sin 4x$

53. $2 \cos^2 15x - 1 = \cos 30x$

54. $\cos\left(\arccos \dfrac{1}{2} - \arcsin \dfrac{1}{2}\right) = \dfrac{\sqrt{3}}{2}$

55–58. *Without using a graphing calculator, match each graph with at least two equations from Column A that represent the graph. Check your answer with a graphing calculator.*

55.

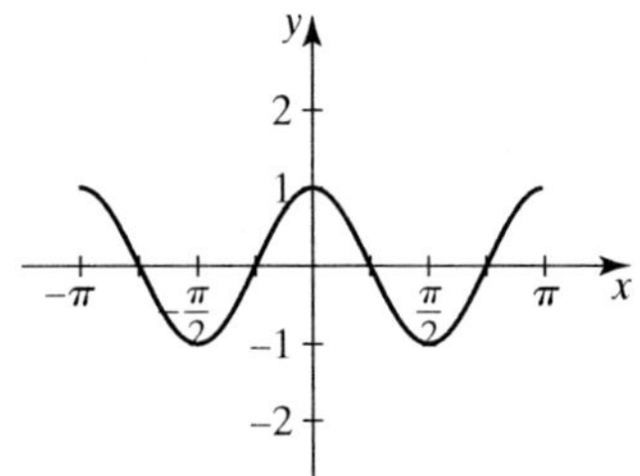

56.

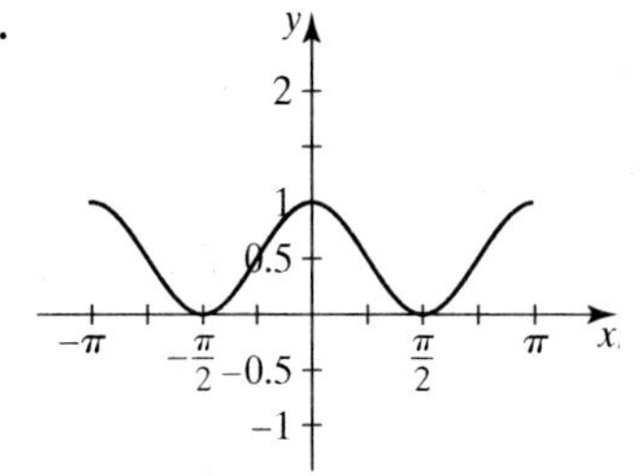

57.

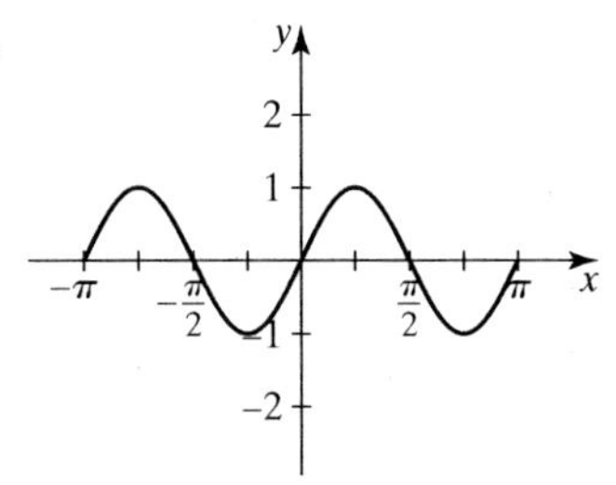

58.

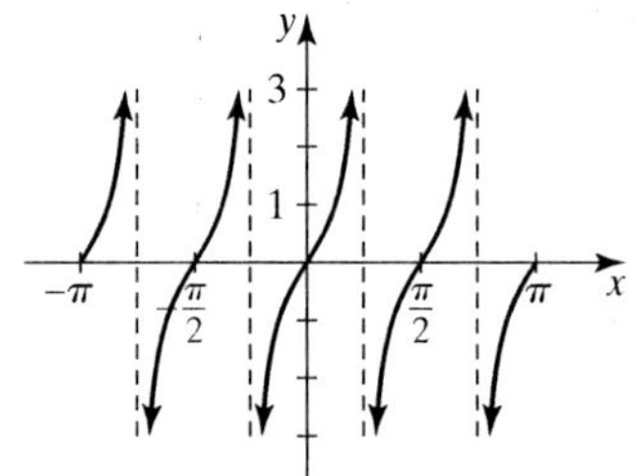

COLUMN A

$y = \cos^2 x$

$y = \cos 2x$

$y = \sin 15x \cos 13x - \cos 15x \sin 13x$

$y = \dfrac{2 \tan x}{1 - \tan^2 x}$

$y = \cos^2 x - \sin^2 x$

$y = \tan 2x$

$y = \dfrac{1}{2} \cos 2x + \dfrac{1}{2} = \dfrac{1 + \cos 2x}{2}$

$y = 2 \sin \dfrac{x}{2} \cos \dfrac{x}{2}$

$y = \sin 2x$

$y = -\cot\left(2x + \dfrac{\pi}{2}\right)$

59–68. *If* $\cos A = \frac{1}{2}$ *and* $\sin B = -\frac{12}{13}$, *where A is in QI and B is in QIII, find the exact value.*

59. $\sin A$

60. $\cos B$

61. $\cos(A + B)$

62. $\sin(A - B)$

63. $\sin 2B$

64. $\tan 2A$

65. $\cos \dfrac{A}{2}$

66. $\sin \dfrac{B}{2}$

67. $\sin^2 \dfrac{A}{2}$

68. $\cos^2 \dfrac{B}{2}$

69–78. *If* $\sin A = \frac{3}{5}$ *and* $\tan B = -\frac{4}{3}$, *where A is in QII and B is in QIV, find the exact value.*

69. $\cos A$

70. $\sin B$

71. $\sin(A + B)$

72. $\cos(A - B)$

73. $\cos 2A$

74. $\tan 2B$

75. $\sin \dfrac{B}{2}$

76. $\cos \dfrac{A}{2}$

77. $\tan \dfrac{A}{2}$

78. $\sin(A - B)$

79. If $\cos 2x = 0.8$ and $2x$ is in QIV $(270° \leq 2x \leq 360°)$, find $\cos x$.

80. If $\sec B = \sqrt{5}$, find $\sin 2B$ if B is in QI.

81–84. *Find the exact value, if defined.*

81. $\cos\left(\arctan\sqrt{3} + \tan^{-1} 1\right)$

82. $\sin\left(\sin^{-1}\frac{\sqrt{3}}{2} + \sin^{-1}(-1)\right)$

83. $\tan\left(\csc^{-1} 2 + \arccos\frac{1}{2}\right)$

84. $\cos(\operatorname{arcsec}(-2) + \operatorname{arccot} 0)$

85. Express $\cos 10x + \cos 3x$ as a product of function values.

86. Express $\cos 5x - \cos 3x$ as a product of function values.

87. Express $\sin 6x \sin 8x$ as the sum or difference of function values.

88. Express $\cos 9x \sin 5x$ as the sum or difference of function values.

Chapter 4 Test

1–8. *Complete the following identities with an equivalent expression. More than one answer may be possible.*

1. $\tan(-x) =$ ________

2. $\cos x \cos y - \sin x \sin y =$ ________

3. $\sin^2 x = 1 -$ ________

4. $1 + \tan^2 x =$ ________

5. $\cos 2x =$ ________

6. $\pm\sqrt{\frac{1 - \cos x}{2}} =$ ________

7. $\cos^2 15x + \sin^2 15x =$ ________

8. $2 \cos x \sin x =$ ________

9–14. *Prove the following identities. Show all your work.*

9. $\sin^2 x\,(1 + \cot^2 x) = 1$

10. $\cot x + \tan x = \sec x \csc x$

11. $\cos 2x - \cos^2 x = -\sin^2 x$

12. $\sin 2x = 2 \sin^2 x \cos x \csc x$

13. $2 \sin^2\frac{x}{2} - 1 = -\cos x$

14. $\cos^2\frac{x}{2} = \frac{\sin^2 x}{2(1 - \cos x)}$

15–16. *If the statement is an identity, prove it. If not, find a counterexample.*

15. $1 - \sin 2x = \frac{1 - \tan^2 x}{1 + \tan^2 x}$

16. $\cos^2 2x + 2 \sin^2 x = 1$

17–22. *Use the appropriate sum or difference formula to find the exact value of the following.*

17. $\cos 105°$

18. $\sin\frac{\pi}{12}$

19. $\sin 12° \cos 48° + \cos 12° \sin 48°$

20. $\cos 75° \cos 75° - \sin 75° \sin 75°$

21. $\frac{\tan 47° - \tan 17°}{1 + \tan 47° \tan 17°}$

22. $\tan(\pi + x)$

23–26. *If* $\cos A = -\frac{1}{2}$, *and* $\sin B = \frac{4}{5}$, *where A is in QIII and B is in QII, find the exact value of the following.*

23. $\sin A$

24. $\cos 2B$

25. $\sin(A + B)$

26. $\sin \dfrac{A}{2}$

27–28. *Identify all the equations from Column A that represent the graph.*

27.

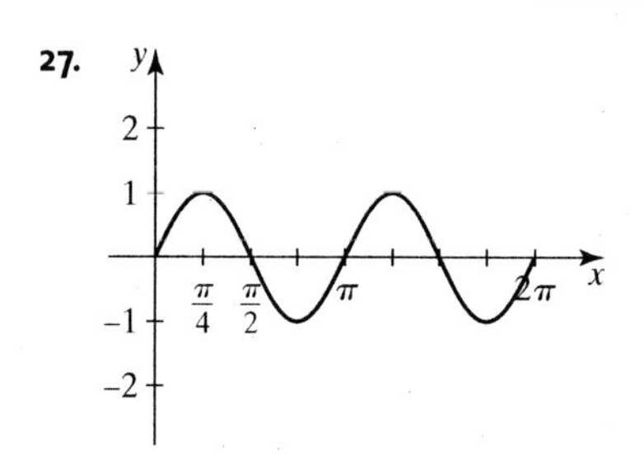

28.

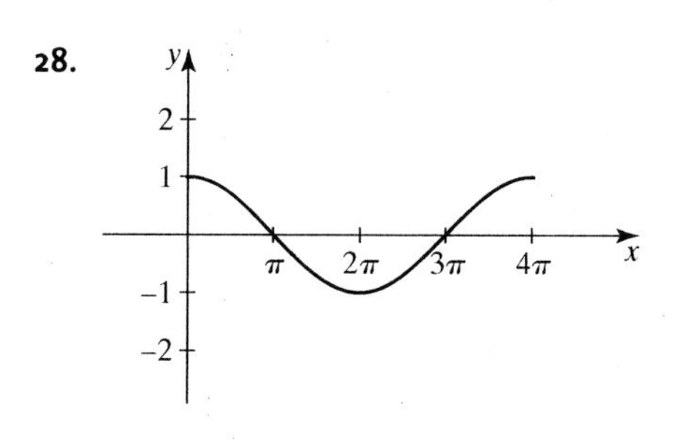

Column A

a. $y = 2 \sin x \cos x$

b. $y = \cos \frac{1}{2}x$

c. $y = \sin 2x$

d. $y = \cos 5x \cos 4x + \sin 5x \sin 4x$

e. $y = \sin x + \cos x \tan x$

f. $y = \sin 5x \cos 3x - \cos 5x \sin 3x$

g. $y = \cos 2x$

h. $y = \cos x \cos \dfrac{x}{2} + \sin x \sin \dfrac{x}{2}$

29. Find the exact value of $\cos\left(\tan^{-1}\sqrt{3} - \sin^{-1}\dfrac{\sqrt{2}}{2}\right)$.

30. Express the sum $\sin 3x + \sin 5x$ as a product.

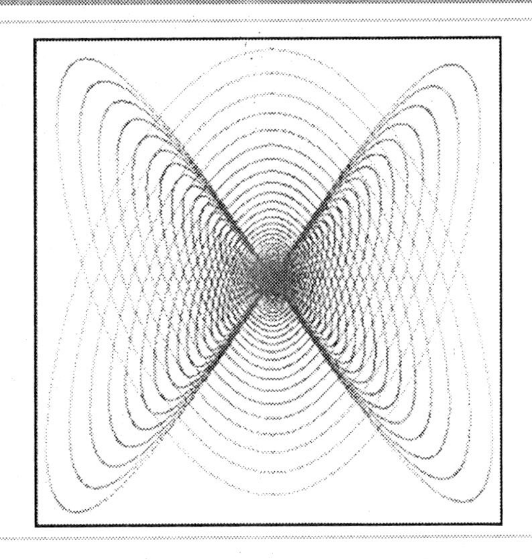

Chapter 5

Trigonometric Equations

When we combine periodic waves moving back and forth with periodic waves moving up and down, we create some interesting patterns that are important to engineers. They are often used in design, logos, and as computer screen savers. These are called Lissajous figures, named after the French physicist Jules Antoine Lissajous (1822–1880). In this chapter we introduce equations that produce these shapes.

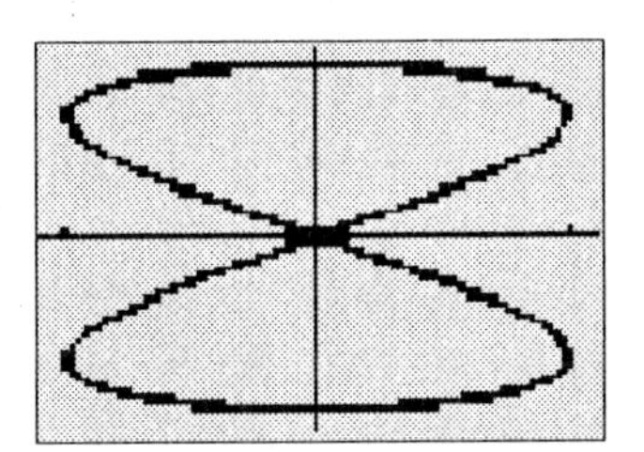

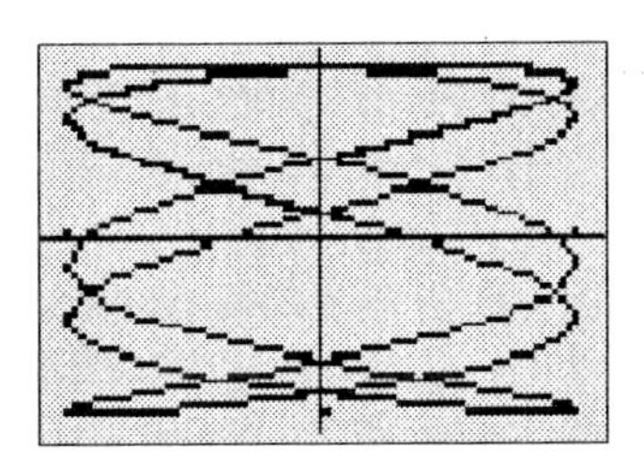

Important Concepts

- solutions to trigonometric equations
- multiple-angle equations
- systems of equations
- parametric equations
- projectile motion

5.1 Solving Conditional Equations I

In Chapter 4 we worked with trigonometric identities, equations that are true for all values for which they are defined. In this chapter we work with **conditional trigonometric equations**, equations that are true for only certain values of a variable for which they are defined. In previous chapters we solved a few conditional trigonometric equations when we found the specific values of a variable (representing degrees, radians, or real numbers) in a restricted interval that made the statements true. For example, in Chapter 2 we solved equations such as the following.

$$\cos x = -\frac{\sqrt{3}}{2} \text{ for } 0 \le x \le \pi \qquad \sin\theta = \frac{1}{2} \text{ for } -90° \le \theta \le 90°$$

$$x = \arccos\left(-\frac{\sqrt{3}}{2}\right) \qquad \theta = \sin^{-1}\left(\frac{1}{2}\right)$$

$$x = \frac{5\pi}{6} \qquad \theta = 30°$$

However, if we use a different interval, such as $0 \le x < 2\pi$ or $0° \le \theta < 360°$, we would have more than one solution as is demonstrated in Examples 1 and 2. For most values in the range of a trigonometric function, there are two solutions for x and θ in these respective intervals. And if we place no restriction on the variable, there can be infinitely many solutions. In this case, we do the following.

- Find all solutions within one period of the function by first establishing the reference angle and then applying it to the appropriate quadrants.
- Then, using the periodic property of each function, express all solutions by adding integer multiples of the period to each solution found within one period.

EXAMPLE 1 Solve $\cos x = -\sqrt{3}/2$ for exact values of x when a) $0 \le x < 2\pi$ and b) $x \in \mathbb{R}$.

SOLUTION

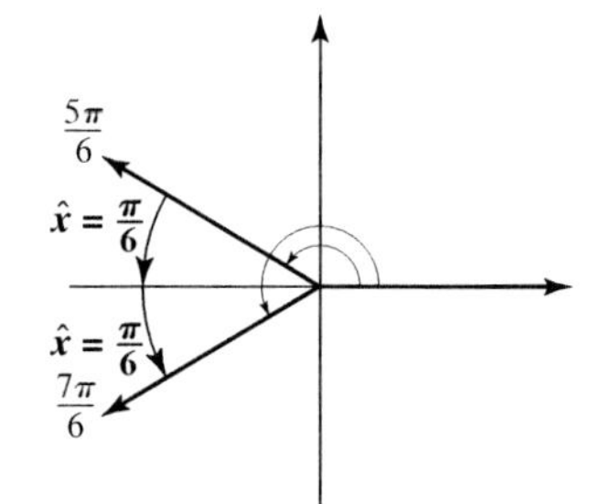

a. When $\cos x = -\sqrt{3}/2$, we know that the reference angle $\hat{x} = \pi/6$ and $\cos x$ is negative in QII and QIII. So,

$$x = \pi - \frac{\pi}{6} = \frac{5\pi}{6} \quad \text{and} \quad x = \pi + \frac{\pi}{6} = \frac{7\pi}{6}.$$

b. The period of the cosine function is 2π, so

$$x = \frac{5\pi}{6} + k \cdot 2\pi, \quad x = \frac{7\pi}{6} + k \cdot 2\pi, \text{ where } k \in \{\ldots,-1, 0, 1, 2, \ldots\}.$$ ■

NOTE: When the answer contains infinitely many values, the solution is an expression that represents all solutions, where k represents an integer ($k \in \{\ldots,-2,-1, 0, 1, 2, \ldots\}$).

EXAMPLE 2 Solve $\sin\theta = \frac{1}{2}$ for exact values of θ when a) $0° \le \theta < 360°$ and b) $\theta \in A°$ (any angle in degrees).

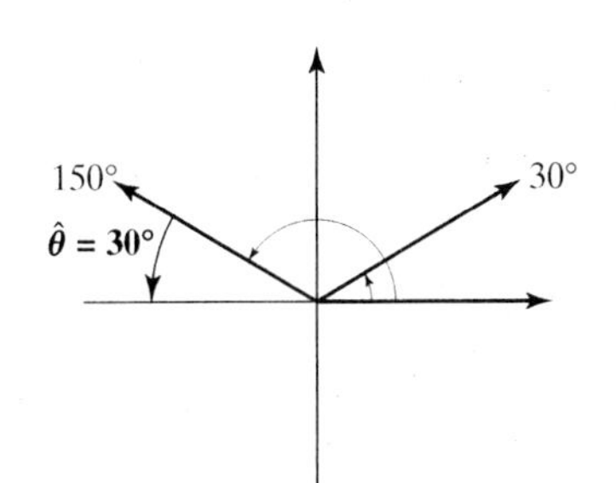

SOLUTION

a. When $\sin\theta = \frac{1}{2}$, we know that $\hat{\theta} = 30°$ and $\sin\theta$ is positive in QI and QII. So,

$$\theta = 30° \quad \text{and} \quad \theta = 180° - 30° = 150°.$$

b. The period of the sine function is 360°, so

$$\theta = 30° + k \cdot 360°, \quad \theta = 150° + k \cdot 360°.$$

Even though the equations in Examples 1 and 2 have an infinite number of solutions, not all values of the variable for which the functions are defined make the statement true. For this reason these are conditional equations, not identities. As with conditional algebraic equations, *you should always check the solutions* to conditional trigonometric equations—but for only one period—which we leave for you to do.

EXAMPLE 3 Solve $\tan\alpha = -\sqrt{3}$ for exact values of α when a) $0° \le \alpha < 180°$ and b) $\alpha \in A°$.

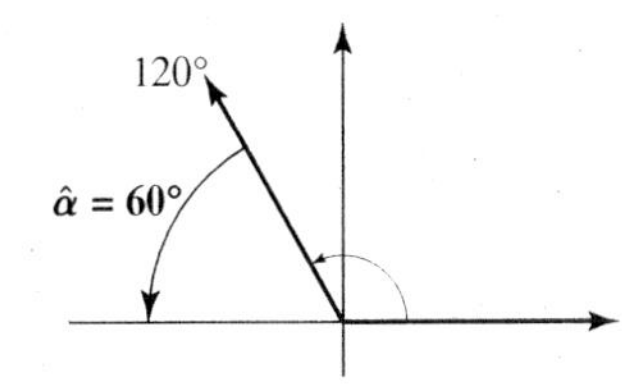

SOLUTION

a. When $\tan\alpha = -\sqrt{3}$, we know that $\hat{\alpha} = 60°$ and $\tan\alpha$ is negative in QII. So,

$$\alpha = 180° - 60° = 120°.$$

b. The period of the tangent function is 180°, so

$$\alpha = 120° + k \cdot 180°.$$

Examples 1–3 involved functional values of common arcs or angles, so we found exact values for the solutions. The next examples demonstrate equations that have approximate solutions found by using a calculator.

EXAMPLE 4 Solve $\sin x = 0.2$ for x when a) $0° \le x < 360°$ and b) $x \in A°$. Approximate the solutions to the nearest tenth of a degree.

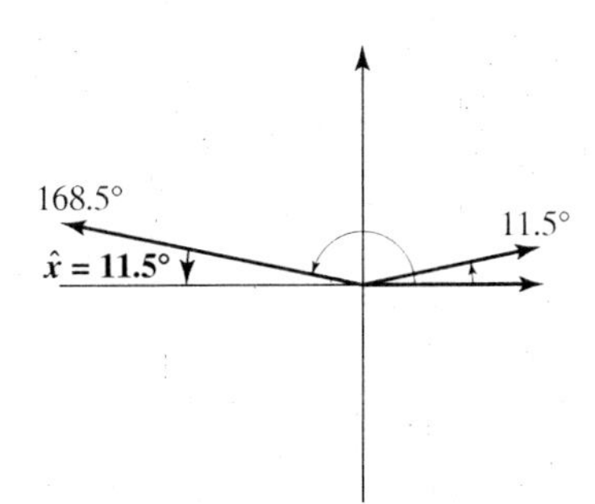

SOLUTION

a. Since 0.2 is not a sine value of a common angle, we find the reference angle using the inverse sine function and a calculator in degree mode.

$$\sin x = 0.2$$
$$\hat{x} = \arcsin(0.2) \approx 11.5°$$

Because the sine is positive in QI and QII,

$$x \approx 11.5° \quad \text{and} \quad x \approx 180° - 11.5° = 168.5°.$$

b. $x \approx 11.5° + k \cdot 360°, \quad x \approx 168.5° + k \cdot 360°.$

EXAMPLE 5 Solve $\tan x = -4.6$ for x when a) $0° \le x < 360°$ and b) $x \in A°$. Approximate the solutions to the nearest tenth of a degree.

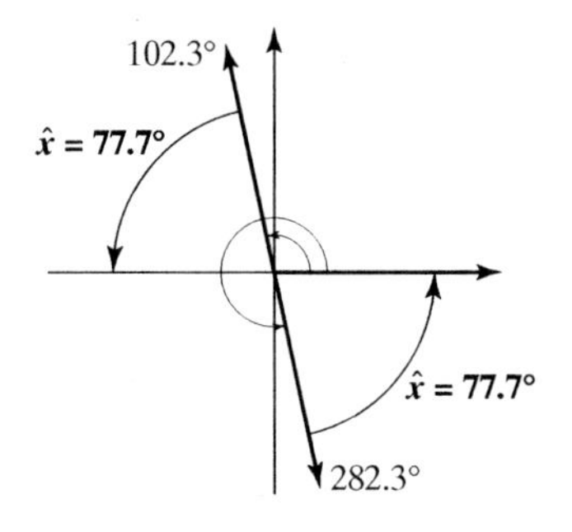

SOLUTION

a. Recall that a reference angle $\hat{x}$ is an angle between 0° and 90°, for which all functional values like $\sin \hat{x}$, $\cos \hat{x}$, $\tan \hat{x}$, and so on, are positive. Therefore, the reference angle is found with a calculator in degree mode by using the following procedure, which applies the inverse function to the positive of the functional value.

$$\tan \hat{x} = |\tan x| = |-4.6|$$
$$\tan \hat{x} = 4.6$$
$$\hat{x} = \tan^{-1}(4.6) \approx 77.7°$$

Since the tangent is negative in QII and QIV,

$$x \approx 180° - 77.7° = 102.3° \quad \text{and} \quad x \approx 360° - 77.7° = 282.3°.$$

b. The period of $\tan x$ is 180°, so we need only the angle between 0° and 180° to find the expression that gives all solutions of x. Therefore,

$$x \approx 102.3° + k \cdot 180°.$$

Note that if $k = 1$, we get $x \approx 102.3° + 1(180°) = 282.3°$. ■

In Example 5 we could have found $x = \tan^{-1}(-4.6) \approx -77.7°$. From $x \approx -77.7°$, we could find the reference angle and then the solutions for $0° \le x < 360°$. But you may find it easier to find the reference angle by taking the inverse function of a positive functional value, which is the procedure we used in the solution of Example 5.

We are able to solve equations for values of x that represent either angles (in degrees or radians) or real numbers. For solutions that require the use of a calculator, check your calculator to make sure you are in the indicated mode (degree or radian).

EXAMPLE 6 Solve $\sec x = 1.5$ for $0 \le x < 2\pi$. Approximate x rounded to four decimal places.

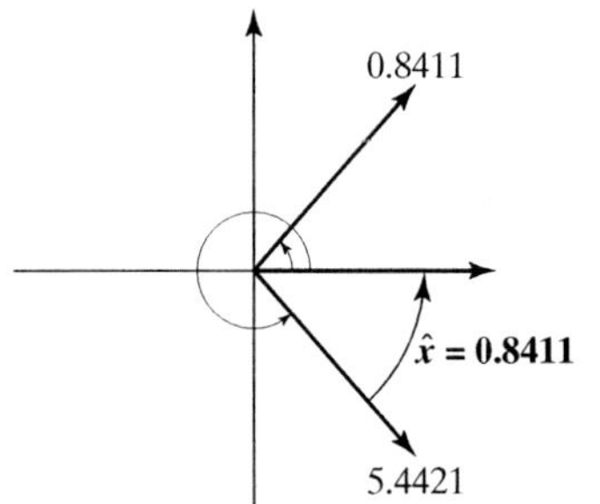

SOLUTION If $\sec x = 1.5 = \frac{3}{2}$, then $\cos x = \frac{2}{3}$. Using a calculator in radian mode, we obtain

$$\hat{x} = \cos^{-1}\left(\tfrac{2}{3}\right) \approx 0.8411.$$

We know $\cos x$ is positive in QI and QIV. Therefore,

$$x \approx 0.8411 \quad \text{and} \quad x \approx 2\pi - 0.8411 = 5.4421.$$ ■

Just as there are equations in algebra that have no solution, the same is true for some equations in trigonometry.

EXAMPLE 7 Explain why the following equations have no solutions.

a. $\cos x = 3$ **b.** $\csc x = -\frac{1}{2}$

SOLUTION

a. The equation $\cos x = 3$ has no solution because $\cos x$ is never larger than 1.

b. The equation $\csc x = -\frac{1}{2}$ has no solution because $\csc x = -\frac{1}{2}$ is equivalent to $\sin x = -2$ and $\sin x$ is never smaller than -1. ■

CONNECTIONS WITH TECHNOLOGY

Graphing Calculator/CAS

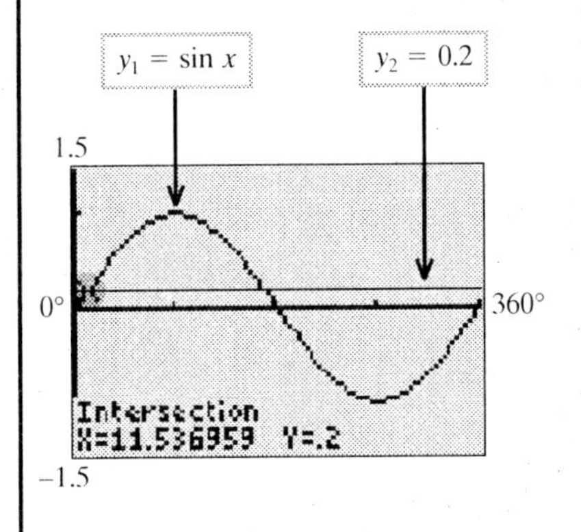

We can verify or approximate real solutions to $f(x) = g(x)$ by using a graphing utility to find the x-coordinates of the points of intersection of the graphs of $y = f(x)$ and $y = g(x)$. We call this the **graph intersection method**. (Check your utility manual to see what INTERSECT features and calculations are available.)

To find or verify the solutions to $\sin x = 0.2$ (Example 4 a), we graph

$$y_1 = \sin x \quad \text{and} \quad y_2 = 0.2.$$

(MODE: degree; WINDOW: Xmin = 0, Xmax = 360, Ymin = −1.5, Ymax = 1.5) The two points of intersection of the graphs for $0° \leq x < 360°$ have x-coordinates 11.5° and 168.5°, rounded to the nearest tenth of a degree. This verifies those found in Example 4. The screen indicates one solution at $x = 11.536959$ (degrees). The other solution can be found similarly using the INTERSECT feature.

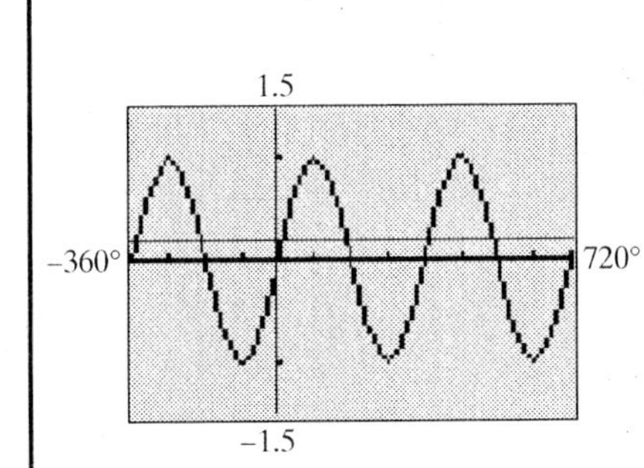

By changing the window so that Xmin = −360 and Xmax = 720, we notice the graphs intersect two times every 360° (period of the sine). Therefore, the solutions for $x \in A°$ (Example 4 b),

$$x \approx 11.5° + k \cdot 360° \quad \text{and} \quad x \approx 168.5° + k \cdot 360°,$$

are graphically demonstrated.

Your ability to solve more difficult equations depends on your ability to solve equations from this section. So let's begin.

Exercise Set 5.1

1–12. *Solve each equation for exact values of x when a) $0 \leq x < 2\pi$ and b) $x \in \mathbb{R}$.*

1. $\cos x = \dfrac{1}{2}$ **2.** $\sin x = -\dfrac{\sqrt{3}}{2}$ **3.** $\sin x = -\dfrac{\sqrt{2}}{2}$

4. $\cos x = \dfrac{\sqrt{3}}{2}$ **5.** $\tan x = \dfrac{\sqrt{3}}{3}$ **6.** $\cot x = \dfrac{\sqrt{3}}{3}$

7. $\sec x = 1$ **8.** $\tan x = -1$ **9.** $\csc x = -2$

10. $\sin x = 0$ **11.** $\cot x = -1$ **12.** $\sec x = -\sqrt{2}$

13–20. *Solve each equation for exact values of α when a) $0° \le \alpha < 360°$ and b) $\alpha \in A°$.*

13. $\tan \alpha = 1$ **14.** $\sin \alpha = \frac{\sqrt{2}}{2}$ **15.** $\csc \alpha = -\frac{2\sqrt{3}}{3}$

16. $\cot \alpha = -\sqrt{3}$ **17.** $\cos \alpha = 0$ **18.** $\sec \alpha = -2$

19. $\cot \alpha$ undefined **20.** $\tan \alpha$ undefined

21–30. *Solve each equation for $0 \le t < 2\pi$. Approximate the solutions rounded to four decimal places.*

21. $\sin t = \frac{7}{25}$ **22.** $\cos t = \frac{3}{5}$ **23.** $\tan t = 5\frac{1}{2}$

24. $\cot t = 2\frac{1}{4}$ **25.** $\cos t = -0.75$ **26.** $\sin t = -0.3445$

27. $\csc t = -1.28766$ **28.** $\sec t = -32.76511$

29. $\cot t = -0.0075$ **30.** $\tan t = 55.8907$

31–40. *Find exact solutions whenever possible for $0° \le x < 360°$. Otherwise approximate the solutions to the nearest tenth of a degree. If there is no solution, explain why.*

EXAMPLE $\tan x = \pm 1$

SOLUTION This is a common angle functional value, and the reference angle is $\hat{x} = 45°$. Since we want the angles for which $\tan x$ is ± 1, we have

$$x = 45°, 135°, 225°, 315°.$$

31. $\sin x = \pm 1$ **32.** $\cos x = \pm\frac{1}{2}$ **33.** $\cos x = -0.45$

34. $\sin x = -0.8989$ **35.** $\tan x = 1.59$ **36.** $\cot x = \frac{12}{13}$

37. $\cot x = \pm\sqrt{3}$ **38.** $\sin x = \pm\frac{\sqrt{3}}{2}$

39. $\sec x = -0.5$ **40.** $\csc x = 0.25$

41–44. *Each graph represents two equations graphed using radians. One point of intersection of these graphs is indicated whose x-coordinate is rounded to four decimal points. Use your knowledge of the graphs of trigonometric and algebraic functions to do the following.*

a. *Determine the equation of each graph.*

b. *Find the equation whose solution can be found using the x-coordinates of the points of intersection of these graphs.*

c. *State the solution for x that is indicated on the screen. Verify that this is a solution of the equation you found in part (b).*

41.

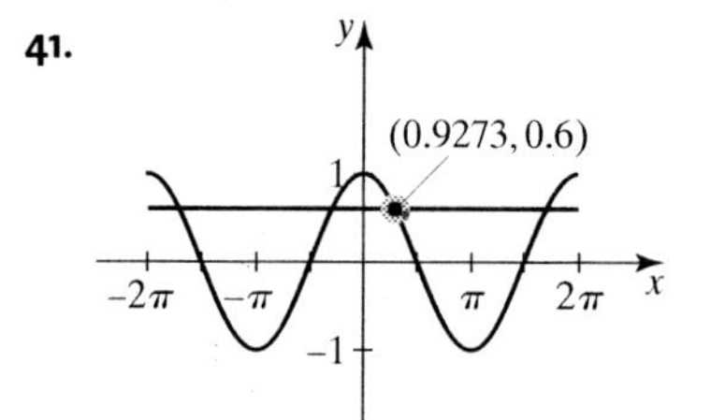

42.

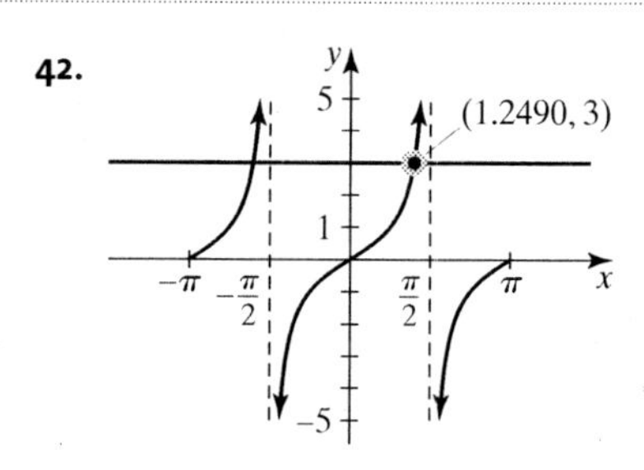

43.

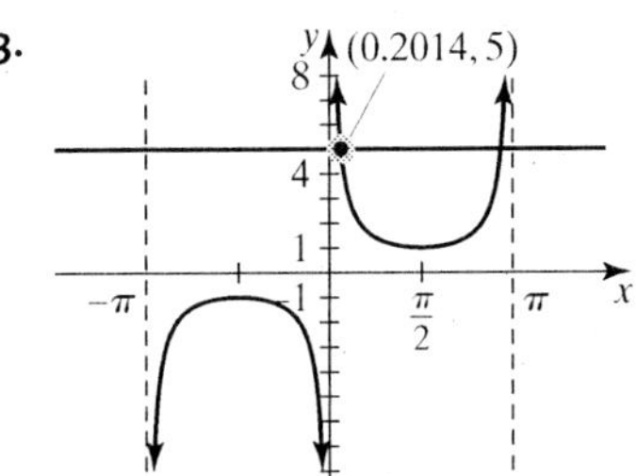

44.

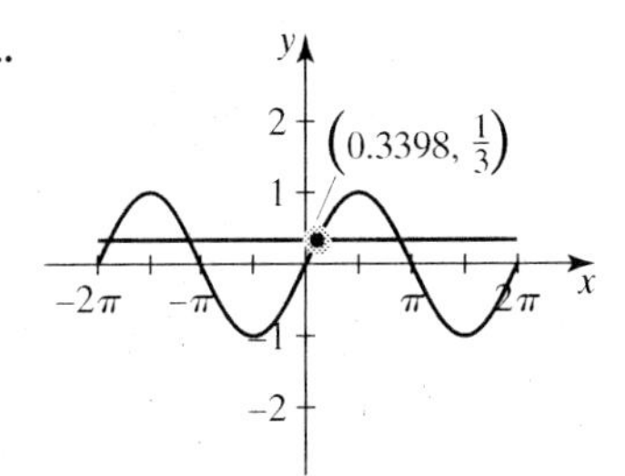

Graphing Calculator/CAS Exercises

45–48. *Find the solutions for $0° \le x \le 360°$ to the nearest tenth of a degree, using the graph intersection method.*

45. $\sec x = 5.60$

46. $\tan x = -0.80$

47. $\cot x = -20$

48. $\csc x = 12$

Discussion

49. Chuck and Kathy are arguing about the answer to Exercise 17. Chuck says that $x = 90° + k \cdot 180°$ are the solutions to the equation $\cos x = 0$. Kathy says his answer is the expression for the x-intercepts of the graph of $y = \cos x$. Are they both correct? Explain why.

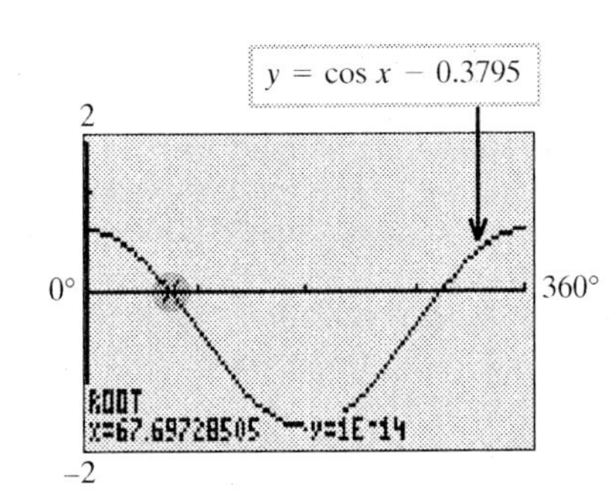

50. Fran is solving $\cos x = 0.3795$ for $0° \le x < 360°$ with a graphing calculator in degree mode. She rewrites the equation as $0 = \cos x - 0.3795$. Then using window settings, Xmin = 0, Xmax = 360, Ymin = −2, and Ymax = 2, she graphs the equation $y = \cos x - 0.3975$. She sees the x-intercepts and uses her graphing utility to find the values of these x-intercepts. She says these values are the solutions to the original equation. Is Fran finding the correct values? Explain your answer.

51. Lindsay and Andy have different expressions for the solution to $\sin x = -1$, where $x \in \mathbb{R}$. Lindsay has $x = -\frac{\pi}{2} + k \cdot 2\pi$, whereas Andy has $x = \frac{3\pi}{2} + k \cdot 2\pi$.

a. Use $k = -1, 0, 1,$ and 2 in *each expression* for the solutions indicated by Lindsay and Andy to find four specific solutions.

b. Do Lindsay's and Andy's expressions describe the same solutions?

c. Is the difference between $-\frac{\pi}{2}$ and $\frac{3\pi}{2}$, which are the first values used in the expression of the solutions, a multiple of 2π (the period of the sine)?

d. If Lindsay has $x = A + k \cdot 2\pi$ and Andy has $x = B + k \cdot 2\pi$ for the solutions to another equation involving $\sin x$ or $\cos x$, what would you need to know about A and B to determine if Lindsay's and Andy's expressions represent the same solutions?

52. Karen and Tim have different expressions for the solution to $\tan x = -1$, where $x \in \mathbb{R}$. Karen has $x = -\frac{\pi}{4} + k \cdot \pi$, whereas Tim has $x = \frac{3\pi}{4} + k \cdot \pi$.

a. Use $k = -1, 0, 1,$ and 2 in *each expression* for the solutions indicated by Karen and Tim to find four specific solutions.

b. Do Karen's and Tim's expressions describe the same solutions?

c. The equation Karen and Tim solved involved the tangent function. Is the difference between $-\frac{\pi}{4}$ and $\frac{3\pi}{4}$ a multiple of π (the period of the tangent)?

(continued)

d. If Karen has $x = A + k\pi$ and Tim has $x = B + k\pi$ for the solutions to another equation involving $\tan x$ or $\cot x$, what would you need to know about A and B to determine if Karen's and Tim's expressions represent the same solutions?

5.2 Solving Trigonometric Conditional Equations II

Equations in Section 5.1, such as

$$\sin x = \frac{1}{2}, \quad \tan x = 5.698, \quad \text{and} \quad \cos x = -0.38764,$$

are in the form of a first-degree trigonometric functional value equal to a real number. These simple equations are in an *immediately solvable form* in that they can be solved immediately by inspection (recognizing a common angle from its functional value) or using inverse functions. The trigonometric equations in this section are not this elementary and require algebraic manipulations to get them into an immediately solvable form. We review how to solve certain algebraic equations as we demonstrate how to solve a comparable trigonometric equation. If you require more review on solving algebraic equations, see Appendix A. You can think of each example as a familiar question with new language.

First Degree (Linear Form)

Example 1 Solve:

a. $2u + 1 = 3$

b. $2 \cos x + 1 = 3, \ 0° \le x < 360°$

Solution

a. Algebra Question

This is a first-degree equation (linear) that involves only one variable. To solve this equation, we use the addition and multiplication properties of equality.

$$2u + 1 = 3$$

$$2u = 2 \qquad \text{Add } -1 \text{ to both sides.}$$

$$u = 1 \qquad \text{Multiply both sides by } \tfrac{1}{2}.$$

b. Algebra Question in Trigonometry

Treating $\cos x$ in the same way as the variable u, we have an equation of the same form as part (a). We solve this equation using the same steps to obtain an immediately solvable form.

$$2 \cos x + 1 = 3$$

$$2 \cos x = 2$$

$$\cos x = 1 \qquad \text{(immediately solvable form)}$$

$$x = 0°$$

■

Example 2 Solve:

a. $4 + 3u = 2u + 5$

b. $4 + 3 \tan x = 2 \tan x + 5, \ 0° \le x < 360°$

Solution

a. Algebra Question

$$4 + 3u = 2u + 5$$

$$u + 4 = 5 \qquad \text{Add } -2u \text{ to both sides.}$$

$$u = 1 \qquad \text{Add } -4 \text{ to both sides.}$$

b. Algebra Question in Trigonometry

$$4 + 3 \tan x = 2 \tan x + 5$$

$$\tan x + 4 = 5$$

$$\tan x = 1$$

$$x = 45°, 225°$$

■

Second Degree (Quadratic Form)

Example 3 Solve:

a. $u^2 - u - 2 = 0$

b. $\sin^2 x - \sin x - 2 = 0,\ 0 \le x < 2\pi$

Solution

a. Algebra Question

This second-degree equation is written in descending powers of the variable and set equal to 0 (standard form). This equation can be solved by the factor method.

$u^2 - u - 2 = 0$	Write in standard form.
$(u+1)(u-2) = 0$	Factor.
$u + 1 = 0$ or $u - 2 = 0$	Set each factor equal to 0.
$u = -1$ or $u = 2$	Solve the resulting equations.

b. Algebra Question in Trigonometry

Treating $\sin x$ in the same way as the variable u, we have an equation in the same form as part (a). Therefore, we use the same steps to obtain an immediately solvable form.

$$\sin^2 x - \sin x - 2 = 0$$
$$(\sin x + 1)(\sin x - 2) = 0$$
$$\sin x + 1 = 0 \quad \text{or} \quad \sin x - 2 = 0$$
$$\sin x = -1 \quad \text{or} \quad \sin x = 2$$
$$x = \frac{3\pi}{2}$$

($\sin x = 2$ has no solution since $\sin x$ is never greater than 1.) ■

Example 4 Solve:

a. $uv = u$

b. $\tan x \cos x = \tan x,\ 0 \le x < 2\pi$

Solution Notice the algebraic equation involves two variables which compares to two functions in the trigonometric equation. Both of these equations can also be solved using the factor method. (Make sure that the equation is set equal to zero before you factor.)

a. Algebra Question

$uv = u$	
$uv - u = 0$	Rewrite in standard form.
$u(v-1) = 0$	Factor.
$u = 0$ or $v - 1 = 0$	Set each factor equal to 0.
$u = 0$ or $v = 1$	Solve the resulting equations.

b. Algebra Question in Trigonometry

$$\tan x \cos x = \tan x$$
$$\tan x \cos x - \tan x = 0$$
$$\tan x(\cos x - 1) = 0$$
$$\tan x = 0 \quad \text{or} \quad \cos x - 1 = 0$$
$$\tan x = 0 \quad \text{or} \quad \cos x = 1$$
$$x = 0, \pi \quad \text{or} \quad x = 0$$

So, the solutions are $x = 0, \pi$. ■

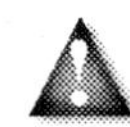

Caution: There is a temptation when solving equations like those in Example 4 to divide both sides of the equation by an unknown value. Recall that the properties of equality state that you can multiply or divide both sides of an equation by an expression that is not equal to 0. In Example 4 it is possible for $u = 0$ and $\tan x = 0$ because these values make their equation true. So an attempt to solve by dividing the respective equations by u or $\tan x$ would be incorrect. By doing so *we wouldn't obtain all the solutions*. Therefore, you should *avoid dividing by a variable expression*.

When you solve trigonometric equations, you are not provided with a comparable algebraic equation. Until you are more familiar with solving trigonometric equations, you may find it helpful to replace the function, such as $\sin x$, with the variable u to help recognize the form and algebraic procedures.

EXAMPLE 5 Solve $\cos^2 t - 3\cos t = 2$, $0 \le t < 2\pi$. Approximate the solutions rounded to four decimal places.

SOLUTION If we let $u = \cos t$, the equation becomes $u^2 - 3u = 2$. This is a quadratic equation, and we first put it in standard form $u^2 - 3u - 2 = 0$. Since it is not factorable, we use the quadratic formula which states that the solutions to $au^2 + bu + c = 0$ are $u = \dfrac{-b \pm \sqrt{b^2 - 4ac}}{2a}$ (see Appendix A). Although *you* can proceed with either the algebraic or trigonometric form, we demonstrate both. In either case, when we use the quadratic formula, the coefficients are $a = 1, b = -3$, and $c = -2$.

ALGEBRA QUESTION

$$u^2 - 3u - 2 = 0$$

$$u = \frac{-(-3) \pm \sqrt{(-3)^2 - 4(1)(-2)}}{2(1)}$$

$$u = \frac{3 \pm \sqrt{17}}{2}$$

$u \approx 3.5616$ or $u \approx -0.5616$

ALGEBRA QUESTION IN TRIGONOMETRY

$$\cos^2 t - 3\cos t - 2 = 0$$

$$\cos t = \frac{-(-3) \pm \sqrt{(-3)^2 - 4(1)(-2)}}{2}$$

$$\cos t = \frac{3 \pm \sqrt{17}}{2}$$

$\cos t \approx 3.5616$ or $\cos t \approx -0.5616$

(no solution, $\cos t \not> 1$)

$\hat{t} \approx \cos^{-1}(0.5616) \approx 0.9745$

Since $\cos t$ is negative in QII and QIII:

$t \approx \pi - 0.9745 \approx 2.1671$

$t \approx \pi + 0.9745 \approx 4.1161$

So, the solutions are $t \approx 2.1671, t \approx 4.1161$. ■

EXAMPLE 6 Solve $3\tan^2 A = 1$ for exact values of A, $0° \le A < 360°$.

SOLUTION Letting $u = \tan A$, we get $3u^2 = 1$. This quadratic equation can be solved using the square root method. (See Appendix A).

ALGEBRA QUESTION

$$3u^2 = 1$$

$$u^2 = \frac{1}{3}$$ Isolate the square term.

$$u = \pm\sqrt{\frac{1}{3}} = \pm\frac{\sqrt{3}}{3}$$ Take the square root of both sides. (If $a^2 = b$, $a = \pm\sqrt{b}$.)

ALGEBRA QUESTION IN TRIGONOMETRY

$$3\tan^2 A = 1$$

$$\tan^2 A = \frac{1}{3}$$

$$\tan A = \pm\sqrt{\frac{1}{3}} = \pm\frac{\sqrt{3}}{3}$$

$$\hat{A} = \tan^{-1}\left(\frac{\sqrt{3}}{3}\right) = 30°$$

Since $\tan A$ is both positive and negative,

$A = 30°, 150°, 210°, 330°$. ■

In Examples 1–6, we found solutions to the equation from 0° to 360° or from 0 to 2π. If we want all the solutions to these equations, we would add multiples of the appropriate period to each solution, as we did in Section 5.1.

EXAMPLE 7 Find all solutions to $\cos^2 t - 3\cos t = 2, t \in \mathbb{R}$ (Example 5).

SOLUTION We add multiples of the cosine's period (2π) to the solutions from Example 5 in order to obtain all solutions. (Had the solutions been in degrees, we would add multiples of 360°.)

Thus, the solutions are

$$t \approx 2.1671 + k \cdot 2\pi, \qquad t \approx 4.1161 + k \cdot 2\pi.$$

Application

In previous sections we found the displacement of an object in simple harmonic motion given a specific time. We are now able to find the time the object is at a given displacement.

EXAMPLE 8 A spring is bobbing up and down such that the displacement at time t is given by the equation $d(t) = 6.2\cos t$, where t is in seconds and $d(t)$ is in inches. At what time t (to the nearest tenth of a second), where $0 \le t < 2\pi$ seconds, will the displacement be 3 inches?

SOLUTION

$$6.2\cos t = d(t)$$

$$6.2\cos t = 3 \qquad \text{Substitute 3 for } d(t).$$

$$\cos t = \frac{3}{6.2} \qquad \text{Divide both sides by 6.2.}$$

$$\hat{t} = \cos^{-1}\left(\frac{3}{6.2}\right) \qquad \text{Use inverse cosine function.}$$

$$t \approx 1.0657, \quad t \approx 2\pi - 1.0657 \qquad \text{The cosine is positive in QI and QIV.}$$

$$t \approx 1.1, \quad t \approx 5.2$$

Therefore, the displacement will be 3 inches at times (to the nearest tenth of a second) 1.1 seconds and 5.2 seconds.

We summarize the *steps to solve trigonometric equations demonstrated in the examples of this section.*

- Determine the form of the equation: first degree (linear), second degree (quadratic), third degree (cubic), or higher. You can use substitution to replace $\sin x$, $\cos x$, and so on, with a single variable such as u or v to help determine the form of the equation.
- Select the appropriate algebraic strategy depending on the form: the factor method, the quadratic formula, the square root method, and so on. If the equation involves more than one function, or if the equation is third degree or higher, try the factor method.

Although the next section will require additional steps for the solution of more difficult trigonometric equations, practicing these steps will provide you with some necessary experience.

Exercise Set 5.2

1–6. *Identify the equations as having first-degree (linear) or second-degree (quadratic) form. Rewrite each equation by substituting u for the trigonometric function. If the equation is quadratic, write it in standard form. Do not solve the equation.*

1. $3 \sin x + 5 = 15 \sin x - 1$

2. $4(\cos x - 1) = 12 \cos x$

3. $3 \cos x - 2 \cos^2 x + 1 = 0$

4. $\sin^2 x - \sin x = 2$

5. $\sin x(\sin x - 5) = 6$

6. $(3 \cos x - 1)(\cos x + 2) = -4$

7–28. *Solve the equations for exact values of x, $0 \le x < 2\pi$. (You may want to use substitution to find a comparable algebraic equation for help with solving.)*

EXAMPLE $\sec^2 x = 2 \sec x$

SOLUTION

$u^2 = 2u$	$\sec^2 x = 2 \sec x$
$u^2 - 2u = 0$	$\sec^2 x - 2 \sec x = 0$
$u(u - 2) = 0$	$\sec x(\sec x - 2) = 0$
$u = 0$ or $u = 2$	$\sec x = 0$ or $\sec x = 2$
	no solution ($\sec x \neq 0$); $\cos x = \frac{1}{2}$
	$x = \frac{\pi}{3}, x = \frac{5\pi}{3}$

7. $2 \sin x = \sqrt{3}$

8. $2 \cos x = 1$

9. $5 \sec x + 8 = 9 \sec x$

10. $8 \cot x - 4 = \cot x + 3$

11. $4 \cos^2 x = 3$

12. $2 \sin^2 x = 1$

13. $15 \tan^2 x = 5$

14. $\csc^2 x = 1$

15. $\cos x \sin x - \cos x = 0$

16. $\sin x \cos x + \sin x = 0$

17. $2 \cos^2 x + \cos x = 1$

18. $2 \sin^2 x + 3 \sin x = 2$

19. $\tan^2 x = \tan x$

20. $\cot^2 x - \cot x = 0$

21. $(\tan^2 x - 3)(\tan x - 1) = 0$

22. $(\sec^2 x - 4)(\sec x - 1) = 0$

23. $4 \sin^3 x - 3 \sin x = 0$

24. $2 \cos^3 x - \cos x = 0$

25. $5 \cos^2 x - 11 \cos x + 6 = 0$

26. $2 \sin^2 x + 3 \sin x = 2$

27. $4 \csc^2 x + 3 \csc x - 1 = 0$

28. $\cot^2 x = 3$

29–36. *Solve the equations for the exact values of θ, $0° \le \theta < 360°$. (You may want to use substitution to find a comparable algebraic equation for help with solving.)*

29. $2 \sin \theta \cos \theta = 0$

30. $1 - 2 \sin^2 \theta = 0$

31. $2 \cos^2 \theta - 1 = 0$

32. $\tan \theta - \sqrt{3} = 2 \tan \theta$

33. $2 \sin^2 \theta + \sin \theta = 1$

34. $2 \cos^2 \theta - 5 \cos \theta = -2$

35. $\sec^2 \theta \tan \theta = \tan \theta$

36. $\csc^2 \theta \cot \theta = 4 \cot \theta$

37–44. *Solve the equations for t, $0 \le t < 2\pi$. When necessary, approximate solutions rounded to four decimal places.*

37. $8 \sin^2 t = 5 \sin t + 2$

38. $15(\cos^2 t - 4) = 12 \cos t - 59$

39. $4 \cos^2 t = \cos(-t)$

40. $\sin^2 t + \sin(-t) = 1$

41. $3 \sin^2 t - 10 \sin(-t) = 8$

42. $4 \cos^2 t + 19 \cos t = 5$

43. $\tan^2 t - 5 \tan t + 3 = 0$

44. $3 \cot^2 t - 4 \cot(-t) = 2$

45–50. *Solve the equations for α, $0° \le \alpha < 360°$. Approximate solutions to the nearest tenth of a degree.*

45. $1 - 4 \cos \alpha = -2 \cos^2 \alpha$

46. $4 \sin \alpha - 1 = 2 \sin^2 \alpha$

47. $2 \sin^2 \alpha - 5 \sin \alpha = 5$

48. $7 \cos^2 \alpha - 3 \cos \alpha - 5 = 0$

49. $5 \tan^2 \alpha = 8(5 - 2 \tan \alpha)$

50. $5 \sec^2 \alpha + 3(5 - \sec \alpha) = 18$

51–56. *Find all solutions to the equation given in the indicated exercise.*

51. Exercise 11

52. Exercise 10

53. Exercise 35

54. Exercise 34

55. Exercise 47

56. Exercise 46

57. The displacement of an oscillating pendulum is given by $d(t) = 3.6 \cos t$, where t is in seconds and $d(t)$ is in cm. Find (to the nearest tenth of a second) the time t for $0 \le t < 6.3$ when the displacement is -2 cm.

58. The instantaneous voltage e in a circuit at time t is given by the equation $e(t) = 1.5 \sin t$, where t is expressed in seconds and e is in volts. Find (to the nearest tenth of a second) the time t for $0 \le t < 6.3$ when the voltage is -1.2 volts.

59. The highest daily body temperature for a particular six-day virus is given by $T(t) = 98.9 + 5 \sin t$, where t is the day into the illness and $T(t)$ is the temperature in degrees Fahrenheit. Find (to the nearest tenth of a day) the following.

a. When will the body temperature be the highest at 103.9° (that is, when will the temperature spike)?

b. When in the illness will the temperature be 101°?

60. (See Exercise 59.) Use the graph of the temperature $T(t) = 98.9 + 5 \sin t$ to verify the solution to the following.

a. Find (to the nearest tenth of a day) the time(s) t when the temperature is 100°.

b. Find (to the nearest tenth of a degree) the temperature at the end of the six-day illness.

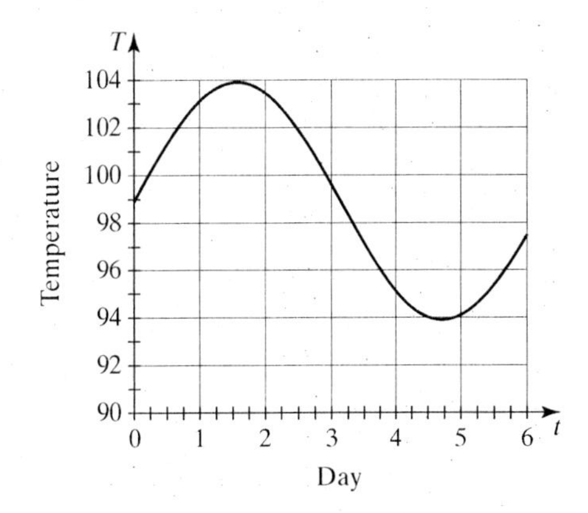

61–62. *The height h of a cannonball fired into the air at time t with an initial velocity of v_0 and with an angle of elevation θ is given by $h = -16t^2 + (v_0 \sin \theta)t$, where t is in seconds, v_0 is in feet per second, and h is in feet. (For -16, units of ft/sec^2 are intended.)*

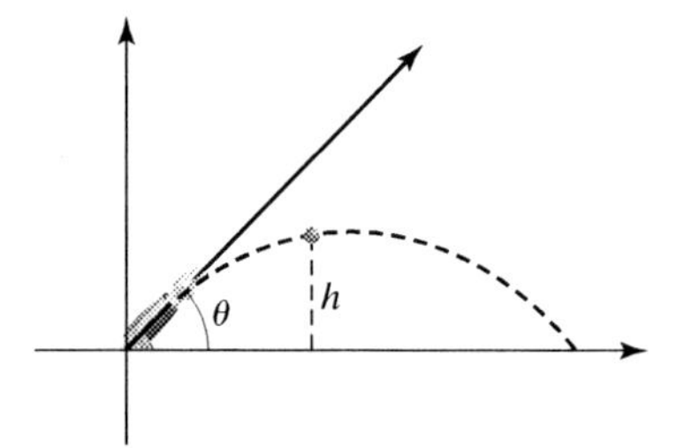

61. **a.** If $v_0 = 200$ ft/sec and the angle of elevation θ is 45°, what is the height (to the nearest foot) of the cannonball after 2 seconds?

b. How many seconds (to the nearest tenth of a second) will it take until the cannonball returns to the ground $(h = 0)$?

c. If we want the cannonball to be at a height of 100 feet after 1 second when $v_0 = 200$ ft/sec, what angle of elevation θ (to the nearest tenth of a degree) should be used?

62. **a.** If $v_0 = 450$ ft/sec and the angle of elevation θ is 30°, what is the height (to the nearest foot) of the cannonball after 2 seconds?

b. How many seconds (to the nearest tenth of a second) will it take until the cannonball returns to the ground $(h = 0)$?

c. If we want the cannonball to be at a height of 300 feet after 1 second when $v_0 = 450$ ft/sec, what angle of elevation θ (to the nearest tenth of a degree) should be used?

Discussion

63. Doug solved the equation $2 \sin x \cos x = \cos x$, $0 \leq x < 360°$ by using an incorrect procedure. Below is the work Doug did to solve the equation.

$$2 \sin x \cos x = \cos x$$
$$2 \sin x = 1$$
$$\sin x = \frac{1}{2}$$

$$x = 30°, 150°$$

Determine the incorrect procedure Doug used and discuss the effect it has on the solution.

5.3 More Trigonometric Equations, Multiple-Angle Equations

In this section we solve trigonometric equations that require trigonometric identities along with algebraic procedures to rewrite the equations into one of the two *solvable forms* discussed in Sections 5.1 and 5.2:

1. an equation with one first-degree trigonometric functional value equal to a real number which is immediately solvable by either recognizing a common angles functional value, or using inverse functions

or

2. an equation involving one, two, or more trigonometric functions that is comparable to an algebraic equation that is solvable using methods of algebra, like the factor method, the square root method, or the quadratic formula, that produce immediately solvable trigonometric equations.

Solving Trigonometric Equations Using Identities and Other Strategies

Let's see what additional strategies are needed to get equations into one of these solvable forms.

EXAMPLE 1 Solve: $3\cos x - 2\sin^2 x = 0, 0 \le x < 2\pi$.

SOLUTION This equation involves two trigonometric functions: $\cos x$ and $\sin x$. We first try to solve this equation with the factor method. But unlike previous equations that involve two trigonometric functions this equation does not factor. We would prefer to solve this equation if it involved only *one trigonometric function*. Is there an identity that will allow us to substitute one of these functions with an equivalent expression in terms of the other function? Yes, because we can use $\sin^2 x = 1 - \cos^2 x$, a form of the Pythagorean identity.

$$3\cos x - 2\sin^2 x = 0$$

$$3\cos x - 2(1 - \cos^2 x) = 0 \qquad \sin^2 x = 1 - \cos^2 x$$

$$2\cos^2 x + 3\cos x - 2 = 0 \qquad \text{Write the equation in standard form (quadratic).}$$

$$(2\cos x - 1)(\cos x + 2) = 0 \qquad \text{Factor.}$$

$$2\cos x - 1 = 0 \quad \text{or} \quad \cos x + 2 = 0 \qquad \text{Set each factor equal to 0.}$$

$$\cos x = \tfrac{1}{2} \quad \text{or} \quad \cos x = -2 \qquad \text{Solve for } \cos x.$$

$$x = \frac{\pi}{3}, \frac{5\pi}{3} \quad \text{(no solution)} \qquad \text{Solve for } x,\ 0 \le x \le 2\pi.$$

EXAMPLE 2 Solve: $6\cos\theta + 7\tan\theta = \sec\theta,\ 0° \le \theta < 360°$

SOLUTION This equation involves *three* trigonometric functions, and we cannot use the factor method. Writing each in terms of $\sin\theta$ and/or $\cos\theta$ provides an idea for the next step.

$$6\cos\theta + 7\tan\theta = \sec\theta$$

$$6\cos\theta + 7\frac{\sin\theta}{\cos\theta} = \frac{1}{\cos\theta} \qquad \text{Rewrite in terms of sine and cosine.}$$

$$6\cos^2\theta + 7\sin\theta = 1 \qquad \text{Multiply both sides by } \cos\theta \text{ to clear fractions. (See Caution.)}$$

$$6(1 - \sin^2\theta) + 7\sin\theta = 1 \qquad \cos^2\theta = 1 - \sin^2\theta$$

$$6\sin^2\theta - 7\sin\theta - 5 = 0 \qquad \text{Write the equation in standard form.}$$

$$(3\sin\theta - 5)(2\sin\theta + 1) = 0 \qquad \text{Factor.}$$

$$3\sin\theta - 5 = 0 \quad \text{or} \quad 2\sin\theta + 1 = 0 \qquad \text{Set each factor equal to zero.}$$

$$\sin\theta = \frac{5}{3} \quad \text{or} \quad \sin\theta = -\frac{1}{2} \qquad \text{Solve for } \sin\theta.$$

$$\text{(no solution)} \qquad \theta = 210°, 330° \qquad \text{Solve for } \theta,\ 0° \le \theta < 360°$$

CAUTION: When we multiplied by $\cos\theta$, we *assumed* that $\cos\theta \ne 0$. Therefore, we must check our answers because if we obtain any solutions for θ in which $\cos\theta = 0$, we will have to eliminate them. Neither 210° nor 330° needs to be eliminated since $\cos 210° \ne 0$ and $\cos 330° \ne 0$.

EXAMPLE 3 Solve: $\sin t = \cos t + 1, 0 \le t < 2\pi$

SOLUTION This equation involves more than one function and again cannot be solved using the factor method. Furthermore, there is no simple identity that reduces the equation to one function, and the equation is already in terms of $\sin t$ and $\cos t$. The Pythagorean identity would help, but it requires either $\sin^2 t$ or $\cos^2 t$ be involved in the equation. If we square both sides of the equation, we can use either $\sin^2 t = 1 - \cos^2 t$ or $\cos^2 t = 1 - \sin^2 t$.

$$
\begin{aligned}
\sin t &= \cos t + 1 && \\
(\sin t)^2 &= (\cos t + 1)^2 && \text{Square both sides. (See Caution.)} \\
\sin^2 t &= \cos^2 t + 2\cos t + 1 && (a+b)^2 = a^2 + 2ab + b^2 \\
1 - \cos^2 t &= \cos^2 t + 2\cos t + 1 && \sin^2 t = 1 - \cos^2 t \\
0 &= 2\cos^2 t + 2\cos t && \text{Write the equation in standard form.} \\
0 &= 2\cos t(\cos t + 1) && \text{Factor.} \\
2\cos t = 0 \quad &\text{or} \quad \cos t + 1 = 0 && \text{Set each factor equal to 0.} \\
\cos t = 0 \quad &\text{or} \quad \cos t = -1 && \text{Solve for } \cos t. \\
t = \frac{\pi}{2}, \frac{3\pi}{2} \quad & \quad\quad t = \pi && \text{Solve for } t,\ 0 \le t < 2\pi.
\end{aligned}
$$

CAUTION: *Squaring both sides of an equation can introduce solutions that are not solutions of the original equation (extraneous solutions).* Therefore, just as with multiplying both sides of an equation by a variable expression, we must check each solution of t.

Check $t = \frac{\pi}{2}$:

$\sin t = \cos t + 1$

$\sin\frac{\pi}{2} \stackrel{?}{=} \cos\frac{\pi}{2} + 1$

$1 = 0 + 1$ ✓

Check $t = \frac{3\pi}{2}$:

$\sin t = \cos t + 1$

$\sin\frac{3\pi}{2} \stackrel{?}{=} \cos\frac{3\pi}{2} + 1$

$-1 \ne 0 + 1$ ✗

Check $t = \pi$:

$\sin t = \cos t + 1$

$\sin\pi \stackrel{?}{=} \cos\pi + 1$

$0 = -1 + 1$ ✓

Therefore, $t = \frac{\pi}{2}, \pi$. ■

Multiple-Angle Equations

Some equations contain trigonometric functions of multiples of an angle. Let's see how to solve this new type of trigonometric equation.

EXAMPLE 4 Solve: $\sin 2x = \frac{\sqrt{3}}{2}, 0° \le x < 360°$

SOLUTION The equation contains a multiple angle, namely, $2x$. Since $0° \le x < 360°$, if we multiply each part of the inequality by 2, we get $0° \le 2x < 720°$. Therefore, the values of $2x$ are between $0°$ and $720°$. So we first find $2x$ between $0°$ and $360°$ and add multiples of $360°$ to each value to get all values of $2x$ between $0°$ and $720°$. Then we solve for x.

$$\sin 2x = \frac{\sqrt{3}}{2}$$
$$2x = 60°, 120°, 60° + 360°, 120° + 360°$$
$$\text{or } 2x = 60°, 120°, 420°, 480°$$
$$x = 30°, 60°, 210°, 240° \qquad \text{Divide each side by 2.}$$

CAUTION $\frac{\sin 2x}{2} \neq \sin x$

If the equation contains *different multiples of the angle* and is not immediately solvable, we first use identities to rewrite the functions in terms of the same multiple angle. Then, if the equation is not in a solvable form, we rewrite the equation to involve only one trigonometric function when possible. The next three examples demonstrate this strategy.

EXAMPLE 5 Solve: $\sin 2x - \sqrt{3} \sin x = 0, 0° \leq x < 360°$

SOLUTION This equation involves different angles: x and $2x$. If we use the sine double-angle identity, then all functions involve only x.

$$\sin 2x - \sqrt{3} \sin x = 0$$
$$2 \sin x \cos x - \sqrt{3} \sin x = 0 \qquad \sin 2x = 2 \sin x \cos x$$
$$\sin x \left(2 \cos x - \sqrt{3}\right) = 0 \qquad \text{Factor.}$$
$$\sin x = 0 \quad \text{or} \quad 2 \cos x - \sqrt{3} = 0 \qquad \text{Set each factor equal to 0.}$$
$$\sin x = 0 \quad \text{or} \quad \cos x = \frac{\sqrt{3}}{2} \qquad \text{Solve for } \sin x \text{ and } \cos x.$$
$$x = 0°, 180° \qquad x = 30°, 330° \qquad \text{Solve for } x, 0° \leq x < 360°.$$

So $x = 0°, 30°, 180°, 330°$.

EXAMPLE 6 Solve: $\cos 4x \cos x + \sin 4x \sin x = 0, 0° \leq x < 360°$

SOLUTION This equation involves different angles: $4x$ and x. We need to rewrite the equation in terms of only one multiple of x. Recognizing the left side of this equation as the expanded side of the cosine difference identity, we rewrite it.

$$\cos 4x \cos x + \sin 4x \sin x = 0$$
$$\cos(4x - x) = 0$$
$$\cos 3x = 0, \text{ for } 0° \leq 3x < 1080°$$
$$3x = 90°, 270°, 90° + 360°, 270° + 360°, 90° + 720°, 270° + 720°$$
$$\text{or } 3x = 90°, 270°, 450°, 630°, 810°, 990°$$
$$x = 30°, 90°, 150°, 210°, 270°, 330° \qquad \text{Divide each side by 3.}$$

EXAMPLE 7 Solve: $\cos^2 2\theta - 2\sin 2\theta + 2 = 0,\ 0 \le \theta < 2\pi$

SOLUTION Since the equation involves only 2θ, we *do not use double-angle identities to rewrite the expressions.* Instead, we notice that the equation involves two different functions, $\sin 2\theta$ and $\cos 2\theta$, so we use identities to rewrite in terms of one function and solve.

$\cos^2 2\theta - 2\sin 2\theta + 2 = 0$	
$(1 - \sin^2 2\theta) - 2\sin 2\theta + 2 = 0$	$\cos^2 2\theta = 1 - \sin^2 2\theta$
$\sin^2 2\theta + 2\sin 2\theta - 3 = 0$	Write the equation in standard form.
$(\sin 2\theta + 3)(\sin 2\theta - 1) = 0$	Factor.
$\sin 2\theta = -3$ or $\sin 2\theta = 1$	Set each factor equal to 0 and solve for $\sin 2\theta$.
(no solution) $\quad 2\theta = \frac{\pi}{2}, \frac{5\pi}{2}$	Solve for 2θ, where $0 \le 2\theta < 4\pi$.
$\theta = \frac{\pi}{4}, \frac{5\pi}{4}$	Divide by 2. $(0 \le \theta < 2\pi)$

It may be possible to solve trigonometric equations in different ways, depending on the strategy or trigonometric identity selected, as we have often demonstrated during our discussion in other areas of trigonometry. The following guidelines are recommended.

Solving Trigonometric Equations

As you proceed with the following steps in order, check before and after each step to see if the equation is in a solvable form outlined at the beginning of this section.

Step 1. If the equation involves different multiples of the angle, use identities to obtain the same multiple.

Step 2. If the equation involves more than one function, if possible, use identities to reduce it to involve one function. If it is not possible to use identities to reduce it to involve one function, try either of the following procedures. *Make sure to check your answers if you use either of these procedures.*

- **a.** Square both sides of the equation.
- **b.** Multiply both sides by the common denominator to eliminate fractions. Then, if possible, use identities to rewrite the equation in terms of one function.

Step 3. Identify the equation as linear or quadratic in form or as an equation that is factorable. Use established algebraic procedures.

Using Technology to Solve Trigonometric Equations

Many trigonometric equations are not solvable using the traditional methods outlined above. Real solutions to trigonometric equations can be found, approximated, or verified, using technology.

CONNECTIONS WITH TECHNOLOGY

Graphing Calculator/CAS

To find the real or approximate real solutions of trigonometric equations, we use a graphing utility and follow either of two methods.

Method 1 Use the graph intersection method. (See Section 5.1 Connections with Technology.)

Method 2 Rewrite the equation $f(x) = g(x)$ as $f(x) - g(x) = 0$ (or $g(x) - f(x) = 0$). Graph $y_1 = f(x) - g(x)$ and find the x-intercepts $(y = 0)$ of the graph over the requested interval. We refer to this as the ***x*-intercept method**. Check your utility manual to see what "solve," "x-intercept," "zero," or "root" functions (calculations) are available.

EXAMPLE Use a graphing utility to solve $x^2 + x - \sin 2x = 1 + \cos 2x$ for $-\pi \leq x \leq \pi$. Approximate solutions to four decimal places.

MODE: Radian; WINDOW: Xmin $= -\pi$, Xmax $= \pi$, Ymin $= -3$, Ymax $= 3$

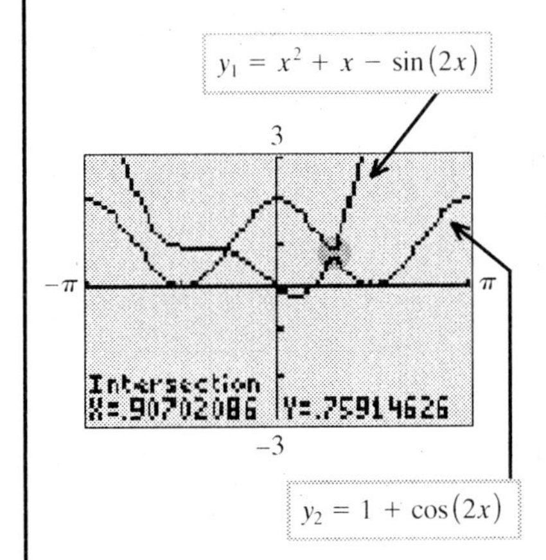

SOLUTION 1 GRAPH INTERSECT METHOD

Input $y_1 = x^2 + x - \sin(2x)$ and $y_2 = 1 + \cos(2x)$. The solutions (the x-coordinates that correspond to the points of intersection) rounded to four decimal places are $x = 0.9070$ and $x = -0.8529$. The screen shown left indicates one solution, and the other solution can be found similarly using the intersect function.

SOLUTION 2 *X*-INTERCEPT METHOD

Input $y_1 = x^2 + x - \sin(2x) - \cos(2x) - 1$. The solutions (the x-intercepts of the graph) rounded to four decimal places are $x = 0.9070$ and $x = -0.8529$. The screen shown below indicates one solution, and the other solution can be found similarly using the zero (root) function.

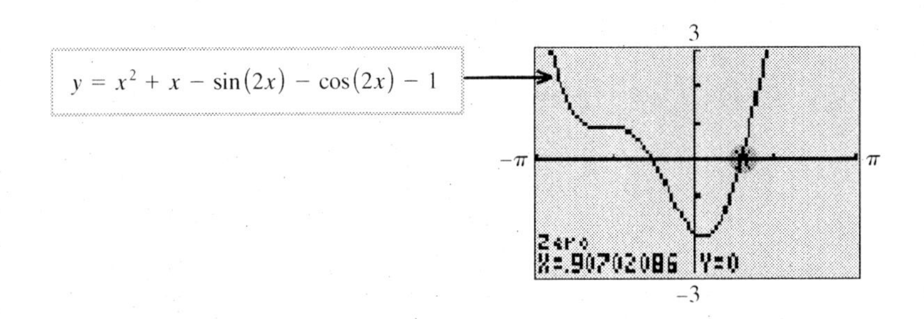

Systems of Equations

We have been solving trigonometric equations of one variable. The solutions are real numbers or measures of angles. If we consider a system of equations in two variables, the solution to the system consists of ordered pairs representing the intersection points of the graphs of the equations in the system.

EXAMPLE 8 Solve the system of equations $y = 4 \cos x$ and $y = 2$, where $0 \le x < 2\pi$.

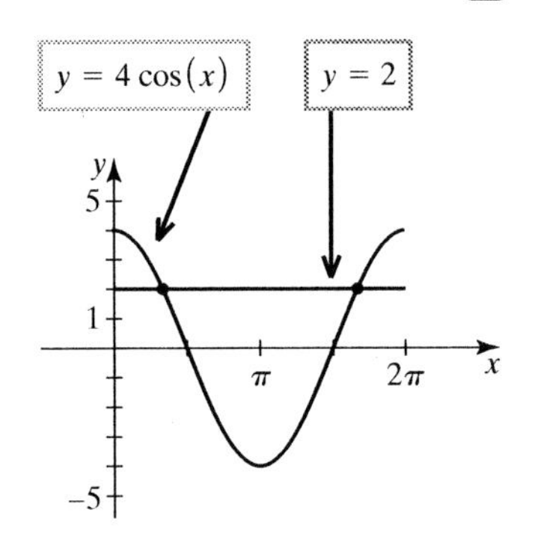

SOLUTION Looking at the graphs of $y = 4 \cos x$ and $y = 2, 0 \le x < 2\pi$, we see two points of intersection, which are the common solutions to the equations. The solution to this system will be the ordered pairs (x, y) representing these common points.

We now find the x-coordinates of these points by solving the system algebraically.

Since both equations are solved for y, we use the substitution method.

$$\left.\begin{array}{l} y = 4 \cos x \\ y = 2 \end{array}\right\} \rightarrow 4 \cos x = 2$$

$$\cos x = \frac{1}{2} \qquad \text{Divide each side by 4.}$$

$$x = \frac{\pi}{3}, \frac{5\pi}{3} \qquad \text{Solve for } x,\ 0 \le x < 2\pi.$$

Using the x-coordinates of the points of intersection, we find the corresponding y-coordinates. Substituting each value of x in either equation $(y = 4 \cos x$ or $y = 2)$ yields $y = 2$. Therefore, the solutions are $(\pi/3, 2)$ and $(5\pi/3, 2)$. ■

When solving systems of equations, it is recommended that you verify your results by graphing the system. (Be careful to set your calculator in the appropriate degree or radian mode.)

EXAMPLE 9 Solve the system of equations $y = \sin x$ and $y = \cos 2x$, where $0° \le x < 360°$. Graph the system to verify your answer.

SOLUTION To find the x-coordinates of the solutions, we use the substitution method.

$$\left.\begin{array}{l} y = \sin x \\ y = \cos 2x \end{array}\right\} \rightarrow \cos 2x = \sin x.$$

This equation involves different angles: x and $2x$. Since we have all the identities from Chapter 4 memorized (*right*?), we know to use the cosine double-angle identity that only involves $\sin x$. This form of the identity not only rewrites the equation to involve only x; it also involves only the sine function.

$$\cos 2x = \sin x$$

$$1 - 2\sin^2 x = \sin x \qquad \cos 2x = 1 - 2\sin^2 x$$

$$2\sin^2 x + \sin x - 1 = 0 \qquad \text{Write in standard form.}$$

$$(2 \sin x - 1)(\sin x + 1) = 0 \qquad \text{Factor.}$$

$$2 \sin x - 1 = 0 \text{ or } \sin x + 1 = 0 \qquad \text{Set each factor equal to 0.}$$

$$\sin x = \tfrac{1}{2} \text{ or } \sin x = -1 \qquad \text{Solve for } \sin x.$$

$$x = 30°, 150°, 270° \qquad \text{Solve for } x,\ 0° \le x < 360°.$$

To find the corresponding y, we replace each x in either $y = \sin x$ or $y = \cos 2x$.

When $x = 30°$:	When $x = 150°$:	When $x = 270°$:
$y = \sin x$	$y = \sin x$	$y = \sin x$
$y = \sin 30°$	$y = \sin 150°$	$y = \sin 270°$
$y = \frac{1}{2}$	$y = \frac{1}{2}$	$y = -1$

Hence, the solutions are $(30°, \frac{1}{2})$, $(150°, \frac{1}{2})$, and $(270°, -1)$.

Notice if we want to find these points using a graphing utility (degree mode), the graph intersection method displays both the x- and y- coordinates of each solution. The first point of intersection (solution) is shown on the following screen.

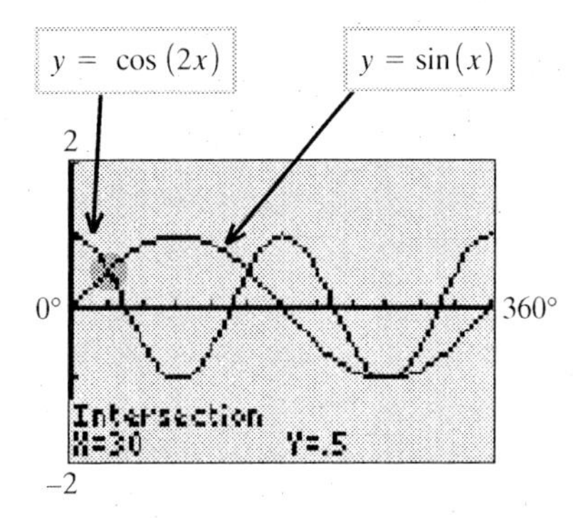

You will find that with experience you will develop insight into solving trigonometric equations. Your recognition of trigonometric identities is as important as your algebraic skills in solving trigonometric equations. And they certainly require practice!

Exercise Set 5.3

1–10. *Solve each equation for exact values of x, where* $0 \le x < 2\pi$.

1. $\sin^2 x - \cos^2 x = 0$

2. $2\cos^2 x - 3 = 3\sin x$

3. $4\cos x = 3\sec x$

4. $4\sin x = 2\csc x$

5. $\cos^2 x + 2\sin x + 2 = 0$

6. $2\sin^2 x - 7\cos x + 2 = 0$

7. $\tan x + \cot x + 2 = 0$

8. $2\cos x + \tan x = \sec x$

9. $\tan^2 x = 1 - \sec x$

10. $\cot^2 x = \csc x - 1$

11–22. *Solve each equation for exact values of x, where* $0° \le x < 360°$.

11. $\cos 2x - \sin x = 0$

12. $\cos x + \cos 2x = 0$

13. $\sin x - \sin 2x = 0$

14. $\sin 2x - \cos x = 0$

15. $\sin 2x = \cos 2x$

16. $\tan x = \cot x$

17. $2\cos^2 \dfrac{x}{2} = 2\cos x$

18. $2\sin^2 \dfrac{x}{2} = \cos x$

19. $\cos 2x = 2 + 3\sin x$

20. $\cos 2x = 2 + \cos x$

21. $\sqrt{3}\cos x = 1 + \sin x$

22. $\sin x - 1 - \sqrt{3}\cos x = 0$

23–32. *Solve each equation for exact values of θ, where $0° \le \theta < 360°$.*

23. $\cos\theta - \sin\theta = 1$
24. $2\cos\theta - \sin 2\theta = 0$
25. $\sin\frac{\theta}{2} + \cos\theta = 0$
26. $\cos\frac{\theta}{2} = \cos\theta$
27. $\cos 2\theta = 1 - 2\sin\theta$
28. $\cos 2\theta + 2 = 3\cos\theta$
29. $\cos^2\frac{\theta}{2} - \cos\theta = 1$
30. $\sin^2\frac{\theta}{2} + \cos\theta = \frac{1}{2}$
31. $\cos 8\theta\cos 6\theta + \sin 8\theta\sin 6\theta = \cos\theta$
32. $\sin 4\theta\cos 2\theta - \cos 4\theta\sin 2\theta = \sqrt{2}\sin\theta$

33–36. *Find all solutions to the equation given in the indicated exercise.*

33. Exercise 9
34. Exercise 8
35. Exercise 21
36. Exercise 22

37–42. *Find x, $0 \le x < 2\pi$, rounded to four decimal places.*

37. $10\cos^2 x + 23\cos x = 5$
38. $4\sin^2 x + 33\sin x - 27 = 0$
39. $\cos^2 x + \sin x = 0$
40. $3\cos x - \sec x = 1$
41. $2\sec^2 x - 7\sec x + 3 = 0$
42. $13\cot x + 11\csc x - 6\sin x = 0$

43–50. *Solve each equation for exact values of x, where $0 \le x < 2\pi$. Verify your answer graphically.*

43. $\sin 2x = \frac{1}{2}$
44. $\cos 3x = -1$
45. $\tan 2x = -\sqrt{3}$
46. $\sec 2x = -2$
47. $\sin 5x\cos 3x - \cos 5x\sin 3x = -\frac{\sqrt{2}}{2}$
48. $\cos 6x\cos 4x + \sin 6x\sin 4x = \frac{\sqrt{2}}{2}$
49. $\cos^2 2x - 8\cos 2x - 9 = 0$
50. $\sin^2 3x + 6\sin 3x + 5 = 0$

51–56. *Solve the system of equations for x, where $0° \le x < 360°$. Verify your answer graphically.*

51. $y = 4\cos^2 x + 2\sin^2 x,\ y = 3$
52. $y = \sin^2 x - \cos^2 x,\ y = 1$
53. $y = \sec x - \tan x,\ y = \cos x$
54. $y = \sec^2 x - 1,\ y = \sqrt{3}\tan x$
55. $y = \sin^2 x + \sin x,\ y = \cos^2 x$
56. $y = \sec x - 1,\ y = -\tan^2 x$

57–58. *The horizontal distance H in feet that a projectile will travel in the air is given by the equation*

$$H = \frac{v_0^2\sin(2\alpha)}{32\text{ ft/sec}^2},$$

where v_0 is the initial velocity of the projectile, and α is the angle of elevation.

57. If you can throw a ball with initial velocity of 110 feet per second, what should be the angle of elevation (to the nearest tenth of a degree) in order to have the ball go a horizontal distance of 350 feet?
58. If you can throw a ball with initial velocity of 140 feet per second, what should be the angle of elevation (to the nearest tenth of a degree) in order to have the ball go a horizontal distance of 600 feet?

Graphing Calculator/CAS Exercises

59–64. *Use a graphing calculator to find the solutions for x, $-\pi \le x \le \pi$, by using either of the two methods described in Connections with Technology in this section. Approximate the solutions rounded to four decimal places where necessary. If the equation has no solution, explain why.*

59. $x + \cos x + \sin x = \tan x$

60. $x + \sin x = \cos x$

61. $x^3 + \tan x + \cos x = 5$

62. $x^2 - 3\cos x + 4\sin x = \tan x$

63. $x^2 + 4 = \sin x$

64. $-2\cos^2 x = x^2 + 2$

Discussion

65. **a.** How many solutions does $\sin x = \frac{1}{2}$ have for $0 \le x < 2\pi$?

b. How many solutions does $\sin 3x = \frac{1}{2}$ have for $0 \le x < 2\pi$?

c. On the basis of your answers for parts (a) and (b), can you find a formula for the number of solutions that $\sin Bx = \frac{1}{2}$ would have for $0 \le x < 2\pi$?

d. Discuss how B affects the number of solutions to this trigonometric equation.

66. Gary is puzzled because his solutions are not correct for the following equation:

$$\cos 2x = 0,\ 0 \le x < 2\pi$$

Below is the work Gary did to solve this equation.

$$\frac{\cos 2x}{2} = \frac{0}{2}$$

$$\cos x = 0$$

$$x = \frac{\pi}{2}, \frac{3\pi}{2}$$

a. Find Gary's error. Explain to him why his work is incorrect. Find the correct solution to this equation.

b. On the basis of your answer to Exercise 63 (d), how would you know without looking at Gary's work that he does not have the correct solution?

67. Emily solved the equation $\csc x \cos x = \csc x$, $0 \le x < 2\pi$, and did not use any steps that *required* her to check for extraneous solutions. Yet her answer is incorrect.

$$\csc x(\cos x - 1) = 0$$

$$\csc x = 0 \text{ or } \cos x = 1$$

$$\text{(no solution)} \qquad x = 0$$

Discuss why her answer is incorrect, and why she needed to check her answer. Does this equation have a solution? What happens when you use either of the two calculator methods to find the solution described in *Connections with Technology* in this section?

5.4 Parametric Equations

In algebra, graphs in the coordinate plane are usually represented by a single equation involving two variables, such as x and y. These equations are called **Cartesian** or **rectangular equations**. We began the discussion of trigonometry in

Chapter 1 with the rectangular equation $x^2 + y^2 = 1$, which represents the unit circle. But suppose an object traveling on the unit circle, or any other plane curve, is starting at a specific point and going in a certain direction that traverses a certain path over a specific period of time. If we just look at the equation of the unit circle, it doesn't tell us any of this information. The equation only indicates the curve on which the object is traveling. It doesn't tell us where the object is on the circle at a given time or in which direction the object is traveling. Could there be equations that would provide this information? We introduce a third variable t, called a **parameter**, to find such equations. If we write both x and y as functions of t, we obtain what are called **parametric equations**.

Parametric Equations

If a plane curve consists of ordered pairs $(x, y) = (f(t), g(t))$, where f and g are continuous on an interval I, the equations $x = f(t)$ and $y = g(t)$, where t is in I, are called **parametric equations** for the curve, and t is called a **parameter**.

A parameter can represent time, the arc length (the distance an object has traveled along its path), or an angle. In Chapter 1 we parametrically defined the x- and y-coordinates on the unit circle as continuous functions of t,

$$x = f(t) = \cos t \quad \text{and} \quad y = g(t) = \sin t,$$

where t represents the length and direction of an arc with initial point $(1, 0)$ on the unit circle. If we let t represent time, these parametric equations tell us the location of an object on the circle at a given time and the direction the object is traveling on the circle. We use these parametric equations in Example 1.

Graphing Parametric Equations

EXAMPLE 1 The equations $x = \cos t$ and $y = \sin t$, where t represents units of time for $0 \le t \le 2\pi$, describe an object moving on the unit circle $x^2 + y^2 = 1$. Determine the following and sketch a graph that describes the motion of the object.

a. Where is the object at $t = 0$ (start time) and at $t = \pi/2$?

b. At what time will the object be at the point $(0, -1)$?

c. What is the direction of the motion?

d. How long will it take the object to return to its starting position?

SOLUTION We know the curve defined by the parametric equations is $x^2 + y^2 = 1$ and that the object is traveling on the unit circle. To determine the location of the object at various times, we set up a table of values for the three variables t, x and y. It is necessary to assign *increasing values of t* in the table, such as $t = 0, \pi/2, \pi, 3\pi/2$, and 2π, to determine the direction of the motion. Then we plot points (x, y) in order of the increasing values of t. Since $0 \le t \le 2\pi$, we refer to $t = 0$ as the start time, and $t = 2\pi$ as the end time.

t	$x = \cos t$	$y = \sin t$
0	1	0 (start)
$\frac{\pi}{2}$	0	1
π	-1	0
$\frac{3\pi}{2}$	0	-1
2π	1	0 (end)

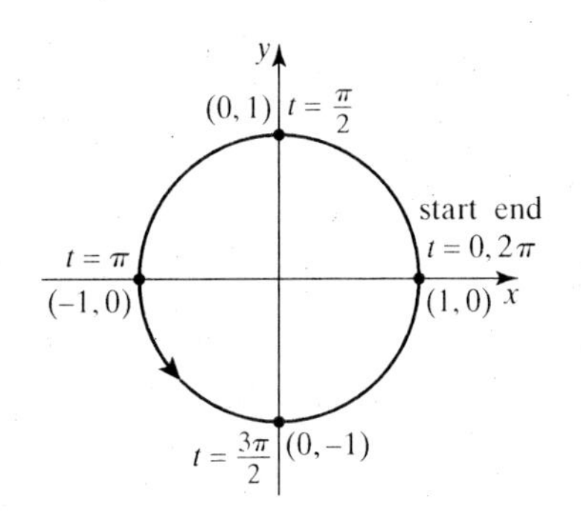

a. When $t = 0$, the object is at $(1, 0)$ (start position), and when $t = \pi/2$, the object is at $(0, 1)$.

b. The object is at $(0, -1)$ when $t = 3\pi/2$.

c. The direction of the motion is counterclockwise around the circle.

d. The object returns to the starting position of $(1, 0)$ when $t = 2\pi$.

■

Along the path we can indicate the t value used to determine the specific point (x, y). On the diagram above, we use an arrow to show the direction of successive points corresponding to increasing values of t, demonstrating the curve is traced out in a specific direction. This direction is called the **orientation** of the curve.

Example 2 demonstrates how it is possible for parametric equations to describe travel on the same curve as the object in Example 1, but in a different direction, with different starting or ending points, and at a different speed.

EXAMPLE 2 Describe the motion of $x = \sin 2t$, $y = \cos 2t$, where t represents time and $0 \le t \le \pi$, by finding the rectangular equation for the curve. Then graph the motion indicating the starting and ending points, the orientation, and find the time it takes to return to the starting position.

SOLUTION This example does not tell us the specific curve of the motion. To determine the rectangular equation of the curve, we *eliminate the parameter t* from the parametric equations to form an equation involving only x and y. Since $x = \sin 2t$ and $y = \cos 2t$, we substitute these into the Pythagorean identity:

$$\cos^2 2t + \sin^2 2t = 1 \qquad \text{Pythagorean identity}$$

$$y^2 + x^2 = 1 \qquad \cos 2t = y,\ \sin 2t = x$$

$$\text{or} \qquad x^2 + y^2 = 1$$

This rectangular equation tells us the curve of the motion is the unit circle. We set up the table for t $(0 \le t \le \pi)$, x, and y, remembering to let the values of t increase, and plot the points.

t	$x = \sin 2t$	$y = \cos 2t$
0	0	1 (start)
$\frac{\pi}{4}$	1	0
$\frac{\pi}{2}$	0	-1
$\frac{3\pi}{4}$	-1	0
π	0	1 (end)

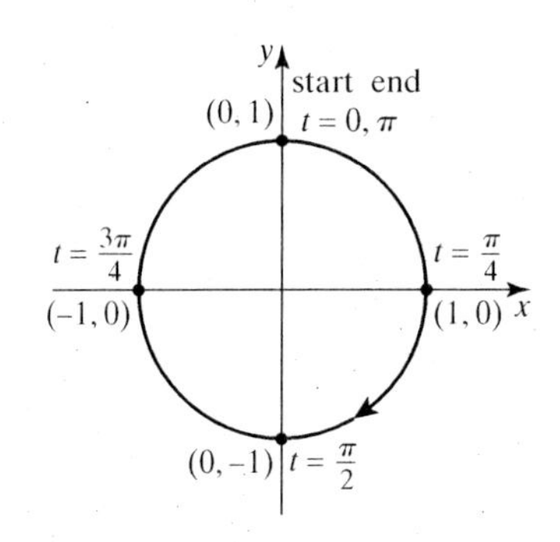

(continued)

The object on the unit circle starts at $(0, 1)$ and moves clockwise. It returns to the starting point making one full wrap of the circle after π units of time. (Does this agree with your knowledge of the period for $\cos 2x$ and $\sin 2x$?) ■

Examples 1 and 2 demonstrate that parametric equations for a particular curve are not unique. Actually there are infinitely many parametric representations of the same curve.

EXAMPLE 3 A particle moves in the plane with its x- and y-coordinates varying with t according to the given parametric equations. For each set of parametric equations, do the following.

i. Eliminate the parameter to find the rectangular equation.

ii. Find the starting point and ending point of the motion on the indicated interval for t.

iii. Sketch a graph that describes the motion and use an arrow to indicate the orientation (the direction of increasing values of t).

a. $x = 2\cos t,\ y = 3\sin t,\ 0 \le t \le 2\pi$ **b.** $x = 3t,\ y = 6t + 4,\ 0 \le t \le 2$

SOLUTION

a. Since the parametric equations involve the cosine and sine, we use the Pythagorean identity to eliminate the parameter. But in order to use this identity, we must solve $x = 2\cos t$ for $\cos t$ and $y = 3\sin t$ for $\sin t$.

$$x = 2\cos t \qquad y = 3\sin t$$

$$\frac{x}{2} = \cos t \qquad \frac{y}{3} = \sin t$$

Now we substitute these values into the Pythagorean identity.

$$\cos^2 t + \sin^2 t = 1$$

$$\left(\frac{x}{2}\right)^2 + \left(\frac{y}{3}\right)^2 = 1$$

$$\frac{x^2}{4} + \frac{y^2}{9} = 1$$

This is the equation of an ellipse with center at the origin, x-intercepts -2 and 2, and y-intercepts -3 and 3. To find the specific motion on the ellipse, we use the following table.

t	$x = 2\cos t$	$y = 3\sin t$
0	2	0 (start)
$\frac{\pi}{2}$	0	3
π	-2	0
$\frac{3\pi}{2}$	0	-3
2π	2	0 (end)

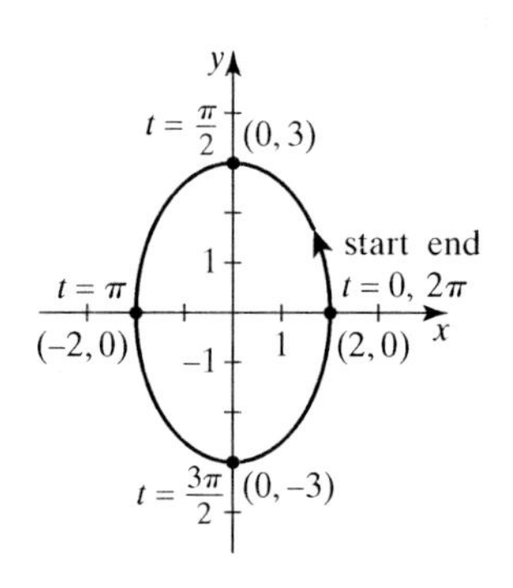

The orientation of the motion is counterclockwise, as the arrow indicates.

NOTE: The rectangular equation can aid us with graphing the curve of the motion. But we still need the parametric equations to tell us the position, direction, and speed of the motion.

b. These parametric equations do not involve trigonometric functions. In order to find a rectangular equation involving only x and y, we solve the first parametric equation for t and then substitute it into the second equation.

$$x = 3t, \ y = 6t + 4, \ 0 \le t \le 2$$

$$x = 3t \qquad \text{Parametric equation for } x.$$

$$\frac{x}{3} = t \qquad \text{Divide both sides by 3 to solve for } t.$$

$$y = 6t + 4 \qquad \text{Parametric equation for } y.$$

$$y = 6\left(\frac{x}{3}\right) + 4 = 2x + 4 \qquad \text{Let } t = \frac{x}{3} \text{ and simplify.}$$

The rectangular equation $y = 2x + 4$ is linear and the graph is a line. To find the motion on the line, we use a table.

t	$x = 3t$	$y = 6t + 4$
0	0	4 (start)
1	3	10
2	6	16 (end)

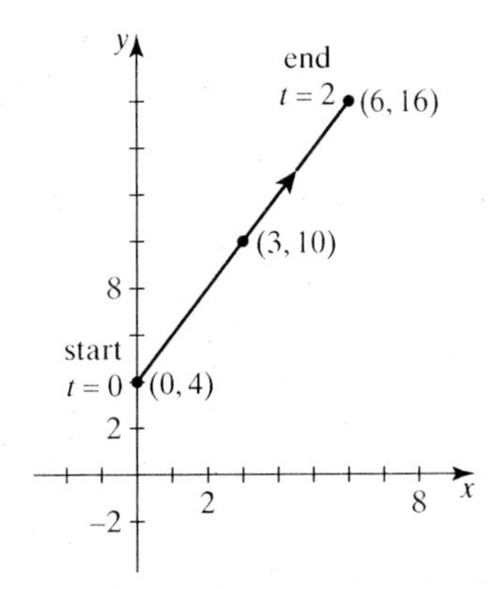

Eliminating the parameter is helpful if the resulting rectangular equation provides us with a more recognizable form of the curve. Although it may not always be possible to eliminate the parameter, it is still possible to obtain a graphical description of the rectangular equation and the motion of the object by using an extensive table of values for t, x, and y. When presented with the parametric equations and any restrictions on the parameter to describe how the curve is traced out, we should use the following steps.

Steps for Sketching Parametric Equations

Step 1. Eliminate the parameter (if possible) to find the rectangular equation. When x is defined in terms of the sine or cosine and y is defined in terms of the other, try using the Pythagorean identity to eliminate the parameter. For other trigonometric functions, try to use an equivalent form of the Pythagorean identity. If x and y are defined otherwise, try the substitution method.

Step 2. Set up a table for t, x, and y, with increasing values of t (the parameter). Where applicable, begin with the smallest and end with the largest value of the parameter. An extensive table may be necessary.

Step 3. Plot several points on the curve in order of the increasing values of the parameter. Indicate the orientation of the curve.

Application of Parametric Equations

EXAMPLE 4 If a projectile is fired with initial velocity v_0 in the direction of the angle of elevation α, then the motion (neglecting air resistance) is given by

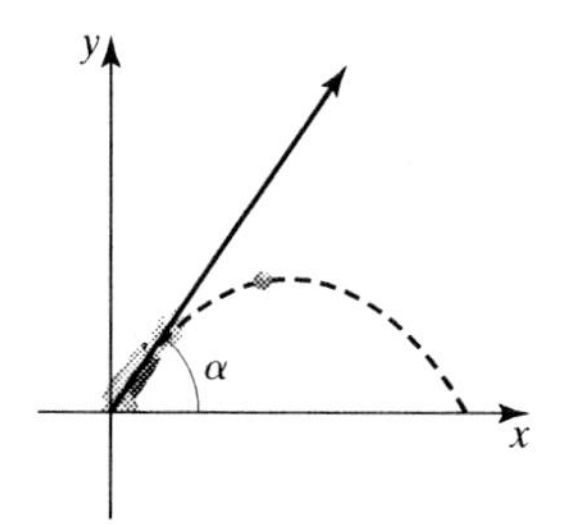

$$x = (v_0 \cos \alpha)t$$
$$y = -16t^2 + (v_0 \sin \alpha)t$$

where t is measured in seconds and x and y are measured in feet. Graph the motion for $\alpha = 30°$ and $v_0 = 80$ feet per second. Determine the time when the projectile strikes the ground and the horizontal distance it traveled.

SOLUTION We substitute the values for v_0 and α into the parametric equations.

$$x = (80 \cos 30°)t = 40\sqrt{3}t$$
$$y = -16t^2 + (80 \sin 30°)t = -16t^2 + 40t$$

Set up a table and use calculator approximations wherever necessary. Then plot the points. Here we let $t \geq 0$.

t	$x = 40\sqrt{3}t$	$y = -16t^2 + 40t$
0	0	0
1	69.3	24
2	138.5	16
3	207.8	−24

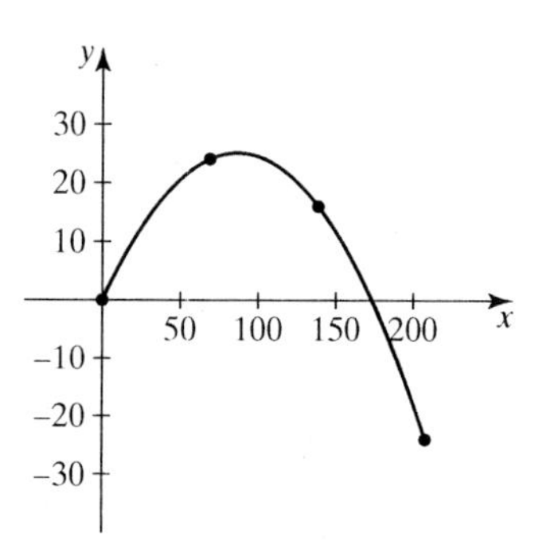

A more extensive table would produce a better description of the motion.

The projectile strikes the ground at a time when $y = 0$. At this time, we then determine the horizontal distance x.

$$y = -16t^2 + 40t$$
$$0 = -8t(2t - 5)$$
$$t = 0, \ t = \tfrac{5}{2} = 2\tfrac{1}{2}$$

Since 0 is the start time, it takes $2\frac{1}{2}$ seconds to strike the ground. We now find x.

$$x = 40\sqrt{3}t = 40\sqrt{3}\left(\frac{5}{2}\right) = 100\sqrt{3}$$

So the horizontal distance traveled is $100\sqrt{3}$ feet.

Notice in this example that instead of eliminating the parameter to find the rectangular equation, we used the parametric equations to determine the curve as well as answering the questions. ■

Using Technology to Graph Parametric Equations

Graphing utilities typically have three graphing modes: function, parametric, and polar. (See Chapter 6.) Most graphing utilities are programmed to accept only rectangular equations that are solved for y. Graphing the circle $x^2 + y^2 = 1$ (see Example 1) in function mode would require the input of two equations that are the result of solving this equation for y: $y = \sqrt{1 - x^2}$ and $y = -\sqrt{1 - x^2}$. In parametric mode, you can represent the graph with $x = \cos t$ and $y = \sin t$. A graphing utility in *parametric mode*, which will easily graph parametric equations and demonstrate their orientation, is particularly helpful when graphing complicated curves.

CONNECTIONS WITH TECHNOLOGY

Graphing Calculator/CAS

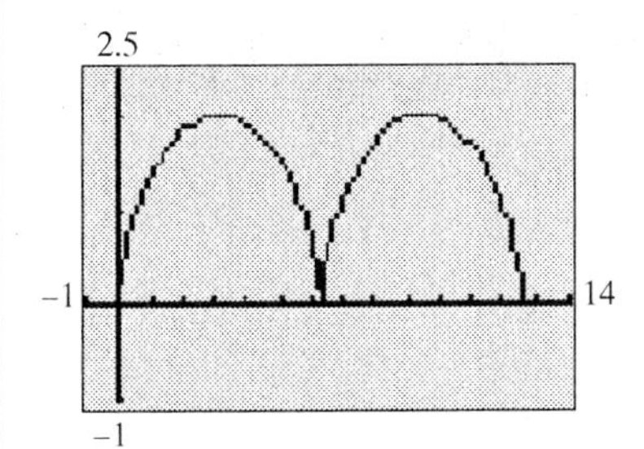

The curve traced by a fixed point P on the circumference of a circle as the circle rolls along a straight line in a plane is called a **cycloid**. The parametric equations for a cycloid traced by a point P on the circumference of a circle of radius a rolling along the x-axis, where one position of P is the origin, are:

$$x = a(t - \sin t)$$
$$y = a(1 - \cos t).$$

To graph the cycloid when $a = 1$, we use the following settings. MODE: radian and parametric; WINDOW: Tmin $= 0$, Tmax $= 4\pi$, Xmin $= -1$, Xmax $= 14$, Ymin $= -1$, Ymax $= 2.5$.

Input $x_1(t) = t - \sin(t)$ and $y_1(t) = 1 - \cos(t)$.

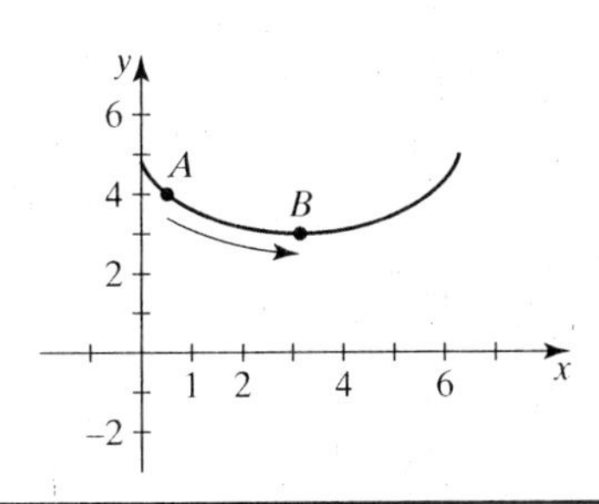

The cycloid is an interesting shape in that when inverted, it presents some perplexing situations. One such situation is that if you place a bead at point A and it falls to point B in a specific time, then placing the bead again at a point anywhere between A and B and letting it again fall to point B takes exactly the same amount of time as it did from point A! Perplexed? The cycloid sparked the interest of Galileo, Newton, Leibniz, and other famous mathematicians.

Finding Parametric Equations

We have not yet been asked to find parametric equations from a rectangular equation. If asked to find a set of parametric equations for a curve of the form $y = f(x)$ with domain $\boldsymbol{I}$, we could simply let $x = t$, and $y = f(x) = f(t)$, for t in $\boldsymbol{I}$.

EXAMPLE 5 Find three different sets of parametric equations for $y = 1 - x^2$.

SOLUTION The domain of $y = 1 - x^2$ is $\mathbb{R}$.

1. If we let $x = t$, then $y = 1 - t^2$, for t in $\mathbb{R}$.
2. If we let $x = 2t$, then $y = 1 - 4t^2$, for t in $\mathbb{R}$.
3. If we let $x = \cos t$, $y = \sin^2 t$, for t in $\mathbb{R}$. ■

Different sets of parametric equations for the same curve can represent different orientations, starting and ending points, speed, and paths on the curve. In fact, we could find many other choices of parametric equations for $y = 1 - x^2$ than the three given in Example 5. More challenging parametric representations of a curve are discussed in advanced mathematics classes.

Exercise Set 5.4

1–12. *Eliminate the parameter t from each set of parametric equations, and find the corresponding rectangular equation. Based on your knowledge of the graph of the rectangular equation, describe the curve represented by the parametric equations.*

1. $x = -\sin t,\ y = \cos t$
2. $x = -\cos t,\ y = \sin t$
3. $x = 4\cos t,\ y = 5\sin t$
4. $x = 2\sin t,\ y = 3\cos t$
5. $x = \cos t,\ y = \cos t + 4$
6. $x = 2\sin t,\ y = 4\sin t - 6$
7. $x = 2t - 4,\ y = 4t$
8. $x = 3t + 1,\ y = 6t$
9. $x = 2t,\ y = 12t^2$
10. $x = \frac{t}{2},\ y = 5t^2 + 2$
11. $x = \sec t,\ y = \tan^2 t$
12. $x = \cot t,\ y = \csc^2 t$

13–24. *The following parametric equations represent the path of a moving object with t in units of time. Eliminate the parameter t from each set of parametric equations, and find the corresponding rectangular equation. Then graph the motion of the object indicating starting and ending points, as well as the orientation of the curve.*

13. $x = -\cos t,\ y = \sin t,\ 0 \le t \le 2\pi$
14. $x = -\sin t,\ y = \cos t,\ 0 \le t \le 2\pi$
15. $x = 3\sin t,\ y = 2\cos t,\ 0 \le t \le 2\pi$
16. $x = 2\sin t,\ y = 3\cos t,\ 0 \le t \le 2\pi$
17. $x = \sin 2t,\ y = -\cos 2t,\ 0 \le t \le \pi$
18. $x = -\sin 2t,\ y = \cos 2t,\ 0 \le t \le \pi$
19. $x = 4t,\ y = 4t - 3,\ 0 \le t \le 2$
20. $x = 3t,\ y = 6 - 6t,\ 0 \le t \le 2$
21. $x = \cos t,\ y = 5\cos t,\ 0 \le t \le \pi$
22. $x = \sin t,\ y = -\sin t,\ 0 \le t \le \pi$
23. $x = 2t,\ y = 4t^2 + 1,\ 0 \le t \le 3$
24. $x = \frac{1}{4}t,\ y = 6 - t^2,\ 0 \le t \le 4$

25–28. *Find the value of t in the indicated interval that determines the given point on the graph of the given parametric equations.*

25. $x = 5 - 2t,\ y = 6t,\ 0 \le t \le 3$

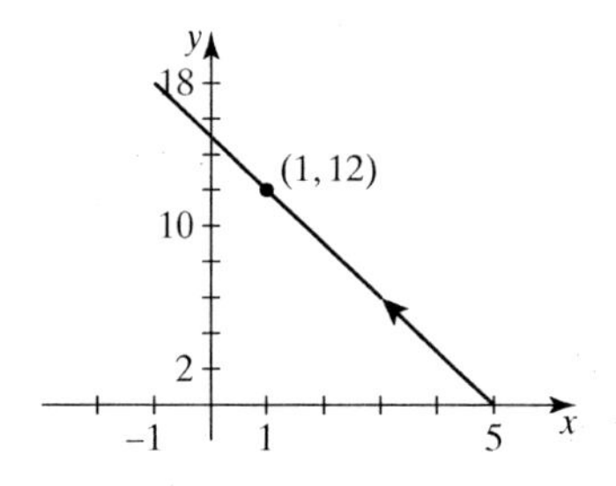

26. $x = \cos 4t,\ y = -\sin 4t,\ 0 \le t \le \frac{\pi}{4}$

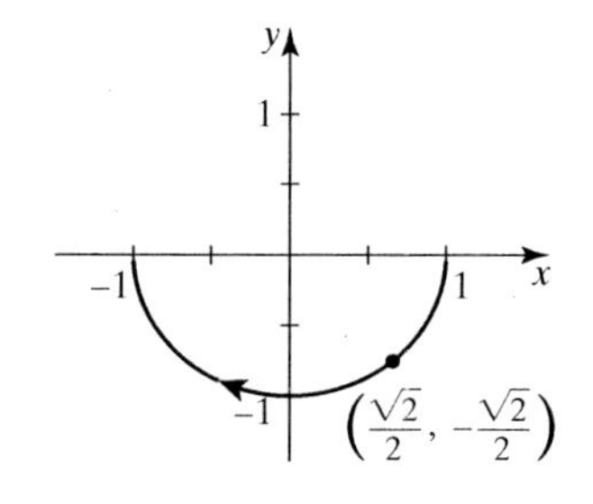

27. $x = 6 - t^2,\ y = 2t,\ 0 \le t \le 7$

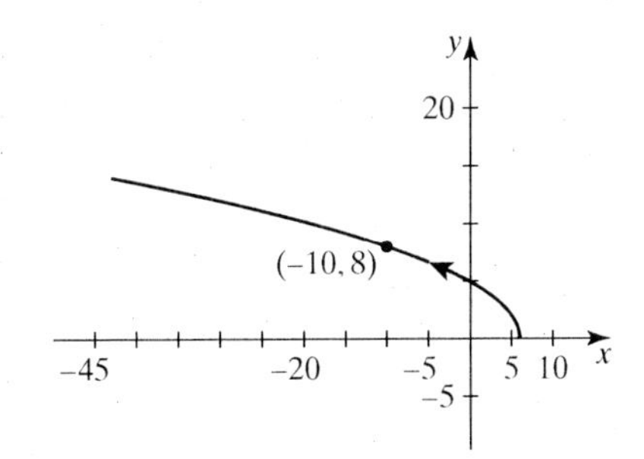

28. $x = 5\cos t,\ y = -4\sin t,\ 0 \le t \le 2\pi$

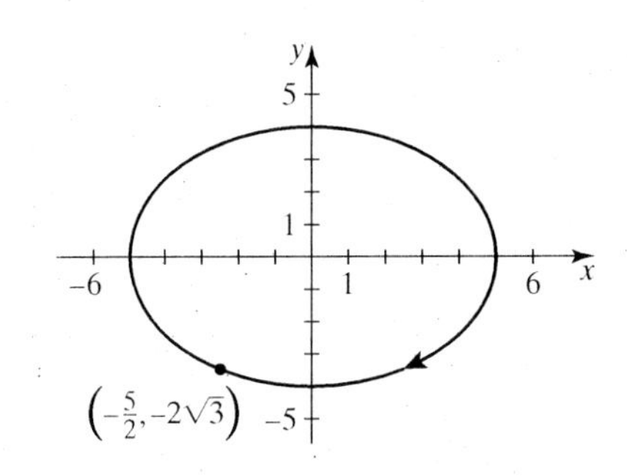

29–32. **a.** *Determine the parametric equations for the path of the projectile by using the given α and v_0, where*

$$x = (v_0 \cos \alpha)t$$

and

$$y = -16t^2 + (v_0 \sin \alpha)t.$$

b. *Graph the motion.*

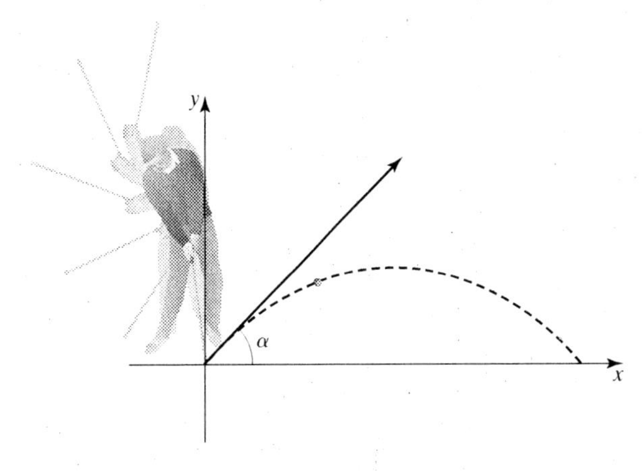

c. *Approximate to the nearest tenth of a second the time it takes for the projectile to return to ground level, and the horizontal distance to the nearest foot it has moved.*

29. A golf ball is hit at an angle of 60° with the ground with initial velocity of 48 feet per second.

30. A model rocket is launched at an angle of 45° with the ground with initial velocity of 100 feet per second.

31. A missile is fired with initial velocity of 120 feet per second at an angle of 75° with the ground.

32. A hockey puck is shot with initial velocity of 60 feet per second at an angle of 15° with the ice (ground).

33–36. *Find two different sets of parametric equations for the given rectangular equation.*

33. $y = 4x + 1$

34. $y = 6x - 12$

35. $x^2 = 1 - 4y^2$

36. $1 - y = x^2$

Graphing Calculator/CAS Exercises

37–40. *Graph the parametric equations over the indicated interval for t. Use parametric and radian mode along with your knowledge of amplitude to estimate window settings for x and y. If necessary, refine your window.*

37. A hypocycloid whose parametric equations are:

$$x = 8\cos t + 2\cos 4t \quad y = 8\sin t - 2\sin 4t \quad (0 \le t \le 2\pi)$$

38. A four-cusped hypocycloid whose parametric equations are:

$$x = 2\cos^3 t \quad y = 2\sin^3 t \quad (0 \le t \le 2\pi)$$

39. A deltoid whose parametric equations are:

$$x = 2\cos t + \cos 2t \quad y = 2\sin t - \sin 2t \quad (0 \le t \le 2\pi)$$

40. The epicycloids whose parametric equations are:

$$x = 4\cos t - \cos 4t \quad y = 4\sin t - \sin 4t \quad (0 \le t \le 2\pi)$$

Discussion

41. Discuss the benefits of parametric equations.

42. Suppose an object is moving along the unit circle described by the parametric equations

$$x = \cos bt \quad \text{and} \quad y = \sin bt.$$

Discuss how different values of b affect the motion. (Make sure to include the possibility of $b < 0$ in your discussion.)

Chapter 5 Summary

5.1–5.3 Solving Trigonometric Equations

Equations involving trigonometric functions that are true only for specific values for which they are defined are called conditional trigonometric equations. To solve equations for these specific values, we use algebraic procedures, trigonometric identities, and either recognize common angle values or use inverse trigonometric functions.

Solvable Form

An equation is in a solvable form if it is in one of the following two forms.

1. an equation with one first-degree trigonometric functional value equal to a real number, which is immediately solvable by either recognizing common angle (arc) functional values or using inverse functions, and
2. an equation involving one, two, or more trigonometric functions that is comparable to an algebraic equation that is solvable using methods of algebra, such as the factor method, the square root method, or the quadratic formula.

To obtain one of these two forms, we use the following steps in order.

Steps to Solve Trigonometric Equations

Always check to see if the equation is in a solvable form before and after trying each step.

Step 1. Use identities to obtain the same multiple angle.

Step 2. If possible, reduce the equation to one function. This may involve the procedures of squaring both sides of the equation or multiplying both sides by an expression to eliminate the denominators, along with the use of identities.

Step 3. Identify the equation (linear, quadratic, and so on) and use established algebraic procedures to solve the equation.

We use the periodic property of the trigonometric functions to write an expression that represents all solutions.

Systems of Equations

The real solutions of systems of two variable trigonometric equations consist of ordered pairs that represent the points of intersection of the graphs of each equation. The two equations are reduced to one equation in one variable, usually by the substitution method. Once the solutions for this equation are found using procedures previously outlined, each of these values are substituted into either of the given equations to find the corresponding second value in the ordered pair.

A graphing utility can be used to find, approximate, or verify real solutions to trigonometric equations as well as systems of trigonometric equations.

5.4 Parametric Equations

If a plane curve is a set of points (x, y) such that $x = f(t)$, $y = g(t)$ are continuous on an interval I, where t is in I, then $x = f(t)$ and $y = g(t)$ are called parametric equations. The parameter, in this case t, can sometimes be eliminated to obtain a rectangular equation. The plane curve can be graphed by selecting values for the parameter t and plotting the corresponding values $(x = f(t), y = g(t))$. The orientation of the curve can be determined by indicating the direction of points corresponding to increasing values of the parameter.

Chapter 5 Review Exercises

5.1–5.3 **1–14.** *Solve the equation for exact values of x, where* $0° \leq x < 360°$.

1. $\sqrt{2}\cos x + 1 = 0$

2. $2\sin(-x) = \sqrt{3}$

3. $\cos x \sin x = 2\sin x$

4. $\tan^3 x - 3\tan x = 0$

5. $\dfrac{1 - \sin x}{1 + \sin x} = 3$

6. $3\sec^2 x - 4 = 0$

7. $2\sin^2 x = 2\cos^2 x - 1$

8. $\sin x + 2\cot x = 2\csc x$

9. $\sin x - \cos x = \sqrt{2}$

10. $4\cos^2 x - 4\sin x - 5 = 0$

11. $\cos^2 x - \cos 2x = \sin^2 x + \sin 2x$

12. $8\sin^4 x - 10\sin^2 x + 3 = 0$

13. $\csc^2 x = \cot x + 1$

14. $\sin\frac{x}{2} = \frac{\sqrt{3}}{2}$

15–22. *Solve the equation for exact values of x, $0 \le x < 2\pi$.*

15. $\tan 2x = 1$

16. $2\cos 3x + 3 = 1$

17. $\cos 2x - \cos x = 0$

18. $2\sin x + \sqrt{2}\sin 2x = 0$

19. $2\sin^2 x - 11\cos x = 7$

20. $\sin 2x - \sqrt{2}\cos x = 0$

21. $\left(\cos\frac{x}{2}\right)^2 + \frac{\sqrt{2}}{2}\cos x = 1 + \frac{1}{2}\cos x$

22. $13\cot x + 11\csc x = 6\sin x$

23–28. *Solve the equation for exact values of θ, where $0° \le \theta < 360°$.*

23. $(\sin\theta + 1)(\sin\theta - 1) = 0$

24. $\left(2\cos\theta + \sqrt{2}\right)\left(\cos\theta + 1\right) = 0$

25. $\sqrt{2}\sin 2\theta\tan 2\theta - \tan 2\theta = 0$

26. $\sin 2\theta + 2\sin^2\theta = 0$

27. $2\cos\theta\sin\theta + \sin 2\theta = 2$

28. $\sqrt{3}\sec^2 2\theta - 2\sec 2\theta = 0$

29–36. *Solve each equation for x, $0° \le x < 360°$, and round to the nearest tenth of a degree. If there is no solution, explain why.*

29. $3\sin^2 x = 1$

30. $\cos^2 x - 2 = 0$

31. $\cos^2 x = 3\sin x + 2$

32. $5\sin^3 x = 2\sin x$

33. $2\sec^2 x - 5\tan x = 0$

34. $6\cot^2 x + 5 = 13\cot x$

35. $\sec^2 x = \frac{1}{2}\sec x$

36. $\csc^2 x = 5$

37–42. *Find all solutions to the equation given in the indicated exercise.*

37. Exercise 5

38. Exercise 6

39. Exercise 19

40. Exercise 20

41. Exercise 29

42. Exercise 28

43–46. *Solve the system of equations for exact values of x, where $0 \le x < 2\pi$. Verify your answer graphically.*

43. $y = 1 + \sin x,\ y = 2\cos^2 x$

44. $y = \sin x,\ y = \csc x$

45. $y = \tan^2 x,\ y = \frac{3}{2}\sec x$

46. $y = \cos^2 x,\ y = \tan x - \sin^2 x$

47–48. *Use a graphing utility to approximate, to four decimal places, the solutions to the following equation for x, where $0 \le x < 2\pi$.*

47. $\cos^3 x = \sin x$

48. $\sin^4 x = \cos x$

5.4

49–54. *Eliminate the parameter t from each set of parametric equations and find the rectangular equation. Then graph the motion of the object, indicating starting and ending points as well as the orientation of the curve.*

49. $x = \sin\frac{1}{2}t,\ y = \cos\frac{1}{2}t,\ 0 \le t \le 2\pi$

50. $x = 3\cos t,\ y = -2\sin t,\ 0 \le t \le 2\pi$

51. $x = \tan t,\ y = \tan^2 t,\ 0 \le t \le \dfrac{\pi}{4}$

52. $x = \cos t,\ y = 1 - \sin^2 t,\ 0 \le t \le \dfrac{\pi}{2}$

53. $x = 4t - 3,\ y = 6 - 8t,\ -2 \le t \le 2$

54. $x = 2 - \frac{1}{2}t,\ y = 6t,\ -4 \le t \le 4$

55. The displacement of an oscillating pendulum is given by $d(t) = 9.5\cos(\pi t)$, where t is in seconds and $d(t)$ is in cm. Find (to the nearest tenth of a second) the time t for $0 \le t < 2$ when the displacement is 0 cm.

56. If a projectile is fired into the air with initial velocity of 1500 ft/sec in the direction of the angle of elevation α, then its height off the ground at time t is given by

$$h = -16t^2 + (1500 \sin \alpha)t,$$

where t is in seconds and h in feet. Find the angle of elevation to the nearest degree if the projectile takes 2 seconds to reach a height of 1200 ft.

Chapter 5 Test

1–6. *Solve the equation for exact values of x, where* $0 \le x < 2\pi$.

1. $\sqrt{2}\sin(-x) - 1 = 0$
2. $2\cos^2 x - \cos x = 0$
3. $\sin^2 x + 8\sin x = 9$
4. $\sin 2x + \sin x = 0$
5. $\cos\dfrac{x}{2} = -\cos x$
6. $4\sin x - \csc x = 0$

7–10. *Solve the equation for exact values of* θ, *where* $0 \le \theta < 360°$.

7. $\dfrac{\tan 2\theta - \tan\theta}{1 + \tan 2\theta \tan\theta} = -\sqrt{3}$
8. $\cos 2\theta \cos 4\theta + \sin 2\theta \sin 4\theta = \frac{1}{2}$
9. $\cos 2\theta + 2 = 3\cos\theta$
10. $2\cos^2\theta - \sqrt{2}\sin\theta = 2$

11–14. *Solve each equation for x,* $0° \le x < 360°$, *rounded to the nearest tenth of a degree.*

11. $\tan^2 x + 4\tan x = 4$
12. $\left(\sin\dfrac{x}{2}\right)^2 - \cos x = 0.4$
13. $10\sin x = 5\csc x + 10$
14. $\cot^3 2x = \dfrac{1}{27}$

15–16. *Solve the system of equations for exact values of x, where* $0 \le x < 2\pi$.

15. $y = 3 - 5\sin x,\ y = \cos 2x$
16. $y = \sqrt{3}\tan x,\ y = \sec^2 x - 1$

17–18. *Solve the equation for exact values of x where a)* $0 \le x < 2\pi$ *and b)* $x \in \mathbb{R}$.

17. $\cos^4 x - \sin^4 x = 0.5$
18. $\cot^2 x = 2\csc^2 x - 2$

19–20. *Eliminate the parameter and find the rectangular equation. Then graph the parametric equations and indicate the starting point, ending point, and orientation.*

19. $x = -4\cos t$, $y = 4\sin t$, $0 \le t \le \pi$
20. $x = t + 1$, $y = 4t^2$, $0 \le t \le 2$

21. The displacement of a spring that is oscillating up and down is given by the equation $d(t) = 2 \sin t + \cos 2t - 1$, where t is in seconds and d is in centimeters. For what values of t, where $0 \leq t \leq 3$, will the displacement be 0 to the nearest tenth of a second?

22. If a projectile is fired with initial velocity v_0 in the direction of the angle of elevation α, then the motion (neglecting air resistance) is given by

$$x = (v_0 \cos \alpha)t$$
$$y = -16t^2 + (v_0 \sin \alpha)t,$$

where t is measured in seconds and x and y are measured in feet. Determine the time (to the nearest tenth of a second) when the projectile strikes the ground if $v_0 = 80$ feet per second and $\alpha = 40°$.

Chapter 6

Vectors, Polar Equations, and Complex Numbers

Pilots that fly jet-powered aircraft are familiar with a technique that produces control forces by redirecting engine exhaust flow, known as thrust vectoring. In 1996, NASA began flight-testing a new thrust-vectoring concept on the F-15 aircraft that should lead to increases in the performance of both civil and military aircraft.

In this chapter we begin the study of vectors and apply them to real-world situations.

Important Concepts

- vectors
- geometric and algebraic vector operations
  - —resultant
  - —dot product
- heading, drift angle, and true course
- projections
- work
- polar coordinates
- polar graphs
- complex numbers
- trigonometric form for complex numbers
- DeMoivre's Theorem

6.1 Geometric Vectors and Applications

Many quantities such as length, area, temperature, speed, and mass are completely described by their magnitudes. These quantities are called **scalars**. Other quantities, called **vectors**, require *both* the magnitude and direction to be completely described. You have probably heard a meteorologist describe wind movement by giving its speed and direction, such as "35 mph southwest." Speed and direction together form a vector quantity called **velocity**. Other examples of vectors are acceleration, displacement, and force since these quantities also involve magnitude and direction.

Vectors Viewed Geometrically

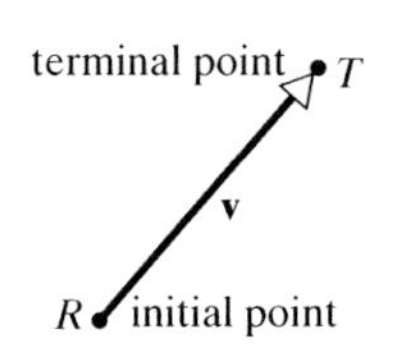

Vectors can be represented geometrically by directed line segments (or arrows). The arrowhead indicates the **direction** of the vector, and the length of the arrow describes the **magnitude** of the vector. The vector with **initial point** R (the tail of the arrow) and **terminal point** T (the tip of the arrowhead) can be represented by

$$\overrightarrow{RT}, \vec{v}, \text{ or } \mathbf{v},$$

read as "vector RT" or "vector v." We use boldface letters to represent vectors in order to distinguish them from scalars. The **magnitude** (or **norm**), designated by $|\mathbf{v}|$, is the length of the vector **v** and is a real number, or scalar. A magnitude's unit of measure depends on the quantity it represents.

Equivalent Vectors

Two vectors are **equivalent** if and only if they have the same magnitude and same direction. If two vectors are equivalent, you can slide one onto the other, keeping the one you are sliding parallel to its original position, so that their initial points coincide and their terminal points coincide. As a result, equivalent vectors are said to be equal even though they may be in different positions.

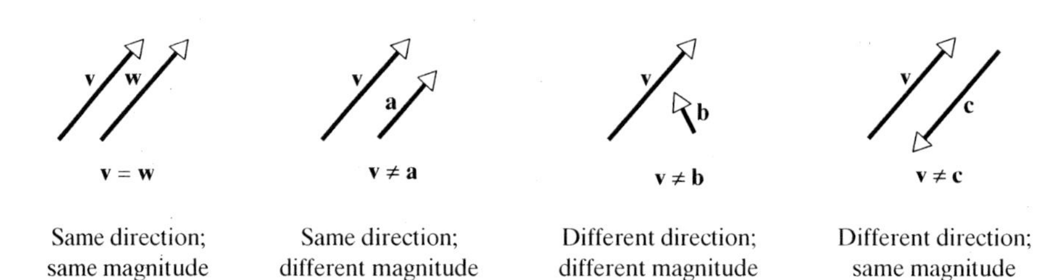

If **v** and **c** have the same magnitude and opposite directions, then **c** is the *opposite* of **v**, or $\mathbf{c} = -\mathbf{v}$.

Vectors can be used in applications of many physical problems. But before we can apply vectors, we need to be able to perform vector operations of multiplication and addition. When dealing with vectors, real numbers are referred to as scalars.

Scalar Multiples of Vectors

For any vector $\mathbf{v}$ and any scalar c, the vector $c\mathbf{v}$ is a vector whose magnitude is $|c|$ times the magnitude of $\mathbf{v}$. The vector $c\mathbf{v}$ is called a **scalar multiple** of $\mathbf{v}$.

If $c > 0$, $c\mathbf{v}$ has the same direction as $\mathbf{v}$.
If $c < 0$, $c\mathbf{v}$ has the opposite direction of $\mathbf{v}$.
If $c = 0$, $c\mathbf{v} = 0\mathbf{v} = \mathbf{0}$.

The vector $\mathbf{0}$, called the **zero vector**, has zero magnitude and arbitrary direction.

The vectors to the left are examples of scalar multiples of $\mathbf{v}$: The geometric vectors $2\mathbf{v}$ and $\frac{1}{2}\mathbf{v}$ have the same direction as $\mathbf{v}$. The magnitude (length) of $2\mathbf{v}$ is $|2\mathbf{v}| = 2|\mathbf{v}|$ and the magnitude of $\frac{1}{2}\mathbf{v}$ is $|\frac{1}{2}\mathbf{v}| = \frac{1}{2}|\mathbf{v}|$. The vector $-3\mathbf{v}$ has the opposite direction as $\mathbf{v}$ with magnitude $|-3\mathbf{v}| = |-3||\mathbf{v}| = 3|\mathbf{v}|$.

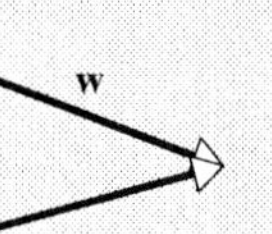

Addition of Vectors Geometrically

For any two vectors $\mathbf{v}$ and $\mathbf{w}$ such that the terminal point of $\mathbf{v}$ is the initial point of $\mathbf{w}$, their sum, denoted by $\mathbf{v} + \mathbf{w}$, is a vector having the same initial point as $\mathbf{v}$ and the same terminal point as $\mathbf{w}$. The vector $\mathbf{v} + \mathbf{w}$ is called the **sum** or **resultant** of $\mathbf{v}$ and $\mathbf{w}$.

EXAMPLE 1 Add vectors $\mathbf{v}$ and $\mathbf{w}$.

a.

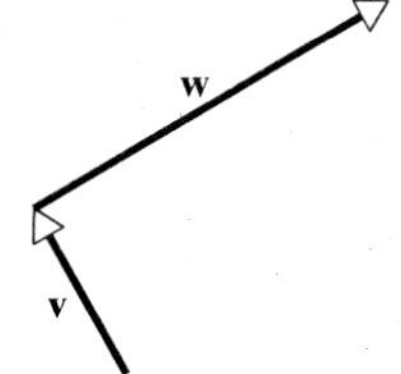

b.

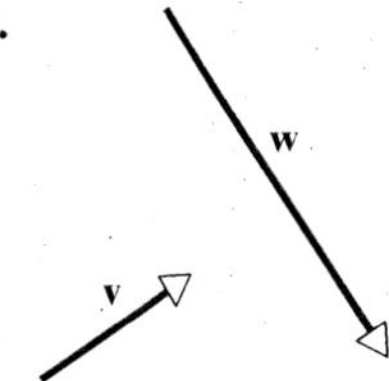

SOLUTION

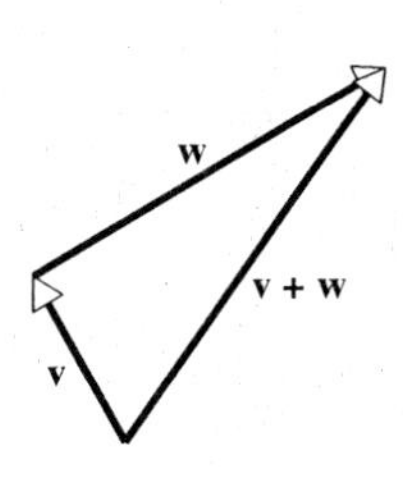

a. The vectors are positioned so that the terminal point of $\mathbf{v}$ coincides with the initial point of $\mathbf{w}$. To obtain the resultant $\mathbf{v} + \mathbf{w}$, we connect the initial point of $\mathbf{v}$ with the terminal point of $\mathbf{w}$. Since a triangle is formed by drawing this resultant, we refer to this procedure as the **triangle method**.

b. Since the terminal point of $\mathbf{v}$ is not the initial point of $\mathbf{w}$, we slide $\mathbf{v}$ (or we could slide $\mathbf{w}$) into this position. To obtain the resultant $\mathbf{v} + \mathbf{w}$, we connect the initial point of $\mathbf{v}$ with the terminal point of $\mathbf{w}$.

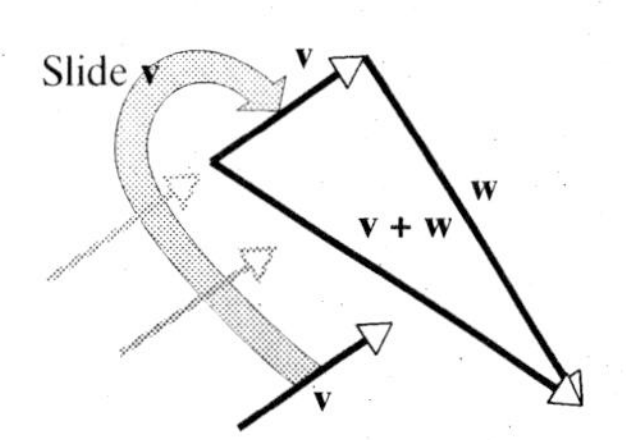

■

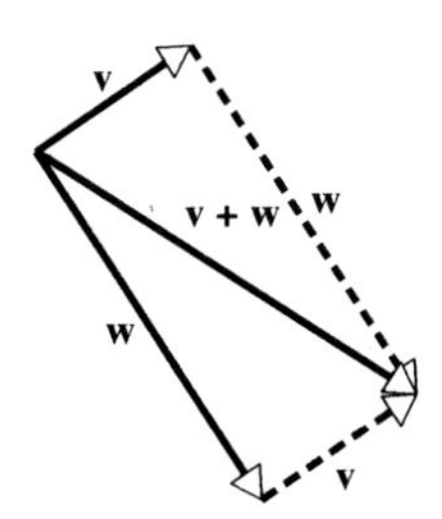

If we position **v** and **w** so that their *initial* points coincide and form a parallelogram, we notice that the resultant is along the diagonal of the parallelogram. We refer to this procedure as the **parallelogram method**. The diagram also demonstrates that vector addition is commutative.

$$\mathbf{v} + \mathbf{w} = \mathbf{w} + \mathbf{v}$$

As a result, we have two ways to geometrically add two vectors **v** and **w**.

Methods For Geometric Addition of Vectors

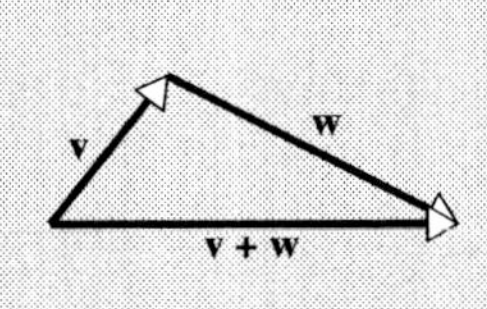

Triangle Method

Position vectors **v** and **w** so that the terminal point of the first vector coincides with the initial point of the second. The resultant is a vector determined by the initial point of the first vector and the terminal point of the second.

Parallelogram Method

Position vectors **v** and **w** so that their initial points coincide. Form a parallelogram using these vectors as the adjacent sides. The resultant is a vector, having the same initial point as **v** and **w**, that coincides with the diagonal of the parallelogram.

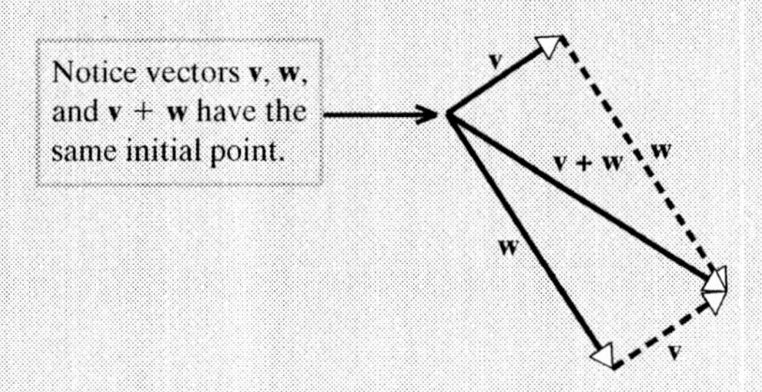

Since subtraction is related to addition, we now can define vector subtraction.

For any two vectors **v** and **w**, **subtraction** of **w** from **v** is defined by

$$\mathbf{v} - \mathbf{w} = \mathbf{v} + (-\mathbf{w}).$$

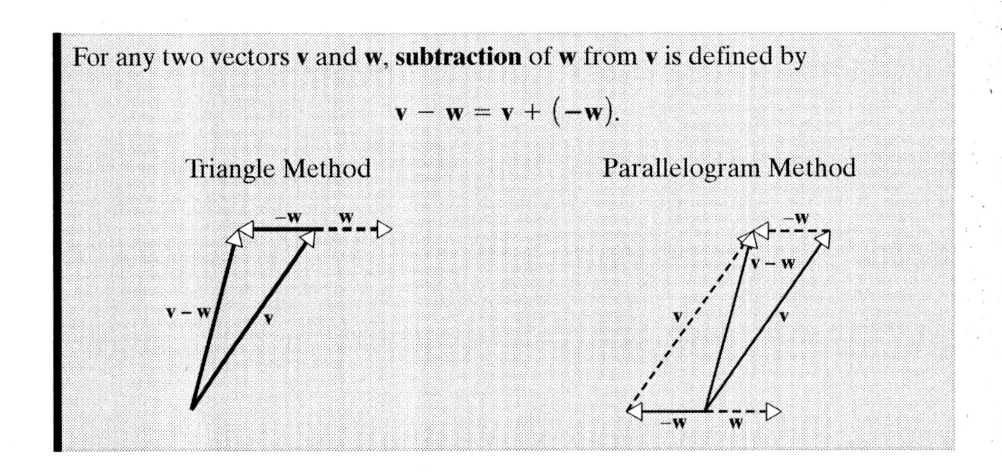

It follows that $\mathbf{v} - \mathbf{v} = \mathbf{0}$ and $\mathbf{v} + \mathbf{0} = \mathbf{v}$.

EXAMPLE 2 Consider the vectors **v**, **w**, and **u**. Sketch the following vectors.

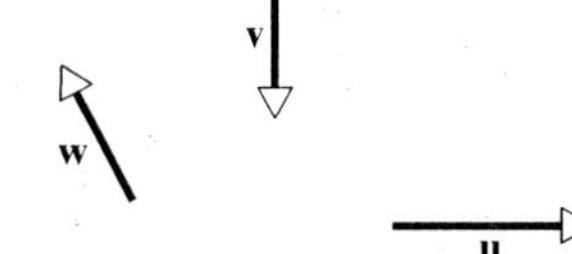

a. $3\mathbf{w}$ b. $\mathbf{v} - 2\mathbf{u}$

c. $\frac{1}{2}\mathbf{u} + \mathbf{w}$ d. $(\mathbf{w} + \mathbf{u}) + \mathbf{v}$

SOLUTION

a. Draw a vector in the same direction as **w** that is 3 times the length of **w**.

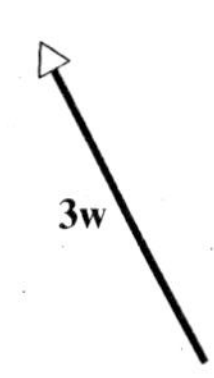

b. Using the triangle method, connect the terminal point of **v** with the initial point of the vector in the direction opposite $2\mathbf{u}$. The resultant is from the initial point of **v** to the terminal point of $-2\mathbf{u}$.

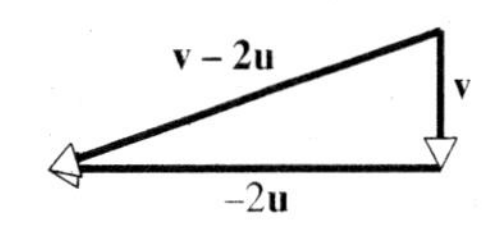

c. Using the parallelogram method, connect the initial points of **w** and $\frac{1}{2}\mathbf{u}$. Form a parallelogram using these vectors as adjacent sides. The resultant has the same initial point as **w** and $\frac{1}{2}\mathbf{u}$ and is along the diagonal.

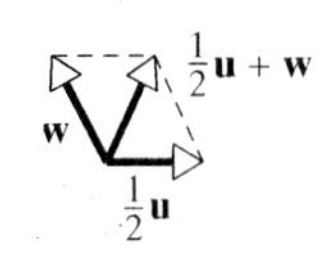

d. First use the triangle method to find $\mathbf{w} + \mathbf{u}$, and then again to obtain the resultant $(\mathbf{w} + \mathbf{u}) + \mathbf{v}$ by connecting the initial point of $\mathbf{w} + \mathbf{u}$ with the terminal point of **v**.

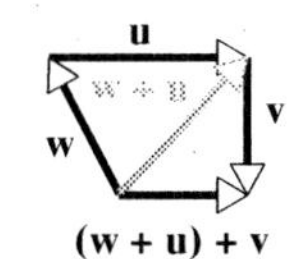

NOTE: Either method for adding vectors involves triangles, so you might want to review Chapter 3. Since we can use parallelograms to perform addition and subtraction of vectors, let's review some of the properties of parallelograms.

Parallelograms

1. Opposite sides are parallel and equal in length.
2. Opposite angles are equal.
3. Adjacent angles are supplementary. (That is, $\alpha + \beta = 180°$.)

Certain applications involving vectors require knowing the measure of the angle that is formed by two vectors.

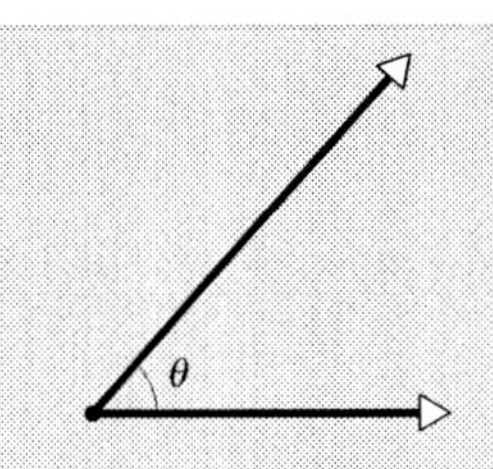

Angle Between Two Vectors

If two non-zero vectors have the same initial point, then the angle θ formed by these vectors is called the angle between the two vectors, where $0° \le \theta \le 180°$.

EXAMPLE 3 Two vectors **v** and **w** have magnitudes of 5 and 7, respectively. The angle between the vectors is 50°. Find the following.

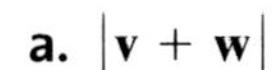

a. $|\mathbf{v} + \mathbf{w}|$

b. The angle the resultant $\mathbf{v} + \mathbf{w}$ makes with vector **w**

SOLUTION

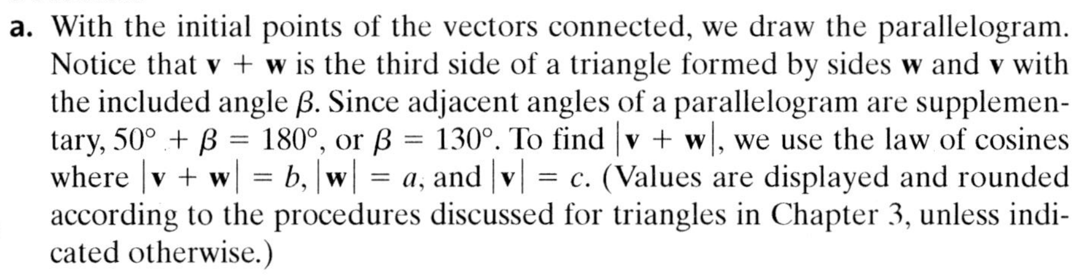

a. With the initial points of the vectors connected, we draw the parallelogram. Notice that $\mathbf{v} + \mathbf{w}$ is the third side of a triangle formed by sides **w** and **v** with the included angle β. Since adjacent angles of a parallelogram are supplementary, $50° + \beta = 180°$, or $\beta = 130°$. To find $|\mathbf{v} + \mathbf{w}|$, we use the law of cosines where $|\mathbf{v} + \mathbf{w}| = b$, $|\mathbf{w}| = a$, and $|\mathbf{v}| = c$. (Values are displayed and rounded according to the procedures discussed for triangles in Chapter 3, unless indicated otherwise.)

$$b^2 = a^2 + c^2 - 2ac\cos\beta$$
$$b^2 = 7^2 + 5^2 - 2(7)(5)\cos 130°$$
$$b^2 \approx 118.9951$$
$$b \approx 10.9085$$

Thus, $|\mathbf{v} + \mathbf{w}| \approx 10.91$.

b. In the triangle diagram, γ is the angle the resultant $\mathbf{v} + \mathbf{w}$ makes with **w**. To find γ, we use the law of sines.

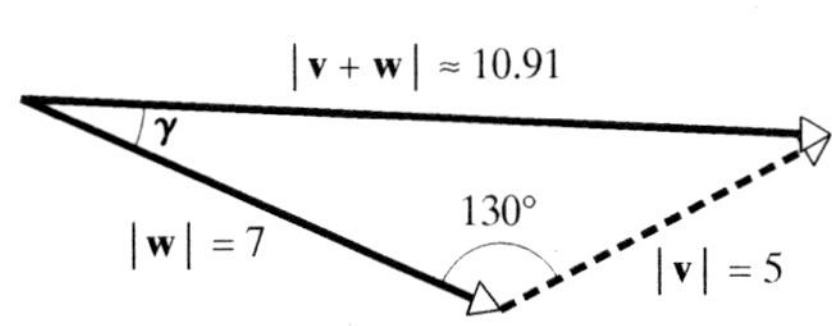

$$\frac{\sin\gamma}{c} = \frac{\sin\beta}{b}$$
$$\frac{\sin\gamma}{5} = \frac{\sin 130°}{10.9085}$$
$$\sin\gamma \approx 0.3511$$
$$\gamma \approx 20.6°$$

Applications of Vectors

Vectors, although relative newcomers to mathematics, can be applied to many real-life situations. We can use a vector to represent **force**, which we can think of as a push or pull of an object. In addition, vectors can be used in navigation to represent the speed and direction of wind, water currents, airplanes, or boats. Before you begin the applications, you may want to review north–south bearing discussed in Section 3.3, and bearing from due north in Section 3.5.

EXAMPLE 4 Two children are pulling on a toy box with forces of 5 pounds and 7 pounds, with an angle of 50° between the forces. Find the magnitude of the resultant force and the measure of the angle the resultant makes with the 7-pound force.

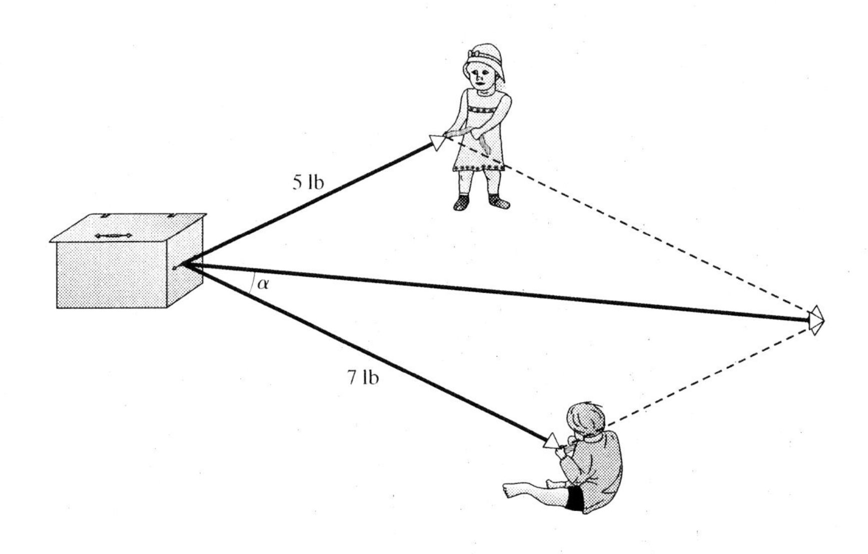

SOLUTION This question is the same problem as Example 3, but given in a physical application. Referring back to Example 3, the *resultant force* is $\mathbf{v} + \mathbf{w}$. So the resultant force has a magnitude of 10.91 pounds and it makes an angle of 20.6° with the 7-pound force. ■

EXAMPLE 5 A boat, heading south, is crossing a river at 15 mph. The river runs west with a current of 3 mph. Find the true course of the boat and the speed relative to the river banks.

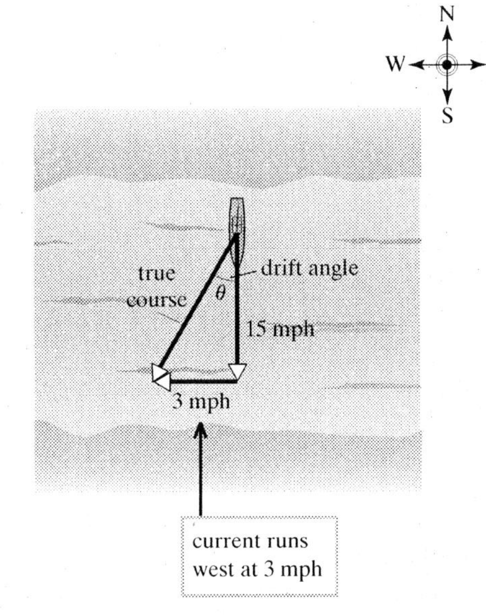

SOLUTION Although the boat is heading south, the current is pushing it on a course that takes it south and slightly west. Since these quantities have direction and magnitude, we use vectors. The **true course** of the boat is the resultant of the two vectors that represent the direction the boat is pointed, called the **heading**, and the current of the water. The magnitude of this resultant represents the speed the boat is traveling relative to banks of the river. Angle θ is called the **drift angle**. Since we have a right triangle, we find θ using the tangent function:

$$\tan\theta = \frac{3}{15} = \frac{1}{5} = 0.2$$

$$\theta = \tan^{-1} 0.2 \approx 11.3°$$

Applying the Pythagorean theorem, the magnitude of the resultant is

$$c = \sqrt{3^2 + 15^2} \approx 15.30.$$

Using the bearing notation discussed in Section 3.3, the true course of the boat is S 11.3° W, with a speed of 15.30 mph relative to the banks of the river. ■

The use of vectors facilitates other problems like air navigation with air currents. The **heading** of an airplane is the direction in which it is pointed, and the **true course** is the direction in which it is actually moving over the ground. The **air speed** is the speed at which it is moving through the air and the **ground speed** is the speed it is moving across the ground below. Wind can cause these two directions and speeds to differ.

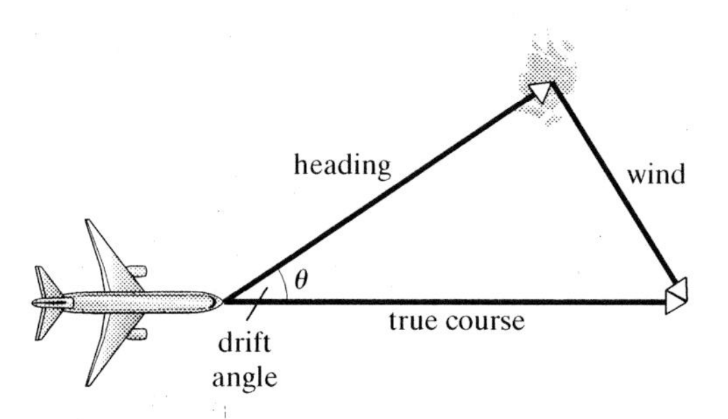

EXAMPLE 6 An airplane with an air speed of 290 mph heads southeast bearing 160° with an east wind (blowing from the east) bearing 270° at a speed of 80 mph. (Recall navigational bearing was discussed in Section 3.5.) Find the drift angle, the ground speed, and the true course of the airplane.

SOLUTION We draw the vectors that represent the velocity of the airplane (**u**) and wind velocity (**w**) using their bearings. We complete the parallelogram and find $\theta = 270° - 160° = 110°$. Therefore, $\alpha = 180° - 110° = 70°$. We draw the resultant, **v**, which represents the true course of the airplane. The ground speed will be represented by $c = |\mathbf{v}|$.

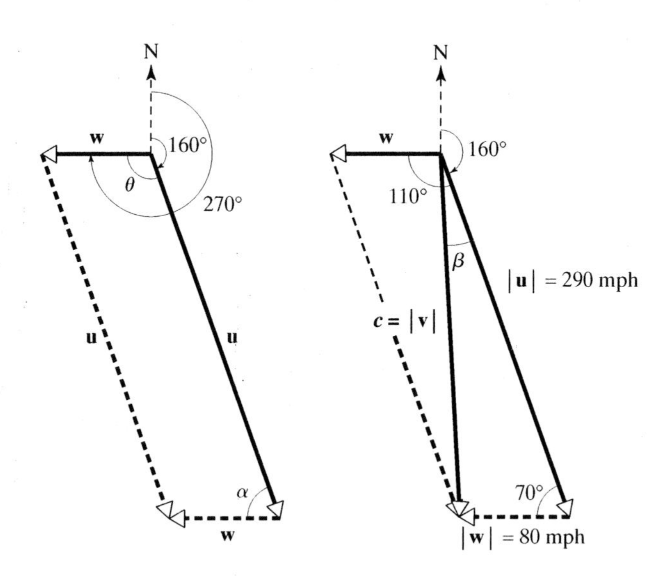

Using the law of cosines we get:

$$c^2 = a^2 + b^2 - 2ab\cos\gamma$$
$$c^2 = 290^2 + 80^2 - 2(290)(80)\cos 70°$$
$$c^2 \approx 74630.2654$$
$$c \approx 273.1854$$

So the airplane has a ground speed of 273.19 mph. We find the drift angle β (the angle the resultant **v** makes with **u**) to determine the true course using the law of cosines.

$$80^2 = 290^2 + (273.1854)^2 - 2(290)(273.1854)\cos\beta$$
$$\cos\beta \approx 0.9614$$
$$\beta \approx 16.0°$$

Therefore **v**, the true course of the airplane, has a bearing of $(160° + 16.0°) = 176.0°$. ■

EXAMPLE 7 Find the magnitude of the force (to the nearest tenth of a pound) required to pull a 100-pound flower pot down a driveway that makes a 25° angle with the horizontal.

SOLUTION The force due to gravity with direction straight down is the resultant of two forces: one parallel to the direction of the incline and the other *perpendicular* to the incline. The force due to gravity can be represented by the 100-pound force **w**. The force **u**, which is perpendicular to the incline (driveway), represents the force the flower pot exerts on the driveway. We want **v**, the force parallel to the direction of the inclined driveway. Notice that **w** = **u** + **v**.

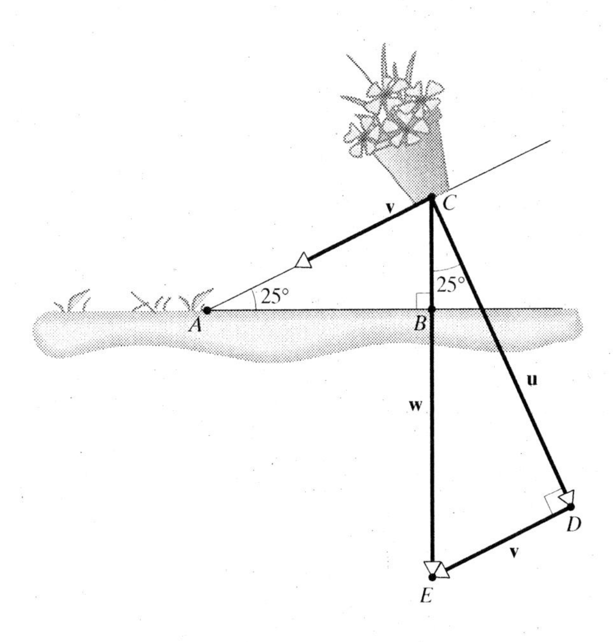

(continued)

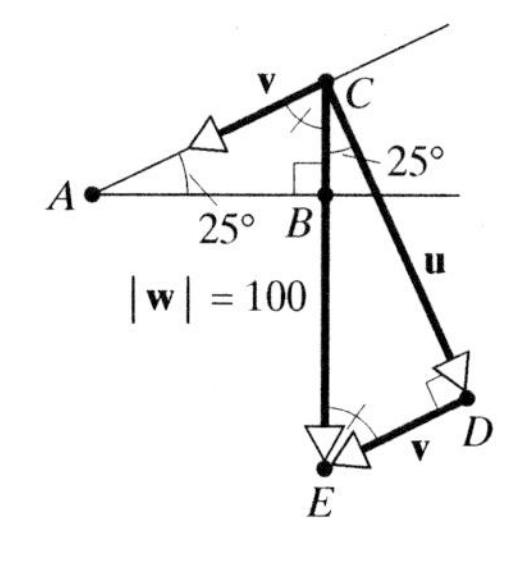

Triangles ABC and CDE are both right triangles, and since $\overline{AC}$ is parallel to $\overline{ED}$, $\angle ACB = \angle CED$ (alternate interior angles of parallel lines are equal). Therefore, since $\triangle ABC$ and $\triangle CDE$ have two pairs of corresponding angles that are equal, the third angles are also equal, or $\angle BAC = \angle DCE$. So both are 25°. Since $\triangle CDE$ is a right triangle we get:

$$\sin \angle DCE = \frac{|\mathbf{v}|}{|\mathbf{w}|}$$

$$\sin 25^\circ = \frac{|\mathbf{v}|}{100}$$

$$|\mathbf{v}| = 100 \sin 25^\circ \approx 42.3.$$

Thus, to the nearest tenth of a pound, a force of 42.3 pounds is necessary to pull the flower pot down the driveway. ■

Problems like these take time. You must carefully read each problem and carefully draw each diagram. Then you need to determine which method you are going to use to solve the triangle. It is easy to get frustrated with word problems due to the fact that you may have to do a lot of steps before you can begin the numerical calculations to determine the result. To quote Albert Einstein, "If I am expected to solve a problem within an hour, I would spend 45 minutes reading and understanding it, 10 minutes developing the right technique, and the last 5 minutes in solving the given problem."

Exercise Set 6.1

1–4. *Label vectors* **a** *and* **a** + **b**.

1.

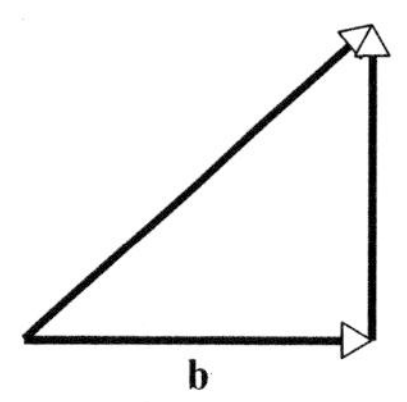

2.

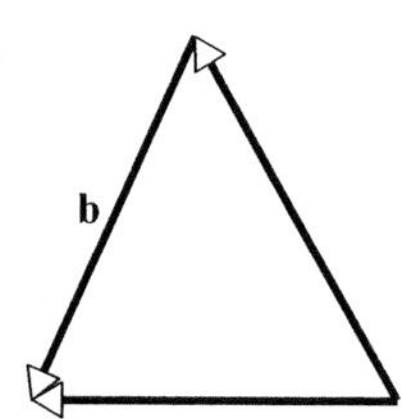

3.

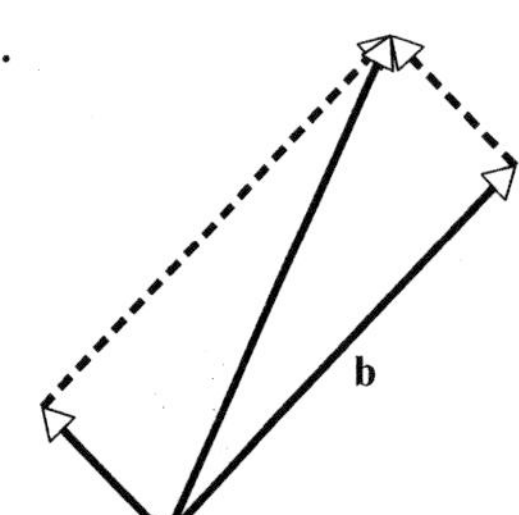

4.

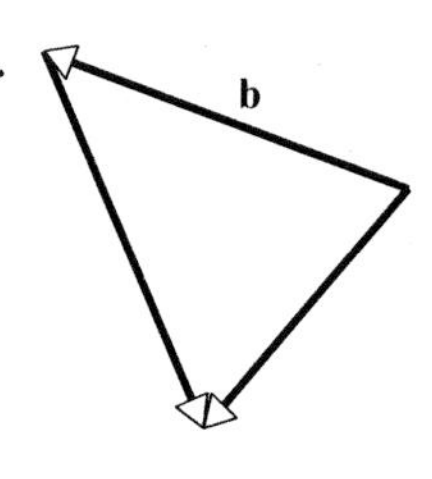

5–10. *By sketching vectors equivalent to* **u**, **v**, *or* **w**, *find the indicated resultant using the triangle method. Label all vectors.*

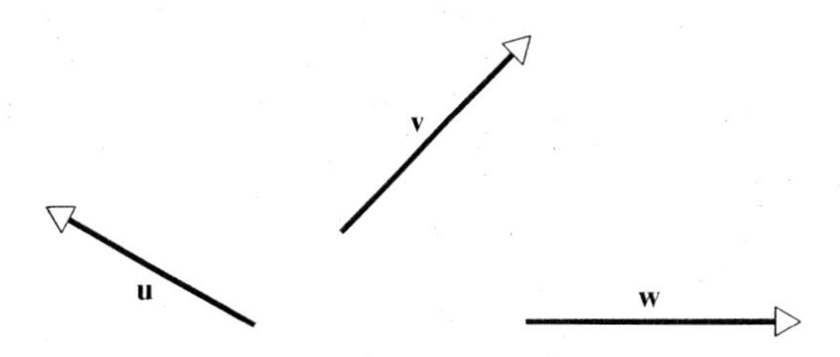

5. $\mathbf{u} + \mathbf{v}$
6. $\mathbf{w} + \mathbf{v}$
7. $2\mathbf{u} - \mathbf{w}$
8. $\mathbf{v} - \frac{1}{2}\mathbf{u}$
9. $(\mathbf{w} + 3\mathbf{u}) + \mathbf{v}$
10. $\mathbf{u} + (2\mathbf{v} + \mathbf{w})$

11–14. *By sketching vectors equivalent to* **a**, **b**, *or* **c**, *find the indicated resultant using the parallelogram method.*

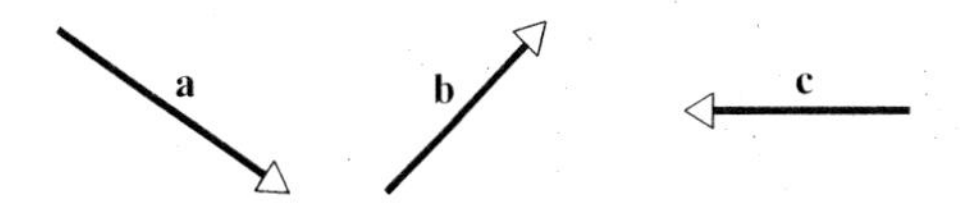

11. $2\mathbf{a} - \mathbf{c}$
12. $\mathbf{b} + 2\mathbf{c}$
13. $\frac{1}{2}\mathbf{b} + \mathbf{c}$
14. $\mathbf{c} - 3\mathbf{a}$

15. Two boys are pulling on a fallen tree branch with forces of 6 and 8 pounds with an angle of 90° between the forces. (See below left.) Find the magnitude of the resultant and the angle the resultant makes with the 6-pound force.

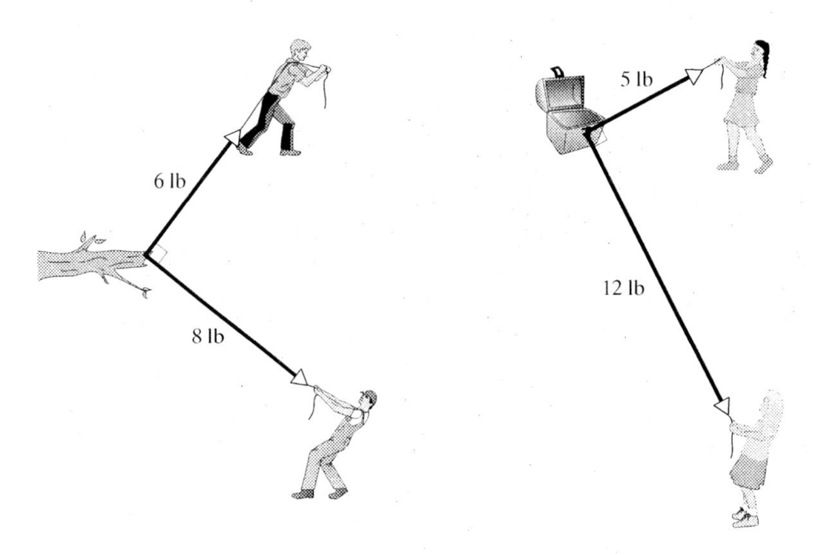

16. Two children are pulling on an object with forces of 5 and 12 pounds with an angle of 90° between the forces. (See above right.) Find the magnitude of the resultant and the angle the resultant makes with the 5-pound force.

17. Two forces of 8 newtons (a measure of force based in the metric system) and 13 newtons have a resultant force of 16.8 newtons. Find the angle between the 8-newton and 13-newton forces. (The angle between the vectors is α, not β.)

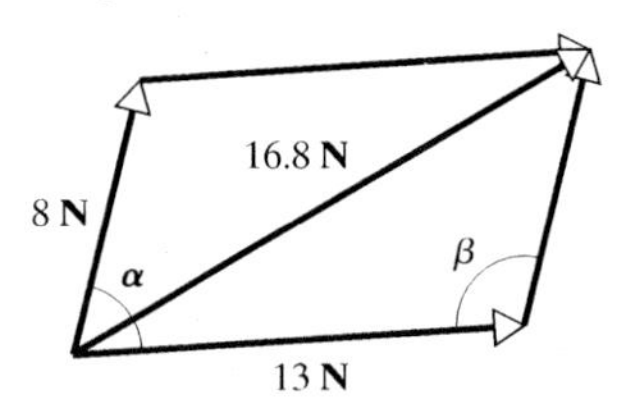

18. Two forces of 10 and 5 newtons have a resultant force of 13.3 newtons. Find the angle between the 10- and 5-newton forces.

19. If two forces with an angle of 60° between them have a resultant of 50 newtons and one of the forces is 30 newtons, find the other force (to the nearest tenth of a newton) and the angle it makes with the resultant.

20. If two forces with an angle of 60° between them have a resultant of 100 newtons and one of the forces is 80 newtons, find the other force (to the nearest tenth of a newton) and the angle it makes with the resultant.

21–26. *Draw the vector* **v** *that represents the given velocity.*

EXAMPLE

45 feet per second (fps) bearing 30°

SOLUTION

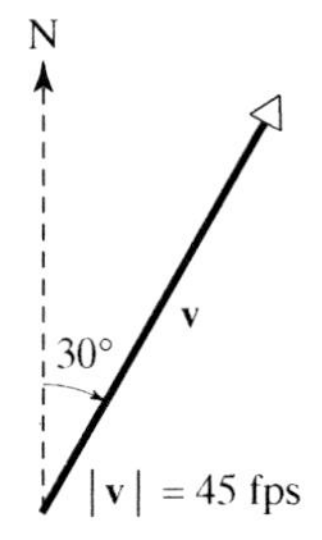

21. 120 miles per hour bearing S 60° E

22. 55 meters per second bearing 120°

23. 25 feet per second bearing 330°

24. 350 miles per hour bearing 210.5°

25. 5 meters per second bearing 137°

26. 100 feet per second bearing N 65° W

27. A boat going through the water at 20 mph is heading across the river at right angles with the current. If the current causes a drift angle of 8°, find the speed of the current.

28. A swimmer heads across the river going 100 feet per minute through the water at right angles to a current of 30 feet per minute. Find the drift angle and how far (to the nearest foot) he traveled if it took him 10 minutes to cross the river.

29. A plane flying with an air speed of 210 mph is headed northeast bearing 78.5°. The wind is blowing 35 mph from the west, bearing 90°. Find the ground speed and the true course of the airplane.

30. A plane flying with an air speed of 195 mph is headed southwest bearing 190°. The ground speed of the airplane is 215 mph with a true course of 225°. Find the drift angle and speed of the wind.

31. A pilot wishes to fly on a course bearing 0° over Acapulco toward Mexico City with a ground speed of 330 mph. If the wind is blowing 20 mph from the southwest bearing 45°, find the heading and the air speed of the airplane.

32. A plane has an airspeed of 210 mph and a heading of 45°. The ground speed of the plane is 195 mph, with a true course bearing 60°. Find the speed (to the nearest mph) and bearing of the wind (to the nearest degree).

33. A new truck, weighing 4200 pounds, is being transported to a car dealer. The transporter stacks trucks on ramps so that each is on an angle of 30° with the horizontal. Find the magnitude of the force that pulls the car down the ramp.

34. Carole's car stalled as she was driving up an incline that makes an angle of 12° with the horizontal. If her car weighs 3400 pounds, find the magnitude of the force that pulls her car down the incline.

35. A force of 75 pounds is needed to keep a 120-pound weight from sliding down a workout bench at the gym. Find the angle the bench makes with the horizontal.

36. A force of 150 pounds is needed to pull a 250-pound sled up the hill. Find the angle the hill makes with the horizontal.

6.2 Algebraic Vectors

Everything in Section 6.1 applies equally to vectors in two or three dimensions. In this section we introduce a coordinate plane and assume that all vectors under discussion are in that plane; that is, they are two-dimensional. We will see how introducing a coordinate system can simplify many vector problems.

Standard Position and Component Form of a Vector

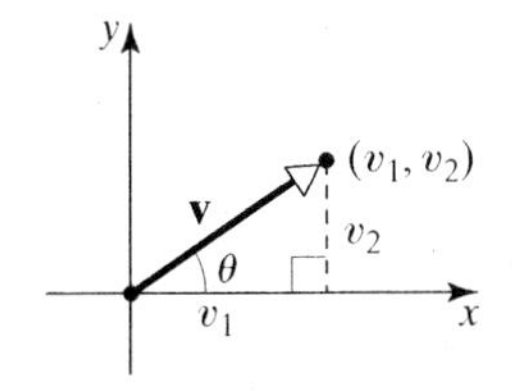

We know that the position of any vector can be changed, provided its magnitude and direction remain the same. If we position **v** so that its initial point is at the origin and label its terminal point (v_1, v_2), **v** is called a **standard position** vector and is represented by $\mathbf{v} = \langle v_1, v_2 \rangle$. The coordinates (v_1, v_2) of the terminal point are called the **components** of **v**. We call $\mathbf{v} = \langle v_1, v_2 \rangle$ the **component form of the vector**. The value v_1 is called the ***x*-component** or the **horizontal component**, and v_2 is called the ***y*-component** or the **vertical component** of **v**. The positive angle (θ) between the vector and the positive side of the *x*-axis is called the **direction angle** for the vector.

Because the zero vector **0** has length zero, its terminal point and initial point coincide at the origin. Thus $\mathbf{0} = \langle 0, 0 \rangle$.

Magnitude, Direction, Horizontal and Vertical Components

Using the preceding diagram, along with the Pythagorean theorem and definitions of the trigonometric ratios, we can find the magnitude, direction angle, and the horizontal and vertical components of a vector.

If vector $\mathbf{v} = \langle v_1, v_2 \rangle$ is in standard position with direction angle θ, we have the following.

- The magnitude of **v** is $|\mathbf{v}| = \sqrt{v_1^2 + v_2^2}$.
- The direction angle of **v** is θ, where $0° \le \theta < 360°$, such that

$$\cos\theta = \frac{v_1}{|\mathbf{v}|} \quad \text{and} \quad \sin\theta = \frac{v_2}{|\mathbf{v}|}.$$

- The horizontal and vertical components of **v** with magnitude $|\mathbf{v}|$ and direction angle θ are

$$v_1 = |\mathbf{v}|\cos\theta \quad \text{and} \quad v_2 = |\mathbf{v}|\sin\theta,$$

  respectively.

Equivalent Vectors

If equivalent vectors are positioned so that their initial points are at the origin, then it is clear that their terminal points must coincide since they would have the same length and direction. The component form of a vector implies that two vectors $\mathbf{v} = \langle v_1, v_2 \rangle$ and $\mathbf{w} = \langle w_1, w_2 \rangle$ are equal if and only if

$$v_1 = w_1 \quad \text{and} \quad v_2 = w_2.$$

EXAMPLE 1 Find the magnitude and direction angle for vector $\mathbf{v}$ with horizontal component 4 and vertical component -5.

SOLUTION Since $v_1 = 4$ and $v_2 = -5$, the magnitude of $\mathbf{v}$ is

$$|\mathbf{v}| = \sqrt{v_1^2 + v_2^2} = \sqrt{4^2 + (-5)^2} = \sqrt{41}.$$

To find the direction angle θ, we get:

$$\cos\theta = \frac{v_1}{|\mathbf{v}|} = \frac{4}{\sqrt{41}}$$

$$\hat{\theta} = \cos^{-1}\left(\frac{4}{\sqrt{41}}\right) \approx 51.3°$$

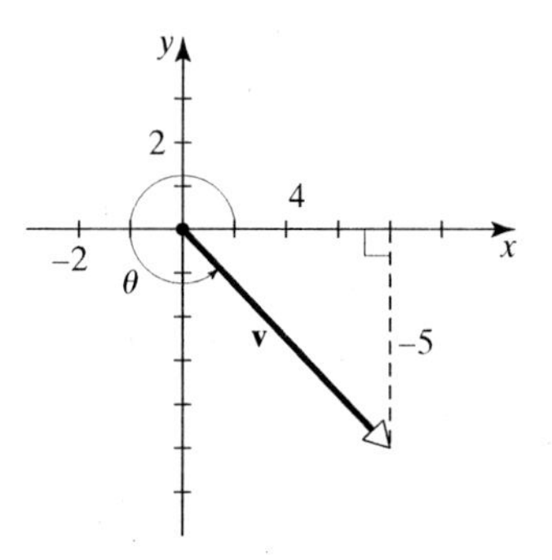

Since $\mathbf{v}$ has a positive x-component and a negative y-component, $\mathbf{v}$ is in QIV. The direction angle θ is defined for $0° \leq \theta < 360°$, so

$$\theta \approx 360° - 51.3° = 308.7°.$$

The component form of vectors can simplify vector addition, subtraction, and scalar multiplication. Let's see how in the next example.

EXAMPLE 2 If $\mathbf{v} = \langle 4, 1 \rangle$ and $\mathbf{u} = \langle 2, 3 \rangle$, find:

a. $\mathbf{v} + \mathbf{u}$ **b.** $2\mathbf{u}$

SOLUTION

a. The diagram demonstrates $\mathbf{v} + \mathbf{u}$ as the diagonal of the parallelogram.

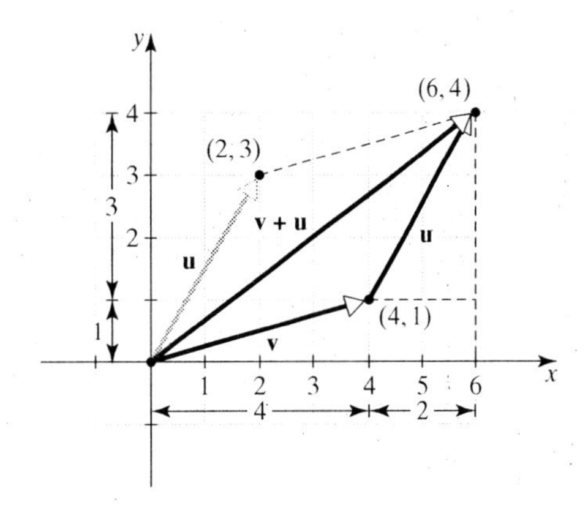

Therefore, $\mathbf{v} + \mathbf{u} = \langle 4, 1 \rangle + \langle 2, 3 \rangle = \langle 4 + 2, 1 + 3 \rangle = \langle 6, 4 \rangle$.

b. The diagram below demonstrates that $2\mathbf{u} = \mathbf{u} + \mathbf{u}$, using triangle addition.

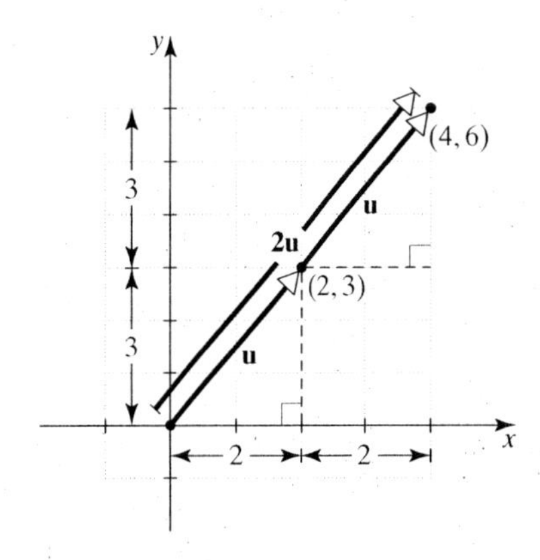

Therefore, $2\mathbf{u} = \langle 2, 3 \rangle + \langle 2, 3 \rangle = \langle 2(2), 2(3) \rangle = \langle 4, 6 \rangle$. ■

We can generalize the results from Example 2 as follows.

Arithmetic Operations for Vectors

For vectors $\mathbf{v} = \langle v_1, v_2 \rangle$ and $\mathbf{w} = \langle w_1, w_2 \rangle$ and any scalar $c \in \mathbb{R}$, the following operations are defined:

Vector Addition $\mathbf{v} + \mathbf{w} = \langle v_1, v_2 \rangle + \langle w_1, w_2 \rangle = \langle v_1 + w_1, v_2 + w_2 \rangle$

Vector Subtraction $\mathbf{v} - \mathbf{w} = \langle v_1, v_2 \rangle - \langle w_1, w_2 \rangle = \langle v_1 - w_1, v_2 - w_2 \rangle$

Scalar Multiplication $c\mathbf{v} = c\langle v_1, v_2 \rangle = \langle cv_1, cv_2 \rangle$

i, j Form of a Vector

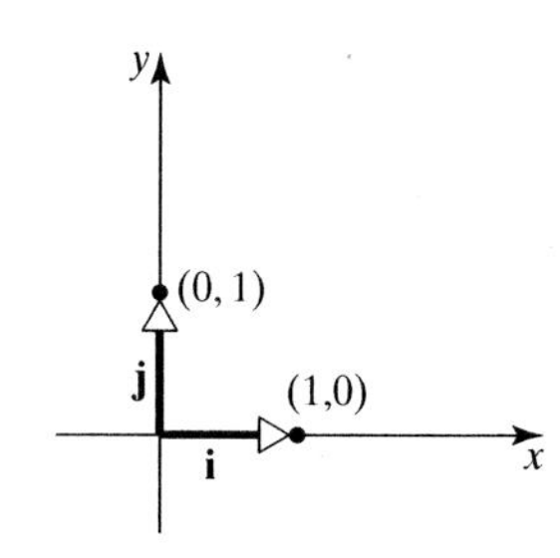

A vector with magnitude 1 is called a **unit vector**. Two special unit vectors are the standard position vectors $\mathbf{i} = \langle 1, 0 \rangle$ and $\mathbf{j} = \langle 0, 1 \rangle$. Using these special vectors, the component form of any vector $\mathbf{v} = \langle v_1, v_2 \rangle$ can be written as $\mathbf{v} = v_1\mathbf{i} + v_2\mathbf{j}$, known as the **i, j form.**

Using this form in Example 2, where $\mathbf{v} = \langle 4, 1 \rangle = 4\mathbf{i} + \mathbf{j}$ and $\mathbf{u} = \langle 2, 3 \rangle = 2\mathbf{i} + 3\mathbf{j}$, we have

$$\mathbf{v} + \mathbf{u} = 6\mathbf{i} + 4\mathbf{j} \quad \text{and} \quad 2\mathbf{u} = 2(2\mathbf{i} + 3\mathbf{j}) = 4\mathbf{i} + 6\mathbf{j}.$$

EXAMPLE 3 Find the magnitude and direction angle of $-2\mathbf{i} - 6\mathbf{j}$.

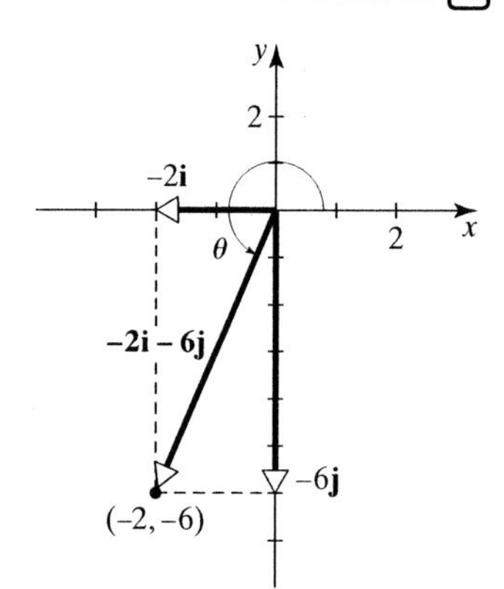

SOLUTION

$$|-2\mathbf{i} - 6\mathbf{j}| = \sqrt{(-2)^2 + (-6)^2} = \sqrt{40} = 2\sqrt{10}$$

For the direction angle θ, we get:

$$\cos\theta = \frac{v_1}{|\mathbf{v}|} = \frac{-2}{2\sqrt{10}} = -\frac{1}{\sqrt{10}}$$

$$\hat{\theta} = \cos^{-1}\left(\frac{1}{\sqrt{10}}\right) \approx 71.6°$$

Since the vector is in QIII, $\theta \approx 180° + 71.6° = 251.6°$. ■

EXAMPLE 4 Express $\mathbf{v}$ in $v_1\mathbf{i} + v_2\mathbf{j}$ form if $|\mathbf{v}| = 2\sqrt{2}$ and $\theta = 315°$. Sketch $\mathbf{v}$.

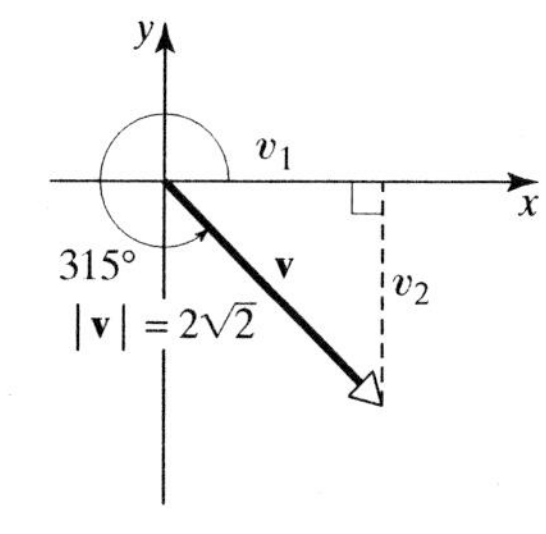

SOLUTION We need the horizontal and vertical components v_1 and v_2, respectively.

$$v_1 = |\mathbf{v}|\cos\theta \qquad v_2 = |\mathbf{v}|\sin\theta$$

$$v_1 = 2\sqrt{2}\cos 315° = 2\sqrt{2}\left(\frac{1}{\sqrt{2}}\right) \qquad v_2 = 2\sqrt{2}\sin 315° = 2\sqrt{2}\left(-\frac{1}{\sqrt{2}}\right)$$

$$v_1 = 2 \qquad v_2 = -2$$

Therefore, $\mathbf{v} = 2\mathbf{i} - 2\mathbf{j}$. ■

Dot Product

So far, we have discussed how to add or subtract two vectors and how to multiply a vector by a scalar. All these operations produce a *vector.* Now we define a special product of two vectors that produces a *scalar,* not a vector. We then see how this product is useful.

Dot Product of Two Vectors

For any two vectors $\mathbf{v} = \langle v_1, v_2 \rangle$ and $\mathbf{w} = \langle w_1, w_2 \rangle$, the **dot product** of $\mathbf{v}$ and $\mathbf{w}$ is

$$\mathbf{v} \cdot \mathbf{w} = \langle v_1, v_2 \rangle \cdot \langle w_1, w_2 \rangle = v_1w_1 + v_2w_2.$$

EXAMPLE 5 Find the dot product $\mathbf{v} \cdot \mathbf{w}$.

a. $\mathbf{v} = \langle 2, -6 \rangle$ and $\mathbf{w} = \langle -1, 5 \rangle$ **b.** $\mathbf{v} = 3\mathbf{i} + 8\mathbf{j}$ and $\mathbf{w} = -9\mathbf{i} + 10\mathbf{j}$

SOLUTION

a. $\mathbf{v} \cdot \mathbf{w} = \langle 2, -6 \rangle \cdot \langle -1, 5 \rangle = 2(-1) + (-6)5 = -32$

b. $\mathbf{v} \cdot \mathbf{w} = (3\mathbf{i} + 8\mathbf{j}) \cdot (-9\mathbf{i} + 10\mathbf{j}) = 3(-9) + 8(10) = -27 + 80 = 53$ ■

Notice that the *dot product produces a scalar,* and this product may also be referred to as the **scalar product**, or **inner product**.

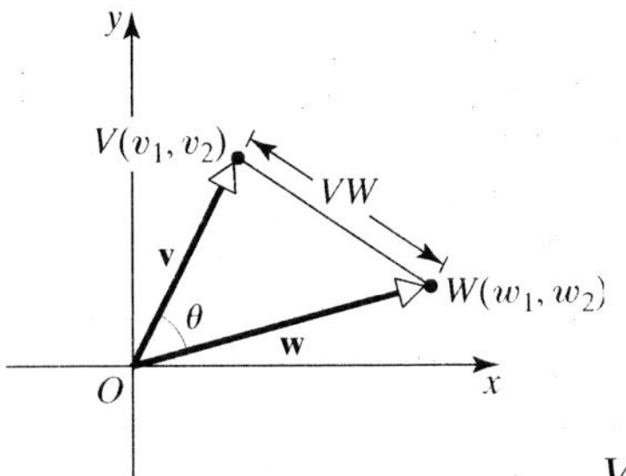

You might wonder why we want to find the dot product or what the scalar result can represent. Let's first look at a connection between the dot product and the angle between two nonzero vectors.

Consider angle θ between vectors $\overrightarrow{OV} = \mathbf{v}$ and $\overrightarrow{OW} = \mathbf{w}$, and let VW be the measure of the line segment between points V and W. Now we apply the law of cosines to $\triangle OVW$.

$$VW^2 = |\mathbf{v}|^2 + |\mathbf{w}|^2 - 2|\mathbf{v}||\mathbf{w}| \cos \theta$$

$$(w_1 - v_1)^2 + (w_2 - v_2)^2 = v_1^2 + v_2^2 + w_1^2 + w_2^2 - 2|\mathbf{v}||\mathbf{w}| \cos \theta \qquad v_1^2 + v_2^2 = |\mathbf{v}|^2,\ w_1^2 + w_2^2 = |\mathbf{w}|^2,\ (w_1 - v_1)^2 + (w_2 - v_2)^2 = VW^2$$

$$w_1^2 - 2w_1v_1 + v_1^2 + w_2^2 - 2w_2v_2 + v_2^2 = v_1^2 + v_2^2 + w_1^2 + w_2^2 - 2|\mathbf{v}||\mathbf{w}| \cos \theta \qquad \text{Simplify.}$$

$$-2v_1w_1 - 2v_2w_2 = -2|\mathbf{v}||\mathbf{w}| \cos \theta \qquad \text{Subtract } v_1^2, v_2^2, w_1^2, \text{ and } w_2^2 \text{ from both sides.}$$

$$v_1w_1 + v_2w_2 = |\mathbf{v}||\mathbf{w}| \cos \theta \qquad \text{Divide each side by } -2.$$

Recognizing that $v_1w_1 + v_2w_2 = \mathbf{v} \cdot \mathbf{w}$, we obtain the following connection between the dot product of two vectors and the cosine of the angle between them.

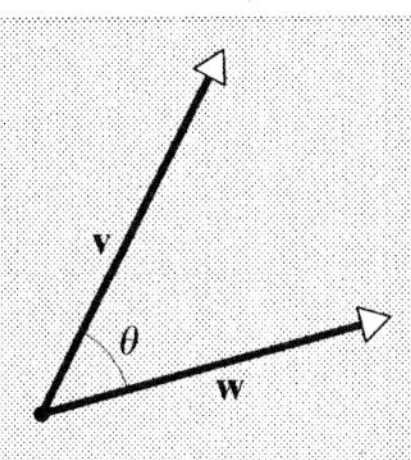

The Angle between Two Vectors and the Dot Product

If θ is the angle between two nonzero vectors $\mathbf{v}$ and $\mathbf{w}$, then

$$\mathbf{v} \cdot \mathbf{w} = |\mathbf{v}||\mathbf{w}| \cos \theta, \qquad \text{or} \qquad \cos \theta = \frac{\mathbf{v} \cdot \mathbf{w}}{|\mathbf{v}||\mathbf{w}|}.$$

As a result, we now have *two ways* to express the dot product of two vectors:

1. as the sum of the products of their respective components, or
2. as the product of their magnitudes and the cosine of the angle between them.

EXAMPLE 6 Find $\mathbf{v} \cdot \mathbf{w}$ if $|\mathbf{v}| = 5$ with $75°$ as the direction angle of $\mathbf{v}$, and $|\mathbf{w}| = 10$ with $135°$ as the direction angle of $\mathbf{w}$.

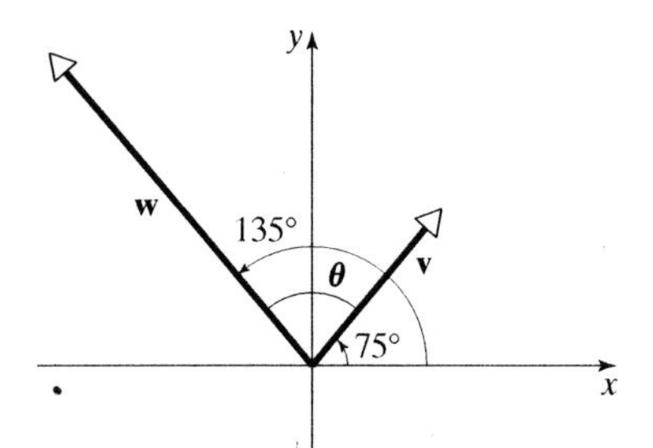

SOLUTION We are given the magnitudes of the vectors, and we can quickly determine the angle between them. So we find their dot product as follows:

$$\mathbf{v} \cdot \mathbf{w} = |\mathbf{v}||\mathbf{w}| \cos \theta$$
$$\mathbf{v} \cdot \mathbf{w} = (5)(10)\cos(135° - 75°)$$
$$= 50 \cos 60°$$
$$= 50\left(\frac{1}{2}\right)$$
$$= 25$$

Finding the Angle between Two Vectors

We can use $\mathbf{v} \cdot \mathbf{w} = |\mathbf{v}||\mathbf{w}| \cos \theta$ in the form $\cos \theta = \frac{\mathbf{v} \cdot \mathbf{w}}{|\mathbf{v}||\mathbf{w}|}$ to find the angle between two vectors $(0° \leq \theta \leq 180°)$.

EXAMPLE 7 Find the angle between $\mathbf{v}$ and $\mathbf{w}$.

a. $\mathbf{v} = 2\mathbf{i} + 3\mathbf{j}$, $\mathbf{w} = 2\mathbf{j}$ **b.** $\mathbf{v} = \langle 1, 1 \rangle$, $\mathbf{w} = \langle 4, -4 \rangle$

SOLUTION

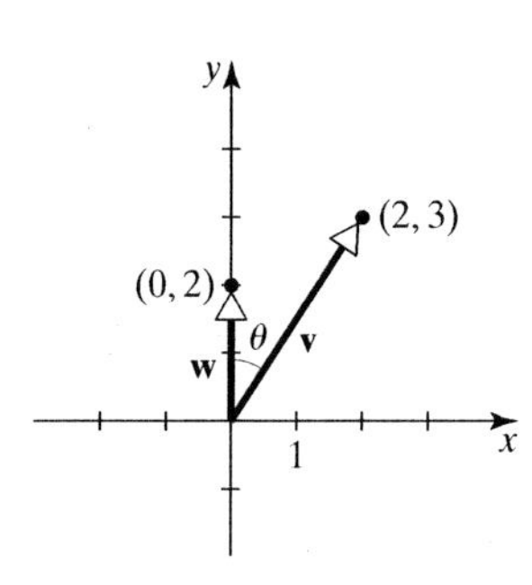

a. Use $\cos \theta = \frac{\mathbf{v} \cdot \mathbf{w}}{|\mathbf{v}||\mathbf{w}|}$ with $\mathbf{v} = 2\mathbf{i} + 3\mathbf{j}$ and $\mathbf{w} = 2\mathbf{j}$.

$$\cos \theta = \frac{(2\mathbf{i} + 3\mathbf{j}) \cdot (0\mathbf{i} + 2\mathbf{j})}{|2\mathbf{i} + 3\mathbf{j}||2\mathbf{j}|}$$
$$\cos \theta = \frac{2(0) + 3(2)}{\sqrt{2^2 + 3^2}\sqrt{0^2 + 2^2}}$$
$$= \frac{6}{2\sqrt{13}} = \frac{3}{\sqrt{13}}$$
$$\theta = \cos^{-1}\left(\frac{3}{\sqrt{13}}\right) \approx 33.7°.$$

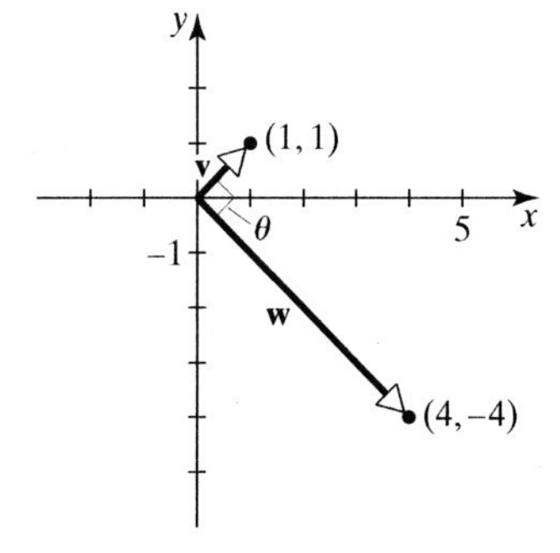

b. Let $\mathbf{v} = \langle 1, 1 \rangle$ and $\mathbf{w} = \langle 4, -4 \rangle$.

$$\cos \theta = \frac{\mathbf{v} \cdot \mathbf{w}}{|\mathbf{v}||\mathbf{w}|}$$
$$\cos \theta = \frac{1(4) + 1(-4)}{\sqrt{1^2 + 1^2}\sqrt{4^2 + (-4)^2}}$$
$$\cos \theta = 0$$
$$\theta = \cos^{-1}(0) = 90°$$

If the dot product of two vectors is zero $(\mathbf{v} \cdot \mathbf{w} = 0)$, then $\cos \theta = 0$. And $\cos \theta = 0$ if and only if $\theta = 90°$, where $0° \leq \theta \leq 180°$. The result in part (b) tells us that two vectors are *perpendicular,* or *orthogonal,* if and only if their dot product is 0.

> **Perpendicular (Orthogonal) Vectors**
> Vectors **v** and **w** are perpendicular, or orthogonal, if and only if $\mathbf{v} \cdot \mathbf{w} = 0$.

Application of Dot Product

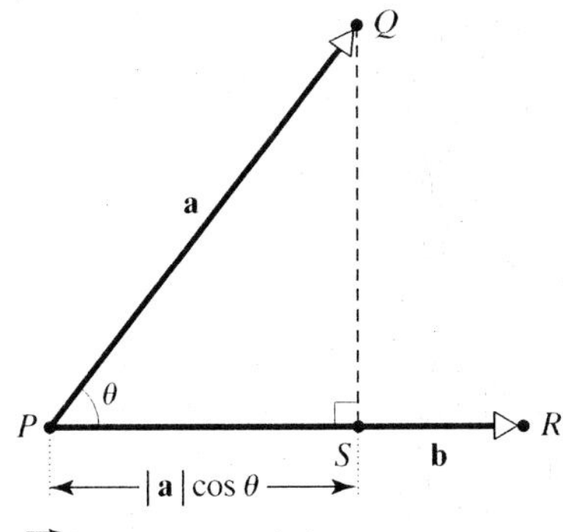

$|\overrightarrow{PS}| = \text{comp}_\mathbf{b}\,\mathbf{a} = |\mathbf{a}| \cos\theta$

Let θ be the angle between two vectors $\mathbf{a} = \overrightarrow{PQ}$ and $\mathbf{b} = \overrightarrow{PR}$. We draw a segment from Q perpendicular to the line containing P and R and label the point of intersection S. The magnitude of $\overrightarrow{PS}$ is called the **component of a along b**, abbreviated $\text{comp}_\mathbf{b}\mathbf{a}$. Using the trigonometric ratio for cosine, we get the following.

$$\cos\theta = \frac{|\overrightarrow{PS}|}{|\mathbf{a}|}$$

$$|\overrightarrow{PS}| = |\mathbf{a}| \cos\theta$$

$$\text{comp}_\mathbf{b}\mathbf{a} = |\mathbf{a}| \cos\theta$$

Now we use the form of the dot product that relates the two vectors to the angle between them.

$$\cos\theta = \frac{\mathbf{a} \cdot \mathbf{b}}{|\mathbf{a}||\mathbf{b}|}$$

$$|\mathbf{a}| \cos\theta = \frac{\mathbf{a} \cdot \mathbf{b}}{|\mathbf{b}|} \qquad \text{Multiply both sides by } |\mathbf{a}|.$$

Since $\text{comp}_\mathbf{b}\mathbf{a} = |\mathbf{a}| \cos\theta$, we get $\text{comp}_\mathbf{b}\mathbf{a} = \dfrac{\mathbf{a} \cdot \mathbf{b}}{|\mathbf{b}|}$.

The notation $\text{comp}_\mathbf{b}\mathbf{a}$ is also referred to as the **scalar projection of a onto b.**

> For any two vectors **a** and **b,** the component of **a** along **b** (or the scalar projection of **a** onto **b**) is given by
>
> $$\text{comp}_\mathbf{b}\mathbf{a} = \frac{\mathbf{a} \cdot \mathbf{b}}{|\mathbf{b}|}.$$

EXAMPLE 8 If $\mathbf{a} = \langle -2, 5 \rangle$ and $\mathbf{b} = \langle 3, 4 \rangle$, find the component of **a** along **b**.

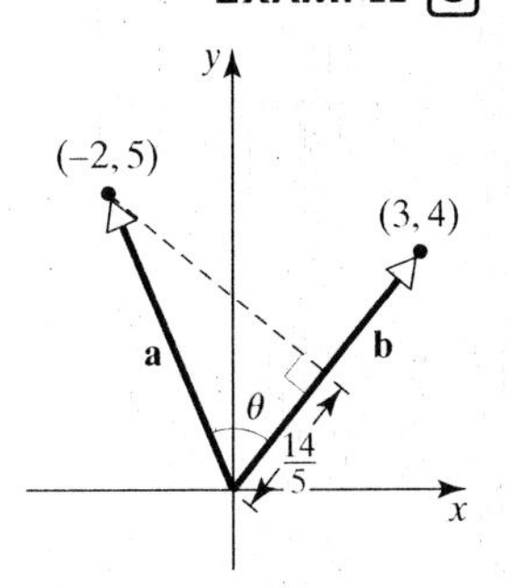

SOLUTION

$$\text{comp}_\mathbf{b}\mathbf{a} = \frac{\mathbf{a} \cdot \mathbf{b}}{|\mathbf{b}|}$$

$$\text{comp}_\mathbf{b}\mathbf{a} = \frac{\langle -2, 5 \rangle \cdot \langle 3, 4 \rangle}{|\langle 3, 4 \rangle|} = \frac{(-2)(3) + (5)(4)}{\sqrt{3^2 + 4^2}}$$

$$= \frac{14}{5}$$

■

Work

One application of dot product is found in the concept of work. If someone pushes a stalled car, **work** W is done when a force $\mathbf{F}$ pushes the car from point A to point B. When a constant force $\mathbf{F}$ points in the same direction as the resulting displacement $\overrightarrow{AB}$, the work W is defined as the magnitude of $\mathbf{F}$ times the magnitude of $\overrightarrow{AB}$, or $W = |\mathbf{F}||\overrightarrow{AB}|$. If the magnitude of $\mathbf{F}$ is in pounds and the magnitude of $\overrightarrow{AB}$ in feet, the unit of work is foot-pounds $(\text{ft} \cdot \text{lb})$. But the constant force $\mathbf{F}$ and the displacement $\overrightarrow{AB}$ do not often point in the same direction. For example, if you are pulling luggage, the force is applied along the handle. Therefore, the force is directed at an angle θ relative to the displacement. In such a case, *only the component of the force along the displacement* is used in defining work. As the diagram indicates, this component is $|\mathbf{F}| \cos \theta$.

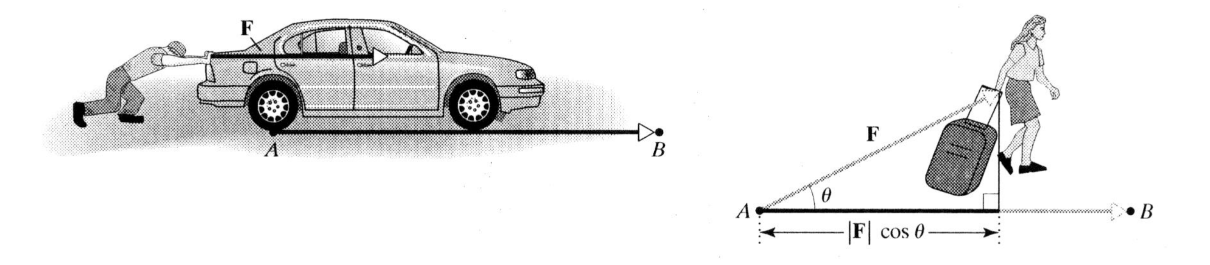

Generalizing this result and using the relationship of dot product, we get the following definition.

Work Done by a Constant Force

The work W done on an object by a constant force $\mathbf{F}$ with displacement $\overrightarrow{AB}$ is the product of the component of $\mathbf{F}$ along $\overrightarrow{AB}$ times the magnitude of the displacement.

$$\text{Thus, } W = (|\mathbf{F}| \cos \theta)|\overrightarrow{AB}| = |\mathbf{F}||\overrightarrow{AB}| \cos \theta,$$
$$\text{or } W = \mathbf{F} \cdot \overrightarrow{AB},$$

where θ is the angle between the force and the displacement.

The last part of the definition, $\mathbf{W} = \mathbf{F} \cdot \overrightarrow{AB}$, tells us that we can determine the work done by finding the dot product of the force vector and the displacement vector.

NOTE: If the force is applied in the direction of the displacement, then $\theta = 0$. So $W = |\mathbf{F}||\overrightarrow{AB}|$.

EXAMPLE 9 Find the work done (to the nearest foot-pound) by a 10-pound force pulling luggage at an angle of 50° for a distance of 120 feet.

SOLUTION

$$\begin{aligned} W &= |\mathbf{F}||\overrightarrow{AB}| \cos \theta \\ &= 10 \text{ lb} \cdot 120 \text{ ft} \cos 50° \\ &\approx 771 \text{ ft} \cdot \text{lb} \end{aligned}$$

We have seen that having vectors in standard position may simplify vector operations of addition, subtraction, and scalar or dot multiplication. But sometimes vectors are not positioned so that their initial points are at the origin. How can we find the equivalent vector in component form? The proof of the following result is left as an exercise.

Equivalent Standard Position Vector

For any vector $\overrightarrow{PQ}$ with initial point $P(p_1, p_2)$ and terminal point $Q(q_1, q_2)$, the equivalent standard position vector is

$$\overrightarrow{PQ} = \langle q_1 - p_1, q_2 - p_2 \rangle$$

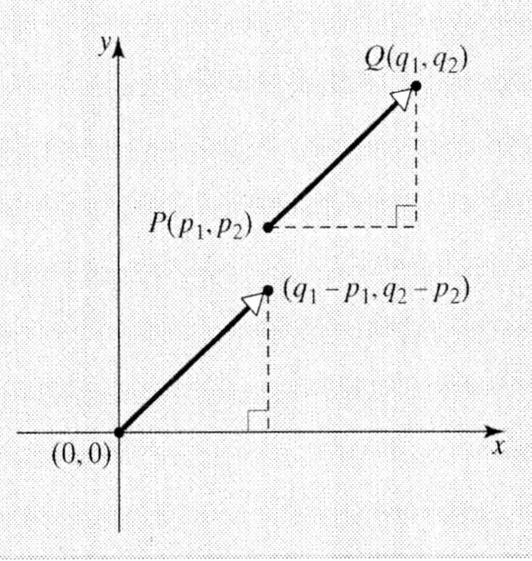

EXAMPLE 10 If $\overrightarrow{PQ}$ has initial point $P(1, -2)$ and terminal point $Q(4, 6)$, find the component form of the standard position vector equivalent to $\overrightarrow{PQ}$.

SOLUTION

$$\overrightarrow{PQ} = \langle 4 - 1, 6 + 2 \rangle = \langle 3, 8 \rangle$$

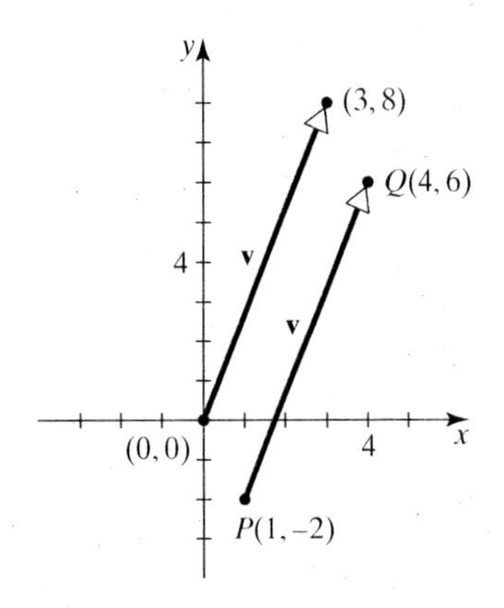

Exercise Set 6.2

1–6. *Draw the standard position vector. Find its magnitude and direction angle, to the nearest tenth of degree, when necessary.*

1. $\langle 1, 1\rangle$
2. $\langle -4, 4\rangle$
3. $-6\mathbf{i} - 3\mathbf{j}$
4. $10\mathbf{i} - 5\mathbf{j}$
5. $\langle -3, 4\rangle$
6. $\langle -6, -2\rangle$

7–10. *Find (to the nearest tenth) the vertical and horizontal components of the vector* **v** *with the given magnitude and direction. Sketch* **v** *and express the vector in component form.*

EXAMPLE $|\mathbf{v}| = 10,\ \theta = 115°$

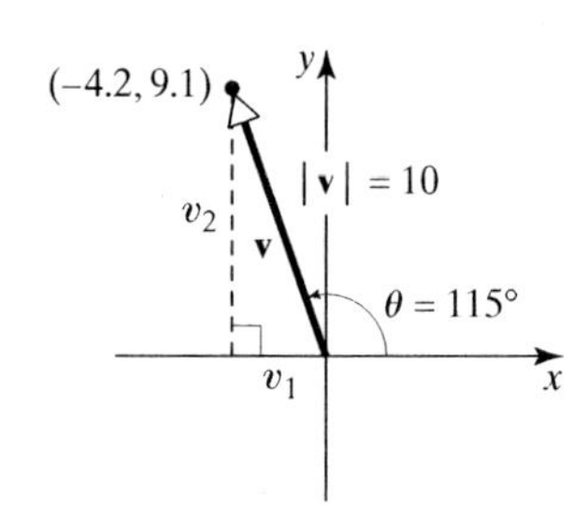

SOLUTION

$v_1 = |\mathbf{v}|\cos\theta$ and $v_2 = |\mathbf{v}|\sin\theta$

$v_1 = 10\cos 115° \approx -4.2$ and $v_2 = 10\sin 115° \approx 9.1$

Therefore, $\mathbf{v} = \langle -4.2, 9.1\rangle$.

7. $|\mathbf{v}| = 3,\ \theta = 50°$
8. $|\mathbf{v}| = 35,\ \theta = 160°$
9. $|\mathbf{v}| = 42,\ \theta = 205°$
10. $|\mathbf{v}| = 17,\ \theta = 305°$

11–16. *Express* **v** *in $v_1\mathbf{i} + v_2\mathbf{j}$ form with the given magnitude and direction.*

11. $|\mathbf{v}| = 8$ and $\theta = 60°$
12. $|\mathbf{v}| = 4\sqrt{3}$ and $\theta = 150°$
13. $|\mathbf{v}| = 11$ and $\theta = 270°$
14. $|\mathbf{v}| = 12$ and $\theta = 135°$
15. $|\mathbf{v}| = 12$ and $\theta = 225°$
16. $|\mathbf{v}| = 13$ and $\theta = 180°$

17–22. *Find* $\mathbf{v} + \mathbf{w}$, $3\mathbf{w} - 2\mathbf{v}$, $|\mathbf{v} - \mathbf{w}|$, *and* $\mathbf{v} \cdot \mathbf{w}$ *for each pair of vectors.*

17. $\mathbf{v} = 2\mathbf{i} + \mathbf{j},\ \mathbf{w} = \mathbf{i} - \mathbf{j}$
18. $\mathbf{v} = \mathbf{i} + 2\mathbf{j},\ \mathbf{w} = -\mathbf{i} + 3\mathbf{j}$
19. $\mathbf{v} = \langle -1, 2\rangle,\ \mathbf{w} = \langle 0, 5\rangle$
20. $\mathbf{v} = \langle -3, 0\rangle,\ \mathbf{w} = \langle -4, -4\rangle$
21. $\mathbf{v} = 6\mathbf{i} - 2\mathbf{j},\ \mathbf{w} = 2\mathbf{i} + 6\mathbf{j}$
22. $\mathbf{v} = 3\mathbf{i} + 7\mathbf{j},\ \mathbf{w} = --7\mathbf{i} + 3\mathbf{j}$

23–26. *Find* $\mathbf{v} \cdot \mathbf{w}$. *Approximate answers to the nearest tenth when necessary.*

23. $|\mathbf{v}| = 6$ and $|\mathbf{w}| = 4$, and the angle θ between **v** and **w** is 30°.
24. $|\mathbf{v}| = 2$ and $|\mathbf{w}| = 9$, and the angle θ between **v** and **w** is 120°.
25. $|\mathbf{v}| = 7$ with the direction angle of **v** as 22° and $|\mathbf{w}| = 14$ with the direction angle of **w** as 130°.
26. $|\mathbf{v}| = \frac{1}{2}$ with the direction angle of **v** as 66° and $|\mathbf{w}| = 50$ with the direction angle of **w** as 40°.

27–34. *Find (to the nearest tenth of a degree) the angle θ between each pair of vectors, where* $0° \le \theta \le 180°$.

27. $\mathbf{v} = \langle 7, 8\rangle,\ \mathbf{w} = \langle -2, 10\rangle$
28. $\mathbf{v} = \langle -1, 4\rangle,\ \mathbf{w} = \langle 1, 3\rangle$
29. $\mathbf{v} = 6\mathbf{i} + 2\mathbf{j},\ \mathbf{w} = 4\mathbf{i}$
30. $\mathbf{v} = -4\mathbf{i} + \mathbf{j},\ \mathbf{w} = 7\mathbf{j}$
31. $\mathbf{v} = \langle 0, -5\rangle,\ \mathbf{w} = \langle 9, 0\rangle$
32. $\mathbf{v} = \langle 7, 2\rangle,\ \mathbf{w} = \langle -2, 7\rangle$
33. $\mathbf{v} = -3\mathbf{i} + 2\mathbf{j},\ \mathbf{w} = 7\mathbf{i} - \mathbf{j}$
34. $\mathbf{v} = \mathbf{i} - 8\mathbf{j},\ \mathbf{w} = 2\mathbf{i} + \mathbf{j}$

35–40. *Determine whether each pair of vectors is orthogonal.*

35. $\langle 2, -1 \rangle, \langle 1, 2 \rangle$

36. $3\mathbf{i} + 6\mathbf{j},\ 2\mathbf{i} - \mathbf{j}$

37. $4\mathbf{i} + 6\mathbf{j},\ -10\mathbf{i}$

38. $\left\langle \frac{1}{2}, -\frac{1}{2} \right\rangle, \left\langle \frac{\sqrt{3}}{3}, 0 \right\rangle$

39. $7\mathbf{i},\ -3\mathbf{j}$

40. $\langle \sqrt{3}, -2\sqrt{3} \rangle, \langle -2, -1 \rangle$

41–44. *Find the component of* $\mathbf{v}$ *along* $\mathbf{w}$ $(\text{comp}_{\mathbf{w}}\mathbf{v})$ *and demonstrate the result graphically.*

41. $\mathbf{v} = -5\mathbf{i} + 2\mathbf{j},\ \mathbf{w} = -2\mathbf{i}$

42. $\mathbf{v} = -7\mathbf{i} + 4\mathbf{j},\ \mathbf{w} = -6\mathbf{i}$

43. $\mathbf{v} = \langle 2, 6 \rangle,\ \mathbf{w} = \langle -5, 4 \rangle$

44. $\mathbf{v} = \langle -10, 9 \rangle,\ \mathbf{w} = \langle 4, 2 \rangle$

45–50. *Find the component of* $\mathbf{w}$ *along* $\mathbf{v}$ $(\text{comp}_{\mathbf{v}}\mathbf{w})$.

45. $\mathbf{v} = -5\mathbf{i} + 2\mathbf{j},\ \mathbf{w} = -2\mathbf{i}$

46. $\mathbf{v} = -7\mathbf{i} + 4\mathbf{j},\ \mathbf{w} = -6\mathbf{i}$

47. $\mathbf{v} = \langle -3, 4 \rangle,\ \mathbf{w} = \langle -1, 3 \rangle$

48. $\mathbf{v} = \langle 8, -6 \rangle,\ \mathbf{w} = \langle 3, -2 \rangle$

49. $\mathbf{v} = 3\mathbf{j},\ \mathbf{w} = 10\mathbf{i}$

50. $\mathbf{v} = 6\mathbf{i},\ \mathbf{w} = 5\mathbf{j}$

51–54. *Find the* $\mathbf{i}, \mathbf{j}$ *form of the standard position vector equivalent to* $\overrightarrow{PQ}$.

51. P(–2, 6) y 1 –4 –1 2 x –6 Q(8, –5)

52.

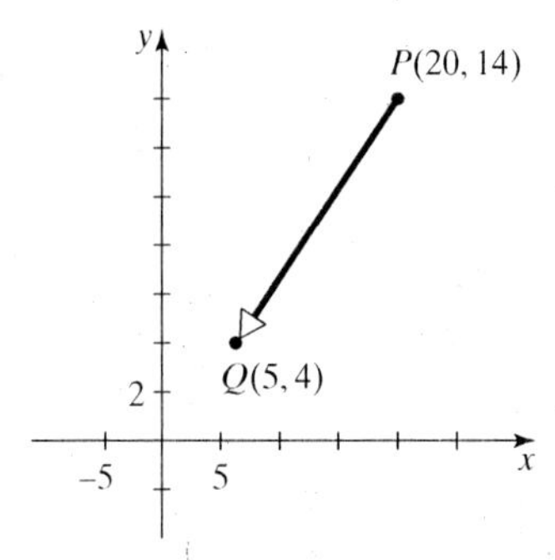

53.

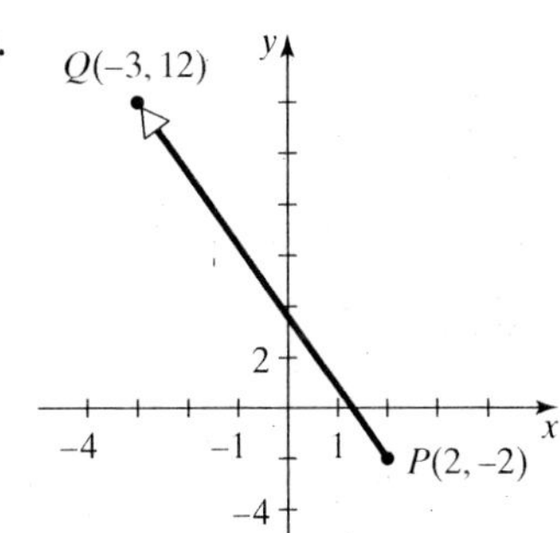

54.

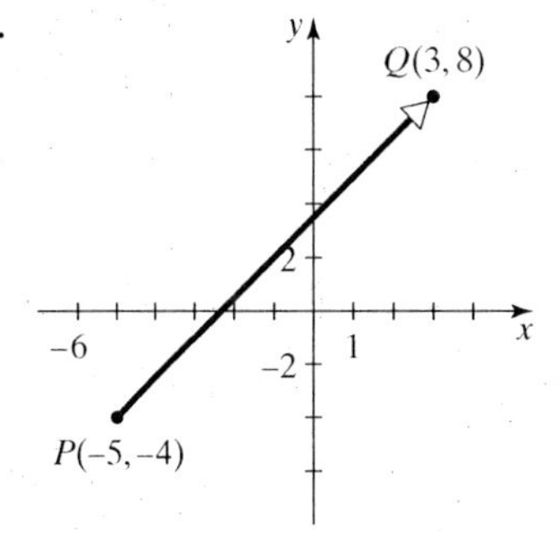

55–68. *Are the following statements true or false? If a statement is false, explain why or give an example that shows why it is false.*

55. The sum of two vectors is a vector.

56. The dot product of two vectors is a vector.

57. The magnitude of a vector is a vector.

58. The product of a scalar and a vector is a vector.

59. Operations with vectors do not always produce vectors.

60. A scalar and a vector can be added.

61. $\mathbf{0} = 0$.

62. A resultant of two vectors is a vector.

63. If two vectors $\mathbf{u}$ and $\mathbf{v}$ are in the same direction, then $\mathbf{u} \cdot \mathbf{v} = |\mathbf{u}||\mathbf{v}|$.

64. Vector subtraction is commutative, or $\mathbf{a} - \mathbf{b} = \mathbf{b} - \mathbf{a}$, for all vectors $\mathbf{a}$ and $\mathbf{b}$.

65. If two vectors are orthogonal, then their dot product is zero.

66. The magnitude of a vector is always nonnegative.

67. If $\mathbf{v} = \langle a, b \rangle$, then $\text{comp}_{\mathbf{i}}\mathbf{v} = a$.

68. If $\mathbf{v} = \langle a, b \rangle$, then $\text{comp}_{\mathbf{j}}\mathbf{v} = b$.

69. Find the work done (to the nearest foot-pound) to pull a boat up a 40-foot ramp that is on a 30° incline if a 600-pound force is used in the direction of the ramp.

70. Find the required work (to the nearest foot-pound) to pull a chair 50 feet by exerting a force of 15 pounds on the arm of the chair that makes a 45° angle with the horizontal.

71. How much work is done by a weight lifter bench pressing a 150-pound barbell $2\frac{1}{2}$ feet vertically above her chest?

72. How much work is done by a 163-pound worker if he climbs a 17 foot vertical ladder?

73. Find the work done moving a box from $P(2, 7)$ to $Q(19, 25)$ with the force $\mathbf{v} = 2\mathbf{i} + 6\mathbf{j}$. Assume the work units are foot-pounds.

74. Find the work done moving an object from $P(1, 3)$ to $Q(-3, 10)$ with force $\mathbf{w} = \langle -5, 6 \rangle$. Assume the work units are foot-pounds.

75. Prove that for any vector $\overrightarrow{PQ}$ with initial point $P(p_1, p_2)$ and terminal point $Q(q_1, q_2)$, $\overrightarrow{PQ} = \langle q_1 - p_1, q_2 - p_2 \rangle$.

76. If a vector $\mathbf{v} = 2\mathbf{i} + 6\mathbf{j}$ is on the line $y = mx + b$, what is the relationship between the slope of the line m and $\tan \theta$, where θ is the direction angle of the vector $\mathbf{v}$?

77–82. *Prove each of the following.*

77. $\mathbf{v} \cdot \mathbf{v} = |\mathbf{v}|^2$

78. $\mathbf{v} + \mathbf{w} = \mathbf{w} + \mathbf{v}$

79. $\mathbf{v} \cdot \mathbf{w} = \mathbf{w} \cdot \mathbf{v}$

80. $\mathbf{i} \cdot \mathbf{j} = 0$

81. $(\mathbf{v} + \mathbf{u}) + \mathbf{w} = \mathbf{v} + (\mathbf{u} + \mathbf{w})$

82. For all vectors $\mathbf{v} \neq \mathbf{0}$, $\dfrac{\mathbf{v}}{|\mathbf{v}|}$ is a unit vector.

Discussion

83. Let θ be the angle between vectors $\mathbf{v}$ and $\mathbf{w}$. What does $\theta = 0°$ tell you about these vectors? What does $\theta = 180°$ tell you about these vectors?

84. Draw two parallel vectors. Do you think that parallel vectors have to point in the same direction? Can parallel vectors point in opposite directions? Discuss what will happen if you slide one vector onto the other.

85. Two vectors are parallel if one is a scalar multiple of the other. Discuss why this statement is true. Draw diagrams to show $\mathbf{v}$ and $\mathbf{w}$ are parallel if $\mathbf{v} = c\mathbf{w}$, where c is a scalar.

86. To find the vertical and horizontal component of $\mathbf{v}$, we use $v_1 = |\mathbf{v}| \cos \theta$ and $v_2 = |\mathbf{v}| \sin \theta$. Verify these relationships using the trigonometric functions defined in Section 3.2.

6.3 Polar Coordinate System

So far we have been graphing plane curves in the rectangular (or Cartesian) coordinate system, using ordered pairs (a, b) to denote a point whose directed distances along the x- and y-axes are a and b, respectively. The equations of these curves have been in either rectangular or parametric form. In this section we consider another method of representing points and equations: the **polar coordinate system**.

Polar Coordinates

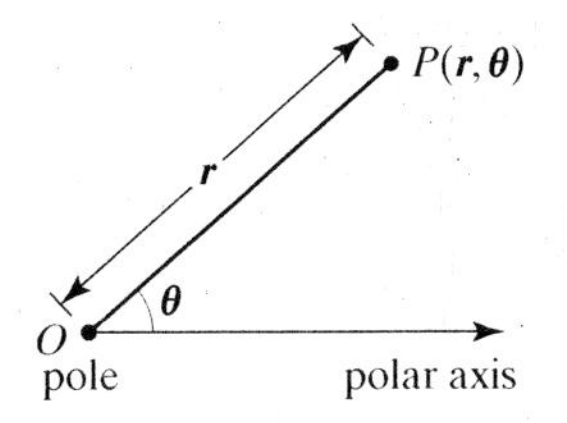

To form the polar coordinate system in the plane, we locate a point O, called the **pole** (or origin). From point O we draw a fixed ray, called the **polar axis**. This fixed ray is usually drawn horizontally to the right and corresponds to the positive x-axis. Then, any point P in the plane can be assigned **polar coordinates** (r, θ) as follows.

$$r = \textit{directed distance} \text{ from } O \text{ to } P$$

$$\theta = \textit{directed angle} \text{ from the polar axis to the segment } OP$$

Either radians or degrees can be used for the measure of θ. The angle θ is considered positive if measured in a counterclockwise direction from the polar axis, and negative if the rotation is clockwise. If $r > 0$, point P is r units from O along the terminal side of θ. If $r < 0$, point P is $|r|$ units along the ray in the *opposite* direction as the terminal side of θ. Since it follows that (r, θ) lies on a circle of radius r, it is convenient to locate points with respect to a grid of concentric circles centered at the pole and rays whose initial points are at the pole. The following are examples of polar coordinates.

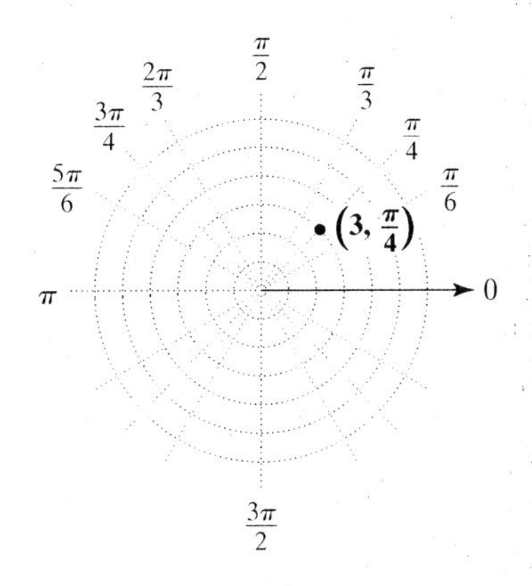

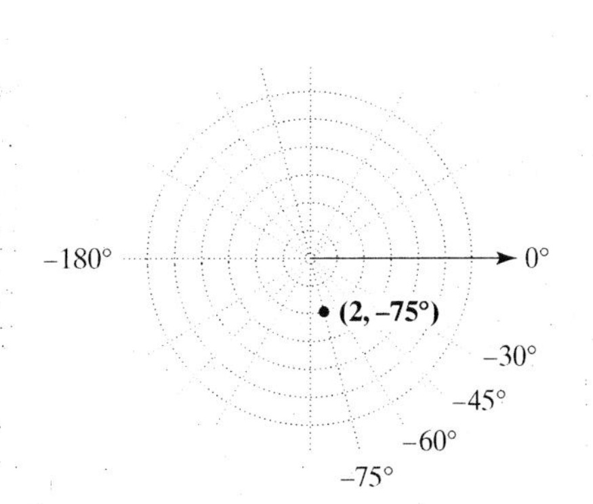

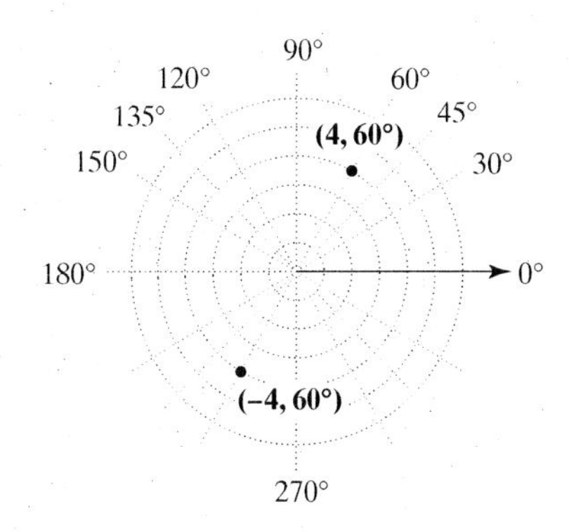

EXAMPLE 1 Graph and label the points $(1, 45°)$, $(4, 90°)$, $(3, 150°)$, and $(-3, 330°)$ in the polar coordinate system.

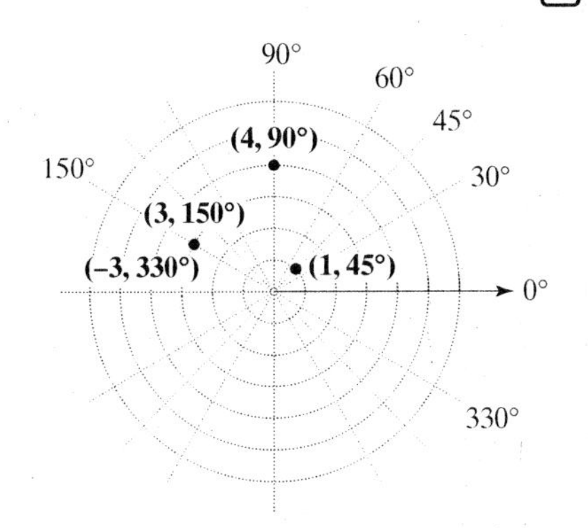

SOLUTION To graph $(1, 45°)$, we locate the point that is 1 unit from the pole along the terminal side of 45°. For $(4, 90°)$, we locate the point that is 4 units from the pole along the terminal side of 90°. Likewise $(3, 150°)$ is 3 units from the pole along the terminal side of 150°. For $(-3, 330°)$, we locate the point that is 3 units from the pole along the ray in the *opposite* direction of 330°. (For negative values of r, you may find it easier to first find the ray that determines the angle. Then locate your point along the ray in the opposite direction of the angle). ■

Notice that $(3, 150°)$ and $(-3, 330°)$ represent the same point. When we use rectangular coordinates, each point P has a *unique* (x, y) representation. This is not the case with polar coordinates, as Example 1 demonstrates. Since the polar axis is the initial side of a given θ, there are infinitely many angles with the same terminal side. We express this idea as follows.

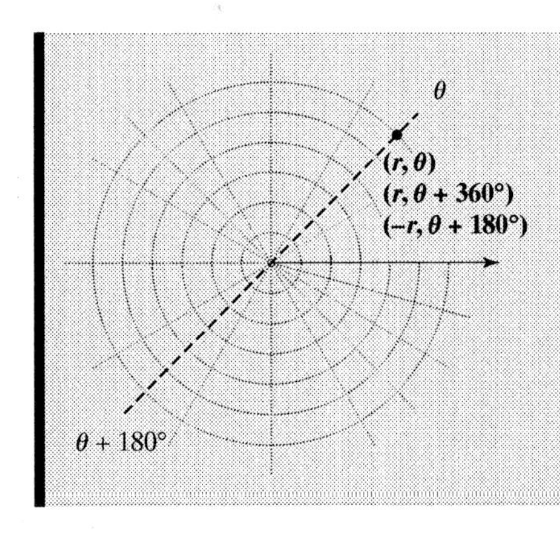

A point with polar coordinates (r, θ) can be written as

$$(r, \theta),\quad (r, \theta + k \cdot 360°),\quad \text{or}\quad (r, \theta) = (-r, \theta + (2k + 1)180°)$$
$$(r, \theta),\quad (r, \theta + k \cdot 2\pi),\quad \text{or}\quad (r, \theta) = (-r, \theta + (2k + 1)\pi),$$

where k is any integer. If point P is at the pole, then its polar coordinates are $(0, \theta)$ for any θ.

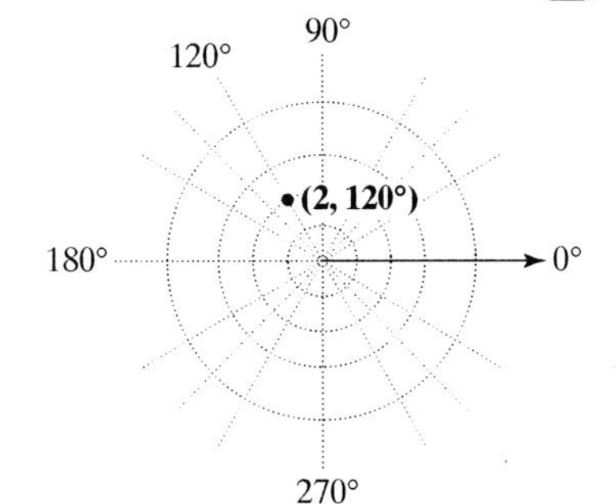

EXAMPLE 2 Find three additional sets of polar coordinates for the point $(2, 120°)$.

SOLUTION For $(2, 120°)$:

$$\underset{k=1}{(2, 120° + (1)360°)},\quad \underset{k=-1}{(2, 120° + (-1)360°)},\quad \underset{k=0}{(-2, 120° + (1)180°)}.$$

Simplifying, we get $(2, 480°)$, $(2, -240°)$, and $(-2, 300°)$. There are infinitely many possible answers. ■

Relationship between Polar Coordinates and Rectangular Coordinates

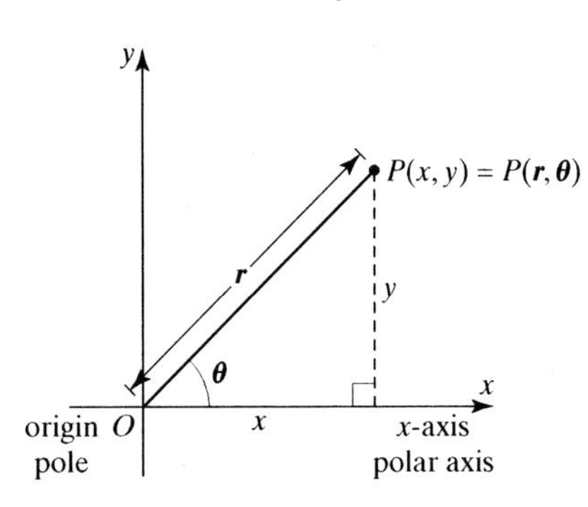

Let's determine the relationship between polar and rectangular coordinates. If we place the pole at the origin and let the polar axis coincide with the positive x-axis, then point P has polar coordinates (r, θ) and rectangular coordinates (x, y). Since (x, y) lies on a circle of radius r, it follows that $x^2 + y^2 = r^2$. Moreover, for $r > 0$, the definition of trigonometric functions from Section 3.2 implies that

$$\cos\theta = \frac{x}{r}, \qquad \sin\theta = \frac{y}{r}, \qquad \text{and} \qquad \tan\theta = \frac{y}{x}.$$

If $r < 0$, we can show that the same relationship applies. For example, consider the point (r, θ), where $r < 0$. Using the previous formula, we know that $(-r, \theta + 180°)$ can represent the same point as (r, θ). And because $r < 0$, we now have $-r > 0$ and therefore are able to use the preceding ratios. Then, by applying the cosine, sine, and tangent sum formulas, we obtain the same results.

$$\cos(\theta + 180°) = \frac{x}{-r} \qquad \sin(\theta + 180°) = \frac{y}{-r} \qquad \tan(\theta + 180°) = \frac{y}{x}$$

$$-\cos\theta = \frac{x}{-r} \qquad -\sin\theta = \frac{y}{-r} \qquad \tan\theta = \frac{y}{x}$$

$$\cos\theta = \frac{x}{r} \qquad \sin\theta = \frac{y}{r}$$

If $r = 0$, then, regardless of θ, the point lies at the pole and the corresponding rectangular coordinates are $(0, 0)$.

These relationships allow us to convert coordinates or equations from one system to the other.

Coordinate Conversion

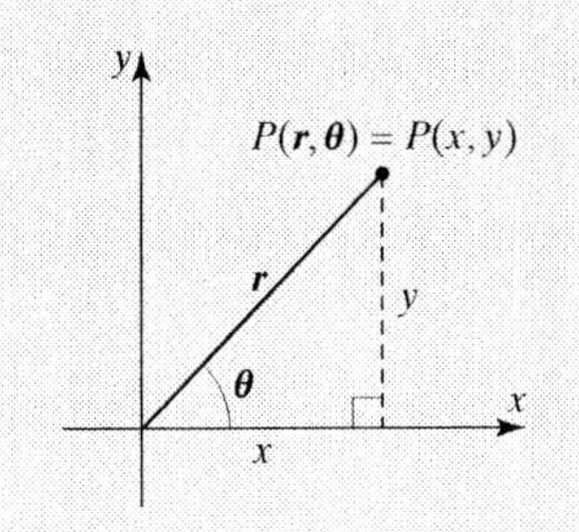

The polar coordinates (r, θ) are related to the rectangular coordinates (x, y) as follows.

$$\tan\theta = \frac{y}{x} \qquad x = r\cos\theta$$

$$r^2 = x^2 + y^2 \qquad y = r\sin\theta$$

These equations allow us to find the polar coordinates of a point when the rectangular coordinates are known.

These equations allow us to find the rectangular coordinates of a point when the polar coordinates are known.

EXAMPLE 3 Find the rectangular coordinates of the point with polar coordinates $(2, 4\pi/3)$.

SOLUTION We have $r = 2$ and $\theta = 4\pi/3$. We use the equations relating the coordinates.

$$x = r\cos\theta \qquad \text{and} \qquad y = r\sin\theta$$

$$x = 2\cos\frac{4\pi}{3} = 2\left(-\frac{1}{2}\right) \qquad y = 2\sin\frac{4\pi}{3} = 2\left(-\frac{\sqrt{3}}{2}\right)$$

$$x = -1 \qquad y = -\sqrt{3}$$

Thus, $(x, y) = (-1, -\sqrt{3})$.

The point and the corresponding ordered pairs are shown.

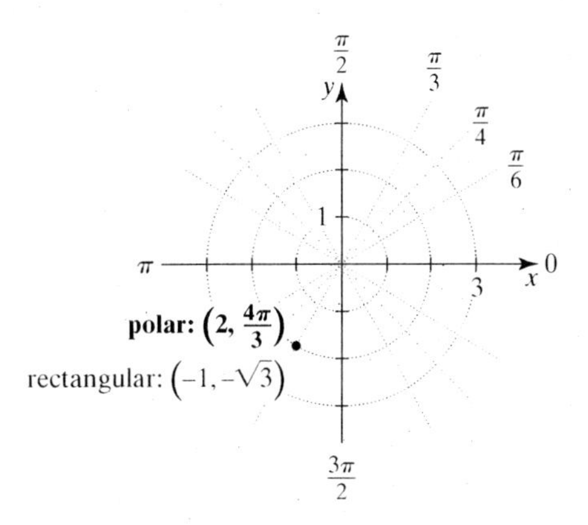

EXAMPLE 4 Find the polar coordinates of each point with the given rectangular coordinates. Assume that $0° \le \theta < 360°$ and that $r > 0$.

a. $(-2, 2)$

b. $(4, -5)$

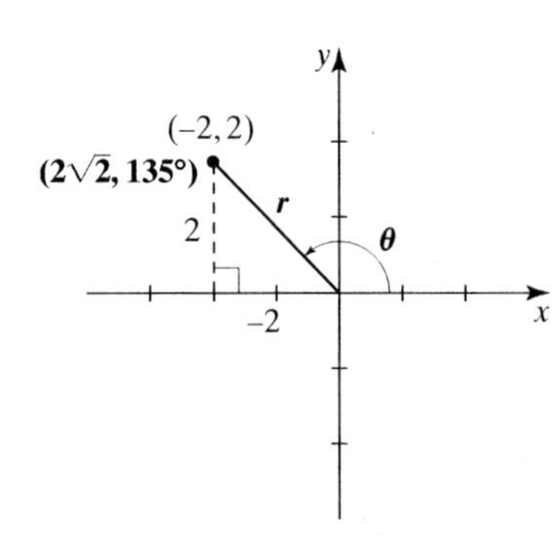

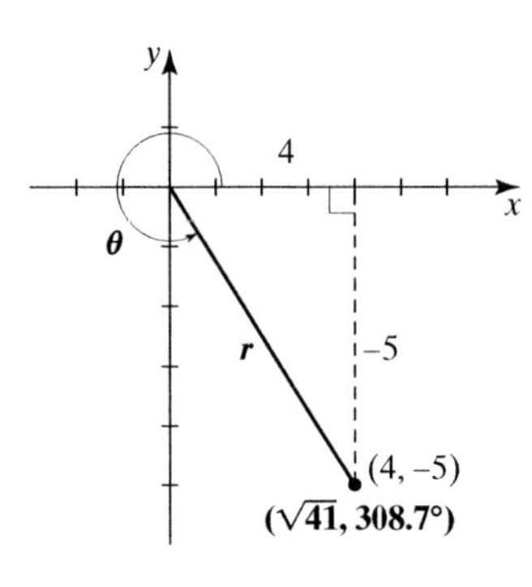

SOLUTION

a. Point $(-2, 2)$ is in QII. Since $r > 0$ and $0° \le \theta < 360°$, we want to make sure that the polar coordinates we select also represent a point in QII.

$$r = \pm\sqrt{(-2)^2 + 2^2} = \pm\sqrt{8} = \pm 2\sqrt{2}$$

We select $r = 2\sqrt{2}$. To find θ, we know

$$\tan\theta = \frac{y}{x} = \frac{2}{-2} = -1,$$

$$\text{and} \qquad \hat{\theta} = \tan^{-1}(1) = 45°.$$

Since the point is in QII and r is positive, $\theta = 135°$. Thus, $(r, \theta) = (2\sqrt{2}, 135°)$.

b. Point $(4, -5)$ is in QIV.

$$r = \pm\sqrt{x^2 + y^2} = \pm\sqrt{4^2 + (-5)^2} = \pm\sqrt{41}.$$

We select $r = \sqrt{41}$. To find θ, we know

$$\tan\theta = \frac{y}{x} = -\frac{5}{4},$$

$$\text{and} \qquad \hat{\theta} = \tan^{-1}\left(\frac{5}{4}\right) \approx 51.3°.$$

Since the point is in QIV and r is positive, $\theta \approx 360° - 51.3° = 308.7°$. Thus, $(r, \theta) = (\sqrt{41}, 308.7°)$. ■

CAUTION: If you compare Examples 3 and 4, you see that converting the polar coordinates of a point to rectangular coordinates is straightforward. But when we change from *rectangular coordinates to polar coordinates,* we must be careful to select polar coordinates that represent the same point by paying attention to the quadrant that contains the given rectangular coordinates.

Polar Equations and Graphs

To sketch curves in polar coordinates, we can proceed as we would with rectangular coordinates by plotting points from a table of values. It is also an option, and sometimes easier, to convert the polar equation to rectangular or the rectangular equation to polar, and then sketch the graph of the converted equation. To convert an equation from rectangular to polar, we just substitute x with $r\cos\theta$ and y with $r\sin\theta$ and simplify where possible. But converting from polar to rectangular can involve some strategies in addition to substitution. Let's take a look at a few conversions between polar and rectangular equations before we graph them.

EXAMPLE 5 Convert the following polar equations to rectangular and describe the graph that both equations represent.

a. $r = 5$ **b.** $\theta = \frac{\pi}{4}$ **c.** $r = \sec\theta$ **d.** $r = 4\sin\theta$

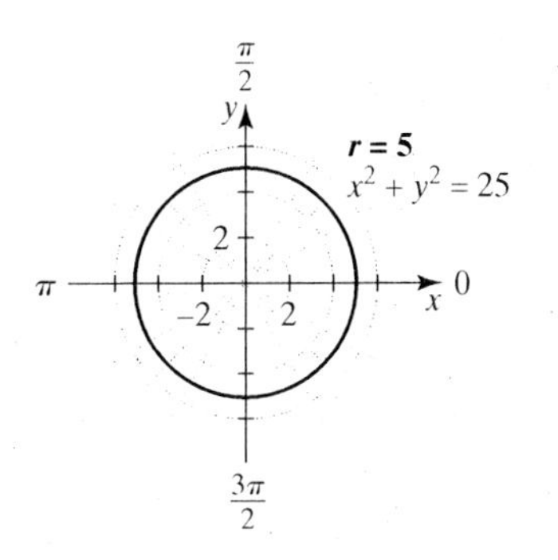

SOLUTION We rewrite each polar equation in order to be able to use the conversions.

a.

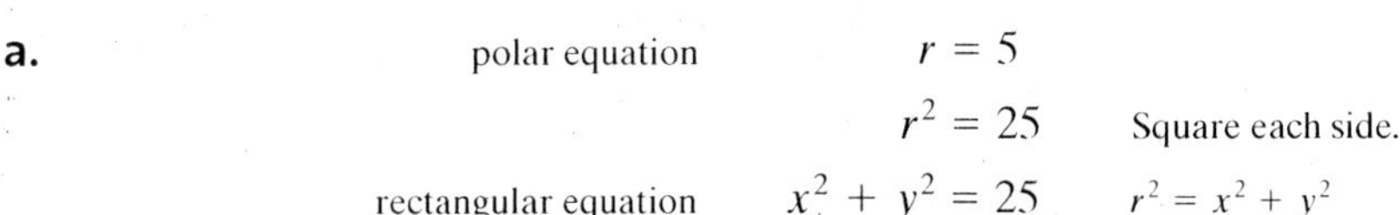

polar equation	$r = 5$	
	$r^2 = 25$	Square each side.
rectangular equation	$x^2 + y^2 = 25$	$r^2 = x^2 + y^2$

The graph of $r = 5$ consists of all points that are 5 units from the pole, where θ can be any angle. Geometrically we have a circle with center at the pole and a radius of 5 units. This circle is the same graph represented by the rectangular equation $x^2 + y^2 = 25$. It would appear that the polar form $r = 5$ is a simpler equation for the circle.

b. For $\theta = \pi/4$, we are not given a specific value for r. Therefore, we use the conversion equation that only involves θ, $\tan\theta = y/x$.

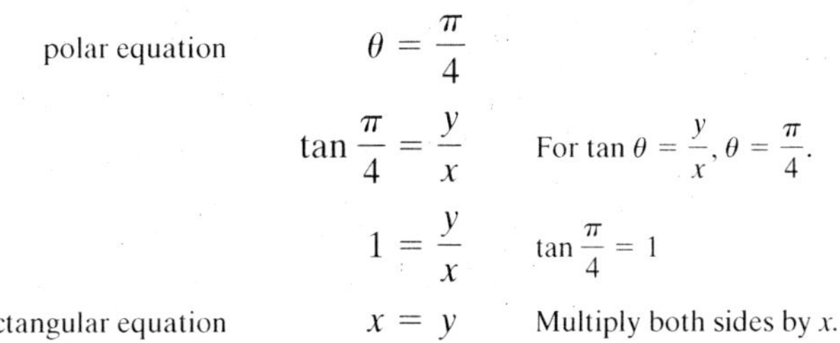

polar equation	$\theta = \dfrac{\pi}{4}$	
	$\tan\dfrac{\pi}{4} = \dfrac{y}{x}$	For $\tan\theta = \dfrac{y}{x}$, $\theta = \dfrac{\pi}{4}$.
	$1 = \dfrac{y}{x}$	$\tan\dfrac{\pi}{4} = 1$
rectangular equation	$x = y$	Multiply both sides by x.

Since r can be any value, the graph of $\theta = \pi/4$ consists of all points on the line that make an angle of $\pi/4$ with the positive x-axis, which is the line $y = x$. It may be debatable which equation is a simpler form for the line.

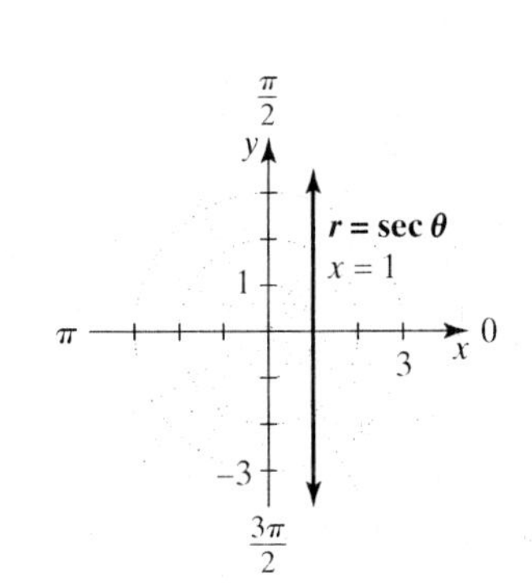

c.

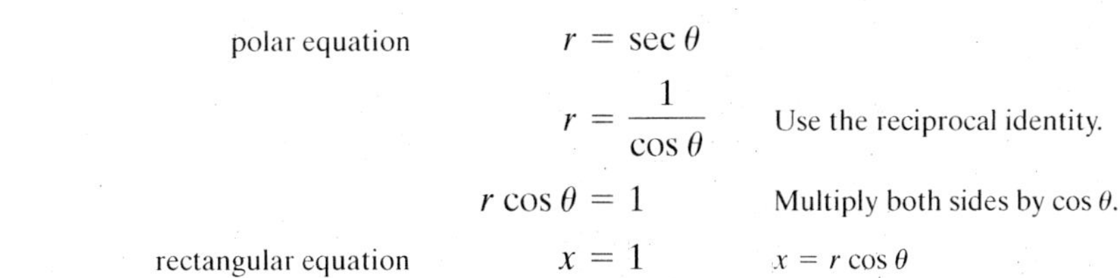

polar equation	$r = \sec\theta$	
	$r = \dfrac{1}{\cos\theta}$	Use the reciprocal identity.
	$r\cos\theta = 1$	Multiply both sides by $\cos\theta$.
rectangular equation	$x = 1$	$x = r\cos\theta$

The graph of $r = \sec\theta$ is not easily determined by its polar equation. But after conversion to rectangular, the graph is recognized as a vertical line. It would appear that the rectangular equation is a simpler form for the vertical line.

d. We multiply each side of $r = 4\sin\theta$ by r in order to substitute $x^2 + y^2$ for r^2 and y for $r\sin\theta$. (We do not square both sides since it would not provide an r on the right side.)

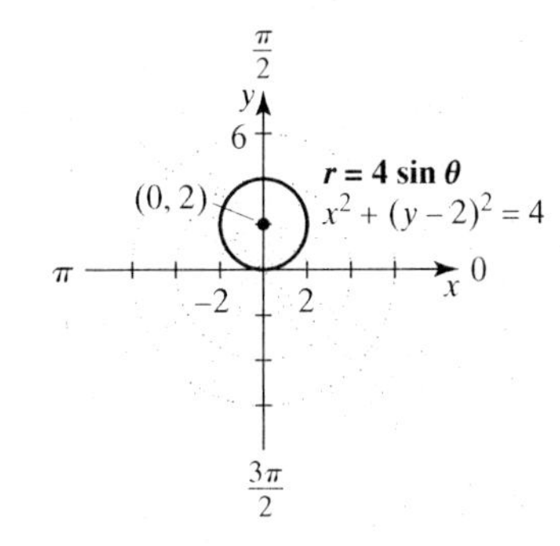

polar equation	$r = 4\sin\theta$	
	$r^2 = 4r\sin\theta$	Multiply both sides by r.
	$x^2 + y^2 = 4y$	$r^2 = x^2 + y^2$, $r\sin\theta = y$
	$x^2 + y^2 - 4y + 4 = 0 + 4$	Complete the square in y.
rectangular equation	$x^2 + (y - 2)^2 = 4$	Factor the completed square.

The equation represents the circle with center at $(0, 2)$ and radius 2. Perhaps the polar equation is a simpler form for this circle. ■

EXAMPLE 6 Find the corresponding polar equation.

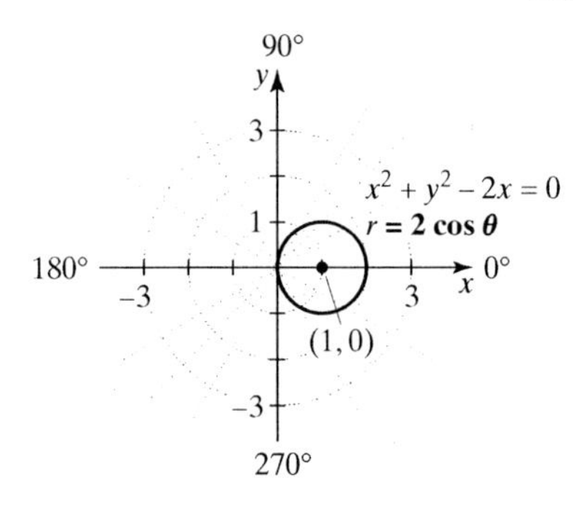

a. $x^2 + y^2 - 2x = 0$ **b.** $y = 9$

SOLUTION

a. rectangular equation

$$x^2 + y^2 - 2x = 0$$

$$r^2 - 2r \cos \theta = 0 \qquad x^2 + y^2 = r^2, x = r \cos \theta.$$

$$r(r - 2 \cos \theta) = 0 \qquad \text{Factor.}$$

$$r = 0 \quad \text{or} \quad r - 2 \cos \theta = 0 \qquad \text{Set each factor equal to 0.}$$

Notice that $r = 0$ is the pole (or origin) and $r = 2 \cos \theta$ includes the origin if $\theta = 90°$. So the corresponding polar equation is

polar equation $$r = 2 \cos \theta.$$

Do you think the polar form is simpler for this circle?

b. rectangular equation

$$y = 9$$

$$r \sin \theta = 9 \qquad y = r \sin \theta$$

$$r = \frac{9}{\sin \theta} \qquad \text{Divide both sides by } \sin \theta.$$

polar equation $$r = 9 \csc \theta$$

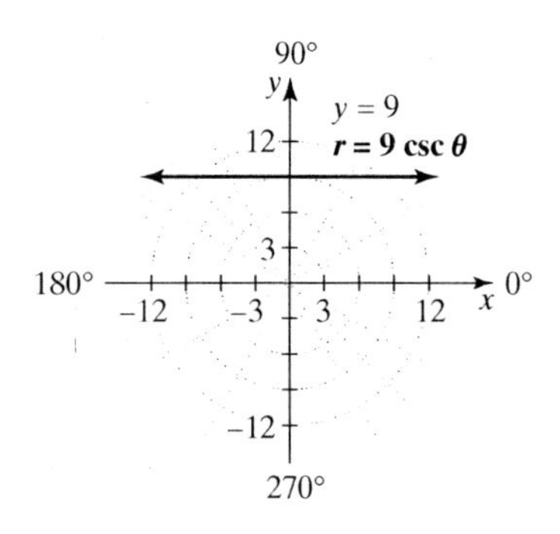

Polar equations are usually solved for r. The rectangular equation would appear to be the simpler equation for this horizontal line. ■

As we can see from the examples, some curves have simpler equations in polar coordinates, whereas some have simpler rectangular equations. And there are equations that do not convert nicely between the two coordinate systems. We therefore need to be able to graph in each coordinate system. As we do when we graph rectangular equations, we make a table of values containing several ordered pair solutions and plot the corresponding points (r, θ) to obtain the graph of a polar equation. We plot the points in the order of increasing values of θ. If points repeat, the curve will simply be traced again.

EXAMPLE 7 Graph $r = 2\sin\theta,\ 0° \le \theta \le 360°$.

SOLUTION

θ	$\sin\theta$	$r = 2\sin\theta$
0°	0	0
30°	$\frac{1}{2}$	1
45°	$\frac{\sqrt{2}}{2}$	$\sqrt{2} \approx 1.4$
60°	$\frac{\sqrt{3}}{2}$	$\sqrt{3} \approx 1.7$
90°	1	2
120°	$\frac{\sqrt{3}}{2}$	$\sqrt{3} \approx 1.7$
180°	0	0
225°	$-\frac{\sqrt{2}}{2}$	$-\sqrt{2} \approx -1.4$
270°	-1	-2
360°	0	0

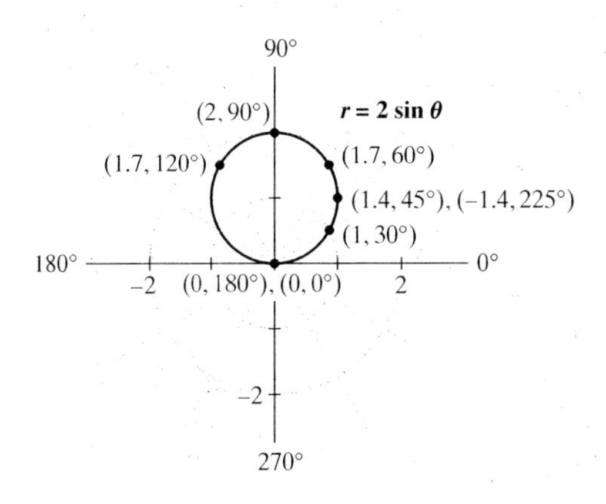

We plot the points and connect them with a smooth curve as θ increases . If we first think about the graph $r = 2\sin\theta$ as θ increases, we would expect that r equals 0 when $\theta = 0°$, r has the largest value when $\theta = 90°$, and r returns to $r = 0$ when $\theta = 180°$. We could also graph this curve by either converting to rectangular coordinates first or using a graphing utility in polar and degree mode with window: θmin = 0, θmax = 360, Xmin = −2, Xmax = 2, Ymin = −2, and Ymax = 2. (Or you could use radian mode and let θ max = 2π.) ■

As we discovered in Chapter 2, once we know the pure form of the graph and what causes variations to these pure forms, we can sketch the graphs quickly. We now give a few common polar equations, along with their names and graphs, and show the variations. Try to mentally visualize these shapes by determining the values of r as θ increases.

Polar graphs, especially those that are not variations of the common ones categorized on this and the following pages, are challenging to do by hand, but are easily done by a graphing utility in polar mode, in the form $r = f(\theta)$.

Circles $(a > 0)$

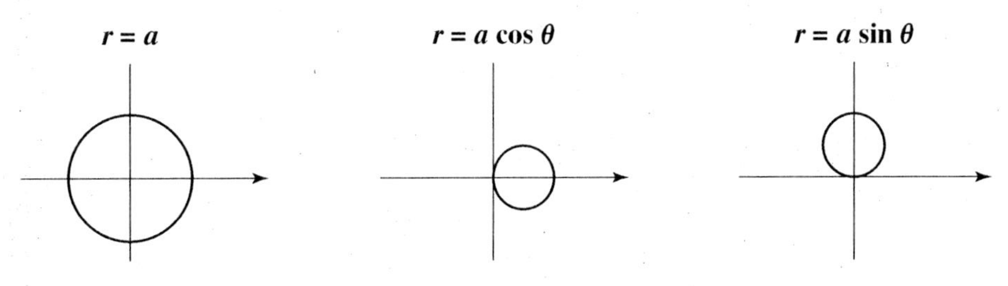

Limaçons $(a > 0, b > 0)$

$r = a + b\cos\theta$

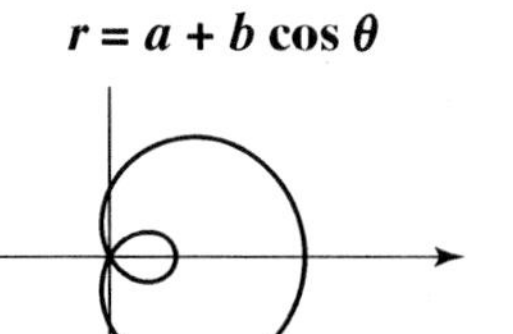

Limaçon $a < b$ (inner loop)

$r = a + b\sin\theta$

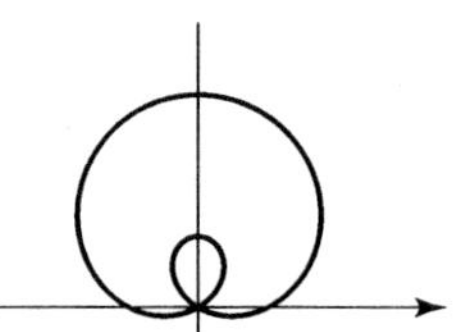

Limaçon $a < b$ (inner loop)

$r = a + b\cos\theta$

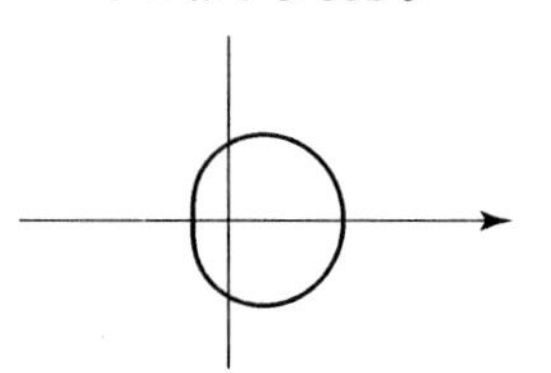

Limaçon $a > b$ (dimpled)

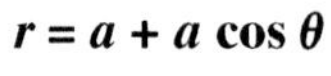

$r = a + a\cos\theta$

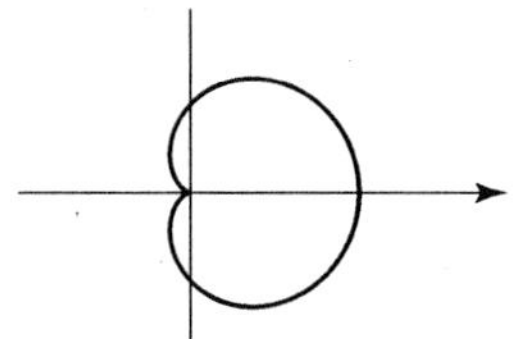

Cardioid $a = b$ (heart-shaped)

$r = a + a\sin\theta$

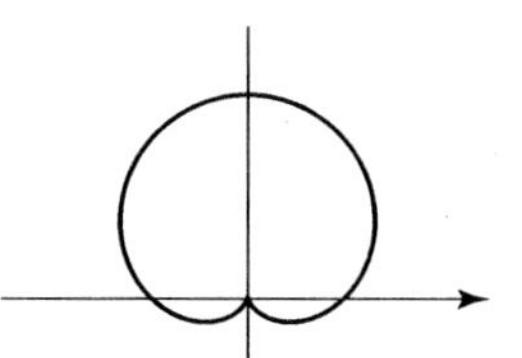

Cardioid $a = b$ (heart-shaped)

Rose Curves $(a > 0)$

n petals if n is odd, $2n$ petals if n is even $(n \geq 2)$.

$r = a\cos n\theta$
$n = 2$ (4 petals)

$r = a\cos n\theta$
$n = 3$ (3 petals)

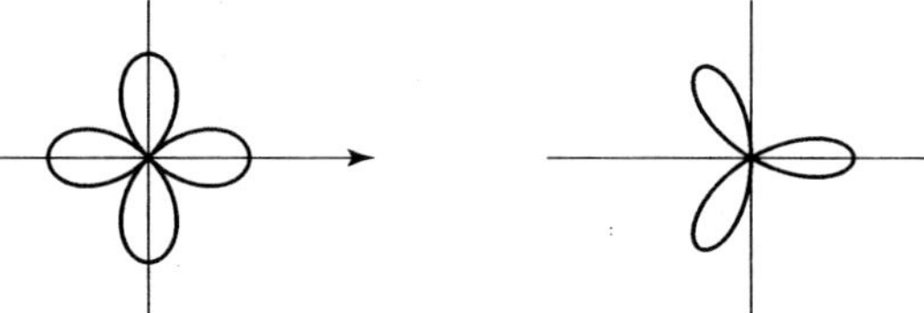

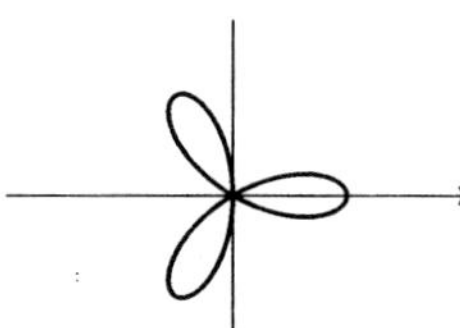

$r = a\sin n\theta$
$n = 4$ (8 petals)

Lemniscates

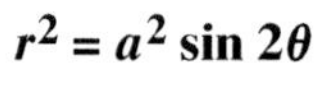

$r^2 = a^2\sin 2\theta$

$r^2 = a^2\cos 2\theta$

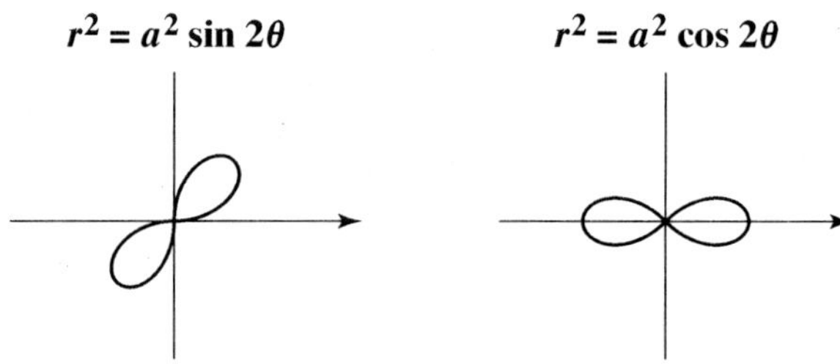

Spirals $(a > 0, k > 0)$

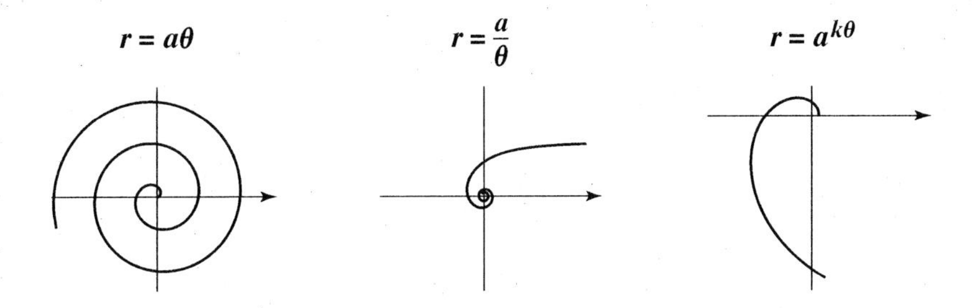

You will be asked in the exercises to investigate what happens to these curves when values of a and b are negative.

CONNECTIONS WITH TECHNOLOGY

Graphing Utility

Many graphing utilities are capable of graphing polar equations. Set the mode to both RADIAN and POLAR (Pol). To investigate what effect $a < 0$ has on $r = a\cos\theta$, we graph $r = 4\cos\theta$ and $r = -4\cos\theta$. Each graph shown here was produced using the window $\theta\text{min} = 0$, $\theta\text{max} = \pi$, Xmin $= -5$, Xmax $= 5$, Ymin $= -4$, Ymax $= 4$. (We could have set the mode to DEGREES and used $\theta\text{max} = 180°$.)

$r = 4\cos\theta$

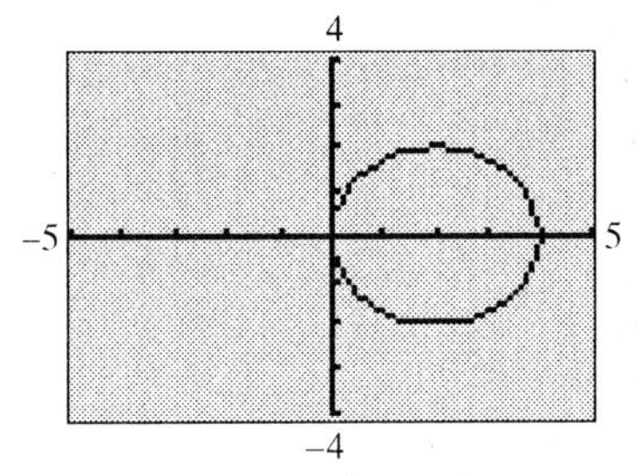

$r = -4\cos\theta$

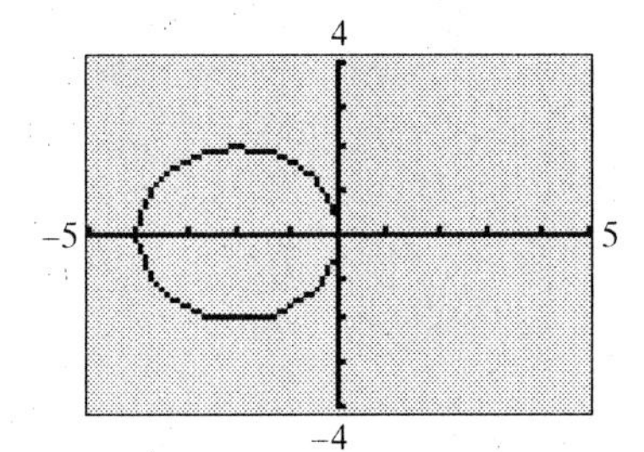

The graph of $r = 4\cos\theta$ and $r = -4\cos\theta$ are circles with radius 2. We see for $y = -a\cos\theta$, the graph of $y = a\cos\theta$ is reflected across the vertical axis.

Exercise Set 6.3

1–6. *Find the polar coordinates* (r, θ) *for the given point such that* $r > 0$ *and* $0° \le \theta < 360°$ *or* $0 \le \theta < 2\pi$.

1.

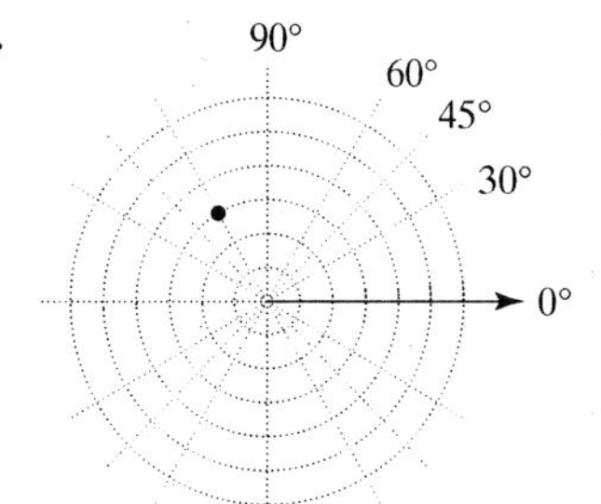

2.

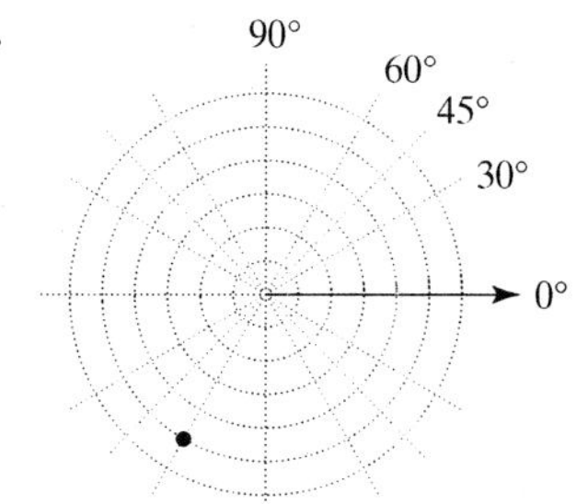

3.

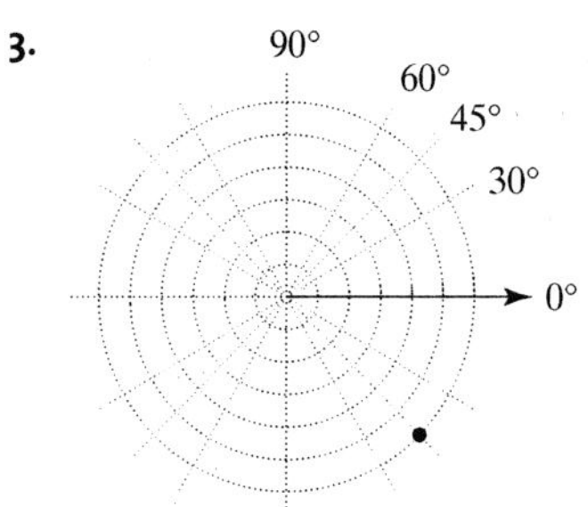

4.

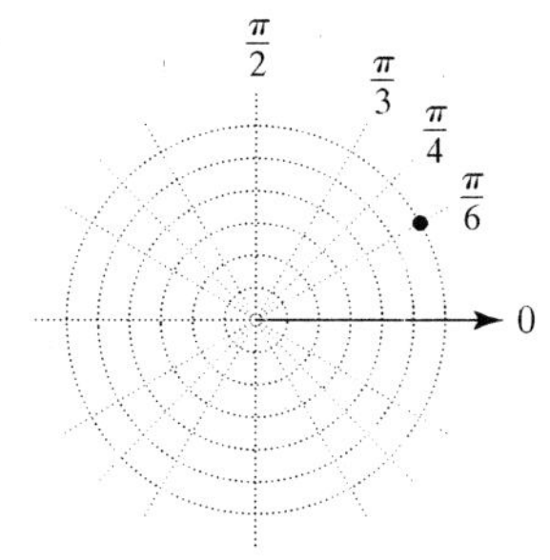

5.

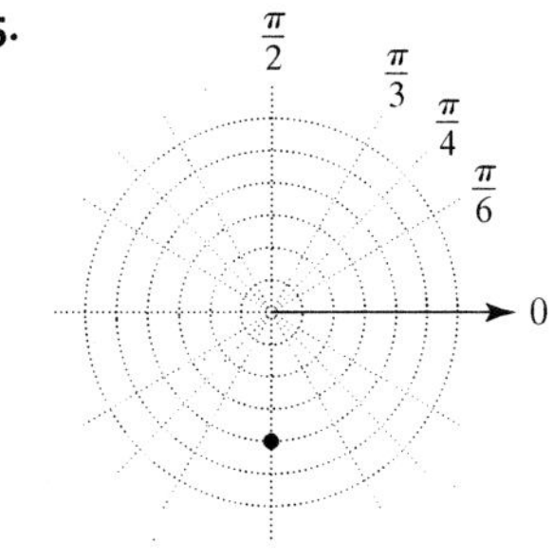

6.

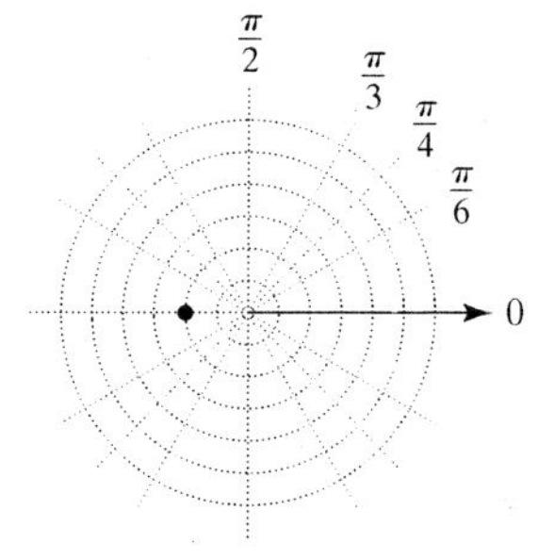

7–12. *On polar graph paper, graph the point with the given polar coordinates and find two additional sets of polar coordinates, where* $-360° < \theta \le 360°$.

7. $(2, 405°)$

8. $(7, 440°)$

9. $(4, -120°)$

10. $(-2, 30°)$

11. $(-5, -200°)$

12. $(-7, -100°)$

13–18. *Convert to rectangular coordinates.*

13. $\left(6, \dfrac{\pi}{6}\right)$

14. $\left(8, \dfrac{7\pi}{6}\right)$

15. $(-5, 120°)$

16. $(-4, 315°)$

17. $(0, 221°)$

18. $(9, 0°)$

19–24. *Convert to polar coordinates, where $r > 0$ and $0 \le \theta < 2\pi$.*

19. $(5, 0)$

20. $(0, -7)$

21. $(7, 7)$

22. $(6, -6)$

23. $(-\sqrt{3}, 1)$

24. $\left(-\frac{1}{2}, -\frac{\sqrt{3}}{2}\right)$

25–30. *Convert to a polar equation.*

25. $x^2 + y^2 = 16$

26. $4x + y = 9$

27. $y = 10$

28. $x = 6$

29. $x^2 + y^2 - 8y = 0$

30. $x^2 - 3y - 9 = 0$

31–36. *Convert to a rectangular equation.*

31. $r \cos \theta = 8$

32. $r = 6$

33. $\theta = \frac{\pi}{6}$

34. $r = 3 \cos \theta$

35. $r^2 = \csc 2\theta$

36. $r = 1 - 2 \sin \theta$

37–42. *Sketch the polar graph of each equation by plotting points. (Check with a graphing calculator.)*

37. $r = 8$

38. $r = 4 \cos \theta$

39. $r = 1 + \cos \theta$

40. $r = 2 + 4 \sin \theta$

41. $\theta = 50°$

42. $r = 2\theta, 0 \le \theta \le 2\pi$

Graphing Calculator/CAS Exercises

43–52. *To investigate what happens to the general form of the polar graphs (see pp. 357–359) when the values of a and b are negative, graph the two polar equations using polar and radian mode, and window: $\theta min = 0$, $\theta max = 2\pi$, $\theta step = 0.1$, $Xmin = -5$, $Xmax = 5$, $Ymin = -4$, $Ymax = 4$. Explain what change, if any, occurs to the polar graph when the value $a > 0$ changes to $a < 0$ or when the value of $b > 0$ changes to $b < 0$.*

43. $r = 2 + 2 \cos \theta$
$r = 2 - 2 \cos \theta$

44. $r = 2 + 2 \sin \theta$
$r = 2 - 2 \sin \theta$

45. $r = 3 \cos 2\theta$
$r = -3 \cos 2\theta$

46. $r = 4 \sin 3\theta$
$r = -4 \sin 3\theta$

47. $r = 4$
$r = -4$

48. $r = 0.8\theta$
$r = -0.8\theta$

49. $r = 2 + 2 \sin \theta$
$r = -2 + 2 \sin \theta$

50. $r = 1 + 3 \cos \theta$
$r = -1 + 3 \cos \theta$

51. $r = 1 + 3 \sin \theta$
$r = 1 - 3 \sin \theta$

52. $r = 3 \cos \theta$
$r = -3 \cos \theta$

53–66. *Match the graph with the correct equation from Column A. Use the information from problems 43–52 to help select the correct equation. Check your answer with a graphing utility.*

COLUMN A

a. $r = 4 - 4 \sin \theta$

b. $r = 5 \cos 3\theta$

c. $r = 5$

d. $r = 4 - 6 \cos \theta$

e. $r = 3 + 3 \cos \theta$

f. $r = 4 \sin 4\theta$

g. $r = 4 - 2 \sin \theta$

h. $r^2 = 25 \sin 2\theta$

i. $r^2 = 16 \cos 2\theta$

j. $r = \dfrac{\theta}{4}$

k. $\theta = \dfrac{\pi}{6}$

l. $\theta = 135°$

m. $r = 3 \sin \theta$

n. $r = 4 \sec \theta$

53.

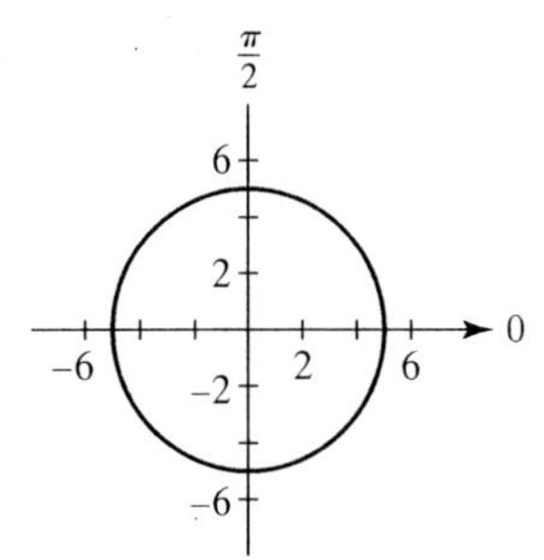

54.

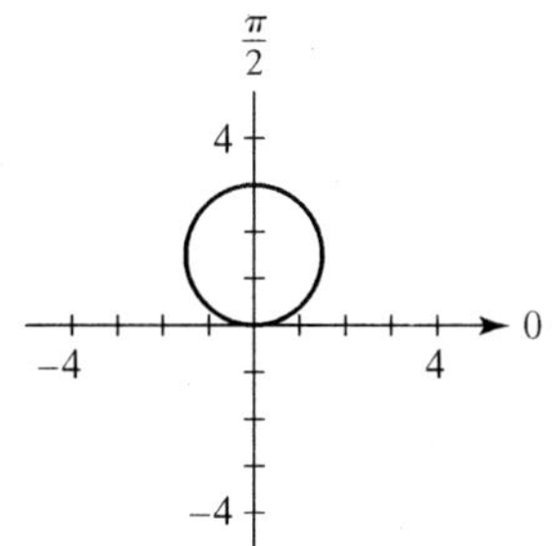

55.

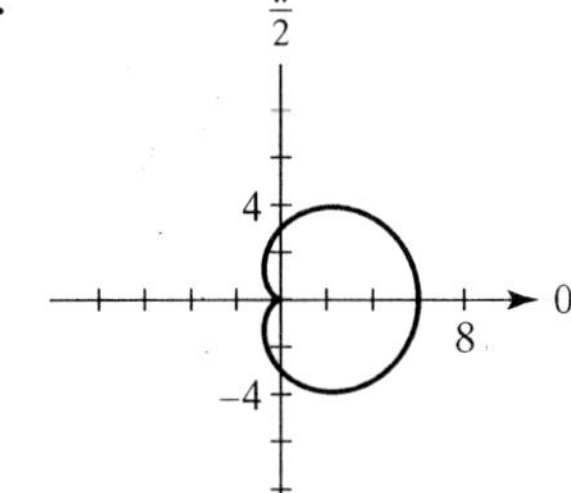

56.

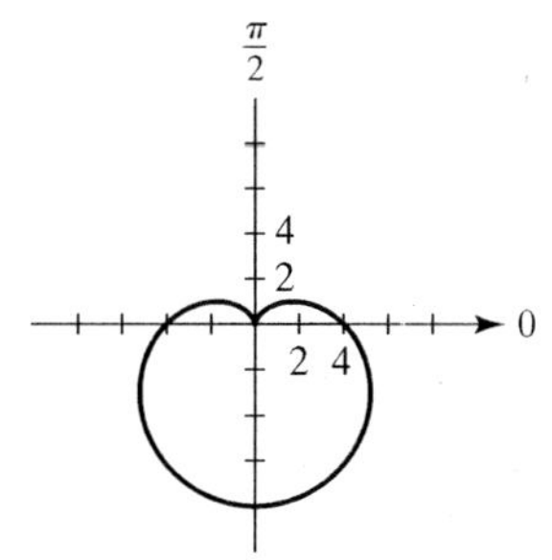

57.

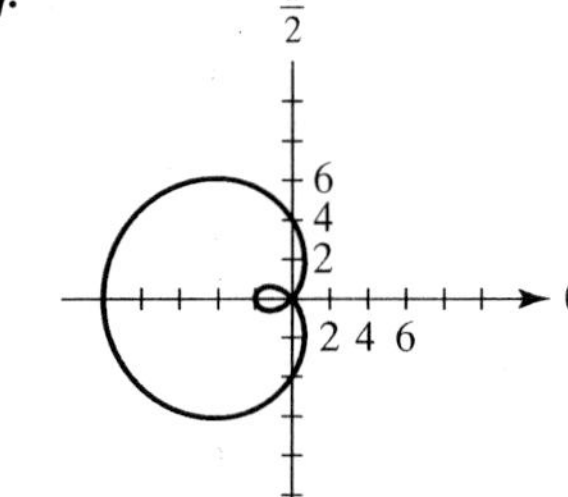

58.

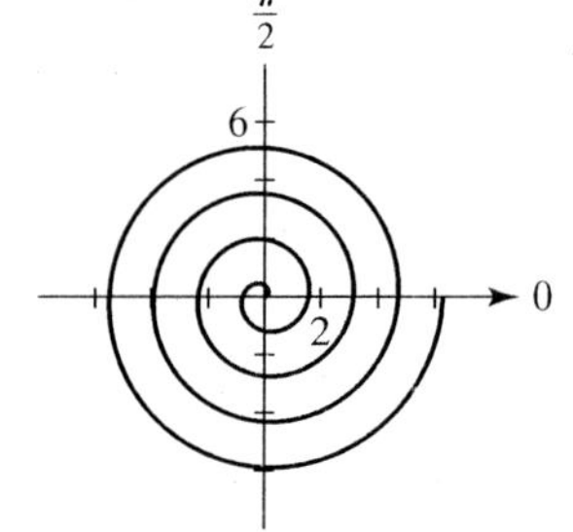

59.

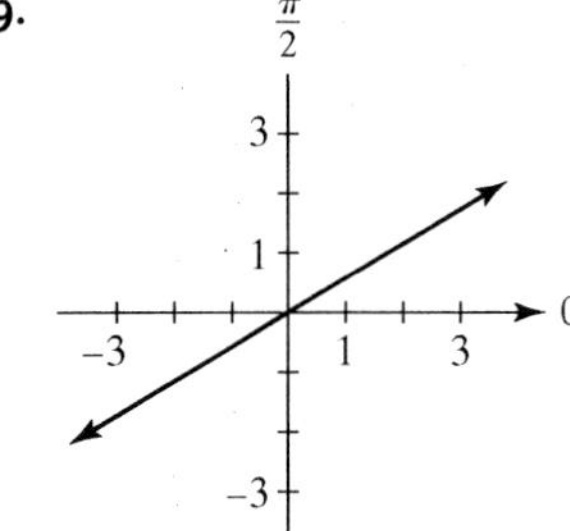

60.

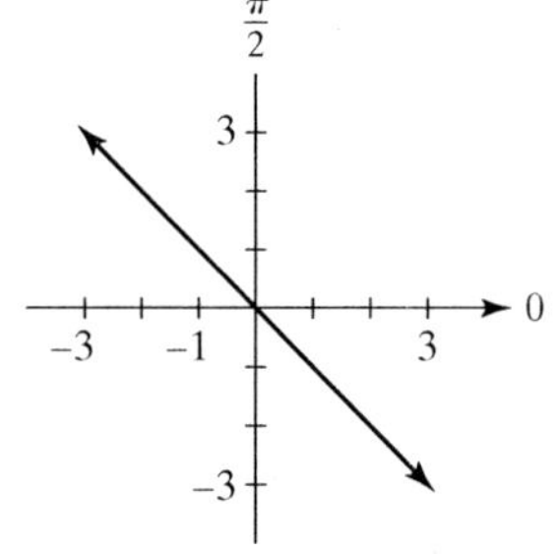

61.

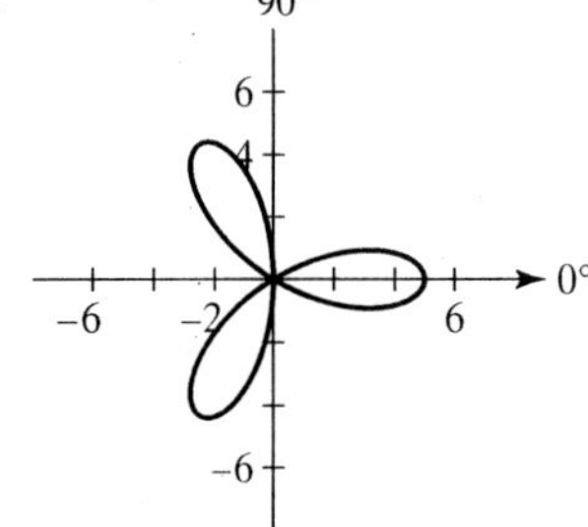

62.

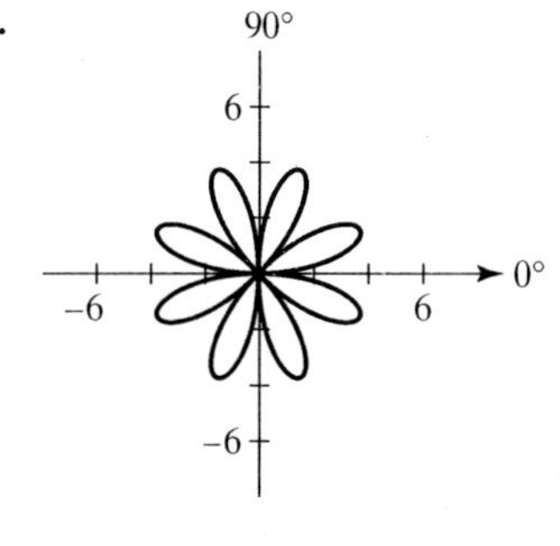

Column A

a. $r = 4 - 4\sin\theta$

b. $r = 5\cos 3\theta$

c. $r = 5$

d. $r = 4 - 6\cos\theta$

e. $r = 3 + 3\cos\theta$

f. $r = 4\sin 4\theta$

g. $r = 4 - 2\sin\theta$

h. $r^2 = 25\sin 2\theta$

i. $r^2 = 16\cos 2\theta$

j. $r = \dfrac{\theta}{4}$

k. $\theta = \dfrac{\pi}{6}$

l. $\theta = 135°$

m. $r = 3\sin\theta$

n. $r = 4\sec\theta$

63.

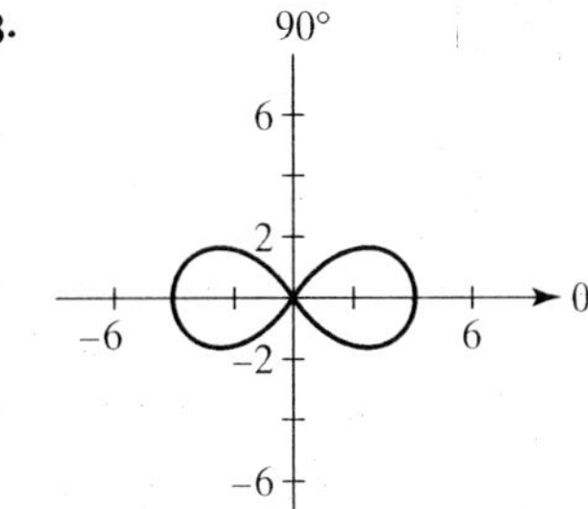

64.

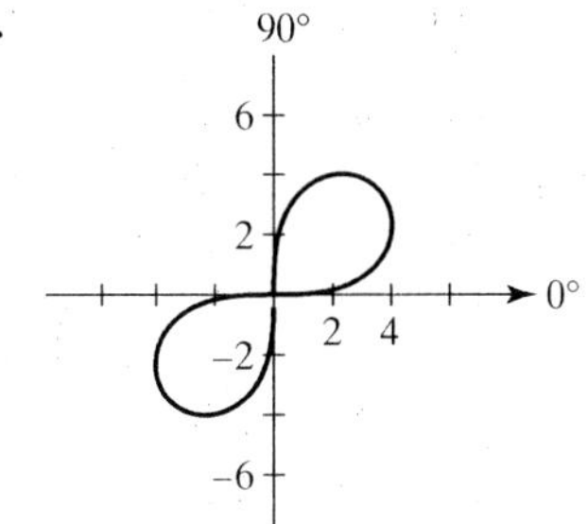

65.

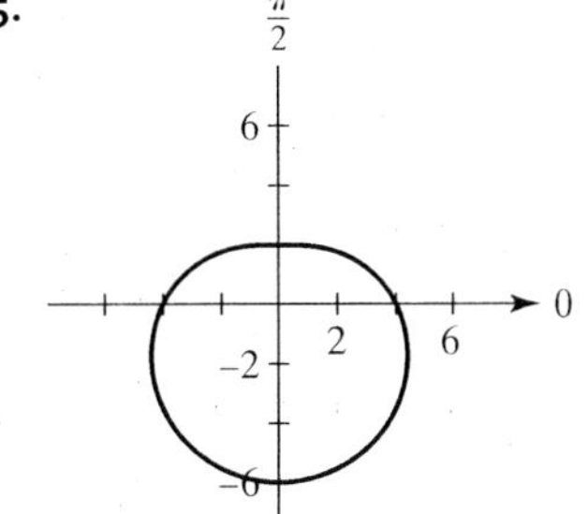

66.

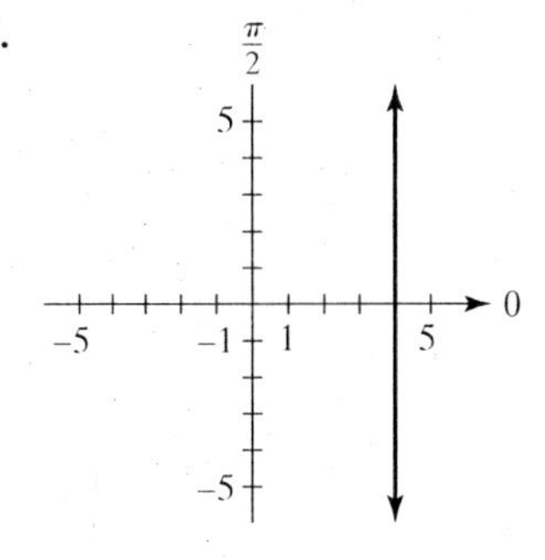

67–68. *Solve each system. Assume that $0 \le \theta < 2\pi$. Use a graphing utility to verify the solutions (points of intersection).*

67. $r = \cos 3\theta$
$r = \dfrac{1}{2}$

68. $r = \cos\theta$
$r = \sin 2\theta$

Discussion

69. The graph of a polar equation is often symmetric to the polar axis $(\theta = 0°)$, the vertical axis $(\theta = 90°)$, and the pole $(r = 0)$. Testing an equation for each of these symmetries involves making certain replacements of r and θ. If the equation with the indicated replacements is equivalent to the original equation, then the graph will have the indicated symmetry.

Symmetry with respect to	Replace (r, θ) with
Polar axis	$(r, -\theta)$, $(-r, \pi - \theta)$
Vertical axis	$(r, \pi - \theta)$, $(-r, -\theta)$
Pole	$(r, \pi + \theta)$, $(-r, \theta)$

For example, Exercise 63 is symmetric to the polar axis, the vertical axis, and the pole.

a. Polar graphs that are symmetric with respect to the polar axis are analogous to graphs in algebra that are symmetric with respect to the x-axis. Which graphs in Exercises 53–66 appear symmetric with respect to the polar axis? If (r, θ) is one point of the graph, is $(r, -\theta)$ also a point of the graph?

b. Polar graphs that are symmetric with respect to the pole are analogous to graphs in algebra that are symmetric with respect to the origin. Which graphs in Exercises 53–66 appear symmetric with respect to the pole? If (r, θ) is one point of the graph, is $(-r, \theta)$ also a point of the graph?

6.4 Complex Numbers

Standard Form of a Complex Number

In Chapter 5 we solved trigonometric equations in which we often used strategies from algebra. One such strategy was the square root method. Solving the following two algebraic equations using this method, we find that the solutions for Equation (1) are real numbers and the solutions for Equation (2) are not real numbers.

1. $u^2 - 9 = 0$

$u^2 = 9$

$u = \pm\sqrt{9}$

$u = \pm 3$

2. $u^2 + 1 = 0$

$u^2 = -1$

$u = \pm\sqrt{-1}$

There is no real number that is the square root of a negative number.

The solution to the second equation requires a definition that extends the number system to include the square root of negative real numbers.

> The number that is the square root of -1 is called i.
>
> $$i = \sqrt{-1} \quad \text{and} \quad i^2 = -1$$
>
> The number i is called the **imaginary unit**.

Thus, the two nonreal solutions to $u^2 + 1 = 0$ (Equation 2) are i and $-i$.

Now that we have defined the imaginary number i, we add a real number to a real multiple of i to get what is called a *complex number*.

> A **complex number** is any number in the form $a + bi$, where $i^2 = -1$ and $a, b \in \mathbb{R}$.

The form $a + bi$ is called the **standard form** of a complex number and the set of all numbers $a + bi$ is called the **set of complex numbers**. Furthermore, a is called the **real part** and b is called the **imaginary part** of $a + bi$. If $b = 0$, then the complex number $a + bi = a$, which is a real number. Therefore, every real number is a complex number. If $b \neq 0$, then the complex number $a + bi$ is called an **imaginary number**. Thus, there are two types of complex numbers (shown in the accompanying diagram): real numbers and imaginary numbers.

Complex Numbers (C)

Real numbers		Imaginary numbers
Rational	Irrational	
$17.5, 0.3\overline{3}$	$e, \sqrt{2}$	$3i, 2 - i,$
$4\frac{1}{2}, -12$	$\pi, \sqrt[3]{5}$	$-6 - 7i$
$0, -\frac{5}{9}, 9$	$\frac{\sqrt{3}}{2}$	$\frac{1}{2} + i\sqrt{3}, \frac{i}{2}$

Operations with Complex Numbers

Now that we have defined the set of complex numbers, we determine when they are equal and how to simplify, add, subtract, multiply, and divide them. In doing so, we again see the use of corresponding algebra skills.

Equality of Complex Numbers

Two complex numbers $a + bi$ and $c + di$, where $a, b, c,$ and $d \in \mathbb{R}$, are equal if and only if

$$a = c \quad \text{and} \quad b = d.$$

That is, two complex numbers are equal if and only if their real parts are equal and their imaginary parts are equal.

EXAMPLE 1 Find x and y if $4x + 16i = 24 + (9y^2 - 9)i$.

SOLUTION

Set the real parts equal.

$$4x = 24$$
$$x = 6 \quad \text{Divide each side by 4.}$$

Set the imaginary parts equal.

$$16 = (9y^2 - 9)$$
$$9y^2 = 25 \quad \text{Add 9 to both sides.}$$
$$y^2 = \frac{25}{9} \quad \text{Divide each side by 9.}$$
$$y = \pm\frac{5}{3} \quad \text{Use the square root method.}$$

So $x = 6$ and $y = \pm\frac{5}{3}$. ■

If we treat the complex numbers $a + bi$ as a binomial in which i is the variable, then we can compute the sum, difference, and product of two complex numbers using familiar rules of algebra. We summarize the results, but you should not have to memorize them.

Sum, Difference, and Product of Complex Numbers

For complex numbers $a + bi$ and $c + di$, the following operations are defined:

Addition	$(a + bi) + (c + di) = (a + c) + (b + d)i$
Subtraction	$(a + bi) - (c + di) = (a - c) + (b - d)i$
Multiplication	$(a + bi)(c + di) = (ac - bd) + (ad + bc)i$

Notice that the multiplication result comes from the fact that $i^2 = -1$.

EXAMPLE 2 Perform the indicated operations with the complex numbers. Leave answers in $a + bi$ form.

a. $(4 + 6i) - (2 - 3i)$ **b.** $(6 - 2i)(4 + 5i)$

c. $(5i)^2$ **d.** $(-5i)^2$

SOLUTION

a. $(4 + 6i) - (2 - 3i) = 4 + 6i - 2 + 3i = (4 - 2) + (6 + 3)i = 2 + 9i$

b. We use rules of algebra for multiplying two binomials.

$$\begin{aligned}(6 - 2i)(4 + 5i) &= 24 + 30i - 8i - 10i^2 && \text{FOIL method}\\ &= 24 + 22i - 10(-1) && i^2 = -1\\ &= 34 + 22i\end{aligned}$$

c. $(5i)^2 = 5^2i^2 = 25(-1) = -25$

d. $(-5i)^2 = (-5)^2i^2 = 25(-1) = -25$ ■

Powers of *i*

When we consider higher integer powers of i, we see a periodic property emerging.

$$\begin{aligned} i^1 &= i & i^5 &= i^4 \cdot i = 1 \cdot i = i\\ i^2 &= -1 & i^6 &= i^4 \cdot i^2 = 1(-1) = -1\\ i^3 &= i^2 \cdot i = -1 \cdot i = -i & i^7 &= i^4 \cdot i^3 = 1(-i) = -i\\ i^4 &= i^2 \cdot i^2 = (-1)(-1) = 1 & i^8 &= i^4 \cdot i^4 = 1,\end{aligned}$$

and so on.

Therefore, any integer power of i can be written as either $1, -1, i$, or $-i$. The pattern suggests that we can simplify higher powers of i by writing them in terms of i^4 since i^4 equals 1.

EXAMPLE 3 Simplify i^{39}.

SOLUTION

$$i^{39} = (i^4)^9 \cdot i^3 = (1)^9(-i) = -i$$ ■

You will be asked in Exercise 61 to find a formula that expresses higher powers of i in terms of powers of i less than 4.

In Examples 2 (c) and 2 (d), we saw that both $(5i)^2 = -25$ and $(-5i)^2 = -25$. This indicates that in the complex number system there are two square roots of -25: $5i$ and $-5i$. So, for any positive real number b there are two square roots of $-b$: $i\sqrt{b}$ and $-i\sqrt{b}$. We call $i\sqrt{b}$ the principal square root of $-b$ and make the following definition.

Square Root of a Negative Number

For any positive real number b,

$$\sqrt{-b} = i\sqrt{b}.$$

EXAMPLE 4 Express the number in terms of i.

a. $\sqrt{-36}$ b. $-\sqrt{-100}$ c. $\sqrt{-17}$ d. $\sqrt{-12}$

SOLUTION Since each number involves the square root of a negative number, we use the preceding definition to express each in terms of i.

a. $\sqrt{-36} = i\sqrt{36} = 6i$

b. $-\sqrt{-100} = -i\sqrt{100} = -10i$

c. $\sqrt{-17} = i\sqrt{17}$ or $\sqrt{17}i$

d. $\sqrt{-12} = i\sqrt{12} = i\sqrt{4}\sqrt{3} = i2\sqrt{3}$ or $2i\sqrt{3}$ ■

NOTE: In Example 4 (c) we see that we can write either $i\sqrt{17}$ or $\sqrt{17}i$. It is customary to write $i\sqrt{a}$ since $\sqrt{a}i$ may be confused with $\sqrt{ai}$ in which i is contained in the radical sign.

In algebra you learned that the **product rule for square roots**, $\sqrt{a}\sqrt{b} = \sqrt{ab}$, can be applied only when $a, b \geq 0$. When working with square roots of negative numbers, (that is, $a, b < 0$), the square root of a product is not equal to the product of the square roots.

$$\sqrt{-2}\sqrt{-8} \neq \sqrt{(-2)(-8)} = \sqrt{16} = 4$$

If we use the definition of the square root of a negative number first, $\sqrt{-b} = i\sqrt{b}$, $b > 0$, we can use the product rule for square roots.

$$\sqrt{-2}\sqrt{-8} = i\sqrt{2} \cdot i\sqrt{8} = i^2\sqrt{16} = (-1)(4) = -4$$

CAUTION: Always write the number $\sqrt{-b}$ as $i\sqrt{b}$, where $b > 0$, before you do any operations with the number. Once complex numbers have been written in this form, they can be simplified, added, subtracted, multiplied, or divided.

EXAMPLE 5 Write each expression in $a + bi$ form.

a. $\sqrt{-32}\sqrt{-2}$ b. $\sqrt{-5}(4 + \sqrt{-5})$ c. $(9 - \sqrt{-3})(9 + \sqrt{-3})$

SOLUTIONS

a. $\sqrt{-32}\sqrt{-2} = i\sqrt{32} \cdot i\sqrt{2} = i^2\sqrt{32 \cdot 2} = i^2\sqrt{64} = (-1)(8) = -8$

b. $\sqrt{-5}(4 + \sqrt{-5}) = i\sqrt{5}(4 + i\sqrt{5})$

$$= 4i\sqrt{5} + i^2\sqrt{5}\sqrt{5}$$
$$= 4i\sqrt{5} + (-1)\sqrt{25}$$
$$= -5 + 4i\sqrt{5}$$

c. $(9 - \sqrt{-3})(9 + \sqrt{-3}) = (9 - i\sqrt{3})(9 + i\sqrt{3})$

$$= 81 + \cancel{9i\sqrt{3}} - \cancel{9i\sqrt{3}} - i^2(3)$$
$$= 81 - (-1)(3)$$
$$= 84$$ ■

The two imaginary numbers in Example 5(c) have a special relationship and are therefore given a special name. The complex numbers $a + bi$ and $a - bi$ are called **complex conjugates** of each other. We notice that *multiplying complex conjugates gives a real number.*

For complex conjugates $a + bi$ and $a - bi$,

$$(a + bi)(a - bi) = a^2 - i^2b^2 = a^2 + b^2$$

We use complex conjugates to divide imaginary numbers. This procedure corresponds to the algebraic procedure of rationalizing denominators.

EXAMPLE 6 Write each quotient in $a + bi$ form.

a. $\dfrac{2 - i}{i}$ **b.** $\dfrac{3 - 2i}{4 + 3i}$

SOLUTION

a. $\dfrac{2 - i}{i} = \dfrac{(2 - i)(-i)}{i(-i)} = \dfrac{-2i + i^2}{-i^2} = \dfrac{-2i - 1}{-(-1)} = -1 - 2i$

b. $\dfrac{3 - 2i}{4 + 3i} = \dfrac{(3 - 2i)(4 - 3i)}{(4 + 3i)(4 - 3i)} = \dfrac{12 - 17i + 6i^2}{4^2 + 3^2}$

$= \dfrac{12 - 17i + 6(-1)}{16 + 9} = \dfrac{6 - 17i}{25} = \dfrac{6}{25} - \dfrac{17}{25}i$ ■

Geometric Representation of a Complex Number

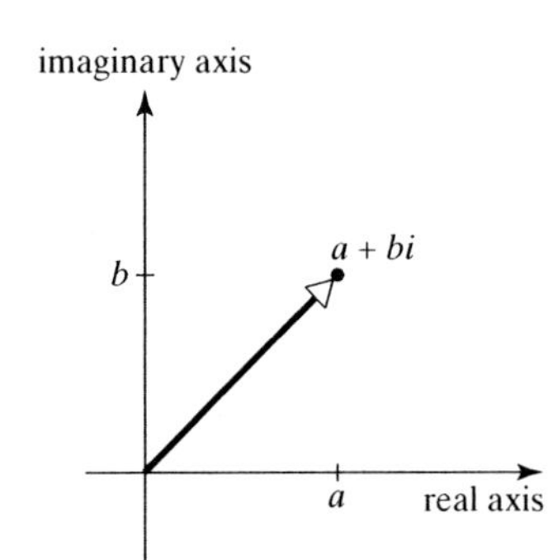

For each ordered pair of real numbers (a, b), there is a unique complex number $a + bi$. Conversely, for each complex number $a + bi$, there is a unique ordered pair of real numbers (a, b). Thus, there is a one-to-one correspondence between the ordered pairs of real numbers and the complex numbers. In addition, there is a one-to-one correspondence between the ordered pairs of real numbers and the points in the rectangular coordinate plane. In other words, $(2, 7)$ corresponds to $2 + 7i$, and $4 - 6i$ corresponds to $(4, -6)$. Therefore the complex numbers and the points in the plane are in a one-to-one correspondence. Thus, we can *geometrically represent the complex number $a + bi$ by a standard position vector with terminal point (a, b)*. The horizontal axis is called the **real axis**, and the vertical axis is called the **imaginary axis** or the ***i*-axis**. The plane in which complex numbers are graphed is called the **complex plane**.

EXAMPLE 7 Graph the complex number in the complex plane.

a. $5 - 2i$ **b.** $2i$ **c.** -6 **d.** $3 + 4i$

SOLUTION

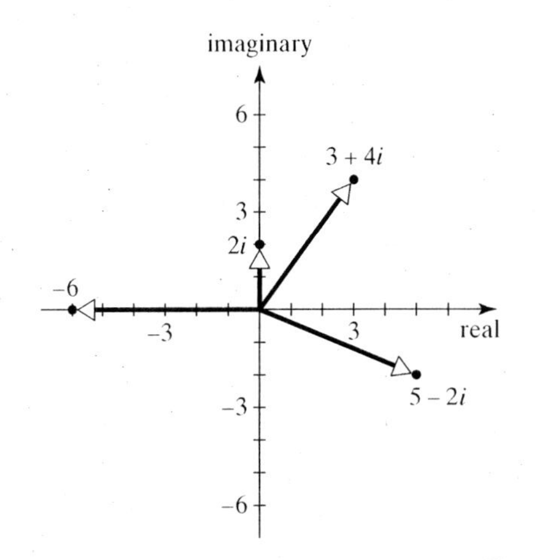

a. $5 - 2i$ corresponds to the standard position vector 5 units along the real axis and -2 units along the imaginary axis.

b. $2i$ corresponds to the standard position vector 0 units along the real axis and 2 units along the imaginary axis.

c. -6 corresponds to the vector -6 units along the real axis.

d. $3 + 4i$ corresponds to the vector 3 units along the real axis and 4 along the imaginary axis. ■

NOTE: The graph of a complex number $a + bi$ can also be represented in the complex plane by just the point (a, b). In this text we will continue to use the vector representation.

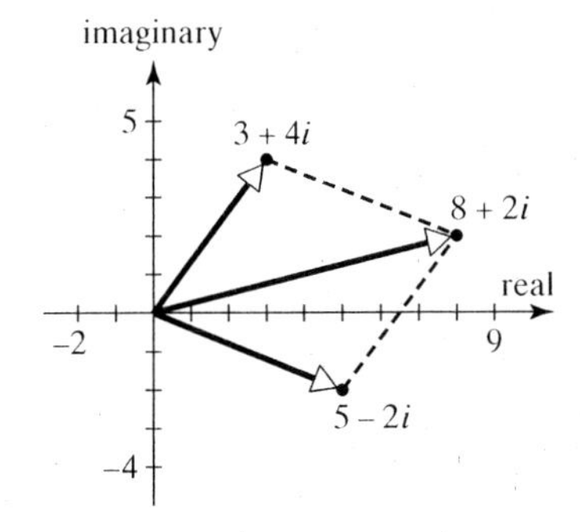

When complex numbers are geometrically represented by vectors, the sum of two complex numbers can be geometrically demonstrated as the diagonal of the parallelogram formed by the vectors. For example, the vector sum

$$(3 + 4i) + (5 - 2i) = 8 + 2i$$

is demonstrated in the accompanying diagram.

The absolute value of a real number x, $|x|$, is the distance from the origin to the point x. In a similar manner, we now define the absolute value of a complex number.

Modulus (Absolute Value) of a Complex Number

The **modulus** (or **absolute value**) of any complex numbers $a + bi$, where $a, b \in \mathbb{R}$, is the distance from the origin to point (a, b). If the distance is represented by r, then

$$r = |a + bi| = \sqrt{a^2 + b^2}.$$

EXAMPLE 8 Find the modulus of the complex number.

a. $5 + 2i$ **b.** 6 **c.** $-4i$

SOLUTION

a. $|5 + 2i| = \sqrt{5^2 + 2^2} = \sqrt{29}$ **b.** $|6| = |6 + 0i| = \sqrt{6^2 + 0^2} = 6$

c. $|-4i| = |0 - 4i| = \sqrt{0^2 + (-4)^2} = 4$ ■

CONNECTIONS WITH TECHNOLOGY

Graphing Calculator

Many graphing calculators have REAL and $a + bi$ mode options. In $a + bi$ mode, operations with complex numbers can be performed.

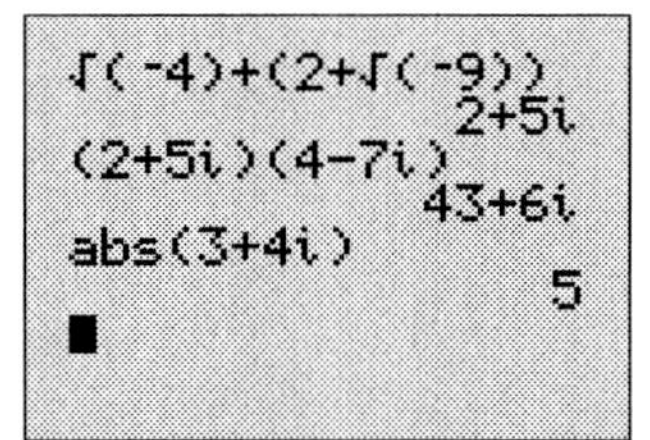

EXAMPLE

a. Express $\sqrt{-4} + \left(2 + \sqrt{-9}\right)$ in standard form.

b. Simplify $(2 + 5i)(4 - 7i)$.

c. Find $|3 + 4i|$.

SOLUTION Using the $a + bi$ mode, we get the following results.

a. $2 + 5i$ **b.** $43 + 6i$ **c.** 5

Note that on some calculators the ordered pair (a, b) represents the complex number $a + bi$.

In this section we review the basics of complex numbers. You may be wondering why we are studying complex numbers in trigonometry. In the previous chapters, starting with the first page of this text, we have used algebra to help develop trigonometry and solve trigonometric problems. In the next section you will see how we use trigonometry to help us with algebraic problems involving complex numbers that might otherwise be time consuming or difficult.

Exercise Set 6.4

1–4. *Find x and y.*

1. $2x + 3yi = (2 + x) - 4i$
2. $7 - 2xi = 14y - 25i$
3. $6x^2 - 15yi = 24 + (3 - 2y)i$
4. $18 - (4 - y^2)i = 9x + 5i$

5–14. *Find the following sums or differences. Leave answers in $a + bi$ form.*

5. $(4 - 6i) + (2 - 5i)$
6. $(4 + 8i) + (-2 + 2i)$
7. $(12 - 2i) - (9 - 3i)$
8. $(-3 + 6i) - (3 + 7i)$
9. $(6 + 6i) - (2 + 9i)$
10. $(8 - i) + (-10 + i)$
11. $(2 + 5i) + (6) + (2 - i)$
12. $(9 - 8i) - (-2i) + (8 - 2i)$
13. $(a + bi) + (a - bi)$
14. $(a + bi) - (a - bi)$

15–26. *Find the following products. Leave answers in $a + bi$ form.*

15. $2i(9 - 6i)$
16. $4i(-2 + 6i)$
17. $(4 - i)(2 + 6i)$
18. $(5 + 6i)(2 - 3i)$
19. $(2 - 5i)^2$
20. $(7 + 3i)^2$

21. $(3 - 2i)(3 + 2i)$
22. $(7 - 9i)(7 + 9i)$
23. $(a + bi)(a - bi)$
24. $(a + bi)(a + bi)$
25. $(1 - i)^4$
26. $(1 + i)^4$

27–32. *Find the following quotients. Leave answers in $a + bi$ form.*

27. $\dfrac{2 + 3i}{4i}$
28. $\dfrac{5 - i}{2i}$
29. $\dfrac{3i}{2 - i}$
30. $\dfrac{5}{6 + 3i}$
31. $\dfrac{2 + i}{2 - i}$
32. $\dfrac{3 + 2i}{4 + 3i}$

33–38. *Simplify the powers of i. Recall that $i^0 = 1, i^1 = i, i^2 = -1$, or $i^3 = -i$.*

33. i^{17}
34. i^{22}
35. i^{19}
36. i^{67}
37. i^{40}
38. i^{4n+1}

39–50. *Write each expression in the form $a + bi$.*

39. $\sqrt{-81}$
40. $\sqrt{-100}$
41. $\sqrt{-49} + \sqrt{-4} + \sqrt{25}$
42. $2\sqrt{-9} - \sqrt{-4} + \sqrt{64}$
43. $\sqrt{-9}(1 - 2\sqrt{-4})$
44. $\sqrt{-8}(2 + 4\sqrt{-2})$
45. $(7 - \sqrt{-4})(3 + 2\sqrt{-4})$
46. $(1 - 5\sqrt{-6})(10 - \sqrt{-6})$
47. $\dfrac{9\sqrt{4}}{\sqrt{-16}}$
48. $\dfrac{-2\sqrt{9}}{\sqrt{-49}}$
49. $\dfrac{6}{2 + \sqrt{-9}}$
50. $\dfrac{3 + \sqrt{-4}}{2 - \sqrt{-9}}$

51–60. *Graph each complex number and find the modulus.*

51. $-1 + 4i$
52. $2 - 6i$
53. $4 + 7i$
54. $-4 - 8i$
55. $8 - i$
56. $3 + 3i$
57. $-8i$
58. $6i$
59. 7
60. -5

61. Find a formula for rewriting i^p as $1, i, i^2$, or i^3, where p is a positive integer.

62. Show for any complex number $a + bi$ and its conjugate $a - bi$,

$$|a + bi||a - bi| = |(a + bi)(a - bi)|.$$

63. Show that $4i$ and $-4i$ are solutions of the equation $x^2 + 16 = 0$.

64. Show that $1 + i$ and $1 - i$ are solutions of the equation $x^2 - 2x + 2 = 0$.

65. Show that $\left(-\dfrac{1}{2} + i\dfrac{\sqrt{3}}{2}\right)^3 = 1.$

66. Show that $\left(-\dfrac{1}{2} - i\dfrac{\sqrt{3}}{2}\right)^3 = 1.$

67–70. *In calculus we use Euler's formula, $e^{i\theta} = \cos\theta + i\sin\theta$, where the irrational number $e = 2.71828\ldots$. Show that the following equations are identities by using Euler's formula. [Some graphing calculators have a mode option of Euler's form (re ^ θi).]*

67. $e^{i(\pi/2)} = i$
68. $e^{i(\pi)} = -1$
69. $\cos\theta = \dfrac{e^{i\theta} + e^{i(-\theta)}}{2}$
70. $\sin\theta = \dfrac{e^{i\theta} - e^{i(-\theta)}}{2i}$

Explore the Pattern

71. Consider the following complex numbers in the forms $a + bi$ and $i(a + bi)$. Simplify the second complex number and then graph both numbers in the complex plane. Determine the geometric relationship between them. Describe the pattern that is the result of multiplying a complex number by i. (*Hint:* Notice the direction and angle of rotation from $a + bi$ to $i(a + bi)$.) Then describe the result of multiplying the complex number $a + bi$ by i^2, i^3, and i^4.

a. $2 + 2i,\ i(2 + 2i)$ **b.** $1 - i,\ i(1 - i)$

c. $4,\ i(4)$ **d.** $2i,\ i(2i)$

72. Consider the following complex numbers in the forms $a + bi$ and $\dfrac{a + bi}{i}$. Simplify the second complex number and then graph both numbers in the complex plane. Determine the geometric relationship between them. Describe the pattern that is the result of dividing a complex number by i. (*Hint*: Notice the direction and angle of rotation from $a + bi$ to $\dfrac{a + bi}{i}$.) Then describe the result of dividing the complex number $a + bi$ by i^2, i^3, and i^4.

a. $2 + 2i,\ \dfrac{2 + 2i}{i}$ **b.** $1 - i,\ \dfrac{1 - i}{i}$

c. $4,\ \dfrac{4}{i}$ **d.** $2i,\ \dfrac{2i}{i}$

Graphing Calculator/CAS Exercises

73–76. *Using a graphing utility in $a + bi$ mode, verify the following.*

73. Exercise 45. **74.** Exercise 50. **75.** Exercise 65. **76.** Exercise 66.

6.5 Trigonometric Form for Complex Numbers

Trigonometric Form

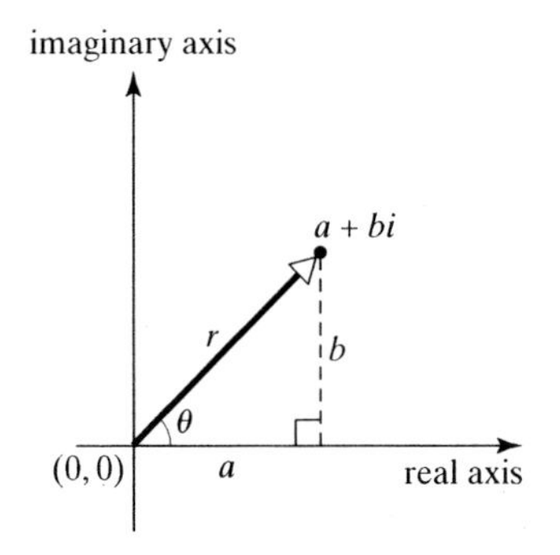

In Section 6.4 we represent each complex number $a + bi$ by a vector from the origin to point (a, b) in the complex plane. This vector along with the positive real axis determine an angle θ in standard position, which is called the **argument** of $a + bi$.

Since the terminal side of θ contains (a, b) and the modulus of $a + bi$ is $r = |a + bi| = \sqrt{a^2 + b^2}$, we obtain the relationships among the complex number $a + bi$, r, and θ:

$$\cos\theta = \frac{a}{r} \quad \text{or} \quad a = r\cos\theta$$

and

$$\sin\theta = \frac{b}{r} \quad \text{or} \quad b = r\sin\theta$$

We now can use these equations to write $a + bi$ in terms of r and θ.

For complex numbers $a + bi$ with modulus r and argument θ, the **trigonometric (or polar) form** is given by

$$a + bi = (r \cos \theta) + (r \sin \theta)i$$
$$\text{or} \quad a + bi = r(\cos \theta + i \sin \theta),$$

where $r \geq 0$.

NOTE: $(\cos \theta + i \sin \theta)$ can be abbreviated as cis θ.

If θ is an argument of $a + bi$, then so is

$$\theta + k \cdot 360^\circ \quad \text{or} \quad \theta + k \cdot 2\pi, \text{ where } k \in \{\ldots, -2, -1, 0, 1, 2, \ldots\}.$$

Therefore, the trigonometric form of a complex number can also be found using either of the following:

$$a + bi = r[\cos(\theta + k \cdot 360^\circ) + i \sin(\theta + k \cdot 360^\circ)]$$
$$a + bi = r[\cos(\theta + k \cdot 2\pi) + i \sin(\theta + k \cdot 2\pi)]$$

We generally select the smallest positive value of θ for the measure of an argument.

NOTE: The trigonometric form of a complex number is also referred to as **polar form**, since this form gives an r and θ in its representation of the complex number.

EXAMPLE 1 Graph $3 - i\sqrt{3}$ and express it in trigonometric form.

SOLUTION The graph of $3 - i\sqrt{3}$ corresponds to the vector that is 3 units along the real axis and $-\sqrt{3}$ units along the imaginary axis. To express $3 - i\sqrt{3}$ in trigonometric form, we need r and θ. Since $a = 3$ and $b = -\sqrt{3}$, we start by finding r.

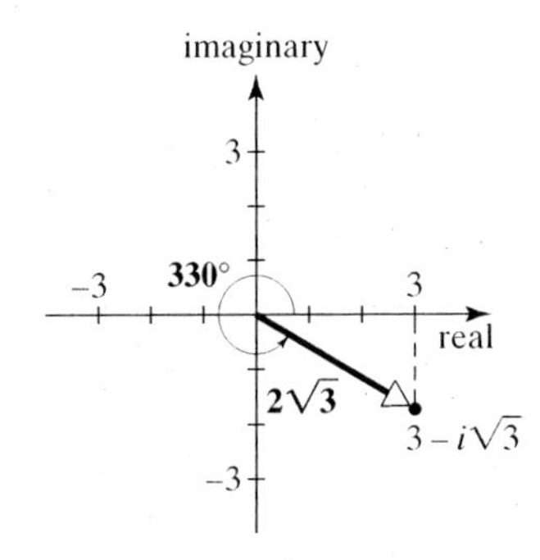

$$r = |a + bi| = \sqrt{a^2 + b^2}$$
$$r = |3 - i\sqrt{3}| = \sqrt{(3)^2 + (-\sqrt{3})^2}$$
$$= \sqrt{9 + 3} = \sqrt{12} = 2\sqrt{3}$$

To find θ, we know that

$$\cos \theta = \frac{a}{r} = \frac{3}{2\sqrt{3}} = \frac{\sqrt{3}}{2} \quad \text{and} \quad \sin \theta = \frac{b}{r} = \frac{-\sqrt{3}}{2\sqrt{3}} = -\frac{1}{2}.$$

Since $\cos \theta$ is positive and $\sin \theta$ is negative, θ is in QIV (as the graph also indicates), and so $\theta = 330^\circ$. Therefore,

$$3 - i\sqrt{3} = r(\cos \theta + i \sin \theta)$$
$$= 2\sqrt{3}(\cos 330^\circ + i \sin 330^\circ).$$

NOTE: Another way to determine θ is to use $\tan \theta = \frac{b}{a} = -\frac{\sqrt{3}}{3}$. Since the graph of $3 - i\sqrt{3}$ is in QIV, we also obtain $\theta = 330^\circ$. ■

CAUTION: If you want the complex number $3 - i\sqrt{3}$ to be in trigonometric form $2\sqrt{3}(\cos 330° + i \sin 330°)$, you must resist the temptation to find the cosine and sine functional values of 330° and then distribute $2\sqrt{3}$. That would put you back into the standard $a + bi$ form!

EXAMPLE 2 Graph the complex number whose trigonometric form is $4(\cos 150° + i \sin 150°)$ and then express it in $a + bi$ form (standard form).

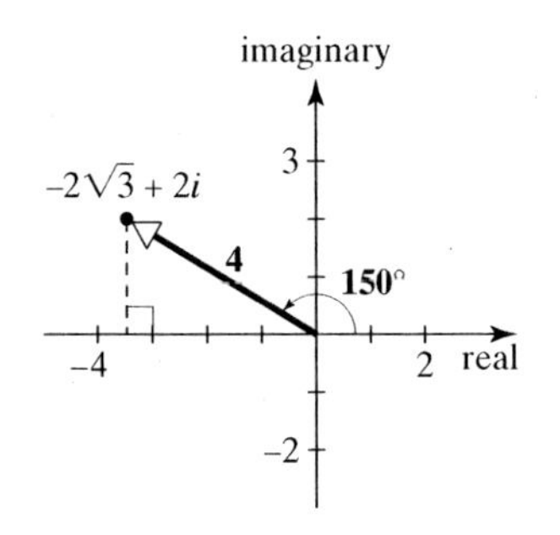

SOLUTION The graph corresponds to a vector 4 units along the terminal side of the standard position angle $\theta = 150°$. Using exact values for cos 150° and sin 150°, we get the standard form.

$$\begin{aligned} a + bi &= 4(\cos 150° + i \sin 150°) \\ &= 4\left(-\frac{\sqrt{3}}{2} + i\frac{1}{2}\right) \\ &= -2\sqrt{3} + 2i \end{aligned}$$

Why would we want to have complex numbers in trigonometric form? One reason is multiplying or dividing complex numbers can be easier when the numbers are in trigonometric form. To see how this is possible, let's consider two complex numbers $a + bi$ and $c + di$. Let r_1 and θ_1 be the modulus and argument for $a + bi$, respectively, and similarly let r_2 and θ_2 be those for $c + di$. Then we have the following.

$$\begin{aligned} (a + bi)(c + di) &= r_1(\cos\theta_1 + i\sin\theta_1) \cdot r_2(\cos\theta_2 + i\sin\theta_2) \\ &= r_1r_2\cos\theta_1\cos\theta_2 + r_1r_2i\cos\theta_1\sin\theta_2 + r_1r_2i\sin\theta_1\cos\theta_2 + r_1r_2i^2\sin\theta_1\sin\theta_2 \\ &= r_1r_2[\cos\theta_1\cos\theta_2 - \sin\theta_1\sin\theta_2] + r_1r_2i[\sin\theta_1\cos\theta_2 + \cos\theta_1\sin\theta_2] \\ &= r_1r_2[\cos(\theta_1 + \theta_2) + i\sin(\theta_1 + \theta_2)] \end{aligned}$$

By using the cosine and sine sum formulas to simplify this expression, we have an alternative way to multiply complex numbers when they are in trigonometric form.

Multiplication of Complex Numbers

If $a + bi = r_1(\cos\theta_1 + i\sin\theta_1)$ and $c + di = r_2(\cos\theta_2 + i\sin\theta_2)$, then

$$(a + bi)(c + di) = r_1r_2[\cos(\theta_1 + \theta_2) + i\sin(\theta_1 + \theta_2)].$$

So to multiply complex numbers in trigonometric form, we multiply the modulus of each number and add the arguments to obtain the modulus and the argument of the result. Using similar reasoning when dividing complex numbers, we get the next result, which you will be asked to verify in the exercise set.

Division of Complex Numbers

If $a + bi = r_1(\cos\theta_1 + i\sin\theta_1)$ and $c + di = r_2(\cos\theta_2 + i\sin\theta_2)$, then

$$\frac{a+bi}{c+di} = \frac{r_1}{r_2}[\cos(\theta_1 - \theta_2) + i\sin(\theta_1 - \theta_2)],$$

where $c \neq 0$ and $d \neq 0$.

To divide complex numbers in trigonometric form, we divide the modulus of the first complex number by the modulus of the second and subtract the arguments in the same order.

EXAMPLE 3 Perform the indicated operation and express the result in $a + bi$ form.

a. $3(\cos 12^\circ + i\sin 12^\circ) \cdot 8(\cos 168^\circ + i\sin 168^\circ)$

b. $\dfrac{3\left(\cos\dfrac{7\pi}{12} + i\sin\dfrac{7\pi}{12}\right)}{6\left(\cos\dfrac{5\pi}{12} + i\sin\dfrac{5\pi}{12}\right)}$

SOLUTION

a. Since the complex numbers are in trigonometric form, we multiply the two modulus values and add the arguments (here in degrees).

$$\begin{aligned}
3(\cos 12^\circ + i\sin 12^\circ) \cdot 8(\cos 168^\circ + i\sin 168^\circ) &= 3 \cdot 8[\cos(12^\circ + 168^\circ) + i\sin(12^\circ + 168^\circ)] \\
&= 24[\cos 180^\circ + i\sin 180^\circ] \qquad \text{trigonometric form} \\
&= 24[(-1) + i(0)] \\
&= -24 \qquad a + bi \text{ form}
\end{aligned}$$

b. Since the complex numbers are in trigonometric form, we divide the modulus of the first complex number by the modulus of the second and subtract the arguments in the same order.

$$\begin{aligned}
\frac{3\left(\cos\frac{7\pi}{12} + i\sin\frac{7\pi}{12}\right)}{6\left(\cos\frac{5\pi}{12} + i\sin\frac{5\pi}{12}\right)} &= \frac{3}{6}\left[\cos\left(\frac{7\pi}{12} - \frac{5\pi}{12}\right) + i\sin\left(\frac{7\pi}{12} - \frac{5\pi}{12}\right)\right] \\
&= \frac{1}{2}\left(\cos\frac{\pi}{6} + i\sin\frac{\pi}{6}\right) \qquad \text{trigonometric form} \\
&= \frac{1}{2}\left(\frac{\sqrt{3}}{2} + i\frac{1}{2}\right) \\
&= \frac{\sqrt{3}}{4} + \frac{1}{4}i \qquad a + bi \text{ form}
\end{aligned}$$

Do you think the problems in Example 3 are quicker in trigonometric form? If so, you are beginning to see some of the advantages of this form of a complex number.

Powers and Roots of Complex Numbers in Trigonometric Form

Section 6.4, Exercise 65, asked you to show that $\left(-\frac{1}{2}+i\frac{\sqrt{3}}{2}\right)^3 = 1$. To complete this exercise, we leave this complex number in the $a + bi$ form and use algebraic methods to multiply this complex number by itself three times! In the next example, let's see if the process can be simplified by first changing the complex number to trigonometric form and then using the procedure for multiplying complex numbers in trigonometric form. (You can use radian or degree measure for θ.)

EXAMPLE 4 Express $-\frac{1}{2}+i\frac{\sqrt{3}}{2}$ in trigonometric form and then show that $\left(-\frac{1}{2}+i\frac{\sqrt{3}}{2}\right)^3 = 1$.

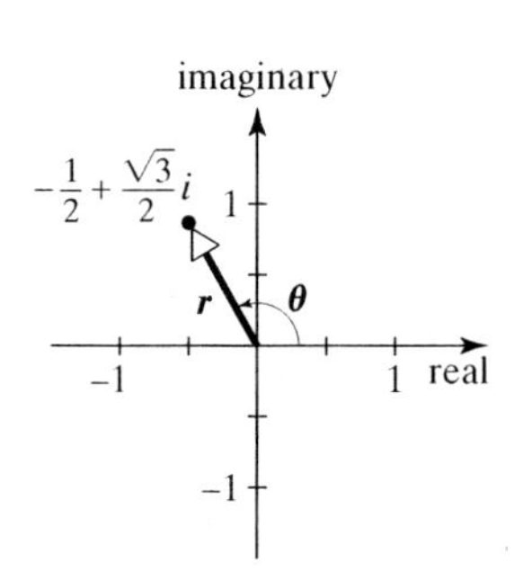

SOLUTION Make a sketch of $-\frac{1}{2}+i\frac{\sqrt{3}}{2}$ in the complex plane. We need r and θ to express this complex number in trigonometric form.

$$a + bi = -\frac{1}{2}+i\frac{\sqrt{3}}{2} \rightarrow a = -\frac{1}{2} \text{ and } b = \frac{\sqrt{3}}{2}$$

$$r = |a+bi| = \sqrt{a^2+b^2} \qquad \text{and} \qquad \tan\theta = \frac{b}{a}$$

$$r = \sqrt{\left(-\tfrac{1}{2}\right)^2+\left(\tfrac{\sqrt{3}}{2}\right)^2} \qquad \tan\theta = \frac{\frac{\sqrt{3}}{2}}{-\frac{1}{2}}$$

$$r = \sqrt{\tfrac{1}{4}+\tfrac{3}{4}} \qquad \tan\theta = \frac{\sqrt{3}}{2}\cdot\left(-\frac{2}{1}\right)$$

$$r = 1 \qquad \tan\theta = -\sqrt{3}$$

$$\hat{\theta} = \tan^{-1}\left(\sqrt{3}\right) = 60^\circ$$

Since θ is in QII, $\theta = 120^\circ$.

Therefore, $\left(-\frac{1}{2}+i\frac{\sqrt{3}}{2}\right) = 1(\cos 120^\circ + i\sin 120^\circ)$. So,

$$\begin{aligned}\left(-\frac{1}{2}+i\frac{\sqrt{3}}{2}\right)^3 &= 1(\cos 120^\circ + i\sin 120^\circ)\cdot 1(\cos 120^\circ + i\sin 120^\circ)\cdot 1(\cos 120^\circ + i\sin 120^\circ)\\ &= 1^2[\cos(120^\circ+120^\circ) + i\sin(120^\circ+120^\circ)]\cdot 1(\cos 120^\circ + i\sin 120^\circ)\\ &= 1^3[\cos(120^\circ+120^\circ+120^\circ) + i\sin(120^\circ+120^\circ+120^\circ)]\\ &= 1^3[\cos(3\cdot 120^\circ) + i\sin(3\cdot 120^\circ)]\\ &= 1[\cos 360^\circ + i\sin 360^\circ]\\ &= 1[1+0]\\ &= 1\end{aligned}$$

If we generalize the procedure from Example 4, we obtain the following statement known as DeMoivre's Theorem.

DeMoivre's Theorem

For any complex number $r(\cos\theta + i\sin\theta)$,

$$[r(\cos\theta + i\sin\theta)]^n = r^n(\cos n\theta + i\sin n\theta),$$

for $n \in \{\ldots, -2, -1, 0, 1, 2, \ldots\}$.

It may be debatable which method of the problem in Example 4 is simpler—using $a + bi$ form or $r(\cos\theta + i\sin\theta)$ form to compute the cube of a complex number. The next example should help convince you of the importance of trigonometric form, which allows you to use DeMoivre's Theorem to find more challenging powers of complex numbers.

EXAMPLE 5 Use DeMoivre's Theorem to find the following and leave the result in $a + bi$ form.

a. $(1 + i)^{10}$ **b.** $\left(-2\sqrt{3} - 2i\right)^{-5}$

SOLUTION To use DeMoivre's Theorem, the complex number must be in trigonometric form.

a. $1 + i \rightarrow a = 1, b = 1$

$$r = |a + bi| = \sqrt{a^2 + b^2} = \sqrt{1^2 + 1^2}$$
$$= \sqrt{2}$$

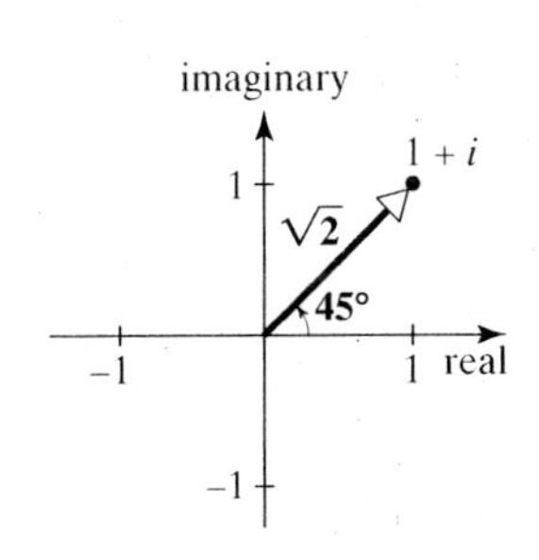

$$\tan\theta = \frac{b}{a} = \frac{1}{1}$$

$$\theta = 45° \qquad \theta \text{ is in QI.}$$

So,

$$(1 + i) = r(\cos\theta + i\sin\theta)$$
$$= \sqrt{2}(\cos 45° + i\sin 45°).$$

Therefore,

$$(1 + i)^{10} = \left(\sqrt{2}\right)^{10}[\cos(10 \cdot 45°) + i\sin(10 \cdot 45°)]$$
$$= 2^5[\cos 450° + i\sin 450°]$$
$$= 32[\cos 90° + i\sin 90°]$$
$$= 32i$$

This method should seem quite a bit easier than doing the problem the algebraic way of multiplying $(1 + i)$ by itself *ten times*!

b. $-2\sqrt{3} - 2i \rightarrow a = -2\sqrt{3}, b = -2$

$$r = |a + bi| = \sqrt{a^2 + b^2} = \sqrt{(-2\sqrt{3})^2 + (-2)^2}$$
$$= 4$$

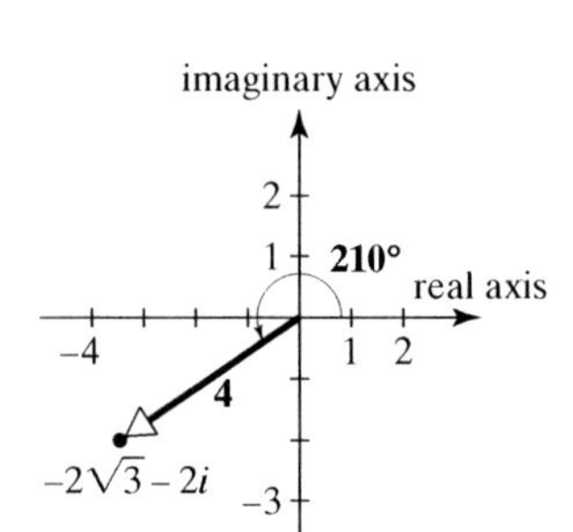

$$\tan\theta = \frac{b}{a} = \frac{-2}{-2\sqrt{3}} = \frac{\sqrt{3}}{3}$$

$$\hat{\theta} = \tan^{-1}\left(\frac{\sqrt{3}}{3}\right) = 30°$$

Since θ is in QIII, $\theta = 210°$.

So, $(-2\sqrt{3} - 2i) = 4(\cos 210° + i \sin 210°)$. Therefore,

$$(-2\sqrt{3} - 2i)^{-5} = 4^{-5}[\cos(-5 \cdot 210°) + i \sin(-5 \cdot 210°)]$$
$$= \frac{1}{4^5}[\cos(-1050°) + i \sin(-1050°)]$$
$$= \frac{1}{1024}[\cos 30° + i \sin 30°]$$
$$= \frac{1}{1024}\left[\frac{\sqrt{3}}{2} + \frac{1}{2}i\right]$$
$$= \frac{\sqrt{3}}{2048} + \frac{1}{2048}i$$

Since using the trigonometric form of a complex number can simplify finding powers of a complex number, you might wonder if it simplifies finding roots of a complex number. The answer is "yes." You may never have found square roots, cube roots, fourth roots, or other roots of complex numbers in your algebra class. It can be shown that all complex numbers have two square roots, three cube roots, four fourth roots, or n nth roots. Let's see how we extend DeMoivre's Theorem for the nth roots of complex numbers, where $n \in \{1, 2, 3, \ldots\}$.

Roots of Complex Numbers

The nth roots of any nonzero complex number $r(\cos\theta + i \sin\theta)$ are given by

$$\sqrt[n]{r}\left[\cos\left(\frac{\theta + k \cdot 360°}{n}\right) + i \sin\left(\frac{\theta + k \cdot 360°}{n}\right)\right],$$

for $k = 0, 1, 2, \ldots, (n - 1)$.

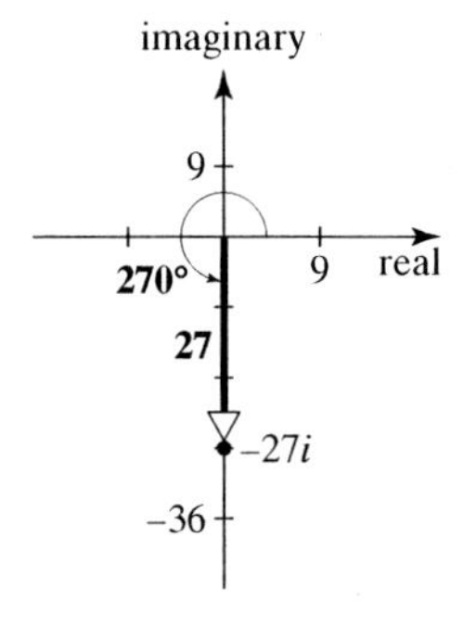

EXAMPLE 6 Find the three cube roots of $-27i$ and graph these roots in the complex plane.

SOLUTION In order to use the extension of DeMoivre's Theorem, we need to put $-27i$ in trigonometric form.

$$r = |0 - 27i| = \sqrt{0^2 + (-27)^2} = 27$$
$$\cos\theta = \tfrac{0}{27} = 0 \quad \text{and} \quad \sin\theta = -\tfrac{27}{27} = -1$$

So, $\theta = 270°$.

Therefore, the three cube roots of $-27i$ are found using $r = 27$, $\theta = 270°$, and $n = 3$.

$$\sqrt[n]{r}\cos\left[\left(\frac{\theta + k \cdot 360°}{n}\right) + i\sin\left(\frac{\theta + k \cdot 360°}{n}\right)\right], \ k = 0, 1, 2, \ldots, (n-1)$$

$$\sqrt[3]{27}\left[\cos\left(\frac{270° + k \cdot 360°}{3}\right) + i\sin\left(\frac{270° + k \cdot 360°}{3}\right)\right], \ k = 0, 1, 2$$

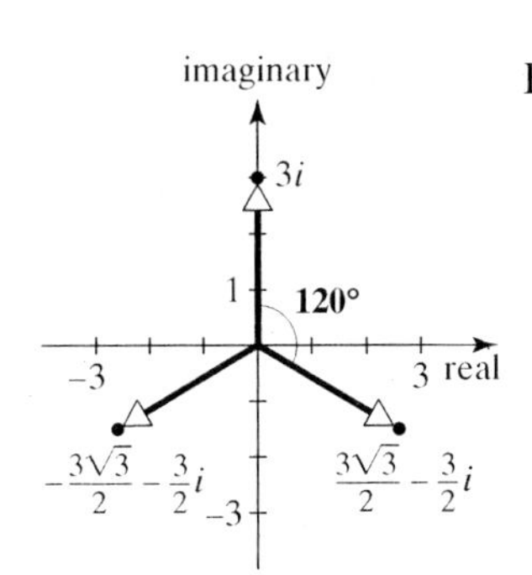

For $k = 0$: $3\left(\cos\dfrac{270° + 0 \cdot 360°}{3} + i\sin\dfrac{270° + 0 \cdot 360°}{3}\right) = 3(\cos 90° + i\sin 90°) = 3i$

$k = 1$: $3(\cos 210° + i\sin 210°) = 3\left(-\dfrac{\sqrt{3}}{2} - \dfrac{1}{2}i\right) = -\dfrac{3\sqrt{3}}{2} - \dfrac{3}{2}i$

$k = 2$: $3(\cos 330° + i\sin 330°) = 3\left(\dfrac{\sqrt{3}}{2} - \dfrac{1}{2}i\right) = \dfrac{3\sqrt{3}}{2} - \dfrac{3}{2}i$

Notice that the trigonometric form and the graph of all three cube roots of $-27i$ demonstrates that each successive vector representation of the complex roots are $360°/3 = 360°/n = 120°$ apart in the complex plane. Would that mean that each successive vector representation of the five fifth roots of a given complex number would be $360°/5 = 72°$ apart? The next example answers this question. ■

EXAMPLE 7 Find the five fifth roots of $\sqrt{2} - \sqrt{2}\,i$. Leave your answers in trigonometric form with θ in degrees.

SOLUTION

$$r = |\sqrt{2} - \sqrt{2}i| = \sqrt{(\sqrt{2})^2 + (-\sqrt{2})^2} = 2$$

$$\cos\theta = \frac{\sqrt{2}}{2} \text{ and } \sin\theta = -\frac{\sqrt{2}}{2}$$

$$\left(\text{or} \quad \tan\theta = \frac{-\sqrt{2}}{\sqrt{2}} = -1\right)$$

The argument θ is in QIV, so $\theta = 315°$. Therefore, the five fifth roots of $\sqrt{2} - \sqrt{2}\,i$ are found as follows:

$$\sqrt[n]{r}\left[\cos\left(\frac{\theta + k \cdot 360°}{n}\right) + i\sin\left(\frac{\theta + k \cdot 360°}{n}\right)\right], \ k = 0, 1, 2, \ldots, (n-1)$$

$$\sqrt[5]{2}\left[\cos\left(\frac{315° + k \cdot 360°}{5}\right) + i\sin\left(\frac{315° + k \cdot 360°}{5}\right)\right], \ k = 0, 1, 2, 3, 4$$

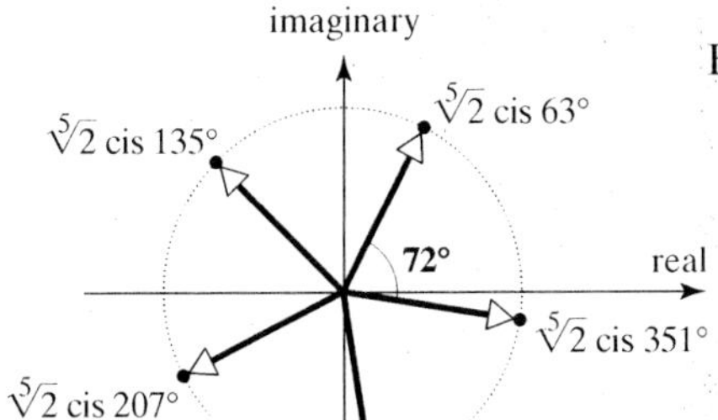

For $k = 0$: $\sqrt[5]{2}(\cos 63° + i\sin 63°)$

$k = 1$: $\sqrt[5]{2}(\cos 135° + i\sin 135°)$

$k = 2$: $\sqrt[5]{2}(\cos 207° + i\sin 207°)$

$k = 3$: $\sqrt[5]{2}(\cos 279° + i\sin 279°)$

$k = 4$: $\sqrt[5]{2}(\cos 351° + i\sin 351°)$

Notice that each successive trigonometric form of the five fifth roots are $360°/5 = 72°$ apart. ■

Solving Algebraic Equations Using Trigonometry

In the next example we demonstrate how trigonometry helps to solve an otherwise difficult algebraic equation by using the ideas developed in this section.

EXAMPLE 8 Find the four solutions to $x^4 + 2x^2 + 4 = 0$, where x is a complex number in standard form.

SOLUTION If we let $u = x^2$, we have the quadratic equation $u^2 + 2u + 4 = 0$. The left side of the equation will not factor so we use the quadratic formula.

$$u = \frac{-2 \pm \sqrt{(2)^2 - 4(1)(4)}}{2}$$

$$u = \frac{-2 \pm \sqrt{-12}}{2}$$

$$u = \frac{-2 \pm 2i\sqrt{3}}{2}$$

$$u = -1 \pm i\sqrt{3}$$

$$x^2 = -1 \pm i\sqrt{3} \qquad u = x^2$$

$$x^2 = -1 + i\sqrt{3} \text{ or } x^2 = -1 - i\sqrt{3} \qquad \text{Rewrite as two equations.}$$

If we write each number in *trigonometric form*, we get:

$$x^2 = 2(\cos 120° + i \sin 120°) \quad \text{or} \quad x^2 = 2(\cos 240° + i \sin 240°)$$

Since each equation has two square roots, we have four solutions to the original equation. For $x^2 = 2(\cos 120° + i \sin 120°)$, we get:

$$x = \sqrt{2}\left(\cos \frac{120° + k \cdot 360°}{2} + i \sin \frac{120° + k \cdot 360°}{2}\right), \quad k = 0, 1$$

Let $k = 0$: $x = \sqrt{2}(\cos 60° + i \sin 60°) = \sqrt{2}\left(\frac{1}{2} + i\frac{\sqrt{3}}{2}\right) = \frac{\sqrt{2}}{2} + i\frac{\sqrt{6}}{2}$

$k = 1$: $x = \sqrt{2}(\cos 240° + i \sin 240°) = \sqrt{2}\left(-\frac{1}{2} - i\frac{\sqrt{3}}{2}\right) = -\frac{\sqrt{2}}{2} - i\frac{\sqrt{6}}{2}$

For $x^2 = 2(\cos 240° + i \sin 240°)$, we get:

$$x = \sqrt{2}\left(\cos \frac{240° + k \cdot 360°}{2} + i \sin \frac{240° + k \cdot 360°}{2}\right), \quad k = 0, 1$$

Let $k = 0$: $x = \sqrt{2}(\cos 120° + i \sin 120°) = \sqrt{2}\left(-\frac{1}{2} + i\frac{\sqrt{3}}{2}\right) = -\frac{\sqrt{2}}{2} + i\frac{\sqrt{6}}{2}$

$k = 1$: $x = \sqrt{2}(\cos 300° + i \sin 300°) = \sqrt{2}\left(\frac{1}{2} - i\frac{\sqrt{3}}{2}\right) = \frac{\sqrt{2}}{2} - i\frac{\sqrt{6}}{2}$

We can either leave the four solutions in standard form using exact values or find decimal approximations to the nearest hundredth.

$$\frac{\sqrt{2}}{2} + i\frac{\sqrt{6}}{2} \approx 0.71 + 1.22i \qquad -\frac{\sqrt{2}}{2} + i\frac{\sqrt{6}}{2} \approx -0.71 + 1.22i$$

$$-\frac{\sqrt{2}}{2} - i\frac{\sqrt{6}}{2} \approx -0.71 - 1.22i \qquad \frac{\sqrt{2}}{2} - i\frac{\sqrt{6}}{2} \approx 0.71 - 1.22i$$

Calculators that solve polynomials for complex solutions can provide these approximate values. ■

Your awareness of the ability to use trigonometry to solve more difficult problems like Example 8 is an invaluable resource that you may need in future math courses. And most of the uses require writing a complex number in trigonometric form. The following exercises provide you with the opportunity to practice this important skill.

Exercise Set 6.5

1–10. *Graph the complex number and then express it in trigonometric form.* $(0° \leq \theta < 360°)$

EXAMPLE $6 + 6i$

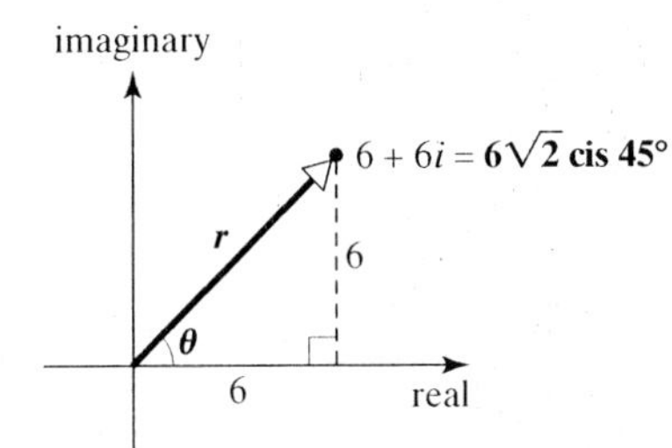

SOLUTION

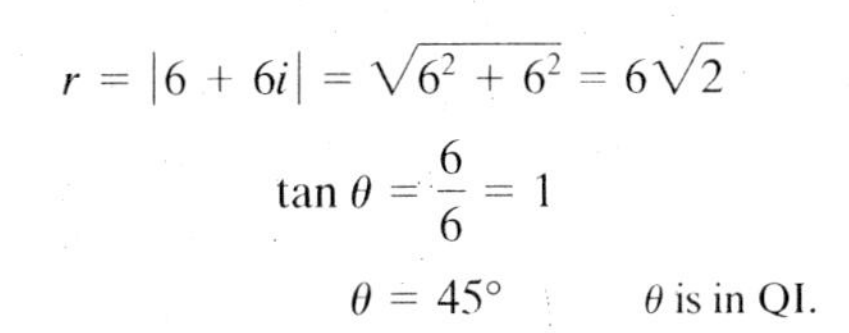

$$r = |6 + 6i| = \sqrt{6^2 + 6^2} = 6\sqrt{2}$$

$$\tan\theta = \frac{6}{6} = 1$$

$$\theta = 45° \qquad \theta \text{ is in QI.}$$

Therefore, $6 + 6i = 6\sqrt{2}(\cos 45° + i\sin 45°)$. ■

1. $\sqrt{3} + i$ **2.** $-2 + 2\sqrt{3}i$ **3.** $-7 + 7i$ **4.** $-1 - i$

5. $-2 - 2\sqrt{3}i$ **6.** $2\sqrt{3} - 2i$ **7.** $-4i$

8. $8i$ **9.** 5 **10.** -6

11–18. *Express each complex number in* $a + bi$ *form.*

EXAMPLE $3(\cos 180° + i\sin 180°)$

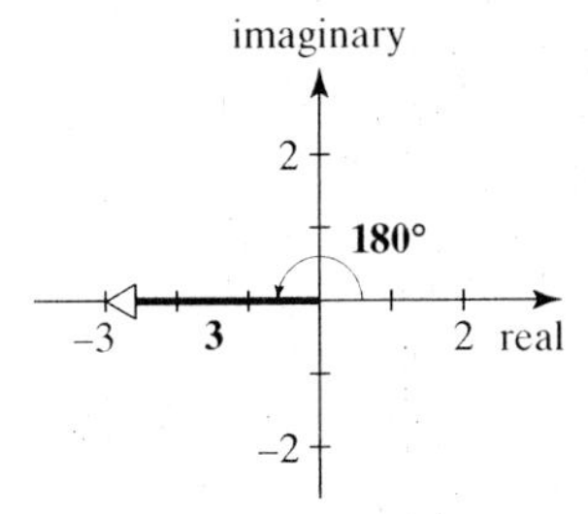

SOLUTION

$$3(-1 + 0i) = -3$$ ■

11. $5(\cos 60° + i\sin 60°)$ **12.** $3(\cos 45° + i\sin 45°)$

13. $4(\cos 135° + i\sin 135°)$ **14.** $8(\cos 210° + i\sin 210°)$

15. $\cos 330° + i\sin 330°$ **16.** $\cos 300° + i\sin 300°$

17. $9(\cos 270° + i\sin 270°)$ **18.** $6(\cos 90° + i\sin 90°)$

19–24. *Find each product and leave the result in a + bi form.*

EXAMPLE $3(\cos 115° + i \sin 115°) \cdot 4(\cos 200° + i \sin 200°)$

SOLUTION

$$\begin{aligned} & 3(\cos 115° + i \sin 115°) \cdot 4(\cos 200° + i \sin 200°) \\ &= 3 \cdot 4[\cos(115° + 200°) + i \sin(115° + 200°)] \\ &= 12[\cos 315° + i \sin 315°] \\ &= 12\left(\frac{\sqrt{2}}{2} - \frac{\sqrt{2}}{2}i\right) \\ &= 6\sqrt{2} - 6\sqrt{2}i \end{aligned}$$

19. $2(\cos 12° + i \sin 12°) \cdot 5(\cos 33° + i \sin 33°)$

20. $4(\cos 8° + i \sin 8°) \cdot 5(\cos 22° + i \sin 22°)$

21. $3(\cos 200° + i \sin 200°) \cdot \frac{1}{3}(\cos 70° + i \sin 70°)$

22. $2(\cos 50° + i \sin 50°) \cdot \frac{1}{2}(\cos 85° + i \sin 85°)$

23. $8(\cos 110° + i \sin 110°) \cdot 2(\cos 130° + i \sin 130°)$

24. $4(\cos 305° + i \sin 305°) \cdot \frac{3}{4}(\cos 55° + i \sin 55°)$

25–30. *Find each quotient and leave the result in a + bi form.*

EXAMPLE $\dfrac{8(\cos 67° + i \sin 67°)}{4(\cos 7° + i \sin 7°)}$

SOLUTION

$$\begin{aligned} \frac{8(\cos 67° + i \sin 67°)}{4(\cos 7° + i \sin 7°)} &= \frac{8}{4}[\cos(67° - 7°) + i \sin(67° - 7°)] \\ &= 2[\cos 60° + i \sin 60°] \\ &= 2\left(\frac{1}{2} + \frac{\sqrt{3}}{2}i\right) \\ &= 1 + \sqrt{3}i \end{aligned}$$

25. $\dfrac{9(\cos 241° + i \sin 241°)}{3(\cos 16° + i \sin 16°)}$

26. $\dfrac{5(\cos 336° + i \sin 336°)}{4(\cos 186° + i \sin 186°)}$

27. $\dfrac{2(\cos 97° + i \sin 97°)}{3(\cos 7° + i \sin 7°)}$

28. $\dfrac{4(\cos 307° + i \sin 307°)}{4(\cos 7° + i \sin 7°)}$

29. $\dfrac{3(\cos 80° + i \sin 80°)}{7(\cos 140° + i \sin 140°)}$

30. $\dfrac{\cos 21° + i \sin 21°}{2(\cos 261° + i \sin 261°)}$

31–40. *Use DeMoivre's Theorem and leave the answer in $a + bi$ form.*

EXAMPLE $(\sqrt{3} + i)^{-2}$

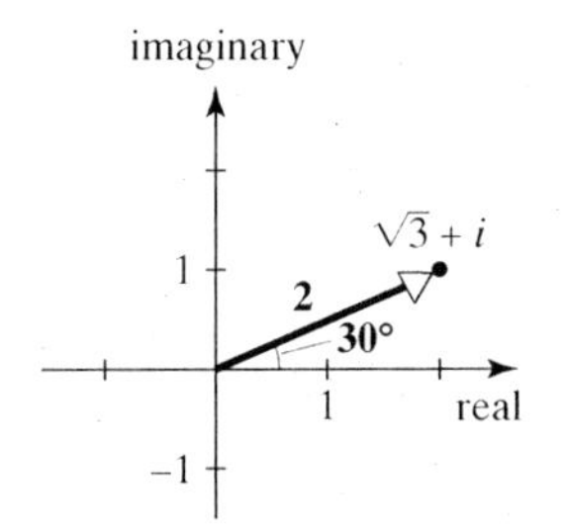

SOLUTION First find the trigonometric form of $\sqrt{3} + i$, where $a = \sqrt{3}$, $b = 1$.

$$r = |\sqrt{3} + i| = \sqrt{(\sqrt{3})^2 + (1)^2} = \sqrt{4} = 2$$

$$\tan\theta = \frac{1}{\sqrt{3}} \rightarrow \theta = 30^\circ$$

$$\begin{aligned}(\sqrt{3} + i)^{-2} &= [2(\cos 30^\circ + i\sin 30^\circ)]^{-2} \\ &= 2^{-2}[\cos(-2 \cdot 30^\circ) + i\sin(-2 \cdot 30^\circ)] \\ &= \frac{1}{2^2}[\cos(-60^\circ) + i\sin(-60^\circ)] \\ &= \frac{1}{4}\left[\frac{1}{2} - \frac{\sqrt{3}}{2}i\right] \\ &= \frac{1}{8} - \frac{\sqrt{3}}{8}i\end{aligned}$$

31. $[2(\cos 12^\circ + i\sin 12^\circ)]^5$

32. $[3(\cos 50^\circ + i\sin 50^\circ)]^3$

33. $(1 - i)^6$

34. $\left(\frac{1}{2} + \frac{\sqrt{3}}{2}i\right)^5$

35. $(1 - \sqrt{3}i)^4$

36. $(-2-2i)^4$

37. $[25(\cos 75^\circ + i\sin 75^\circ)]^{-2}$

38. $[3(\cos 110^\circ + i\sin 110^\circ)]^{-3}$

39. $(-2 + 2i)^{-4}$

40. $(-\sqrt{3} - i)^{-3}$

41–52. *Find the requested number of roots of the given complex number. Leave the answers in trigonometric form.*

EXAMPLE The two square roots of $49(\cos 180^\circ + i\sin 180^\circ)$

SOLUTION Using the extension of DeMoivre's Theorem, the two square roots are:

$$\sqrt{49}\left[\cos\left(\frac{180^\circ + k \cdot 360^\circ}{2}\right) + i\sin\left(\frac{180^\circ + k \cdot 360^\circ}{2}\right)\right], \quad \text{for } k = 0, 1$$

For $k = 0$: $7[\cos 90^\circ + i\sin 90^\circ]$
$k = 1$: $7[\cos 270^\circ + i\sin 270^\circ]$

41. The two square roots of $16(\cos 90^\circ + i\sin 90^\circ)$

42. The two square roots of $36(\cos 60^\circ + i\sin 60^\circ)$

43. The two square roots of $\frac{1}{2} + \frac{\sqrt{3}}{2}i$

44. The two square roots of $25i$

45. The three cube roots of $8(\cos 135^\circ + i\sin 135^\circ)$

46. The three cube roots of $27(\cos 180^\circ + i\sin 180^\circ)$

47. The five fifth roots of $\frac{1}{2} - \frac{\sqrt{3}}{2}i$

48. The four fourth roots of $-\frac{1}{2} - \frac{\sqrt{3}}{2}i$

49. The two square roots of $100(\cos 100° + i \sin 100°)$

50. The three cube roots of $8(\cos 75° + i \sin 75°)$

51. The four fourth roots of $\frac{1}{2} - \frac{\sqrt{3}}{2}i$

52. The five fifth roots of $\frac{\sqrt{2}}{2} - \frac{\sqrt{2}}{2}i$

53–56. *Find the standard form for the complex numbers found in the given exercise. Where exact values are not possible, approximate values to the nearest hundredth. Then graph the complex numbers.*

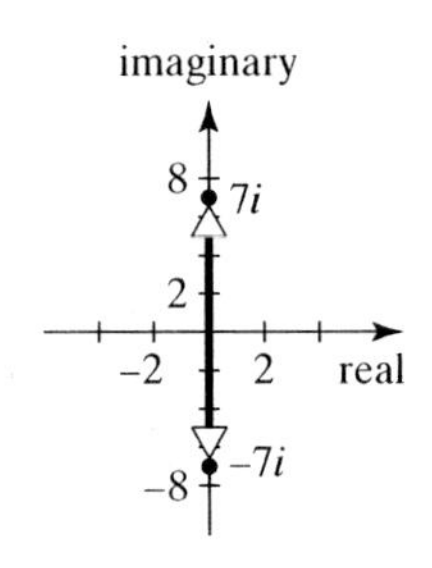

EXAMPLE a. $7[\cos 90° + i \sin 90°]$
b. $7[\cos 270° + i \sin 270°]$

SOLUTION

a. $7[\cos 90° + i \sin 90°] = 7[0 + i] = 7i$

b. $7[\cos 270° + i \sin 270°] = 7[0 + i(-1)] = -7i$

53. Exercise 41 **54.** Exercise 42 **55.** Exercise 45 **56.** Exercise 50 ■

57–60. *Solve each equation for x, where x is a complex number. Leave answers in trigonometric form unless you are able to find the standard form with exact values.*

EXAMPLE $x^3 + 1 = 0$

SOLUTION

$$x^3 + 1 = 0$$
$$x^3 = -1 = -1 + 0i = 1(\cos 180° + i \sin 180°)$$

We use the extension of DeMoivre's Theorem to find the three cube roots of $1(\cos 180° + i \sin 180°)$.

$$x = \sqrt[3]{1}\left[\cos\left(\frac{180° + k \cdot 360°}{3}\right) + i \sin\left(\frac{180° + k \cdot 360°}{3}\right)\right], \quad \text{for } k = 0, 1, \text{and } 2$$

For $k = 0$: $x = 1(\cos 60° + i \sin 60°) = \frac{1}{2} + \frac{\sqrt{3}}{2}i$

$k = 1$: $x = 1(\cos 180° + i \sin 180°) = -1$

$k = 2$: $x = 1(\cos 300° + i \sin 300°) = \frac{1}{2} - \frac{\sqrt{3}}{2}i$ ■

57. $x^4 + 1 = 0$

58. $x^3 + 8 = 0$

59. $x^4 - 2x^2 + 4 = 0$

60. $x^4 - 2\sqrt{3}x^2 + 4 = 0$

61–70. *Are the following statements true or false? If a statement is false, explain why or give an example that shows why it is false.*

61. The complex number $a + bi$ is in trigonometric form.

62. Every complex number has a unique argument.

63. Every complex number has a modulus.

64. Every complex number has a vector representation in the complex plane.

65. The complex number $0 + 0i$ does not have trigonometric form.

66. The trigonometric form of complex numbers can be useful when adding or subtracting the complex numbers.

67. The trigonometric form of complex numbers can be useful when multiplying or dividing the complex numbers.

68. The trigonometric form of a complex number can be helpful when finding roots of the complex number.

69. The trigonometric form of a complex number can be helpful when finding integer powers of complex numbers.

70. The successive roots of a complex number have arguments whose measures are the same number of degrees (radians) apart.

71. Prove if $a + bi = r_1(\cos\theta_1 + i\sin\theta_1)$ and $c + di = r_2(\cos\theta_2 + i\sin\theta_2)$, then

$$\frac{a + bi}{c + di} = \frac{r_1}{r_2}[\cos(\theta_1 - \theta_2) + i\sin(\theta_1 - \theta_2)],$$

where $c \neq 0$ and $d \neq 0$.

72. Rewrite the rules given in this section that involve trigonometric form using the notation $r(\text{cis}\ \theta)$.

Graphing Calculator/CAS Exploration

73–76. *You can use a graphing utility or CAS to find the product, quotient, or power of complex numbers, usually in approximate form. Find the power of a complex number in the given exercises using a calculator or computer.*

73. Exercise 33 74. Exercise 36 75. Exercise 37 76. Exercise 40

77–78. *Investigate if your graphing utility or CAS can find either of the two square roots and any of the three cube roots of the complex number in the given exercise.*

77. Exercise 43 78. Exercise 46

Chapter 6 Summary

6.1 Geometric Vectors and Applications

A vector is a quantity with both direction and magnitude. A vector is geometrically represented by an arrow and can be denoted by a boldface letter. We can perform various operations with vectors.

Scalar Multiple of a Vector

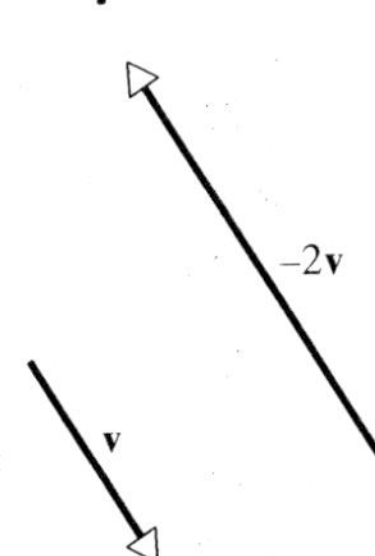

For any scalar c, $c\mathbf{v}$ is a vector whose magnitude is $|c|$ times the magnitude of $\mathbf{v}$.

If $c > 0$, $c\mathbf{v}$ points in the same direction as $\mathbf{v}$.
If $c < 0$, $c\mathbf{v}$ points in the opposite direction as $\mathbf{v}$.

The zero vector, $\mathbf{0}$, has magnitude 0 and arbitrary direction.

Addition of Vectors

Vectors can be added geometrically using the triangle or parallelogram method. The sum of two vectors is a vector called the **resultant.**

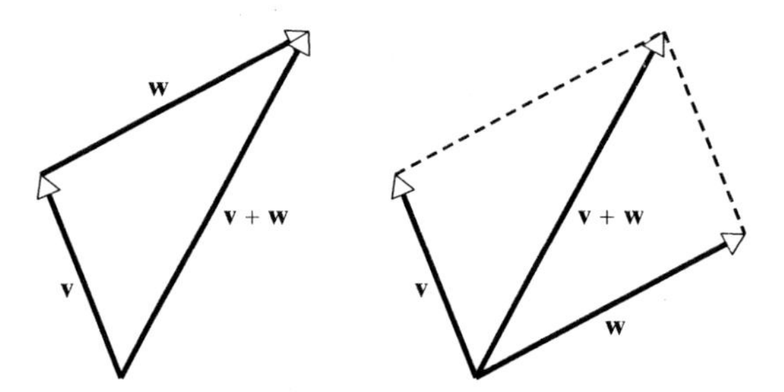

Geometric vectors can be used to help solve physical problems involving force, velocity, and other quantities that have both direction and magnitude.

6.2 Algebraic Vectors

A two-dimensional vector positioned so that the initial point is at the origin is said to be in standard position. The vector $\mathbf{v}$ can be algebraically described by the ordered pair at the terminal point (v_1, v_2) of the vector using the notation $\mathbf{v} = \langle v_1, v_2 \rangle$. This component form of a vector has v_1 as the horizontal component and v_2 as the vertical component. The angle the vector makes with the positive side of the x-axis is called the direction angle. Two vectors in component form are equivalent if their corresponding components are equal. Two unit vectors, $\langle 1, 0 \rangle = \mathbf{i}$ and $\langle 0, 1 \rangle = \mathbf{j}$, allow us to write the component form of a vector as $\mathbf{v} = \langle v_1, v_2 \rangle = v_1\mathbf{i} + v_2\mathbf{j}$.

For any vector $\mathbf{v} = \langle v_1, v_2 \rangle$,

- the magnitude of $\mathbf{v} = |\mathbf{v}| = \sqrt{v_1^2 + v_2^2}$;
- the direction angle θ of $\mathbf{v}$ is such that $\cos\theta = \dfrac{v_1}{|\mathbf{v}|}$ and $\sin\theta = \dfrac{v_2}{|\mathbf{v}|}$, $0° \le \theta < 360°$; and
- the horizontal and vertical components, respectively, of $\mathbf{v}$ with magnitude $|\mathbf{v}|$ and direction angle θ are

$$v_1 = |\mathbf{v}|\cos\theta, \quad v_2 = |\mathbf{v}|\sin\theta.$$

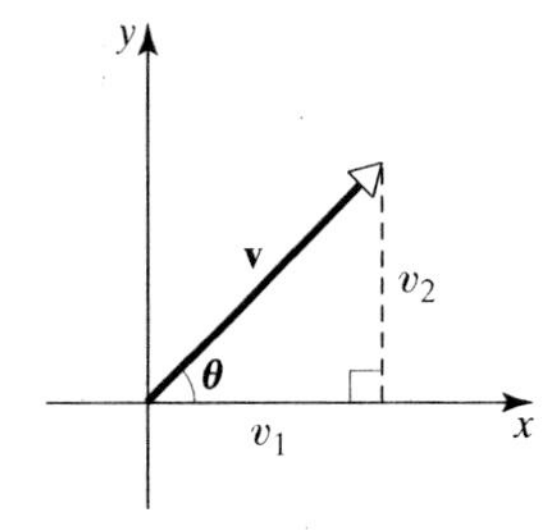

For any vector $\overrightarrow{PQ}$ with initial point $P(p_1, p_2)$ and terminal point $Q(q_1, q_2)$, its equivalent standard position vector is $\overrightarrow{PQ} = \langle q_1 - p_1, q_2 - p_2 \rangle$.

Vector Operations

For vectors $\mathbf{v} = \langle v_1, v_2 \rangle = v_1\mathbf{i} + v_2\mathbf{j}$ and $\mathbf{u} = \langle u_1, u_2 \rangle = u_1\mathbf{i} + u_2\mathbf{j}$: (produces a vector)

Addition $\mathbf{v} + \mathbf{u} = \langle v_1, v_2 \rangle + \langle u_1, u_2 \rangle = \langle v_1 + u_1, v_2 + u_2 \rangle$

or $\mathbf{v} + \mathbf{u} = (v_1\mathbf{i} + v_2\mathbf{j}) + (u_1\mathbf{i} + u_2\mathbf{j}) = (v_1 + u_1)\mathbf{i} + (v_2 + u_2)\mathbf{j}$

Subtraction $\mathbf{v} - \mathbf{u} = \langle v_1, v_2 \rangle - \langle u_1, u_2 \rangle = \langle v_1 - u_1, v_2 - u_2 \rangle$

or $\mathbf{v} - \mathbf{u} = (v_1\mathbf{i} + v_2\mathbf{j}) - (u_1\mathbf{i} + u_2\mathbf{j}) = (v_1 - u_1)\mathbf{i} + (v_2 - u_2)\mathbf{j}$

Scalar Multiplication $c\mathbf{v} = c\langle v_1, v_2 \rangle = \langle cv_1, cv_2 \rangle$

or $c\mathbf{v} = c(v_1\mathbf{i} + v_2\mathbf{j}) = cv_1\mathbf{i} + cv_2\mathbf{j}$

produces a vector

Dot Product $\mathbf{v} \cdot \mathbf{u} = v_1u_1 + v_2u_2$

$\mathbf{v} \cdot \mathbf{u} = |\mathbf{u}||\mathbf{v}| \cos\theta$, where θ is the angle between $\mathbf{u}$ and $\mathbf{v}$.

produces a scalar

Applications of Dot Product

- The angle θ between two nonzero vectors $\mathbf{u}$ and $\mathbf{v}$ is found by

$$\cos\theta = \frac{\mathbf{u} \cdot \mathbf{v}}{|\mathbf{u}||\mathbf{v}|}.$$

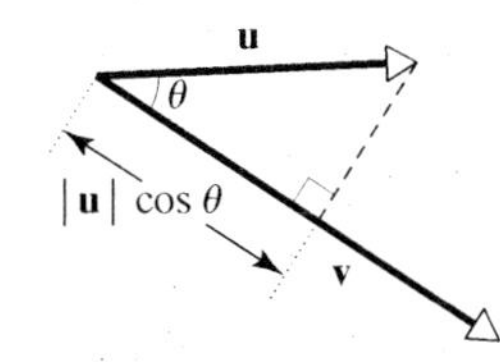

Two vectors are orthogonal (perpendicular) if their dot product is 0.

- The projection of one vector ($\mathbf{u}$) on another ($\mathbf{v}$), called $\text{comp}_{\mathbf{v}}\mathbf{u}$, is the scalar

$$\text{comp}_{\mathbf{v}}\mathbf{u} = |\mathbf{u}| \cos\theta = \frac{\mathbf{u} \cdot \mathbf{v}}{|\mathbf{v}|}.$$

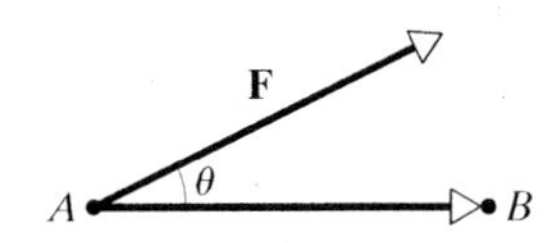

- The work W done by a constant force $\mathbf{F}$ that displaces an object from A to B, can be found using $W = \mathbf{F} \cdot \overrightarrow{AB} = |\mathbf{F}||\overrightarrow{AB}| \cos\theta$, where θ is the angle between the force and displacement.

6.3 Polar Coordinates

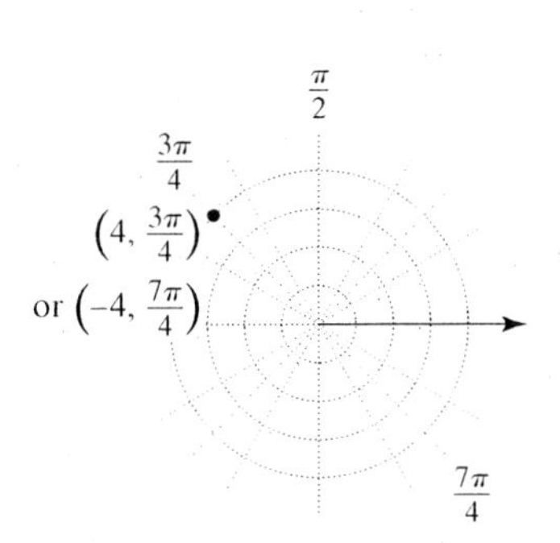

The ordered pair (r, θ) names a point that is the directed distance of r units from the origin along the ray (or opposite ray) containing the terminal side of angle θ in standard position. The coordinates r and θ are called the polar coordinates and specify a unique point in the plane. However, each point in the plane corresponds to infinitely many ordered pairs of polar coordinates.

The following relationships convert between polar coordinates (r, θ) and rectangular coordinates (x, y).

$$\tan\theta = \frac{y}{x} \qquad x = r\cos\theta$$

$$r^2 = x^2 + y^2 \qquad y = r\sin\theta$$

These relationships are the same ones used to convert back and forth between equations given in polar coordinates and those in rectangular coordinates.

Polar equations can be graphed by making a table of values containing ordered pair solutions and plotting the corresponding points (r, θ) as the value of θ increases. They can also be graphed using a graphing utility in polar and radian (or degree) mode.

6.4 Complex Numbers

A complex number is any number that can be written in the standard form $a + bi$, $a, b \in \mathbb{R}$ and $i^2 = -1$. The number a is called the real part, and b is called the imaginary part of $a + bi$.

Two complex numbers are equal if and only if their real parts are equal and their imaginary parts are equal.

$$a + bi = c + di \leftrightarrow a = c \text{ and } b = d$$

If we treat the complex numbers as binomials in which i is the variable, then sums and products can be computed using familiar rules of algebra and replacing i^2 with -1. For any integer p, then i^p can be simplified to $1, -1, i$, or $-i$.

For any positive real number b, $\sqrt{-b} = i\sqrt{b}$. Always write square roots of negative real numbers $(\sqrt{-b}, b > 0)$ in $i\sqrt{b}$ form *before* performing any operations. The conjugate of $a + bi$ is $a - bi$. The product $(a + bi)(a - bi) = a^2 + b^2$, a real number. Conjugates are used to simplify expressions that involve division by a complex number.

$$\frac{\sqrt{-9}}{1 - i} = \frac{3i}{1 - i} = \frac{3i(1 + i)}{(1 - i)(1 + i)} = \frac{3i + 3i^2}{2} = -\frac{3}{2} + \frac{3}{2}i$$

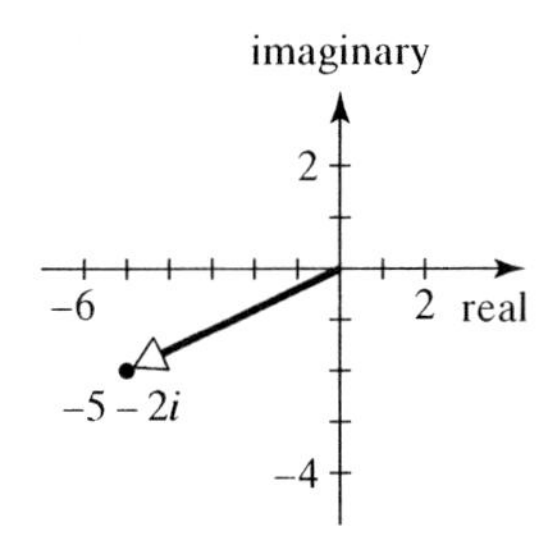

The geometric representation of a complex number $a + bi$ is a vector with initial point at the origin and terminal point (a, b) graphed in the complex plane in which the x-axis is called the real axis, and the y-axis is called the imaginary axis. Complex numbers can be added graphically using the parallelogram method for vectors.

The modulus or absolute value of a complex number $a + bi$ is the distance from the origin to the point (a, b). If r represents this distance, then

$$r = |a + bi| = \sqrt{a^2 + b^2}.$$

6.5 Trigonometric Form For Complex Numbers

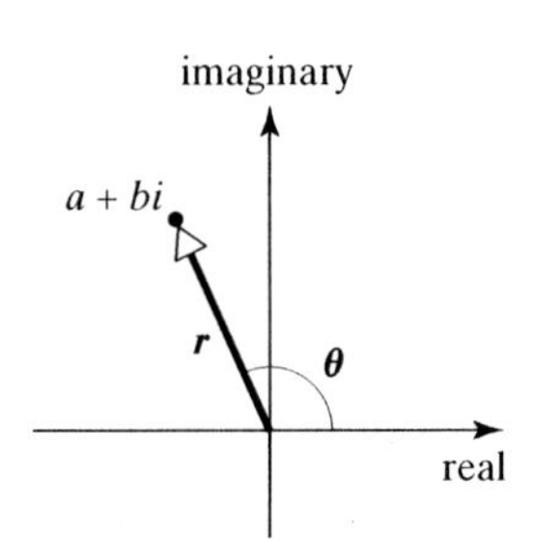

The argument of $a + bi$ is the least positive angle θ from the real axis to the graph of $a + bi$. The following relates $a + bi$, r and θ.

$$a = r\cos\theta \qquad b = r\sin\theta$$

$$\text{and } \tan\theta = \frac{b}{a}$$

The trigonometric (or polar) form for the complex number $a + bi$ is given by

$$a + bi = (r\cos\theta) + (r\sin\theta)\,i$$

$$\text{or} \quad a + bi = r(\cos\theta + i\sin\theta),$$

where $r = \sqrt{a^2 + b^2}$ is the modulus and θ is the argument of $a + bi$.

If $a + bi = r_1(\cos\theta_1 + i\sin\theta_1)$ and $c + di = r_2(\cos\theta_2 + i\sin\theta_2)$, then

$$(a + bi)(c + di) = r_1 r_2[\cos(\theta_1 + \theta_2) + i\sin(\theta_1 + \theta_2)]$$
$$\frac{a + bi}{c + di} = \frac{r_1}{r_2}[\cos(\theta_1 - \theta_2) + i\sin(\theta_1 - \theta_2)],$$

where $c \neq 0$ and $d \neq 0$.

DeMoivre's Theorem is used to find powers of complex numbers. For any complex number $r(\cos\theta + i\sin\theta)$,

$$[r(\cos\theta + i\sin\theta)]^n = r^n(\cos n\theta + i\sin n\theta),$$

for $n \in \{\ldots, -2, -1, 0, 1, 2, \ldots\}$.

For $n \in \{1, 2, \ldots\}$, the nth roots of any nonzero complex number $r(\cos\theta + i\sin\theta)$ are given by

$$\sqrt[n]{r}\left[\cos\left(\frac{\theta + k \cdot 360°}{n}\right) + i\sin\left(\frac{\theta + k \cdot 360°}{n}\right)\right],$$

for $k = 0, 1, 2, \ldots, (n - 1)$.

Chapter 6 Review Exercises

6.1

1–6. *By sketching vectors equivalent to* **u**, **v**, *or* **w**, *find the indicated resultant using the triangle or parallelogram method.*

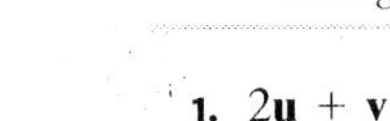

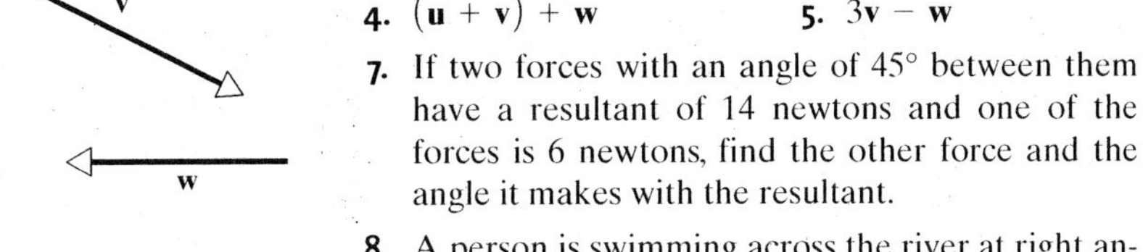

1. $2\mathbf{u} + \mathbf{v}$ **2.** $\mathbf{v} - 3\mathbf{w}$ **3.** $\mathbf{u} + (\mathbf{v} + \mathbf{w})$

4. $(\mathbf{u} + \mathbf{v}) + \mathbf{w}$ **5.** $3\mathbf{v} - \mathbf{w}$ **6.** $4\mathbf{u} + 3\mathbf{w}$

7. If two forces with an angle of 45° between them have a resultant of 14 newtons and one of the forces is 6 newtons, find the other force and the angle it makes with the resultant.

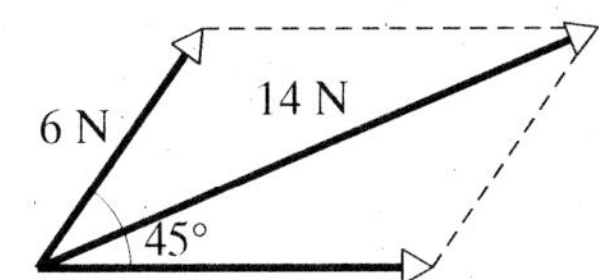

8. A person is swimming across the river at right angles with the current. The current drifts the swimmer 15° off course, and causes her to go 7 miles per hour relative to the banks of the river. What is the speed of the swimmer (to the nearest tenth of a mile per hour) in still water?

9. An airplane is headed on a bearing of 160° with an air speed of 400 kilometers per hour. The true course has a bearing of 172°. The ground speed is 430 kilometers per hour. Find the wind speed and the bearing of the wind.

6.2

10–13. *Find the exact value of the magnitude and direction angle to the nearest tenth of a degree for each vector.*

10. $\langle 7, -7\rangle$ **11.** $\langle 6\sqrt{2}, 6\sqrt{2}\rangle$ **12.** $-3\mathbf{i} + 6\mathbf{j}$ **13.** $5\mathbf{i} - 12\mathbf{j}$

14–17. *Express* **v** *in* $v_1\mathbf{i} + v_2\mathbf{j}$ *form with the given magnitude and direction. Approximate* v_1 *and* v_2 *to the nearest tenth when necessary.*

14. $|\mathbf{v}| = 9, \theta = 100°$ **15.** $|\mathbf{v}| = 12, \theta = 310°$

16. $|\mathbf{v}| = 1, \theta = 90°$ **17.** $|\mathbf{v}| = 7, \theta = 180°$

18–21. *Find* $2\mathbf{v} + \mathbf{w}$, *and* $|2\mathbf{v} + \mathbf{w}|$ *for each pair of vectors.*

18. $\mathbf{v} = \langle 2, -5 \rangle$, $\mathbf{w} = \mathbf{i} + \mathbf{j}$

19. $\mathbf{v} = 6\mathbf{i} + 2\mathbf{j}$, $\mathbf{w} = \langle -3, 4 \rangle$

20. $\mathbf{v} = \mathbf{i} + 4\mathbf{j}$, $\mathbf{w} = 9\mathbf{i} - 9\mathbf{j}$

21. $\mathbf{v} = -3\mathbf{i} + \mathbf{j}$, $\mathbf{w} = \mathbf{i} - 7\mathbf{j}$

22–25. *Find the inner product* $\mathbf{v} \cdot \mathbf{w}$. *Approximate answers to the nearest tenth when necessary.*

22. $\mathbf{v} = \langle 3, -3 \rangle$, $\mathbf{w} = \langle 10, -8 \rangle$

23. $\mathbf{v} = -2\mathbf{i} - 4\mathbf{j}$, $\mathbf{w} = 5\mathbf{i} + 7\mathbf{j}$

24. $|\mathbf{v}| = 10$ and the direction angle of $\mathbf{v}$ is 55°, $|\mathbf{w}| = 4$ and the direction angle of $\mathbf{w}$ is 107°.

25. $|\mathbf{v}| = 5$, $|\mathbf{w}| = 26$, and the angle θ between $\mathbf{v}$ and $\mathbf{w}$ is 78°.

26–27. *Find to the nearest tenth of a degree the angle* θ *between each pair of vectors, where* $0° \le \theta \le 180°$.

26. $\mathbf{v} = \langle 7, -3 \rangle$, $\mathbf{w} = \langle -1, -3 \rangle$

27. $\mathbf{v} = -6\mathbf{i} - 2\mathbf{j}$, $\mathbf{w} = 10\mathbf{i} - \mathbf{j}$

28–29. *Find the component of* $\mathbf{u}$ *along* $\mathbf{v}$ $(\text{comp}_{\mathbf{v}}\mathbf{u})$ *and the component of* $\mathbf{v}$ *along* $\mathbf{u}$ $\text{comp}_{\mathbf{u}}\mathbf{v}$.

28. $\mathbf{u} = -5\mathbf{i} + 3\mathbf{j}$, $\mathbf{v} = -6\mathbf{i} - 4\mathbf{j}$

29. $\mathbf{u} = 10\mathbf{i} + 4\mathbf{j}$, $\mathbf{v} = 8\mathbf{i} - 3\mathbf{j}$

30–31. *Find the* $a\mathbf{i} + b\mathbf{j}$ *form for* $\overrightarrow{RS}$.

30. $R = (-2, 4)$, $S = (7, 12)$

31. $R = (8, -2)$, $S = (-6, -3)$

32. Find the work Carole has done by pulling a 50-pound suitcase 200 feet if the handle makes an angle of 35° with the horizontal.

6.3

33–36. *Plot each point given in polar coordinates and find its rectangular coordinates.*

33. $\left(5, -\frac{\pi}{2}\right)$

34. $(8, 120°)$

35. $(-2, 225°)$

36. $\left(-2, -\frac{\pi}{6}\right)$

37–40. *For each point given in rectangular coordinates, find two sets of polar coordinates where* $r > 0$, $-360° < \theta \le 360°$.

37. $(-3, \sqrt{3})$

38. $(1, -1)$

39. $(7, 0)$

40. $(0, -9)$

41–42. *Convert each rectangular equation into an equation in polar form.*

41. $3x^2 + 3y^2 = 12y$

42. $y = -2$

43–44. *Convert each polar equation into an equation in rectangular form.*

43. $\theta = \frac{7\pi}{4}$

44. $r \cos \theta = 7r \sin \theta + 12$

45–48. *Sketch the graph of each polar equation.*

45. $r = 4 \sin \theta$

46. $r = 3 - 3 \cos \theta$

47. $r = 2 + \cos \theta$

48. $r = 1 - 2 \sin \theta$

6.4

49–54. *Perform the indicated operations and leave each expression in* $a + bi$ *form.*

49. $(-3 + 4i) + 6(2 - 5i)$

50. $\frac{6 - i}{2 + 2i}$

51. $(1 + i)^2$

52. $(\sqrt{-4} + \sqrt{25})(\sqrt{-9} + 3)$

53. $i^{33} + i^2 - i^{11}$

54. $\frac{-4}{1 + \sqrt{-2}}$

55–60. *Find the modulus of each complex number.*

55. $1 - i$ **56.** $-\sqrt{3} + i$ **57.** $-3 - 4i$

58. $-5 + 12i$ **59.** $6 + \sqrt{2}i$ **60.** $3 - \sqrt{5}i$

6.5

61–64. *Express each complex number in trigonometric form.*

61. $1 - i$ **62.** $-\sqrt{3} + i$ **63.** $-2 - 2\sqrt{3}i$ **64.** $-8i$

65–68. *Express each complex number in $a + bi$ form.*

65. $7(\cos 150° + i \sin 150°)$ **66.** $\cos 240° + i \sin 240°$

67. $6(\cos 315° + i \sin 315°)$ **68.** $2(\cos 180° + i \sin 180°)$

69–70. *Find the product and leave the result in $a + bi$ form.*

69. $3(\cos 88° + i \sin 88°) \cdot 5(\cos 32° + i \sin 32°)$

70. $7(\cos 40° + i \sin 40°) \cdot 8(\cos 350° + i \sin 350°)$

71–72. *Find the quotient and leave the result in $a + bi$ form.*

71. $\dfrac{15(\cos 209° + i \sin 209°)}{25(\cos 29° + i \sin 29°)}$ **72.** $\dfrac{34(\cos 315° + i \sin 315°)}{17(\cos 270° + i \sin 270°)}$

73–76. *Express the power of each complex number in $a + bi$ form.*

73. $[3(\cos 15° + i \sin 15°)]^6$ **74.** $(\sqrt{3} - i)^6$

75. $[5(\cos 120° + i \sin 120°)]^{-2}$ **76.** $(6 + 6i)^{-3}$

77–78. *Find the requested number of roots of the given complex number. Leave answers in trigonometric form.*

77. The three cube roots of $216(\cos 27° + i \sin 27°)$.

78. The two square roots of $\frac{1}{2} - \frac{\sqrt{3}}{2}i$.

79–80. *Solve the equations for x, where x is a complex number, and leave your answers in the indicated form.*

79. $x^2 + 4i = 0$, trigonometric form **80.** $x^3 = 1$, $a + bi$ form

Chapter 6 Test

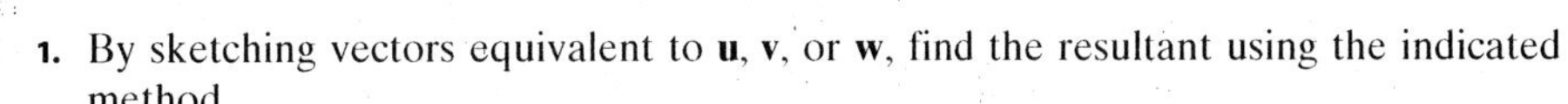

1. By sketching vectors equivalent to **u**, **v**, or **w**, find the resultant using the indicated method.

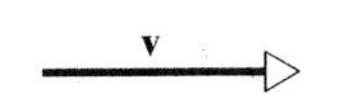

a. $\mathbf{u} + \mathbf{w}$, triangle method **b.** $\mathbf{v} - 2\mathbf{w}$, parallelogram method

2. Draw the standard position vector and find its magnitude and direction angle to the nearest tenth of a degree.

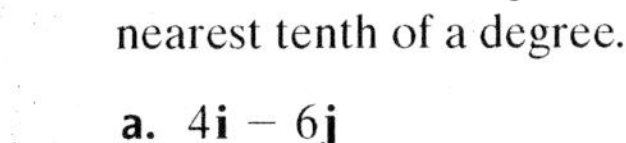

a. $4\mathbf{i} - 6\mathbf{j}$ **b.** $\langle -3, 6 \rangle$

3. Find to the nearest tenth the vertical and horizontal components of the vector **v** with the given magnitude and direction. Sketch **v** and express the vector in $v_1\,\mathbf{i} + v_2\,\mathbf{j}$ form.

   a. $|\mathbf{v}| = 1,\ \theta = \dfrac{\pi}{6}$ b. $|\mathbf{v}| = 5,\ \theta = 138°$

4. If $\mathbf{v} = 2\mathbf{i} - \mathbf{j}$, $\mathbf{u} = 4\mathbf{i} + 8\mathbf{j}$, and $\mathbf{w} = -3\mathbf{j}$, find the following.

   a. $2\mathbf{v} - 4(\mathbf{u} + \mathbf{w})$ b. $\mathbf{v} \cdot \mathbf{w}$

   c. The angle (to the nearest tenth of a degree) between **u** and **w**.

   d. $\text{comp}_{\mathbf{v}}\mathbf{w}$ e. $|\mathbf{v} - \mathbf{u}|$

5. Find the dot product of **v** and **w** if $|\mathbf{v}| = 8$, $|\mathbf{w}| = 3$ and the angle between **v** and **w** is 120°.

6. Two forces of 3 and 12 pounds have an angle of 20° between them. Find the magnitude (to the nearest tenth of a pound) of the resultant and the angle (to the nearest tenth of a degree) the resultant makes with the 3-pound force.

7. An airplane is headed southwest on a bearing of 210° with an air speed of 500 miles per hour, with the wind blowing from the north (bearing 180°) at a speed of 100 miles per hour. Find (to the nearest mile per hour) the ground speed and (to the nearest tenth of a degree) the course of the airplane.

8. Find the rectangular coordinates for the point whose polar coordinates are given.

   a. $(3, 120°)$ b. $(-4, 45°)$

9. For each point given in rectangular coordinates, find two sets of polar coordinates $(-360° < \theta \le 360°)$.

   a. $(10, -10)$ b. $(5\sqrt{3}, 5)$

10. Convert each rectangular equation to an equation in polar form.

    a. $x^2 + y^2 = 25$ b. $y = -1$

11. Convert each polar equation to an equation in rectangular form.

    a. $r = 3\cos\theta$ b. $r = 5\sec\theta$

12. Express the result in $a + bi$ form.

    a. $(-4 + 2i) - 3(2 + i)$ b. $\sqrt{-25}(4 - 2i)$

13. Express each complex number in trigonometric form.

    a. $2 - 2i$ b. $6i$

14. Express in $a + bi$ form.

    a. $4(\cos 45° + i\sin 45°)$ b. $\sqrt{3}(\cos 60° + i\sin 60°)$

    c. $9(\cos 90° + i\sin 90°) \cdot \frac{1}{3}(\cos 180° + i\sin 180°)$

    d. $\dfrac{12(\cos 217° + i\sin 217°)}{6(\cos 67° + i\sin 67°)}$

15. Express each power in $a + bi$ form.

    a. $[2(\cos 30° + i\sin 30°)]^6$ b. $(1 + \sqrt{3}i)^9$

16. Solve the equation for x, where x is a complex number, and leave your answer in trigonometric form.

    a. $x^2 + 9i = 0$ b. $x^4 + 1 = 0$

Chapters 1–6 Cumulative Review

1.a. Copy the diagram and complete the entries.

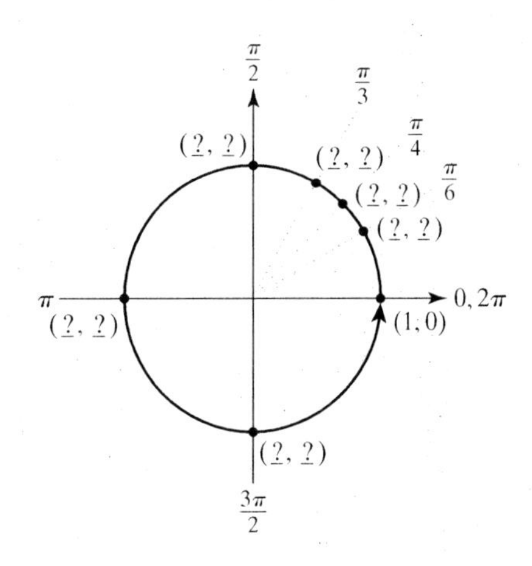

b. Copy the following table and complete the entries.

x	$\cos x$	$\csc x$	$\tan x$
0			
$\frac{\pi}{6}$			
45°			
60°			
$\frac{\pi}{2}$			
π			
$\frac{3\pi}{2}$			
360°			

2. If $\sin A = -\frac{4}{5}$ with $\cos A > 0$ and $\cos B = \frac{\sqrt{2}}{2}$ with B in QIV, find each of the following. In the case of finding a functional value, give the exact value.

a. The quadrant that A is in

b. $\sec A$

c. $\tan B$

d. $\cos 2A$

(continued)

e. $\tan 2A$

f. $\sin(A + B)$

g. The quadrant that $\dfrac{A}{2}$ is in

h. $\cos \dfrac{A}{2}$

3. Graph at least two periods of each trigonometric equation, indicating the period, amplitude, and asymptotes if appropriate. Label the critical values for one cycle.

a. $y = -3\cos\left(\frac{1}{2}x\right)$

b. $y = \tan(2x)$

4. Give two trigonometric equations that represent each graph.

a. 

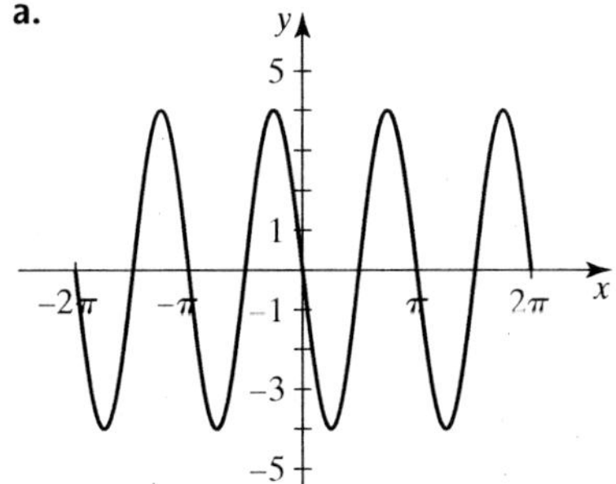

b.

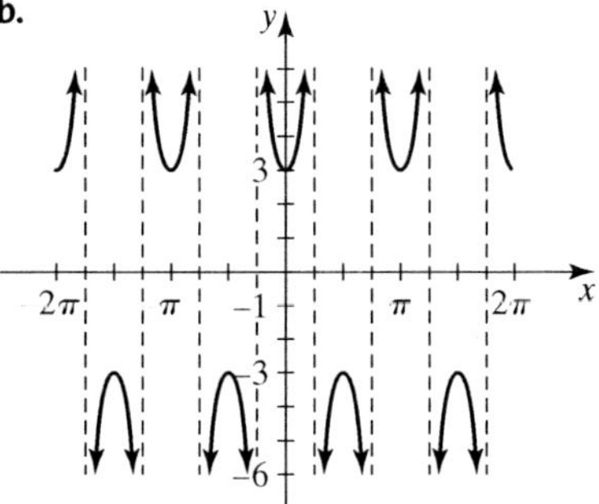

5. Identify each inverse trigonometric graph.

a.

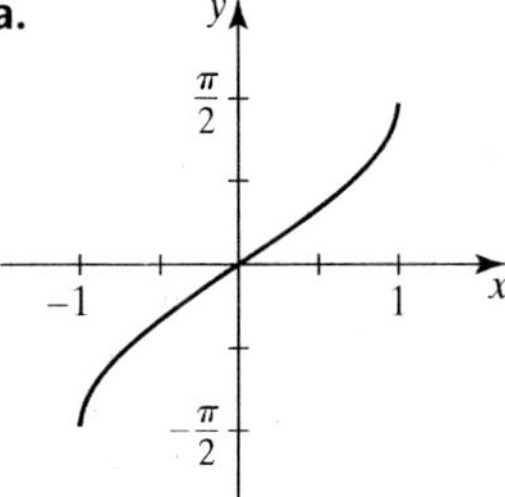

b.

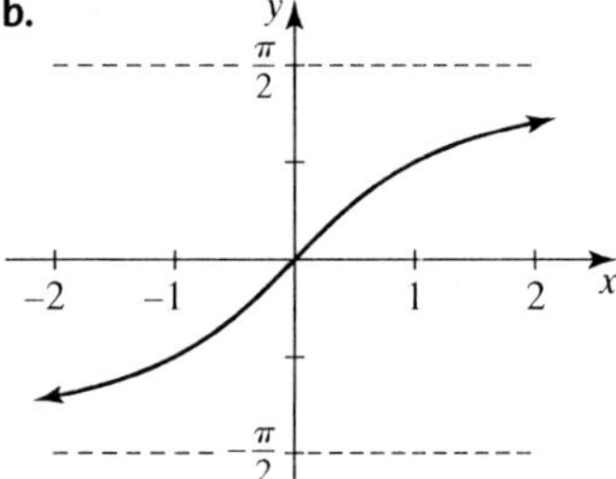

c.

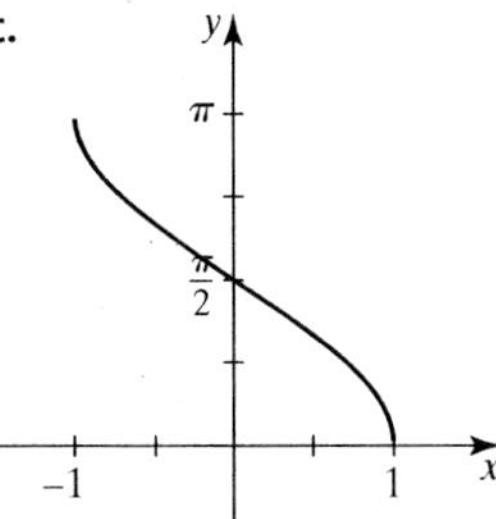

6. Find the exact value unless otherwise indicated.

a. $\sin(-225°)$

b. $\cos 570°$

c. $\cos\left(\dfrac{3\pi}{4}\right)$

d. $\csc\left(-\dfrac{7\pi}{3}\right)$

e. $\tan\left(\dfrac{9\pi}{4}\right)$

f. $\arctan(-1)$ in radians

g. $\cos\left(\arcsin -\dfrac{\sqrt{3}}{2}\right)$

h. $\operatorname{arcsec}(5.2)$ to the nearest tenth of a degree

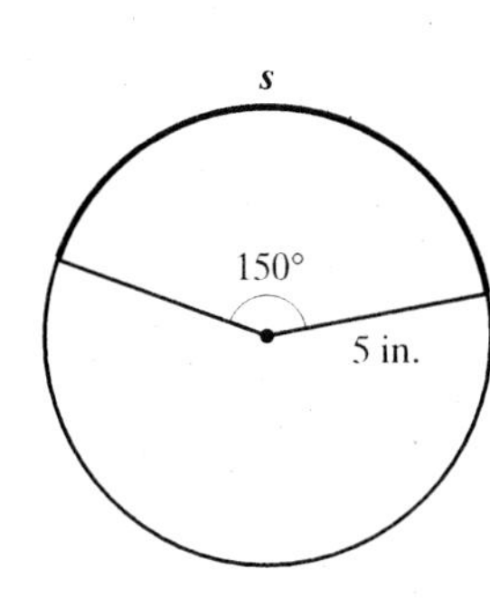

i. $2\sin^2\frac{\pi}{3}\tan\frac{\pi}{3} + \sec\frac{5\pi}{4}$

j. $\sin(\arctan\frac{4}{3})$

k. $\cos^{-1}(-\frac{1}{2})$

l. $\operatorname{arccsc}\sqrt{2}$

7. Find the length of the arc s with $\theta = 150°$ and $r = 5$ inches.

8. The displacement of a pendulum is given by $d(t) = -5\cos(0.2\pi t)$, where t is in seconds and $d(t)$ is in centimeters. (Approximate answers to the nearest tenth.)

   a. Find the displacement if $t = \frac{3}{4}$.

   b. Find t if $0 < t < 4$ when $d(t) = -2$.

9. Prove the following identity using an appropriate sum or difference identity.

$$\tan(360° - x) = -\tan x$$

10. Find the exact value of $\cos(-15°)$.

11. Complete the identity. More than one answer is possible.

   a. $\sin^2 x = 1 - \underline{\ ?\ }$

   b. $\sec^2 x = 1 + \underline{\ ?\ }$

   c. $\cos^2\frac{x}{2} = \frac{?}{2}$

   d. $\sin 2x = 2\underline{\ ?\ }$

   e. $2\cos^2 x - 1 = \underline{\ ?\ }$

   f. $\tan(-x) = \underline{\ ?\ } = \frac{1}{?}$

12. If the statement is an identity, prove it. If not, find a counterexample.

   a. $2\left(\cos\frac{x}{2}\right)^2 - \cos x = 1$

   b. $\frac{\sin 5x\cos 3x - \cos 5x\sin 3x}{\cos 2x + 1} = \sin x\sec x$

   c. $\sec^2\theta + \csc^2\theta = \sec^2\theta\csc^2\theta$

   d. $\sin x + \cos x = 1$

13. Solve the equation for x, $0° \le x < 360°$. Approximate to nearest tenth of a degree when necessary.

   a. $4\cos x = 2$

   b. $\tan x = 0$

   c. $\cos^2 2x = 1$

   d. $3\sin^2 x + 5\sin x = 4$

   e. $\sin\frac{x}{2} = \cos x$

   f. $\sin x = \cos x + 1$

   g. $\csc^2 x + \csc x = 2$

14. Solve the equation for x, $0 \le x < 2\pi$. Approximate to the nearest hundredth of a radian when necessary.

   a. $\sec x = -\sqrt{2}$

   b. $\sin 2x = \cos 2x$

   c. $3\sin x - \cos^2 x = 3$

15. On the top of a 25-foot building, Nina determines the angle of depression to a 10-dollar bill on the ground below is 75°. How far is the 10-dollar bill from a point on the ground directly below Nina?

16. In a triangle, $a = 5$, $b = 10$, and $\alpha = 30°$. Solve the triangle(s).

17. By sketching vectors equivalent to **a** and **b**, find the resultant **a** + **b** and state which method you are using.

18. A geometric vector **u** has direction angle $\alpha = 135°$ and magnitude $5\sqrt{2}$.

a. Find the components of **u** and find the $a\mathbf{i} + b\mathbf{j}$ form for **u**.

b. Graph **u** in standard position.

c. Does the vector $\overrightarrow{PQ} = \mathbf{u}$ if $P = (7, -2)$ and $Q = (2, 3)$? Explain your answer.

19. Consider $\mathbf{u} = 3\mathbf{i} - 4\mathbf{j}$, $\mathbf{v} = 4\mathbf{i} + 3\mathbf{j}$, and $\mathbf{w} = \mathbf{i} - \mathbf{j}$. Find the following.

a. $2\mathbf{v} - 4\mathbf{w}$ 　　**b.** $\mathbf{u} \cdot \mathbf{v}$

c. The angle between **u** and **v**.

d. The angle to the nearest degree between **u** and **w**.

e. $|\mathbf{v}|$ 　　**f.** $\text{comp}_{\mathbf{w}}\mathbf{v}$

20. Two forces of 8 pounds and 12 pounds act on a point with an angle of 120° between the forces. Find the magnitude of the resultant force and the angle it makes with the 12-pound force. (Approximate answers to the nearest tenth.)

21. An airplane is headed southeast, bearing 135°, with an air speed of 300 mph and a wind blowing from the northwest, bearing 160° at a speed of 50 mph. Find the ground speed and the drift angle. (Approximate answers to the nearest tenth.)

22. **a.** Graph and label the point $(2, -45°)$ in the polar coordinate system.

b. Find two other sets of polar coordinates for the polar coordinates $(2, -45°)$.

c. Find the rectangular coordinates for $(2, -45°)$

23. **a.** Find the polar form for the rectangular coordinates $(-6, 6\sqrt{3})$.

b. Find the polar form for the rectangular equation $x^2 + y^2 + 6y = 7$.

24. Find the rectangular form for the polar equation $r = 3\cos\theta$.

25. Graph the following equations in polar form.

a. $r = 1 + \sin\theta$ 　　**b.** $r = \sin 2\theta$

26. Find a polar equation that represents the graph.

a.

b.

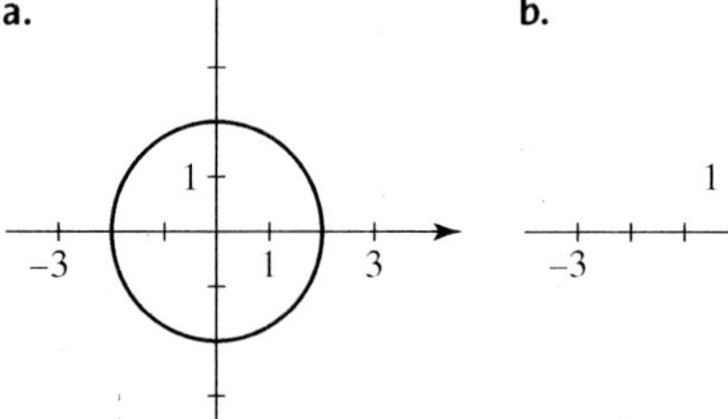

c.

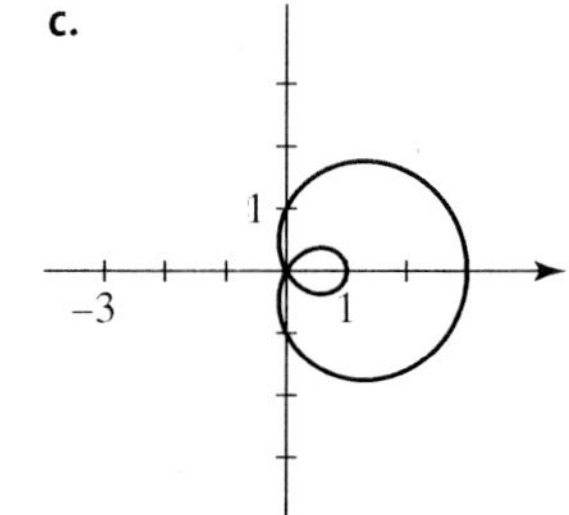

27. Find the rectangular equation for the parametric equations $x = 2\cos t$, $y = 3\sin t$ and graph the curve represented by these parametric equations for $0 \le t \le \pi$. Make sure to indicate the orientation.

28. Find the following.

a. $(1 + i)(1 + \sqrt{-4})$ 　　**b.** $\dfrac{3i}{2 - i}$

c. $|8 - 2i|$ 　　**d.** The trigonometric form of $3 + 3i$.

e. $(3 + 3i)^4$ in $a + bi$ form.

f. The two square roots of $3 + 3i$ in trigonometric form.

29–70. *Are the following statements true or false? If a statement is false, explain why or give an example that shows why it is false.*

29. Angle $\theta = \dfrac{11\pi}{6}$ terminates in QIII.

30. If $\sin t < 0$, $\cos t > 0$, the t is in QII.

31. $\cos^2 x + \sin^2 y = 1$.

32. $\csc x = \dfrac{1}{\cos x}$

33. If $\sin t = \dfrac{1}{2}$, then $\cos t = \dfrac{\sqrt{3}}{2}$.

34. $\cos(-x) = \cos x$

35. $\cos x = \sin y$ if $x + y = 90°$.

36. The period of the tangent function is π.

37. The tangent is undefined at $\dfrac{\pi}{2}$.

38. $\tan^2 x = 1 - \sec^2 x$.

39. $\sin 2x = 2 \sin x$

40. $\dfrac{\sin 2x}{2} = \sin x$

41. $(\sin x)^2 = \sin^2 x$

42. $(\sin x)^{-1} = \sin^{-1} x$

43. $\cos^2 x = 1 - \sin^2 x$

44. $\cos 2x = 1 - 2\sin^2 x$

45. $\cos 6x = 1 - 2\sin^2 3x$

46. If $\sin\theta = y$, $\theta = \arcsin y$ for any θ.

47. The period of $y = 2\sin 3\left(x + \dfrac{\pi}{2}\right)$ is $\dfrac{\pi}{3}$.

48. $\sin(\arccos x) = \sqrt{1 - x^2}$

49. $\tan x = \dfrac{\sin 2x}{\cos 2x}$

50. For any triangle whose sides are a, b, and c and angle α across from side a, $a^2 = b^2 + c^2 - 2bc\cos\alpha$.

51. For a triangle whose angles in degrees are 30–60–90, the relationship between the sides are as indicated on the diagram shown below on the left.

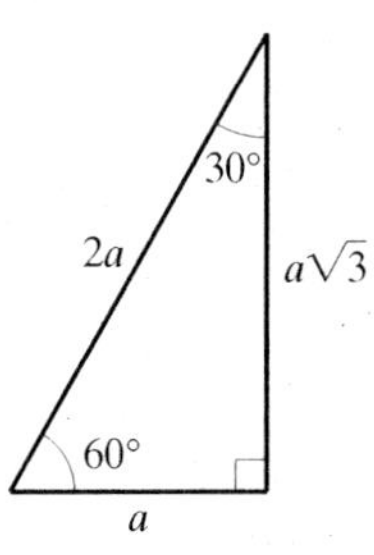

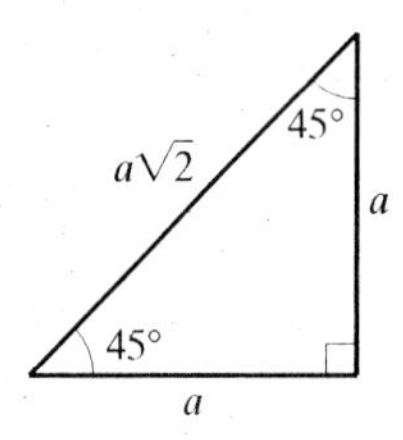

52. For a triangle whose angles in degrees are 45–45–90, the relationships between the sides are as indicated on the diagram shown above on the right.

53. $\cot(-x) = -\cot x$

54. In trigonometry, answers can be expressed in only *one* way.

55. $\sin(x + y) = \sin x + \sin y$

56. $\cos 110° \cos 20° + \sin 110° \sin 20° = 0$

57. $\tan(360° - x) = -\tan x$

58. $\cos^2 x - \sin^2 x = 1$

59. $\cos 4x = 1 - 8\sin^2 x \cos^2 x$

60. $\cot 4x = -1$ has 8 solutions.

61. The parametric equations $x = \cos^2 t$, $y = \sin t$ represent a parabola.

62. Polar coordinates are unique.

63. $r = 7 \cos \theta$ is a cardioid.

64. The sum of two vectors is a scalar.

65. The product of a scalar and a vector is a vector.

66. Vector addition is commutative, or $\mathbf{a} + \mathbf{b} = \mathbf{b} + \mathbf{a}$, for all vectors **a** and **b**.

67. If two vectors are orthogonal then their dot product is zero.

68. The dot product of two vectors produces a scalar.

69. To find the trigonometric form of the complex number $a + bi$, you need the modulus and argument θ.

70. $|1 - 6i| = 1 + 6i$

Appendix A Algebra Review

A.1 Real Number System

In Appendix A we review fundamental concepts of algebra that are important for learning trigonometry. We begin with the sets of numbers that make up the real number system.

Natural Numbers:	$\{1, 2, 3, 4, \dots\}$
Whole Numbers:	$\{0, 1, 2, 3, 4, \dots\}$
Integers:	$\{\dots, -2, -1, 0, 1, 2, 3, \dots\}$
Rational Numbers:	$\left\{\frac{a}{b}, \text{ where } a, b \text{ are integers and } b \neq 0\right\}$

There are also numbers, such as π, that are considered to be not rational; that is, they cannot be written as the ratio of two integers. These numbers are called the **irrational numbers**. The rational and the irrational numbers together make up the set of **real numbers**, denoted by the symbol $\mathbb{R}$. The following diagram demonstrates the relationship between the sets of numbers.

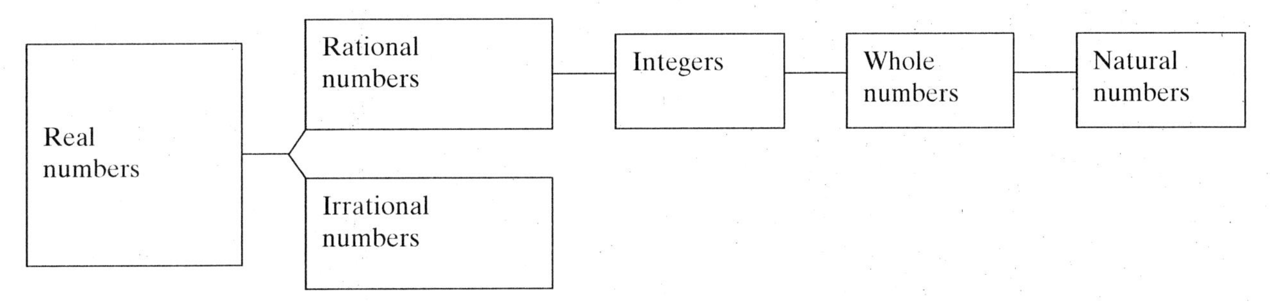

Each real number has a decimal representation. Rational numbers correspond to decimals that either end (terminate) or repeat a sequence of digits (indicated by a bar above the digits that repeat indefinitely), such as the following.

$$\frac{1}{2} = 0.5 \qquad -\frac{4}{3} = -1.33333\ldots = -1.\overline{3}$$

$$\frac{3}{11} = 0.272727\ldots = 0.\overline{27} \qquad \frac{15}{1} = 15.0$$

Irrational numbers correspond to decimals that do not terminate and do not repeat. The following are two examples of irrational numbers.

$$\pi = 3.141592653589793\ldots \qquad \sqrt{2} = 1.414213562373095\ldots$$

When working with irrational numbers, it is advisable to retain exact values (like π or $\sqrt{2}$) during calculations rather than obtain a decimal approximation, unless otherwise indicated. If a decimal approximation is needed, approximate the answer to the specified number of decimal places.

Real Number Line

The real numbers can be geometrically represented by points on a line. To do this, we select on a line an arbitrary point, called the **origin**, to correspond to the number 0. Then we select a point to the right of 0 and call it 1. The distance between 0 and 1 provides a unit of measure in order to locate the positions of the positive numbers to the right of 0 and the negative numbers to the left of 0. The line is called a **real number line**, or a **number line**. The number line geometrically demonstrates the order of the real numbers. We notice that a is less than b, $a < b$, since a is to the left of b. The number that corresponds to a point on the number line is called the **coordinate** of the point.

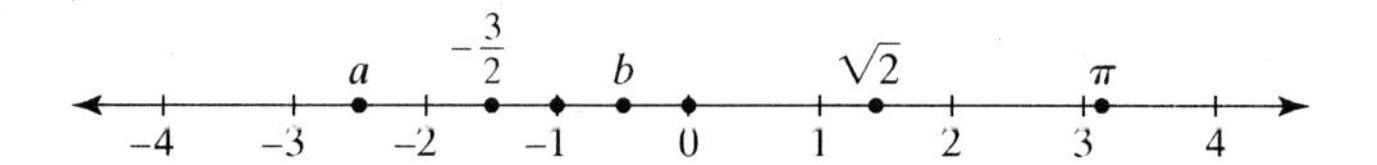

Absolute Value of a Number

The **absolute value** of a real number a, denoted by $|a|$, is the distance on a number line between 0 and the point with coordinate a. Since distance is always nonnegative, the absolute value of a number is always a positive value or zero.

> If a is a real number, then the **absolute value** of a is
>
> $$|a| = \begin{cases} a & \text{if } a \geq 0 \ (a \text{ is positive or zero.}) \\ -a & \text{if } a < 0 \ (a \text{ is negative, so } -a \text{ is positive.}) \end{cases}$$

EXAMPLE 1 Simplify the following expressions.

a. $|7.2|$ **b.** $|-9|$ **c.** $|0|$

SOLUTION

a. $|7.2| = 7.2$ **b.** $|-9| = -(-9) = 9$ **c.** $|0| = 0$ ■

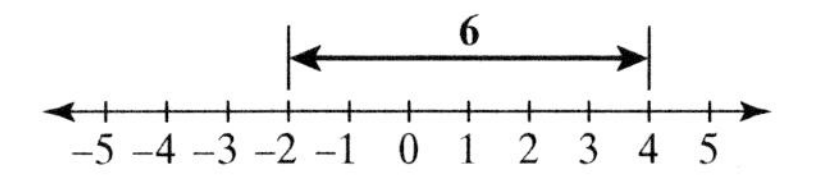

How do we find the distance between -2 and 4 on the real number line? We see the distance is 6. We arrive at this answer by applying the absolute value to the difference of these numbers getting either $|(-2) - 4| = 6$ or $|4 - (-2)| = 6$. We generalize this result in the next statement.

> ### Distance between Points on the Real Number Line
>
> If a and b are real numbers, then the distance between the points a and b on the real number line is
>
> $$d(a, b) = |a - b| = |b - a|.$$

Properties of the Real Numbers

The following properties apply for real numbers a, b, and c.

Commutative Properties for Addition and Multiplication

$$a + b = b + a \qquad ab = ba$$

Associative Properties for Addition and Multiplication

$$(a + b) + c = a + (b + c) \qquad (ab)c = a(bc)$$

Distributive Property of Multiplication over Addition or Subtraction

$$a(b + c) = ab + ac \qquad (b - c)a = ab - ac$$

Zero-Product Property

If $ab = 0$, then either $a = 0$ or $b = 0$, or both equal 0.

CAUTION: The commutative and associative properties *do not apply to subtraction or division.*

$$7 - 5 \neq 5 - 7 \qquad 7/(5/2) \neq (7/5)/2$$

Order of Operations

When problems involve more than one operation, we perform the operations in the following order.

Order of Operations

Calculations are done using the following order of operations.

1. If there are any symbols of inclusion like parentheses, braces, brackets, bars, absolute value signs, radicals, and so on, in the expression, then the inclusive part is evaluated first, starting with the innermost symbol.
2. Any evaluation always proceeds in three ordered steps.
   - **Step 1:** Powers (exponents) and roots are done in any order.
   - **Step 2:** Multiplication and division are done in order from left to right.
   - **Step 3:** Addition and subtraction are done in order from left to right.

EXAMPLE 2 Simplify the following expressions.

a. -9^2 **b.** $\left(-\frac{1}{2}\right)^2 + \left|-\frac{1}{4}\right|$ **c.** $8 \div 4 \cdot 2 + (9 - 6)^2$

d. $\sqrt{3^2 + 4^2}$ **e.** $\dfrac{5^2 - 3^2 - 6^2}{-2 \cdot 3 \cdot 6}$

Solution

a. $-9^2 = -81$

b. $\left(-\frac{1}{2}\right)^2 + \left|-\frac{1}{4}\right| = \frac{1}{4} + \frac{1}{4}$

$= \frac{2}{4} = \frac{1}{2}$

c. $8 \div 4 \cdot 2 + (9 - 6)^2 = 8 \div 4 \cdot 2 + 3^2$

$= 2 \cdot 2 + 9$

$= 4 + 9 = 13$

d. $\sqrt{3^2 + 4^2} = \sqrt{9 + 16}$

$= \sqrt{25} = 5$

e. $\frac{5^2 - 3^2 - 6^2}{-2 \cdot 3 \cdot 6} = \frac{25 - 9 - 36}{-36}$

$= \frac{-20}{-36} = \frac{5}{9}$ ■

A.2 Exponents and Radicals

Exponential Notation

If x is any real number and n is a positive integer, then the nth power of x is

$$x^n = \underbrace{x \cdot x \cdot x \cdot \cdots \cdot x}_{n \text{ factors}},$$

where x is called the base and n is called the exponent.

Since exponents are evaluated first in the order of operations, it is important that we can work with them efficiently and correctly. This can be done by using the rules of exponents.

Rules of Exponents

1. $a^m a^n = a^{m+n}$ $\qquad 8^5 \cdot 8 \cdot 8^9 = 8^{5+1+9} = 8^{15}$

2. $(a^m)^n = a^{mn}$ $\qquad (9^2)^7 = 9^{14}$

3. $(ab)^m = a^m b^m$ $\qquad (-2x)^4 = (-2)^4 x^4$

4. $\left(\frac{a}{b}\right)^m = \frac{a^m}{b^m}, \quad b \neq 0$ $\qquad \left(\frac{3}{5}\right)^3 = \frac{3^3}{5^3}$

5. $a^0 = 1, \quad a \neq 0$ $\qquad 150^0 = 1$

6. $\frac{a^m}{a^n} = a^{m-n}, \quad a \neq 0$ $\qquad \frac{6^{14}}{6^3} = 6^{14-3} = 6^{11}$

7. $a^{-n} = \frac{1}{a^n}, \quad a \neq 0$ $\qquad 12^{-3} = \frac{1}{12^3}$

8. $\left(\frac{a}{b}\right)^{-n} = \left(\frac{b}{a}\right)^n \quad a, b \neq 0$ $\qquad \left(\frac{7}{3}\right)^{-8} = \left(\frac{3}{7}\right)^8$

EXAMPLE 1 Simplify the following expressions.

a. $\left(\frac{1}{2}\right)^{-2}$ b. $\frac{x^7x^{-2}}{x^{-5}}$ c. $\frac{(a^2b^{-3})^2}{(2ab)^4}$ d. $\left(\frac{x}{y}\right)^2\left(\frac{y^3z^2}{x^{-2}}\right)^4$

SOLUTION

a. $\left(\frac{1}{2}\right)^{-2} = \left(\frac{2}{1}\right)^2 = 4$ Rule 8, followed by Rule 4.

b. $\frac{x^7x^{-2}}{x^{-5}} = \frac{x^5}{x^{-5}}$ Rule 1

$= x^{5-(-5)}$ Rule 6

$= x^{10}$

c. $\frac{(a^2b^{-3})^2}{(2ab)^4} = \frac{a^4b^{-6}}{2^4a^4b^4}$ Rule 3, Rule 2

$= \frac{a^{4-4}b^{-6-4}}{2^4}$ Rule 6

$= \frac{a^0b^{-10}}{16}$

$= \frac{1}{16b^{10}}$ Rule 5, Rule7

d. $\left(\frac{x}{y}\right)^2\left(\frac{y^3z^2}{x^{-2}}\right)^4 = \frac{x^2}{y^2}\cdot\frac{y^{12}z^8}{x^{-8}}$ Rule 4, Rule 3, Rule 2

$= \frac{x^{2-(-8)}y^{12-2}z^8}{1}$ Rule 6

$= x^{10}y^{10}z^8$ ■

Radicals

Often we must find a number whose square is a particular value a. If such a number can be found, it is called a **square root of a**. For example,

$$4 \text{ is a square root of 16 because } 4^2 = 16,$$

$$\text{and} \quad -4 \text{ is a square root of 16 because } (-4)^2 = 16.$$

Hence, there are two square roots of 16. However, we use the symbol $\sqrt{\ }$, called a **radical sign**, to indicate **the positive square root**, or simply the **principal square root**. Therefore, $\sqrt{16} = 4$. The number under the radical sign, which is 16 in this case, is called the **radicand**.

Principal Square Root

If $a > 0$, the **principal square root** of a is the positive square root of a, denoted as $\sqrt{a}$.

The following important facts come from this definition.

- $\sqrt{a^2} = |a|$ The principal square root of a positive number is positive.
- Negative numbers do not have square roots in the real number system because the square of every real number is *nonnegative.*
- $\sqrt{0} = 0$

Sometimes we need to find a number whose cube is a particular value a. Although we can have two real square roots of a particular value, we have only one real cube root of a particular value. For example,

$$\sqrt[3]{27} = 3, \qquad \sqrt[3]{-8} = -2, \qquad \sqrt[3]{125x^3} = 5x.$$

Since we may want to find fourth roots, fifth roots, and so on, we make the following definition for an nth root.

nth Root

If n is any positive integer and a and b are real numbers such that $a = b^n$, then the principal nth root of a is b, or

$$\sqrt[n]{a} = b \qquad \text{if } b^n = a.$$

If n is even, then $a \geq 0$ and $b \geq 0$.

EXAMPLE 2 Simplify the radical expressions where defined.

a. $\sqrt[4]{81}$ **b.** $\sqrt[5]{-32}$ **c.** $\sqrt{-9}$

SOLUTION

a. $\sqrt[4]{81} = 3$ because $3^4 = 81$.

b. $\sqrt[5]{-32} = -2$ because $(-2)^5 = -32$.

c. $\sqrt{-9}$ is not defined since there is no real number whose square is negative.

■

Rules for Radicals

Just as there are rules for exponents, there are rules for radicals. Because we frequently use square roots in trigonometry, we also state the rules specific to square roots.

For any positive integer n and real numbers a and b, the following is true.

	General Rule	Square Root
Product Rule for Radicals	$\sqrt[n]{a}\sqrt[n]{b} = \sqrt[n]{ab}$	$\sqrt{a}\sqrt{b} = \sqrt{ab}, \quad a, b \geq 0$
Quotient Rule for Radicals	$\sqrt[n]{\frac{a}{b}} = \frac{\sqrt[n]{a}}{\sqrt[n]{b}}, \quad b \neq 0$	$\sqrt{\frac{a}{b}} = \frac{\sqrt{a}}{\sqrt{b}}, \quad a \geq 0, b > 0$

provided the nth roots are defined.

The following important facts come from this definition for $a \geq 0$ and $b \geq 0$.

- $(\sqrt{a})^2 = \sqrt{a}\sqrt{a} = \sqrt{a^2} = a$ Example: $(\sqrt{5})^2 = 5$
- $(\sqrt{a} + \sqrt{b})(\sqrt{a} - \sqrt{b}) = \sqrt{a}(\sqrt{a} - \sqrt{b}) + \sqrt{b}(\sqrt{a} - \sqrt{b})$

$$= a - \sqrt{ab} + \sqrt{ab} - b$$
$$= a - b$$

Example: $(\sqrt{6} + \sqrt{2})(\sqrt{6} - \sqrt{2}) = 6 - 2 = 4$

Simplified Form of a Radical

We use the rules for radicals to simplify radical expressions. A radical is in simplified or proper form when the following three conditions are met.

1. All nth power factors of the radicand of $\sqrt[n]{}$ are removed. In doing so, we obtain the smallest possible number under the radical sign.
2. The radicand contains no fractions.
3. No denominator contains a radical.

EXAMPLE 3 Simplify the radical expressions.

a. $\sqrt{8}$ **b.** $\sqrt{18} + \sqrt{50}$ **c.** $\sqrt[3]{\dfrac{54}{8}}$

d. $\dfrac{1}{\sqrt{2}}$ **e.** $\dfrac{2}{\sqrt{3}}$ **f.** $\dfrac{2}{\sqrt{2} + 1}$

SOLUTION

a.
$$\begin{aligned} \sqrt{8} &= \sqrt{4 \cdot 2} && \text{Factor out the largest square.} \\ &= \sqrt{4}\sqrt{2} && \text{Product Rule} \\ &= 2\sqrt{2} \end{aligned}$$

b.
$$\begin{aligned} \sqrt{18} + \sqrt{50} &= \sqrt{9 \cdot 2} + \sqrt{25 \cdot 2} && \text{Factor out the largest squares.} \\ &= \sqrt{9}\sqrt{2} + \sqrt{25}\sqrt{2} && \text{Product Rule} \\ &= 3\sqrt{2} + 5\sqrt{2} && \text{Simplify.} \\ &= 8\sqrt{2} && \text{Distributive Property} \end{aligned}$$

c.
$$\begin{aligned} \sqrt[3]{\frac{54}{8}} &= \frac{\sqrt[3]{54}}{\sqrt[3]{8}} && \text{Quotient Rule} \\ &= \frac{\sqrt[3]{27 \cdot 2}}{2} && \text{Factor out the largest cube.} \\ &= \frac{3\sqrt[3]{2}}{2} && \text{Simplify.} \end{aligned}$$

d. $\dfrac{1}{\sqrt{2}} = \dfrac{1}{\sqrt{2}} \cdot \dfrac{\sqrt{2}}{\sqrt{2}}$ Rationalize the denominator by multiplying by 1 in the form $\dfrac{\sqrt{2}}{\sqrt{2}}$.

$= \dfrac{\sqrt{2}}{2}$

e. $\dfrac{2}{\sqrt{3}} = \dfrac{2}{\sqrt{3}}\dfrac{\sqrt{3}}{\sqrt{3}}$ Rationalize the denominator by multiplying by 1 in the form $\dfrac{\sqrt{3}}{\sqrt{3}}$.

$= \dfrac{2\sqrt{3}}{3}$

f. $\dfrac{2}{\sqrt{2}+1} = \left(\dfrac{2}{\sqrt{2}+1}\right)\left(\dfrac{\sqrt{2}-1}{\sqrt{2}-1}\right)$ Rationalize the denominator by multiplying by $1 = \dfrac{\sqrt{2}-1}{\sqrt{2}-1}$.

$= \dfrac{2(\sqrt{2}-1)}{2-1}$

$= 2\sqrt{2} - 2$ ■

Rational Exponents

The radical sign $\sqrt[n]{}$ has an equivalent exponent form. To find the connection, we use the rules of exponents and obtain the following.

$$\sqrt{25} = 5^1 = 5^{2/2} = 5^{2(1/2)} = (5^2)^{1/2} = (25)^{1/2}$$

This calculation suggests that the fractional exponent $\frac{1}{2}$ is equivalent to the square root. We generalize this result with the following definition:

Rational Exponents

If m and n are positive integers, then

$$\sqrt[n]{a} = a^{1/n},$$

and $$(\sqrt[n]{a})^m = (a^{1/n})^m = a^{m/n},$$

provided that $a^{1/n}$ is defined.

EXAMPLE 4 Simplify the expressions.

a. $100^{1/2}$ **b.** $-16^{1/2}$ **c.** $(-125)^{1/3}$

d. $27^{-2/3}$ **e.** $x^{2/3}x^{4/3}$ **f.** $\left(\dfrac{-32y^5}{x^{20}}\right)^{1/5}$

SOLUTION

a. $100^{1/2} = \sqrt{100} = 10$

b. $-16^{1/2} = -\sqrt{16} = -4$

c. $(-125)^{1/3} = \sqrt[3]{-125} = -5$

d. $27^{-2/3} = \dfrac{1}{27^{2/3}} = \dfrac{1}{(27^{1/3})^2} = \dfrac{1}{(\sqrt[3]{27})^2} = \dfrac{1}{3^2} = \dfrac{1}{9}$

e. $x^{2/3}x^{4/3} = x^{2/3+4/3} = x^{6/3} = x^2$

f. $\left(\dfrac{-32y^5}{x^{20}}\right)^{1/5} = \dfrac{(-32)^{1/5}(y^5)^{1/5}}{(x^{20})^{1/5}} = \dfrac{-2y}{x^4}$ ■

A.3 Algebraic Expressions

Expressions

An algebraic expression in the real numbers is any combination of letters (called **variables**), such as x, y, or a, numbers (called **constants**), such as 2, 0.5, or -4, and the operations addition, subtraction, multiplication, division, and raising to a power. The **domain** of a variable is the set of numbers that can replace the variable without making the expression undefined. So numbers that produce a zero in a denominator or a negative number under an even-root radical are not permitted to be in the domain of the variable. The following are two examples of expressions with restricted domains:

$\dfrac{x+2}{x-9}$ The domain is all real numbers x *except* $x = 9$, that is, $\{x \mid x \neq 9\}$.

$\sqrt{x}$ The domain is all nonnegative real numbers x, that is, $\{x \mid x \geq 0\}$.

Evaluating Expressions

If we replace the variables of a particular expression with specified values, we are **evaluating the expression**.

EXAMPLE 1 Evaluate the expression if $a = \frac{1}{2}$, $b = 3$, and $c = 5$.

a. $\sqrt{1-a^2}$ **b.** $\dfrac{b-c}{1+bc}$ **c.** $|3b - 10ac|$

SOLUTION

a. $\sqrt{1-a^2} = \sqrt{1-\left(\dfrac{1}{2}\right)^2}$

$= \sqrt{1 \cdot \dfrac{4}{4} - \dfrac{1}{4}}$

$= \sqrt{\dfrac{3}{4}} = \dfrac{\sqrt{3}}{2}$

b. $\dfrac{b-c}{1+bc} = \dfrac{(3)-(5)}{1+(3)(5)}$

$= \dfrac{-2}{16} = -\dfrac{1}{8}$

c. $|3b - 10ac| = \left|3(3) - 10\left(\dfrac{1}{2}\right)(5)\right|$

$= |9 - 25|$

$= |-16| = 16$ ■

Simplifying Algebraic Expressions

When we add and subtract expressions, we combine *like* terms. An **algebraic term** is a constant, a variable, or the product or quotient of constants and variables raised to powers. Some examples of terms are:

$$-5y \qquad 6z^2p \qquad \frac{x}{9} \qquad \frac{19a^2}{b}$$

The **numerical coefficients** of these terms are $-5, 6, \frac{1}{9}$, and 19, respectively. In algebraic expressions, terms are separated by addition and subtraction signs. For example, $11y^2 - 10y + 5$ has three terms, and $4x^2 - y^2$ has two terms. Terms are called **like terms** if they have the same variables with the same exponents.

Like terms	Unlike terms
$5a^2, \; -17a^2$	$5a^2, \quad 17a$
$9xyz, \; -\frac{1}{2}xyz$	$9xyz, \; -\frac{1}{2}yz$

We combine like terms by using the distributive property, which suggests that to combine like terms we add or subtract their numerical coefficients.

EXAMPLE 2 Simplify by combining like terms.

a. $5a^2 - 17a^2$ **b.** $6x + 8y + 2x$ **c.** $-2a^2b + 5 + 9a^2b - 10$

SOLUTION

a. $5a^2 - 17a^2 = (5 - 17)a^2$
$= -12a^2$

b. $6x + 8y + 2x = 6x + 2x + 8y$
$= (6 + 2)x + 8y$
$= 8x + 8y$

c. $-2a^2b + 5 + 9a^2b - 10 = (-2 + 9)a^2b + (5 - 10) = 7a^2b - 5$ ■

Polynomials

A **polynomial** is the sum of one or more algebraic terms whose variables have whole-number exponents. A polynomial with one term is called a **monomial**, with two terms is called a **binomial**, and with three terms is called a **trinomial**.

Monomials	Binomials	Trinomials
$7x^2$	$6t^2 + 2$	$a^2 + 2a - 3$
v^3z	$-5m^6n^7 - c^2$	$-12x^5yz + 24x^3p - 2n$

The **degree of a term** of a polynomial is the sum of the exponents on the variables. And the **degree of a polynomial** is the highest degree of any of its nonzero terms. For example,

$$7x^2 \text{ is of degree 2,}$$

and

$$-5m^6n^7 - c^2 \text{ is of degree 13.}$$

EXAMPLE 3 Simplify the expression $(x^2 - 3x + 6) - (x - 7x^2)$. Then, for the resulting polynomial, state the degree and whether it is a monomial, binomial, or trinomial.

SOLUTION

$$\begin{aligned}(x^2 - 3x + 6) - (x - 7x^2) &= x^2 - 3x + 6 - x + 7x^2 && \text{Distributive property}\\ &= (x^2 + 7x^2) + (-3x - x) + 6 && \text{Group like terms.}\\ &= 8x^2 - 4x + 6 && \text{Combine like terms.}\end{aligned}$$

The trinomial is of degree 2. ■

Product of Algebraic Expressions

We find the product of expressions that involve more than one term by using the distributive property.

Multiplying a Monomial and a Polynomial

EXAMPLE 4 Use the distributive property to find the product.

a. $5y^2(9y^5 + 2)$ **b.** $-3x^5(-x^3 + 4x^2 + 3x - 10)$

SOLUTION

a. $$\begin{aligned}5y^2(9y^5 + 2) &= (5y^2)(9y^5) + (5y^2)(2)\\ &= 45y^7 + 10y^2\end{aligned}$$

b. $$\begin{aligned}&-3x^5(-x^3 + 4x^2 + 3x - 10)\\ &= -3x^5(-x^3) - 3x^5(4x^2) - 3x^5(3x) - 3x^5(-10)\\ &= 3x^8 - 12x^7 - 9x^6 + 30x^5\end{aligned}$$ ■

Multiplying Binomials by the FOIL Method

We use the distributive property to multiply two binomials.

$$(a + b)(c + d) = (a + b)c + (a + b)d = ac + bc + ad + bd$$

A shortcut to this process is the **FOIL method**, which comes from the abbreviation for *First, Outer, Inner, Last.* To use this method, we identify and multiply certain terms of the binomial as follows.

$$(a + b)(c + d) = ac + ad + bc + bd$$

First: ac (F), Outer: ad (O), Inner: bc (I), Last: bd (L)

The result is the sum of the products of the first, outer, inner, and last terms.

EXAMPLE 5 Find the product of the binomials $3x - 7$ and $4x + 2$.

SOLUTION Use the FOIL method to find the product of the binomials.

$$\begin{aligned}(3x - 7)(4x + 2) &= \overset{\text{F}}{12x^2} + \overset{\text{O}}{6x} - \overset{\text{I}}{28x} - \overset{\text{L}}{14} \\ &= 12x^2 - 22x - 14 \qquad \text{Combine like terms.}\end{aligned}$$

Special Products

Using the FOIL method, you can verify the following special products.

$$(a + b)^2 = (a + b)(a + b) = a^2 + 2ab + b^2 \qquad (1)$$
$$(a - b)^2 = (a - b)(a - b) = a^2 - 2ab + b^2 \qquad (2)$$
$$(a + b)(a - b) = a^2 - b^2 \qquad (3)$$

The trinomials produced in Equations (1) and (2) are called **perfect squares**. Equation (3) produces a binomial called the **difference of two squares**.

EXAMPLE 6 Find the product.

a. $(3x + 4)^2$ **b.** $(5y - 2)(5y + 2)$

SOLUTION

a. $(3x + 4)^2 = (3x)^2 + 2(3x)(4) + 4^2 = 9x^2 + 24x + 16$ Perfect square

b. $(5y - 2)(5y + 2) = (5y)^2 - 2^2 = 25y^2 - 4$ Difference of two squares

Factoring Polynomials

To factor a polynomial means to rewrite the polynomial as a product, which involves reversing the distributive property. The *first step* in factoring is to *look for a greatest common factor (GCF) among all the terms.*

EXAMPLE 7 Factor the polynomial by taking out the greatest common factor.

a. $36t^2 - 72$ **b.** $24x^3y - 6x^2y^2 + 12xy$

SOLUTION

a. $36t^2 - 72 = 36(t^2 - 2)$

b. $24x^3y - 6x^2y^2 + 12xy = 6xy(4x^2 - xy + 2)$

To factor a binomial that is the difference of two squares, we reverse the special product in Equation (3).

EXAMPLE 8 Factor the polynomial.

a. $100x^2 - 49$ **b.** $2 - 8y^2$

SOLUTION

a. $100x^2 - 49 = (10x)^2 - 7^2$
$= (10x - 7)(10x + 7)$

b. $2 - 8y^2 = 2(1 - 4y^2)$ Take out the GCF 2.
$= 2(1^2 - (2y)^2)$
$= 2(1 - 2y)(1 + 2y)$ ■

We can factor trinomials that are **quadratic** in the form $x^2 + bx + c$ if we can find two integers m and n whose product is c and whose sum is b. Then we get

$$x^2 + \underset{\substack{\uparrow \\ m+n}}{b}x + \underset{\substack{\uparrow \\ mn}}{c} = (x + m)(x + n).$$

EXAMPLE 9 Factor the trinomial $x^2 + 11x - 12$.

SOLUTION We are looking for two integers whose product is -12. The possibilities are:

$$12(-1) \quad 6(-2) \quad 4(-3) \quad -12(1) \quad -6(2) \quad -4(3)$$

Since the sum of the integers must be 11, we select -1 and 12.

Therefore, the factorization is

$$x^2 + 11x - 12 = (x - 1)(x + 12).$$ ■

To factor a trinomial in quadratic form $ax^2 + bx + c$, $a \neq 1$, as the product of two binomials, that is,

$$ax^2 + bx + c = (px + q)(rx + s)$$

(factors of a: p and r; factors of c: q and s)

we use the method of **trial and error**. This method tries possible integer factors of a as the coefficients of the first terms of the binomials in combination with possible integer factors of c as the second terms of the binomials until the product of the binomials produces the given trinomial.

EXAMPLE 10 Factor each trinomial.

a. $20x^2 - 11x - 4$ **b.** $9x^2 + 6x + 1$

SOLUTION

a. The factors of 20 are $20 \cdot 1$, $10 \cdot 2$, or $5 \cdot 4$, and the factors of -4 are $-4 \cdot 1$, $4(-1)$, or $2(-2)$. After trying these possibilities, we get the factorization

$$20x^2 - 11x - 4 = (5x - 4)(4x + 1).$$

b. $9x^2 + 6x + 1 = (3x + 1)^2$ Perfect square ■

Rational Expressions

The quotient of two polynomials is called a **rational expression**.

EXAMPLE 11 Simplify the rational expression $\dfrac{2x^2 + 4x + 2}{x^2 + 4x + 3}$.

SOLUTION We factor the polynomials in the numerator and denominator and divide out the common factors.

$$\frac{2x^2 + 4x + 2}{x^2 + 4x + 3} = \frac{2(x^2 + 2x + 1)}{(x + 3)(x + 1)} = \frac{2(x + 1)\cancel{(x + 1)}}{(x + 3)\cancel{(x + 1)}} = \frac{2x + 2}{x + 3}$$

A.4 Solving Equations

An **equation in one variable** is a statement in which two expressions, with at least one containing the variable, are equal. The expressions are called the **sides** of the equation. For example, $3x + 4 = 10$ is an equation in the single variable x.

Any value of the variable that makes the equation true is called a **solution** of the equation. To **solve an equation** means to use a method or procedure to find all the solutions of the equation. Many equations have more than one solution, many have no solution, and many have infinitely many solutions. For example:

$3x + 4 = 10$ is true when $x = 2$ and false for all other values of x (one solution).

$x^2 - 7 = 2$ is true for $x = 3$ and $x = -3$ and false for all other values of x (two solutions).

$x + 6 = x + 7$ is never true (no solution).

$2x + 1 = x + x + 1$ is true for all values of x (infinitely many solutions).

$\dfrac{1}{x + 1} = \dfrac{1}{x + 1}$ is true for all values except $x = -1$, since the equation is not defined at this value.

The last two equations are examples of an **identity**, a statement that is true for all values of the variable for which both sides are defined.

Finding the solutions of an equation uses procedures that result in equations having the same solutions, or **equivalent equations**. The following procedures that produce equivalent equations are called the **properties of equality**.

Properties of Equality

If A, B, and C are algebraic expressions, then the following are equivalent to $A = B$.

Addition Property of Equality

$$A + C = B + C$$

Subtraction Property of Equality

$$A - C = B - C$$

Multiplication Property of Equality

$$AC = BC \quad (C \neq 0)$$

Division Property of Equality

$$\frac{A}{C} = \frac{B}{C} \quad (C \neq 0)$$

First-Degree (Linear) Equations

The simplest type of equation, $ax + b = 0$, is a **first-degree equation**, or **linear equation**, in which the variable is of the first degree. To solve a first-degree equation, we proceed with the following steps.

Step 1: Simplify each side of the equation.

Step 2: If the variable is on both sides of the equation, use the addition or subtraction property of equality to get the variable to one side, and the constants to the other side.

Step 3: Use the multiplication or division property of equality to obtain a coefficient of one on the variable, hence obtaining the form

$$x = \textit{some number} \text{ (the solution to the equation).}$$

Step 4: Check the solution by replacing the variable in the *original equation.*

EXAMPLE 1 Solve the linear equation.

a. $7(x + 3) = 5(x - 3)$ **b.** $5^2 = 3^2 + 6^2 - 2(3)(6)x$

SOLUTION

a. $7(x + 3) = 5(x - 3)$

$7x + 21 = 5x - 15$ Distribute.

$2x + 21 = -15$ Subtract $5x$ from each side.

$2x = -36$ Subtract 21 from each side.

$x = -18$ Divide both sides by 2.

Check the answer.

$$7(-18 + 3) \stackrel{?}{=} 5(-18 - 3)$$

$$7(-15) \stackrel{?}{=} 5(-21)$$

$$-105 = -105 \checkmark$$

b. $5^2 = 3^2 + 6^2 - 2(3)(6)x$

$25 = 45 - 36x$ Simplify each side.

$25 - 45 = -36x$ Subtract 45 from each side.

$-20 = -36x$

$\frac{5}{9} = x$ Divide both sides by -36.

Check the answer.

$$5^2 \stackrel{?}{=} 3^2 + 6^2 - 2(3)6(\tfrac{5}{9})$$

$$25 \stackrel{?}{=} 9 + 36 - 36(\tfrac{5}{9})$$

$$25 \stackrel{?}{=} 45 - 20$$

$$25 = 25 \checkmark$$

■

Second-degree (Quadratic) Equations

A **second-degree**, or **quadratic equation**, is an equation that can be written in the **standard form**

$$ax^2 + bx + c = 0,$$

where a, b, and c are real numbers with $a \neq 0$.

Methods to Solve a Quadratic Equation

We now look at three methods for solving a quadratic equation. Answers should be checked in the original equation, which we leave for you to do.

Factor Method

If a quadratic equation is written in standard form and the nonzero side can be factored, we use the zero-product property of the real numbers to solve the equation. This first method is known as the **factor method**.

EXAMPLE 2 Solve the quadratic equation $x^2 - 6x = 16$.

SOLUTION

$x^2 - 6x = 16$	
$x^2 - 6x - 16 = 0$	Standard form
$(x - 8)(x + 2) = 0$	Factor.
$x - 8 = 0$ or $x + 2 = 0$	Zero-product property (Set each factor equal to zero.)
$x = 8$ or $x = -2$	Solve each linear equation.

■

Square Root Method

If a quadratic equation has no middle term $(b = 0)$, it is in the form $ax^2 + c = 0$. In this case, we can use the **square root method** to isolate the square term and then apply the following property.

Square Root Property for Quadratic Equations

If k is a positive number and $x^2 = k$, then the solutions to the equation are

$$x = \sqrt{k} \quad \text{or} \quad x = -\sqrt{k}.$$

We can also write the solutions as $x = \pm\sqrt{k}$.

EXAMPLE 3 Use the square root method to solve the quadratic equation $4x^2 - 3 = 0$.

SOLUTION

$4x^2 - 3 = 0$	
$4x^2 = 3$	To isolate the square term, add 3 to both sides.
$x^2 = \frac{3}{4}$	Then divide both sides by 4.
$x = \pm\frac{\sqrt{3}}{2}$	Square root property

■

Quadratic Formula

To solve any quadratic equation, we can use the **quadratic formula**.

Quadratic Formula

The solutions of the quadratic equation $ax^2 + bx + c = 0$, $a \neq 0$, are

$$x = \frac{-b + \sqrt{b^2 - 4ac}}{2a} \quad \text{and} \quad x = \frac{-b - \sqrt{b^2 - 4ac}}{2a},$$

or

$$x = \frac{-b \pm \sqrt{b^2 - 4ac}}{2a}.$$

Whether you use the factor method or the quadratic formula to solve a quadratic equation, the equation first must be set equal to 0 (standard form).

EXAMPLE 4 Solve the quadratic equation $5x^2 - 3x - 1 = 0$.

SOLUTION Since the equation is already in standard form, we identify $a = 5$, $b = -3$, and $c = -1$, and substitute these values into the formula.

$$\begin{aligned} x &= \frac{-b \pm \sqrt{b^2 - 4ac}}{2a} \\ &= \frac{-(-3) \pm \sqrt{(-3)^2 - 4(5)(-1)}}{2(5)} \\ &= \frac{3 \pm \sqrt{9 + 20}}{10} \\ &= \frac{3 \pm \sqrt{29}}{10} \end{aligned}$$

Approximating these answers to four decimal places using a calculator, we get

$$x = \frac{3 + \sqrt{29}}{10} \approx 0.8385 \quad \text{and} \quad x = \frac{3 - \sqrt{29}}{10} \approx -0.2385.$$

Equations that have quadratic form can be solved using the same methods as those used to solve quadratic equations.

EXAMPLE 5 Solve the equation $x^4 - 5x^2 + 4 = 0$.

SOLUTION First we show that the equation has quadratic form.

$$\begin{aligned} x^4 - 5x^2 + 4 &= 0 & \\ (x^2)^2 - 5(x^2) + 4 &= 0 & \text{Quadratic form} \\ y^2 - 5y + 4 &= 0 & \text{Let } y = x^2. \\ (y - 4)(y - 1) &= 0 & \text{Factor.} \end{aligned}$$

(continued)

$$y - 4 = 0 \quad \text{or} \quad y - 1 = 0 \qquad \text{Zero-product property}$$
$$y = 4 \qquad y = 1$$
$$x^2 = 4 \qquad x^2 = 1 \qquad \text{Let } x^2 = y.$$
$$x = \pm 2 \qquad x = \pm 1 \qquad \text{Square root property}$$

Rational and Radical Equations

Equations that involve rational expressions are called **rational equations**, and equations that involve radical expressions are called **radical equations**. Solving rational and radical equations can produce solutions that do not satisfy the original equation, which are called **extraneous solutions**.

Solving Rational Equations

- Multiply each side of the equation by the least common denominator (LCD) to eliminate the fractions. Then proceed with the method required by the resulting equation.

Solving Radical Equations

- Isolate the radical and raise each side of the equation to the power necessary to eliminate the radical. (If more than one radical was in the original equation, you may have to repeat this step until all radicals are eliminated). Then proceed with the method required by the resulting equation.

As a result of either

1. multiplying each side of the equation by a *variable expression,* or
2. squaring each side of the equation (or raising each side to the same power)

we must check for extraneous solutions.

EXAMPLE 6 Solve the equation.

a. $\dfrac{x+1}{2(x-1)} = \dfrac{2}{(x-1)(x+1)}$ **b.** $\sqrt{2x-3} = x - 3$

SOLUTION

a. To solve the given rational equation, we begin by multiplying each side by the least common denominator.

$$\frac{x+1}{2(x-1)} = \frac{2}{(x-1)(x+1)}$$

$$\cancel{2}(x+1)\cancel{(x-1)}\left(\frac{x+1}{\cancel{2}\cancel{(x-1)}}\right) = \left(\frac{2}{\cancel{(x-1)}\cancel{(x+1)}}\right)2\cancel{(x+1)}\cancel{(x-1)} \qquad \text{Multiply by LCD } 2(x+1)(x-1) \text{ and simplify.}$$

$$(x+1)(x+1)=4$$

$$x^2+2x-3=0 \quad \text{Simplify and put in standard form.}$$

$$(x+3)(x-1)=0 \quad \text{Factor.}$$

$$x+3=0 \quad \text{or} \quad x-1=0 \quad \text{Zero-product property}$$

$$x=-3 \qquad x=1$$

Because we multiplied by a variable expression, we must check our solutions. We see that the solution $x=1$ makes the original equation undefined since the denominator becomes zero, so $x=1$ must be discarded. Substituting x with -3 makes the statement true, so the only solution to the equation is

$$x=-3.$$

b. Since the radical is already isolated we begin solving this radical equation by squaring both sides.

$$\sqrt{2x-3}=x-3$$

$$(\sqrt{2x-3})^2=(x-3)^2 \quad \text{Square both sides.}$$

$$2x-3=x^2-6x+9 \quad \text{Simplify.}$$

$$0=x^2-8x+12 \quad \text{Standard form}$$

$$0=(x-6)(x-2) \quad \text{Factor.}$$

$$x-6=0 \quad \text{or} \quad x-2=0 \quad \text{Zero-product property}$$

$$x=6 \qquad x=2$$

Check $x=6$:

$$\sqrt{2(6)-3} \stackrel{?}{=} (6)-3$$

✓ $\sqrt{9}=3$

Check $x=2$:

$$\sqrt{2(2)-3} \stackrel{?}{=} (2)-3$$

✗ $\sqrt{1} \neq -1$

The only solution is $x=6$. ■

A.5 The Coordinate Plane

In a plane, we draw two perpendicular real number lines, one horizontal and one vertical, that intersect at 0 on each line. The horizontal line with the positive values to the right is called the ***x*-axis**; the vertical line with positive values upward is called the ***y*-axis**. The point where the lines intersect is called the **origin**, and the lines divide the plane into four regions called **quadrants**, denoted by QI, QII, QIII, and QIV. Points can be identified with ordered pairs of real numbers to form the **Cartesian plane**, or the **rectangular coordinate plane**, by assigning each ordered pair of numbers (a,b) the point P that is a units along the x-axis followed by b units along the y-axis. The first value a is called the ***x*-coordinate**, and the second value b is called the ***y*-coordinate**.

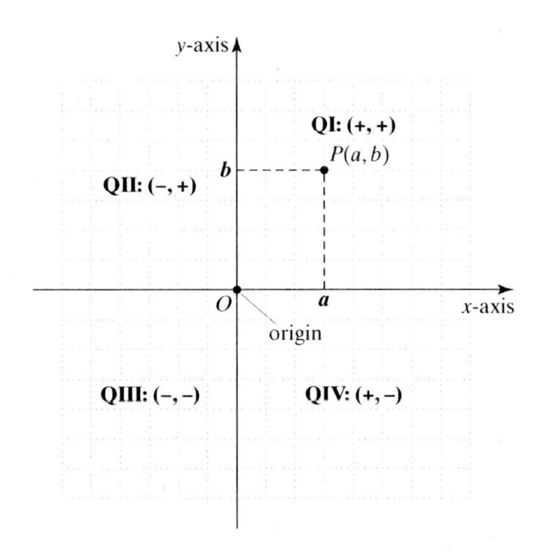

EXAMPLE 1 Plot the following points and indicate which quadrant they are in.

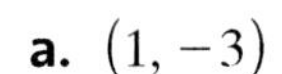

a. $(1, -3)$ **b.** $(-2, -1)$ **c.** $(6, 0)$

d. $(1, 4)$ **e.** $(0, -4)$ **f.** $(-4, 5)$

SOLUTION

a. $(1, -3)$ is in QIV. **b.** $(-2, -1)$ is in QIII

c. $(6, 0)$ is not in a quadrant **d.** $(1, 4)$ is in QI

e. $(0, -4)$ is not in a quadrant. **f.** $(-4, 5)$ is in QII ■

To find the distance between two points in the plane, we use the following formula.

Distance Formula

The distance between two points $P(x_1, y_1)$ and $Q(x_2, y_2)$ in the plane is given by the formula

$$d(P, Q) = \sqrt{(x_2 - x_1)^2 + (y_2 - y_1)^2}.$$

EXAMPLE 2 Find the distance between the points $P(-2, 3)$, and $Q(4, -5)$.

SOLUTION Let $P(-2, 3) = (x_1, y_1,)$ and $Q(4, -5) = (x_2, y_2)$.

$$\begin{aligned} d(P, Q) &= \sqrt{(x_2 - x_1)^2 + (y_2 - y_1)^2} \\ &= \sqrt{(4 - (-2))^2 + (-5 - 3)^2} \\ &= \sqrt{6^2 + (-8)^2} \\ &= \sqrt{36 + 64} \\ &= \sqrt{100} = 10 \end{aligned}$$

■

The Graph of an Equation

If we have an equation in at most two variables, such as $x^2 + 6 = y$, the point (x, y) is called a solution if the equation is true when the coordinates of the point are substituted into the equation. The point $(2, 10)$ is a solution to $x^2 + 6 = y$ since $2^2 + 6 = 10$. The **graph** of an equation is the set of all points (x, y) in the coordinate plane that are solutions to the equation. There are times when we can identify the graph from the form of the equation.

Graphs of First-degree (Linear) Equations

The graph of every linear equation $Ax + By = C$ $(A, B$ not both zero) is a line.

EXAMPLE 3 Graph $6x - 3y = 12$.

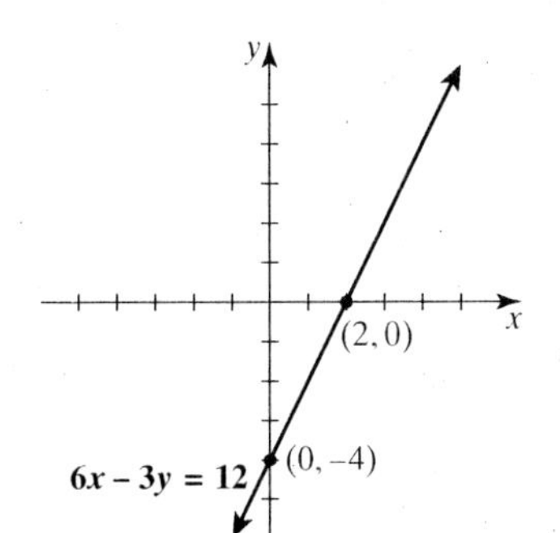

SOLUTION The graph is a line since the equation is linear. We make a table of values to find two points necessary to graph the line. We use the point where the line crosses the x-axis, called the ***x*-intercept**, by letting $y = 0$, and the point where the line crosses the y-axis, called the ***y*-intercept**, by letting $x = 0$.

x	y
2	0
0	-4

$6x - 3(0) = 12 \rightarrow x = 2$

$6(0) - 3y = 12 \rightarrow y = -4$

The graphs of nonlinear equations can produce curves. Two such curves are the *parabola* and the *circle*. ■

Graphs of Second-degree (Quadratic) Equations

The graph of the quadratic equation $y = ax^2 + bx + c$ $(a \neq 0)$ is a parabola.

EXAMPLE 4 Graph $y = 2x^2 + 1$.

SOLUTION To graph the parabola, we use a table to find ordered pairs and then plot the points.

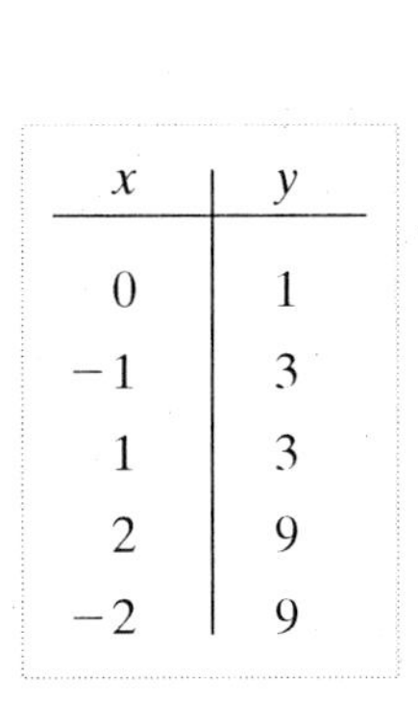

x	y
0	1
-1	3
1	3
2	9
-2	9

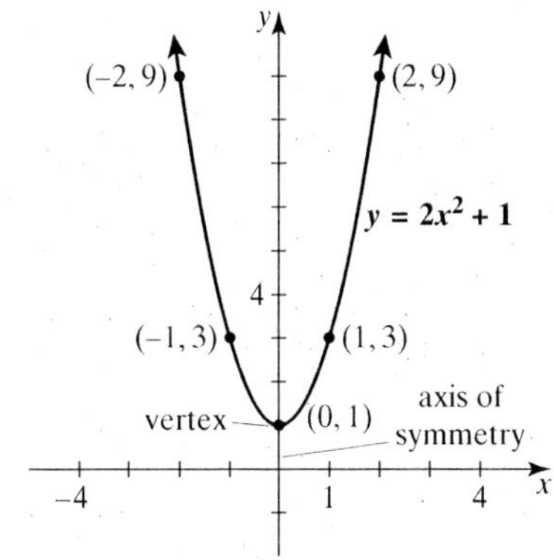

(continued)

The lowest point of the parabola is called the **vertex**, which in this example is $(0, 1)$, The line that passes through the vertex and divides the parabola into two equal parts is called the **axis of symmetry**. In this case, the axis of symmetry is the y-axis, or $x = 0$. ■

Circles

A **circle** is the set of all points in the plane that are a fixed distance r from a fixed point called the **center**. The fixed distance r is called the **radius**. The equation of a circle with center (h, k) and radius r is

$$(x - h)^2 + (y - k)^2 = r^2.$$

EXAMPLE 5 Graph:

a. $x^2 + y^2 = 25$ **b.** $(x - 2)^2 + y^2 = 9$

SOLUTION

a. The graph of $x^2 + y^2 = 25$ is a circle whose center is $(0, 0)$ and whose radius is $r = 5$.

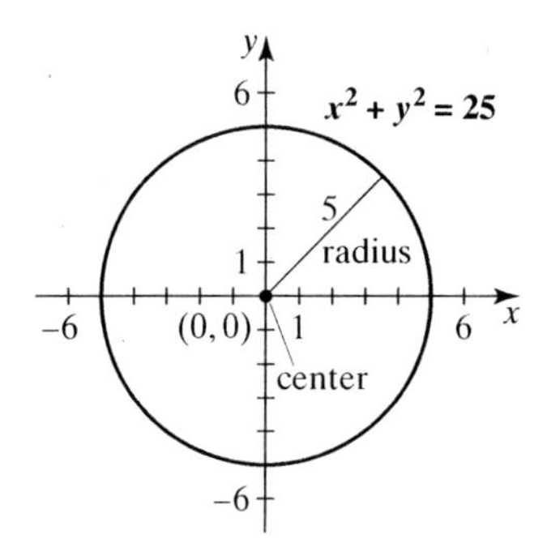

b. The graph of $(x - 2)^2 + y^2 = 9$ is a circle with center $(2, 0)$ and $r = 3$. ■

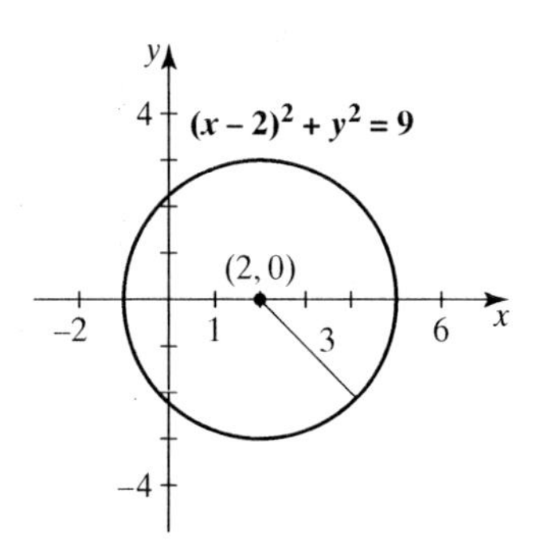

A.6 Relations and Functions

A **relation** is a set of ordered pairs $\{(x, y)\}$, with x as the first coordinate and y as the second coordinate. Coordinates can also be referred to as **components**. If the set of ordered pairs are such that each first component x is associated with *exactly* *one* second component y, we call the relationship a **function** and say that y is a function of x. The set consisting of all the first components is called the **domain**, and the **range** is the set consisting of all the second components of the ordered pairs.

Functional Notation

If y is a function of x, we can use the notation $f(x)$, rather than y, for the range value. The notation $f(x)$ is read "f of x," or "the value of f at x." So both y and $f(x)$ are symbols for the second component when the first component is x, that is, $y = f(x)$.

When using functional notation, x can be interpreted as a "place-holder." For example, the function $f(x) = 5x^2 - x + 1$ can be interpreted as

$$f(\blacksquare) = 5 \cdot \blacksquare^2 - \blacksquare + 1,$$

where $\blacksquare$ holds the place for any value or expression. To evaluate the function at a particular value (or expression), the value is placed in each position of the placeholder $\blacksquare$. It is recommended that you use parentheses around the value or expression that you are inserting into the placeholder position.

EXAMPLE 1 If $f(x) = 5x^2 - x + 1$, find:

a. $f(2)$ **b.** $f(-1)$ **c.** $f(a + b)$

SOLUTION

a. $f(2) = 5 \cdot (2)^2 - (2) + 1$
$= 5 \cdot 4 - 2 + 1 = 19$

b. $f(-1) = 5 \cdot (-1)^2 - (-1) + 1$
$= 5 \cdot 1 + 1 + 1 = 7$

c. $f(a + b) = 5 \cdot (a + b)^2 - (a + b) + 1$
$= 5 \cdot (a^2 + 2ab + b^2) - (a + b) + 1$
$= 5a^2 + 10ab + 5b^2 - a - b + 1$ ■

Vertical Line Test

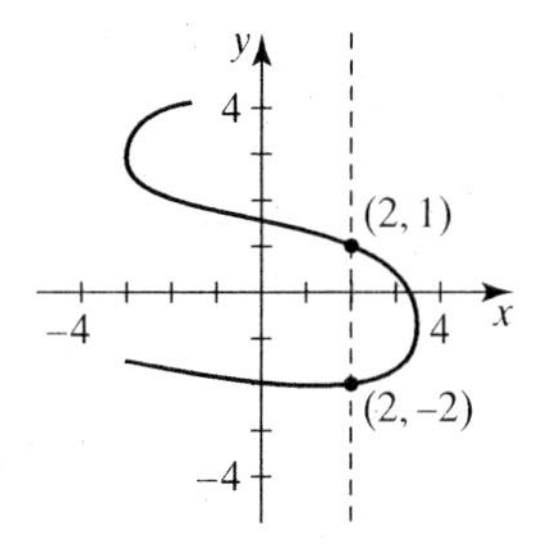

The **vertical line test** is used to determine whether the graph of an equation (a set of ordered pairs) represents a function. Any vertical line intersecting the graph in more than one place indicates that for a specific x there is *more than one* y. In this case, the graph cannot be the graph of a function.

Vertical Line Test

A graph represents a function if and only if every vertical line that intersects the graph does so in exactly one point.

EXAMPLE 2 Determine whether the following graphs represent functions.

a.

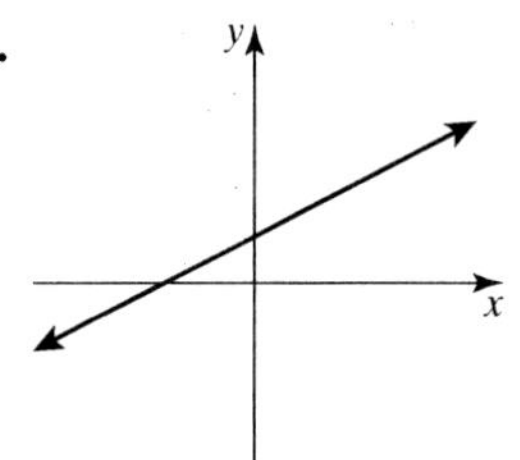

b.

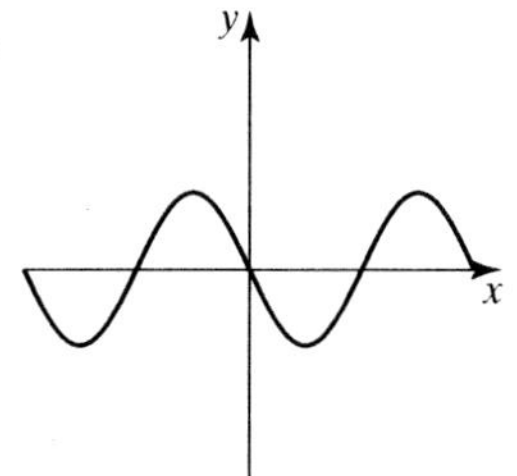

c.

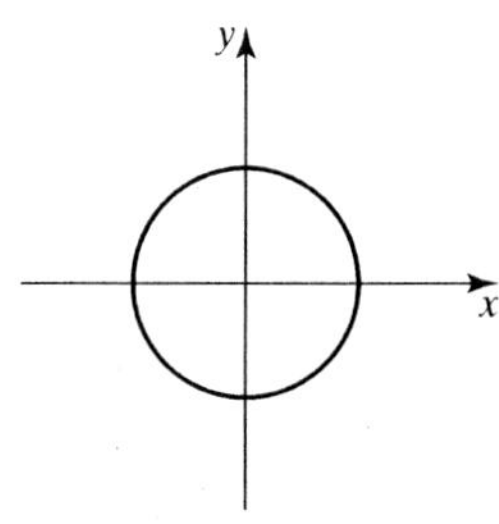

SOLUTION

a. function

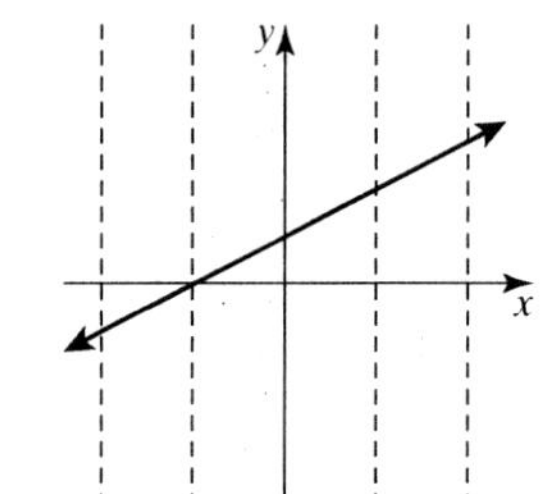

b. function

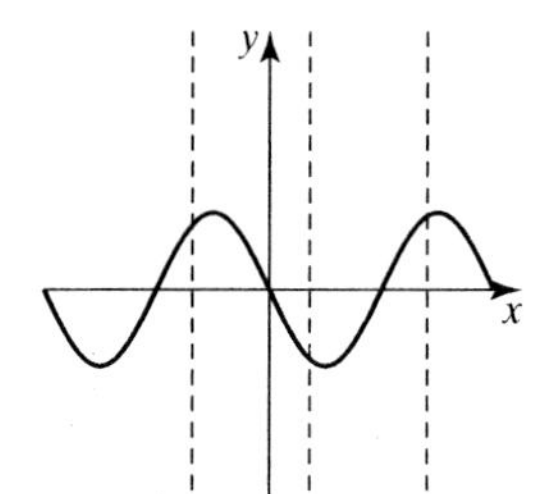

c. not a function

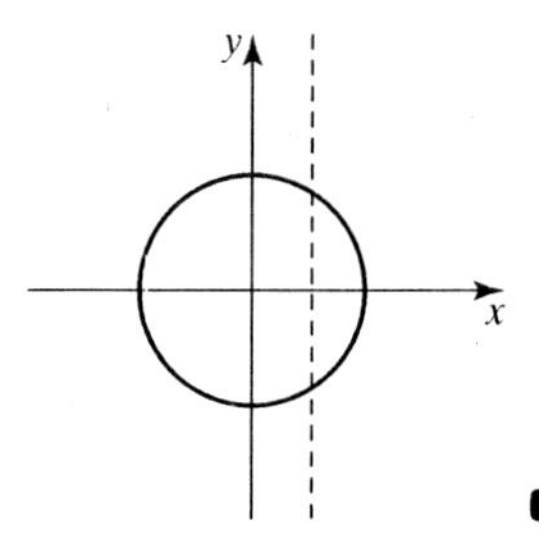

If $f(x)$ is a function, we know that for every x there is only one y. And if it is also the case that for every y there is only one x, the function is said to be **one-to-one**.

One-to-One Functions

A function is said to be **one-to-one** if for every two elements in the domain x_1 and x_2 such that $x_1 \neq x_2$, then

$$f(x_1) \neq f(x_2).$$

Horizontal Line Test

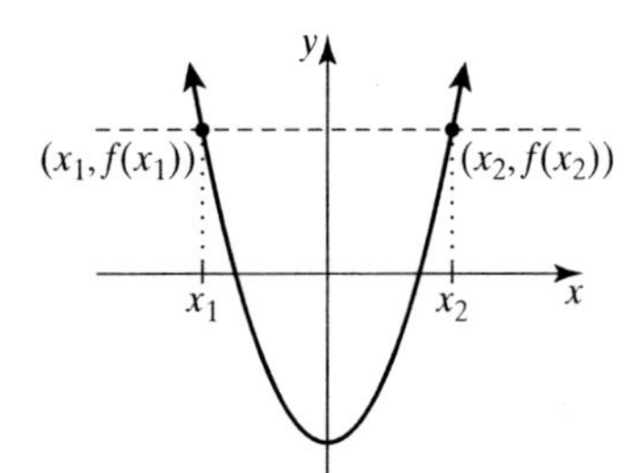

If a horizontal line intersects the graph of a function $f(x)$ in more than one point, it indicates that for $x_1 \neq x_2$ we have $f(x_1) = f(x_2)$. Hence, the graph does not represent a one-to-one function. We use this geometric method to determine whether a function is one-to-one.

Horizontal Line Test

A function $f(x)$ is said to be one-to-one if and only if every horizontal line that intersects the graph does so in exactly one point.

Inverse Functions

To obtain the inverse relation of a function, we interchange the components of all the ordered pairs of the function. For example, if (a, b) is an ordered pair of the function, then (b, a) is an ordered pair of the inverse relation. When we interchange the components for a one-to-one function, the resulting ordered pairs are such that every first component has only one second component. In other words, the inverse relation is a function.

EXAMPLE 3 Determine by their graphs whether the functions are one-to-one and thus have inverse functions.

a.

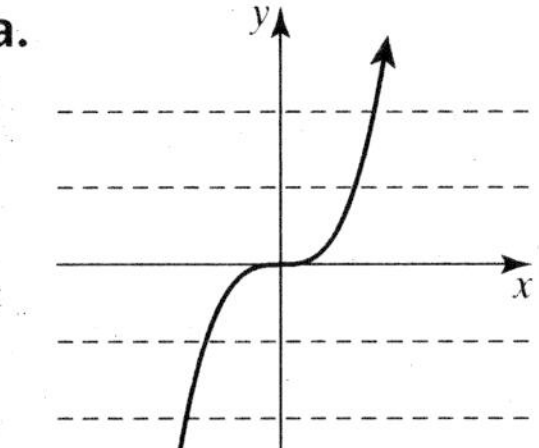

b.

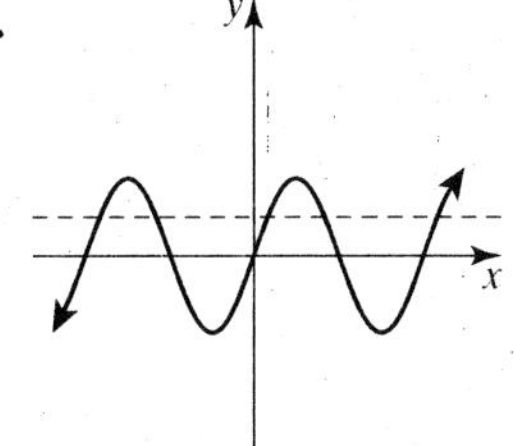

c.

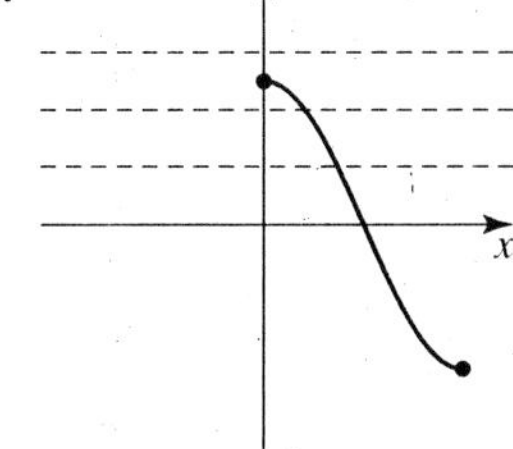

SOLUTION

a. Yes, the function is one-to-one and has an inverse function.

b. No the function is not one-to-one.

c. Yes, the function is one-to-one and has an inverse function. ■

Inverse Notation

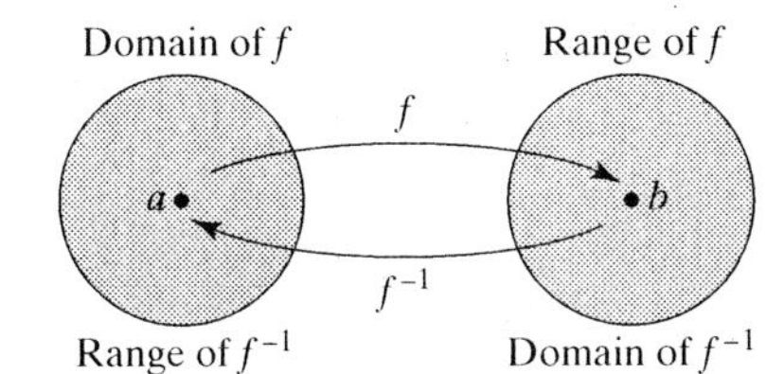

If $f(x)$ is a one-to-one function with domain A and range B, then the inverse function, denoted by $f^{-1}(x)$, has domain B and range A such that

$$f^{-1}(b) = a \text{ is equivalent to } f(a) = b,$$

for any a in A and b in B.

Finding the Inverse Function

If $f(x)$ is one-to-one, we can find the inverse function $f^{-1}(x)$ as follows.

Step 1: If needed, replace $f(x)$ with y.

Step 2: Interchange x and y in the equation.

Step 3: Solve the equation for y, then replace y with $f^{-1}(x)$.

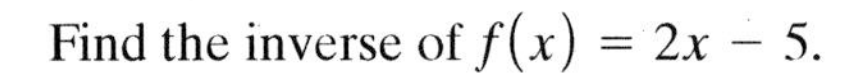

EXAMPLE 4 Find the inverse of $f(x) = 2x - 5$.

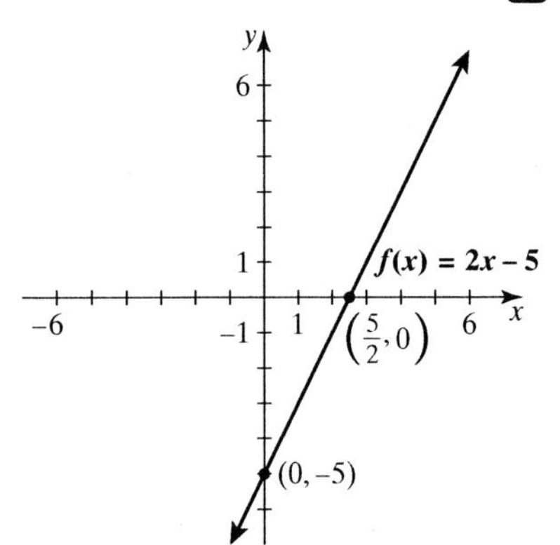

SOLUTION We know $f(x)$ is one-to-one using the horizontal line test on its graph. To find the inverse, we follow the steps shown.

$$2x - 5 = f(x)$$

$$2x - 5 = y \qquad \text{Replace } f(x) \text{ with } y.$$

$$2y - 5 = x \qquad \text{Interchange } x \text{ and } y.$$

$$2y = x + 5 \qquad \text{Solve for } y.$$

$$y = \tfrac{1}{2}x + \tfrac{5}{2}$$

$$f^{-1}(x) = \tfrac{1}{2}x + \tfrac{5}{2} \qquad \text{Replace } y \text{ with } f^{-1}(x).$$

■

Graphing the inverse function, we see that the graph of $f^{-1}(x)$ is the reflection of the graph of $f(x)$ about the line $y = x$.

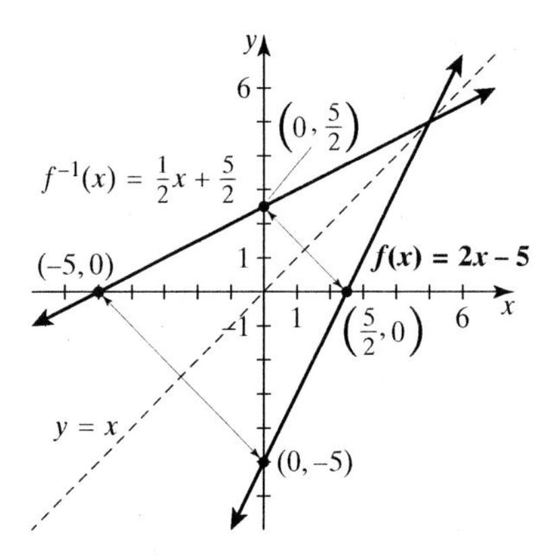

Symmetry

The graphs below demonstrate symmetry with respect to the y-axis, origin, and x-axis.

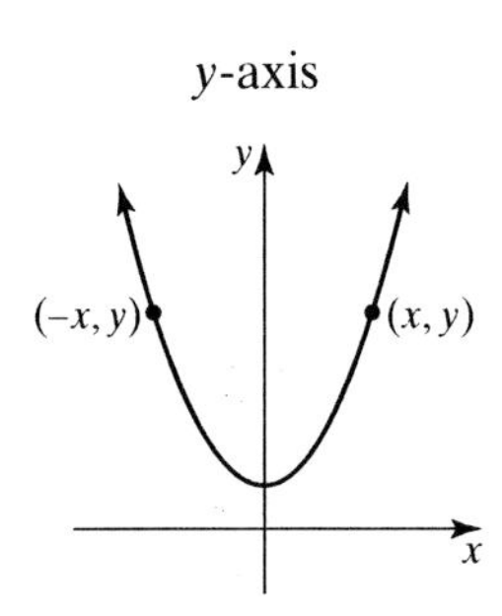

The graph is unchanged when rotated 180° about the y-axis.

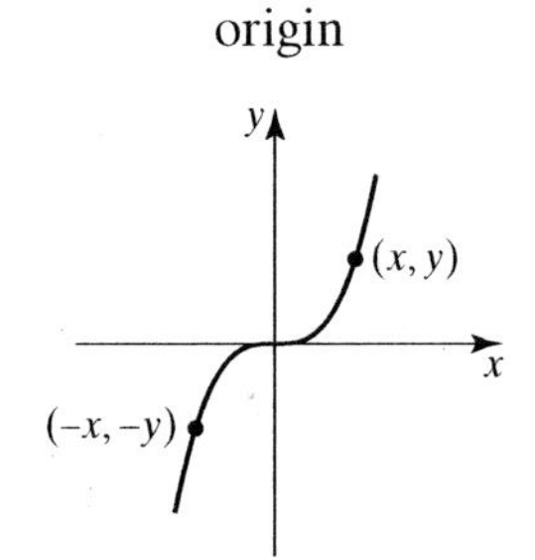

The graph is unchanged when rotated 180° about the origin.

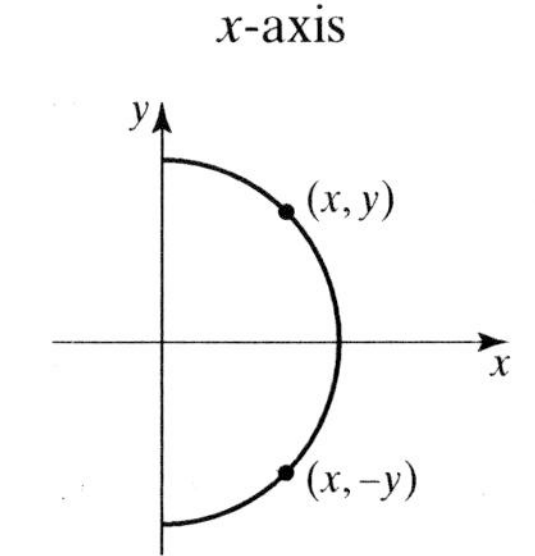

The graph is unchanged when rotated 180° about the x-axis.

Appendix B Geometry Review

In Appendix B we review the concepts of geometry that are critical for the study of trigonometry, and then list some important formulas. Although there is an obvious difference between a side of a triangle and the length of the side as there is a difference between an angle and the measure of the angle, we may not always make the distinction in this text. For the sake of brevity, the length of a side of a triangle and measure of an angle may simply be referred to as "side" or a letter such as c, and "angle" or a letter such as α, respectively.

B.1 Special Triangles

A triangle in which each angle has a measure less than 90° is called an **acute triangle**. If the triangle has an angle whose measure is greater than 90°, it is called an **obtuse triangle**, and if one of the angles measures 90°, then it is called a **right triangle**. In a right triangle, the side across from the right angle is called the **hypotenuse**, and the other two sides are called **legs**.

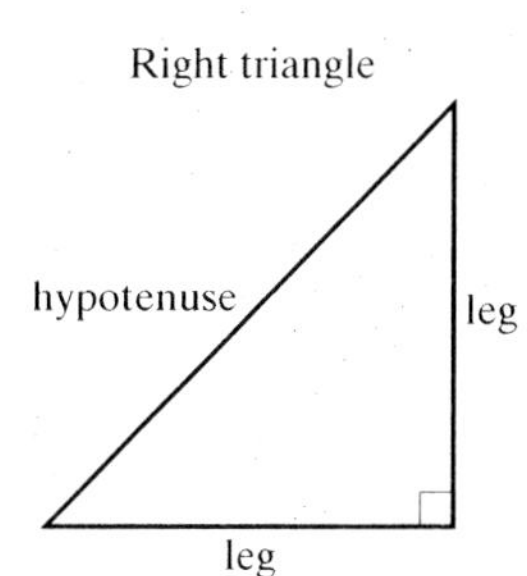

If two sides of a triangle are equal, the triangle is called an **isosceles triangle**. The two angles across from the two equal sides are equal.

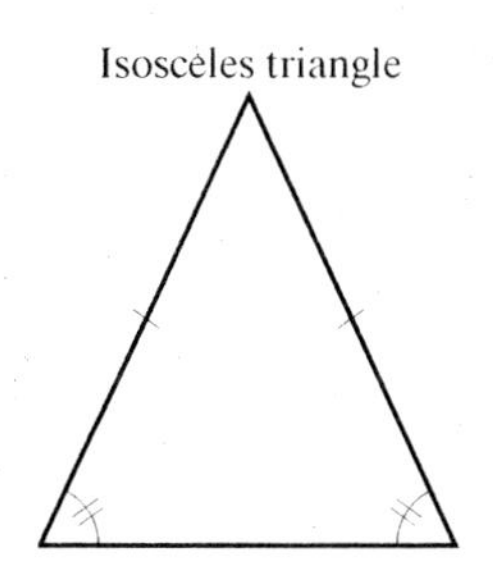

If all three sides of a triangle are equal, the triangle is called **equilateral**. Each angle of an equilateral triangle is 60°.

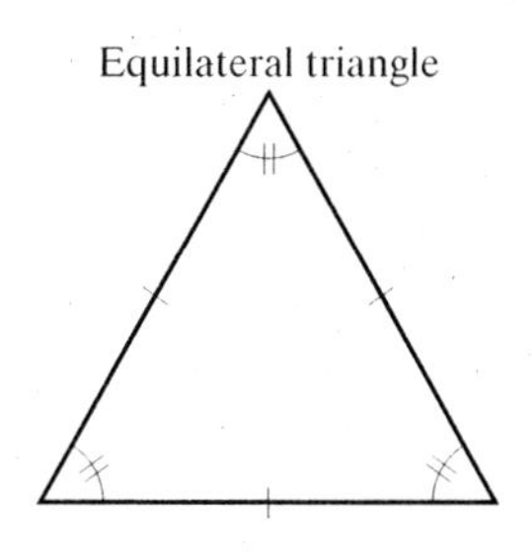

The sum of the measures of the angles of any triangle is 180°.

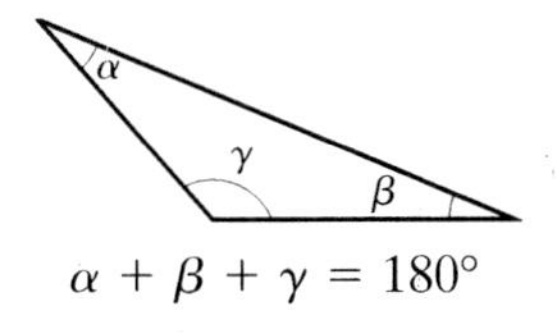

$\alpha + \beta + \gamma = 180°$

An **altitude** h of a triangle is the line segment drawn from the vertex of an angle of a triangle perpendicular to the line containing the opposite side.

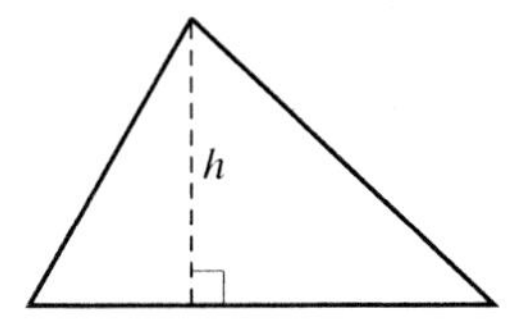

Congruent Triangles

Two triangles are **congruent** (denoted by $\cong$) if they have the same shape and same size. When triangles ABC and DEF are congruent, we write $\triangle ABC \cong \triangle DEF$, which means the corresponding angles and corresponding sides are equal.

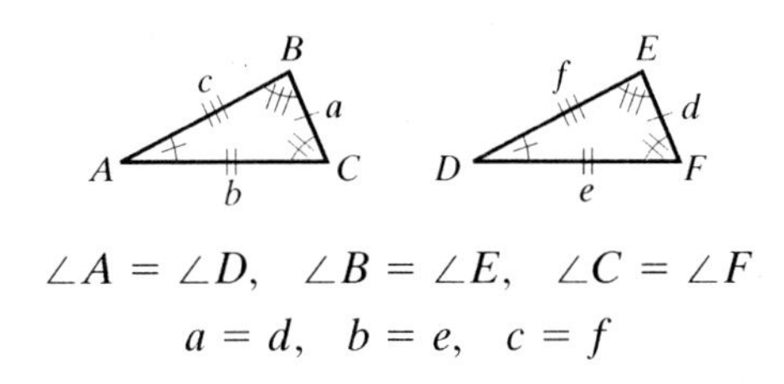

$$\angle A = \angle D, \quad \angle B = \angle E, \quad \angle C = \angle F$$
$$a = d, \quad b = e, \quad c = f$$

Since the measures of the angles of any triangle add to 180°, if two angles of one triangle are equal to two angles of a second triangle, then their third angles are equal.

When establishing that two triangles are congruent, it is not necessary to show that all three corresponding sides and all three corresponding angles are equal. Two triangles will be congruent if any of the following conditions are satisfied.

SSS **Side-Side-Side**

If each side of one triangle is equal to the corresponding side of a second triangle, the triangles are congruent.

AAS, ASA **Angle-Angle-Side, Angle-Side-Angle**

If two angles and a side of one triangle are equal to the corresponding parts of a second triangle, the triangles are congruent.

SAS **Side-Angle-Side**

If two sides and the included angle of one triangle are equal to the corresponding two sides and the included angle of a second triangle, the triangles are congruent.

EXAMPLE 1 Determine whether the two triangles are congruent.

SOLUTION

a. $\triangle ABC \cong \triangle DEF$ by AAS.

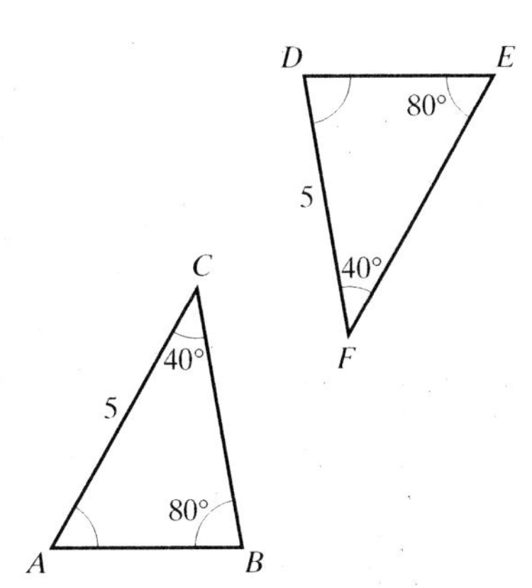

b. $\triangle RST \cong \triangle OPQ$ by SAS.

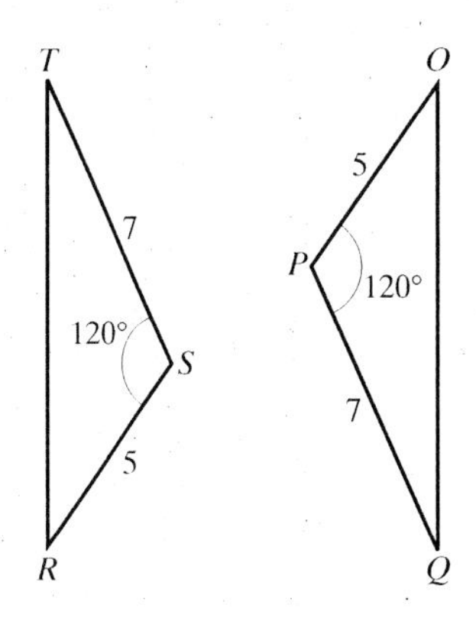

c. We cannot conclude congruence since side-side-angle does not guarantee congruence.

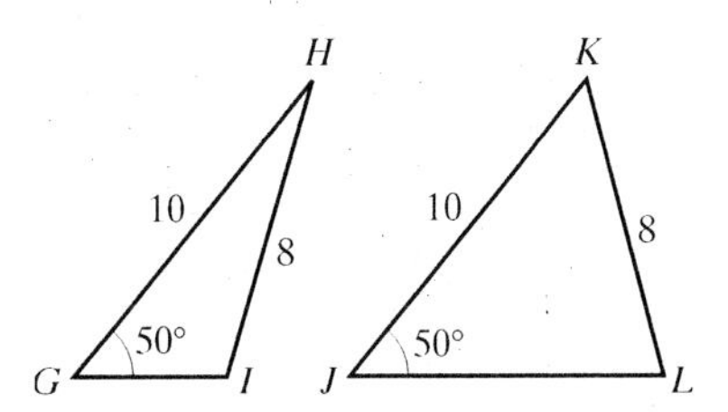

Similar Triangles

Two triangles are **similar** (denoted by ~) if they have the same shape, but not necessarily the same size. If triangles RST and UVW are similar, we write $\triangle RST \sim \triangle UVW$, which means the corresponding angles are equal and the corresponding sides are proportional.

$$\angle R = \angle U, \quad \angle S = \angle V, \quad \angle T = \angle W$$

$$\frac{RS}{UV} = \frac{ST}{VW} = \frac{RT}{UW}$$

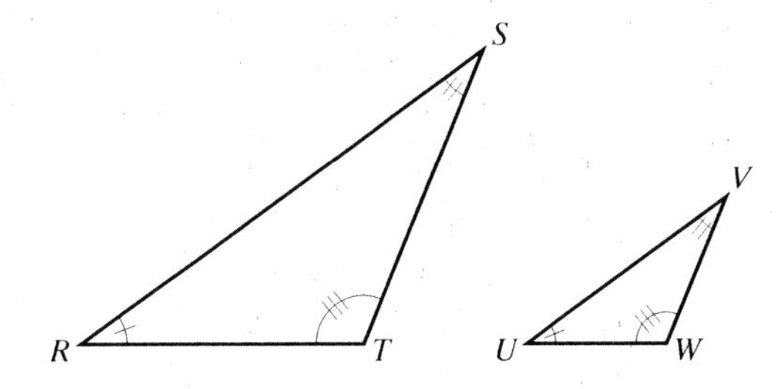

When establishing that two triangles are similar, it is not necessary to show that all three angles are equal and that corresponding sides are proportional. Two triangles will be similar if the following conditions are satisfied.

AA If two angles of one triangle are equal to two angles of a second triangle, the triangles are similar. (Recall that if two pairs of corresponding angles are equal, then the third pair of corresponding angles must also be equal.)

SSS If the three corresponding sides are proportional, the triangles are similar.

SAS If one angle of one triangle is equal to one angle of a second triangle and the adjacent sides of the first triangle are proportional to the adjacent sides of the second, the triangles are similar.

EXAMPLE 2 If $\triangle XYZ \sim \triangle RST$, state which angle is equal to $\angle YZX$ and find x and y.

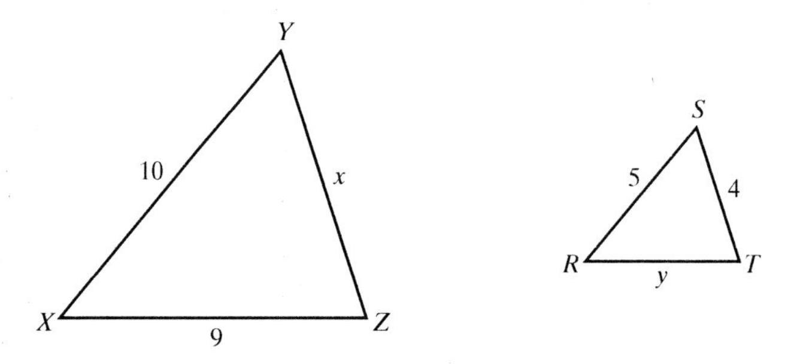

SOLUTION Since $\triangle XYZ \sim \triangle RST$, the corresponding angles are equal, so $\angle YZX = \angle STR$. Also, the corresponding sides are proportional. So, $\frac{x}{4} = \frac{10}{5} \rightarrow x = 8$ and $\frac{y}{9} = \frac{5}{10} \rightarrow y = 4.5$. ■

Parallel Lines

Lines that lie in a plane and do not intersect are called **parallel**. If two parallel lines are intersected by a third line, angles of equal measure are formed.

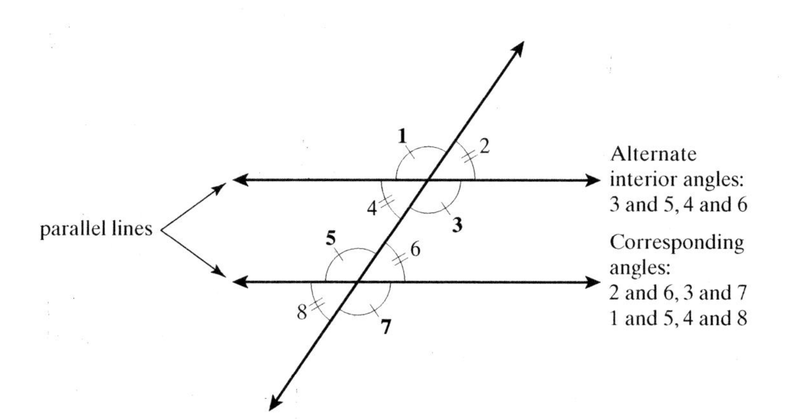

B.2 Geometry Formulas

Supplementary Angles

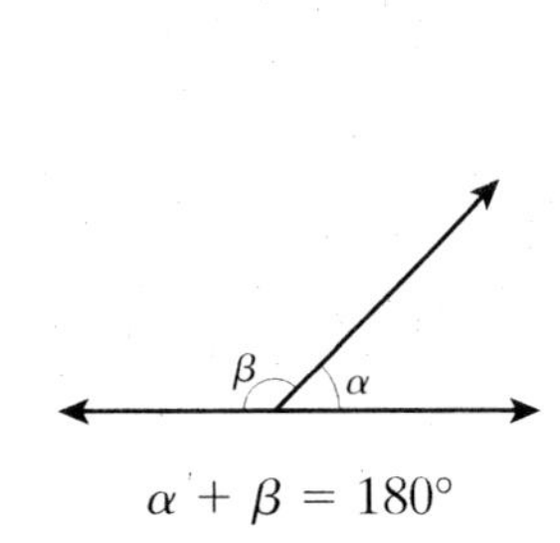

$\alpha + \beta = 180°$

Complimentary Angles

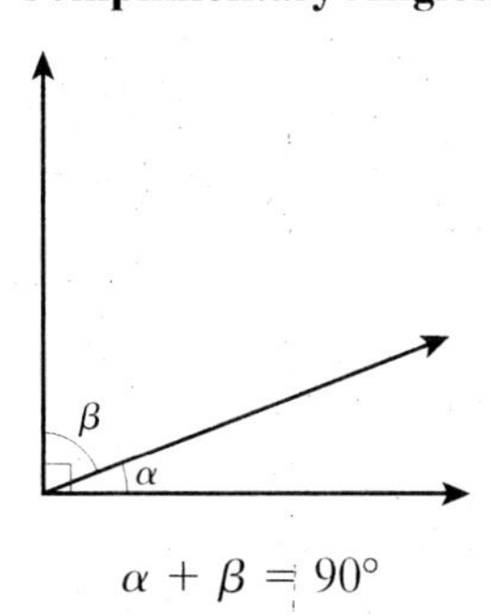

$\alpha + \beta = 90°$

Circle

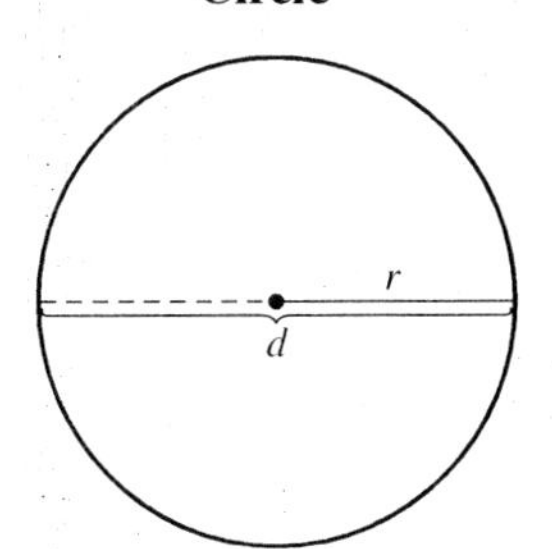

Circumference: $C = \pi d = 2\pi r$

Area: $A = \pi r^2$

Rectangle

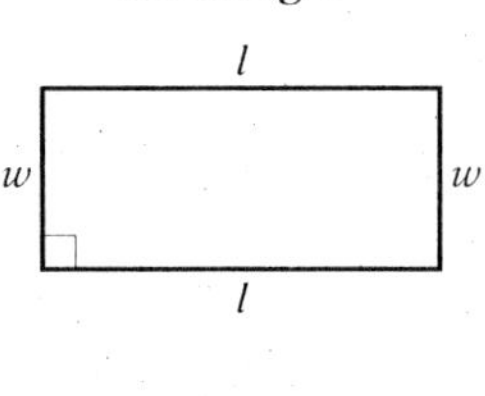

Perimeter: $P = 2l + 2w$

Area: $A = lw$

Square

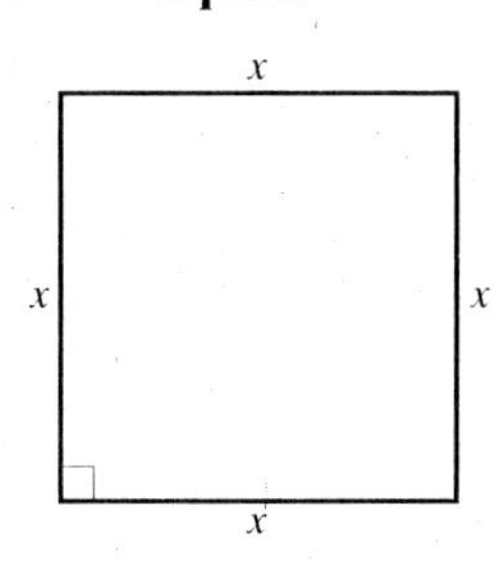

Perimeter: $P = 4x$

Area: $A = x^2$

Parallelogram

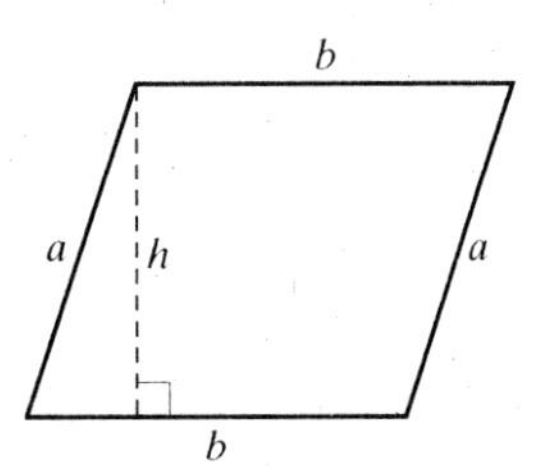

Perimeter: $P = 2a + 2b$

Area: $A = bh$

Triangle

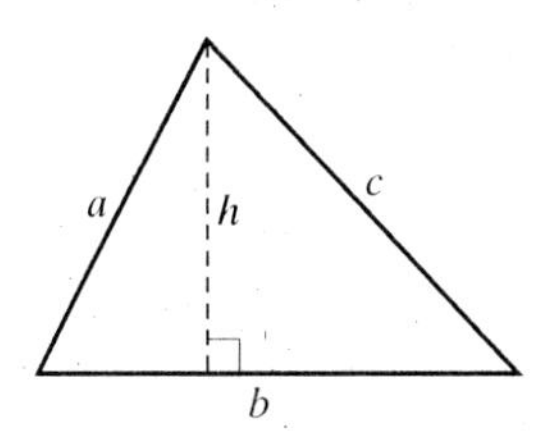

Perimeter: $P = a + b + c$

Area: $A = \frac{1}{2}bh$

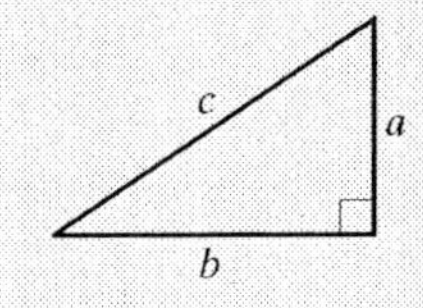

Pythagorean Theorem

In any right triangle, the square of the length of the longest side (hypotenuse) is equal to the sum of the squares of the lengths of the other two sides (legs).

$$a^2 + b^2 = c^2$$

To understand why the Pythagorean Theorem is true, consider the four congruent right triangles drawn to form a square with each side of length $a + b$. The angles marked in the diagram as α, β, and ϕ are such that $\alpha + \beta + \phi = 180°$. And since α and β are the acute angles of the right triangle, we know that $\alpha + \beta = 90°$. So we can conclude that $\phi = 90°$. Therefore, the inner figure is a square.

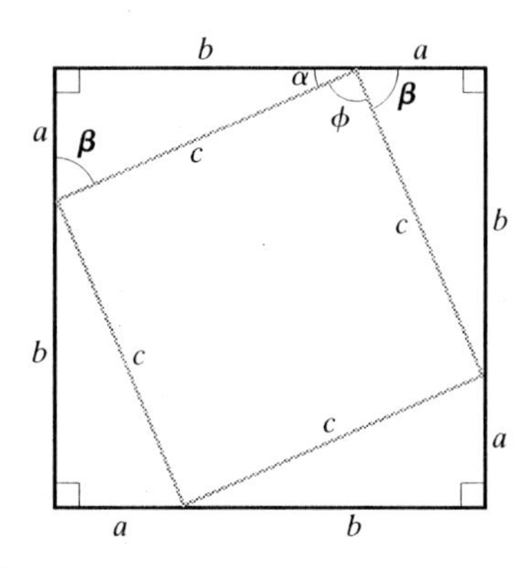

We notice that the area of the large square is equal to the sums of the areas of the four triangles and the inner square.

Area of large square		Area of the four triangles		Area of the inner square	
$(a + b)^2$	$=$	$4(\frac{1}{2}ab)$	$+$	c^2	
$a^2 + 2ab + b^2$	$=$	$2ab + c^2$			Simplify each side.
$a^2 + b^2$	$=$	c^2			Subtract $2ab$ from each side.

Three-dimensional Geometric Figures

Rectangular Box

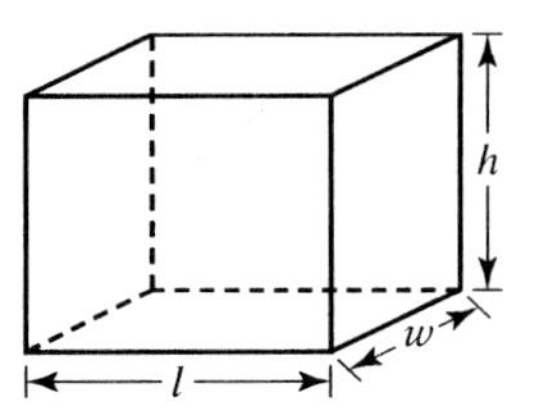

Volume: $V = lwh$

Cylinder

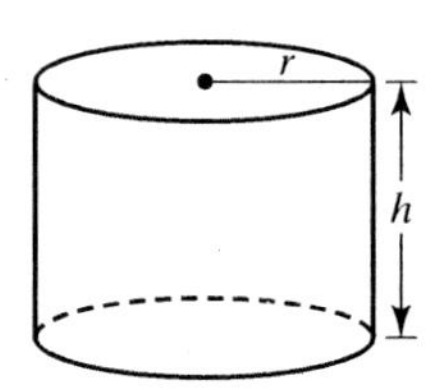

Volume: $V = \pi r^2 h$

Sphere

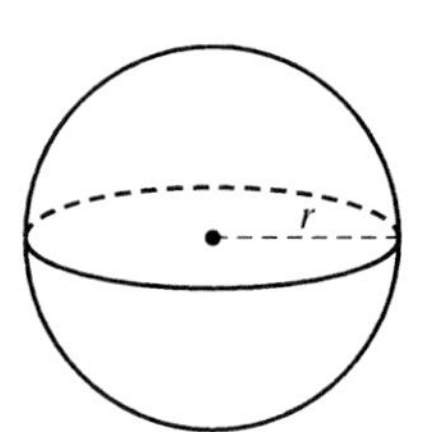

Volume: $V = \frac{4}{3}\pi r^3$

Answers to Selected Exercises

Chapter 1

Exercise Set 1.2, p. 13

1. a. number line b. unit c. direction
   d. $(1, 0)$ e. counterclockwise f. negative
3. a. y-coordinate b. t c. 1 d. -1
5. a. $\cos t$ b. $\sin^2 t$
7. a. 0 b. $\frac{\pi}{6}$ c. $\frac{\pi}{4}$ d. $\frac{\pi}{3}$
   e. $\frac{\pi}{2}$ f. π g. $\frac{3\pi}{2}$
9. QIII 11. QIII 13. QIV 15. QII
17. QI 19. QII 21. QI 23. QII
25. QI 27. QIII 29. QIV 31. QI
33. QIV 35. $\frac{12}{13}$ 37. $-\frac{\sqrt{2}}{2}$
39. a. QII b. $\frac{15}{17}$ c. $-\frac{17}{8}$
   d. $\frac{17}{15}$ e. $-\frac{15}{8}$ f. $-\frac{8}{15}$
41. a. QIV b. $\frac{1}{2}$ c. 2
   d. $-\frac{2}{\sqrt{3}} = -\frac{2\sqrt{3}}{3}$ e. $-\sqrt{3}$ f. $-\frac{1}{\sqrt{3}} = -\frac{\sqrt{3}}{3}$
43. $\sin t = -\sqrt{1 - x^2}$

Exercise Set 1.3, p. 23

1. $t = \frac{\pi}{4}$, $\sin t = \frac{1}{\sqrt{2}} = \frac{\sqrt{2}}{2}$
3. $\sin t = -1, \cos t = 0$
5.

x	$\cos x$
0	1
$\frac{\pi}{4}$	$\frac{1}{\sqrt{2}} = \frac{\sqrt{2}}{2}$
π	-1
$\frac{\pi}{3}$	$\frac{1}{2}$
$\frac{3\pi}{2}$	0
$\frac{\pi}{2}$	0
$\frac{\pi}{6}$	$\frac{\sqrt{3}}{2}$
2π	1

7. $\frac{3}{4}$ 9. $\frac{\sqrt{2}}{4}$ 11. $\frac{6 - \sqrt{3}}{2}$ 13. 1
15. $\frac{\sqrt{3}}{3}$ 17. undefined 19. $\frac{3}{2}$ 21. 2
23. T 25. F 27. T 29. F
31. 0.9320 33. -0.9902 35. 0.4647 37. 0.7344
39. 3.5326 41. -8.8188

Exercise Set 1.4, p. 36

1.

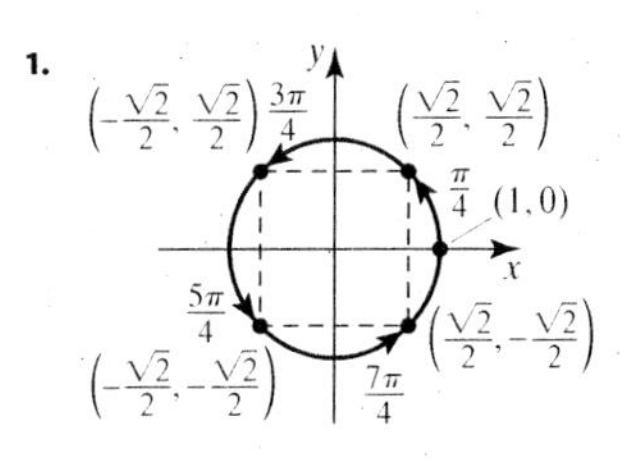

3.

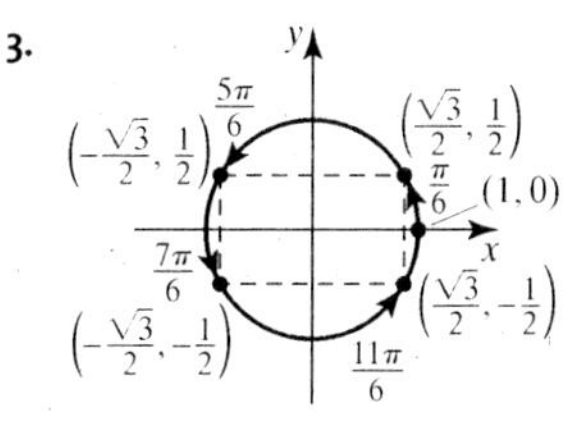

5. $x = -\frac{\pi}{3}$ 7. $x = -\frac{5\pi}{4}$

9. $\hat{x} = \frac{\pi}{3}$

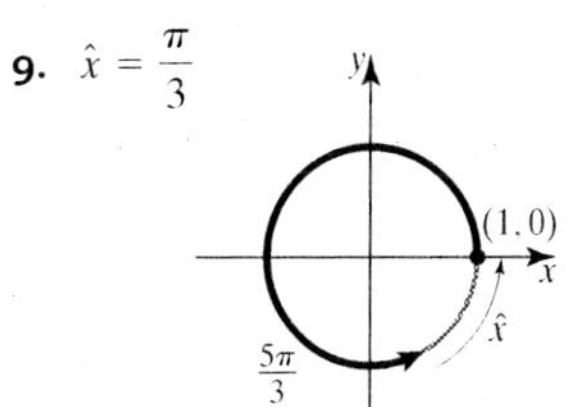

11. $\hat{x} = \frac{\pi}{6}$

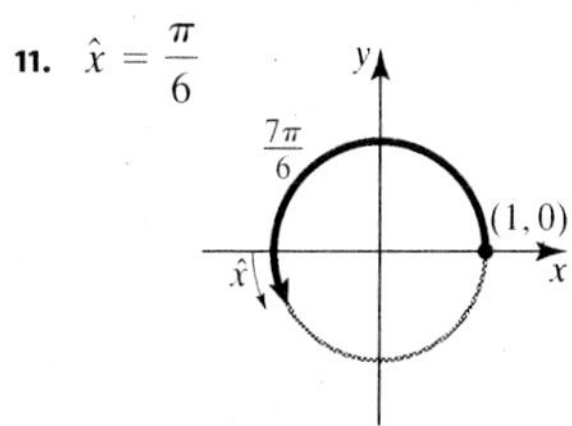

13. $\hat{x} = \frac{\pi}{4}$

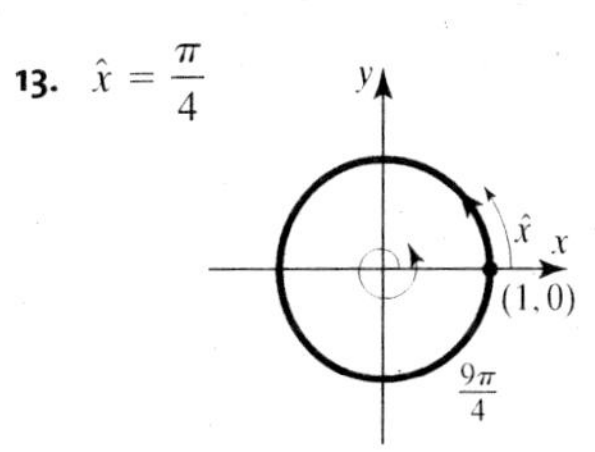

15. $\hat{x} = \frac{\pi}{3}$

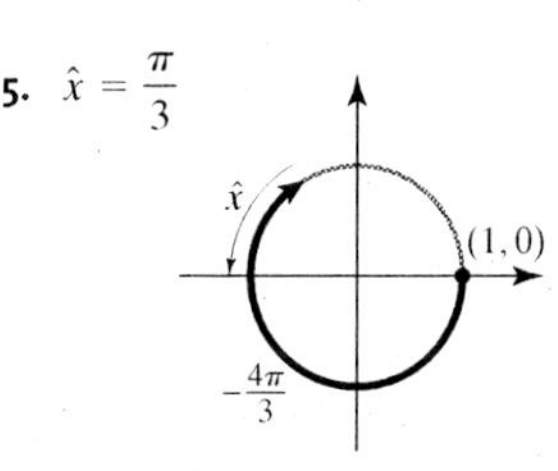

17. $\hat{x} = \frac{\pi}{6}$

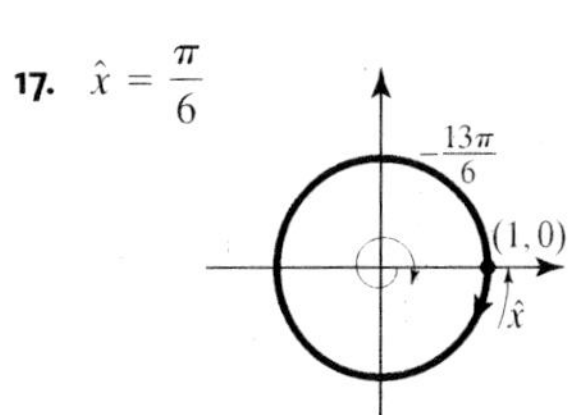

19. $\hat{x} = \frac{\pi}{6}$

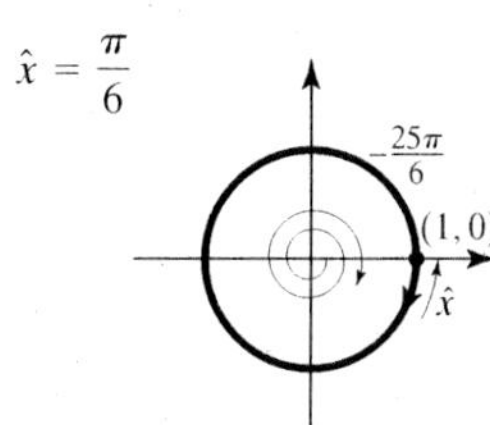

21. $\frac{\sqrt{3}}{2}$ 23. $-\frac{\sqrt{3}}{2}$ 25. $\frac{\sqrt{2}}{2}$ 27. $\frac{1}{2}$
29. $-\frac{\sqrt{2}}{2}$ 31. $\frac{1}{2}$ 33. $\frac{\sqrt{3}}{2}$ 35. $-\sqrt{2}$

37. -0.7240 **39.** 8.3553 **41.** 0.2817 **43.** $\frac{\sqrt{2}-1}{2}$

45. 5.3806 **47.** F **49.** T **51.** F

53. T **55.** F **57.** F **59.** T

61. T **63.** $\frac{5\pi}{3}$ **65.** $\frac{3\pi}{2}$

67. $-\frac{\pi}{4}$ **69.** $\frac{5\pi}{6}$

Exercise Set 1.5, p. 48

1. $\tan 0 = 0, \tan\frac{\pi}{6} = \frac{\sqrt{3}}{3}, \tan\frac{\pi}{4} = 1, \tan\frac{\pi}{3} = \sqrt{3}, \tan\frac{\pi}{2}$ undefined

3. $\sec 0 = 1, \sec\frac{\pi}{6} = \frac{2\sqrt{3}}{3}, \sec\frac{\pi}{3} = 2, \sec\frac{3\pi}{4} = -\sqrt{2}, \sec\pi = -1,$ $\sec\frac{3\pi}{2}$ undefined, $\sec\left(-\frac{\pi}{4}\right) = \sqrt{2}$

5. $\frac{\pi}{4}$ **7.** $\frac{\pi}{6}$ **9.** QIV **11.** QI

13. QIII **15.** QII **17.** 1 **19.** $-\frac{2\sqrt{3}}{3}$

21. 2 **23.** $\sqrt{2}$ **25.** $-\frac{\sqrt{3}}{3}$ **27.** $-\frac{\sqrt{3}}{3}$

29. $\sqrt{2}$ **31.** $-\sqrt{3}$ **33.** undefined **35.** $-\frac{2\sqrt{3}}{3}$

37. 6 **39.** $\frac{1-\sqrt{3}}{2}$ **41.** -0.8391 **43.** 1.0640

45. 2.3212 **47.** $65{,}532.7566$ **49.** -14.2299 **51.** 1.3600

53. $\frac{5}{12}$

55. a. $\frac{24}{25}$ **b.** $-\frac{25}{7}$ **c.** $-\frac{24}{7}$
d. $\frac{25}{24}$ **e.** $-\frac{7}{24}$

57. 5 **59.** F **61.** T **63.** T

65. F **67.** F **69.** T **71.** Identity

73. Identity **75.** $\frac{5\pi}{4}$ **77.** $\frac{2\pi}{3}$ **79.** $\frac{3\pi}{2}$

81. For $k = -1, 0, 1, 2,\ x \neq -\pi/2, \pi/2, 3\pi/2, 5\pi/2$, respectively.

Exercise Set 1.6, p. 61

1. cosine, sine, secant, cosecant

3. T **5.** T **7.** T **9.** F

11. T **13.** $\frac{\sqrt{3}}{2}$, iii, vii **15.** $-\frac{\sqrt{3}}{2}$, iv, v, vi

17. a. $x = \frac{\pi}{3}, x = \frac{5\pi}{3}; x = \frac{\pi}{3} + k2\pi, x = \frac{5\pi}{3} + k2\pi$
b. $x = \frac{5\pi}{4}, x = \frac{7\pi}{4}; x = \frac{5\pi}{4} + k2\pi, x = \frac{7\pi}{4} + k2\pi$
c. $x = \frac{\pi}{4}, x = \frac{5\pi}{4}; x = \frac{\pi}{4} + k\pi$

19. 0 in. **21.** 0 in. **23.** 15.0 cm **25.** -12.6 cm

27. 161.9 volts **29.** -41.7 volts

31. 0 psf (pounds per square foot)

33. 0.001 psf

35. Since $-1 \le \cos(\pi t) \le 1$, we know $5.2 \ge -5.2\cos(\pi t) \ge -5.2$, which indicates the most it will bounce above or below is 5.2 in.

37. A: all functions are positive in QI, S: $\sin x$ and $\frac{1}{\sin x}$ are positive in QII, T: $\tan x$ and $\frac{1}{\tan x}$ are positive in QIII, and C: $\cos x$ and $\frac{1}{\cos x}$ are positive in QIV.

Chapter 1 Review Exercises, p. 66

1. a. x **b.** terminal **c.** t
d. $(1, 0)$ **e.** unit **f.** length

2. 1 **3.** -1

4. a. positive **b.** clockwise

5. $\cos^2 t + \sin^2 t = 1$

6. $\mathbb{R}$ (the real numbers)

7. a. QIV **b.** QI **c.** QIII
d. QII **e.** QIV **f.** QIII
g. QIII **h.** QII

8. a. QIV **b.** QIII

9. a. QII **b.** $-\frac{1}{2}$ **c.** $\frac{2\sqrt{3}}{3}$
d. -2 **e.** $-\sqrt{3}$ **f.** $-\frac{\sqrt{3}}{3}$

10. a. IV **b.** $-\frac{4}{5}$ **c.** $-\frac{5}{4}$
d. $\frac{5}{3}$ **e.** $-\frac{4}{3}$ **f.** $-\frac{3}{4}$

11. a. $\cos t = -1, \sin t = 0$ **b.** $t = \frac{\pi}{3}, \sin t = \frac{\sqrt{3}}{2}$
c. $t = \frac{\pi}{4}, \cos t = \frac{\sqrt{2}}{2}$

12. a. $\frac{\sqrt{3}+2}{2}$ **b.** 2 **c.** $\frac{1}{2}$
d. $\sqrt{3}$ **e.** $-\frac{1}{4}$ **f.** $\sqrt{2}$

13. a. 1.7969 **b.** -3.9938 **c.** 2.8356
d. 3.9558 **e.** 100,000,000.3333

14. **a.** T **b.** F **c.** T **d.** T
e. T **f.** F **g.** T **h.** F
i. F **j.** F

15. **a.** $\frac{\pi}{3}$ **b.** $\frac{\pi}{4}$

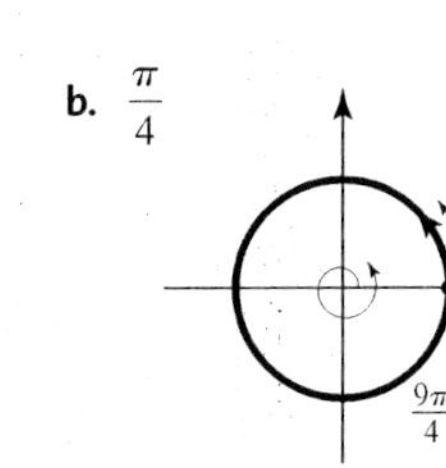

c. $\frac{\pi}{6}$ **d.** $\frac{\pi}{4}$

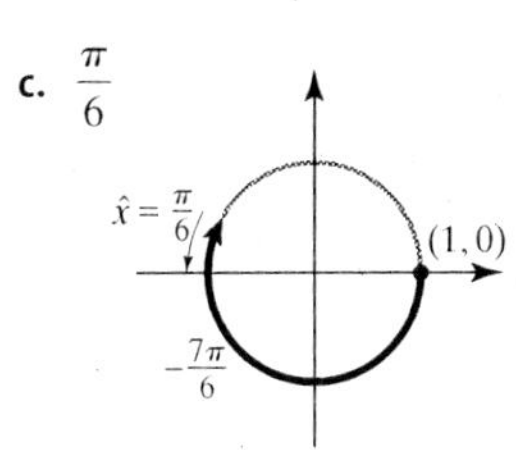

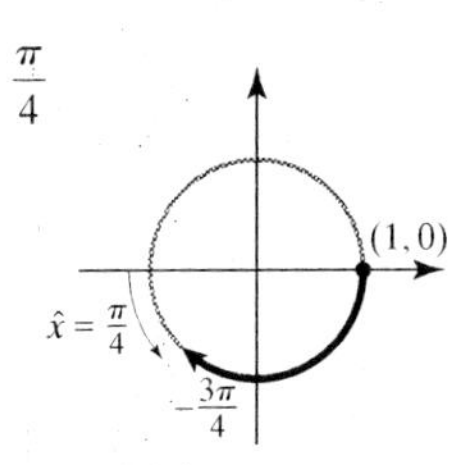

16. **a.** $-\frac{\sqrt{3}}{2}$ **b.** $\frac{\sqrt{2}}{2}$ **c.** $-\frac{\sqrt{3}}{2}$ **d.** $-\frac{\sqrt{2}}{2}$

17. **a.** 1 **b.** 1 **c.** 0.0689 **d.** -2

18. **a.** $\frac{7\pi}{6}$ **b.** $\frac{3\pi}{4}$ **c.** $\frac{3\pi}{2}$ **d.** $\frac{\pi}{3}$

19. **a.** $x = \frac{5\pi}{6}, \csc x = 2$ **b.** $x = -\frac{\pi}{3}, \tan x = -\sqrt{3}$

c. $x = \frac{5\pi}{4}, \sec x = -\sqrt{2}$

20. **a.** QIV **b.** QIII **c.** not possible
d. QIV **e.** QI **f.** QII **g.** QI or QIII

21. **a.** $-\sqrt{3}$ **b.** $-\sqrt{2}$ **c.** $-\frac{2\sqrt{3}}{3}$ **d.** $-\frac{\sqrt{3}}{3}$

e. $\frac{1}{3}$ **f.** undefined **g.** 4 **h.** undefined

i. $\frac{3 - 2\sqrt{3}}{3}$

22. **a.** -4.4609 **b.** -0.3511 **c.** -10.6371 **d.** 52.7594

23. **a.** $\cos x = -\frac{15}{17}, \tan x = -\frac{8}{15}$ **b.** $\sin x = -\frac{\sqrt{2}}{2}, \cot x = -1$

c. $\sin x = -\frac{5}{13}, \csc x = -\frac{13}{5}, \sec x = -\frac{13}{12}$

d. $\cos x = \frac{\sqrt{5}}{3}, \tan x = -\frac{2\sqrt{5}}{5}$

24. **a.** $\frac{\pi}{3}$ **b.** $\frac{\pi}{4}$ **c.** $\frac{\pi}{6}$

25. **a.** $\frac{3\pi}{4}$ **b.** $-\frac{\pi}{6}$ **c.** $-\frac{\pi}{4}$

26. **a.** T **b.** T **c.** F **d.** F
e. T

27. **a.** T **b.** F **c.** T **d.** F
e. T **f.** T

28. **a.** $x = \frac{\pi}{3}, x = \frac{2\pi}{3}; x = \frac{\pi}{3} + k2\pi, x = \frac{2\pi}{3} + k2\pi$

b. $x = \frac{\pi}{4}, x = \frac{7\pi}{4}; x = \frac{\pi}{4} + k2\pi, x = \frac{7\pi}{4} + k2\pi$

c. $x = \frac{5\pi}{6}, x = \frac{11\pi}{6}; x = \frac{5\pi}{6} + k\pi$

29. 3.7082 cm **30.** 1.0353 in.

31. **a.** The arc $4\pi/11$ is not a common arc.
b. Elka **c.** Tom input $1/\sin(4\pi/11)$.

Chapter 1 Test, p. 71

1. **a.** y-coordinate **b.** unit **c.** direction
d. $\cos^2 x + \sin^2 x = 1$ **e.** $\mathbb{R}$ **f.** $-1 \leq \sin x \leq 1$
g. 2π **h.** $x \neq \frac{\pi}{2} + k\pi, x \in \mathbb{R}$
i. $\mathbb{R}$ **j.** π

2.

x	$\sin x$	$\tan x$	$\sec x$
0	0	0	1
$\frac{\pi}{6}$	$\frac{1}{2}$	$\frac{1}{\sqrt{3}}$ or $\frac{\sqrt{3}}{3}$	$\frac{2}{\sqrt{3}}$ or $\frac{2\sqrt{3}}{3}$
$\frac{\pi}{4}$	$\frac{1}{\sqrt{2}}$ or $\frac{\sqrt{2}}{2}$	1	$\sqrt{2}$
$\frac{\pi}{3}$	$\frac{\sqrt{3}}{2}$	$\sqrt{3}$	2
$\frac{\pi}{2}$	1	undefined	undefined
π	0	0	-1
$\frac{3\pi}{2}$	-1	undefined	undefined

3. **a.** QII **b.** QIV

4. **a.** $-\frac{\sqrt{2}}{2}$ **b.** $-\frac{2\sqrt{3}}{3}$

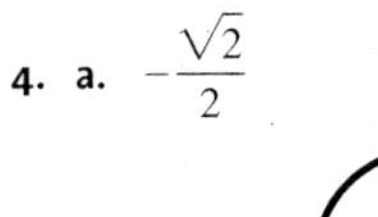

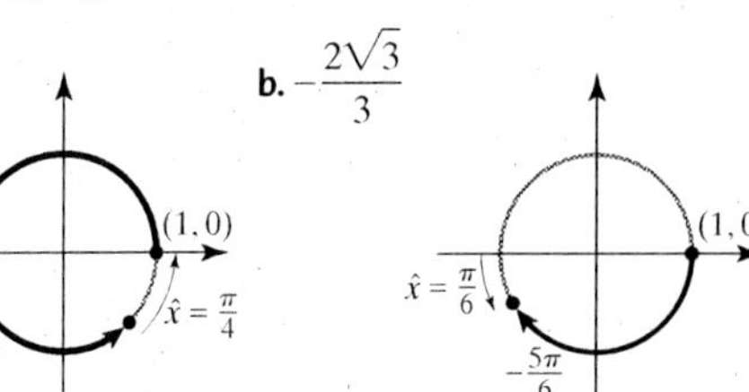

5. a. $-\frac{\sqrt{2}}{2}$ b. $\frac{3}{4}$ c. $-\sqrt{3}$ d. 2
e. $-\frac{1}{2}$ f. 1 g. $-\frac{3\sqrt{3}}{2}$

6. a. $-\frac{\sqrt{2}}{2}$ b. 1 c. $-\sqrt{2}$

7. a. QII b. $-\frac{\sqrt{7}}{4}$ c. $-\frac{3\sqrt{7}}{7}$

8. a. QIV b. $\frac{5\pi}{3}$

9. a. 0.7510 b. 0.9391 c. 1.4702
d. −2.6988 e. −21.7523

10. a. T b. F c. T d. F
e. F f. F g. T h. T
i. F j. F

11. a. $x = \frac{\pi}{6}, x = \frac{11\pi}{6}; x = \frac{\pi}{6} + k2\pi, x = \frac{11\pi}{6} + k2\pi$
b. $x = \frac{2\pi}{3}, x = \frac{5\pi}{3}; x = \frac{2\pi}{3} + k\pi$

12. 2.1631 cm

Chapter 2

Exercise Set 2.1, p. 85

1. a. $y = \cos x$; labeled points $(0, 1)$, $(\frac{\pi}{2}, 0)$, $(\frac{3\pi}{2}, 0)$, $(2\pi, 1)$, $(\pi, -1)$

b.

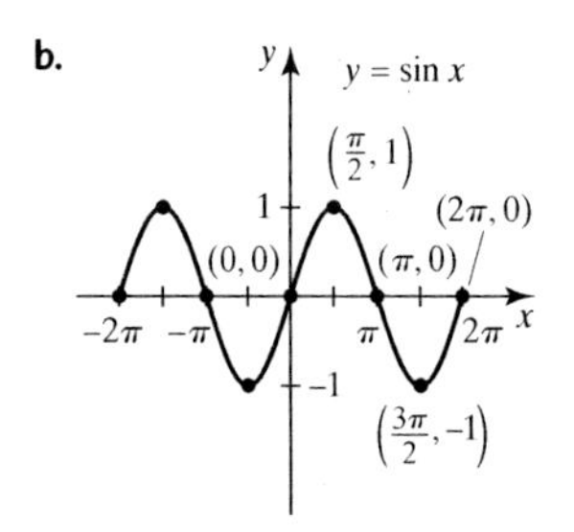

3.

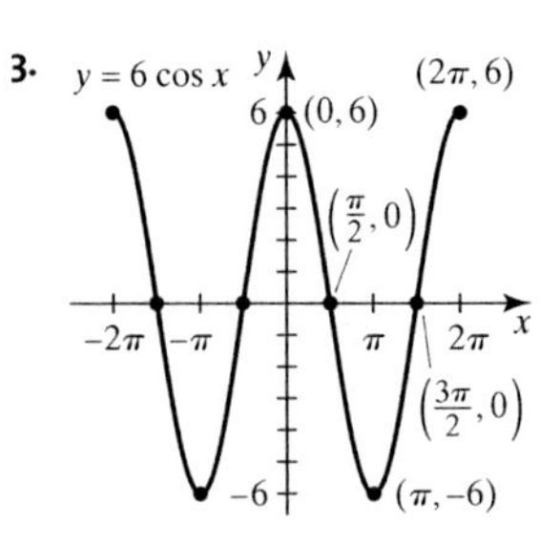

range: $|y| \le 6$, x-intercepts: $-3\pi/2$, $-\pi/2$, $\pi/2$, $3\pi/2$ or $x = (\pi/2) + k\pi$

5.

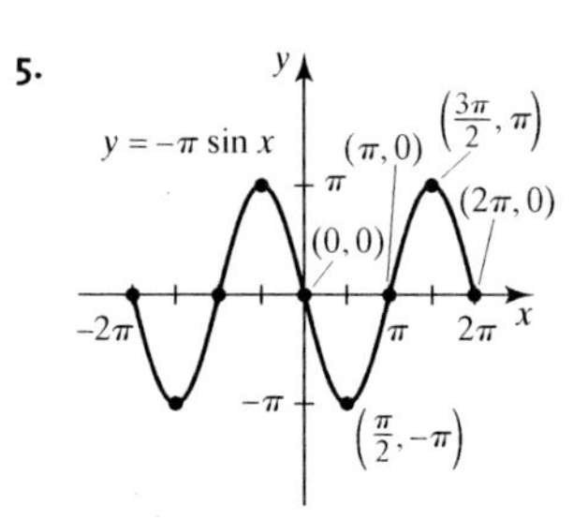

range: $|y| \le \pi$, x-intercepts: $-2\pi, -\pi, 0, \pi, 2\pi$ or $x = k\pi$

7.

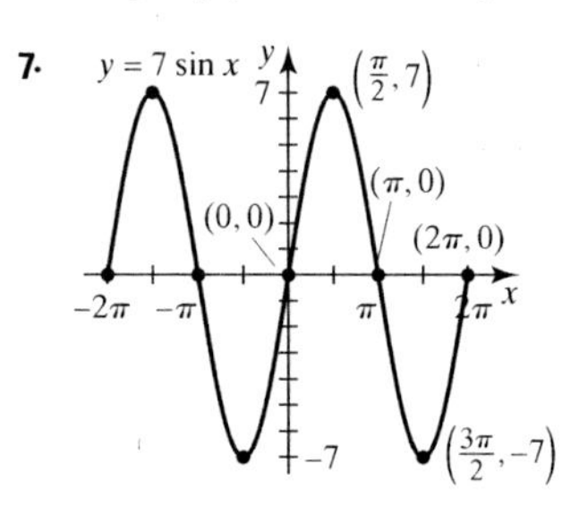

range: $|y| \le 7$, x-intercepts: $-2\pi, -\pi, 0, \pi, 2\pi$ or $x = k\pi$

9.

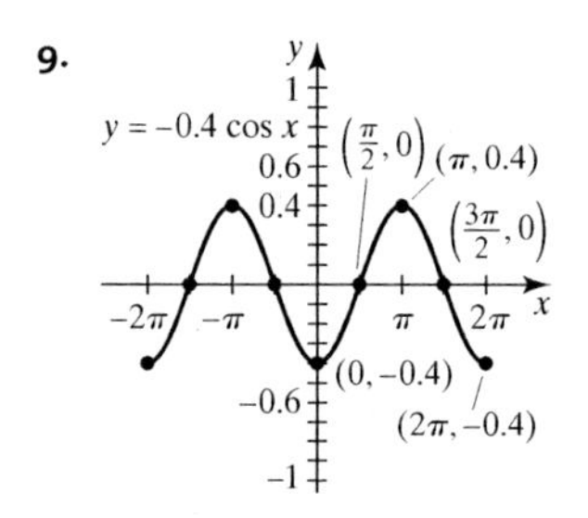

range: $|y| \le 0.4$, x-intercepts: $-3\pi/2, -\pi/2, \pi/2, 3\pi/2$ or $x = (\pi/2) + k\pi$

11.

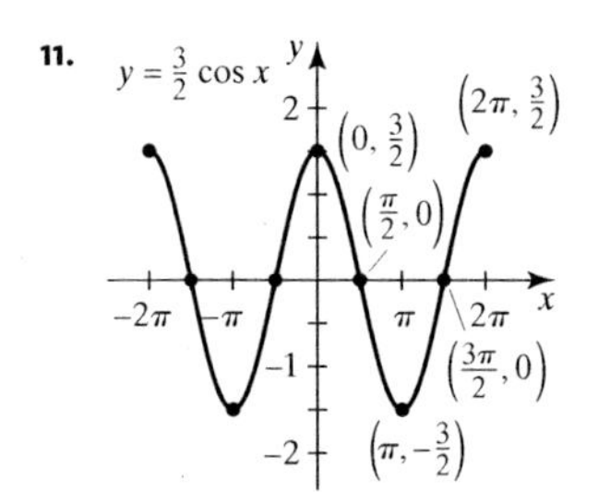

range: $|y| \le 3/2$, x-intercepts: $-3\pi/2, -\pi/2, \pi/2, 3\pi/2$ or $x = (\pi/2) + k\pi$

13. $y = -2 \sin x$ 15. $y = 3 \cos x$ 17. $y = 0.3 \sin x$

19. No, the period should be 2π.

21. No, graph should be arc shaped.

23. No, there should be four equal arcs; the period should be 2π.

25. a. 5 b. 2

27. negative identities: $\cos(-x) = \cos x$

29. a.

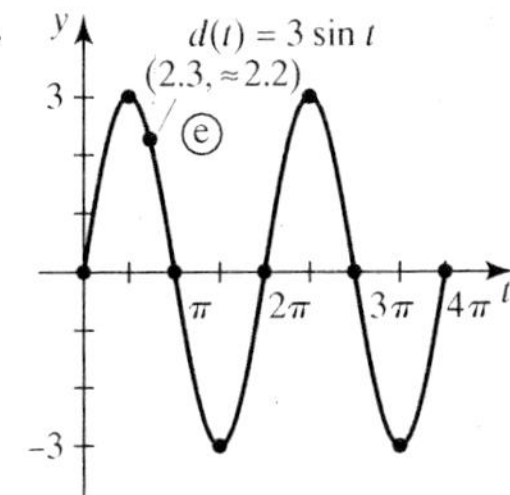

b. $t = 0, \pi, 2\pi, 3\pi, 4\pi$ **c.** 3 in.

d. The maximum displacement is equal to the amplitude.

e. $d \approx 2.2$ in. (see graph)

31. a. B **b.** F

33. T **35.** T **37.** T **39.** F

Exercise Set 2.2, p. 104

1.

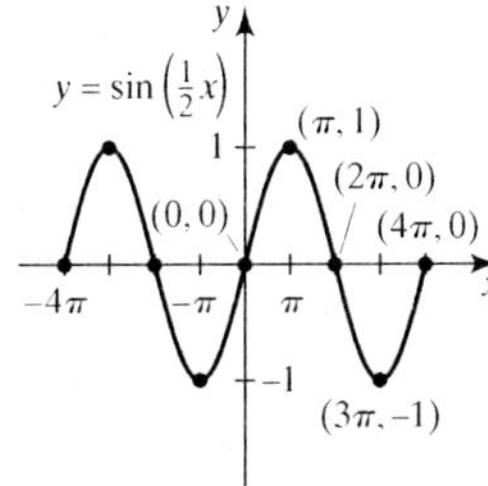

amplitude: 1, period: 4π, x-intercepts: -4π, -2π, 0, 2π or $x = k \cdot 2\pi$

3.

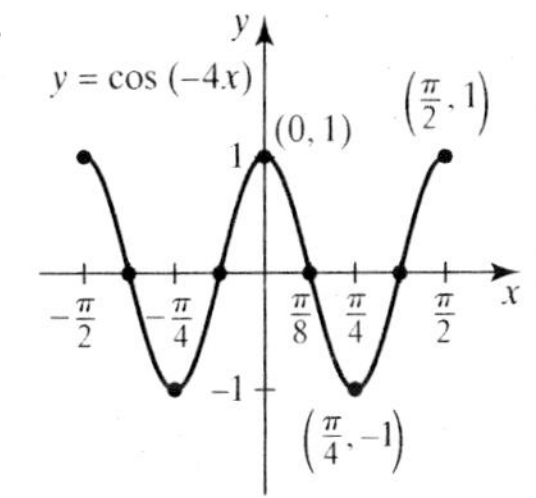

amplitude: 1, period: $\pi/2$, x-intercepts: $-3\pi/8$, $-\pi/8$, $\pi/8$, $3\pi/8$ or $x = (\pi/8) + k \cdot \pi/4$

5.

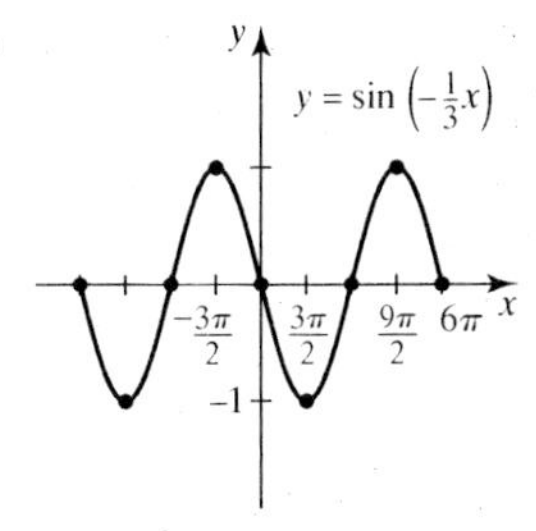

amplitude: 1, period: 6π, x-intercepts: -6π, -3π, 0, 3π, 6π or $x = k \cdot 3\pi$

7.

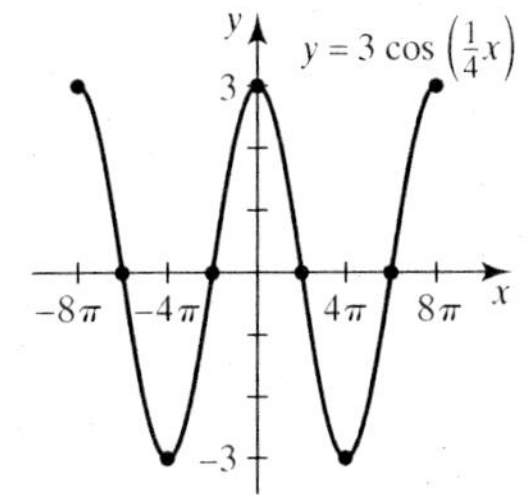

amplitude: 3, period: 8π, x-intercepts: -6π, -2π, 2π, 6π or $x = 2\pi + k \cdot 4\pi$

9.

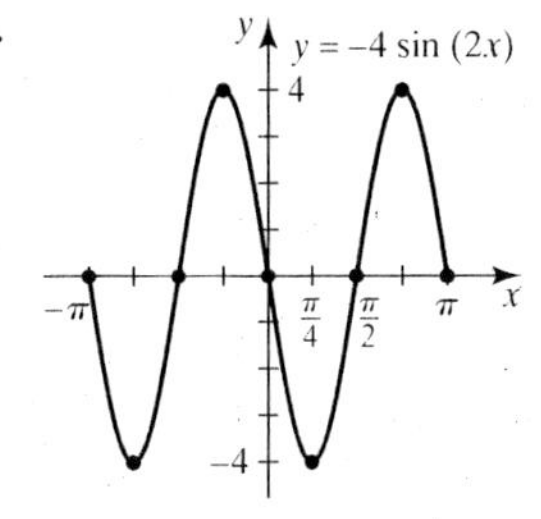

amplitude: 4, period: π, x-intercepts: $-\pi$, $-\pi/2$, 0, $\pi/2$, π or $x = k \cdot \pi/2$

11.

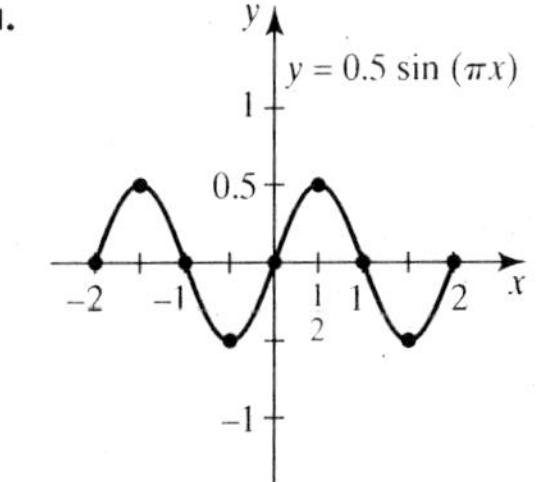

amplitude: 0.5, period: 2, x-intercepts: $-2, -1, 0, 1, 2$ or $x = k$

13.

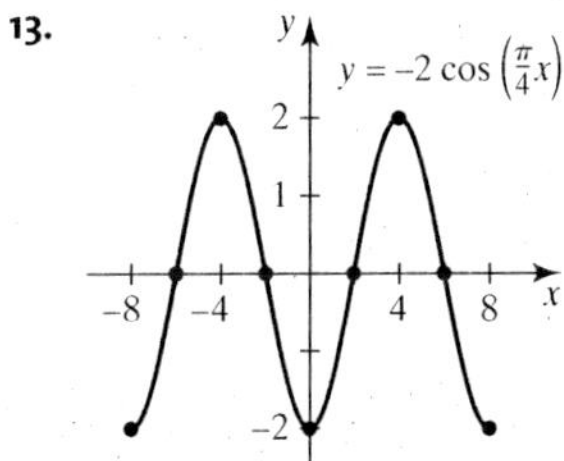

amplitude: 2, period: 8, x-intercepts: $-6, -2, 2, 6$ or $x = 2 + k \cdot 4$

15.

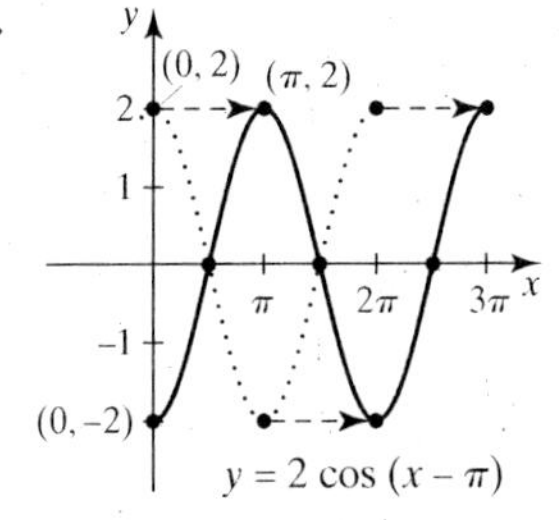

amplitude: 2, period: 2π, phase shift: right π, range: $|y| \leq 2$

Chapter 2

17.

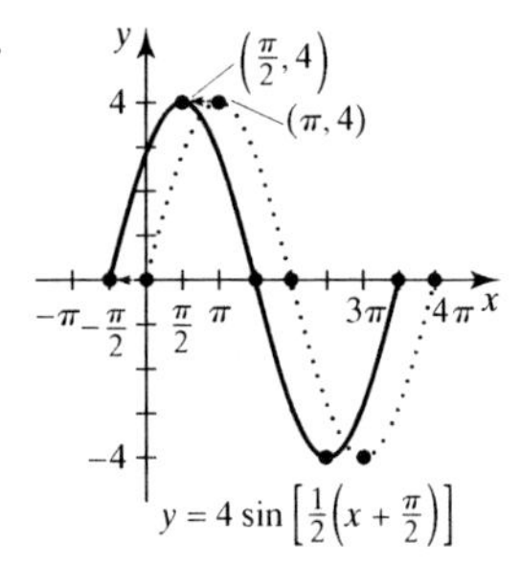

amplitude: 4, period: 4π, phase shift: left $\pi/2$, range: $|y| \le 4$

19.

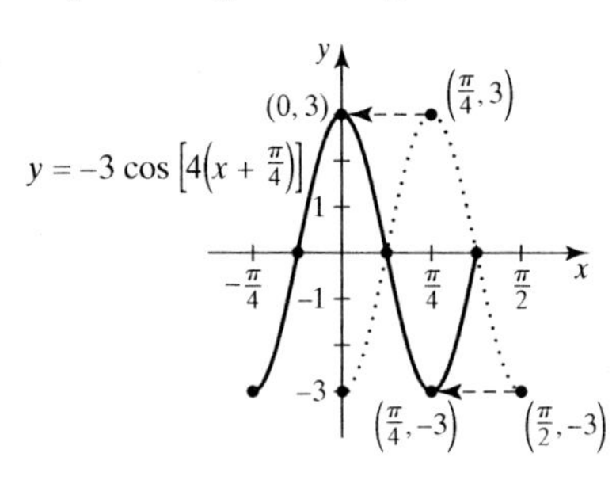

amplitude: 3, period: $\pi/2$, phase shift: left $\pi/4$, range: $|y| \le 3$

21.

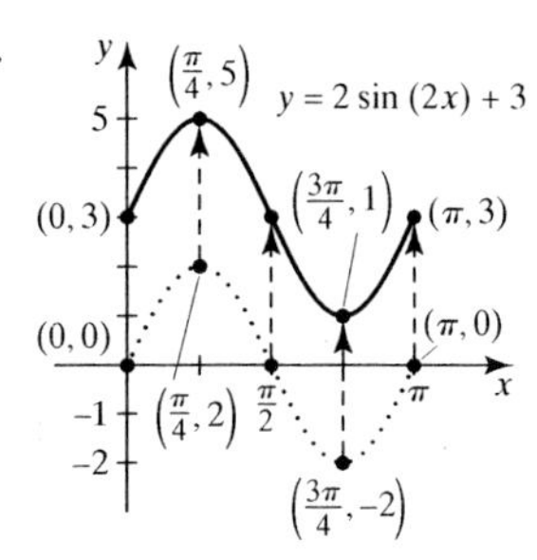

amplitude: 2, period: π, vertical shift: up 3, range: $1 \le y \le 5$

23. b, g

25. d, h

27. $y = 2\cos(2x)$

29. $y = 2\sin x + 3$

31. $y = -4\cos\left(\frac{1}{2}x\right)$

33. $y = \sin\left[2\left(x + \dfrac{\pi}{4}\right)\right]$, or $y = \sin\left[2\left(x - \dfrac{3\pi}{4}\right)\right]$

35. $y = \cos\left[\dfrac{1}{2}\left(x - \dfrac{\pi}{2}\right)\right]$, or $y = \sin\left[\dfrac{1}{2}\left(x + \dfrac{\pi}{2}\right)\right]$

37. $y = \sin\left(x - \dfrac{\pi}{2}\right)$

39. $y = -6\sin\left(\frac{1}{2}x\right)$, $y = 6\cos\left[\frac{1}{2}(x + \pi)\right]$

41. amplitude $= \frac{1}{4}$, $P = \pi/2$, phase shift: right $\frac{1}{2}$

43. amplitude $= 2.3$, $P = 12\pi$, vertical shift: up π

45. **a.** $y = \sin x + 2$ shifts the graph of $y = \sin x$ up 2 units; $y = \sin(x + 2)$ shifts the graph of $y = \sin x$ left 2 units.

b. $y = \sin(2x)$ has an amplitude of 1 and a period of π. $y = 2\sin x$ has an amplitude of 2 and a period of 2π.

c. $y = \frac{1}{2}\sin 2x$ has an amplitude of $\frac{1}{2}$ and a period of π; $y = \sin x$ has an amplitude of 1 and a period of 2π.

d. $y = \sin(x + (\pi/2))$ shifts $y = \sin x$ left $\pi/2$ units; $y = \sin x + \sin(\pi/2)$ is the same as $y = \sin x + 1$, which shifts $y = \sin x$ up 1 unit.

Exercise Set 2.3, p. 115

1. **a.** max = 88.7, min = 80.1

b. amplitude = 4.3

3. **a.** max = 74.5, min = 58.5

b. amplitude = 8

5. Yes

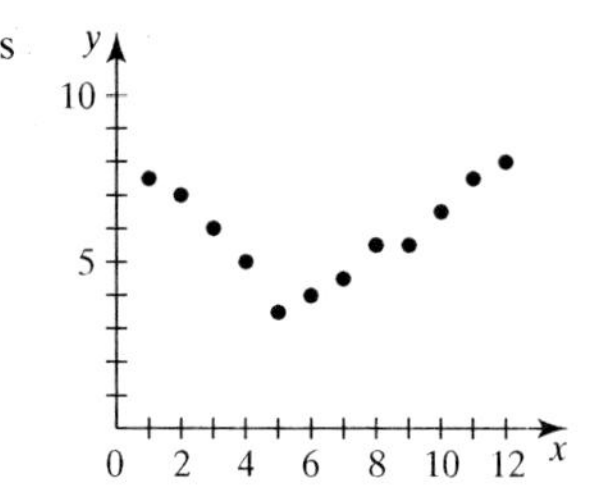

7. No

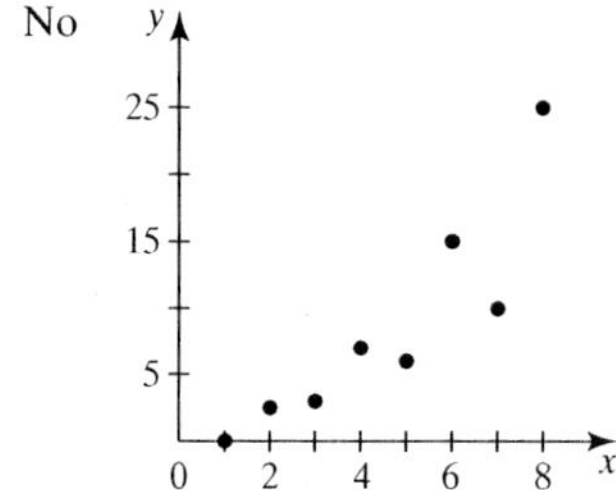

9. Yes

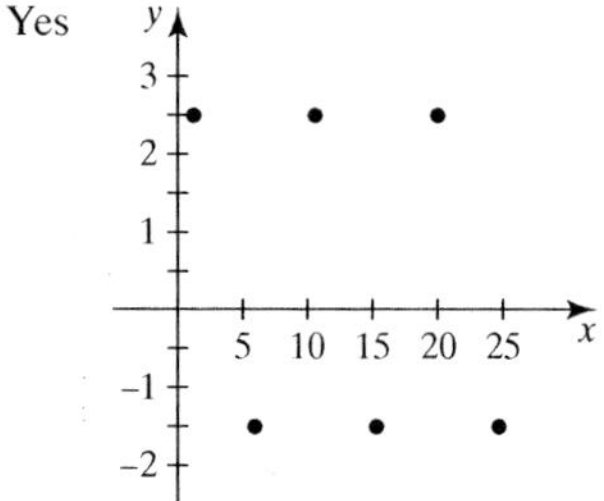

11. $y = 5.25\sin\left(\dfrac{2\pi}{15}(x + 1.75)\right) + 15.75$

13. $y = 9.5\sin\left(\dfrac{\pi}{6}(x - 5)\right) + 80.5$

15. $a \approx 5.1767$, $b \approx 0.3748$, $c \approx 1.4498$, $d \approx 16.1445$

17. **a.** $a \approx 1.4346$, $b \approx 0.5664$, $c \approx -1.8332$, $d \approx 5.3372$

b. At 5 AM, $x = 5$ and depth $\approx$ 6.5 ft; at 4 PM, $x = 16$ and depth $\approx$ 6.5 ft.

c. At 9 PM, $x = 21$ and depth $\approx$ 4.5 ft; the depth is not enough to bring the boat in.

19. **a.** $P_1 = \cos(20\pi x)$, $P_2 = -\cos(24\pi x)$

b. No

c. period $= \frac{1}{2}$, frequency $= 2$

21. Graph (b) is correct, period too small in graph (a), minimum height below ground level in graph (c)

Exercise Set 2.4, p. 129

	Function	Domain	Range	Period	Asymptotes	x-intercepts	y-intercept	Graph
1.	$y = \sin x$	$x \in \mathbb{R}$	$\|y\| \leq 1$	2π	—	$x = k\pi$	$(0, 0)$	
3.	$y = \tan x$	$x \neq \dfrac{\pi}{2} + k\pi$	$y \in \mathbb{R}$	π	$x = \dfrac{\pi}{2} + k\pi$	$x = k\pi$	$(0, 0)$	
5.	$y = \sec x$	$x \neq \dfrac{\pi}{2} + k\pi$	$\|y\| \geq 1$	2π	$x = \dfrac{\pi}{2} + k\pi$	—	$(0, 1)$	

7. $y = \sin x$, $y = \tan x$

9. $y = \tan x$, $y = \cot x$, $y = \csc x$, $y = \sin x$

11. all

13. $y = \tan x$, $y = \cot x$, $y = \sec x$, $y = \csc x$

15. asymptotes: $x = \pi + k \cdot 2\pi$, x-intercepts: $x = k \cdot 2\pi$

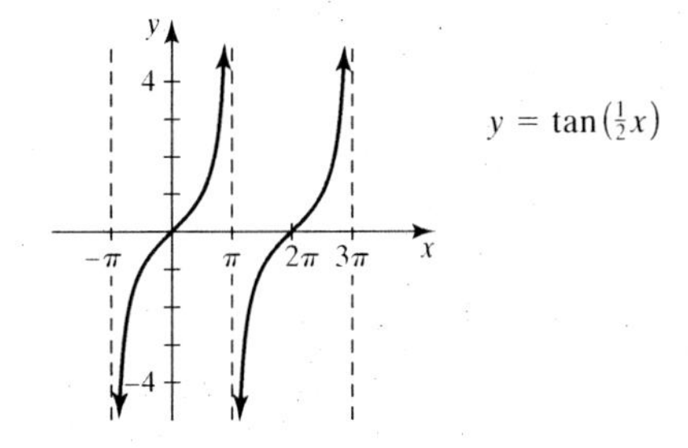

$y = \tan(\frac{1}{2}x)$

17. asymptotes: $x = k\pi$, x-intercepts: $x = (\pi/2) + k\pi$

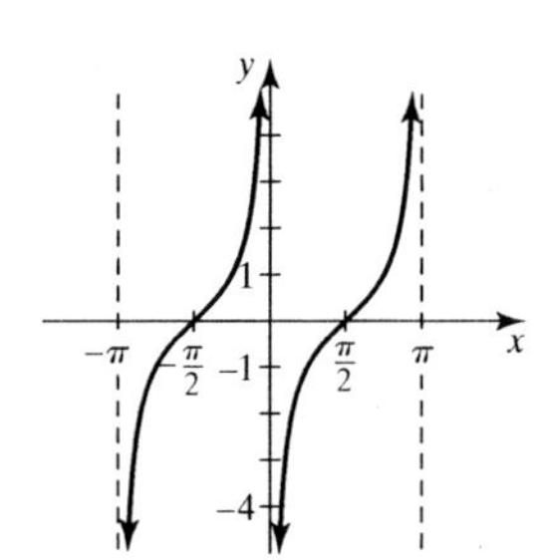

$y = \tan\left(x + \frac{\pi}{2}\right)$

19. asymptotes: $x = k \cdot \pi/2$, x-intercepts: $x = (\pi/4) + k \cdot \pi/2$

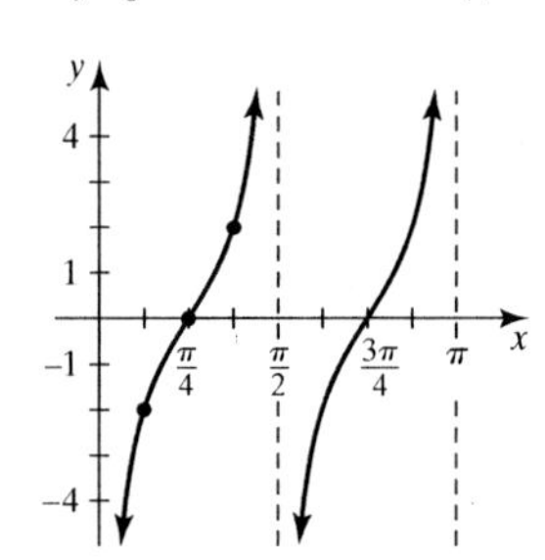

$y = -2\cot(2x)$

21. asymptotes: $x = k\pi$, x-intercepts: $x = (\pi/4) + k\pi$

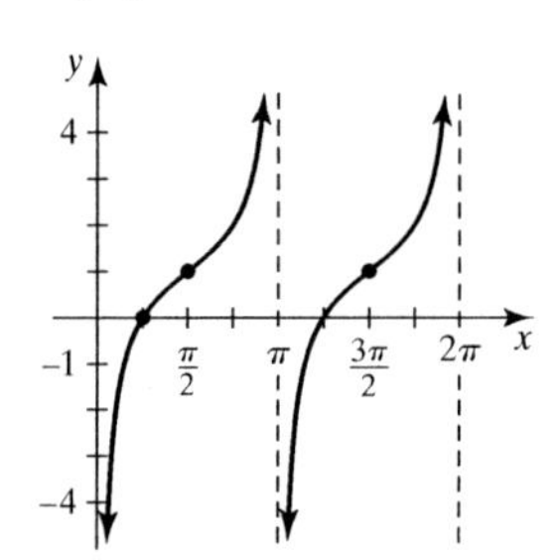

$y = -\cot x + 1$

23. $P = 2\pi$, range: $|y| \geq 4$

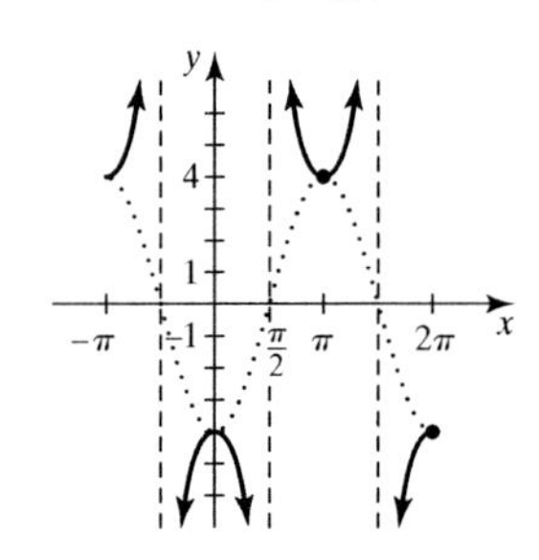

$y = -4\sec x$

25. $P = \pi$, range: $|y| \geq 1$

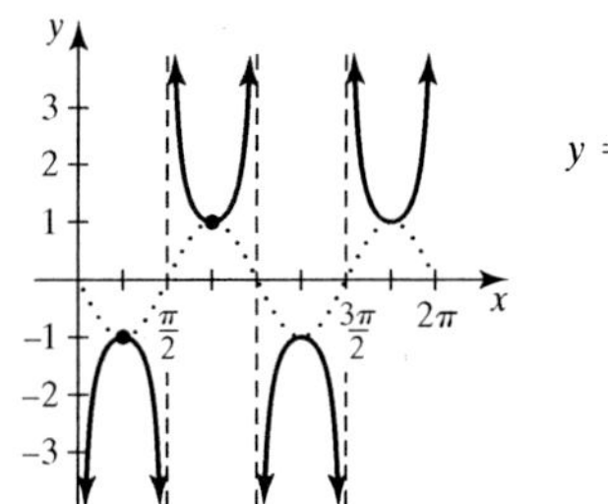

$y = \csc(-2x)$

27. $P = 2\pi$, range: $|y| \geq 4$

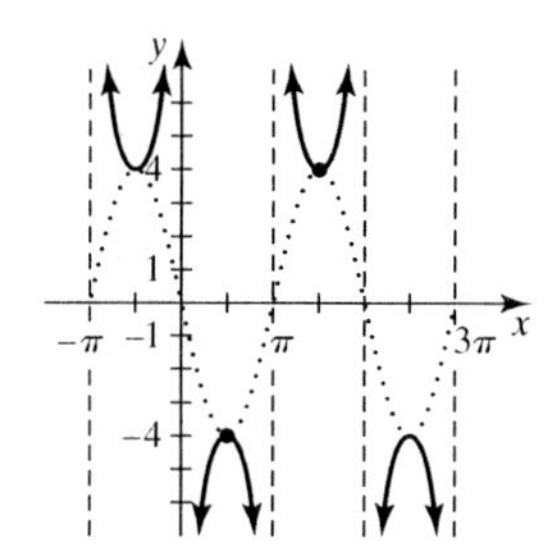

$y = 4\sec\left(x + \frac{\pi}{2}\right)$

29. c

31. a

33. Graph does not represent a function, should not cross the asymptotes.

35. One branch is missing (incomplete cycle)

37. a, c

39–41. *Variations on* Ymin *and* Ymax *are possible.*

39. $P = 10$, Range: $\mathbb{R}$ WINDOW Xmin $= -10$, Xmax $= 10$, Ymin $= -20$, Ymax $= 20$

41. $P = \pi$, Range: $|y| \geq 5$, WINDOW Xmin $= -\pi$, Xmax $= \pi$, Ymin $= -20$, Ymax $= 20$

Exercise Set 2.5, p. 143

1. $\sin y = \sqrt{3}/2$, $y = \pi/3$

3. $\cos y = 1$, $y = 0$

5. $\csc y = -\sqrt{2}$ or $\sin y = \frac{-1}{\sqrt{2}}$, $y = -\pi/4$

7. $\pi/4$	**9.** $\pi/3$	**11.** 0	**13.** π
15. $\pi/4$	**17.** $-\pi/3$	**19.** $5\pi/6$	**21.** 2.3005
23. 0.9273	**25.** 2.5201	**27.** 0.3948	**29.** 0.7004
31. -1.4048	**33.** 1.3221	**35.** $\sqrt{3}/2$	**37.** 1
39. $\sqrt{3}$	**41.** 0	**43.** 2	**45.** $\sqrt{2}/2$
47. $-\sqrt{3}/3$	**49.** $-\pi/3$	**51.** 0	**53.** $\pi/6$
55. π	**57.** $\frac{5}{13}$	**59.** $\frac{15}{17}$	**61.** 0.2756
63. 0.8840	**65.** -0.6261	**67.** -1.8000	**69.** 1.3149

71. **a.** $y = \tan x$, **ii.** $y = \arctan x$

b. $y = \sin x$, **i.** $y = \arcsin x$

c. $y = \cos x$, **iii.** $y = \arccos x$

73. $x = \frac{1}{6}\sin^{-1}(-1) = -\pi/12$

75. $x = \frac{1}{9}\tan^{-1}\left(\frac{7}{4}\right) \approx 0.1169$

77. $\sin^{-1} x$ is the inverse function of $\sin x$, whereas $(\sin x)^{-1}$ is the reciprocal function of $\sin x$. Another name for $\sin^{-1} x$ is $\arcsin x$, and another name for $(\sin x)^{-1}$ is $\csc x$.

Chapter 2 Review Exercises, p. 151

1. amplitude = 2

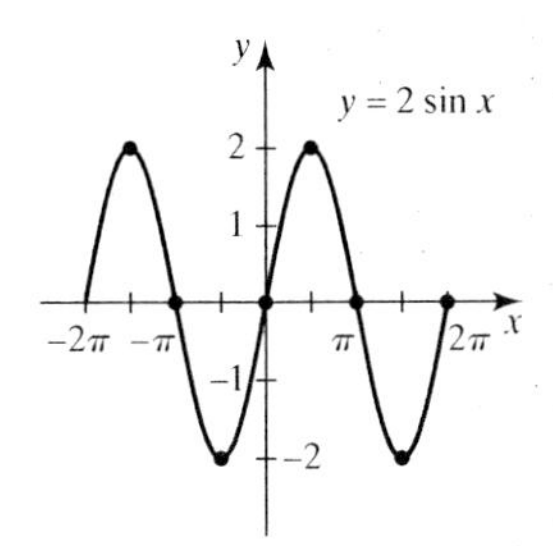

2. amplitude = 3

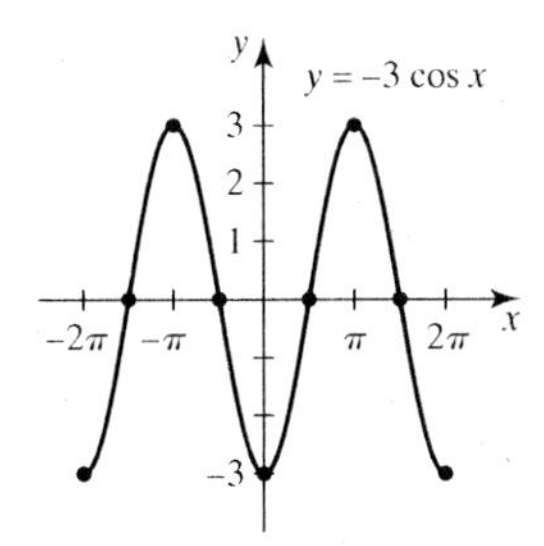

3. amplitude = 0.7

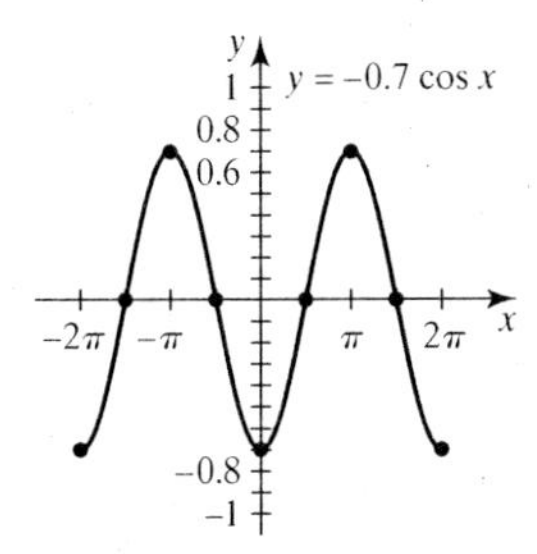

4. amplitude = $\frac{1}{2}$

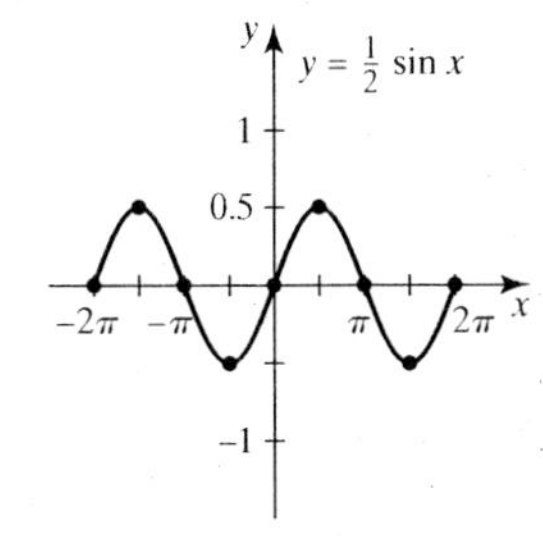

5. amplitude = $\frac{3}{2}$

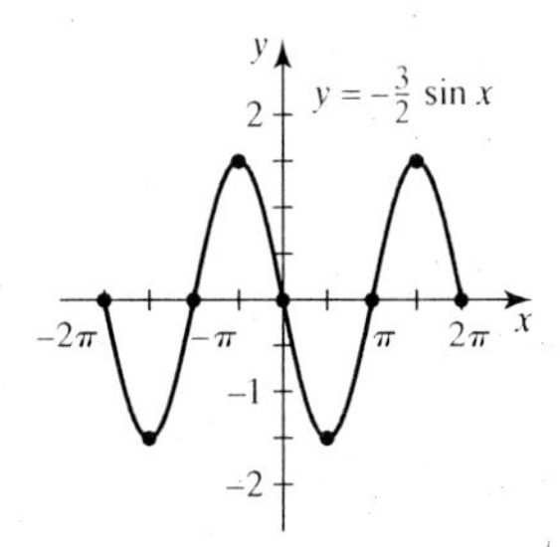

6. amplitude = 2π

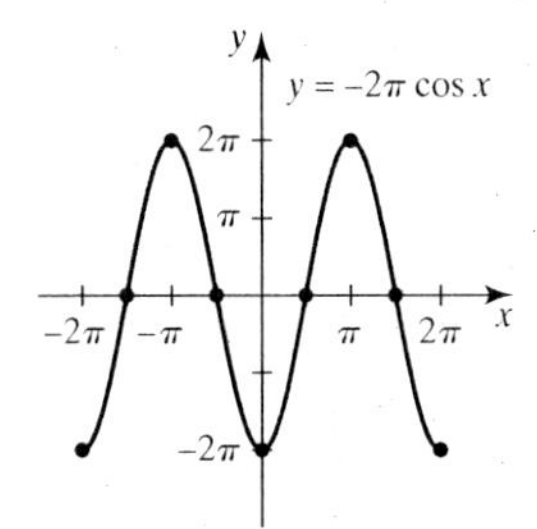

7. period: 4π, range: $|y| \le 4$, x-intercepts: $x = \pi + k \cdot 2\pi$

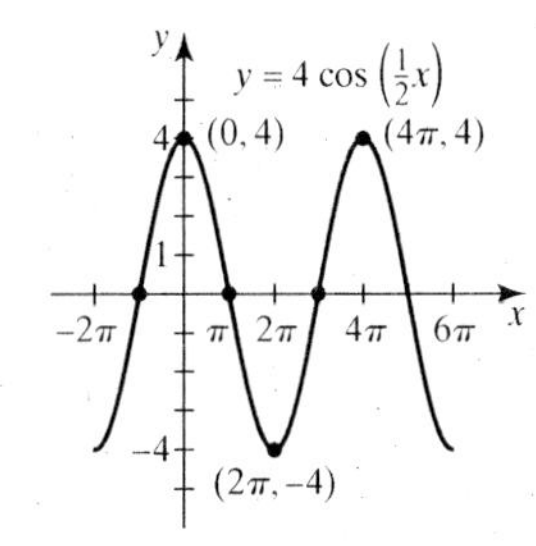

8. period: π, range: $|y| \le 1.5$, x-intercepts: $x = k \cdot \pi/2$

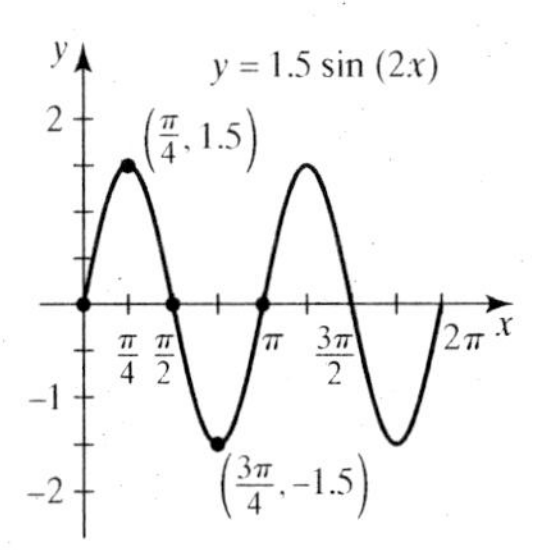

9. period: 8π, range: $|y| \le 3$, x-intercepts: $x = k \cdot 4\pi$

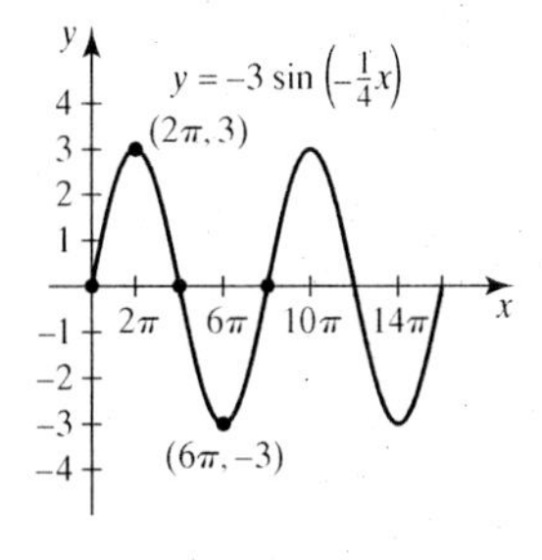

10. period: 4π, range: $|y| \le 2$, x-intercepts: $x = \pi + k \cdot 2\pi$

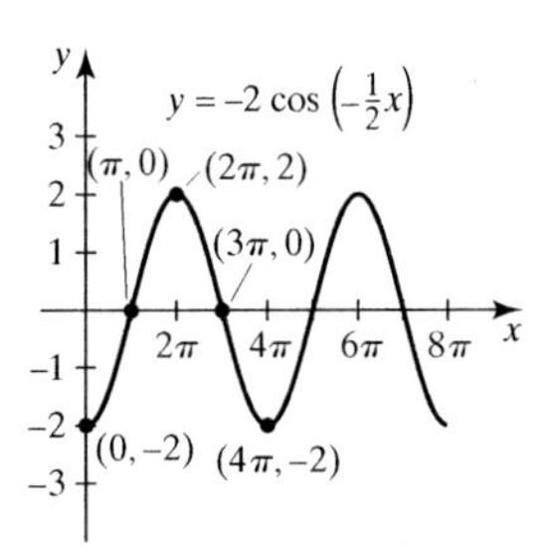

11. period: 2, range: $|y| \le 1$, x-intercepts: $x = \frac{1}{2} + k$

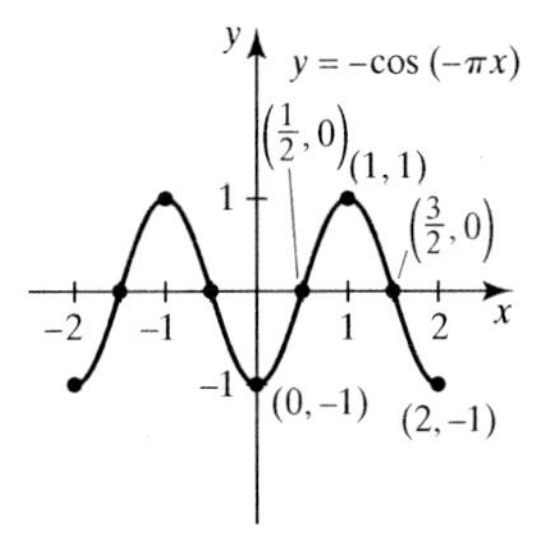

12. period: 5, range: $|y| \le 5$, x-intercepts: $x = k \cdot \frac{5}{2}$

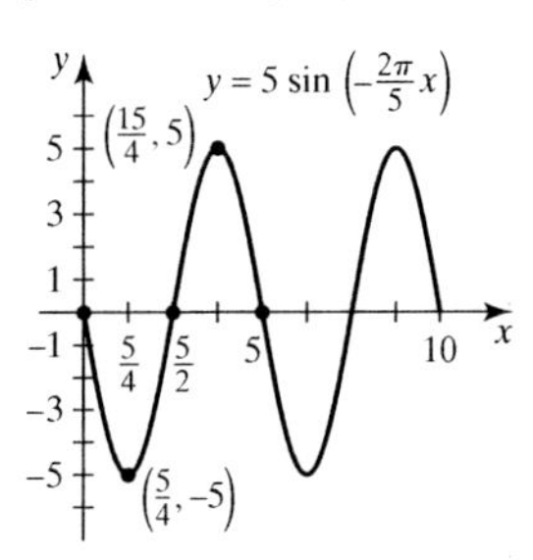

13. period: 2π, phase shift: right $\pi/2$

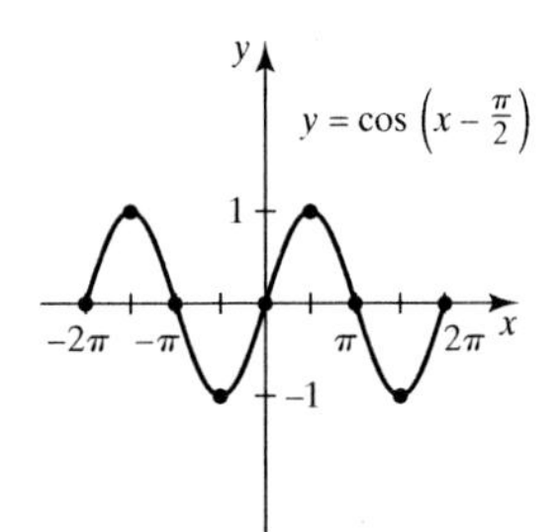

14. period: 2π, phase shift: left $\pi/2$

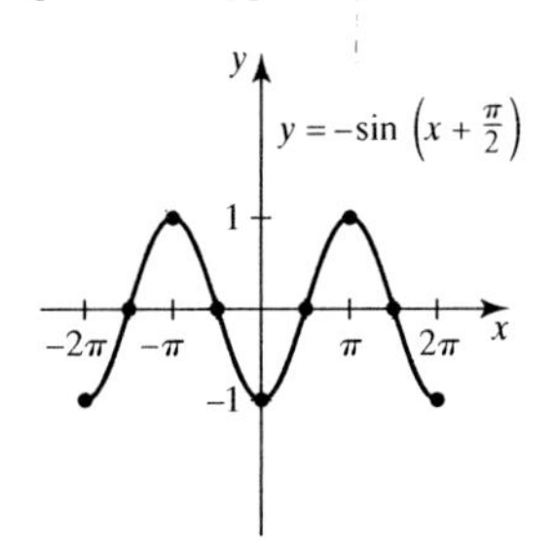

15. period: 4π, phase shift: left π

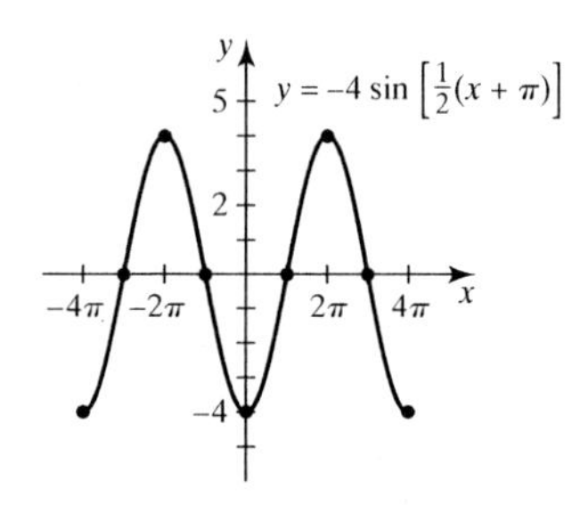

16. period: π, phase shift: right $\pi/4$

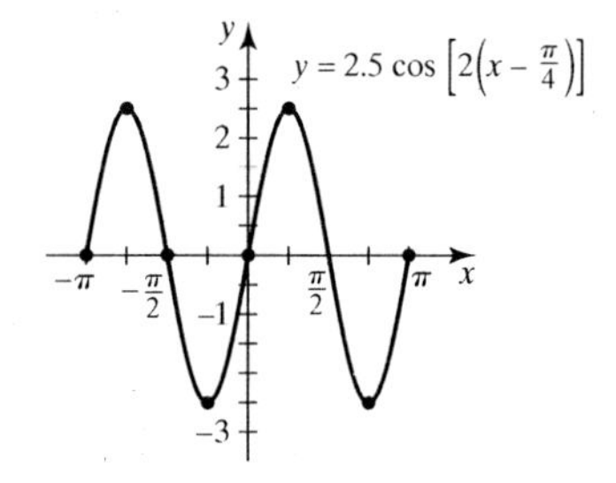

17. period: 6π, vertical shift: up 4

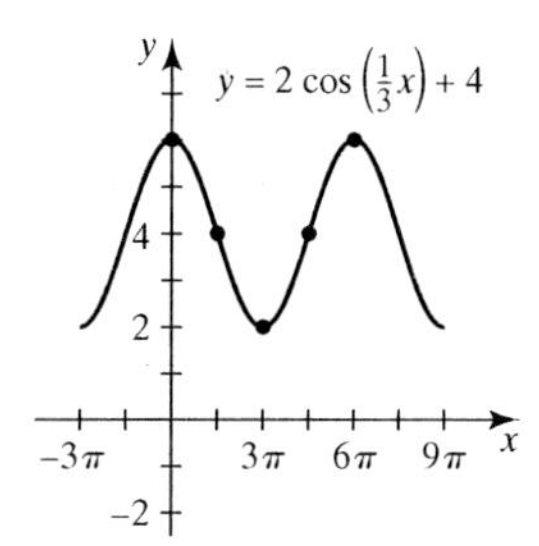

18. period: π, vertical shift: down $\frac{1}{2}$

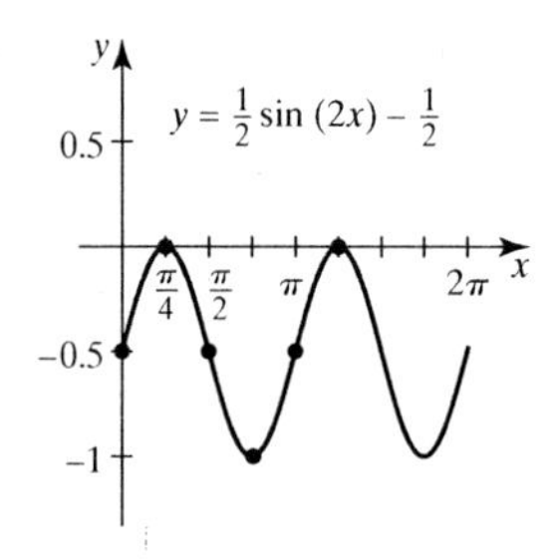

19. $y = 5\sin x$

20. $y = -4\cos x$

21. $y = 0.6\cos x$

22. $y = -\frac{1}{2}\sin x$

23–24. *There are many possible answers.*

23. $y = 4\cos\left[2\left(x + \frac{\pi}{4}\right)\right]$, $y = 4\sin\left[2\left(x + \frac{\pi}{2}\right)\right]$,

$y = 4\sin\left[2\left(x - \frac{\pi}{2}\right)\right]$

24. $y = \sin\left[2\left(x + \frac{\pi}{8}\right)\right]$, $y = \cos\left[2\left(x - \frac{\pi}{8}\right)\right]$, $y = \sin\left[2\left(x - \frac{7\pi}{8}\right)\right]$, $y = \cos\left[2\left(x + \frac{7\pi}{8}\right)\right]$

25. See answers for Exercise 23 or check with a graphing utility.

26. See answers for Exercise 24 or check with a graphing utility.

27. a.

b. $t = \frac{1}{2}, \frac{3}{2}, \frac{5}{2}, \frac{7}{2}$

c. See graph in part (a).

28. a.

b. See graph in part (a).

29. a. $y = -2\frac{5}{8}\cos\left(\frac{\pi}{6}x\right) + 12\frac{1}{8}$, $y = 2\frac{5}{8}\sin\left(\frac{\pi}{6}(x - 3)\right) + 12\frac{1}{8}$

c. For $y = a\sin(bx + c) + d$: $a = 2.6327$, $b = 0.5073$, $c = -1.4729$, $d = 12.1466$

30. a. $y = -5\frac{1}{4}\cos\left(\frac{\pi}{6}(x - 1)\right) + 12\frac{1}{4}$,

$y = 5\frac{1}{4}\sin\left(\frac{\pi}{6}(x - 4)\right) + 12\frac{1}{4}$

c. $a = 5.1516$, $b = 0.5094$, $c = -1.8199$, $d = 12.1474$

31. period: 3π, range: $\mathbb{R}$

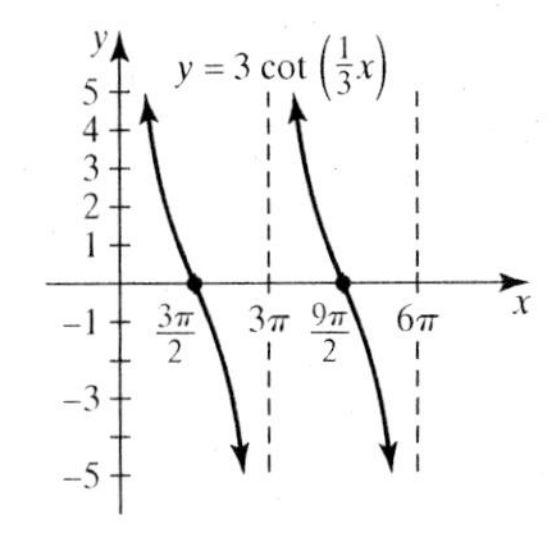

32. period: $\pi/2$, range: $\mathbb{R}$

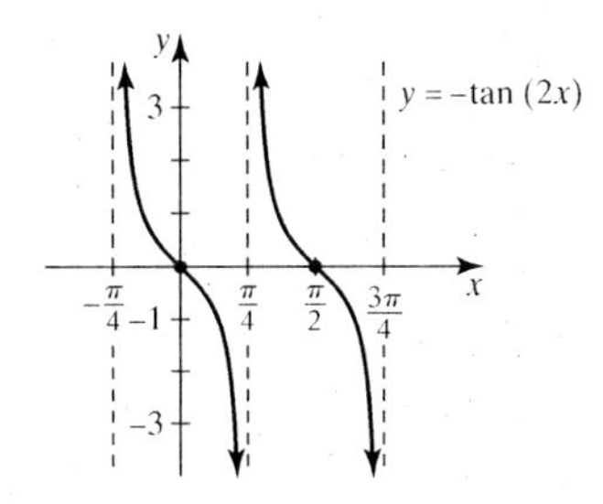

33. period: 4π, range: $\mathbb{R}$

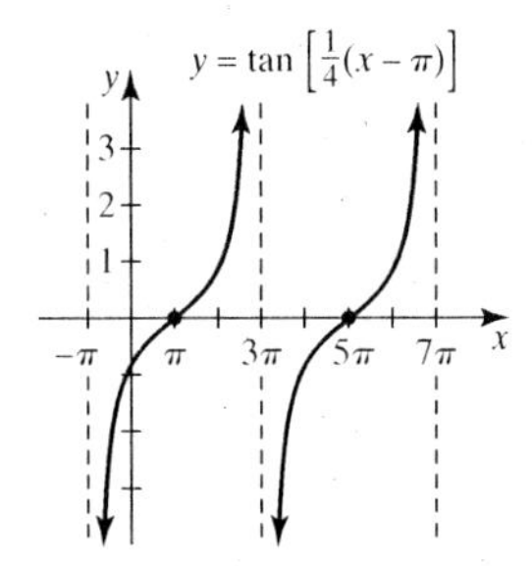

34. period: 2π, range: $\mathbb{R}$

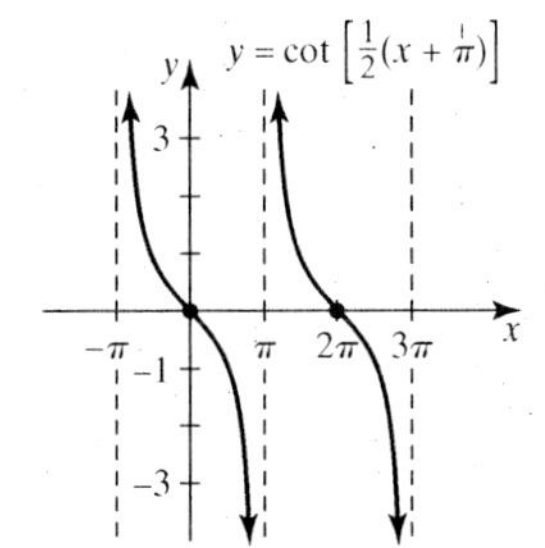

35. period: 4π, range: $|y| \geq \frac{1}{2}$

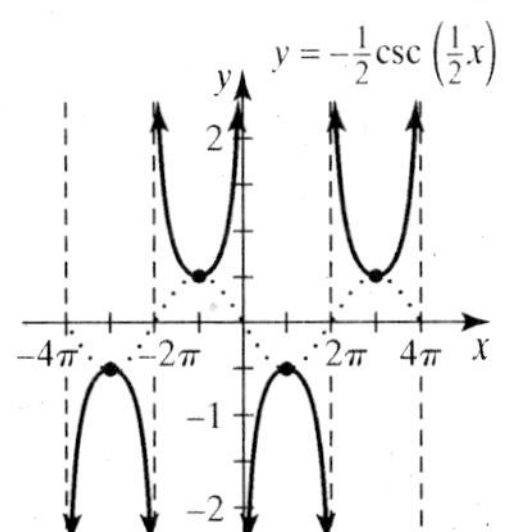

36. period: π, range: $|y| \geq 3$

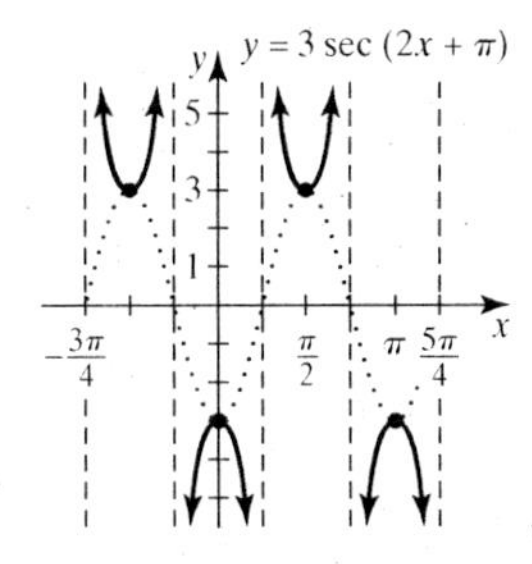

37. period: π, range: $y \geq 5$ or $y \leq 3$

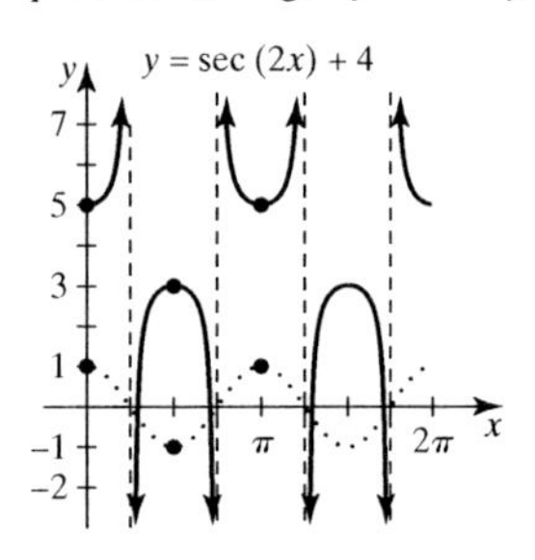

38. period: 2π, range: $y \geq 0$ or $y \leq -4$

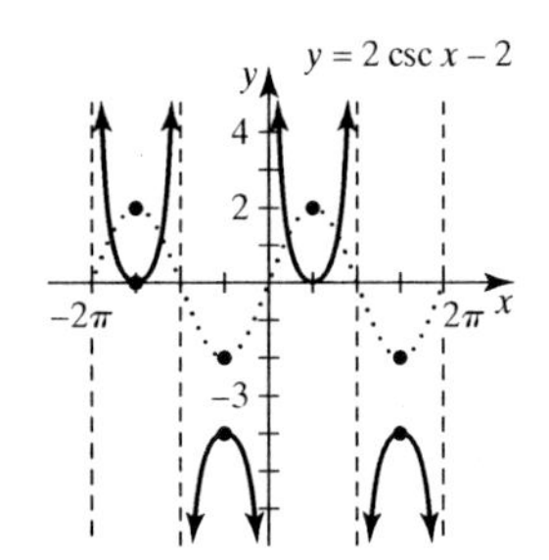

39–42. *Many answers are possible.*

39. $y = -\tan\left(\frac{1}{2}x\right)$

40. $y = 3\csc(2x)$

41. $y = 2\sec\left(\frac{1}{4}x\right)$

42. $y = \cot\left(x + \frac{\pi}{4}\right)$

43. $\sin y = \frac{\sqrt{2}}{2}, y = \frac{\pi}{4}$

44. $\cos y = -1, y = \pi$

45. $\tan y = -\sqrt{3}, y = -\frac{\pi}{3}$

46. $\sin y = -\frac{\sqrt{3}}{2}, y = -\frac{\pi}{3}$

47. $\cos y = -\frac{1}{\sqrt{2}}, y = \frac{3\pi}{4}$

48. $\cot y = 0$, or $\tan y$ undefined, $y = \frac{\pi}{2}$

49. $\frac{\sqrt{2}}{2}$ **50.** $\frac{\pi}{6}$ **51.** $\sqrt{2}$ **52.** 0

53. $-\frac{\pi}{2}$ **54.** $\frac{2\pi}{3}$ **55.** $\frac{12}{13}$ **56.** $\frac{8}{15}$

57. 3.1195 **58.** 0.2570 **59.** 0.9897 **60.** 1.2470

61. 0.7850 **62.** 0.3309 **63.** $y = \sin^{-1} x$

64. $y = \tan^{-1} x$

65. $x = \frac{1}{2}\arcsin\left(\frac{0.5}{4}\right) \approx 0.0627$

66. $x = 2\arccos\left(\frac{0.9}{3.2}\right) \approx 2.5714$

67. $x = \frac{1}{4}\arctan\left(-\frac{y}{3}\right)$

68. $x = -\frac{1}{4}\arcsin\left(\frac{y}{\pi}\right)$

69. F **70.** F **71.** F **72.** T

73. T **74.** F **75.** F **76.** T

77. T **78.** F **79.** T **80.** F

81. F **82.** F **83.** T **84.** F

85. Jo is correct. If you use the guide Jack suggests, you will not have any asymptotes since $y = \cos(2x) + 4$ has no x-intercepts.

Chapter 2 Test, p. 155

1. period: 4π, amplitude: 5, x-intercepts: $x = k \cdot 2\pi$

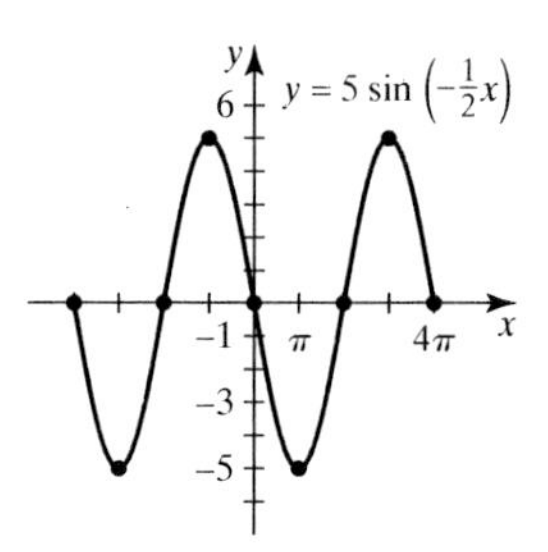

2. period: π, phase shift: right $\pi/2$, x-intercepts: $x = (\pi/4) + k \cdot \pi/2$

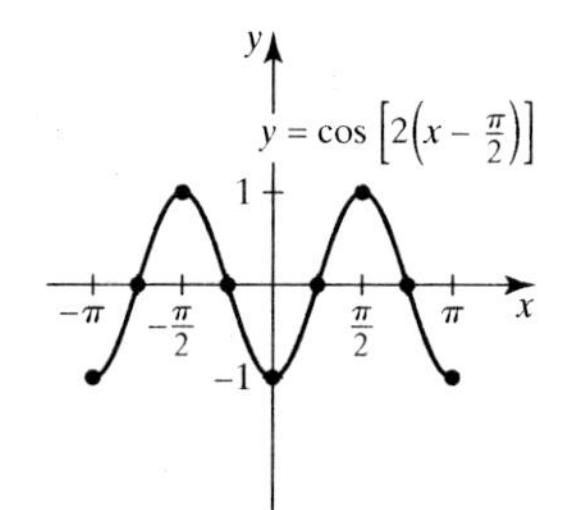

3. period: 2π, asymptotes: $x = \pi + k \cdot 2\pi$, range: $\mathbb{R}$

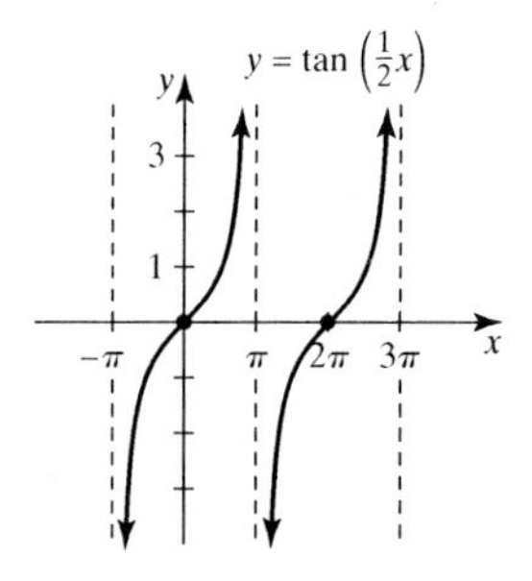

4. period: 2π, asymptotes: $x = (\pi/2) + k \cdot \pi$, range: $|y| \geq 3$

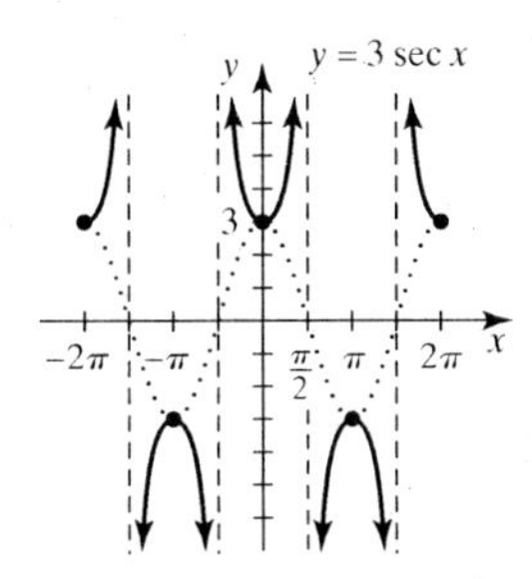

5. period: 2, x-intercepts: $x = \frac{3}{2} + k \cdot 2$, range: $0 \leq y \leq 4$

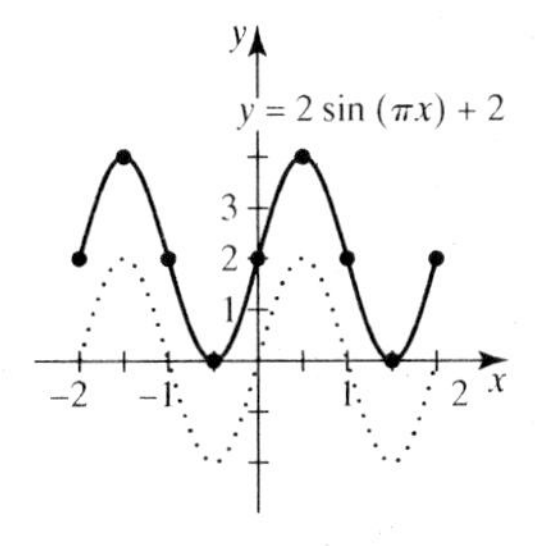

6. range: $|y| \geq 5$, asymptotes: $x = (\pi/2) + k\pi$

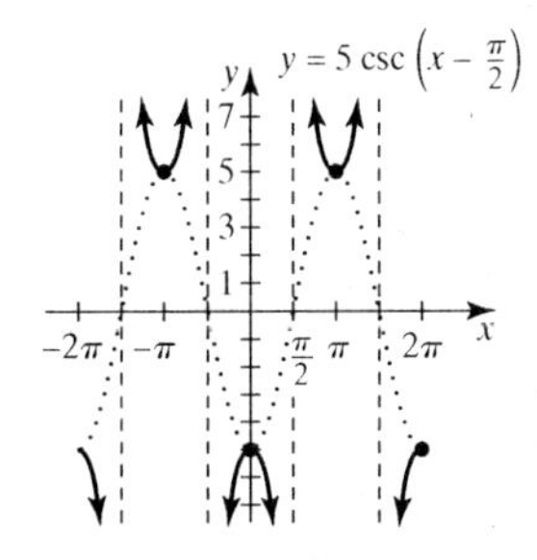

7–8. *There are other possible answers.*

7. a. $y = -2\cos x$ b. $y = \tan(2x)$
   c. $y = 4\sec(\frac{1}{2}x)$ d. $y = \sin\left(x + \frac{\pi}{4}\right)$

8. $y = 2\sin\left(x - \frac{\pi}{2}\right)$

9. a.

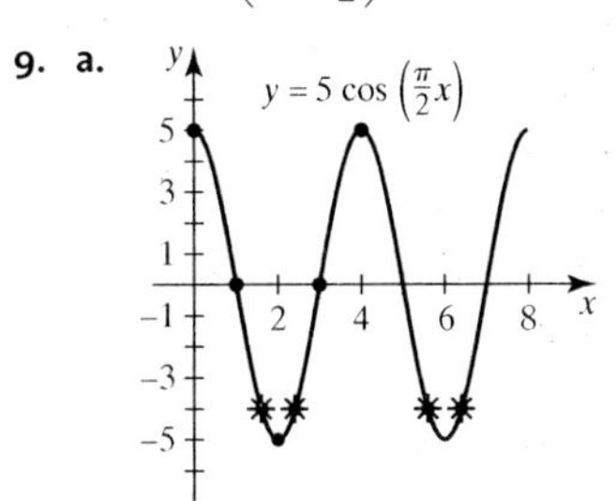

   b. 5 cm c. See graph.

10. a. $\pi/3$ b. 0 c. $\pi/4$ d. $\pi/6$
    e. $3\pi/4$ f. $-\pi/4$ g. π h. $\pi/3$
    i. $\frac{4}{5}$ j. π

11. a. 1.4963 b. 2.6681 c. 2.1394 d. 0.5944
    e. −3.6700 f. 0.9776

12. a. $x = \frac{1}{8}\sin^{-1}(\frac{1}{3}) = \frac{\pi}{48} \approx 0.0654$ b. $x = \frac{1}{2}\tan^{-1}(y/4)$

13. a. $y = \arctan x$ b. $y = \arccos x$

14. a. F b. T c. T
    d. F e. F f. T
    g. T h. F i. T

15. a. 6, February b. April to December c. 18, August

Chapter 3

Exercise Set 3.1, p. 169

1. $\frac{180}{\pi}$

3. Values go from 0° counterclockwise: 0, 30°, $\frac{\pi}{4}$, $\frac{\pi}{3}$, 90°, 120°, $\frac{3\pi}{4}$, $\frac{5\pi}{6}$, 180°, 210°, 225°, $\frac{4\pi}{3}$, $\frac{3\pi}{2}$, $\frac{5\pi}{3}$, 315°, 330°

5. 45° 7. 150° 9. 480° 11. −210°

13. −540° 15. 57° 17. $\frac{5\pi}{12}$ 19. $\frac{5\pi}{4}$

21. $-\frac{\pi}{15}$ 23. $-\frac{2\pi}{3}$ 25. 0.28 27. 4.16

29. a. $\frac{2\pi}{3}$ b. $\frac{2\pi}{3} + k \cdot 2\pi$

31. a. 52° b. $52° + k \cdot 360°$

33. 1.07, $1.07 + k \cdot 2\pi$ 35. 10°, $10° + k \cdot 360°$

37. 18.33 in. 39. 10.5 cm 41. 33.51 m 43. $r = 8$ in.

45. a. $\frac{\pi}{4}$ b. $\frac{\pi}{3}$ c. π d. $\frac{\pi}{2}$

47. $v = \frac{s}{t} = \frac{r\theta}{t} = r\frac{\theta}{t} = r\omega$, or $\frac{v}{r} = \omega$ 49. $v \approx 0.79$ ft/hr

51. $\omega \approx 0.16$ rad/sec 53. $v = 8$ ft/min 55. $s = 140$ ft

57. $s = 62.4$ ft 59. F 61. T 63. F

65. 5000π rad/min ≈ 15,707.96 rad/min 67. 1,600,921 mi

69. 11,206,448 mi/wk 71. 1.16 mi 73. 2356 in./min

75. a. 49 sq in. b. 4π sq ft c. $\frac{25\pi}{3}$ sq cm

Chapter 3

Exercise Set 3.2, p. 182

	$\cos\theta$	$\sin\theta$	$\tan\theta$
1.	$\frac{8}{17}$	$\frac{15}{17}$	$\frac{15}{8}$
3.	$\frac{\sqrt{2}}{2}$	$-\frac{\sqrt{2}}{2}$	-1
5.	$-\frac{4}{5}$	$\frac{3}{5}$	$-\frac{3}{4}$
7.	$-\frac{\sqrt{3}}{2}$	$-\frac{1}{2}$	$\frac{1}{\sqrt{3}} = \frac{\sqrt{3}}{3}$
9.	-1	0	0

11. QIV

13. QII

	$\cos\beta$	$\sin\beta$	$\tan\beta$	$\sec\beta$	$\csc\beta$	$\cot\beta$
15.	$\frac{1}{4}$	$-\frac{\sqrt{15}}{4}$	$-\sqrt{15}$	4	$-\frac{4\sqrt{15}}{15}$	$-\frac{1}{\sqrt{15}} = -\frac{\sqrt{15}}{15}$
17.	$\frac{2}{3}$	$\frac{\sqrt{5}}{3}$	$\frac{\sqrt{5}}{2}$	$\frac{3}{2}$	$\frac{3\sqrt{5}}{5}$	$\frac{2\sqrt{5}}{5}$
19.	$-\frac{\sqrt{15}}{4}$	$-\frac{1}{4}$	$\frac{\sqrt{15}}{15}$	$-\frac{4\sqrt{15}}{15}$	-4	$\sqrt{15}$
21.	$-\frac{2\sqrt{5}}{5}$	$\frac{\sqrt{5}}{5}$	$-\frac{1}{2}$	$-\frac{\sqrt{5}}{2}$	$\sqrt{5}$	-2

23. $\frac{\sqrt{3}}{2}$ **25.** -2 **27.** 1 **29.** -1

31. 0 **33.** $\frac{\sqrt{3}}{3}$ **35.** undefined **37.** 0.4663

39. 0.9848 **41.** 0.5962 **43.** -1.7116 **45.** -5.6632

47. 0.8352 **49.** $30°$ **51.** $-30°$ **53.** $90°$

55. $77.0°$ **57.** $-27.1°$ **59.** $-24.6°$ **61.** $\frac{\sqrt{15}}{4}$

63. $\frac{4}{3}$ **65.** $\frac{13}{5}$ **67.** $\frac{\sqrt{4 - x^2}}{2}$

Exercise Set 3.3, p. 195

1. $\sin 60°$ **3.** $\csc 6.3°$ **5.** $\cot 76°43'$

7. a. y **b.** x **c.** $\frac{x}{z}$ **d.** $\frac{x}{y}$ **e.** T

9. $\cos 70° = \frac{3}{x}$ **11.** $\sin x = \frac{3}{5}$ **13.** not possible

15. $\sin\alpha = \frac{4}{5}$ $\cos\alpha = \frac{3}{5}$ $\tan\alpha = \frac{4}{3}$

17. $\sin\alpha = \frac{1}{2}$ $\cos\alpha = \frac{\sqrt{3}}{2}$ $\tan\alpha = \frac{\sqrt{3}}{3}$

19. $46.4°$ **21.** 9.34 **23.** $3\sqrt{10} \approx 9.49$

25. 3.09 **27.** $19.5°$

29.

$a = 7$	$\alpha = 74.8°$
$b \approx 1.90$	$\beta = 15.2°$
$c \approx 7.25$	$\gamma = 90°$

31.

$a = 3$	$\alpha \approx 23.2°$
$b = 7$	$\beta \approx 66.8°$
$c = \sqrt{58} \approx 7.62$	$\gamma = 90°$

33.

$a = 21$	$\alpha = 80°$
$b \approx 3.70$	$\beta = 10°$
$c \approx 21.32$	$\gamma = 90°$

35.

$a = \sqrt{27.75} \approx 5.27$	$\alpha \approx 31.8°$
$b = 8.5$	$\beta \approx 58.2°$
$c = 10$	$\gamma = 90°$

37.

$a \approx 4.70$	$\alpha = 11.4°$
$b \approx 23.33$	$\beta = 78.6°$
$c = 23.8$	$\gamma = 90°$

39. Each trigonometric ratio would have two unknown values.

41. a. No, since there are two adjacent sides to γ: a and b.
b. $\tan\gamma = \tan 90°$, which is undefined.
c. $\cos\gamma = \cos 90° = 0 = \frac{\text{adj}}{\text{hyp}}$, would imply the adjacent side is 0.

43. a. 1, 3, 8, 11 **b.** 4, 5, 9, 10
c. Neither side of the angle is horizontal.

45. 33 ft **47.** 31.3 m **49.** $26.6°$

51. $h = 2\sqrt{3}$ in., $A = 4\sqrt{3}$ sq in. ≈ 6.93 sq in. **53.** 15 ft

55. a. 1 **b.** 4

57. a. 2.01 mi **b.** N 28° W, 4.27 mi

59. $h \approx 27.57$ **61.** $h \approx 468.35$ ft **63.** $6.8°$

65. a. $65°$
b. Between two points the angle of depression is equal to the angle of elevation.

Exercise Set 3.4, p. 214

1. AAS, law of sines
3. SSA, law of sines (ambiguous case)
5. SAS, not law of sines
7. SSS, not law of sines
9. AAS;

$a = 7$	$\alpha = 104°$
$b \approx 1.75$	$\beta = 14°$
$c \approx 6.37$	$\gamma = 62°$

11. AAS;

$a = 6$	$\alpha = 15.6°$
$b \approx 21.97$	$\beta = 100°$
$c \approx 20.12$	$\gamma = 64.4°$

13. SSA (ambiguous case);

$a \approx 14.36$	$\alpha \approx 127.1°$
$b = 9$	$\beta = 30°$
$c = 7$	$\gamma \approx 22.9°$

15. SSA (ambiguous case);

$a \approx 4.82$	$\alpha \approx 13.8°$
$b = 16.8$	$\beta \approx 56.2°$
$c = 19$	$\gamma = 110°$

17. SSA (ambiguous case); no solution.
19. SSA (ambiguous case);

SOLUTION 1

$a \approx 8.28$	$\alpha \approx 85.6°$
$b = 7$	$\beta \approx 57.4°$
$c = 5$	$\gamma = 37°$

SOLUTION 2

$a' \approx 2.90$	$\alpha' \approx 20.4°$
$b = 7$	$\beta' \approx 122.6°$
$c = 5$	$\gamma = 37°$

21. SSA (ambiguous case)

$a = 1.6$	$\alpha = 30°$
$b \approx 2.77$	$\beta = 60°$
$c = 3.2$	$\gamma = 90°$

23. 21.10 in.
25. $h \approx 18.58$, $A \approx 297$ sq units
29. 342 mi
31. 60 ft
33. 272.98 ft
35. 7 ft

Exercise Set 3.5, p. 224

1. SAS, law of cosines
3. SSS, law of cosines
5. SSA, law of sines or law of cosines
7. AAS, law of sines
9.

$a = 15$	$\alpha \approx 26.6°$
$b = 24$	$\beta \approx 134.2°$
$c = 11$	$\gamma \approx 19.2°$

11.

$a = 6$	$\alpha \approx 38.4°$
$b = 9$	$\beta \approx 68.6°$
$c \approx 9.24$	$\gamma = 73°$

13.

$a = 16$ ft	$\alpha \approx 57.3°$
$b \approx 18.95$ ft	$\beta = 94.4°$
$c = 9$ ft	$\gamma \approx 28.3°$

15.

$a = 7$ cm	$\alpha \approx 108.7°$
$b = 5$ cm	$\beta \approx 42.6°$
$c \approx 3.55$ cm	$\gamma = 28.7°$

17. No solution; not enough information.
19. No solution; $a + c \not> b$.
21.

$a = 5$	$\alpha \approx 55.8°$
$b = 6$	$\beta \approx 82.8°$
$c = 4$	$\gamma \approx 41.4°$

23.

$a \approx 9.92$	$\alpha = 82°$
$b \approx 5.6$	$\beta = 34°$
$c = 9$	$\gamma = 64°$

25. 90°
27. 513 mi
29. a. 222.49 mi b. 14.9°
31. a. 57.1° b. 1470 sq ft
33. 32.4°
35. 180 ft
37. a. 41.1°, 138.9° b. longer
39.

$a = 7$	$\alpha = 60°$
$b = 3$	$\beta \approx 21.8°$
$c = 8$	$\gamma \approx 98.2°$

41. **SOLUTION 1**

$a \approx 4.61$	$\alpha \approx 81.0°$
$b = 3$	$\beta = 40°$
$c = 4$	$\gamma \approx 59.0°$

SOLUTION 2

$a' \approx 1.52$	$\alpha' \approx 19.0°$
$b = 3$	$\beta = 40°$
$c = 4$	$\gamma' \approx 121.0°$

Chapter 3 Review Exercises, p. 230

1. 150°, QII 2. 229°, QIII 3. $\frac{11\pi}{36}$ 4. $-\frac{7\pi}{9}$

5. a. 273° b. $273° + k360°$

6. a. $\frac{\pi}{4}$ b. $\frac{\pi}{4} + k2\pi$

7. 12.6 ft 8. 20.1 ft/sec

9. 333.3 degrees/sec 10. 1.5 rad

11. $\cos\theta = -\frac{15}{17}, \sin\theta = \frac{8}{17}, \tan\theta = -\frac{8}{15}$

12. $\cos\theta = \frac{5}{13}, \sin\theta = \frac{12}{13}, \tan\theta = \frac{12}{5}$

13. $\cos\theta = \frac{\sqrt{3}}{2}, \sin\theta = -\frac{1}{2}, \tan\theta = -\frac{\sqrt{3}}{3}$

14. $\cos\theta = -\frac{5\sqrt{29}}{29}, \sin\theta = -\frac{2\sqrt{29}}{29}, \tan\theta = \frac{2}{5}$

15. a. QIV b. QII

16. Values going from point $(1,0)$ counterclockwise: $(1,0)$, 0, 0°; $\left(\frac{\sqrt{3}}{2}, \frac{1}{2}\right), \frac{\pi}{6}$, 30°; $\left(\frac{\sqrt{2}}{2}, \frac{\sqrt{2}}{2}\right), \frac{\pi}{4}$, 45°; $\left(\frac{1}{2}, \frac{\sqrt{3}}{2}\right), \frac{\pi}{3}$, 60°; $(0,1), \frac{\pi}{2}$, 90°; $(-1,0), \pi$, 180°; $(0,-1), -\frac{\pi}{2}$, $-90°$

17. $-\frac{1}{2}$ 18. $\frac{1}{2}$ 19. undefined

20. $\frac{1}{4}$ 21. -1 22. $-\frac{\sqrt{3}}{3}$ 23. $-\frac{1}{2}$

24. $-\frac{\sqrt{2}}{2}$ 25. 1 26. 1 27. 2

28. 0 29. 135° 30. $-90°$ 31. 60°

32. 60° 33. not possible 34. 0 35. 150°

36. 30° 37. 30° 38. $-45°$ 39. not possible

40. not possible 41. $2\sqrt{6}$ 42. $\frac{\sqrt{17}}{17}$ 43. cos 45°

44. cot 30° 45. csc 74° 25′ 46. sin 18°

47.

$a = 5$	$\alpha = 27°$
$b \approx 9.81$	$\beta = 63°$
$c \approx 11.01$	$\gamma = 90°$

48.

$a \approx 8.62$	$\alpha = 38°$
$b \approx 11.03$	$\beta = 52°$
$c = 14$	$\gamma = 90°$

49.

$a = 5$	$\alpha \approx 38.7°$
$b \approx 6.24$	$\beta \approx 51.3°$
$c = 8$	$\gamma = 90°$

50.

$a = 93.1$	$\alpha \approx 62.2°$
$b = 49$	$\beta \approx 27.8°$
$c \approx 105.21$	$\gamma = 90°$

51. N 3.7° W

52.

$a = 10$	$\alpha = 68°$
$b \approx 2.24$	$\beta = 12°$
$c \approx 10.62$	$\gamma = 100°$

53. no solution

54. **Solution 1**

$a = 3$	$\alpha = 10°$
$b = 6$	$\beta \approx 20.3°$
$c \approx 8.72$	$\gamma \approx 149.7°$

Solution 2

$a = 3$	$\alpha = 10°$
$b = 6$	$\beta' \approx 159.7°$
$c' \approx 3.09$	$\gamma' \approx 10.3°$

55.

$a = 50$	$\alpha = 122.5°$
$b = 35$	$\beta \approx 36.2°$
$c \approx 21.54$	$\gamma \approx 21.3°$

56. 1156 ft 57. 83.1°

58.

$a = 3$	$\alpha \approx 49.5°$
$b \approx 3.88$	$\beta = 100°$
$c = 2$	$\gamma \approx 30.5°$

59. No solution. $(a + c \not> b)$

60.

$a = 5$	$\alpha \approx 51.3°$
$b = 2$	$\beta \approx 18.2°$
$c = 6$	$\gamma \approx 110.5°$

61.

$a = 2.3$	$\alpha \approx 90.4°$
$b = 1.2$	$\beta \approx 31.4°$
$c \approx 1.95$	$\gamma \approx 58.2°$

62. 123.46 mi, 30.9°

63. a. law of cosines b. law of cosines
c. not possible d. law of sines or law of cosines
e. right triangle f. not possible
g. law of sines

64. a. T b. F c. F d. T
e. F f. T g. T h. F
i. T j. T k. F

Chapter 3 Test, p. 233

1. a. 157.5°, QII b. 57.3°, QI

2. a. $-\frac{\pi}{12}$ b. $\frac{7\pi}{6}$

3. a. 576 ft b. 96 ft/sec c. 1 rad/sec

4. a. $\frac{3\sqrt{13}}{13}$ **b.** $-\frac{2\sqrt{13}}{13}$

c. $-\frac{3}{2}$

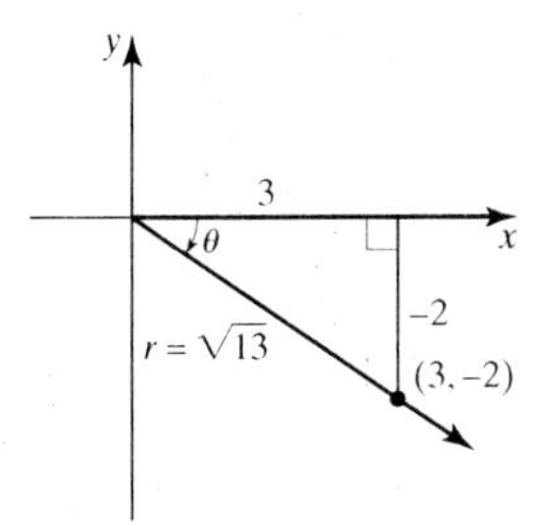

5. a. $-\frac{3}{5}$

b. $\frac{5}{4}$

6. a. $-\frac{1}{2}$ **b.** $-\sqrt{3}$ **c.** $\sqrt{3}$

d. $-\sqrt{2}$ **e.** -1 **f.** $-\frac{\sqrt{2}}{2}$

7. a. $180°$ **b.** $30°$ **c.** $60°$

d. Not possible since $\sin x$ is never greater than 1.

8. a. 0.8473 **b.** -1.5399 **c.** 1.0642

d. 3.8905 **e.** $64.2°$ **f.** 1.8605

9. a. $24.4°$ **b.** 4.10

10. a. law of cosines **b.** law of sines

c. not enough information **d.** law of cosines

e. right triangle

11. a. 8.92 **b.** $14.9°$

12. 9386.5 ft **13.** 41 mi

14. No, it is not long enough. These measurements do not produce a triangle; using the law of sines (SSA) we obtain $\sin \alpha > 1$.

15. a. $90°$

b. $\cos \alpha = b/c$ and $\sin \beta = b/c \rightarrow \cos \alpha = \sin \beta$, or $\cos \alpha = \sin \beta$ since $\alpha + \beta = 90°$.

Chapter 4

The solutions given provide one proof for an identity. Many others are possible.

Exercise Set 4.1, p. 245

1. $\sin^2 x$ **3.** $\cot^2 x$ **5.** $\cos^2 x$

7. $\tan^2 x$ **9.** $\cos^2 x$ **11.** $\pm\sqrt{1 - \sin^2 x}$

13. 1 **15.** 1 **17.** $\sin^2 \alpha$

19. $1 + 2\cos A \sin A$ **21.** $\csc \beta$ **23.** $\sin x$

25. $\cos^2 B - \sin^2 B$ **27.** $1 + \sin x$

29.
$$\cos A \tan A = \sin A$$
$$\cos A \cdot \frac{\sin A}{\cos A} =$$
$$\sin A = \quad \downarrow$$

31.
$$\sin^2 x \sec x \csc x = \tan x$$
$$\sin^2 x \cdot \frac{1}{\cos x} \cdot \frac{1}{\sin x} =$$
$$\frac{\sin x}{\cos x} =$$
$$\tan x = \quad \downarrow$$

33.
$$\sec t - \cos t = \tan t \sin t$$
$$\frac{1}{\cos t} - \cos t \cdot \frac{\cos t}{\cos t} =$$
$$\frac{1 - \cos^2 t}{\cos t} =$$
$$\frac{\sin^2 t}{\cos t} =$$
$$\frac{\sin t}{\cos t} \cdot \frac{\sin t}{1} =$$
$$\tan t \sin t = \quad \downarrow$$

35.
$$1 + \cot x = \frac{\cos x + \sin x}{\sin x}$$
$$\downarrow \quad = \frac{\cos x}{\sin x} + \frac{\sin x}{\sin x}$$
$$= \cot x + 1$$

37.
$$\cos^2 x - \sin^2 x = 2\cos^2 x - 1$$
$$\cos^2 x - (1 - \cos^2 x) =$$
$$2\cos^2 x - 1 = \quad \downarrow$$

Chapter 4

39.

$$\tan^2\beta - \sin^2\beta = \tan^2\beta\sin^2\beta$$
$$\frac{\sin^2\beta}{\cos^2\beta} - \sin^2\beta\cdot\frac{\cos^2\beta}{\cos^2\beta} =$$
$$\frac{\sin^2\beta - \sin^2\beta\cos^2\beta}{\cos^2\beta} =$$
$$\frac{\sin^2\beta(1-\cos^2\beta)}{\cos^2\beta} =$$
$$\frac{\sin^2\beta}{\cos^2\beta}\cdot\sin^2\beta =$$
$$\tan^2\beta\sin^2\beta =$$

41.

$$\frac{\sin^4 x - \cos^4 x}{\sin^2 x - \cos^2 x} = 1$$
$$\frac{(\sin^2 x - \cos^2 x)(\sin^2 x + \cos^2 x)}{\sin^2 x - \cos^2 x} =$$
$$\sin^2 x + \cos^2 x =$$
$$1 =$$

43.

$$\frac{1}{1+\cos A} = \csc^2 A - \csc A\cot A$$
$$= \frac{1}{\sin^2 A} - \frac{1}{\sin A}\cdot\frac{\cos A}{\sin A}$$
$$= \frac{1-\cos A}{\sin^2 A}$$
$$= \frac{1-\cos A}{(1+\cos A)(1-\cos A)}$$
$$= \frac{1}{1+\cos A}$$

45.

$$\frac{1-\cos\alpha}{\sin\alpha} + \frac{\sin\alpha}{1-\cos\alpha} = 2\csc\alpha$$
$$\frac{(1-\cos\alpha)}{(1-\cos\alpha)}\cdot\frac{(1-\cos\alpha)}{\sin\alpha} + \frac{\sin\alpha}{(1-\cos\alpha)}\cdot\frac{\sin\alpha}{\sin\alpha} =$$
$$\frac{1-2\cos\alpha+\cos^2\alpha}{(1-\cos\alpha)\sin\alpha} + \frac{\sin^2\alpha}{(1-\cos\alpha)\sin\alpha} =$$
$$\frac{2-2\cos\alpha}{(1-\cos\alpha)\sin\alpha} =$$
$$\frac{2(1-\cos\alpha)}{(1-\cos\alpha)\sin\alpha} =$$
$$\frac{2}{\sin\alpha} =$$
$$2\csc\alpha =$$

47.

$$\sec^2\theta\csc^2\theta = \sec^2\theta + \csc^2\theta$$
$$= \frac{1}{\cos^2\theta}\cdot\frac{\sin^2\theta}{\sin^2\theta} + \frac{1}{\sin^2\theta}\cdot\frac{\cos^2\theta}{\cos^2\theta}$$
$$= \frac{\sin^2\theta + \cos^2\theta}{\cos^2\theta\sin^2\theta}$$
$$= \frac{1}{\cos^2\theta\sin^2\theta}$$
$$= \frac{1}{\cos^2\theta}\cdot\frac{1}{\sin^2\theta}$$
$$= \sec^2\theta\csc^2\theta$$

49.

$$(\sin x + \cos x)^2 - 1 = 2\sin x\cos x$$
$$\sin^2 x + 2\sin x\cos x + \cos^2 x - 1 =$$
$$2\sin x\cos x + (\sin^2 x + \cos^2 x) - 1 =$$
$$2\sin x\cos x =$$

51.

$$\frac{\csc^2 x - \cot^2 x}{\sec^2 x} = \cos^2 x$$
$$\frac{(1+\cot^2 x) - \cot^2 x}{\sec^2 x} =$$
$$\frac{1}{\sec^2 x} =$$
$$\cos^2 x =$$

53.

$$2\cos^2 B - 1 = 1 - 2\sin^2 B$$
$$2(1-\sin^2 B) - 1 =$$
$$2 - 2\sin^2 B - 1 =$$
$$1 - 2\sin^2 B =$$

55.

$$\sec(-A)\cot(-A)\sin(-A) = 1$$
$$\frac{1}{\cos(-A)}\cdot\frac{\cos(-A)}{\sin(-A)}\cdot\sin(-A) =$$
$$1 =$$

57. The identity is true for $x \neq (\pi/2) + k\pi,\ x \in \mathbb{R}$.

59. The identity is true for $x \neq k\pi,\ x \in \mathbb{R}$.

61. Krystn has an equivalent form of the correct answer.

$$-\tan x + \frac{\sin x}{\cos x - 1} =$$
$$\frac{-\sin x}{\cos x}\left(\frac{\cos x - 1}{\cos x - 1}\right) + \left(\frac{\sin x}{\cos x - 1}\right)\left(\frac{\cos x}{\cos x}\right) =$$
$$\frac{-\sin x\cos x + \sin x + \sin x\cos x}{\cos x(\cos x - 1)} =$$
$$\frac{\sin x}{\cos x(\cos x - 1)} =$$
$$\tan x\left(\frac{1}{\cos x - 1}\right)$$

63–71. *Different counterexamples can be used to verify statements that are not identities.*

63. The statement is an identity.

$$\sin x\cot x\sec x = 1$$
$$\sin x\cdot\frac{\cos x}{\sin x}\cdot\frac{1}{\cos x} =$$
$$1 =$$

65. The graphs do not coincide; not an identity. For a possible counterexample, let $x = \pi/4$.

$$\left(\tan\frac{\pi}{4} + \cot\frac{\pi}{4}\right)^2 \neq \left(\sec\frac{\pi}{4}\right)^2$$
$$(1+1)^2 \neq (\sqrt{2})^2$$
$$4 \neq 2$$

67. The statement is an identity.

$$\frac{\tan x \sin x}{\tan x + \sin x} = \frac{\tan x - \sin x}{\tan x \sin x}$$

$$\frac{\frac{\sin x}{\cos x} \cdot \sin x}{\frac{\sin x}{\cos x} + \sin x} =$$

$$\frac{\frac{\sin^2 x}{\cos x}}{\frac{\sin x}{\cos x} + \frac{\sin x \cos x}{\cos x}} =$$

$$\frac{\frac{\sin^2 x}{\cos x}}{\frac{\sin x(1 + \cos x)}{\cos x}} =$$

$$\frac{\sin^2 x}{\cos x} \cdot \frac{\cos x}{\sin x(1 + \cos x)} =$$

$$\frac{\sin x}{(1 + \cos x)} \frac{(1 - \cos x)}{(1 - \cos x)} =$$

$$\frac{\sin x(1 - \cos x)}{\sin^2 x} =$$

$$\left(\frac{1 - \cos x}{\sin x}\right) \cdot \frac{\tan x}{\tan x} =$$

$$\frac{\tan x - \sin x}{\tan x \sin x} =$$

69. The graphs do not coincide; not an identity. For a possible counterexample, let $x = \pi$.

$$2 \cos \pi \neq \cos 2\pi$$
$$2(-1) \neq 1$$
$$-2 \neq 1$$

71. The statement is an identity.

$$\frac{1 - \cos^4 x}{\sin x} + \sin^3 x = 2 \sin x$$

$$\frac{1 - \cos^4 x}{\sin x} + \sin^3 x \cdot \frac{\sin x}{\sin x} =$$

$$\frac{1 - (\cos^4 x - \sin^4 x)}{\sin x} =$$

$$\frac{1 - (\cos^2 x + \sin^2 x)(\cos^2 x - \sin^2 x)}{\sin x} =$$

$$\frac{1 - (1)((1 - \sin^2 x) - \sin^2 x)}{\sin x} =$$

$$\frac{1 - 1 + 2 \sin^2 x}{\sin x} = \frac{2 \sin x}{1} =$$

73. **a.** Yes, on this interval the graphs appear to coincide.

b. Yes, he is correct that the statement is not an identity.

$$\left(\sin \frac{\pi}{6} + 0.01\right)^2 \neq \left(\sin \frac{\pi}{6}\right)^2 + (0.01)^2$$
$$(0.5 + 0.01)^2 \neq (0.5)^2 + (0.01)^2$$
$$(0.51)^2 \neq 0.25 + 0.0001$$
$$0.2601 \neq 0.2501$$

75. Roger is not working each side independently. When proving an identity, performing the same operation on both sides, like squaring both sides, is not permitted. As this example demonstrates, such operations can make a false statement erroneously appear true.

77. The statement is an identity if it is true for *all* values for which it is defined, not just *one* value. The value $\theta = \pi/4$ could be used as a counterexample to show it is not an identity.

Exercise Set 4.2, p. 254

1. $\frac{\sqrt{6} + \sqrt{2}}{4}$ **3.** $-\frac{\sqrt{6} + \sqrt{2}}{4}$ **5.** $\frac{\sqrt{2} - \sqrt{6}}{4}$

7. $\sqrt{6} - \sqrt{2}$ **9.** -1 **11.** 0

13. $-\frac{\sqrt{2}}{2}$ **15.** $\frac{\sqrt{3}}{2}$ **17.** T

19. T **21.** F **23.** F

25. T **27.** d, g **29.** e, f

31. $-\sin x$ **33.** $\sin x$ **35.** $-\cos x$

37. $\sin x$ **39.** $\cos(A + B) = 0, \cos(A - B) = \sqrt{3}/2$

41. $\cos(A + B) = -33/65, \cos(A - B) = -63/65$

43. $\cos(A + B) = -3\sqrt{10}/10, \cos(A - B) = -3\sqrt{10}/10$

45. $\cos(A + B) = 24/25, \cos(A - B) = 0$

47. $(\sqrt{6} - \sqrt{2})/4$ **49.** $(\sqrt{2} - \sqrt{6})/4$ **51.** $(-\sqrt{6} - \sqrt{2})/4$

53.
$$\cos(90° - \alpha) = \sin \alpha$$
$$\cos 90° \cos \alpha + \sin 90° \sin \alpha =$$
$$0 \cdot \cos \alpha + 1 \cdot \sin \alpha =$$
$$+\sin \alpha =$$

55.
$$\cos(2\pi - A) = \cos A$$
$$\cos 2\pi \cos A + \sin 2\pi \sin A =$$
$$1 \cdot \cos A + 0 \cdot \sin A =$$
$$\cos A =$$

57. Let $A = \pi, B = \pi/2$.
$$\cos\left(\pi + \frac{\pi}{2}\right) \neq \cos \pi + \cos \frac{\pi}{2}$$
$$\cos\left(\frac{3\pi}{2}\right) \neq (-1) + 0$$
$$0 \neq -1$$

59.
$$\cos(2x) = \cos^2 x - \sin^2 x$$
$$\cos(x + x) =$$
$$\cos x \cos x - \sin x \sin x =$$
$$\cos^2 x - \sin^2 x =$$

61. **a.** $y = \cos x$

$$\begin{aligned}\cos 2x \cos 3x + \sin 2x \sin 3x &= \cos(2x - 3x)\\ &= \cos(-x)\\ &= \cos x\end{aligned}$$

b. $y = \cos 3x \cos x + \sin 3x \sin x$

$$\begin{aligned} &= \cos(3x - x)\\ &= \cos 2x\end{aligned}$$

$y = \cos 2x$

Exercise Set 4.3, p. 263

1. $\dfrac{\sqrt{6} + \sqrt{2}}{4}$ **3.** $2 - \sqrt{3}$ **5.** $\dfrac{\sqrt{2} - \sqrt{6}}{4}$

7. $2 + \sqrt{3}$ **9.** $\sin 180° = 0$ **11.** $\sin \dfrac{3\pi}{4} = \dfrac{\sqrt{2}}{2}$

13. $\tan 45° = 1$ **15.** $\tan 120° = -\sqrt{3}$

17. F **19.** T **21.** T **23.** F **25.** f **27.** d

29. $\sin x$ **31.** $-\cos x$ **33.** $\tan x$

35. $-56/65$ **37.** $-117/125$

39. **a.** $-63/65$ **b.** $-33/56$ **c.** $-56/65$ **d.** QII

41. $(\sqrt{2} - \sqrt{6})/4$ **43.** $-2 - \sqrt{3}$ **45.** 0 **47.** 1

49.
$$\frac{\sin(A - B)}{\cos A \cos B} = \tan A - \tan B$$
$$\frac{\sin A \cos B - \cos A \sin B}{\cos A \cos B} =$$
$$\frac{\sin A \cos B}{\cos A \cos B} - \frac{\cos A \sin B}{\cos A \cos B} =$$
$$\frac{\sin A}{\cos A} - \frac{\sin B}{\cos B} =$$
$$\tan A - \tan B =$$

51.
$$\sin 2x = 2 \sin x \cos x$$
$$\sin(x + x) =$$
$$\sin x \cos x + \cos x \sin x =$$
$$2 \sin x \cos x =$$

53.
$$\tan\left(\frac{\pi}{2} - x\right) = \cot x$$
$$\frac{\sin\left(\frac{\pi}{2} - x\right)}{\cos\left(\frac{\pi}{2} - x\right)} =$$
$$\frac{\cos x}{\sin x} =$$
$$\cot x =$$

55.
$$\begin{aligned}\sin x + \cos x &= \sqrt{2} \sin\left(x + \frac{\pi}{4}\right)\\ &= \sqrt{2}\left[\sin x \cos \frac{\pi}{4} + \cos x \sin \frac{\pi}{4}\right]\\ &= \sqrt{2}\left[\sin x \cdot \frac{1}{\sqrt{2}} + \cos x \cdot \frac{1}{\sqrt{2}}\right]\\ &= \sin x + \cos x\end{aligned}$$

57. Substitute $A = \pi/2$ and $B = \pi/2$ in $\sin(A + B) = \sin A + \sin B$:

$$\begin{aligned}\sin\left(\frac{\pi}{2} + \frac{\pi}{2}\right) &\neq \sin \frac{\pi}{2} + \sin \frac{\pi}{2}\\ \sin(\pi) &\neq \sin \frac{\pi}{2} + \sin \frac{\pi}{2}\\ 0 &\neq 1 + 1\\ 0 &\neq 2\end{aligned}$$

59. Donald is using the cofunction identity $\sin 75° = \cos 15°$, since $75° + 15° = 90°$.

Exercise Set 4.4, p. 270

1. $\cos 2x$ **3.** $2 \cos 2x$ **5.** $3 \sin 6x$

7. $\cos 16x$ **9.** $\tan 2x$ **11.** $1/2$

13. $-\sqrt{3}/2$ **15.** $-\sqrt{3}/3$ **17.** $3\sqrt{3}/2$

19. $8\sqrt{2}$ **21.** $-24/25$ **23.** $-527/625$

25. $-24/7$ **27.** $2\sqrt{13}/13$ **29.** $-169/119$

31.
$$\begin{aligned}\sin 2x &= \frac{2 \tan x}{1 + \tan^2 x}\\ &= \frac{2 \tan x}{\sec^2 x}\\ &= \frac{\dfrac{2 \sin x}{\cos x}}{\dfrac{1}{\cos^2 x}}\\ &= \frac{2 \sin x}{\cos x} \cdot \frac{\cos^2 x}{1}\\ &= 2 \sin x \cos x\\ &= \sin 2x\end{aligned}$$

33.
$$\begin{aligned}\cot \theta &= \frac{\sin 2\theta}{1 - \cos 2\theta}\\ &= \frac{2 \sin \theta \cos \theta}{1 - (1 - 2 \sin^2 \theta)}\\ &= \frac{2 \sin \theta \cos \theta}{2 \sin^2 \theta}\\ &= \frac{\cos \theta}{\sin \theta}\\ &= \cot \theta\end{aligned}$$

35. $\cos^2 A = \dfrac{1 + \cos 2A}{2}$

$= \dfrac{1 + (2\cos^2 A - 1)}{2}$

$= \dfrac{2\cos^2 A}{2}$

$= \cos^2 A$

37. $1 = 2\sin^2 x + \cos 2x$

$= 2\sin^2 x + (1 - 2\sin^2 x)$

$= 1$

39.
$$\begin{aligned} \sin 3x &= 3\sin x - 4\sin^3 x \\ \sin(2x + x) &= \\ \sin 2x \cos x + \cos 2x \sin x &= \\ (2\sin x \cos x)\cos x + (1 - 2\sin^2 x)\sin x &= \\ 2\sin x \cos^2 x + \sin x - 2\sin^3 x &= \\ 2\sin x(1 - \sin^2 x) + \sin x - 2\sin^3 x &= \\ 3\sin x - 4\sin^3 x &= \end{aligned}$$

41.
$$\begin{aligned} (\sin x - \cos x)^2 &= 1 - \sin 2x \\ \sin^2 x - 2\sin x \cos x + \cos^2 x &= \\ 1 - 2\sin x \cos x &= \\ 1 - \sin 2x &= \end{aligned}$$

43. F 45. F 47. T 49. F

51.
$$\begin{aligned} \tan 2x &= \frac{2\tan x}{1 - \tan^2 x} \\ \tan(x + x) &= \\ \frac{\tan x + \tan x}{1 - \tan x \tan x} &= \\ \frac{2\tan x}{1 - \tan^2 x} &= \end{aligned}$$

53. 14.4 ft

55. $A = lw,\ w = d\sin\theta,\ l = d\cos\theta$

$A(\theta) = d^2 \sin\theta\cos\theta$

$= d^2\,\dfrac{2\sin\theta\cos\theta}{2}$

$= \dfrac{d^2}{2}\sin 2\theta$

57.
$$\begin{aligned} y_2 &= \cos 2x \\ \cos^4 x - \sin^4 x &= \cos 2x \\ (\cos^2 x + \sin^2 x)(\cos^2 x - \sin^2 x) &= \\ 1(\cos^2 x - \sin^2 x) &= \\ \cos 2x &= \end{aligned}$$

59.
$$\begin{aligned} y_2 &= \cos 2x \\ \frac{1 - \tan^2 x}{1 + \tan^2 x} &= \cos 2x \\ \frac{1 - \tan^2 x}{\sec^2 x} &= \\ (1 - \tan^2 x)(\cos^2 x) &= \\ \cos^2 x - \sin^2 x &= \\ \cos 2x &= \end{aligned}$$

Exercise Set 4.5, p. 280

1. $(\sqrt{2 + \sqrt{2}})/2$ 3. $-(\sqrt{2 - \sqrt{3}})/2$ 5. $-\sqrt{7 - 4\sqrt{3}}$ or $-2 + \sqrt{3}$ 7. $(\sqrt{2 + \sqrt{2}})/2$ 9. $(2 + \sqrt{3})/4$ 11. $\cos \pi/14$ 13. $\sin 244°$ 15. $\cos(x/2)$ 17. $(1 - \cos 2x)/2$ 19. $\tan 10°$ 21. a. $\sqrt{10}/10$ b. $3\sqrt{10}/10$ c. 3 23. a. $-(\sqrt{2 - \sqrt{3}})/2$ b. $(2 + \sqrt{3})/4$ c. $-2 - \sqrt{3}$ or $-\sqrt{7 + 4\sqrt{3}}$ 25. $(\sqrt{50 + 10\sqrt{5}})/10$ 27. 1/3 29. $-4/5$ 31. $3\sqrt{10}/10$ 33. 3 35. 7/25 37. 3/4 39. $(3 + 4\sqrt{3})/10$ 41. $(-4 - 3\sqrt{3})/10$ 43. $(48 - 25\sqrt{3})/11$

45. $\sin^2 \dfrac{\theta}{2} = \dfrac{\csc\theta - \cot\theta}{2\csc\theta}$

$= \dfrac{\dfrac{1}{\sin\theta} - \dfrac{\cos\theta}{\sin\theta}}{\dfrac{2}{\sin\theta}}$

$= \dfrac{1 - \cos\theta}{\sin\theta} \cdot \dfrac{\sin\theta}{2}$

$= \dfrac{1 - \cos\theta}{2}$

$= \sin^2 \dfrac{\theta}{2}$

47.
$$\begin{aligned} \left(\cos\frac{\beta}{2} - \sin\frac{\beta}{2}\right)^2 &= 1 - \sin\beta \\ \cos^2\frac{\beta}{2} - 2\sin\frac{\beta}{2}\cos\frac{\beta}{2} + \sin^2\frac{\beta}{2} &= \\ 1 - 2\sin\frac{\beta}{2}\cos\frac{\beta}{2} &= \\ 1 - \sin\beta &= \end{aligned}$$

49. $\tan \dfrac{x}{2} = \csc x - \cot x$

$= \dfrac{1}{\sin x} - \dfrac{\cos x}{\sin x}$

$= \dfrac{1 - \cos x}{\sin x}$

$= \tan \dfrac{x}{2}$

Chapter 4

51.
$$1 - \tan^2 \frac{A}{2} = \frac{2\cos A}{1 + \cos A}$$
$$1 - \frac{1 - \cos A}{1 + \cos A} =$$
$$\frac{1 + \cos A}{1 + \cos A} - \frac{1 - \cos A}{1 + \cos A} =$$
$$\frac{1 + \cos A - (1 - \cos A)}{1 + \cos A} =$$
$$\frac{2\cos A}{1 + \cos A} =$$

53.
$$\sin^2 \frac{x}{2} = \frac{\sec x - 1}{2\sec x}$$
$$= \frac{\frac{1}{\cos x} - 1}{\frac{2}{\cos x}}$$
$$= \left(\frac{1 - \cos x}{\cos x}\right)\left(\frac{\cos x}{2}\right)$$
$$= \frac{1 - \cos x}{2}$$
$$= \sin^2 \frac{x}{2}$$

55. $\frac{1}{2}(\sin 60° - \sin 28°)$ **57.** $\frac{1}{2}(\cos 8x - \cos 12x)$

59. $5(\cos 2° + \cos 8°)$ **61.** $2\cos 45° \cos 30° = \sqrt{6}/2$

63. $2\sin 3y \cos y$ **65.** $-2\sin 3x \sin 2x$

67. a.
$$\tan \frac{x}{2} = \sqrt{\frac{1 - \cos x}{1 + \cos x}}$$
$$\frac{\sin \frac{x}{2}}{\cos \frac{x}{2}} =$$
$$\frac{\sqrt{\frac{1 - \cos x}{2}}}{\sqrt{\frac{1 + \cos x}{2}}} =$$
$$\sqrt{\frac{1 - \cos x}{2} \cdot \frac{2}{1 + \cos x}} =$$
$$\sqrt{\frac{1 - \cos x}{1 + \cos x}} =$$

b. Using $\tan \frac{x}{2} = \frac{\sin x}{1 + \cos x}$, let $\frac{x}{2} = A$ and $x = 2A$.
$$\tan A = \frac{\sin 2A}{1 + \cos 2A}$$
$$= \frac{2\sin A \cos A}{1 + (2\cos^2 A - 1)}$$
$$= \frac{2\sin A \cos A}{2\cos^2 A}$$
$$= \frac{\sin A}{\cos A}$$
$$= \tan A$$
$$= \tan \frac{x}{2}$$

c. Using part (b)
$$\tan \frac{x}{2} = \frac{1 - \cos x}{\sin x}$$
$$\frac{\sin x}{1 + \cos x} =$$
$$\left(\frac{\sin x}{1 + \cos x}\right)\frac{1 - \cos x}{1 - \cos x} =$$
$$\frac{\sin x\,(1 - \cos x)}{\sin^2 x} =$$
$$\frac{1 - \cos x}{\sin x} =$$

69.
$$\sin A \cos B = \tfrac{1}{2}[\sin(A + B) + \sin(A - B)]$$
$$= \tfrac{1}{2}[(\sin A \cos B + \cos A \sin B) + (\sin A \cos B - \cos A \sin B)]$$
$$= \tfrac{1}{2}[2\sin A \cos B]$$
$$= \sin A \cos B$$

71. $x = \left(\frac{A + B}{2}\right)$ and $y = \left(\frac{A - B}{2}\right) \longrightarrow (x + y) = A$ and $(x - y) = B$

$$\cos A + \cos B = 2\cos\left(\frac{A + B}{2}\right)\cos\left(\frac{A - B}{2}\right)$$
$$= 2\cos x \cos y$$
$$= 2\left[\tfrac{1}{2}(\cos(x + y) + \cos(x - y))\right]$$
$$= \cos A + \cos B$$

73. $x = \left(\frac{A + B}{2}\right)$ and $y = \left(\frac{A - B}{2}\right) \longrightarrow (x + y) = A$ and $(x - y) = B$

$$\cos A - \cos B = -2\sin\left(\frac{A + B}{2}\right)\sin\left(\frac{A - B}{2}\right)$$
$$= -2\sin x \sin y$$
$$= -2\left[\tfrac{1}{2}(\cos(x - y) - \cos(x + y))\right]$$
$$= -(\cos B - \cos A)$$
$$= \cos A - \cos B$$

75. $y = (\cos x - 1)(\cos x + 2)$

$y = \cos x + \frac{1}{2}\cos 2x - \frac{3}{2}$

$$(\cos x - 1)(\cos x + 2) = \cos x + \tfrac{1}{2}\cos 2x - \tfrac{3}{2}$$
$$= \cos x + \tfrac{1}{2}(2\cos^2 x - 1) - \tfrac{3}{2}$$
$$= \cos^2 x + \cos x - 2$$
$$= (\cos x - 1)(\cos x + 2)$$

77.

$$y_2 = \cot\frac{x}{2}$$

$$2\cot x + \tan\frac{x}{2} = \cot\frac{x}{2}$$

$$\frac{2\cos x}{\sin x} + \frac{1-\cos x}{\sin x} =$$

$$\frac{1+\cos x}{\sin x} =$$

$$\frac{1}{\left(\frac{\sin x}{1+\cos x}\right)} =$$

$$\frac{1}{\tan\frac{x}{2}} =$$

$$\cot\frac{x}{2} =$$

79.

$$y_2 = \cos x$$

$$\cos^4\frac{x}{2} - \sin^4\frac{x}{2} = \cos x$$

$$\left(\cos^2\frac{x}{2} + \sin^2\frac{x}{2}\right)\left(\cos^2\frac{x}{2} - \sin^2\frac{x}{2}\right) =$$

$$1\left(\cos^2\frac{x}{2} - \sin^2\frac{x}{2}\right) =$$

$$\cos x =$$

Chapter 4 Review Exercises, p. 286

1. $\cos^2 x$ **2.** $\cos 2x$ **3.** $\sec^2 x$ **4.** $\sin 2x$ **5.** $\frac{1}{\cos x}$ **6.** $\cot x$ **7.** $\cos 2x$ **8.** 1 **9.** $\cos x$ **10.** $\sin x$ **11.** $\cos x\cos y + \sin x\sin y$ **12.** 1 **13.** $\sin x\cos y + \cos x\sin y$ **14.** $\csc^2 x$ **15.** $-\cos x$ **16.** $\cos x$ **17.** $4\cos x$ **18.** $4\sin x$ **19.** $\cos^2 x$ **20.** $-\csc x$ **21.** $\sin 10x$ **22.** $\cos 20x$ **23.** $\cos x$ **24.** $\sin 8x$ **25.** $\sin 4x$ **26.** $\tan 2x$

27.

$$\cos^2 x\left(\sec^2 x - 1\right) = \sin^2 x$$

$$\cos^2 x\left(\tan^2 x\right) =$$

$$\cos^2 x\left(\frac{\sin^2 x}{\cos^2 x}\right) =$$

$$\sin^2 x =$$

28.

$$\tan^2 A\csc^2 A - \tan^2 A = 1$$

$$\tan^2 A\left(\csc^2 A - 1\right) =$$

$$\tan^2 A\left(\cot^2 A\right) =$$

$$1 =$$

29.

$$\frac{\cos y\tan y}{\sin y} = 1$$

$$\frac{\cos y\,\frac{\sin y}{\cos y}}{\sin y} =$$

$$\frac{\sin y}{\sin y} =$$

$$1 =$$

30.

$$\cot\alpha - \tan\alpha = \frac{\cos 2\alpha}{\sin\alpha\cos\alpha}$$

$$= \frac{\cos^2\alpha - \sin^2\alpha}{\cos\alpha\sin\alpha}$$

$$= \frac{\cos^2\alpha}{\cos\alpha\sin\alpha} - \frac{\sin^2\alpha}{\cos\alpha\sin\alpha}$$

$$= \frac{\cos\alpha}{\sin\alpha} - \frac{\sin\alpha}{\cos\alpha}$$

$$= \cot\alpha - \tan\alpha$$

31.

$$\frac{1+\sin\theta}{\cos\theta} = \frac{\cos\theta}{1-\sin\theta}$$

$$= \left(\frac{\cos\theta}{1-\sin\theta}\right)\cdot\left(\frac{1+\sin\theta}{1+\sin\theta}\right)$$

$$= \frac{\cos\theta\left(1+\sin\theta\right)}{\cos^2\theta}$$

$$= \frac{1+\sin\theta}{\cos\theta}$$

32.

$$\csc t = 2\cos t\csc 2t$$

$$= 2\cos t\,\frac{1}{\sin 2t}$$

$$= 2\cos t\,\frac{1}{2\sin t\cos t}$$

$$= \frac{1}{\sin t}$$

$$= \csc t$$

33.

$$\sin 2x = 2\sin^2 x\cos x\csc x$$

$$= 2\sin^2 x\cos x\,\frac{1}{\sin x}$$

$$= 2\sin x\cos x$$

$$= \sin 2x$$

34.

$$\cos^4\beta - \sin^4\beta = \cos 2\beta$$

$$(\cos^2\beta - \sin^2\beta)(\cos^2\beta + \sin^2\beta) =$$

$$(\cos^2\beta - \sin^2\beta)\cdot 1 =$$

$$\cos 2\beta =$$

35. $$\tan y + \sec y = \frac{1}{\sec y - \tan y} = \left(\frac{1}{\sec y - \tan y}\right)\left(\frac{\sec y + \tan y}{\sec y + \tan y}\right) = \frac{\sec y + \tan y}{\sec^2 y - \tan^2 y} = \frac{\sec y + \tan y}{1} = \tan y + \sec y$$

36. $$\frac{\cos x}{1 - \tan x} + \frac{\sin x}{1 - \cot x} = \cos x + \sin x$$
$$\frac{\cos x}{1 - \frac{\sin x}{\cos x}} + \frac{\sin x}{1 - \frac{\cos x}{\sin x}} =$$
$$\left(\frac{\cos x}{\cos x}\right)\left(\frac{\cos x}{1 - \frac{\sin x}{\cos x}}\right) + \left(\frac{\sin x}{\sin x}\right)\left(\frac{\sin x}{1 - \frac{\cos x}{\sin x}}\right) =$$
$$\frac{\cos^2 x}{\cos x - \sin x} + \left(\frac{\sin^2 x}{\sin x - \cos x}\right)\left(\frac{-1}{-1}\right) =$$
$$\frac{\cos^2 x - \sin^2 x}{\cos x - \sin x} =$$
$$\frac{(\cos x - \sin x)(\cos x + \sin x)}{\cos x - \sin x} =$$
$$\cos x + \sin x =$$

37. $$\cos(90° + \theta) = -\sin\theta$$
$$\cos 90° \cos\theta - \sin 90° \sin\theta =$$
$$0 \cdot \cos\theta - 1 \cdot \sin\theta =$$
$$-\sin\theta =$$

38. $$\sin(90° + x) = \cos x$$
$$\sin 90° \cos x + \cos 90° \sin x =$$
$$1 \cdot \cos x + 0 \cdot \sin x =$$
$$\cos x =$$

39. $$\tan\left(\alpha - \frac{\pi}{4}\right) = \frac{\tan\alpha - 1}{1 + \tan\alpha}$$
$$\frac{\tan\alpha - \tan\frac{\pi}{4}}{1 + \tan\alpha\tan\frac{\pi}{4}} =$$
$$\frac{\tan\alpha - 1}{1 + \tan\alpha} =$$

40. $$\tan\frac{A}{2} = \frac{\sec A \sin A}{\sec A + 1} = \left(\frac{\sec A \sin A}{\sec A + 1}\right)\left(\frac{\cos A}{\cos A}\right) = \frac{\sin A}{1 + \cos A} = \tan\frac{A}{2}$$

41. $A = \pi/4$
$$\tan\frac{\pi}{4}\cos\frac{\pi}{4}\sin\frac{\pi}{4} \neq \sec\frac{\pi}{4}$$
$$1 \cdot \frac{1}{\sqrt{2}} \cdot \frac{1}{\sqrt{2}} \neq \sqrt{2}$$
$$\frac{1}{2} \neq \sqrt{2}$$

42. $x = \pi/2$
$$\cos 2\left(\frac{\pi}{2}\right) \neq 1 + 2\left(\cos\frac{\pi}{2}\right)^2$$
$$\cos\pi \neq 1 + 2 \cdot 0$$
$$-1 \neq 1$$

43. The statement is not an identity. Let $x = \pi$:
$$\left(\sin\frac{\pi}{2}\right)^2 \neq \frac{(\cos\pi)^2}{2 - 2\cos\pi}$$
$$1 \neq \frac{1}{4}$$

44. The statement is not an identity. Let $x = \pi/4$:
$$\left(\cot\frac{\pi}{4}\right)^2 \neq \frac{1 - \left(\cos 2\left(\frac{\pi}{4}\right)\right)^2}{\sin\frac{\pi}{4}}$$
$$1 \neq \sqrt{2}$$

45. $(\sqrt{6} - \sqrt{2})/4$ **46.** $(\sqrt{6} + \sqrt{2})/4$ **47.** F, $\cos 109°$
48. T **49.** F, $(\sqrt{2} - \sqrt{6})/4$ **50.** T
51. T **52.** F, $2\sin 16x$ **53.** T
54. T **55.** $y = \cos 2x$, $y = \cos^2 x - \sin^2 x$
56. $y = \dfrac{1 + \cos 2x}{2}$, $y = \cos^2 x$
57. $y = \sin 2x$, $y = \sin 15x \cos 13x - \cos 15x \sin 13x$
58. $y = \tan 2x$, $y = \dfrac{2\tan x}{1 - \tan^2 x}$, $y = -\cot 2\left(x + \dfrac{\pi}{4}\right)$
59. $\sqrt{3}/2$ **60.** $-5/13$ **61.** $(-5 + 12\sqrt{3})/26$
62. $(-5\sqrt{3} + 12)/26$ **63.** $120/169$ **64.** $-\sqrt{3}$
65. $\sqrt{3}/2$ **66.** $3\sqrt{13}/13$ **67.** $1/4$
68. $4/13$ **69.** $-4/5$ **70.** $-4/5$
71. 1 **72.** $-24/25$ **73.** $7/25$
74. $24/7$ **75.** $\sqrt{5}/5$ **76.** $\sqrt{10}/10$
77. 3 **78.** $-7/25$ **79.** $-3\sqrt{10}/10$
80. $4/5$ **81.** $(\sqrt{2} - \sqrt{6})/4$ **82.** $-1/2$
83. undefined **84.** $-\sqrt{3}/2$ **85.** $2\cos\dfrac{13x}{2}\cos\dfrac{7x}{2}$
86. $-2\sin 4x \sin x$ **87.** $\frac{1}{2}(\cos 2x - \cos 14x)$
88. $\frac{1}{2}(\sin 14x - \sin 4x)$

Chapter 4 Test, p. 289

1. $-\tan x$
2. $\cos(x + y)$
3. $\cos^2 x$
4. $\sec^2 x$
5. $\cos^2 x - \sin^2 x, 2\cos^2 x - 1$, or $1 - 2\sin^2 x$
6. $\sin \frac{x}{2}$
7. 1
8. $\sin 2x$
9.
$$\begin{aligned} \sin^2 x(1 + \cot^2 x) &= 1 \\ \sin^2 x(\csc^2 x) &= \\ \sin^2 x\left(\frac{1}{\sin^2 x}\right) &= \\ 1 &= \end{aligned}$$
10.
$$\begin{aligned} \cot x + \tan x &= \sec x \csc x \\ \frac{\cos x}{\sin x} + \frac{\sin x}{\cos x} &= \\ \left(\frac{\cos x}{\sin x}\right)\left(\frac{\cos x}{\cos x}\right) + \left(\frac{\sin x}{\cos x}\right)\left(\frac{\sin x}{\sin x}\right) &= \\ \frac{\cos^2 x + \sin^2 x}{\cos x \sin x} &= \\ \frac{1}{\cos x} \cdot \frac{1}{\sin x} &= \\ \sec x \csc x &= \end{aligned}$$
11.
$$\begin{aligned} \cos 2x - \cos^2 x &= -\sin^2 x \\ (\cos^2 x - \sin^2 x) - \cos^2 x &= \\ -\sin^2 x &= \end{aligned}$$
12.
$$\begin{aligned} \sin 2x &= 2\sin^2 x \cos x \csc x \\ &= 2\sin^2 x \cos x \frac{1}{\sin x} \\ &= 2\sin x \cos x \\ &= \sin 2x \end{aligned}$$
13.
$$\begin{aligned} 2\sin^2 \frac{x}{2} - 1 &= -\cos x \\ 2\left(\frac{1 - \cos x}{2}\right) - 1 &= \\ 1 - \cos x - 1 &= \\ -\cos x &= \end{aligned}$$
14.
$$\begin{aligned} \cos^2 \frac{x}{2} &= \frac{\sin^2 x}{2(1 - \cos x)} \\ &= \left(\frac{\sin^2 x}{2(1 - \cos x)}\right)\left(\frac{1 + \cos x}{1 + \cos x}\right) \\ &= \frac{\sin^2 x(1 + \cos x)}{2\sin^2 x} \\ &= \frac{1 + \cos x}{2} \\ &= \cos^2 \frac{x}{2} \end{aligned}$$
15. The statement is not an identity. Let $x = \pi/3$:
$$1 - \sin 2\left(\frac{\pi}{3}\right) \neq \frac{1 - \left(\tan \frac{\pi}{3}\right)^2}{1 + \left(\tan \frac{\pi}{3}\right)^2}$$
$$1 - \frac{\sqrt{3}}{2} \neq \frac{1 - (\sqrt{3})^2}{1 + (\sqrt{3})^2}$$
$$1 - \frac{\sqrt{3}}{2} \neq -\frac{1}{2}$$
16. The statement is not an identity. Let $x = \pi/2$:
$$\left(\cos 2\left(\frac{\pi}{2}\right)\right)^2 + 2\left(\sin \frac{\pi}{2}\right)^2 \neq 1$$
$$(-1)^2 + 2 \neq 1$$
$$3 \neq 1$$
17. $(\sqrt{2} - \sqrt{6})/4$
18. $(\sqrt{6} - \sqrt{2})/4$
19. $\sqrt{3}/2$
20. $-\sqrt{3}/2$
21. $\sqrt{3}/3$
22. $\tan x$
23. $-\sqrt{3}/2$
24. $-7/25$
25. $(3\sqrt{3} - 4)/10$
26. $\sqrt{3}/2$
27. a, c, f
28. b, h
29. $(\sqrt{2} + \sqrt{6})/4$
30. $2\sin 4x \cos x$

Chapter 5

Exercise Set 5.1, p. 295

1. **a.** $\frac{\pi}{3}, \frac{5\pi}{3}$ **b.** $\frac{\pi}{3} + k \cdot 2\pi, \frac{5\pi}{3} + k \cdot 2\pi$
3. **a.** $\frac{5\pi}{4}, \frac{7\pi}{4}$ **b.** $\frac{5\pi}{4} + k \cdot 2\pi, \frac{7\pi}{4} + k \cdot 2\pi$
5. **a.** $\frac{\pi}{6}, \frac{7\pi}{6}$ **b.** $\frac{\pi}{6} + k \cdot \pi$
7. **a.** 0 **b.** $k \cdot 2\pi$
9. **a.** $\frac{7\pi}{6}, \frac{11\pi}{6}$ **b.** $\frac{7\pi}{6} + k \cdot 2\pi, \frac{11\pi}{6} + k \cdot 2\pi$
11. **a.** $\frac{3\pi}{4}, \frac{7\pi}{4}$ **b.** $\frac{3\pi}{4} + k \cdot \pi$
13. **a.** 45°, 225° **b.** $45° + k \cdot 180°$
15. **a.** 240°, 300° **b.** $240° + k \cdot 360°, 300° + k \cdot 360°$
17. **a.** 90°, 270° **b.** $90° + k \cdot 180°$
19. **a.** 0°, 180° **b.** $k \cdot 180°$
21. 0.2838, 2.8578
23. 1.3909, 4.5325
25. 2.4189, 3.8643
27. 4.0308, 5.3939
29. 1.5783, 4.7199
31. 90°, 270°
33. 116.7°, 243.3°
35. 57.8°, 237.8°
37. 30°, 150°, 210°, 330°
39. No solution, since $|\sec x| \geq 1$, or $|\cos x| \leq 1$.
41. **a.** $y = 0.6, y = \cos x$ **b.** $\cos x = 0.6$ **c.** $x = 0.9273$
43. **a.** $y = 5, y = \csc x$ **b.** $\csc x = 5$ **c.** $x = 0.2014$
45. 79.7°, 280.3°
47. 177.1°, 357.1°

Exercise Set 5.2, p. 302

1. $3u + 5 = 15u - 1$, first degree

3. $2u^2 - 3u - 1 = 0$, second degree

5. $u^2 - 5u - 6 = 0$, second degree

7. $x = \frac{\pi}{3}, \frac{2\pi}{3}$ **9.** $x = \frac{\pi}{3}, \frac{5\pi}{3}$ **11.** $x = \frac{\pi}{6}, \frac{5\pi}{6}, \frac{7\pi}{6}, \frac{11\pi}{6}$

13. $x = \frac{\pi}{6}, \frac{5\pi}{6}, \frac{7\pi}{6}, \frac{11\pi}{6}$ **15.** $x = \frac{\pi}{2}, \frac{3\pi}{2}$ **17.** $x = \frac{\pi}{3}, \pi, \frac{5\pi}{3}$

19. $x = 0, \frac{\pi}{4}, \pi, \frac{5\pi}{4}$ **21.** $x = \frac{\pi}{4}, \frac{\pi}{3}, \frac{2\pi}{3}, \frac{5\pi}{4}, \frac{4\pi}{3}, \frac{5\pi}{3}$

23. $x = 0, \frac{\pi}{3}, \frac{2\pi}{3}, \pi, \frac{4\pi}{3}, \frac{5\pi}{3}$ **25.** $x = 0$ **27.** $x = \frac{3\pi}{2}$

29. 0°, 90°, 180°, 270° **31.** 45°, 135°, 225°, 315°

33. 30°, 150°, 270° **35.** 0°, 180°

37. 1.1247, 2.0169, 3.4224, 6.0024 **39.** $\frac{\pi}{2}, \frac{3\pi}{2}$, 1.3181, 4.9651

41. 0.7297, 2.4119 **43.** 0.6089, 1.3424, 3.7505, 4.4840

45. 73.0°, 287.0° **47.** 230.0°, 310.0°

49. 58.8°, 101.7°, 238.8°, 281.7° **51.** $\frac{\pi}{6} + k \cdot \pi, \frac{5\pi}{6} + k \cdot \pi$

53. $k \cdot 180°$ **55.** $230.0° + k \cdot 360°, 310.0° + k \cdot 360°$

57. 2.2 sec, 4.1 sec

59. **a.** 1.6 days **b.** 0.4 days, 2.7 days

61. **a.** 219 ft **b.** 8.8 sec **c.** 35.5°

Exercise Set 5.3, p. 311

1. $\frac{\pi}{4}, \frac{3\pi}{4}, \frac{5\pi}{4}, \frac{7\pi}{4}$ **3.** $\frac{\pi}{6}, \frac{5\pi}{6}, \frac{7\pi}{6}, \frac{11\pi}{6}$ **5.** $\frac{3\pi}{2}$

7. $\frac{3\pi}{4}, \frac{7\pi}{4}$ **9.** $0, \frac{2\pi}{3}, \frac{4\pi}{3}$ **11.** 30°, 150°, 270°

13. 0°, 60°, 180°, 300° **15.** 22.5°, 112.5°, 202.5°, 292.5°

17. 0° **19.** 210°, 270°, 330° **21.** 30°, 270°

23. 0°, 270° **25.** 180° **27.** 0°, 90°, 180°

29. 180° **31.** 0°, 120°, 240°

33. $k \cdot 2\pi, \frac{2\pi}{3} + k \cdot 2\pi, \frac{4\pi}{3} + k \cdot 2\pi$

35. $30° + k \cdot 360°, 270° + k \cdot 360°$

37. 1.3694, 4.9137 **39.** 3.8078, 5.6169 **41.** 1.2310, 5.0522

43. $\frac{\pi}{12}, \frac{5\pi}{12}, \frac{13\pi}{12}, \frac{17\pi}{12}$ **45.** $\frac{\pi}{3}, \frac{5\pi}{6}, \frac{4\pi}{3}, \frac{11\pi}{6}$

47. $\frac{5\pi}{8}, \frac{7\pi}{8}, \frac{13\pi}{8}, \frac{15\pi}{8}$ **49.** $\frac{\pi}{2}, \frac{3\pi}{2}$

51. (45°, 3), (135°, 3), (225°, 3), (315°, 3)

53. (0°, 1), (180°, −1) **55.** $(30°, \frac{3}{4}), (150°, \frac{3}{4}), (270°, 0)$

57. 33.9° or 56.1° **59.** 1.1888

61. −1.6727, 1.2288, 1.9766

63. No solution: $\sin x$ is never greater than 1 and $x^2 + 4$ is always greater than 1.

Exercise Set 5.4, p. 320

1. $x^2 + y^2 = 1$, circle **3.** $\frac{x^2}{16} + \frac{y^2}{25} = 1$, ellipse

5. $y = x + 4$, line **7.** $y = 2x + 8$, line

9. $y = 3x^2$, parabola **11.** $y = x^2 - 1$, parabola

13. $x^2 + y^2 = 1$

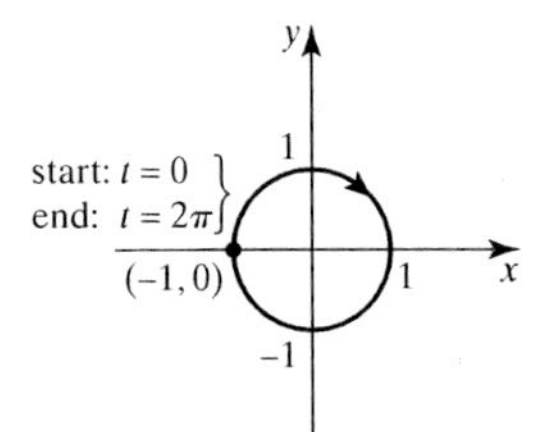

15. $\frac{x^2}{9} + \frac{y^2}{4} = 1$

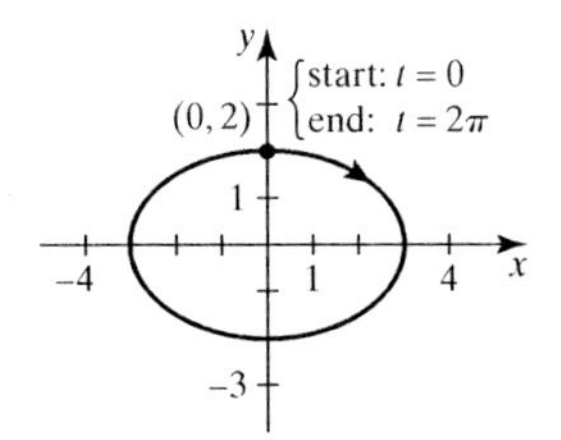

17. $x^2 + y^2 = 1$

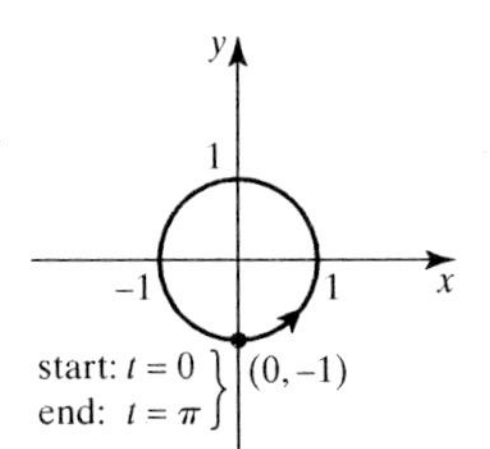

19. $y = x - 3$

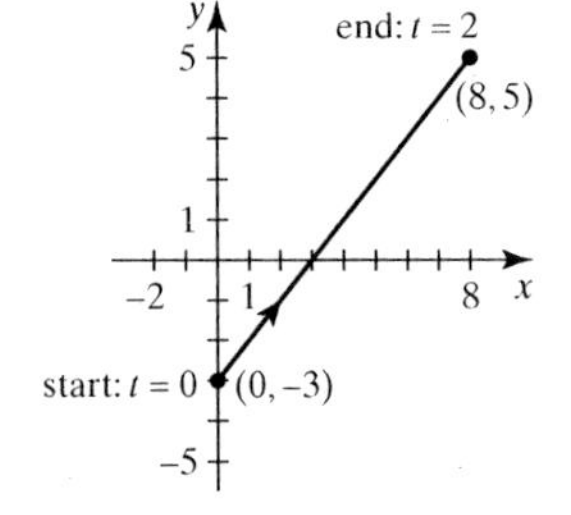

Chapter 5

21. $y = 5x$

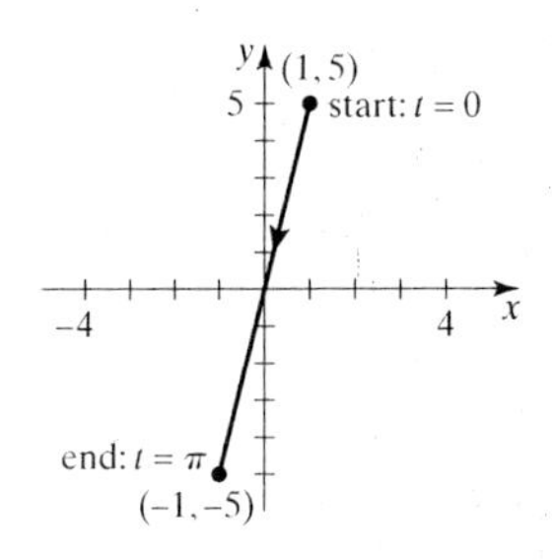

23. $y = x^2 + 1$

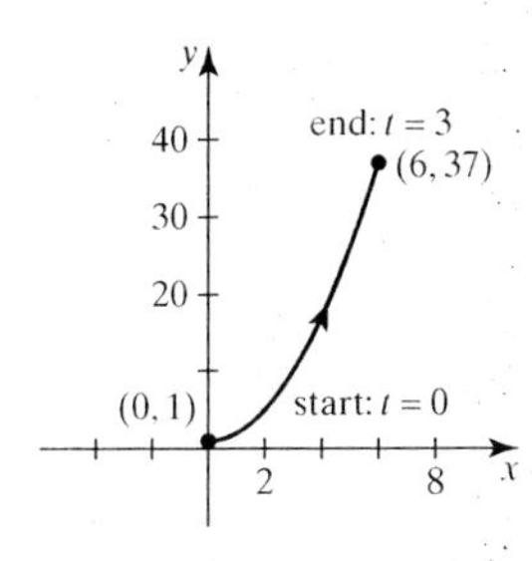

25. $t = 2$ **27.** $t = 4$

29. **a.** $x = 24t,\ y = -16t^2 + 24\sqrt{3}\,t$ **b.**

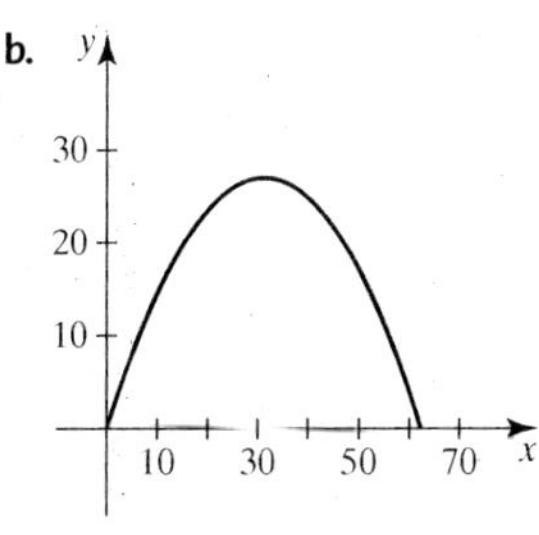

c. 2.6 sec, 62 ft

31. **a.** $x = (120\cos 75°)t,\ y = -16t^2 + (120\sin 75°)t$

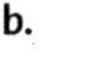

b.

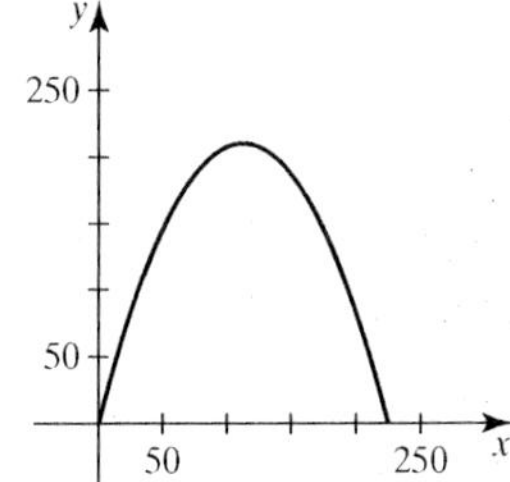

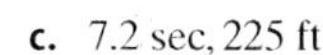

c. 7.2 sec, 225 ft

33–36. *Other answers are possible.*

33. $x = t,\ y = 4t + 1;\ x = \cos t,\ y = 4\cos t + 1$

35. $x = \cos t,\ y = \frac{1}{2}\sin t;\ x = \sin t,\ y = \frac{1}{2}\cos t$

37.

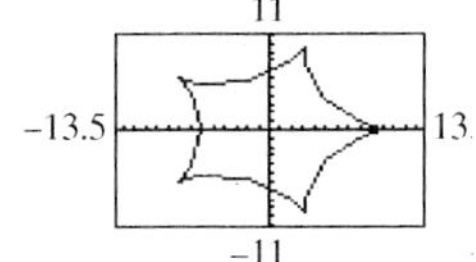

39.

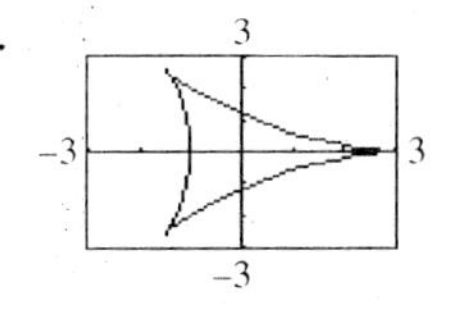

Chapter 5 Review Exercises, p. 323

1. 135°, 225° **2.** 240°, 300° **3.** 0°, 180°

4. 0°, 60°, 120°, 180°, 240°, 300° **5.** 210°, 330°

6. 30°, 150°, 210°, 330° **7.** 30°, 150°, 210°, 330°

8. no solution **9.** 135°

10. 210°, 330° **11.** 0°, 90°, 180°, 270°

12. 45°, 60°, 120°, 135°, 225°, 240°, 300°, 315°

13. 45°, 90°, 225°, 270°

14. 120°, 240° **15.** $\frac{\pi}{8}, \frac{5\pi}{8}, \frac{9\pi}{8}, \frac{13\pi}{8}$ **16.** $\frac{\pi}{3}, \pi, \frac{5\pi}{3}$

17. $0, \frac{2\pi}{3}, \frac{4\pi}{3}$ **18.** $0, \frac{3\pi}{4}, \pi, \frac{5\pi}{4}$ **19.** $\frac{2\pi}{3}, \frac{4\pi}{3}$

20. $\frac{\pi}{4}, \frac{\pi}{2}, \frac{3\pi}{4}, \frac{3\pi}{2}$ **21.** $\frac{\pi}{4}, \frac{7\pi}{4}$ **22.** $\frac{2\pi}{3}, \frac{4\pi}{3}$

23. 90°, 270° **24.** 135°, 180°, 225°

25. 0°, 22.5°, 67.5°, 90°, 180°, 202.5°, 247.5°, 270°

26. 0°, 135°, 180°, 315° **27.** 45°, 225°

28. 15°, 165°, 195°, 345° **29.** 35.3°, 144.7°, 215.3°, 324.7°

30. No solution **31.** 202.5°, 337.5°

32. 0°, 39.2°, 140.8°, 180°, 219.2°, 320.8°

33. 26.6°, 63.4°, 206.6°, 243.4° **34.** 31.0°, 63.4°, 211.0°, 243.4°

35. No solution **36.** 26.6°, 153.4°, 206.6°, 333.4°

37. $210° + k \cdot 360°, 330° + k \cdot 360°$

38. $30° + k \cdot 180°, 150° + k \cdot 180°$

39. $\frac{2\pi}{3} + k \cdot 2\pi, \frac{4\pi}{3} + k \cdot 2\pi$

40. $\frac{\pi}{2} + k \cdot \pi, \frac{\pi}{4} + k \cdot 2\pi, \frac{3\pi}{4} + k \cdot 2\pi$

41. $35.3° + k \cdot 180°, 144.7° + k \cdot 180°$

42. $15° + k \cdot 180°, 165° + k \cdot 180°$

43. $\left(\frac{\pi}{6}, \frac{3}{2}\right), \left(\frac{5\pi}{6}, \frac{3}{2}\right), \left(\frac{3\pi}{2}, 0\right)$ **44.** $\left(\frac{\pi}{2}, 1\right), \left(\frac{3\pi}{2}, -1\right)$

45. $\left(\frac{\pi}{3}, 3\right), \left(\frac{5\pi}{3}, 3\right)$ **46.** $\left(\frac{\pi}{4}, \frac{1}{2}\right), \left(\frac{5\pi}{4}, \frac{1}{2}\right)$

47. 0.5988, 3.7404 **48.** 1.0182, 5.2650

49. $x^2 + y^2 = 1$

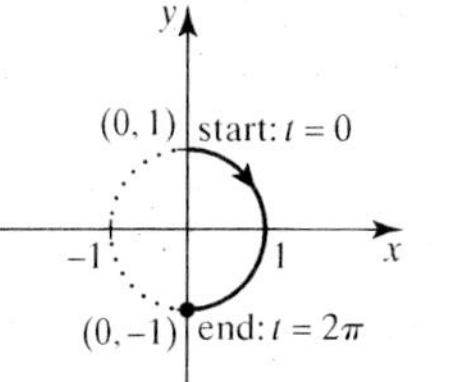

50. $\frac{x^2}{9} + \frac{y^2}{4} = 1$

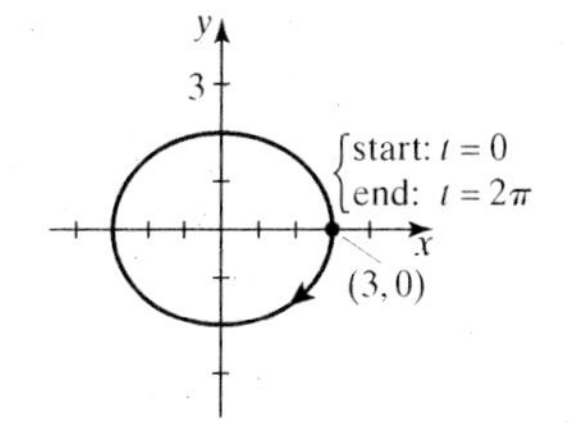

51. $y = x^2$

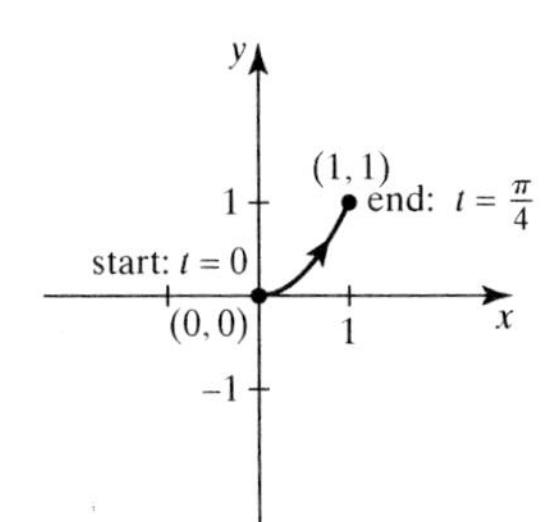

52. $y = x^2$

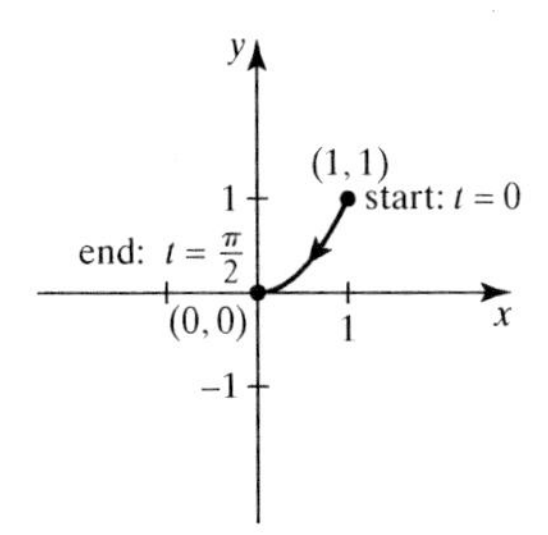

53. $y = -2x$

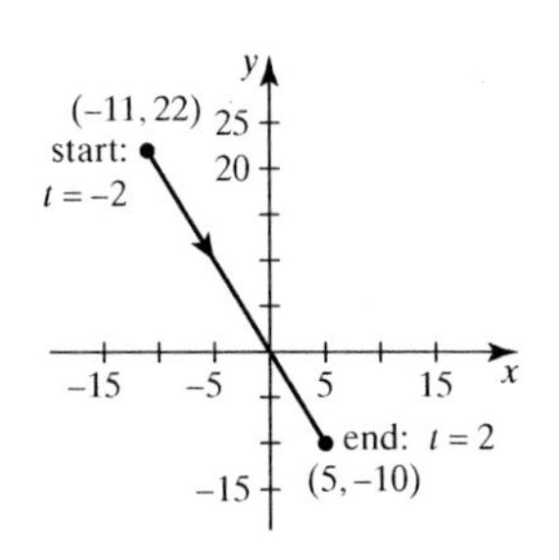

54. $y = -12x + 24$

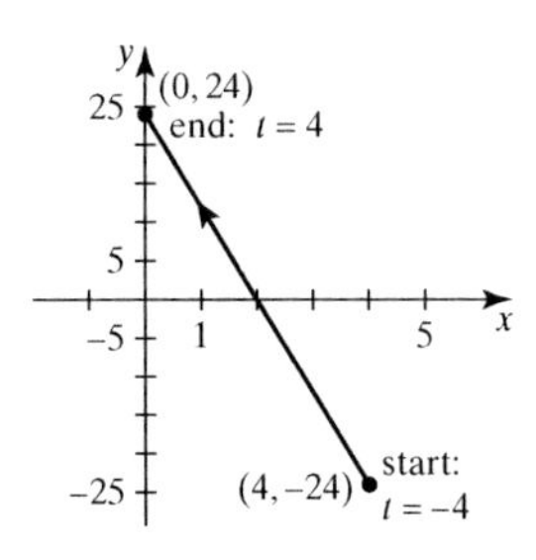

55. $\frac{1}{2}$ sec, $\frac{3}{2}$ sec

56. 25°

Chapter 5 Test, p. 325

1. $\frac{5\pi}{4}, \frac{7\pi}{4}$

2. $\frac{\pi}{3}, \frac{\pi}{2}, \frac{3\pi}{2}, \frac{5\pi}{3}$

3. $\frac{\pi}{2}$

4. $0, \frac{2\pi}{3}, \pi, \frac{4\pi}{3}$

5. $\frac{2\pi}{3}$

6. $\frac{\pi}{6}, \frac{5\pi}{6}, \frac{7\pi}{6}, \frac{11\pi}{6}$

7. 120°, 300°

8. 30°, 150°, 210°, 330°

9. 0°, 60°, 300°

10. 0°, 180°, 225°, 315°

11. 39.6°, 101.7°, 219.6°, 281.7°

12. 86.2°, 273.8°

13. 201.5°, 338.5°

14. 35.8°, 125.8°, 215.8°, 305.8°

15. $\left(\frac{\pi}{6}, \frac{1}{2}\right), \left(\frac{5\pi}{6}, \frac{1}{2}\right)$

16. $(0, 0), (\pi, 0), \left(\frac{\pi}{3}, 3\right), \left(\frac{4\pi}{3}, 3\right)$

17. **a.** $\frac{\pi}{6}, \frac{5\pi}{6}, \frac{7\pi}{6}, \frac{11\pi}{6}$ **b.** $\frac{\pi}{6} + k\pi, \frac{5\pi}{6} + k\pi$

18. **a.** $\frac{\pi}{2}, \frac{3\pi}{2}$ **b.** $\frac{\pi}{2} + k\pi$

19. $x^2 + y^2 = 16$

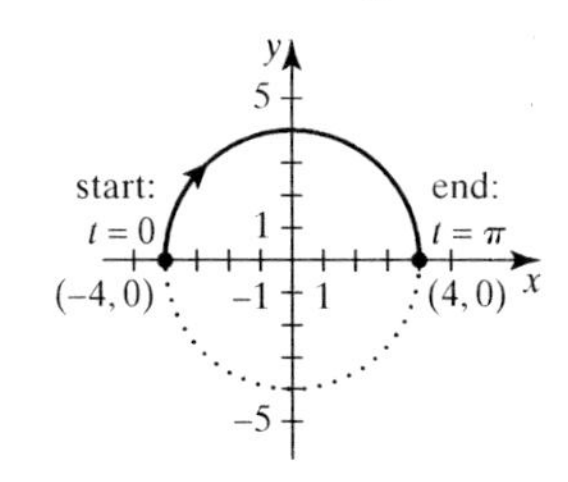

20. $y = 4(x - 1)^2$

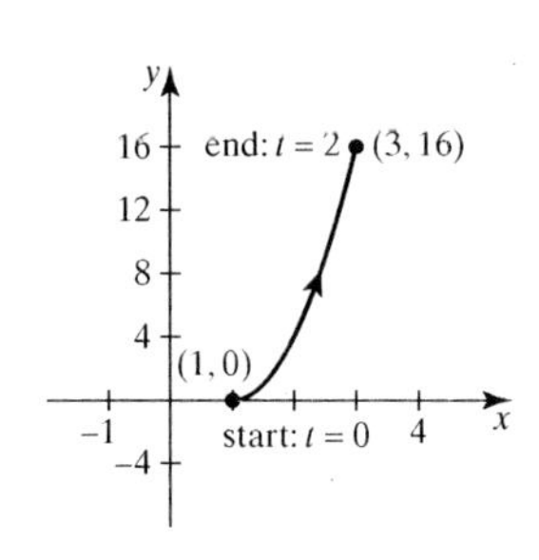

21. 0 sec, 1.6 sec

22. 3.2 sec

Chapter 6

Exercise Set 6.1, p. 336

1.

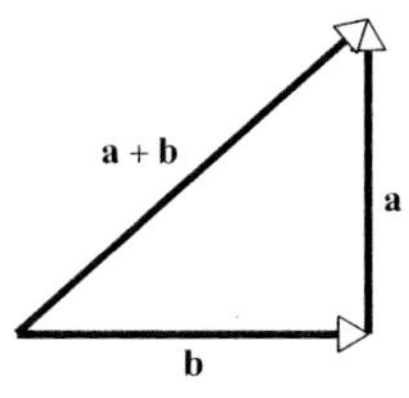

3.

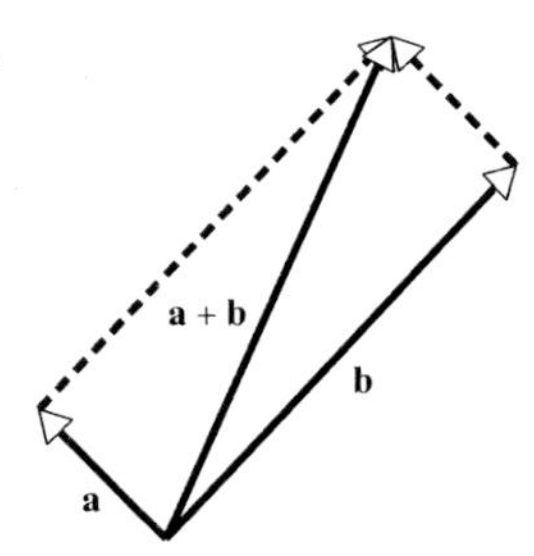

5.

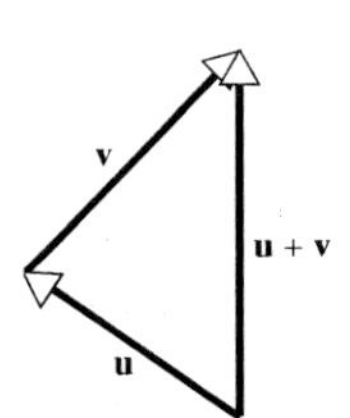

7.

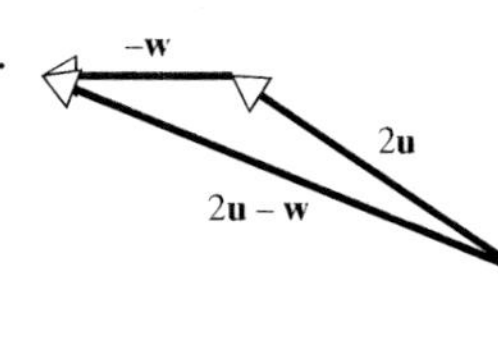

9.

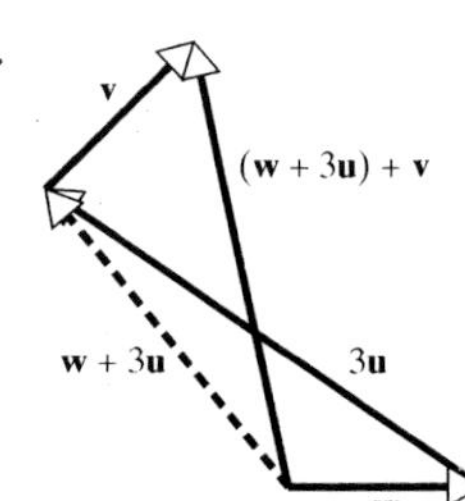

11.

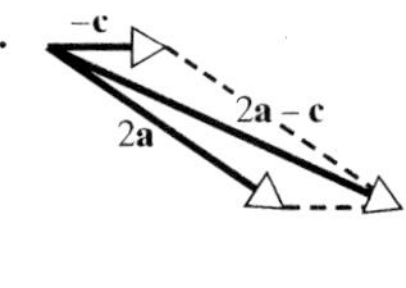

13. 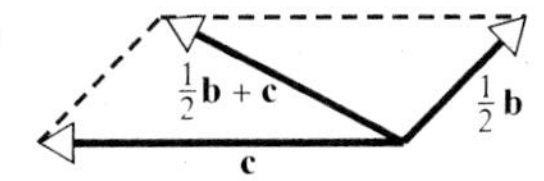

15. 10 lb, 53.1° 17. 76.3° 19. 27.7 newtons, 31.3°

21. 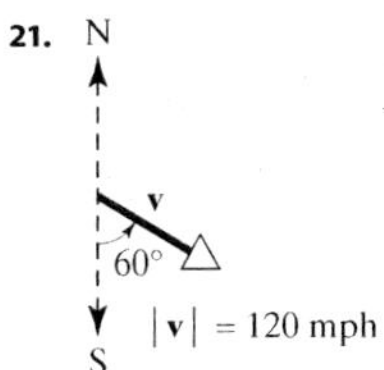

23. 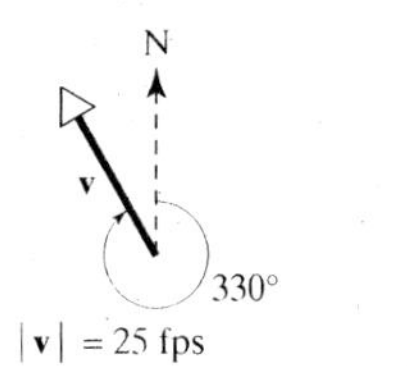

25. 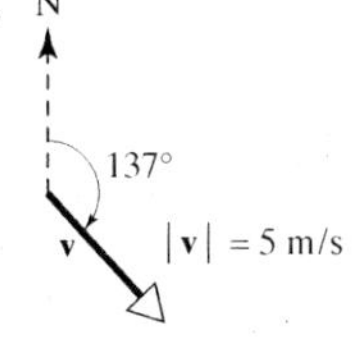

27. 2.8 mph 29. 244.4 mph, 80.1° 31. 316.2 mph, 357.4°
33. 2100 lb 35. 38.7°

Exercise Set 6.2, p. 348

1. $\sqrt{2}$, 45°

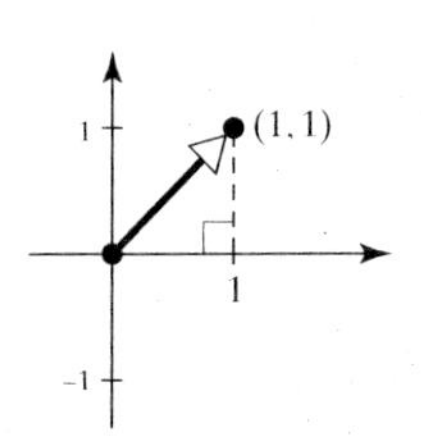

3. $3\sqrt{5}$, 206.6°

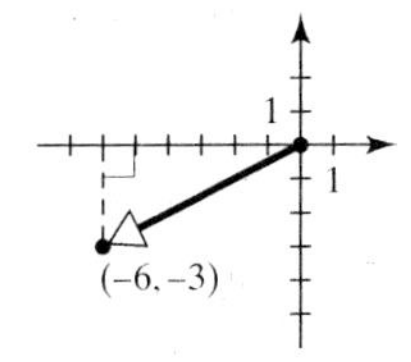

5. 5, 126.9°

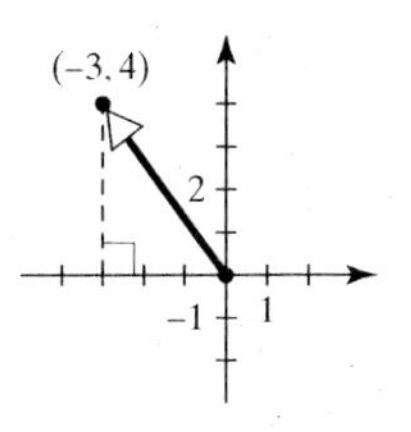

7. $\langle 1.9, 2.3\rangle$ 9. $\langle -38.1, -17.7\rangle$ 11. $4\mathbf{i} + 4\sqrt{3}\mathbf{j}$
13. $-11\mathbf{j}$ 15. $-6\sqrt{2}\mathbf{i} - 6\sqrt{2}\mathbf{j}$

	$\mathbf{v} + \mathbf{w}$	$3\mathbf{w} - 2\mathbf{v}$	$\lvert\mathbf{v} - \mathbf{w}\rvert$	$\mathbf{v} \cdot \mathbf{w}$
17.	$3\mathbf{i}$	$-\mathbf{i} - 5\mathbf{j}$	$\sqrt{5}$	1
19.	$\langle -1, 7\rangle$	$\langle 2, 11\rangle$	$\sqrt{10}$	10
21.	$8\mathbf{i} + 4\mathbf{j}$	$-6\mathbf{i} + 22\mathbf{j}$	$4\sqrt{5}$	0

23. $12\sqrt{3} \approx 20.8$ 25. −30.3 27. 52.5°
29. 18.4° 31. 90° 33. 154.4°
35. yes 37. no 39. yes
41. 5

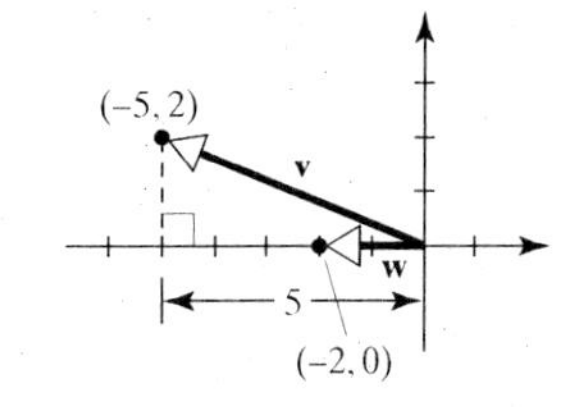

43. $\frac{14\sqrt{41}}{41}$

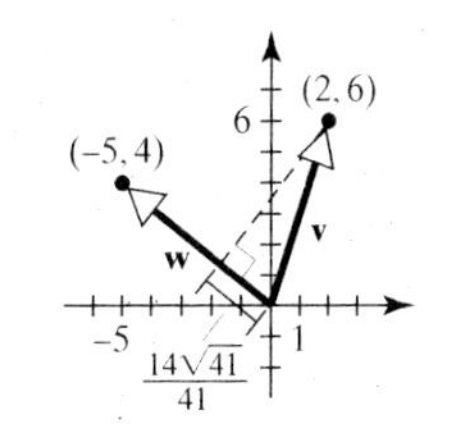

45. $\frac{10\sqrt{29}}{29}$ 47. 3 49. 0 51. $10\mathbf{i} - 11\mathbf{j}$
53. $-5\mathbf{i} + 14\mathbf{j}$ 55. T 57. F 59. T
61. F 63. T 65. T 67. T
69. 20,785 foot-pounds 71. 375 ft · lb 73. 142 ft · lb

Exercise Set 6.3, p. 360

1. (3, 120°) 3. (6, 315°) 5. $\left(4, \frac{3\pi}{2}\right)$

7–12. *Other answers are possible.*

7. (2, 45°), (2, −315°)

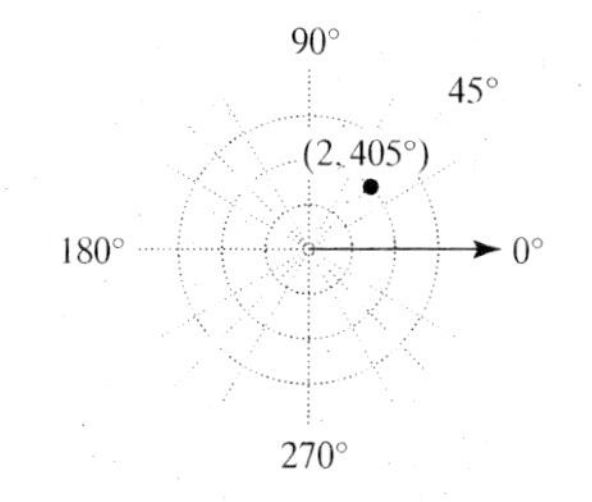

9. (4, 240°), (−4, 60°)

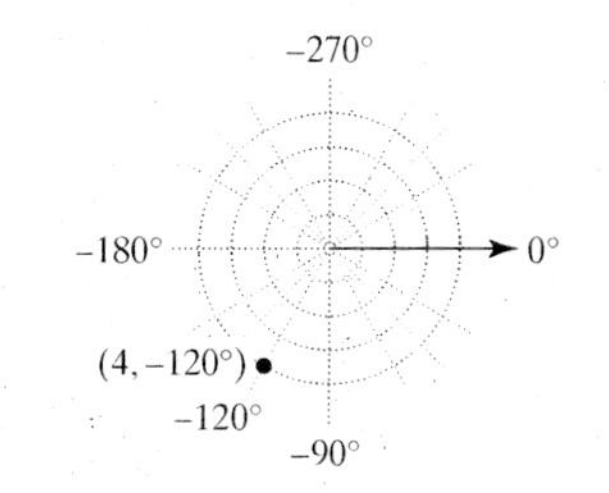

11. $(5, 340^\circ), (5, -20^\circ)$

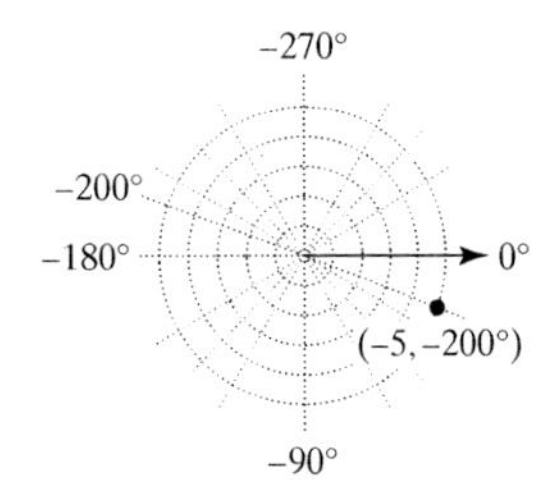

13. $(3\sqrt{3}, 3)$ **15.** $\left(\frac{5}{2}, -\frac{5\sqrt{3}}{2}\right)$ **17.** $(0, 0)$

19. $(5, 0)$ **21.** $\left(7\sqrt{2}, \frac{\pi}{4}\right)$ **23.** $\left(2, \frac{5\pi}{6}\right)$

25. $r = 4$ **27.** $r = 10 \csc \theta$ **29.** $r = 8 \sin \theta$

31. $x = 8$ **33.** $y = \frac{\sqrt{3}}{3}x$ **35.** $y = \frac{1}{2x}$

37.

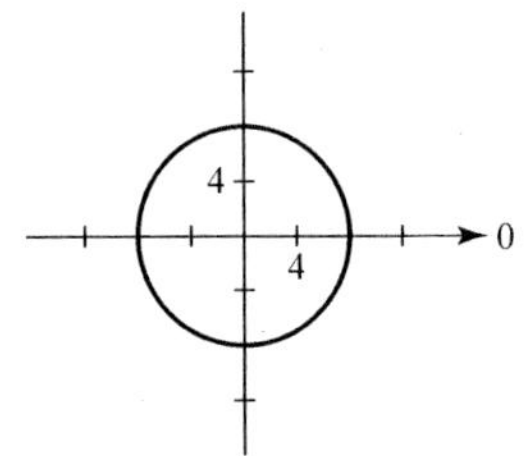

39.

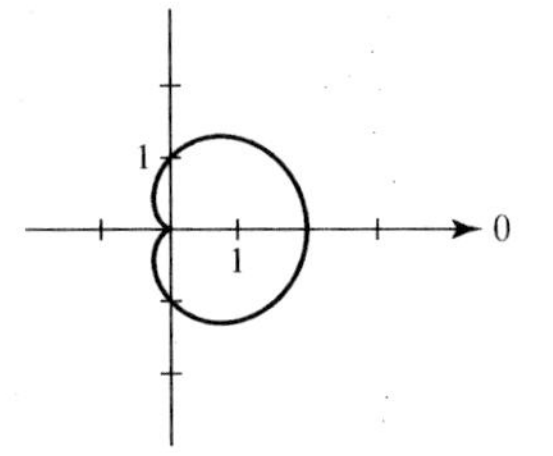

41.

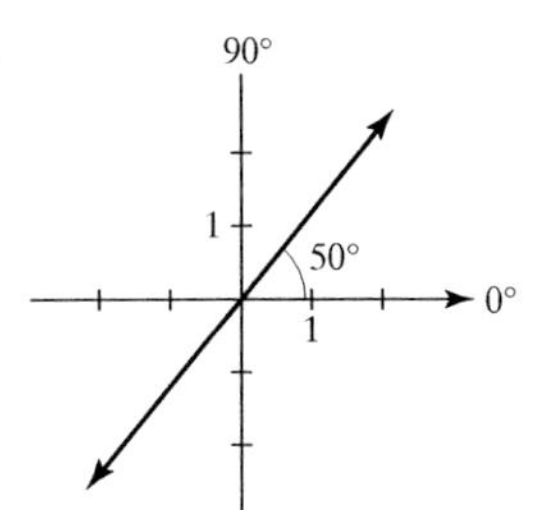

53. c **55.** e **57.** d **59.** k

61. b **63.** i **65.** g

67. $\left(\frac{1}{2}, \frac{\pi}{9}\right), \left(\frac{1}{2}, \frac{5\pi}{9}\right), \left(\frac{1}{2}, \frac{7\pi}{9}\right), \left(\frac{1}{2}, \frac{11\pi}{9}\right), \left(\frac{1}{2}, \frac{13\pi}{9}\right), \left(\frac{1}{2}, \frac{17\pi}{9}\right)$

Exercise Set 6.4, p. 370

1. $x = 2, y = -\frac{4}{3}$ **3.** $x = \pm 2, y = -\frac{3}{13}$ **5.** $6 - 11i$

7. $3 + i$ **9.** $4 - 3i$ **11.** $10 + 4i$ **13.** $2a$

15. $12 + 18i$ **17.** $14 + 22i$ **19.** $-21 - 20i$ **21.** 13

23. $a^2 + b^2$ **25.** -4 **27.** $\frac{3}{4} - \frac{1}{2}i$ **29.** $-\frac{3}{5} + \frac{6}{5}i$

31. $\frac{3}{5} + \frac{4}{5}i$ **33.** i **35.** $-i$ **37.** 1

39. $9i$ **41.** $5 + 9i$ **43.** $12 + 3i$ **45.** $29 + 22i$

47. $-\frac{9}{2}i$ **49.** $\frac{12}{13} - \frac{18}{13}i$

51. $\sqrt{17}$

imaginary

−1 + 4i

−1 1 real

53. $\sqrt{65}$

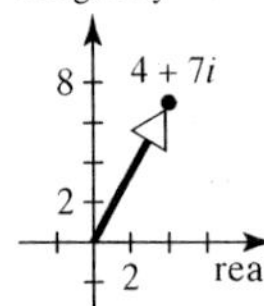

55. $\sqrt{65}$

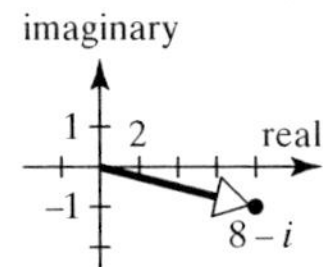

57. 8

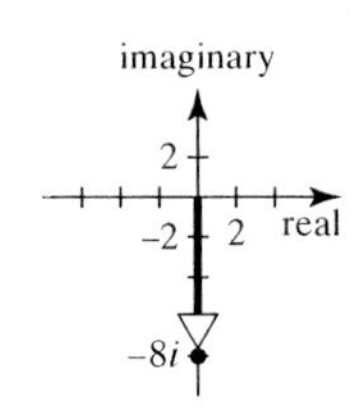

59. 7

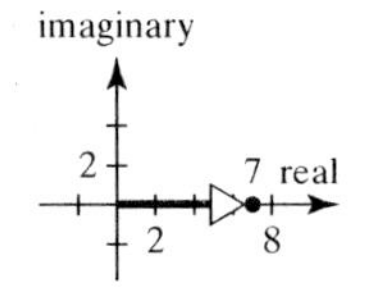

Exercise Set 6.5, p. 381

1. $2(\cos 30^\circ + i \sin 30^\circ) = 2 \text{ cis } 30^\circ$

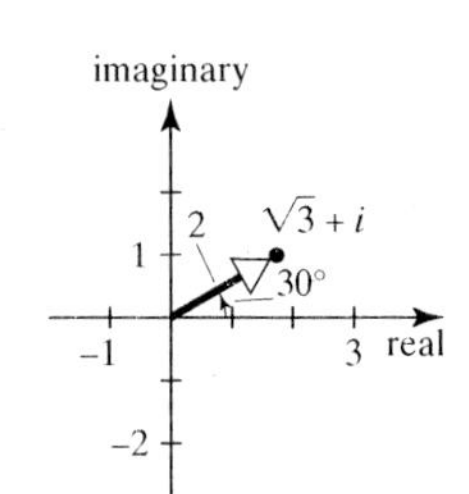

3. $7\sqrt{2} \text{ cis } 135^\circ$

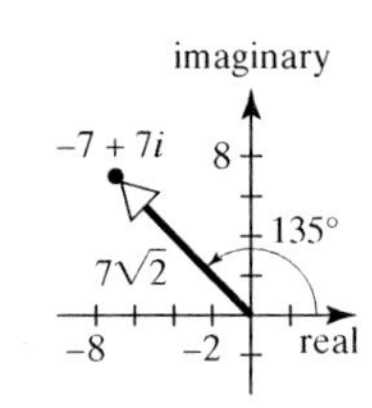

5. $4 \text{ cis } 240^\circ$

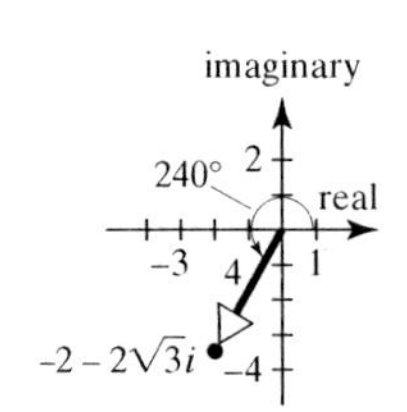

7. $4 \text{ cis } 270^\circ$

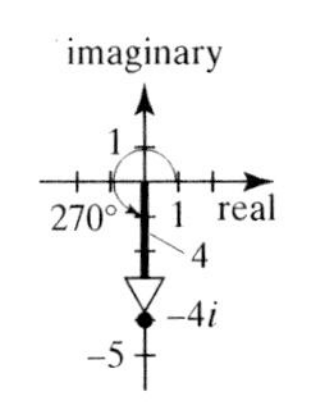

9. $5 \text{ cis } 0^\circ$

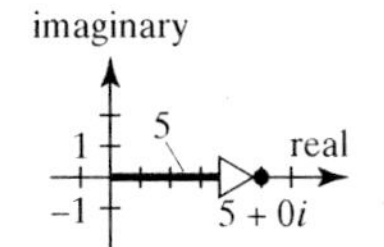

11. $\frac{5}{2} + \frac{5\sqrt{3}}{2}i$ **13.** $-2\sqrt{2} + 2\sqrt{2}i$ **15.** $\frac{\sqrt{3}}{2} - \frac{1}{2}i$

17. $-9i$ **19.** $5\sqrt{2} + 5\sqrt{2}i$ **21.** $-i$

23. $-8 - 8\sqrt{3}\,i$ **25.** $-\frac{3\sqrt{2}}{2} - \frac{3\sqrt{2}}{2}i$ **27.** $\frac{2}{3}i$

29. $\frac{3}{14} - \frac{3\sqrt{3}}{14}i$ **31.** $16 + 16\sqrt{3}\,i$ **33.** $8i$

35. $-8 + 8\sqrt{3}\,i$ **37.** $-\frac{\sqrt{3}}{1250} - \frac{1}{1250}i$ **39.** $-\frac{1}{64}$

41. 4 cis 45°, 4 cis 225° **43.** cis 30°, cis 210°

45. 2 cis 45°, 2 cis 165°, 2 cis 285°

47. cis 60°, cis 132°, cis 204°, cis 276°, cis 348°

49. 10 cis 50°, 10 cis 230°

51. cis 75°, cis 165°, cis 255°, cis 345°

53. $2\sqrt{2} + 2\sqrt{2}\,i, -2\sqrt{2} - 2\sqrt{2}\,i$

55. $\sqrt{2} + \sqrt{2}\,i, -1.93 + 0.52i, 0.52 - 1.93i$

57. $\frac{\sqrt{2}}{2} + \frac{\sqrt{2}}{2}i, \frac{\sqrt{2}}{2} - \frac{\sqrt{2}}{2}i, -\frac{\sqrt{2}}{2} + \frac{\sqrt{2}}{2}i, -\frac{\sqrt{2}}{2} - \frac{\sqrt{2}}{2}i$

59. $\frac{\sqrt{6}}{2} + \frac{\sqrt{2}}{2}i, \frac{\sqrt{6}}{2} - \frac{\sqrt{2}}{2}i, -\frac{\sqrt{6}}{2} + \frac{\sqrt{2}}{2}i, -\frac{\sqrt{6}}{2} - \frac{\sqrt{2}}{2}i$

61. F **63.** T **65.** F **67.** T **69.** T

Chapter 6 Review Exercises, p. 389

1.

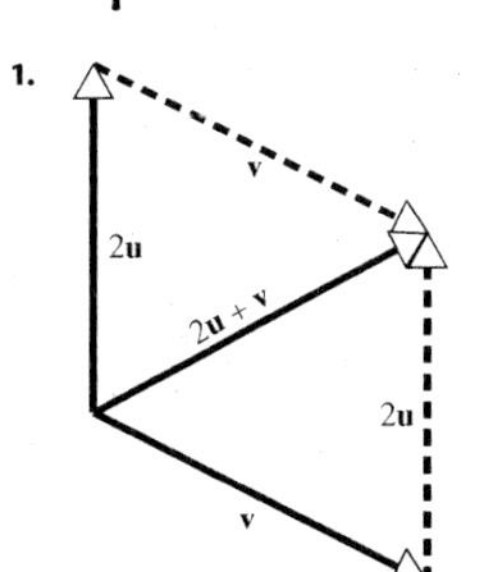

2.

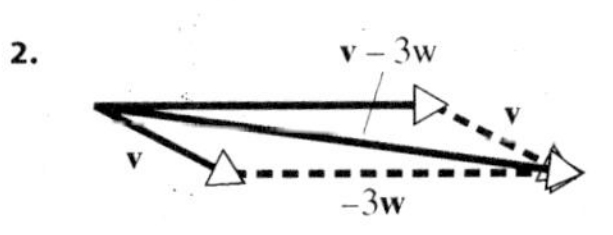

3.

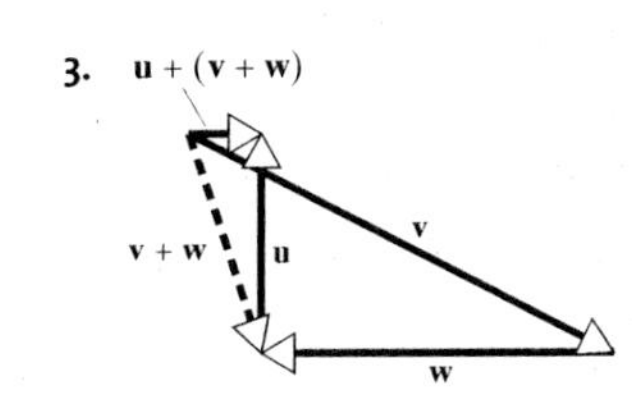

4.

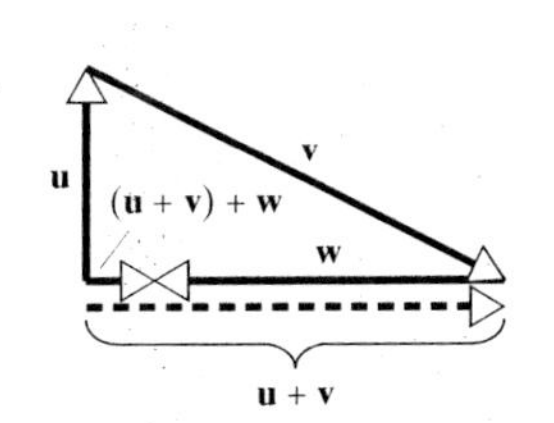

5.

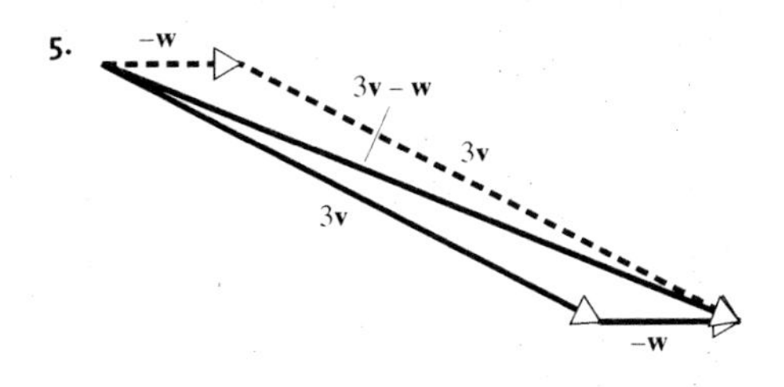

6.

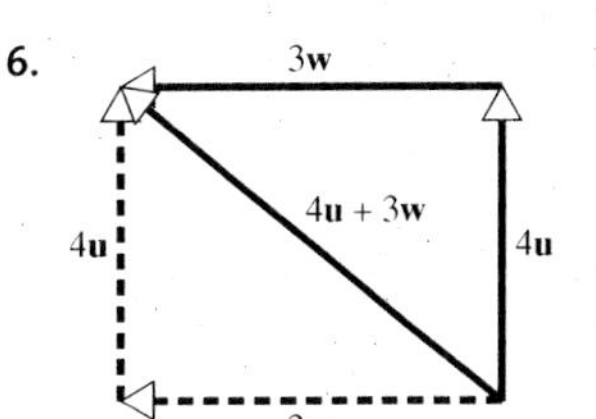

7. 9.10 newtons, 17.6° **8.** 6.8 mph **9.** 91.75 mph, 237.0°

10. $7\sqrt{2}$, 315° **11.** 12, 45° **12.** $3\sqrt{5}$, 116.6°

13. 13, 292.6° **14.** $-1.6\mathbf{i} + 8.9\mathbf{j}$ **15.** $7.7\mathbf{i} - 9.2\mathbf{j}$

16. $\mathbf{j}$ **17.** $-7\mathbf{i}$ **18.** $5\mathbf{i} - 9\mathbf{j}, \sqrt{106}$

19. $9\mathbf{i} + 8\mathbf{j}, \sqrt{145}$ **20.** $11\mathbf{i} - \mathbf{j}, \sqrt{122}$ **21.** $-5\mathbf{i} - 5\mathbf{j}, 5\sqrt{2}$

22. 54 **23.** −38 **24.** 24.6

25. 27.0 **26.** 85.2° **27.** 155.9°

28. $\text{comp}_{\mathbf{v}}\mathbf{u} = \frac{9\sqrt{13}}{13}, \text{comp}_{\mathbf{u}}\mathbf{v} = \frac{9\sqrt{34}}{17}$

29. $\text{comp}_{\mathbf{v}}\mathbf{u} = \frac{68\sqrt{73}}{73}, \text{comp}_{\mathbf{u}}\mathbf{v} = \frac{34\sqrt{29}}{29}$

30. $9\mathbf{i} + 8\mathbf{j}$ **31.** $-14\mathbf{i} - \mathbf{j}$ **32.** 8191.52 ft · lb

33. $(0, -5)$

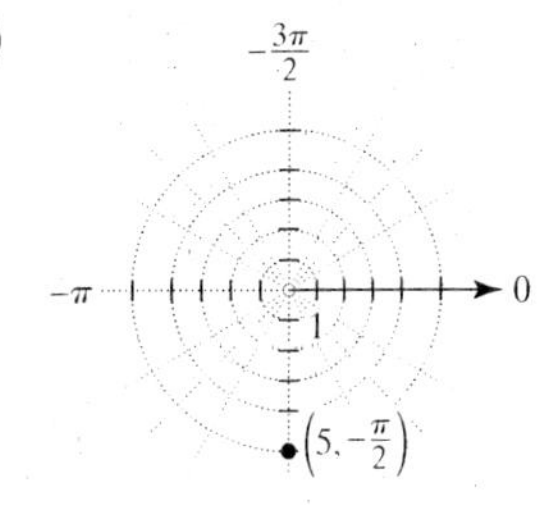

34. $(-4, 4\sqrt{3})$

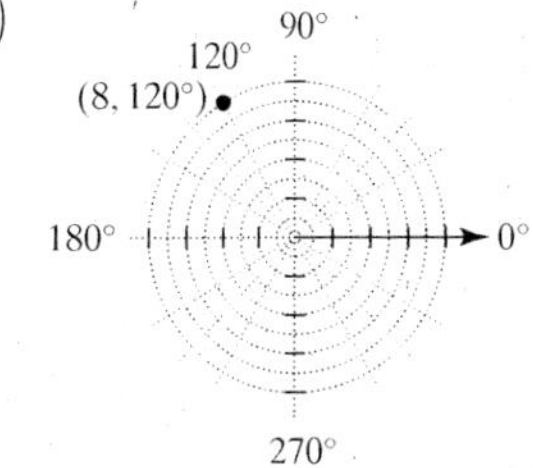

35. $(\sqrt{2}, \sqrt{2})$

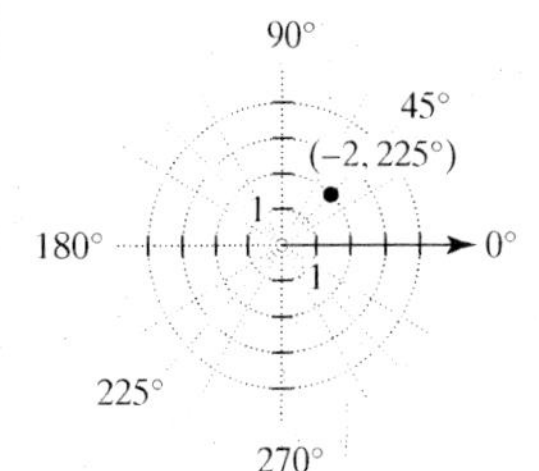

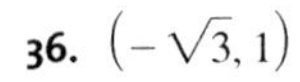

36. $(-\sqrt{3}, 1)$

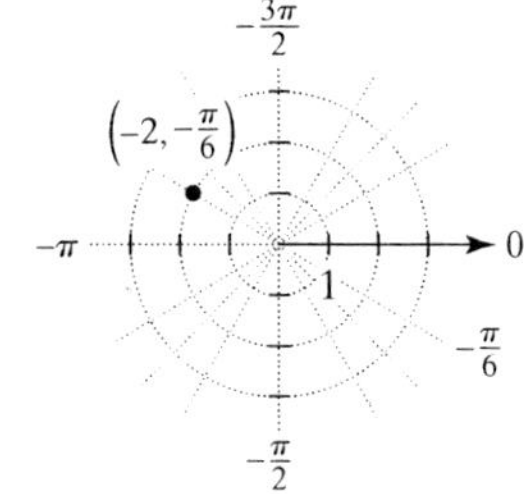

37. $(2\sqrt{3}, 150°), (2\sqrt{3}, -210°)$ **38.** $(\sqrt{2}, -45°), (\sqrt{2}, 315°)$

39. $(7, 0°), (7, 360°)$ **40.** $(9, 270°), (9, -90°)$

41. $r = 4 \sin \theta$ **42.** $r = -2 \csc \theta$

43. $y = -x$ **44.** $x - 7y - 12 = 0$

45. $r = 4 \sin \theta$ **46.** $r = 3 - 3 \cos \theta$

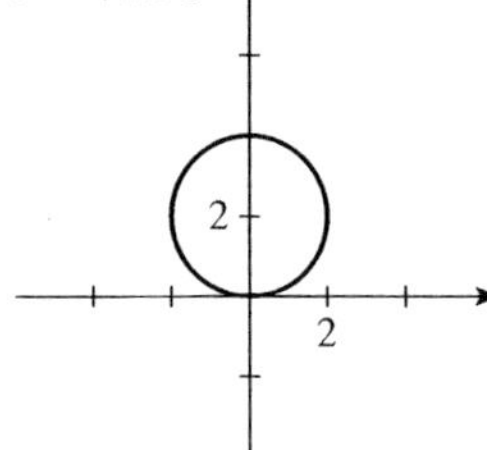

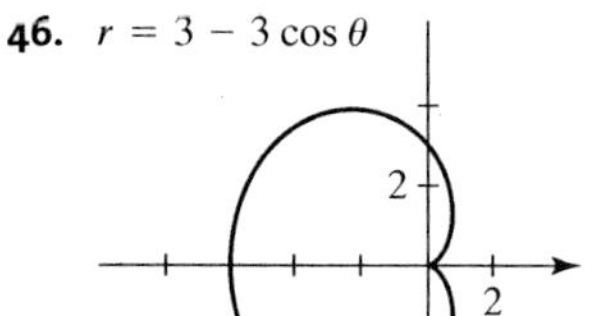

47. $r = 2 + \cos \theta$

48. $r = 1 - 2 \sin \theta$

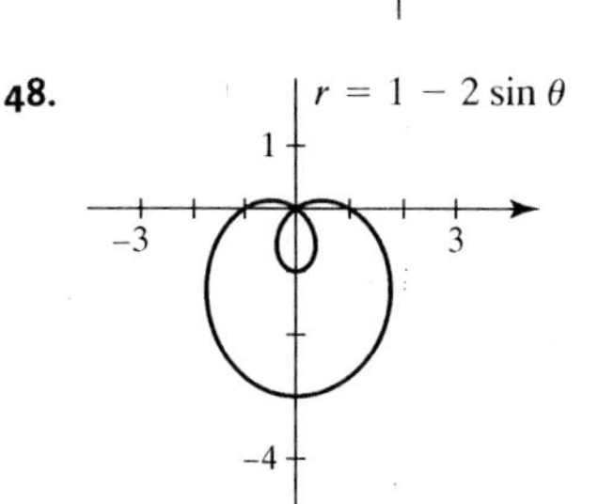

49. $9 - 26i$ **50.** $\frac{5}{4} - \frac{7}{4}i$ **51.** $2i$

52. $9 + 21i$ **53.** $-1 + 2i$ **54.** $-\frac{4}{3} + \frac{4\sqrt{2}}{3}i$

55. $\sqrt{2}$ **56.** 2 **57.** 5

58. 13 **59.** $\sqrt{38}$ **60.** $\sqrt{14}$

61. $\sqrt{2}$ cis 315° **62.** 2 cis 150° **63.** 4 cis 240°

64. 8 cis 270° **65.** $-\frac{7\sqrt{3}}{2} + \frac{7}{2}i$ **66.** $-\frac{1}{2} - \frac{\sqrt{3}}{2}i$

67. $3\sqrt{2} - 3\sqrt{2}i$ **68.** -2 **69.** $-\frac{15}{2} + \frac{15\sqrt{3}}{2}i$

70. $28\sqrt{3} + 28i$ **71.** $-\frac{3}{5}$ **72.** $\sqrt{2} + \sqrt{2}i$

73. $729i$ **74.** -64 **75.** $-\frac{1}{50} + \frac{\sqrt{3}}{50}i$

76. $-\frac{1}{864} - \frac{1}{864}i$ **77.** 6 cis 9°, 6 cis 129°, 6 cis 249°

78. cis 150°, cis 330° **79.** 2 cis 135°, 2 cis 315°

80. $1, -\frac{1}{2} + \frac{\sqrt{3}}{2}i, -\frac{1}{2} - \frac{\sqrt{3}}{2}i$

Chapter 6 Test, p. 391

1. a.

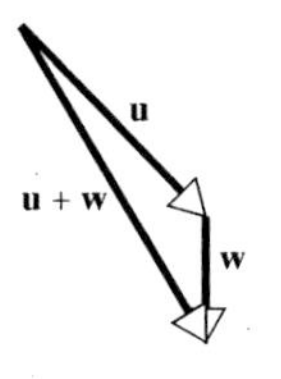

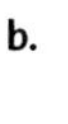

b.

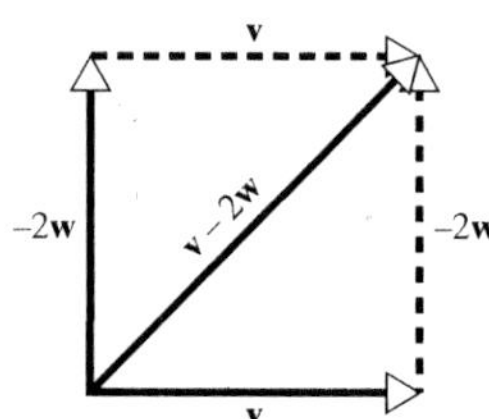

2. a. $2\sqrt{13}$, 303.7°

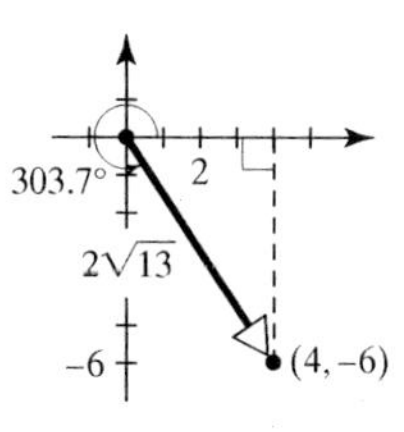

b. $3\sqrt{5}$, 116.6°

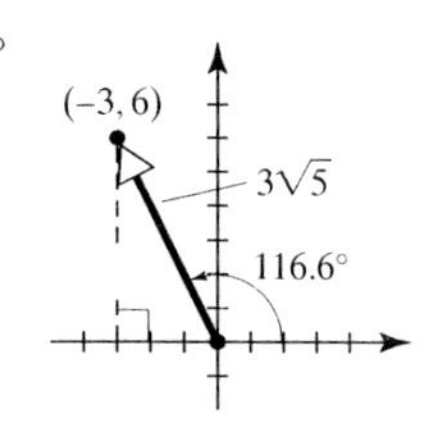

3. a. $0.9\mathbf{i} + 0.5\mathbf{j}$ **b.** $-3.7\mathbf{i} + 3.3\mathbf{j}$

4. a. $-12\mathbf{i} - 22\mathbf{j}$ **b.** 3 **c.** 153.4°

d. $\frac{3\sqrt{5}}{5}$ **e.** $\sqrt{85}$

5. -12 **6.** 14.9 lb, 16.0° **7.** 589 mph, 205.1°

8. a. $\left(-\frac{3}{2}, \frac{3\sqrt{3}}{2}\right)$ **b.** $(-2\sqrt{2}, -2\sqrt{2})$

9. a. $(10\sqrt{2}, 315°), (10\sqrt{2}, -45°)$ **b.** $(10, 30°), (10, -330°)$

10. a. $r = 5$ **b.** $r = -\csc \theta$

11. a. $x^2 + y^2 = 3x$ **b.** $x = 5$

12. a. $-10 - i$ **b.** $10 + 20i$

13. a. $2\sqrt{2}$ cis 315° **b.** 6 cis 90°

14. a. $2\sqrt{2} + 2\sqrt{2}i$ **b.** $\frac{\sqrt{3}}{2} + \frac{3}{2}i$

c. $-3i$ **d.** $-\sqrt{3} + i$

15. a. -64 **b.** -512

16. a. 3 cis 135°, 3 cis 315° **b.** cis 45°, cis 135°, cis 225°, cis 315°

Chapters 1–6

Cumulative Review, p. 393

1. a. Values are given counterclockwise from $(1, 0)$:
$\left(\frac{\sqrt{3}}{2}, \frac{1}{2}\right), \left(\frac{\sqrt{2}}{2}, \frac{\sqrt{2}}{2}\right), \left(\frac{1}{2}, \frac{\sqrt{3}}{2}\right), (0, 1), (-1, 0), (0, -1)$

b.

x	$\cos x$	$\csc x$	$\tan x$
0	1	undefined	0
$\frac{\pi}{6}$	$\frac{\sqrt{3}}{2}$	2	$\frac{\sqrt{3}}{3}$
$45°$	$\frac{\sqrt{2}}{2}$	$\sqrt{2}$	1
$60°$	$\frac{1}{2}$	$\frac{2\sqrt{3}}{3}$	$\sqrt{3}$
$\frac{\pi}{2}$	0	1	undefined
π	-1	undefined	0
$\frac{3\pi}{2}$	0	-1	undefined
$360°$	1	undefined	0

2. a. QIV **b.** $\frac{5}{3}$ **c.** -1 **d.** $-\frac{7}{25}$

e. $\frac{24}{7}$ **f.** $-\frac{7\sqrt{2}}{10}$ **g.** QII **h.** $-\frac{2\sqrt{5}}{5}$

3. a. $P = 4\pi$, $A = 3$

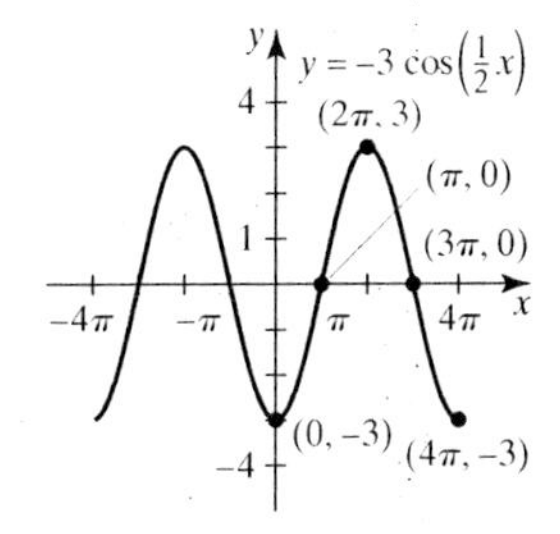

b. $P = \frac{\pi}{2}$, Asymptotes: $x = \frac{\pi}{4} + k \cdot \frac{\pi}{2}$

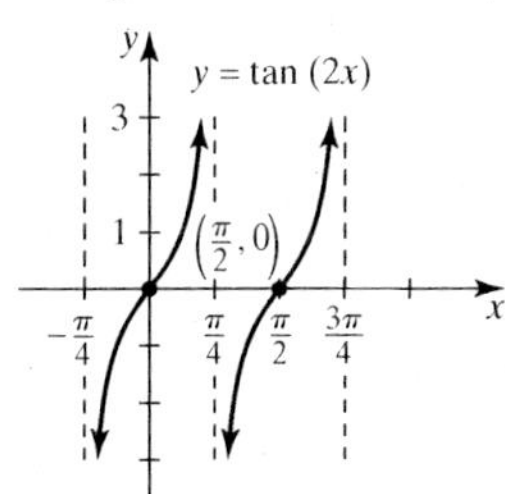

4. a. $y = -4\sin(2x)$, $y = 4\sin\left[2\left(x - \frac{\pi}{2}\right)\right]$

b. $y = 3\sec(2x)$, $y = 3\csc\left[2\left(x + \frac{\pi}{4}\right)\right]$

5. a. $y = \arcsin x$ **b.** $y = \tan^{-1} x$ **c.** $y = \arccos x$

6. a. $\frac{\sqrt{2}}{2}$ **b.** $-\frac{\sqrt{3}}{2}$ **c.** $-\frac{\sqrt{2}}{2}$ **d.** $-\frac{2\sqrt{3}}{3}$

e. 1 **f.** $-\frac{\pi}{4}$ **g.** $\frac{1}{2}$ **h.** $78.9°$

i. $\frac{3\sqrt{3} - 2\sqrt{2}}{2}$ **j.** $\frac{4}{5}$ **k.** $120°$ or $\frac{2\pi}{3}$ **l.** $45°$ or $\frac{\pi}{4}$

7. 13.1 in.

8. a. -4.5 cm **b.** 1.8 sec

9.
$$\frac{\tan 360° - \tan x}{1 + \tan 360° \tan x} = -\tan x$$
$$\frac{0 - \tan x}{1 + 0} =$$
$$-\tan x =$$

10. $\frac{\sqrt{2 + \sqrt{3}}}{2}$ or $\frac{\sqrt{6} + \sqrt{2}}{4}$

11. a. $\sin^2 x = 1 - \cos^2 x$ **b.** $\sec^2 x = 1 + \tan^2 x$

c. $\cos^2 \frac{x}{2} = \frac{1 + \cos x}{2}$ **d.** $\sin 2x = 2\sin x \cos x$

e. $2\cos^2 x - 1 = \cos 2x$ **f.** $\tan(-x) = -\tan x = \frac{-1}{\cot x}$

12. a.
$$2\left(\cos\frac{x}{2}\right)^2 - \cos x = 1$$
$$2\left(\frac{1 + \cos x}{2}\right) - \cos x =$$
$$1 + \cos x - \cos x =$$
$$1 =$$

b.
$$\frac{\sin 5x \cos 3x - \cos 5x \sin 3x}{\cos 2x + 1} = \sin x \sec x$$
$$\frac{\sin(5x - 3x)}{(2\cos^2 x - 1) + 1} =$$
$$\frac{2\sin x \cos x}{2\cos^2 x} =$$
$$\frac{\sin x}{\cos x} =$$
$$\sin x \sec x =$$

c.
$$\sec^2\theta + \csc^2\theta = \sec^2\theta\csc^2\theta$$
$$\frac{1}{\cos^2\theta} + \frac{1}{\sin^2\theta} =$$
$$\frac{\sin^2\theta + \cos^2\theta}{\sin^2\theta\cos^2\theta} =$$
$$\frac{1}{\sin^2\theta\cos^2\theta} =$$
$$\frac{1}{\sin^2\theta} \cdot \frac{1}{\cos^2\theta} =$$
$$\sec^2\theta\csc^2\theta =$$

d. Let $x = \pi$.
$$\sin x + \cos x = 1$$
$$\sin \pi + \cos \pi \neq 1$$
$$0 + (-1) \neq 1$$

13. a. $60°, 300°$ **b.** $0°, 180°$ **c.** $0°, 90°, 180°, 270°$

d. $36.2°, 143.8°$ **e.** $60°, 300°$ **f.** $90°, 180°$

g. $90°, 210°, 330°$

14. **a.** $\frac{3\pi}{4}, \frac{5\pi}{4}$ **b.** $\frac{\pi}{8}, \frac{5\pi}{8}, \frac{9\pi}{8}, \frac{13\pi}{8}$ **c.** $\frac{\pi}{2}$

15. 6.70 ft

16. $a = 5, b = 10, c = 5\sqrt{3}, \alpha = 30°, \beta = 90°, \gamma = 60°$

17. Triangle method

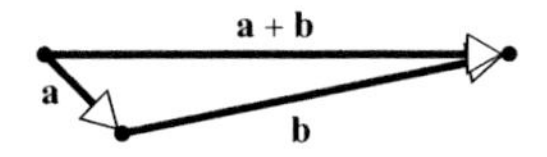

Parallelogram method

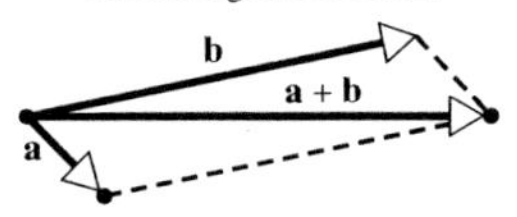

18. **a.** $-5\mathbf{i} + 5\mathbf{j}$

b.

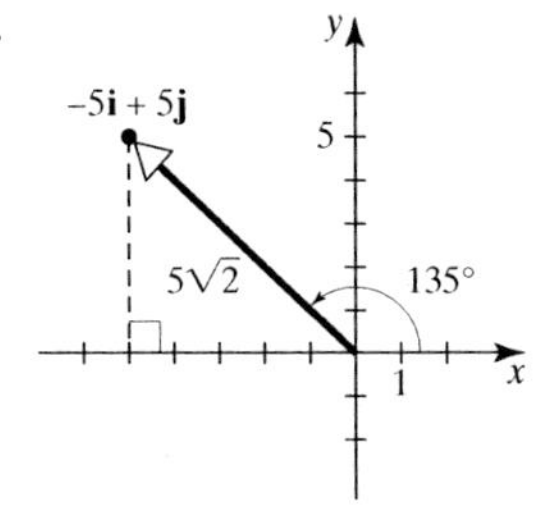

c. Yes

19. **a.** $4\mathbf{i} + 10\mathbf{j}$ **b.** 0 **c.** 90°

d. 8.1° **e.** 5 **f.** $\frac{\sqrt{2}}{2}$

20. 10.6 lb, 40.9°

21. 346.0 mph, 3.5°

22. **a.**

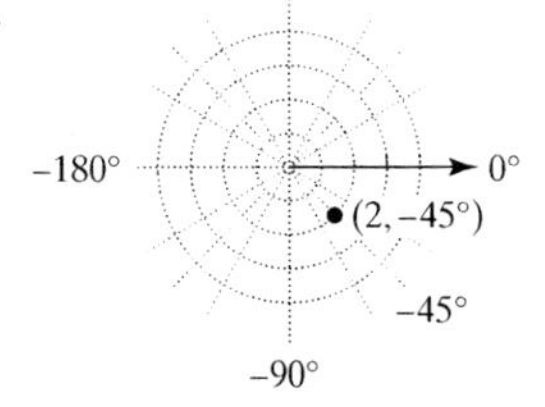

b. $(2, 315°), (-2, 135°)$ **c.** $(\sqrt{2}, -\sqrt{2})$

23. **a.** $(12, 120°)$ **b.** $r^2 + 6r\sin\theta = 7$

24. $x^2 + y^2 - 3x = 0$

25. **a.**

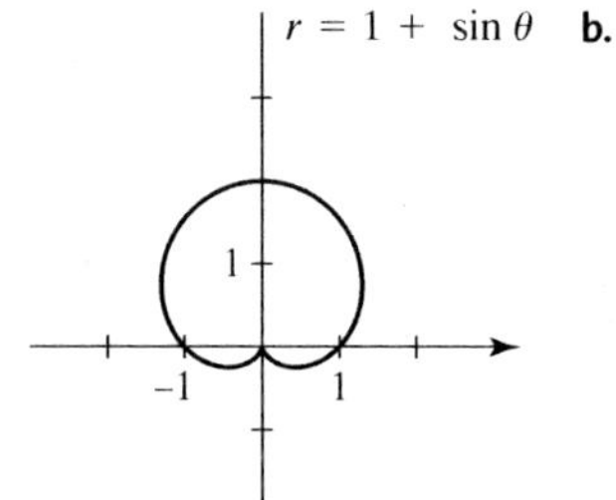

b.

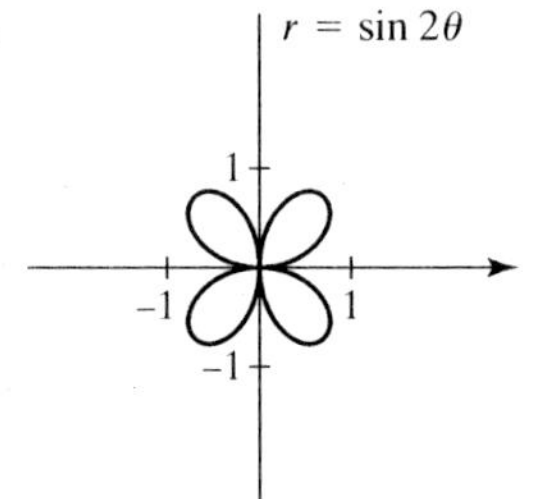

26. **a.** $r = 2$ **b.** $r = 2\cos\theta$ **c.** $r = 1 + 2\cos\theta$

27. $\left(\frac{x}{2}\right)^2 + \left(\frac{y}{3}\right)^2 = 1$

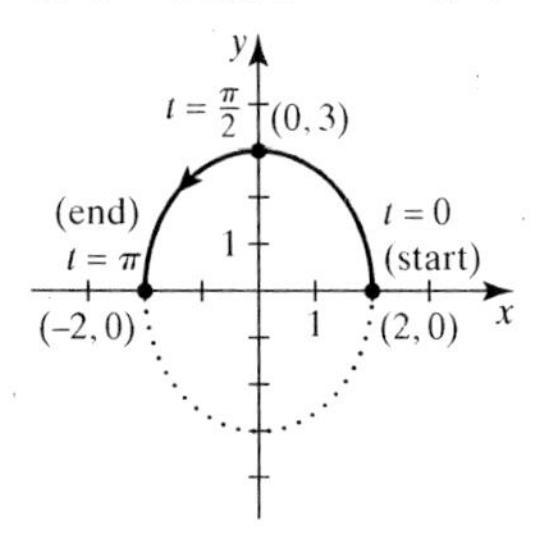

28. **a.** $-1 + 3i$ **b.** $-\frac{3}{5} + \frac{6}{5}i$ **c.** $2\sqrt{17}$ **d.** $3\sqrt{2}\text{ cis }45°$

e. -324 **f.** $\sqrt[4]{18}\text{ cis }22.5°, \sqrt[4]{18}\text{ cis }202.5°$

29. F	**30.** F	**31.** F	**32.** F
33. F	**34.** T	**35.** T	**36.** T
37. T	**38.** F	**39.** F	**40.** F
41. T	**42.** F	**43.** T	**44.** T
45. T	**46.** F	**47.** F	**48.** T
49. F	**50.** T	**51.** T	**52.** T
53. T	**54.** F	**55.** F	**56.** T
57. T	**58.** F	**59.** T	**60.** F
61. T	**62.** F	**63.** F	**64.** F
65. T	**66.** T	**67.** T	**68.** T
69. T	**70.** F		

Index

A

B

C

D

Q

R

S

T

Trigonometric Ratios for Right Triangles

$$\sin \alpha = \frac{\text{side opposite } \alpha}{\text{hypotenuse}} = \frac{a}{c}$$

$$\cos \alpha = \frac{\text{side adjacent } \alpha}{\text{hypotenuse}} = \frac{b}{c}$$

$$\tan \alpha = \frac{\text{side opposite } \alpha}{\text{side adjacent } \alpha} = \frac{a}{b}$$

$$\alpha + \beta = 90°$$

Pythagorean Theorem: $a^2 + b^2 = c^2$

Oblique Triangles

Law of Sines (AAS, SSA)

$$\frac{\sin \alpha}{a} = \frac{\sin \beta}{b} = \frac{\sin \gamma}{c}$$

Law of Cosines (SAS, SSS, SSA)

$$c^2 = a^2 + b^2 - 2ab \cos \gamma$$

$$b^2 = a^2 + c^2 - 2ac \cos \beta$$

$$a^2 = b^2 + c^2 - 2bc \cos \alpha$$

Basic Identities

Pythagorean Identity

$$\cos^2 x + \sin^2 x = 1$$

$$\cos^2 x = 1 - \sin^2 x, \qquad \sin^2 x = 1 - \cos^2 x$$

$$1 + \tan^2 x = \sec^2 x, \qquad \cot^2 x + 1 = \csc^2 x$$

Cofunction Identities

$$\cos\left(\frac{\pi}{2} - x\right) = \sin x$$

$$\sin\left(\frac{\pi}{2} - x\right) = \cos x$$

$$\tan\left(\frac{\pi}{2} - x\right) = \cot x$$

Sum and Difference Identities

$$\cos(A + B) = \cos A \cos B - \sin A \sin B$$

$$\sin(A + B) = \sin A \cos B + \cos A \sin B$$

$$\tan(A + B) = \frac{\tan A + \tan B}{1 - \tan A \tan B}$$

$$\cos(A - B) = \cos A \cos B + \sin A \sin B$$

$$\sin(A - B) = \sin A \cos B - \cos A \sin B$$

$$\tan(A - B) = \frac{\tan A - \tan B}{1 + \tan A \tan B}$$

Double-Angle Identities

$$\cos 2x = \cos^2 x - \sin^2 x$$

$$= 2\cos^2 x - 1$$

$$= 1 - 2\sin^2 x$$

$$\sin 2x = 2 \sin x \cos x$$

$$\tan 2x = \frac{2 \tan x}{1 - \tan^2 x}$$

Half-Angle Identities

$$\cos^2 x = \frac{1 + \cos 2x}{2} \qquad \cos \frac{x}{2} = \pm\sqrt{\frac{1 + \cos x}{2}}$$

$$\sin^2 x = \frac{1 - \cos 2x}{2} \qquad \sin \frac{x}{2} = \pm\sqrt{\frac{1 - \cos x}{2}}$$

$$\tan \frac{x}{2} = \pm\sqrt{\frac{1 - \cos x}{1 + \cos x}} = \frac{\sin x}{1 + \cos x} = \frac{1 - \cos x}{\sin x}$$